D0085226

INTRODUCTORY SOIL MECHANICS AND FOUNDATIONS: GEOTECHNICAL ENGINEERING

INTRODUCTORY

George F. Sowers

Regents Professor of Civil Engineering, Georgia Institute of Technology; Senior Vice President and Consultant, Law Engineering Testing Company

SOIL MECHANICS AND FOUNDATIONS: GEOTECHNICAL ENGINEERING

FOURTH EDITION

MACMILLAN PUBLISHING CO., INC.
New York
COLLIER MACMILLAN PUBLISHERS
London

Printed in the United States of America

Macmillan Publishing Co., Inc.
866 Third Avenue, New York, New York 10022
Collier Macmillan Canada, Ltd.

Library of Congress Cataloging in Publication Data

Sowers, George F.
Introductory soil mechanics and foundations.
Includes bibliographies and index.
1. Soil mechanics. 2. Foundations.
TA710.S67 1979 624'.15 78-19069
ISBN 0-02-413870-3

Printing: 1 2 3 4 5 6 7 8 Year: 9 0 1 2 3 4 5

Preface to the Fourth Edition

Nearly thirty years have passed since the first edition appeared in 1951. There have been many changes during that period—changes that became challenges that generated new textbooks. Among these was the second edition in 1961 and the third in 1970.

One major change has been the accumulation of geotechnical experience that has made this fourth edition necessary: more than a century in the Sowers' family, spanning three generations. George B. Sowers, coauthor of the first three editions, died after sixty years of intensive practice. The present author adds forty years. The third generation is already making its contributions to the earth sciences.

A second change and challenge is the continuing growth of the fundamental sciences and their applications to the solution of problems. We have more elegant theories, more sophisticated techniques for field and laboratory measurement, and more powerful computational tools—from the electronic calculator that has replaced the traditional slide rule to the advanced computers that even present their results in the form of graphs and drawings. Paradoxically, these advances have not always been helpful. We have some solutions to problems that are insignificant, but none to some of the more pressing problems. For example, we can analyze the ground motion generated by an earthquake, but we cannot predict when and where the earthquake will occur. We can measure the movement in a landslide and evaluate its mechanism, but we are not sure why one mountainside moves and a similar adjacent one remains stable. Science and engineering have become so intoxicated by the elegant tools and techniques for analysis that they forget that the ultimate aim of engineering is to solve problems—the problems of society that are related to both the natural and man-made environment.

A third challenge is the public's reaction to the failure to solve real problems. Contemporary "consumerism" demands performance and even perfection. Engineers are being held accountable for their mistakes.

Unfortunately, they are also blamed for the effects of misused technology brought on by the nearsighted demands of the public and politicians for instant solutions to the problems. For example, the billion-dollar Teton Dam failure has been rightly blamed on errors in engineering judgment. However, public demand and political expediency determined that the dam should be built, and economic pressures from established bureaucracy made it impossible to build the dam in a safe manner. The engineer must be aware of the limitations of technological solutions to problems: both those limitations of the state of scientific knowledge and engineering experience and the limitations imposed by political expediency, economic inadequacy, and public misinformation. The engineer is obligated to resist misapplied political and bureaucratic pressure and to inform the public of the risks of the project as well as its benefits.

The students remain a fourth challenge. They have been exposed to a more sophisticated education and have at their command more powerful tools for solving problems. However, these tools can be the student's enemy because they focus the student's attention on the techniques for problem solving instead of on the problem. The result is that a problem is solved with great precision, but sometimes the wrong problem!

It is the aim of this fourth edition to emphasize the real world of soil and rock with which the engineer must cope. It emphasizes physical concepts of soil and rock behavior under the environmental changes induced by nature and man. It provides analytical tools for the solution of real problems, not the most complex and precise tools, but tools that the student can understand and whose limitations he or she can appreciate. More sophisticated tools are left to more advanced texts after the student engineer understands the use of the basic concepts.

The author hopes that the users of the text will be intrigued by solving real problems, using not only the tools of analysis, but also their intuition and growing experience where no analyses are available. Creating new solutions for old problems and for those of the future using knowledge and ingenuity is the real challenge of engineering and its greatest reward. The author hopes that this text will contribute a challenge, promote ingenuity, and add enjoyment to the practice of engineering.

The author is indebted to the many users of the previous editions who have made helpful suggestions for improvements. These users include students, professors, and practicing engineers and geologists. Particularly, Clay Sams, Ralph Brown, C. Fogle, Steve Collins, Don Miller, R. Wells, and R. White, of the Law Engineering Testing Company; Professors C. E. Weaver, R. D. Barksdale, B. B. Mazanti, and Q. Robinett, of the Georgia Institute of Technology; Professors H. B. Seed and J. M. Duncan of the University of California; G. W. Clough, of Stanford University; S. P. Clemence, of Syracuse University; W. C. Lovell, of Purdue University; L. C. Reese, of the University of Texas; and A. Casco, of the National University of Honduras, made specific suggestions. The work of Professors Jose Menendez Menendez, of the University of Havana, and Alfanso Rico Rodriguez, of the National University of Mexico, in translating the third edition into Spanish, and the Asian edition emphasized the need for a worldwide view of geotechnical problems, which has been given greater emphasis in this revision.

Finally, the work would not have been completed without the help of Frances L. Sowers, who reviewed the entire text.

G.F.S.

Contents

Nomenclature and Symbols

The following symbols have been used throughout this text. They generally conform to those adopted by the International Society for Soil Mechanics and Foundation Engineering at its Ninth Conference in Tokyo, 1977; and to the earlier D-653-67, "Terms and Symbols Relating to Soil Mechanics." Some symbols are used in more than one way in geotechnical engineering. In a few cases, there is no standard—more than one symbol is used. Other usages conflict with applied mechanics. These important conflicts that are often confusing are noted.* In geotechnical engineering, depths below the ground or water surface are frequently positive and above negative. Compressive stresses are positive and tensile negative, similar to hydraulic engineering usage.

			Customary Units		
Symbol	Nomenclature	Dimensions	English (Force)	Metric (Force)	SI
a	acceleration	L/T^2	ft/sec^2	cm/sec^2, m/sec^2	m/sec^2
A	area	L^2	ft^2, in.^{2}	cm^2, m^2	mm^2, m^2
a_v	coefficient of compressibility	L^2/F	ft^2/lb	m^2/kg	m^2/kN
B	width	L	ft	cm, m	mm, m
C_c	compression index	dimensionless			
C_u	uniformity coefficient	dimensionless			
c	shear strength in undrained or quick shear or apparent cohesion	F/L^2, M/LT^2	lb/ft^2 t/ft^2	kg/cm^2 kg/m^2	N/mm^2 kN/m^2
c_a	adhesion of soil to a surface	F/L^2	lb/ft^2	kg/m^2	kN/m^2
c_v	coefficient of consolidation	L^2/T	ft^2/min	cm^2/sec	m^2/sec
C_α	coefficient of secondary consolidation	dimensionless			
d	diameter of tube, crack depth	L	in., ft	mm, m	mm, m
D	diameter of soil grain	L	in.	mm	mm

Symbol	Nomenclature	Dimensions	Customary Units		
			English (Force)	Metric (Force)	SI
D_r^*	relative density (also I_D, R_D)	dimensionless			
D_{10}	diameter of grain of which 10% by weight smaller	L	in.	mm	mm
D_{50}	median grain diameter	L	in.	mm	mm
E^*	modulus of elasticity (also modulus of deformation M)	F/L^2, M/LT^2	lb/ft^2	kg/cm^2	kN/m^2
e	mechanical efficiency of a pile hammer	dimensionless			
e	void ratio (express as decimal)	dimensionless			
e_0	initial void radio before loading or deformation	dimensionless			
F	force	F, ML/T^2	lb	kg	N, kN
F_s	factor of safety	dimensionless			
G_s	specific gravity of solids	dimensionless			
G_E^*	shear modulus of elasticity	F/L^2, M/LT^2	lb/ft^2	kg/cm^2	kN/m^2
H	vertical depth of cut, height of wall, thickness of soil stratum	L	ft	cm, m	mm, m
h	head	L	ft	cm, m	mm, m
i	hydraulic gradient	dimensionless			
K	coefficient of earth pressure	dimensionless			
K_0	coefficient of earth pressure at rest	dimensionless			
K_A	coefficient of active earth pressure	dimensionless			
K_P	coefficient of passive earth pressure	dimensionless			
k	coefficient of permeability	L/T	ft/min	cm/sec	mm/sec, m/sec, m/year
k_s	coefficient of subgrade reaction	F/L^3	lb/in.^{3}	kg/cm^3	kN/m^3
L	length	L	ft	cm, m	mm, m
LL	liquid limit (L_w, w_L)	dimensionless			
M	mass	FT^2/L, M	slug	kgm	g, kg
m	stability number	dimensionless			
n	number of piles in a group	dimensionless			
n	porosity (percentage)	dimensionless			
n	coefficient of restitution	dimensionless			
P	resultant force of earth pressure over an area, of unit length	F, ML/T^2	lb, kip	kg	N, kN
P_0	resultant force of at-rest earth pressure per unit length of wall	F, ML/T^2	lb, kip	kg	N, kN

Symbol	Nomenclature	Dimensions	Customary Units		
			English (Force)	Metric (Force)	SI
P_A	resultant force of active earth pressure per unit length of wall	$F, ML/T^2$	lb, kip	kg	N, kN
P_P	resultant force of passive earth pressure per unit length of wall	$F, ML/T^2$	lb, kip	kg	N, kN
p	normal stress or pressure of soil or rock against a structure	$F/L^2, M/LT^2$	lb/ft^2	kg	N, kN
Q	total load	$F, ML/T^2$	lb, kip, ton	kg, ton (metric)	N, kN
Q_L	load/unit of length	$F/L, M/T^2$	lb, kip, ft	kg/m	kN/m
q	flow of fluid	L^3/T	ft^3/sec	cm^3/sec, m^3/sec	m^3/sec
q	foundation pressure on soil/rock	$F/L^2, M/LT^2$	lb/ft^2, kip/ft^2	kg/cm^2	kN/m^2
q_a	allowable pressure	$F/L^2, M/LT^2$	lb/ft^2, kip/ft^2	kg/cm^2	kN/m^2
q_0	ultimate or failure pressure	$F/L^2, M/LT^2$	lb/ft^2, kip/ft^2	kg/cm^2 t/m^2	kN/m^2
q_r	compressive strength	$F/L^2, M/LT^2$	lb/ft^2, kip/ft^2	kg/cm^2 t/m^2	kN/m^2
q_q	surcharge loading	$F/L^2, M/LT^2$	lb/ft^2, kip/ft^2	kg/cm^2 t/m^2	kN/m^2
q_s	safe bearing capacity	$F/L^2, M/LT^2$	lb/ft^2, kip/ft^2	kg/cm^2, t/m^2	kN/m^2
q_u	unconfined compressive strength	$F/L^2, M/LT^2$	lb/ft^2, kip/ft^2	kg/cm^2 t/m^2	kN/m^2
r	radius	L	ft	cm, m	mm, m
S	degree of saturation	dimensionless			
s	displacement of a structural element	L	in., ft	cm, m	mm, m
T	time factor	dimensionless			
T_0	surface tension	$F/L, M/T^2$	lb/ft	g/cm	N/mm, N/m
t	time	T	sec, min, year	sec, min, year	sec, min, year
$U\%$	percentage consolidation	dimensionless			
U	uplift force	$F, ML/T^2$	lb, kip	kg	N, kN
u	neutral stress	$F/L^2, M/LT^2$	lb/ft^2, kip/ft^2	kg/cm^2	kN/m^2
V	volume	L^3	ft^3	cm^3, m^3	mm^3, m^3
W	weight	$F, ML/T^2$	lb, kip, ton	gm, kg	N, kN
z	depth (positive down)	L	ft	cm, m	mm, m

Symbol	Nomenclature	Dimensions	Customary Units		
			English (Force)	Metric (Force)	SI
α	rate of secondary consolidation	dimensionless			
β	angle between a slope (earth dam face, wall backfill, or excavation face) and a horizontal plane	deg, radian	deg	deg	deg, radian**
γ*	shear strain	radian			
γ*	unit weight (see density)	F/L^3, M/L^2T^2	lb/ft^3	g/cm^2, kg/m^3	kN/m^3
γ_w	unit weight of water	F/L^3, M/L^2T^2	lb/ft^3	g/cm^2, kg/m^3	kN/m^3
γ'	unit weight of soil submerged in water	F/L^3, M/L^2T^2	lb/ft^3	g/cm^2, kg/m^3	kN/m^3
γ_d	weight of soil solids in unit volume of soil	F/L^3, M/L^2T^2	lb/ft^3	g/cm^2, kg/m^3	kN/m^3
Δ	change or increment				
δ	angle of wall friction		deg	deg	deg, radian*
ε	linear strain	L/L (dimensionless)			
θ	angle between planes	dimensionless	deg	deg	deg, radian**
ν	Poisson's ratio	dimensionless			
ρ	density (also see γ)	FT^2/L, M/L^3	slug/ft^3	g/cm^2 (mass)	kg/m^3
σ	normal stress	F/L^2, M/LT^2	lb/ft^2, kip/ft^2	kg/cm^2	kN/m^2
σ', $\bar{\sigma}$	effective normal stress	F/L^2, M/LT^2	lb/ft^2, kip/ft^2	kg/cm^2	kN/m^2
σ_{ff}	normal stress on failure surface at time of failure	F/L^2, M/LT^2	lb/ft^2, kip/ft^2	kg/cm^2	kN/m^2
σ_1, σ_2, σ_3	principal stresses	F/L^2, M/LT^2	lb/ft^2, kip/ft^2	kg/cm^2	kN/m^2
σ'_c	preconsolidation stress	F/L^2, M/LT^2	lb/ft^2, kip/ft^2	kg/cm^2	kN/m^2
σ_v, (σ_z)	vertical normal stress	F/L^2, M/LT^2	lb/ft^2 kip/ft^2	kg/cm^2	kN/m^2
σ_h, (σ_x, σ_y)	horizontal normal stress	F/L^2, M/LT^2	lb/ft^2, kip/ft^2	kg/cm^2	kN/m^2
σ_θ	normal stress on plane making angle of θ with major principal plane	F/L^2, M/LT^2	lb/ft^2, kip/ft^2	kg/cm^2	kN/m^2
τ	shear stress	F/L^2, M/LT^2	lb/ft^2, kip/ft^2	kg/cm^2	kN/m^2
τ_{ff}	shear strength (shear stress at time of failure on failure plane)	F/L^2, M/LT^2	lb/ft^2 kip/ft^2	kg/cm^2	kN/m^2

Symbol	Nomenclature	Dimensions	Customary Units		
			English (Force)	Metric (Force)	SI
τ_Θ	shear stress on plane making angle of with major principal plane	F/L^2, M/LT^2	lb/ft^2, kip/ft^2	kg/cm^2	kN/m^2
ϕ	angle of internal friction		deg	deg	deg, radian**
ψ	angle of asperities on rock surfaces		deg	deg	deg, radian**

* Conflicting usage of symbols.

** Although the radian is the correct SI angular unit, geotechnical engineers continue to employ degrees.

Useful Equivalents

Three systems for expressing the units of measurement are in use among engineers: the English, based on the foot of length and pound of force; the metric, based on the meter of length and kilogram of force; and the SI, based on the meter of length and the kilogram of mass. Most of the world has officially adopted the SI system; however, practicing engineers continue to think in terms of pounds or kilograms of force. Moreover, many of the outstanding technical publications to which engineers refer employ the older units. Therefore, it is essential that engineers keep in mind a few useful approximations, that they learn to think in both the older and SI system, and that they make their computations and presentation in whichever of the three forms the situation demands.

Dimension	Equivalent	Useful Approximation
Length, L	1 ft = 305 mm	10 ft = 3 m
	1 in. = 25.4 mm	
	1 mile = 1.61 km	1 km = 0.6 mile
	1 m = 3.28 ft	
Area, L^2	1 ft^2 = 0.0929 m^2	1 ft^2 = 0.1 m^2
	1 $in.^2$ = 645 mm^2 = 6.45×10^{-4} m	
	1 acre = 4047 m^2 = 0.4047 hectare	1 acre = 0.4 ha
	1 mi^2 = 2.59 km^2 = 2.59×10^6 m^2	1 km^2 = 0.4 mi^2
Velocity, L/T	1 ft/sec = 0.305 m/sec	
	1 ft/min = 5.08 mm/sec	2 ft/min = 10 mm/sec
	1 ft/year = 9.67×10^{-6} mm/sec	1×10^{-5} mm/sec
	1 m/year = 3.17×10^{-5} mm/sec	3×10^{-5} mm/sec
Volume, L^3	1 ft^3 = 0.0283 m^3 = 28.3 liter = 28.3×10^3 ml	
	1 yd^3 = 0.765 m^3 1 liter = 1×10^{-3} m^3	1 yd^3 = $\frac{3}{4}$ m^3
	1 gal = 0.0038 m^3 = 3.8 liter	1 m^3 = 263 gal
	1 acre-foot = 1233 m^3	
	1 m^3 = 10^6 ml	

Dimension	Equivalent	Useful Approximation
Flow, L^3/T	1 gpm $= 6.31 \times 10^{-5}$ m^3/sec	
	1 ft^3/sec $= 0.0283$ m^3/sec	
Mass, M	1 lbm = 454 g	
	1 ton* = 907.2 kg	
	1 kg = 2.20 lb	
Force, F, ML/T^2***	1 lb (1 lbf) = 4.45 newtons (N)	
	1 kgf = 9.81 N	1 kgf = 10 N
	1 kip = 4.45 kN	
	1 ton f* = 8900 N = 8.9 kN	1 tf* = 9 kN
	1 ton f** = 9.81 kN	1 tf** = 10 kN
Energy, Torque FL, ML^2/T^2	1 ft-lb = 1.356 N·m	
Stress, F/L^2, M/LT^2	1 lb/ft^2 = 47.9 N/m^2	1 lb/ft^2 = 50 N/m^2
	1 kip/ft^2 = 47.9 kN/m^2	1 kip/ft^2 = 50 kN/m^2
	1 kip/ft^2 = 0.488 kgf/cm^2	1 kip/ft^2 = $\frac{1}{2}$ kgf/cm^2
	1 lb/in.^{2} = 6.89 kN/m^2***	1 lb/ft^2 = 7 kN/m^2
	1 kgf/cm^2 = 98.1 kN/m^2***	1 kgf/cm^2 = 100 kN/m^2
	1 ton*/ft^2 = 95.8 kN/m^2	1 t/ft^2 = 100 kN/m^2
	1 tonf**/m^2 = 9.81 kN/m^2	1 tf/m^2 = 10 kN/m^2
	1 Pascal = 1 N/m^2 (The term "Pascal" for stress is seldom seen in U.S. practice)	
	1 bar = 100 kN/m^2 (The term "bar" for stress is sometimes used in Europe)	1 bar = 1 atmosphere
Density (Unit Mass), M/L^3	1 lbm/ft^3 = 0.016 g/cm^3 = 16 kg/m^3	
	62.4 lbm/ft^3 = 1 g/cm^3 = 1000 kg/m^3 = 1 g/ml = 1×10^6 g/m^3	
Unit weight, F/L^3, M/L^2T^2	1 lb/ft^3 = 0.157 kN/m^3	
	1 lb/in.^{3} = 271 kN/m^3	
	62.4 lb/ft^3 = 1 gf/cm^3 = 9.81 kN/m^3 = 1 gf/ml = 1 gf/cm^3	62.4 lb/ft^3 = 10 kN/m^3
Acceleration, L/T^2	32 ft/sec^2 = 9.81 m/sec^2 (avg)	g = 10 m/sec^2
Angle	57.3 deg = 1 radian	

* U.S. or short ton.

** Metric ton.

*** lbf = lb (force); gf = g (force); kgf = kg (force). In this text English units imply lbf or tf; however, consistent with engineering practice, the abbreviation "lb" will be used. The text also uses SI units that include g (mass) or kg (mass) but angles in degrees. Forces are denoted by different terms: newtons or kilonewtons. In practice, the metric system use of g (mass) or g (force) is ambiguous; therefore, when g (force) or kg (force) is intended, it will be noted "gf" or "kgf" in this text. In the English units lb (mass) is lbm. Although prefixes in the SI system are assigned in the ratio of 10, the ASTM guide suggests using only those in the ratio of 1000: nano, n− = 10^{-9}; micro, μ− = 10^{-6}; milli, m− = 10^{-3}; kilo, k− = 10^3; Mega, M− = 10^6. When derived dimensions are employed, such as stress, the prefix is preferably attached to the numerator only: MN/m^2 is preferred to N/mm^2. An exception is g/ml (grams per milliliter).

CHAPTER 1
The Nature of Soils and Rocks

The site for a convenience grocery was strategically located at the intersection of two main streets although most of the property was a hillside. In order to provide level space for the building and for the parking of automobiles, a wide cut was made at the toe of the slope. Of course, this resulted in a steeper hillside, but because the soil appeared to be very stiff, the builder assumed that the cut would be safe. A few months later the owner of the property noticed that the rear corner of his new building, 6 m (20 ft) from the toe of the cut, was rising. At the same time the driveway between the building and the hill was narrowing. The builder diagnosed the case as sliding of the earth at the toe of the slope and he erected a concrete retaining wall to restrain the movement. Instead of stopping the slide, the wall moved with the hill toward the building. Frantically, the contractor drove steel sheet piling between the building and the hillside to support the concrete wall and the hill. The movement continued at the same rate. Finally, in desperation, he constructed a horizontal reinforced concrete beam against the sheet piling and supported the beam with steel H-piles driven at an angle and bearing on rock. The hillside, the wall, the sheeting, and the beam continued to advance on the building.

An investigation of soil conditions disclosed that the stiff clay in the hillside absorbed water and expanded when the weight on the soil was reduced by cutting at the toe. The expansion took place slowly, and initially the freshly excavated slope was stable. The expanded soil became much weaker than in its original state and eventually was incapable of supporting itself on the new-cut slope. The retaining wall and the sheeting were designed on the basis of the ordinary formulas and were incapable of resisting the unsupported mass in motion on the hillside.

The project was not large and the cost of the excavation and the building was just over $200,000. The cost of the retaining wall, the sheet piling, and the concrete beam was $150,000, or nearly as much as the original project, but they were of no value in correcting the difficulty. At that point the entire project was a financial disaster because correction of the unstable hillside would be more expensive than the value of the building.

The owner decided that the only alternative to bankruptcy was to try to operate the store. He rented a power shovel and a truck, and each week removed the sliding soil from the toe of the slope. He hauled it to the top of the slide area and dumped it on the moving mass. His objective was to fill the depression formed at the top of the slide and to protect a city street and some houses higher on the hill. The refilling only aggravated the movement and failed to support the land above the slide. First the street, then a gas main, and finally a house beyond were destroyed. The store owner ultimately paid for them. The business at his location was so good and his increasing damages so high that he could not afford to stop. Finally, after three years he secured professional advice and installed a drainage system that gradually reduced the water in the hillside. The movement virtually ceased. The cost of the engineered solution to the problem was a fraction of the amount he had spent.

The project described was in Tennessee with construction commencing in the 1950s. The author has received a number of letters from widely scattered areas in the United States inquiring how he learned of that failure in their city.

Such failures are not uncommon, and they illustrate the need for careful, scientific soils engineering on even small projects. Although the soil conditions encountered in the preceding example were unusual, they could have been detected by an investigation costing less than $2000, and a safe design for the hillside and the structure could have been prepared within the economic limits of the project.

Traditionally, soil problems have not received the attention they deserve from engineers and constructors. Too often, designs have been based on handbooks written years ago, experiences with other sites that are not representative, or even on guesses about the soil properties. Only extremely generous safety factors (as high as 20 in some cases) have prevented serious failures. Too often, construction operations involving soil have been based on blind trial and error, but costly failures and dead workmen are a high price to pay for experience.

1:1 Definition of Soil and Rock

The reason for the lack of a rational approach to problems of design and construction has been the misunderstanding of the complex nature and behavior of these materials. Until the twentieth century engineers considered them to be a sort of witch's brew—a mysterious mixture that was incapable of scientific study. Advances in the techniques of studying soils and rocks have led to a better understanding of their nature and to rational methods in design and construction.

DEFINITION OF SOIL / To a farmer soil is the substance that supports plant life, whereas to the geologist it is an ambiguous term meaning the material that supports life plus the unconsolidated rock from which it was derived. To the engineer the term *soil* has a broader meaning.

Earth, or *soil*, in the engineering sense is defined as "any unconsolidated material composed of discrete solid particles with gases or liquids between." The maximum particle size that qualifies as soil is not fixed but is defined by the engineering function involved. For small excavations and trenches where hand excavation is employed, and for the construction of fills in layers, the limiting size is 0.3 m (12 in.) in diameter or about 40 kg (85 lb), the maximum size a man can lift. Where power shovels are used for excavation, the limit is sometimes given as $\frac{1}{2}$ to 1 m^3 ($\frac{1}{2}$ to 1 yd^3) or 2000 to 4000 kg (1 to 2 tons).

Soils include a wide variety of materials, such as the gravel, sand, and clay mixtures deposited by glaciers, the alluvial sands and silts and clays of the flood plains of rivers, the soft marine clays and beach sands of the coast, the badly weathered rocks of the tropics, and even the cinders, bedsprings, tin cans, and ashes of a city dump. Soils can be well-defined mixtures of a few specific minerals or chaotic mixtures of almost anything.

DEFINITION OF ROCK / Rock is defined by the engineer as any indurated material that requires drilling and blasting or similar methods of brute force for excavation. The minimum degree of induration that qualifies as rock has sometimes been defined by a compressive strength of 1500 kN/m^2 (about 200 $lb/in.^2$). In all cases the dividing line between soil and rock is not definite; there is a continuous series of materials from the loosest soil to the hardest rock, and any division into two categories is arbitrary. In preparing engineering documents, such as specifications, the engineer must define the limit so that all who are affected will be in agreement.

The definition of rock from an engineering or functional viewpoint is complicated by structures and defects. A rock that is hard but fractured may be easier to excavate than a softer but more coherent material. Although the fractured rock is easier to excavate, it may require bracing to support it in a deep excavation, whereas the soft rock may stand without support.

Problems in soil and rock engineering, therefore, can seldom be solved by blind reliance on either empirical data gathered on past projects or on the most sophisticated computerized analyses. Each situation is unique; it requires careful investigation and thorough scientific analysis tempered with engineering judgment based on varied experience. Most important, soil and rock engineering requires imagination, intuition, initiative, and courage: *imagination* to visualize the three-dimensional interplay of forces and reactions in the complex materials; *intuition* to sense what cannot be deduced from scientific knowledge or past experience; *initiative* to devise new solutions for old and new problems; and *courage* to carry the work to completion despite skeptics and in the face of the ever-present risk of the unknown. It is this continuing challenge that makes real engineering intriguing and rewarding.

1:2 Development of Soil and Rock Engineering

Construction involving the earth began before the dawn of history. Some of the first construction operations were the digging of holes to bury the dead or to dispose of excrement and the building of earth mounds for worship and burial. Earth, in the form of sun-dried brick or of mud daubed on interlaced sticks or reeds, was used in building houses.

The builders of the ancient civilizations, such as those of India and Babylon, have left numerous examples of their ability to handle soil problems. Some earth dams in India have been storing water for more than 2000 years. The cities of Babylon were placed on fills to raise them above the flood plains, and the buildings were erected on stone mats that spread the loads to the weak soils below. The walls were jointed to allow for settlement without cracking. These builders were skilled artisans who learned from their own bitter experience or from the success and failure of others. During the Middle Ages craftsmen improved the art of construction involving soils by the process of trial and error. They received little help from the early scientists, who felt that problems involving the earth were beneath the dignity of a gentleman.

Since the eighteenth century, however, the need for better structures led scientists and engineers to study soil problems and to try to analyze them like other problems in structural design. Such eminent investigators as Coulomb and Rankine, who are well known in the fields of physics and applied mechanics, turned their attention to the mechanics of soil masses. They started from mathematical expressions of soil strength or from crude experiments on piles of sand, and from them developed expressions for earth pressure on walls and the bearing capacity of foundations. This procedure was logical, and when applied to other problems in mechanics, led to theories that are still in good repute. The theories they developed for soils, however, often proved to be dismal failures. Retaining walls failed, buildings settled, and excavations caved in when designed in accordance with these simple theories.

The tremendous increase in the size of the structures in the early twentieth century and the need for greater economy in their construction forced many brilliant engineers to reevaluate the work of the earlier investigators and to develop new and more realistic methods of analyzing soil masses. The work of Fellenius in Sweden, Kogler in Germany, Hogentogler in the United States, and above all the contributions of Karl Terzaghi in both Europe and the United States brought about the birth of a new phase of civil engineering—*soil mechanics*, and its application to practical problems, *soil engineering*. Since the middle 1930s soil mechanics has become an indispensable tool to the planner and designer and an aid to the builder who must work with the ground.

Rock mechanics as an applied science has had a parallel development. Shafts and tunnels for water supply were excavated through rock during the early days of Greece and Rome. Quarry techniques were evolved from the demand for accurately shaped rock to be used in constructing the architectural masterpieces and the marble sculpture of that era. Knowledge was gained by experience and not by scientific reasoning until the late nineteenth century. It is interesting that many of the early mechanical engineering advances, including pumps and the steam engine, were generated by the needs of the mines; however, the rock itself was given little scientific study. Modern scientific rock mechanics has developed from the needs of both the mining and the construction industries. Although soil mechanics and rock mechanics outwardly appear different, the differences are in emphasis and in terminology. The art of rock mechanics was initially mine-oriented. The scientific analysis has emphasized the behavior of relatively rigid masses laced with cracks. Soil mechanics has been construction-oriented, and has emphasized the behavior of weak, compressible materials. In the overall view there is little difference scientifically and practically. The civil engineer, the builder, and the miner are concerned with the full spectrum of materials, from the softest soil to the hardest rock; any separation between soil and rock mechanics is in the point of view of the engineers and artisans involved.

Mechanics is a science involving masses and their response to forces. Soil and rock mechanics describe the response of soil and rock masses to force systems. Engineering is the use of science, plus experience and ingenuity, to solve problems economically. Thus engineering involving soil and rocks must blend the scientific analyses of these materials with the experience of the centuries and the imagination of the engineer to find practical solutions to real-life problems of planning, design, construction, and operation. The terms *geotechnique* or *geotechnical engineering* describe this multidisciplinary effort.

TYPES OF PROBLEMS IN SOIL AND ROCK ENGINEERING / Two distinct types of problems are involved in geotechnical engineering. The first type deals with soils and rocks as they actually occur in nature. Buildings ordinarily are founded on undisturbed materials; excavations and highway cuts are made through natural deposits; and drainage networks are constructed to remove water from existing soil masses. The second type of problem involves soils or rocks as raw materials for construction. Fills for highways and railroads, earth dams and levees, and airport and highway subgrades use the earth as a source of construction materials. The characteristics of the soil and rock are altered to form new materials similar to the way sand, cement, and stone are made into concrete.

In handling either type of problem, the engineer must keep one thought uppermost—he or she is dealing with highly complex materials and with variable ingredients that at times will appear to defy all laws of nature. With careful study based on scientific analysis and sound judgment, even the most difficult problems can be analyzed.

1:3 Phases of Soil and Rock Composition

THREE-PHASE COMPOSITION / Because soils by definition include all unconsolidated materials, we may expect them to be composed of many different ingredients in all three states or phases of matter—solid, liquid, and gaseous. The same applies to many rocks. Although they are consolidated and indurated, most contain some liquid and gaseous matter.

The interrelations of the masses (or weights) and volumes of the different phases are important because they help define the condition or the physical makeup of the material. The definitions and terms attached to these relations must be clearly understood before the engineer can gain an understanding of the properties of soils and rocks.[1:1]

The volumes and masses or weights of the different phases of matter in a soil can be represented by a block diagram or graph such as Fig. 1.1. The

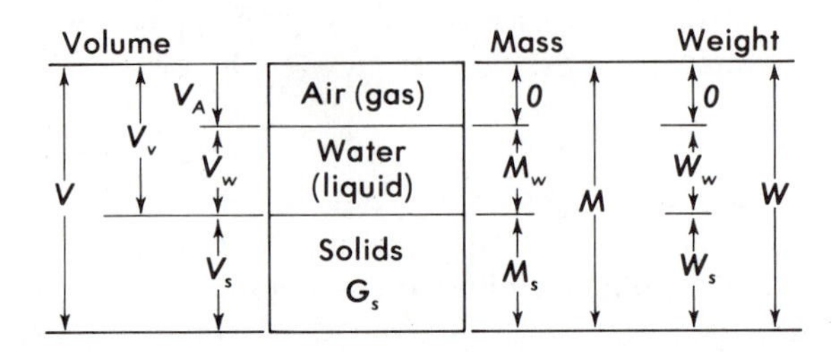

Figure 1.1 *Block diagram showing relationships of weights, masses, and volumes of solids, water, and air in soil or rock.*

total volume, mass, or weight is denoted by the entire block; the solids, by the lower section; the liquids, by the middle section; and the gases, by the upper section. For all practical purposes the gases are considered to be air (although methane is occasionally found in some soils containing decaying organic matter), and the liquid is ordinarily water (although in some instances the water contains small quantities of dissolved salts). The composition of the solids, however, varies considerably and will be discussed in detail in Section 1:4.

VOLUME RELATIONSHIPS / The volume of the solids in an intact piece of soil or rock is denoted as V_s, the volume of the water V_w, and the volume of the air V_a. The total volume, including air, water, and solids is V. The spaces between the solid particles that are occupied by air and water are the *voids*, and their volume is denoted by V_v. The ratio of the volume of voids to the volume of solids is expressed by the *void ratio*, e, as

$$e = \frac{V_v}{V_s} \tag{1:1}$$

The void ratio is always expressed as a decimal. It can exceed 1. A second way of expressing the relation between voids and solids is the *porosity*, n, which is defined by

$$n = \frac{V_v}{V} \times 100\% \tag{1:2}$$

and is always expressed as a percentage. Porosity cannot exceed 100%.

The voids in rock are in two forms: discrete voids like tiny bubbles completely surrounded by solids; and interconnected or open voids, similar to those in soils. The changes that occur in the interconnected voids play a dominant role in physical behavior. The *apparent void ratio* or *apparent porosity* of the connected, open voids is less than the *true* or *total porosity* that includes all voids. The total porosity is difficult to measure and is of less significance from the viewpoint of engineering behavior. In this text total and apparent void ratio are considered to be the same unless otherwise noted.

The *degree of saturation*, S, expresses the relative volume of water in the voids and is always expressed as a percentage,

$$S = \frac{V_w}{V_v} \times 100\% \tag{1:3}$$

A soil is said to be *saturated* if $S = 100\%$.

MASS OR WEIGHT RELATIONSHIPS / In the English and metric systems, in which force is a fundamental unit, the weight of the soil or rock is employed; in the SI system, where mass is a fundamental unit, the mass is employed. The mass of solids in an intact piece of soil or rock is denoted M_s; the weight is W_s. The mass of water is M_w; the weight is W_w. The mass and weight of the total are, respectively, M and W. The ratio of the mass or weight of the solid phase to the water is termed the *water content*, w, and is expressed by

$$w = \frac{M_w}{M_s} \times 100\% = \frac{W_w}{W_s} \times 100\% \tag{1:4}$$

The density or unit mass is the mass per unit of volume, ρ. The corresponding *unit weight* is γ.

$$\rho = \frac{M}{V} \tag{1:5a}$$

$$\gamma = \frac{W}{V} \tag{1:5b}$$

The density of water, ρ_w, is 1000 kg/m^3 or 1 g/ml; the unit weight of water, γ_w, is 62.4 lb/ft^3 or 1 g/ml. In the SI system unit weight is expressed in kN/m^3; the unit weight of water is 9.81 kN/m^3.

The *specific gravity* of a substance is the ratio of its weight to the weight of an equal volume of water. The specific gravity of an intact piece of soil or rock (including air, water, and solids) is termed *mass specific gravity* or *apparent specific gravity*. It is denoted by G_m and may be expressed by the formula

$$G_m = \frac{\rho}{\rho_w} = \frac{M}{V\rho_w} = \frac{\gamma}{\gamma_w} = \frac{W}{V\gamma_w} \tag{1:6}$$

The *specific gravity of the solids*, G_s (excluding air and water), is expressed by

$$G_s = \frac{M_s}{V_s\rho_w} = \frac{W_s}{V_s\gamma_w} \tag{1:7}$$

It is the weighted average of the soil minerals.

The specific gravity of rock is expressed in two ways: including the discrete voids and any gas or liquid that fills them, or the specific gravity of the solid matter alone. It is usually expedient to include the effect of the discrete voids in the specific gravity of the rock. The specific gravity of a rock, therefore, may be somewhat less than that of the weighted average of its minerals. Specific gravity is a dimensionless ratio and therefore has no units. It is the same in all systems of measurement.

COMPUTATIONS INVOLVING RELATIONSHIPS / The relations between volumes and masses are very important in many types of calculations, such as those to determine the stability of a soil body, to estimate building settlements, or to specify the amount of compaction necessary to construct an earthfill. These calculations are the arithmetic of soil mechanics and must be mastered before proceeding further. In making each calculation, a block diagram showing the relation of the different phases should be drawn and the different data entered on it as shown in the following examples.

Example 1:1

Compute the density, unit weight, void ratio, porosity, and degree of saturation of an undisturbed prism of moist soil, Fig. 1.2a. Measurement found a volume of 0.0112 m^3 and a mass of 19.8 kg. After oven-drying at 105°C, the mass was 17 kg. The average specific gravity of solids was 2.69.

1. $V = 0.0112$ m^3, $M = 19.8$ kg, $G_s = 2.69$, $M_s = 17$ kg, Fig. 1.2a.
2. $M_w = 19.8 - 17 = 2.8$ kg.

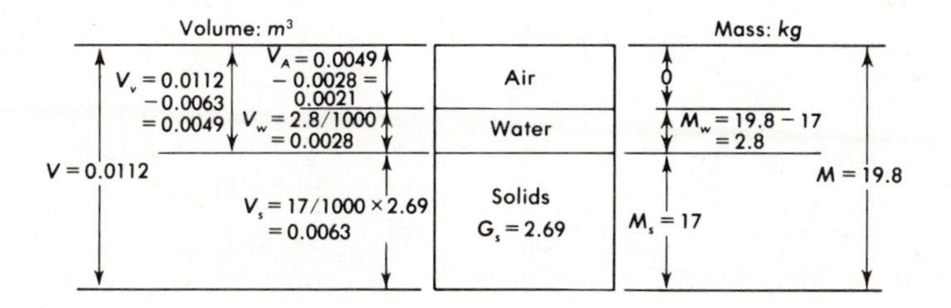

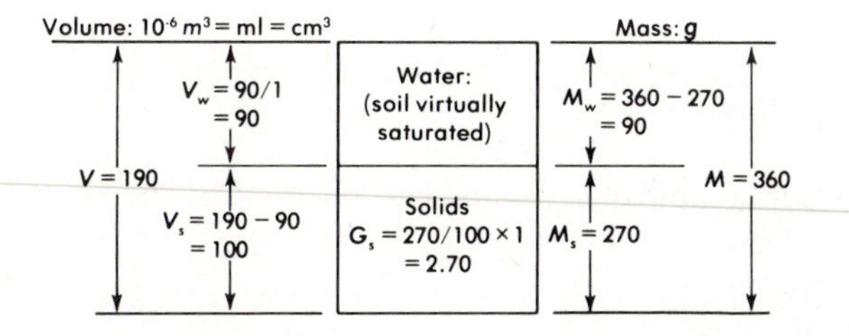

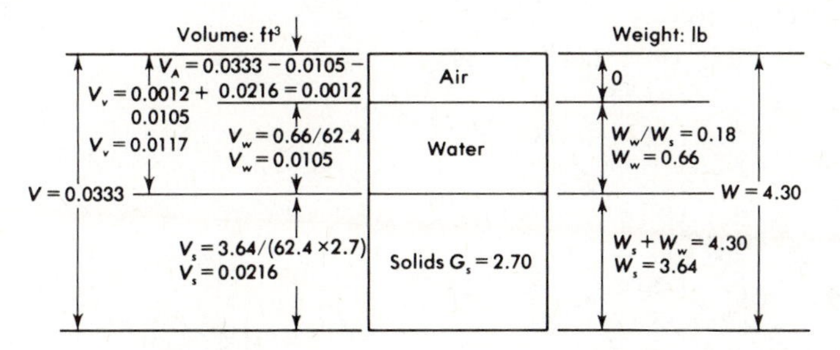

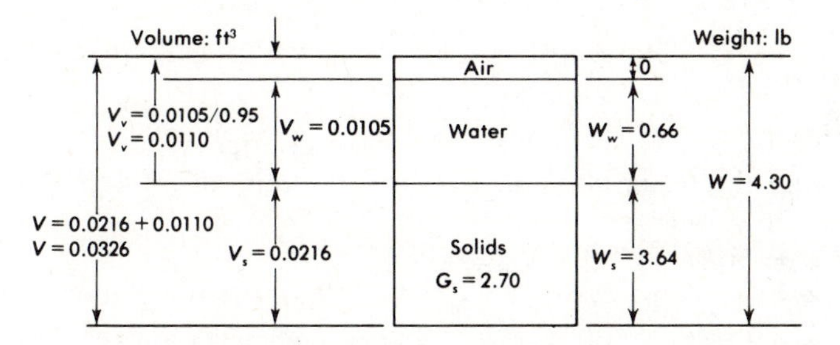

Figure 1.2 *Computing phase relationships with the aid of block diagrams*: (*a*) *Example 1.1*, (*b*) *Examples 1.2 and 1.3 with* $G_s = 2.80$, (*c*) *Example 1.4 after initial compaction, and* (*d*) *Example 1.4 after added compaction.*

3. $\rho = 19.8/0.0122 = 1768\ \text{kg/m}^3$.
4. $\gamma = 1768 \times 9.81 = 17{,}340\ \text{N/m}^3 = 17.3\ \text{kN/m}^3 = 1.768\ \text{g/cm}^3$ $= 110.3\ \text{lb/ft}^3$.
5. $V_w = 2.8 \div 1000 = 0.0028\ \text{m}^3$.
6. $V_s = 17/(1000 \times 2.69) = 0.0063\ \text{m}^3$.
7. $V_v = 0.0112 - 0.0063 = 0.0049\ \text{m}^3$.

8. $V_A = 0.0049 - 0.0028 = 0.0021\ \text{m}^3$.
9. $e = 0.0049/0.0063 = 0.78$.
10. $n = (0.0049/0.0122) \times 100\% = 4.4\%$.
11. $S = (0.0028/0.0049) \times 100\% = 57\%$.
12. $w = (2.8/17) \times 100\% = 16.5\%$.

NOTE: Void ratio is usually rounded out to nearest 0.01 or sometimes 0.001; porosity to nearest percent or 0.1%; saturation to nearest percent or 0.1%; and water content to nearest percent or 0.1%. The accuracy of the measurements seldom justifies more than three significant figures.

Example 1:2

An undisturbed sample of saturated clay from below the groundwater table had a mass of 360 g and a volume of 190 cm^3, Fig. 1.2b. Dried, its mass was 270 g. Find the density, unit weight, water content, void ratio, porosity, and specific gravity of solids.

1. $V = 190\ \text{cm}^3 = 190 \times 10^{-6}\ \text{m}^3$, $M = 360$ g, $M_s = 270$ g.
2. $\rho = 360/190 \times 10^{-6} = 1{,}895{,}000\ \text{g/m}^3 = 1895\ \text{kg/m}^3$.
3. $M_w = 360 - 270 = 90$ g.
4. $w = 90/270 \times 100\% = 33\%$.
5. $V_w = 90/10^6 = 90 \times 10^{-6}\ \text{m}^3 = V_v$ (saturated).
6. $V_s = 190 \times 10^{-6} - 90 \times 10^{-6} = 100 \times 10^{-6}\ \text{m}^3$.
7. $e = 90/100 = 0.90$.
8. $n = (90/190) \times 100\% = 47\%$.
9. $G_s = (M_s/V_s\rho_w) = 270/(100 \times 10^{-6} \times 1 \times 10^6) = 2.70$.
10. $\gamma = 18.59\ \text{kN/m}^3 = 1.895\ \text{gf/ml} = 118\ \text{lb/ft}^3$.

Example 1:3

If the specific gravity of solids in Example 1:2, Fig. 1.2b, were 0.1 higher, what would the water content, density, and unit weight be, assuming no difference in volumes;

1. $V_s = 100 \times 10^{-6}\ \text{m}^3$, $V_w = 90 \times 10^{-6}\ \text{m}^3$, $G_s = 2.8$.
2. $M_s = 100 \times 10^{-6} \times 2.8 \times 1 \times 10^{-6} = 280\ \text{g} = 0.28\ \text{kg}$.
3. $w = 90/280 \times 100\% = 32\%$ (increased 3%).
4. $\rho = (280 + 90)/190 \times 10^{-6} = 1{,}947{,}000\ \text{g/m}^3 = 1947\ \text{kg/m}^3$ (increased 2%).
5. $\gamma = 19.03\ \text{kN/m}^3$ (increased 2%).

NOTE: Small changes in G_s do not influence the other properties. For this reason, estimates of G_s are adequate for many computations.

Example 1:4

A clayey soil is being compacted in a fill. Its water content is 18% and the specific gravity of solids is 2.70. An undisturbed sample of the soil, 1/30 ft^3, weighed 4.30 lb, Fig. 1.2c. Compute the void ratio, porosity, degree of saturation, and weight of solids in a unit volume of soil.*

1. $V = 0.0333\ \text{ft}^3$; $W = 4.30$ lb; $w = 18\%$, $G_s = 2.70$.
2. $W_w/W_s = 0.18$; $W_w + W_s = 4.30$; $W_s + 0.18W_s = 4.30$; $W_s = 3.64$ lb; $W_w = 0.66$ lb.
3. $V_w = 0.65/62.4 = 0.0105$; $V_s = 3.64/(62.4 \times 2.7) = 0.0216\ \text{ft}^3$.
4. $V_A = 0.0333 - 0.0216 - 0.0105 = 0.0012\ \text{ft}^3$.
5. $S = 0.0105/(0.0012 + 0.0105) \times 100\% = 90\%$.
6. $e = (0.0012 + 0.0105)/0.0216 = 0.54$.
7. $\gamma = 4.3/0.0333 = 129\ \text{lb/ft}^3 = 2.07\ \text{g/cm}^3 = 20.3\ \text{kN/m}^3$ $\rho = 2070\ \text{kg/m}^3$.
8. $\gamma_d^* = 3.64/0.0333 = 109.3\ \text{lb/ft}^3 = 1.75\ \text{g/cm}^3 = 17.2\ \text{kN/m}^3$ $\rho_d^* = 1750\ \text{kg/m}^3$.

If further compaction can increase the degree of saturation to 95%, compute the increased density and reduced void ratio, assuming that the water content does not change, Fig. 1.2d.

1. $W = 4.30$ lb; $W_s = 3.64$ lb; $W_w = 0.66$ lb; $S = 95\%$; $V_W = 0.0105\ \text{ft}^3$; $V_s = 0.0216\ \text{ft}^3$.
2. $S = V_w/V_v$; $0.95 = 0.0105/V_v$; $V_v = 0.0110\ \text{ft}^3$.
3. $V = 0.0110 + 0.0216 = 0.0326\ \text{ft}^3$.
4. $\gamma = 4.30/0.0326 = 132\ \text{lb/ft}^3 = 2.11\ \text{gf/cm}^3 = 20.7\ \text{kN/m}^3$ $\rho = 2110\ \text{kg/m}^3$.
5. $\gamma_d = 3.64/0.0326 = 112\ \text{lb/ft}^3 = 1.79\ \text{gf/cm}^3 = 17.6\ \text{kN/m}^3$ $\rho_d = 1790\ \text{kg/m}^3$.
6. $e = 0.0110/0.0216 = 0.51$.

1:4 Solids: Rock Minerals and Weathering

The solid phase plays a major part in determining the engineering behavior of soil and the dominant part in rock. According to the engineering definition, almost anything on the earth's crust is included in the definition of soil and rock. The most important solids fall into three classes: (1) minerals, (2) the products of organic synthesis and decay, and (3) man-made materials. By far the most important constituents of soil and rock are minerals: naturally occurring chemical compounds of definite composition and crystal structure.

PREDOMINANT MINERALS / Although hundreds of different minerals are listed in mineralogy manuals, a relatively few make up the greater part of rocks and soils. These minerals and some of their pertinent properties related to civil engineering are listed in Table 1:1.

ROCK WEATHERING / Rock weathering is the breakdown of intact masses of rock into smaller pieces by mechanical, chemical, or

* The density and weight of solids per unit of total volume are ρ_d and γ_d; the terms *dry density* and *dry unit weight* are used frequently.

TABLE 1:1 / MAJOR MINERALS IN ROCK AND SOILS

Mineral Group	Variety	Hardness*	Color	Cleavage	Specific Gravity
Silica	Quartz	7	Colorless-white	None	2.66
	Chert	7	Light to black	None	2.66
Feldspar	Orthoclase, microcline	6	White-pink	Right-angle	2.56
	Plagioclase	6	White-gray	Right-angle (striated surface)	2.6–2.75
Mica	Muscovite	2–2.5	Silvery	Thin platy	2.75–3.0
	Biotite	2.5–3	Dark	Thin platy	
Ferromagnesian	Pyroxene: augite	5–6	Black	Right-angle	3.1–3.6
	Amphibole: hornblende	5–6	Black	Oblique-angle	2.9–3.8
	Olivene	6–5.7	Greenish		3.3
Iron oxides	Limonite, magnetite	5, 6	Red, yellow, black		5.4
Carbonate	Calcite†	3	White, gray	3 parallelogram faces	2.7
	Dolomite‡	4	White, gray	3 parallelogram faces	2.8
Clay minerals	Kaolinite, illite, montmorillonite	2 1	White, gray	Platy, earthy texture	2.2–2.6
Cellulose (not a mineral)				Fibrous	1.5–2

* Hardness: fingernail = 2; copper coin = 3; pocket knife = 5; glass = 5.5.
† Will effervesce with cold hydrochloric acid. (1 part commercial acid to 10 parts water.)
‡ Weak effervesce with cold acid. (1 part commercial acid to 10 parts water.)

solution processes.[1:2] *Mechanical weathering*, or *disintegration*, is a combination of grinding, shattering, and breaking that reduces the rock to smaller and smaller fragments that have the same mineral composition as the original rock. It is caused by the freezing of water in cracks and pores, the impact of water, the abrasion of gravel and boulders carried by mountain streams and rivers, the pounding of water waves on beaches or cliffs, the sand blast of sand-laden desert winds, the expansion and contraction of rock by violent temperature changes, and by the plowing action of glaciers. *Chemical weathering*, or *decomposition*, is a chemical alteration of the rock minerals to form new minerals that usually have chemical and physical properties completely different from their parent materials. It is caused by the reaction of the minerals with water, dissolved carbon dioxide and oxygen from the air, organic acids from plant decay, and dissolved salts present in the water. *Solution* is the dissolving of soluble minerals from the rock, leaving the insoluble minerals behind as a residue. All three processes occur simultaneously but at different rates, depending on the climate, topography, and composition of the original rock. In general, decomposition predominates in warm, humid regions and in areas with flat topography, and disintegration predominates in dry regions and areas with rugged topography. Solution obviously is predominant in humid regions underlain by soluble rocks.

MINERALS AND WEATHERING / *Silica* (silicon dioxide) is one of the most important constituents of many rocks and most soils. It occurs in two forms: crystalline (quartz) and cryptocrystalline (chert, flint, and chalcedony). It is inert in chemical weathering and insoluble in water, although it is slightly soluble in a basic environment. In the crystalline form and in most cases of the cryptocrystalline it is hard and tough with no cleavage, and resists mechanical weathering better than the other important rock minerals. It eventually breaks into tough, angular, irregular fragments that resist abrasion.

Feldspars constitute a second important group of rock-forming minerals that consist of potassium, sodium, calcium, or similar aluminum silicate. They are brittle, with pronounced planes of cleavage, and they break easily to form small prismlike particles. The feldspars are very susceptible to chemical breakdown, and the mechanical disintegration accelerates the chemical processes to such an extent that feldspar fragments are rarely found in soils in humid regions. The decomposition products of feldspars are exceedingly variable, depending on the type of feldspar and on the weathering conditions, but they can be described in three groups: complex hydrous aluminum silicates, soluble or semisoluble carbonates of sodium and similar metals, and silica (usually in a colloidal suspension). The hydrous aluminum silicates comprise a family called the *clay minerals* that are physically very different from the feldspars from which they came.

Micas comprise a second family of silicate minerals that often contain iron and magnesium in addition to their potassium. The mica flakes are soft and resilient, with a pronounced cleavage. They split easily and break to form still smaller, thinner flakes. Their chemical breakdown is similar to that of the feldspars, producing the clay minerals, carbonates, and silica, but in addition various oxides of iron are formed from those containing iron. The chemical weathering of the micas is not so rapid as that of feldspar; thus micas are often present in soils in humid regions.

The *ferromagnesian* family of minerals (including horneblende, olivine, and augite) are complex aluminum silicates that contain both iron and magnesium. They are moderately hard and tough, some with no pronounced cleavage, and break mechanically into irregular dark-colored fragments. They alter chemically to form iron oxides, clay minerals, and the other products of silicate decomposition.

Iron oxides and hydroxides occur in various crystalline and noncrystalline forms and in both the ferrous and ferric state. They may be present in the original rock, but are also produced by the weathering of iron-bearing minerals such as biotite micas or the ferromagnesian group. Iron is responsible for much of the coloration, from the greenish hues of ferrous iron in deeply submerged formations, through the yellows of surface soils in temperate zones to the bright reds and purples of the highly oxidized ferric materials of the tropics.

The carbonate minerals, calcite and dolomite, break down mechanically into both irregular and prismlike fragments, depending on the degree of crystallization of the rock. Carbonate fragments, particularly the smaller sizes, are found most frequently in arid and glaciated regions. In humid regions, chemical weathering takes the form of solution. Weak acids from organic decay and plant roots, but principally from carbon dioxide dissolved in water, produce the following reactions:

$$H_2O + CO_2 = H_2CO_3$$

$$2H_2CO_3 + CaCO_3 = Ca(HCO_3)_2 + H_2O$$

The soluble bicarbonate is leached away with the ground water leaving behind any insoluble portions of the original rock, such as chert, quartz, clay minerals, and iron oxides.

REPRECIPITATION / The soluble products of weathering, carbonates and bicarbonates, as well as silica, iron oxides, and iron hydroxides in the colloidal state, are carried away from their point of origin by percolating moisture. In a new remote environment, further chemical and physical changes can take place and cause precipitation of the materials, either in the colloidal or crystalline form. Frequently, this occurs in soil voids or rock fissures, filling them and sometimes cementing the mass into a new material.

The weathering process is dynamic: Rock minerals are broken physically, changed chemically, and dissolved in water. The end products, soils, consist of a relatively small group: predominantly quartz and clay minerals, with varying amounts of mica, ferromagnesian minerals, iron oxides, and carbonates. The alteration continues, with changes in environment that occur naturally and that are induced by drainage, excavation, flooding, filling, and the weight of structures.

1:5 Clay Minerals[1:3, 1:4]

The decomposition of the feldspars, micas, and ferromagnesian minerals, all of which are complex aluminum silicates, occurs in many ways. The major factors are moisture, temperature, oxidation or reducing conditions, the ions present in solution (including those released by weathering), pressure, and time. The reactions are varied, but an oversimplified form of decomposition of feldspar illustrates how they might occur.

$$2KAlSi_3O_8 + 2CO_2 + 3H_2O = 4SiO_2 + 2KHCO_3 + Al_2O_3 \cdot 2SiO_2 \cdot 2H_2O$$

The first product, silica, is in the form of a colloidal gel or suspension; the second product, potassium bicarbonate, is in solution; and the third product is a hydrous aluminum silicate—a simplified clay mineral.

There are many forms of clay minerals, with some similarities and wide differences in composition, structure, and behavior. All are extremely fine grained, with large surface areas per unit of mass. Probably all have definite crystal structures that include large numbers of atoms arranged in complex three-dimensional patterns. All are electrically active.

SHEET STRUCTURE / Most clay crystals consist of atomic sheets, principally of two types: silica and alumina. The silica sheet is a repeating two-dimensional linkage of silicons with valences of 4 and oxygens with valences of 2. Each silicon is surrounded by four oxygens, each of which contributes one valence link to the central silicon. Some of the remaining oxygen valence bonds are linked with adjoining silicons, as shown in Fig. 1.3a, but the oxygens on one side of the sheet are unsatisfied. The geometric shape is that of tetrahedrons with oxygens at the points, a silicon in each center, and the tetrahedrons arranged in a plane to form repeating hexagons (Fig. 1.3b). The height of the tetrahedron, and therefore the sheet thickness, is 0.22 nm or 2.2×10^{-10} m.

The alumina sheet (Fig. 1.3c) is more complex. It consists of units of one aluminum surrounded by oxygens and hydroxyls (OH) that form octahedrons. Adjacent aluminums share oxygens and also OH groups, by alternating one to another. The thickness of this unit is likewise 0.22 nm. The sheet formed by linking unit octahedra does not balance the unit valences as does that of silica, so 1 of 3 unit octahedra will contain no

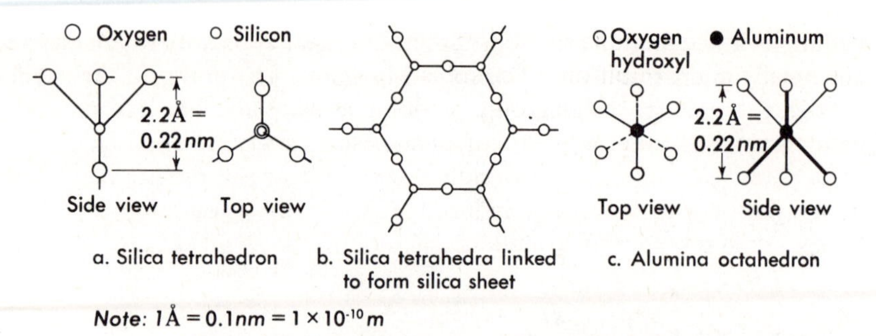

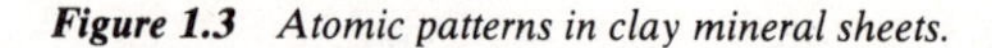

Figure 1.3 *Atomic patterns in clay mineral sheets.*

aluminum. This makes the sheet nonsymmetrical, and nonuniform, but helps to balance valence bonds.

The complexity of the sheet is enhanced by *isomorphous substitution*, the replacement of one or more aluminums by magnesium, which occupy the vacant spaces in the octahedra. The substitution of magnesium with a valence of 2 for aluminum with a valence of 3 locally aggravates imbalance. Similarly, iron and other atoms can substitute for aluminum provided they physically fit the space; even aluminum can substitute for silicon in the tetrahedral sheet.

Most clay minerals consist of silica and alumina sheets stacked together (Fig. 1.4) to form plates. The unsatisfied oxygens of the silica sheet are

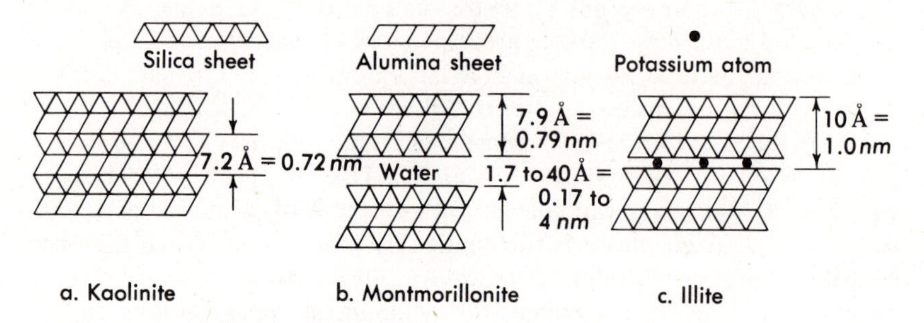

Figure 1.4 *Sheet structure of major clay mineral families.*

shared with the alumina sheet, to form a more or less balanced whole. Any remaining unbalance may be satisfied by cations supplied by salts in the surrounding water. In some instances cations are shared by adjacent plates. Similarly, hydrogen atoms may switch back and forth between plates. The shared attraction, known as the *hydrogen bond*, can provide linkage to hold the plates together in stacks.

KAOLINITE / Kaolinites consist of alternating alumina and silica sheets that form a clay mineral whose nominal unit plate thickness is 0.72 nm. There are several members of the kaolinite family, depending on variations in the alumina sheet. In general they are relatively well balanced electrically and exhibit only limited isomorphous substitution. The sheets are tightly bonded in the plates; moreover, they stack like leaves in a book to form stacks as thick as 0.01 mm. Such a stack is shown in Fig. 1.5.

Figure 1.5 *Electron photomicrograph of kaolinite showing overlapping plates forming a stack, magnified 7400 diameter. (Courtesy Electron Microscope Laboratory, Engineering Experiment Station, Georgia Institute of Technology.)*

HALLOYSITE / Halloysite is a peculiar member of the kaolinite family that contains a sheet of water between adjacent clay units. The clay unit thickness is 1.0 nm rather than 0.72 nm. Halloysite can be dehydrated by drying, and reduced in thickness; it may not revert to the hydrated form upon rewetting. Its engineering behavior is completely changed by dehydration. Serious technical problems in embankment construction have been traced to halloysite that was tested in the laboratory after drying but used moist as a fill material.

MONTMORILLONITE / Montmorillonites or smectites are clay minerals consisting of one alumina sheet sandwiched between two silica sheets. The unit plate is 0.79 nm thick, but can be as wide as 1 μm. The units do not stack together readily, and when stacked, break apart easily. Sheets of water or organic molecules occur between the unit plates, varying their spacing from 0.17 to 4 nm or more. The montmorillonites are characterized by extensive isomorphous substitution, with each substitution theoretically producing a different mineral. The variety of cations that compensate for each substitution further multiplies the variety of montmorillonites. Many different minerals, such as nontronite, sauconite, and saponite are included in this group, but from the engineering point of view differentiation has little value. Generally, montmorillonites form in regions rich in ferromagnesian rocks (the source of magnesium) such as volcanics, and particularly in areas of high temperature and intense rainfall.

ILLITE / The illites, like montmorillonites, consist of an alumina sheet between silicas but 1 nm thick. However, adjacent illite units are linked by their sharing of potassium ions and thus stack together rather tightly. There is only limited isomorphous substitution in the alumina sheet, but there is some substitution of aluminum for silicon in the silica sheet. The illites are often present in shales and other deposits that have been subject to a changing environment; they appear to be the alteration products of potassium feldspars or other clay minerals.

Other clay mineral groups include *chlorite* and *vermiculite* that form sheets like mica. Vermiculite is similar to montmorillonite and is sometimes considered to be a part of that family. However, like halloysite, it contains sheets of water between the vermiculite unit sheets. Both minerals exhibit isomorphous substitution.

The *attapulgite* or polygorskite minerals are different in that silica tetrahedrons form a double-layer chain with magnesium and aluminum atoms providing the link between the silica sheets. The crystal is in the form of a long fiber or curled ribbon that can contain water molecules. *Sepiolite* is similar. It appears to form from montmorillonite in a saline environment.

IDENTIFICATION / Because of the extremely small size, identification of the clay minerals is difficult. Most soils contain several, so fine grained and so similar in their size and weight that separation is virtually impossible. One method of identification of mixed clay minerals is *differential thermal analysis*. The clay is heated slowly. At different temperatures, depending on the clay mineral, water is released. The heat required to release the water produces a characteristic change in the rate of heating. The clay is identified empirically and semiquantitatively by comparing the rate of heating with standard curves for the pure minerals. The *X-ray diffraction* method subjects the clay to an X-ray beam at varying angles. The reflection produced by the atoms in the structure create a characteristic pattern from which the atomic arrangement can be deduced. This is probably the most reliable method for identification and can give both the clay mineral and the approximate amount present.

The electron microscope makes a shadow photograph of the mineral. The well-stacked kaolin particles can be easily identified as in Fig. 1.5, but the montmorillonites, whose unit plates separate easily, sometimes produce no identifiable pattern.

Identification of the clay minerals and appreciation of their structure are helpful in gaining an understanding of the behavior of the soils that contain them. The character of the clay greatly influences a soil's drainage, strength, compressibility, and particularly its reaction to changes in moisture. However, these properties cannot be predicted quantitatively from mineral analyses; they must be found from empirical tests (Chapters 3, 4, and 5).

1:6 Organic, Precipitation, Volcanic, and Man-Made Solids

Other forms of solids, although scattered in occurrence and often only a small percentage of the total weight, can have an influence on the ultimate behavior of soil and rock mass that is far greater than their relative proportion suggests.

Organic materials are present in many surface soils, particularly when the environment is not conducive to rapid decomposition. Fibrous mats of roots and partially decayed vegetation accumulate in swampy regions where the water is stagnant, or where the materials are buried in soils that limit the circulation of water, oxygen, and nutrients. The partial decay produces hydrogen sulfide gas that is a serious hazard to life in excavations and which induces corrosion of construction materials. The soluble organic products leach away as brown swamp water.

As decay continues, the individual pieces lose their identity and a nearly structureless, fibrous peat is produced. This is largely cellulose but is often mixed with mineral matter that is deposited simultaneously. The peat continues to decompose if it has some access to circulating ground water and nutrients producing methane gas. This presents an explosion hazard in excavations and tunnels. Soils containing large amounts of root matter or peat are so compressible that they are avoided or discarded if possible. However, it is essential to understand them and know how to use them because there may be no alternatives.

Organic decay produces humic acids that reduce ferric iron to ferrous and aid in the decomposition of rock minerals and the development of kaolin-type clays. Concentrations of organic cations also play a part in clay mineral structure; organic clays behave differently from those that are inorganic. These humic acids often percolate deeply into soil and rock, causing local alteration at great depth and peculiar discontinuities in engineering properties.

Organic growth produces soil and rock. The shells, shell fragments, and coral formations of the sea are well-known examples of biologic calcareous deposits. The beach sands of some islands and even of mainlands, such as Florida, consist largely of shell coral fragments and finer biologic calcareous particles. Diatoms, minute sea- and freshwater organisms, have skeletons of silica, like tiny snowflakes; accumulations of these are occasionally found in sedimentary soils. Iron-fixing organisms similarly are believed to be responsible for some accumulations of iron ore as well as iron cementation of other sediments.

Precipitation of carbonates and silica in the sea form great accumulations of limesands and nodular silica. The Bahama Island chain is still growing by this process, in addition to coral reef accumulations. Replacement of calcite by less soluble silica entirely alters the character of the lime deposit, forming nodules of chert in the softer calcite mass.

Volcanic action generates soil directly when the lava explodes into *volcanic ash* or *tephra*. This is a porous, cinderlike material. It rapidly decomposes into soil, cements itself into a volcanic sandstone or *tuff*, or remains in its initial form, depending on the environment.

MAN-MADE MATERIALS / Man-made materials are becoming increasingly important soil and rock solids. The major source is waste of àll kinds, ranging from uniform products of an industrial process to the heterogeneous accumulations of garbage, building debris, and metallic spoil from domestic and varied industrial sources. Each presents its special problems. Some industrial by-products, such as slag, are excellent soil-like raw materials for embankment construction. Some even harden into rocklike masses that obstruct future work. Others, such as old mine tailings, are hazards, as was the old pile of coal mine waste that demolished a Welsh school in the 1960s. Some can be utilized as foundations, but some wastes, like most sanitary landfills, must be kept covered but vented for years to minimize health and explosion hazards. It is particularly important to realize that most manmade materials are likely to be far more active chemically than natural minerals and react more quickly and more dramatically to changes in environment.

1:7 Grain Size[1:6]

The range in the sizes of soil particles or grains is almost limitless; the largest grains are by definition those that can be moved by hand, while the finest grains are so small they cannot be identified by an ordinary microscope. The particles produced by mechanical weathering are rarely smaller than 0.001 mm in diameter and are usually much larger, for nature's grinding processes are not very efficient, and the small grains often escape further punishment by slipping through the voids between the larger grains. The products of chemical weathering, including iron oxides and the clay minerals, are tiny crystals that occasionally are larger than 0.005 mm in diameter but are usually very much finer.

GRAIN SIZE TESTS / Two methods are commonly used to determine the grain sizes present in a soil. Calibrated sieves or screens having openings as large as 100 mm (4 in.) and as small as 0.075 mm (U.S. Standard No. 200) are used for separating the coarser grains, Table 1:2. Sieves with

TABLE 1:2 / STANDARD SIEVE SIZES (ASTM)[1:6]

Size (in.)		Opening (mm)	Size No.		Opening (mm)	Size No.		Opening (mm)
4		100				40		0.425
3	(1) (2)	75	4	(1) (2)	4.75	50	(2)	0.300
2	(1)	50	8	(2)	2.36	60	(1)	0.250
$1\frac{1}{2}$	(1) (2)	37.5	10	(1)	2.00	70		0.212
1	(1)	25	16	(2)	1.18	100	(2)	0.150
$\frac{3}{4}$	(1) (2)	19	20	(1)	0.850	140	(1)	0.106
$\frac{1}{2}$		12.5	30	(2)	0.600	200	(1) (2)	0.075
$\frac{3}{8}$	(1) (2)	9.5						

(1) A standard series for ASTM D422 "Grain size test."
(2) Alternate series for uniformly spaced points on log diameter graph.

smaller openings are available but are impractical for soil work. The portions finer than 0.1 mm can be measured by sedimentation. This is based on the principle that the smaller the size of a particle, the more slowly it settles through water. This method is unsatisfactory for grains smaller than 0.0005 mm because such particles are kept in suspension indefinitely through molecular agitation. For particles that have a near-spherical shape, both the sieve and the sedimentation tests give the same results in the size range in which they overlap. For flat particles, however, the sieve measures the intermediate dimension, or width, as 1.4 times the sieve opening, while the sedimentation indicates the diameter of a sphere that settles at the same rate through water as the soil particle. This equivalent diameter is approximately the grain thickness. Particles smaller than 0.0005 mm can be measured by an electron microscope, but the data are of little use in soil engineering. Particles larger than 100 mm are measured by calipers or special sieves.

GRAIN SIZE SCALES / Because of the extreme range in grain sizes, scientists and engineers have attempted to break the entire scale into smaller divisions. Many methods have been proposed, but all are arbitrary and no one method is better than another. A convenient scale, adopted by the ASTM, is shown in Fig. 1.6. The coarsest division is *gravel*, which includes all soil grains larger than a No. 4 sieve. *Sand* includes all particles smaller than the No. 4 sieve and coarser than a No. 200. The grains smaller than the No. 200 sieve are the *fines*. This fraction is sometimes subdivided

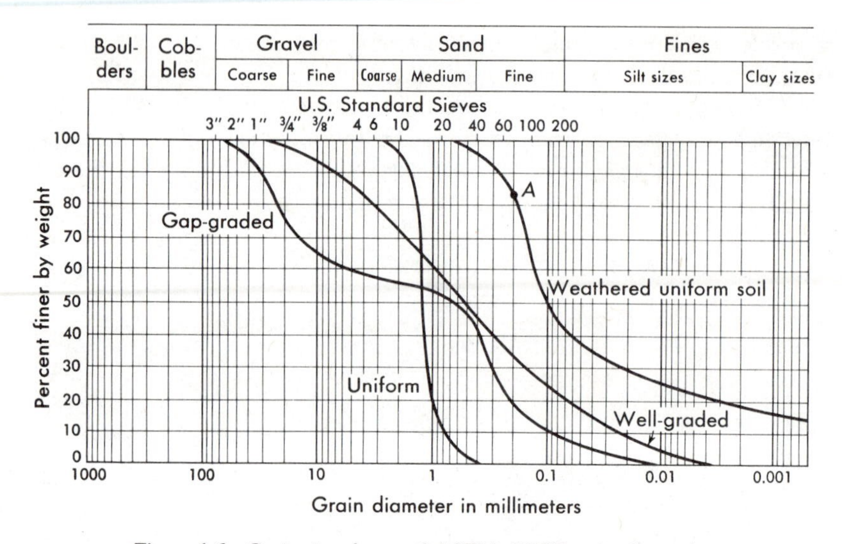

Figure 1.6 *Grain size chart and ASTM–ASCE grain size scale.*

into *silt sizes*, the particles coarser than 0.002 mm, and *clay sizes*, the particles finer than 0.002 mm. Unfortunately, any attempt to designate clay by particle size is misleading, for some soils finer than 0.002 mm contain no clay minerals and some clay mineral grains are larger than 0.002 mm.

GRAIN SIZE CHART / A better method of representing the different grain sizes in a soil is the logarithmic chart shown in Fig. 1.6. The entire range in grain sizes, plotted on a logarithmic scale, forms the horizontal divisions, while the percentages by weight of the soil grains that are finer than a given size form the vertical divisions. For example, point *A* on Fig. 1.6 means that 83% by weight of that soil is finer than 0.2 mm. A curve drawn through all the points that represent a single soil is known as the *grain size curve* of that soil. The interrelation of the different grain sizes in a soil can be seen from the shape of grain size curve. A steep curve indicates that the grains are nearly the same size, and such a soil is termed *uniform*. A flat curve shows a wide range in grain sizes; such a soil is termed *well graded*. Humps in the curve indicate that a soil is composed of a mixture of two or more uniform soils. Such a soil is *gap-graded*. A steep curve in the sand range that gradually becomes a long, flat curve in the fines range may indicate a soil that was formed by mechanical weathering and later altered by chemical weathering.

EFFECTIVE SIZE AND UNIFORMITY / The *effective size* of the grains is defined as the size corresponding to 10% on the grain size curve. It is given the symbol D_{10}.

Other sizes statistically defined that are useful include the *median*, D_{50}, the *finest quartile*, D_{25}, and D_{15}. The effective size, as well as D_{15}, has been found to be a major factor in the effective pore diameters and is related empirically to drainage and seepage (Chapter 3).

The uniformity of the soil can be statistically defined in a number of ways. An old but useful index is the *uniformity coefficient* C_u. It is defined by the relation

$$C_u = \frac{D_{60}}{D_{10}} \tag{1:8}$$

Soils with C_u less than 4 are said to be uniform; soils with C_u greater than 6 are well graded, provided that the grain size curve is smooth and reasonably symmetrical. Another measure of uniformity, frequently encountered in geologic work, is the sorting coefficient S_0. It is defined by the relation

$$S_0 = \sqrt{\frac{D_{75}}{D_{25}}} \tag{1:9}$$

1:8 Particle Shape[1:7]

The shape of the particles is fully as important as the size in determining the engineering behavior of soil and clastic rock. However, because shape is more difficult to measure and describe quantitatively, it is often neglected. Three classes of grain shapes have been defined: bulky grains, flaky or scalelike grains, and needlelike grains. The first two are far more important, but all three are significant in their difference in physical behavior.

When the length, width, and thickness of the particles are of the same order of magnitude, the shape is *bulky*. Bulky grains are formed by the mechanical disintegration of rocks and minerals, by precipitation, and by volcanic action. They are rarely finer than 0.001 mm in diameter, and most can be readily examined with a good magnifying glass or microscope (Fig. 1.7). A binocular-stereoscopic microscope is better, because of the three-dimensional representation.

SPHERICITY / Two aspects of the bulky shape are significant: the *sphericity* and the *angularity* or *roundness*. The sphericity describes the differences between length L, width B, and thickness H. The equivalent diameter of the particle D_e is the diameter of the sphere of equal volume.

$$D_e = \sqrt{\frac{6V}{\pi}} \tag{1:10a}$$

The *sphericity*, X, more correctly *operational sphericity*, is defined as follows:

$$X = \frac{D_e}{L} \tag{1:10b}$$

Figure 1.7 *Photomicrographs of sand particles*: (Left) *Subangular to subrounded beach sand, 0.3 to 0.8 mm in diameter*; *and* (right) *subrounded to rounded dune sand 0.2 to 0.5 mm in diameter.*

A sphere has sphericity of 1, while flat or elongated particles have lower values. A second index is the *flatness* defined by:

$$F = \frac{B}{H} \tag{1:10c}$$

The elongation, E, is defined by:

$$E = \frac{L}{B} \tag{1:10d}$$

The ease of handling soil or broken rock, the ability to remain stable when subject to shock, and the resistance to breakdown under load are related to sphericity. The higher the sphericity or the less the flatness or elongation, the less the tendency of the particles to fracture or degrade into smaller particles under loading. Such materials are easier to manipulate in construction. Flat and elongated particles tend to orient themselves so they are parallel when used in embankment or subgrade construction, forming planes of weakness. On the other hand, once interlocked by compaction they form a relatively stable mass.

ANGULARITY / Angularity of roundness, R, is a measure of the sharpness of the corners. It is quantitatively defined by:

$$R = \frac{\text{average radius of corners and edges}}{\text{radius of maximum inscribed sphere}} \tag{1:11}$$

Because it is difficult to measure, it is usually described qualitatively as shown in Fig. 1.8. When bulky grains are first formed by crushing or grinding

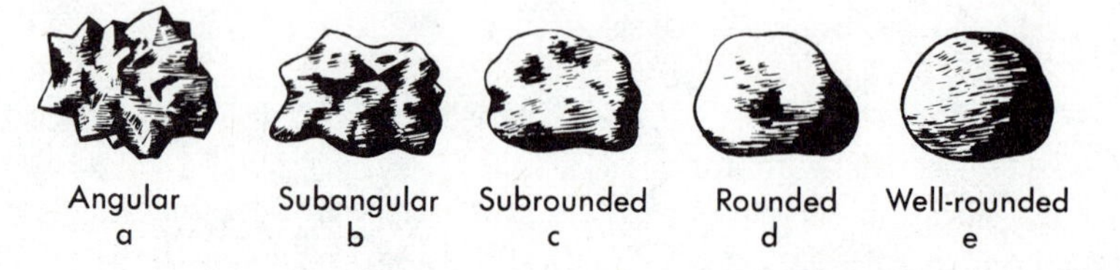

Figure 1.8 *Roundness or angularity of bulky particles.*

of rocks, they are *angular.* After the sharpest edges become smoothed, they are *subangular.* When areas between the edges are somewhat smoothed and the corners begin to be worn away, the particle is *subrounded.* It is *rounded* when all the irregularities are nearly smooth, but the original shape can still be seen. When all traces of the original shape disappear, it is *well rounded.* Small sand particles close to their point of origin tend to be very angular, while the gravel and boulder sizes in the same environment are subrounded to rounded. Beach sands tossed by wind and wave are subangular to rounded, depending on the minerals and the distance to their origin.

Windblown sands that roll and tumble in dunes become well rounded, while the same sands in water are more angular. The microscopic examination of soils is a fascinating experience in which a common river sand is found to be shining quartz fragments, and a dirty beach sand is found to be a collection of tiny shells, and jewellike garnets and zircons, on a background of crystal quartz.

The angularity has a profound influence on the engineering behavior. Under load the angular corners break and crush, but the particles tend to resist displacement. The smoother, more rounded particles are less resistant to displacement, but they are less likely to crush.

A more subtle surface quality is *roughness.* These are the micro irregularities on the particle as seen by a microscope or scanning electron microscope. It consists of surface pitting and small irregularities that contribute to resistance to displacement.

FLAKY GRAINS / Flaky grains have very low sphericity (typically less than 0.01); they are thin but not necessarily elongated. Generally, they resemble a sheet of paper in their relative dimensions. They form from the mechanical weathering or disintegration of the micas, but the predominant flaky grains are the clay minerals. Compared to the bulky grains, they are limber and resilient, like dry leaves. If the particles are randomly oriented, they may resist displacement; if stacked parallel, they resist displacement and shock perpendicular to their planes but are easily displaced parallel to their surfaces. Small amounts of flaky mica can change the behavior of a predominantly bulky grain soil. The flakes act like springs, separating the bulky grains and making the soil resilient and fluffy.

NEEDLELIKE GRAINS / Extremely elongated particles (E greater than 100) occur in some coral deposits and in the attapulgite clays. The particles are resilient and break easily under load.

EFFECT OF GRAIN SHAPE / Soils composed of bulky grains behave like loose bricks or broken stone. They are capable of supporting heavy, static loads with little deformation, especially if the grains are angular. Vibration and shock, however, cause them to be displaced easily. Soils composed of flaky grains tend to be compressible and deform easily under static loads, like dried leaves or loose paper in a basket. They are relatively stable when subjected to shock and vibration. The presence of only a small percentage of flaky particles is required to change the character of the soil and to produce the typical behavior of a flaky material.

1:9 Interaction of Water and Solid Phase[1:4, 1:8, 1:9]

If equal volumes of dry sand and water are mixed together in a container, some of the water will fill the voids in the sand, and the excess will merely rise above the sand surface. If more or less water is used, the effect on the sand will be the same (provided that there is enough to fill the soil voids), and the only difference will be the amount of the excess water that covers the sand. The sand feels gritty, either dry or wet, and does not appear to be affected by the water. If equal volumes of dry montmorillonite clay (such as commercial bentonite) and water are mixed, the water will disappear and a soft, sticky, greasy-feeling solid will be formed. If two volumes of water to one of montmorillonite are used, the result will be similar: The water will disappear and a sticky solid will be formed. The only difference is that the second solid will be somewhat softer than the first. In the case of the clay mineral, there is a reaction between the solid and the water, which results in a change in the characteristics of both. This phenomenon is termed *adsorption*, the binding of the water to the solid surface, and it has a profound effect on the physical properties of any soils containing minerals which exhibit it.

SURFACE CHARGE / The causes of adsorption are not fully understood, but they are definitely related to the electrical charges in the surface of the material. A number of mechanisms are responsible: (1) unsatisfied valence bonds at the edges of clay minerals and broken edges of other particles; (2) unbalance caused by isomorphous substitution of low-valence atoms ($Mg = 2^+$) for one of higher valence ($Al = 3^+$) in the sheets of the clay minerals; (3) nonuniform distribution of atoms and nonuniformity of the electrical charge in the surface; and (4) disassociation of ions such as hydrogen, hydroxyl, and cations from the clay surface in water. Both the electrical sign and the charge intensity vary with position. Clay mineral faces are generally negative, due to isomorphous substitution, and the edges are positive or negative. There are also localized areas of high and low charge. The total charge per unit of mass varies with the charger per unit of area and the ratio of area to mass. In most mineral particles with high sphericities and small intensities, the total charges are small. In the flaky clay minerals with relative intense surface charges, the total charges are high. The resulting electrostatic field is most intense close to the clay surface but decreases rapidly with distance, as shown in Fig. 1.9a.

ADSORPTION OF WATER / Water is a peculiar molecule because of the nonsymmetrical distribution of the positive hydrogen atoms about the negative oxygen. The molecule, although neutral, is polar, with a positive charge on one side and a negative on the other. This polar molecule or *dipole*, is attracted to the surface of solids and particularly clay minerals with their large electrostatic surface charges. It is held by a number of mechanisms:

1. The dipole—electrostatic attraction.
2. Hydrogen bonding—the sharing of hydrogen atoms with the clay.
3. Hydration of the cations that are attracted to the clay surface to compensate for isomorphous substitution.

The water collects close to the clay surface almost as iron filings are attracted to a magnet (Fig. 1.9b). The water closest to the surface is tightly held, and the molecules are oriented in the electrostatic field. The water

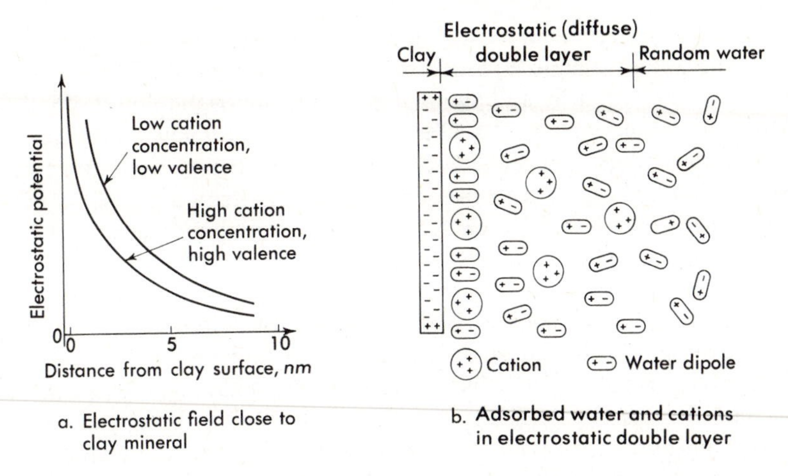

Figure 1.9 *Force field of a clay mineral and the resulting adsorbed water.*

closest to the clay surface appears denser than ordinary water. The attractive forces hold it to the clay, and movement of the water is largely restricted to slow migration parallel to the surface. In this way the adsorbed water appears to have a higher viscosity than ordinary water. Further from the surface the attraction is less and the degree of orientation and apparent viscosity increase are smaller. The thickness of the innermost layer of water is probably less than 1 nm. The total thickness of water that is attracted to the clay may approach 40 nm. This oriented water zone is termed the *electrostatic* or *diffuse double layer.*

EXCHANGEABLE CATIONS / Cations from soil moisture are attracted to the surface of clay minerals to balance the negative electrical charge produced by isomorphous substitution and possibly to balance the unsatisfied valence bonds at the particle edges. These are not fully integrated in the clay minerals; instead they can be replaced by other cations so long as the total valence balance is maintained. The cations present are those that were derived from the soluble products of the weathering plus any that may be brought by percolating water. Sodium and potassium predominate, but magnesium, calcium, aluminum, hydrogen, and even organic cations may be present, depending on the environment.

The cations play a part in determining the clay behavior. Their positive screen reduces the effect of the negative clay charge and the behavior of the electrostatic or diffuse double layer. The higher the valence of the cation, the less the total attraction of the clay for water. The higher valence ions form a thinner layer with higher density positive charge than do the lower valence ions, and thus a more effective screen (Fig. 1.9a). The order of increasing cation effect is as follows:

$$\mathrm{Li}(1^+) < \mathrm{Na}(1^+) < \mathrm{H}(1^+) \ll \mathrm{Mg}(2^+) < \mathrm{Ca}(2^+) \ll \mathrm{Al}(3^+)$$

The cations are somewhat mobile, changing places with one another within the diffuse double layer. If the concentration of one cation, such as calcium, is increased by adding lime to a wet clay containing sodium, the calcium ions will replace the sodium ions in proportion to the relative concentrations. This is termed *base exchange*, and the cations that take part are *exchangeable ions*. The exchange capability can be determined by leaching an ionic solution of a different cation through the clay and measuring the change. The amount that is exchangeable depends on the clay mineral. The kaolinites have low exchange capacity; the montmorillonites have a high exchange capacity. (In warm regions when the soil moisture has a pH exceeding 8, anion exchange also can occur.)

CLAY PARTICLE INTERACTION / Clay particles in the presence of moisture exhibit greatly different behavior from that of other minerals because of the interaction of the electrostatic fields and the diffuse double layers. Arranged face-to-face, they are held apart by their like electrical charges, with the diffuse double layers occupying the space between. At the same time they are locally attracted by unlike charges (such as at the edges), and the sharing of some hydrogens (*hydrogen bonding*) and possibly sharing of cations. A powerful attractive mechanism is the Van der Waals forces. This is essentially the dipole effect in which neutral molecules attract others because of their nonsymmetrical charges. These forces decrease rapidly with increasing spacing. Finally, there may be cementing by other minerals such as calcium carbonate or iron oxide between the grains.

The force system around the clay particles, including any external force, is in equilibrium. A large external force can move the particles closer to one another, squeezing some of the water of the diffuse double layer away. A reduction in water content by evaporation can reduce the diffuse double-layer thickness and move the particles closer. The closer spacing increases the particle attraction and reduces potential movement between the particles. This gives rise to the phenomenon of *plasticity* in soils containing clay. A clay soil can deform plastically without cracking at varying water contents: the greater the moisture, the greater the particle spacing; the less the attraction, the greater the particle mobility. At low water contents the same clay particles are closer, exhibit more attraction, and form a more rigid body.

COHESIVE AND COHESIONLESS SOILS / Soils in which the adsorbed water and particle attraction work together to produce a body which holds together and deforms plastically at varying water contents are known as *cohesive soils* or *clays* (largely because this cohesive quality results from some proportion of clay minerals). Those soils that do not exhibit this cohesion are termed *cohesionless*. Soils composed of bulky grains are cohesionless regardless of the fineness of their particles. Many soils are mixtures of bulky grains and clay minerals and exhibit some degree of varying consistency with changes in moisture. These, too, are termed *cohesive soils* if the effect is significant. Obviously, there is no sharp dividing line between cohesionless and cohesive, but it is often convenient to divide soils into these two groups for the purpose of study.

1:10 Plasticity and the Atterberg Limits[1:6c,d]

The Swedish soil scientist Atterberg developed a method of describing quantitatively the effect of varying water content on the consistency of fine-grained soils. He established stages of soil consistency and defined definite but arbitrary limits for each.

Each boundary or limit is defined by the water content that produces a specified consistency; the difference between the limits represents the range in water content for which the soil is in a certain stage or state.

TABLE 1:3 / ATTERBERG LIMITS

Stage	Description	Boundary or Limit
Liquid	A slurry; pea soup to soft butter; a viscous liquid	
		Liquid limit (LL)
Plastic	Soft butter to stiff putty; deforms but will not crack	
		Plastic limit (PL)
Semisolid	Cheese; deforms permanently but cracks	
		Shrinkage limit (SL)
Solid	Hard candy; breaks up upon deformation	

The *liquid limit* (LL) is defined as the water content at which a trapezoidal groove of specified shape, cut in moist soil held in a special cup, is closed after 25 taps on a hard rubber plate. The *plastic limit* (PL) is the water content at which the soil begins to break apart and crumble when rolled by hand into threads 3 mm ($\frac{1}{8}$ in.) in diameter. The *shrinkage limit* (SL) is the water content at which the soil reaches its theoretical minimum volume as it dries out from a saturated condition. This is described in Section 4.5. The limits are customarily expressed as whole numbers of percentage, omitting the percent sign.

In themselves the Atterberg limits mean little, but as indexes to the significant properties of a soil they are very useful. The liquid limit has been found to be directly proportional to the compressibility of a soil. The difference between the liquid and plastic limits, termed the *plasticity index* (PI), represents the range in water contents through which the soil is in the plastic state.

The most important use of the Atterberg limits is in classifying fine-grained soils. In addition, a number of relationships involving the Atterberg limits are useful in correlating soil behavior with simple test data. The activity, A, is the ratio of plasticity index to percentage of clay sizes (finer than 0.002 mm)

$$A = \frac{\text{PI}}{(\% < 0.002 \text{ mm})} \tag{1:12a}$$

The water–plasticity ratio or liquidity index, I_L, relates the water content of the soil to the liquid and plastic limits.

$$R_w = I_L = \frac{(w/100\%) - \text{PL}}{\text{PI}} \tag{1:12b}$$

The Atterberg limits and their associated relationships are simple empirical expressions of the water adsorbing and absorbing ability of soils containing clay. They thus express both the clay–water behavior and also how that is diluted by the nonclay particles. The tests are usually standardized on the portion of the soil finer than 0.42 mm (passing No. 40 sieve: fine sand sizes and smaller). However, if little coarse sand is present, the tests are often made on the total sample, but with some differences in the results. The liquid limit is related to the total moisture potentially held in the diffuse double layer plus any water held by absorption within the solids; the plastic limit is related to the innermost moisture plus absorption. The plasticity index, the difference, is thus related to the potential water content changes of the diffuse double layer. The activity expresses the plasticity of the finest fraction, which is largely clay minerals. This is a measure of the water-holding ability of the clay minerals and also suggests whether the clay is a kaolinite (low activity, <1), a montmorillonite (high activity, >4), or illite (intermediate activity, 1–2).

1:11 Microstructure or Fabric

The mineral particles, water, and air are arranged in many different ways to form the materials we know as soils and rocks. In soil mechanics the term *structure* (more properly *microstructure*) is used to describe the geometry of the particle-void formation. In rock mechanics the petrographic term *fabric* is used to denote the arrangement of the mineral grains, and structure is reserved for the larger features of the entire formation. The latter in soil and foundation engineering is sometimes termed *mass structure*. In agricultural engineering the structure means the pattern of the layering, cracking, and agglomeration exhibited by soils close to the ground surface. In soil engineering this is sometimes termed *macrostructure*.

The confusion of terminology merely adds to the problem of describing the endless variety of fabric or microstructure possible, depending on the grain shapes, the interparticle forces, and the manner in which the soil or rock formed. For the purposes of study, most can be placed in four groups: cohesionless, cohesive, composite, and crystalline. The first three apply to soils and sedimentary rocks; the last applies to certain sediments and to igneous and metamorphic rocks. Many materials are encountered, however, that do not fit these simple basic patterns. Therefore, each soil or rock must be evaluated individually, and unwarranted conclusions regarding engineering behavior should not be based on the description of the microstructure.

COHESIONLESS STRUCTURES / The cohesionless soils are composed largely of bulky grains that can be represented by spheres or similar regular, equidimensional bodies. The simplest arrangement of such particles is similar to oranges stacked on a grocer's shelf; each grain is in contact with those surrounding it. Such a structure is termed *single-grain* and is typical of sands and gravels.

Depending on the relative positions of the grains, it is possible to have a wide range in void ratios. If we pack uniform, rounded grains in a box, with each directly above the one below as shown in Fig. 1.10a, a structure with a void ratio of about 0.90 is formed. If we place them so that each succeeding layer falls into the depression between the spheres in the layer below, as in Fig. 1.10b, a structure with a void ratio of about 0.35 is formed. The

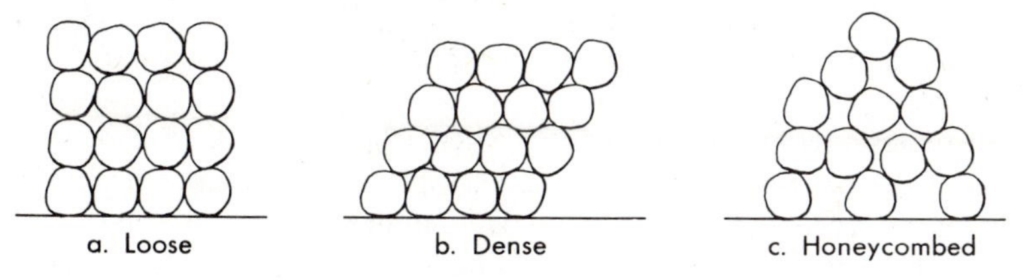

Figure 1.10 *Cohesionless soil structures.*

arrangement corresponding to the higher void ratio is described as *loose*, and that corresponding to the lower is described as *dense*. Various arrangements of the same grains could be made to produce any void ratio between these two limits.

Similar variations in void ratio are possible in cohesionless soils having irregular grain shapes and mixed sizes. The highest void ratio possible for a given soil (and still have each particle touching its neighbors) is the *maximum void ratio*, e_{max}. The smallest void ratio is the *minimum void ratio*, e_{min}. The approximate minimum void ratio of a soil is determined by compacting it by combined tamping and vibration until no further densification is possible. The tamping must not be so vigorous that the soil grains are fractured, however. The approximate maximum void ratio is found by pouring the dry soil through a funnel into a graduated cylinder.

For uniform spheres, $e_{max} = 0.90$ and $e_{min} = 0.35$, and the range in e between the limits is 0.55. Soils with angular grains tend to have both higher maximum and minimum void ratios than spheres, but the range is usually somewhat smaller. Soils with mixed grain sizes, on the other hand, usually have lower values for e_{max} and e_{min}, and the range in e is also smaller than for uniform spheres. Typical ranges in void ratio and density are in Table 1:4.

TABLE 1:4 / TYPICAL VOID RATIOS AND DENSITIES OF SINGLE-GRAINED STRUCTURES

						Unit Weight			
		Void Ratio		Density (kg/m³)		(kN/m³)		(lb/ft³)	
Soil Description	Moisture	max	min	min	max	min	max	min	max
Uniform subangular	Dry	0.85	0.45	1443	1841	14.1	18.0	90	115
sand ($G_s = 2.67$)	Saturated	0.85	0.45	1903	2152	18.6	21.1	119	134
Well-graded subangular	Dry	0.75	0.35	1526	1978	14.9	19.4	95	123
sand ($G_s = 2.67$)	Saturated	0.75	0.35	1954	2237	19.1	21.9	122	140
Well-graded silty sandy	Dry	0.65	0.25	1606	2120	15.7	20.8	100	132
rounded gravel	Saturated	0.65	0.25	2000	2320	19.6	22.7	125	145
($G_s = 2.65$)									
Micaceous silty sand	Dry	1.25	0.80	1200	1500	11.8	14.7	75	94
($G_s = 2.70$)	Saturated	1.25	0.80	1756	1944	17.2	19.0	110	121

Particles having low sphericity, particularly the slablike fragments from layered rock, do not form simple cohesionless structures. The slabs may bridge over large voids, yet be wedged tightly in a stable mass. When manipulated the slabs tend to orient themselves parallel to the direction of their movements. Such an *oriented structure* is *anisotropic* in its properties with entirely different behavior perpendicular to the orientation than parallel to it. Maximum and minimum void ratios probably have little significance in such materials.

Flaky particles, such as mica, with extremely low sphericity, similarly may form oriented structures. In soils formed entirely by decomposition and not transported from their point of origin, the orientation is a relic of the original rock fabric. Orientation in micaceous soils can also develop during sedimentation and from movements generated by shear or high pressure. Generally, oriented flaky structures have low void ratios. Nonoriented flaky grains may cause high void ratios when the flakes bridge over large voids. Maximum and minimum void ratios may be of little significance in micaceous soils. The heterogeneous arrangement of mica flakes wedged between bulky grains causes high void ratios but rather stable arrangement. The more dense, oriented mica is highly anisotropic with a rather low resistance to displacement parallel to the planes of the particles. The maximum void ratio for soils containing even small amounts of mica is much higher than for the same soil without mica. The minimum void ratio depends on orientation; it is often impossible to determine with consistency.

The relation between the actual void ratio of a soil and its limiting values, e_{max} and e_{min}, is expressed by the *relative density* or relative void ratio D_r:

$$D_r = \frac{e_{max} - e}{e_{max} - e_{min}} \times 100\% = \frac{\gamma - \gamma_{min}}{\gamma_{max} - \gamma_{min}} \times 100\% \qquad (1:13)$$

A natural soil is said to be loose if its relative density is less than about 50%; dense, if it is higher.

The properties of loose, single-grain structures are greatly different from those of dense soil. The loose soil, with its grains perched directly on top of one another, is inherently unstable. Shock and vibration cause the grains to move and shift into a more dense, stable arrangement. The rounded particles are particularly unstable when loose, but even angular grains exhibit instability if their void ratios are sufficiently high. Dense, single-grain structures are inherently stable and are only slightly affected by shock and vibration. Both loose and dense structures are capable of supporting static loads with little distortion.

HONEYCOMBED STRUCTURE / Under some conditions it is possible to arrange cohesionless bulky grains in crude arches so that the void ratio exceeds the maximum for the single-grain arrangement. Such a structure has a negative relative density and is termed *honeycombed* (Fig. 1.10c). Honeycombed structures can develop when fine sand or cohesionless silt particle settle out of still water. Because of their small size they settle slowly and wedge between each other without rolling into more stable positions, as do the larger particles. The structure also may develop when damp, fine sand is dumped into a fill or a pile without densification, a condition often termed *bulked.*

The honeycombed structure is usually able to support static loads with little distortion, similar to the way in which a stone arch carries its load without deflection. Under vibration and shock, however, the structure may collapse. In some cases this merely results in rapid settlement of the soil. In others the collapse sets off a chain reaction of soil failure that converts the entire formation momentarily into a heavy liquid capable of filling an excavation or swallowing a bulldozer. Fortunately, such structures are not common and usually occur in lenses and pockets of limited extent. Because of the hazards involved, however, the engineer should view all water-deposited silts and very fine sands with suspicion until void-ratio tests prove them to be stable.

COHESIVE STRUCTURE—DISPERSON AND FLOCCULATION / In cohesive soils the structure is determined largely by the clay minerals and the forces acting between them. The clay particles in water are acted upon by a complex series of forces, some of which, including the universal attractive forces and mutual attraction to individual cations, tend to pull the particles together; others, such as the electric charge on each grain and the electric charges on the adsorbed cations, cause the particles to repel one another. The forces of both attraction and repulsion increase, but at different rates, as the distance between particles decreases. In a dilute suspension with wide particle spacings, the total repulsion usually exceeds the attraction. The particles remain apart and stay in suspension or settle very slowly while bounding about from the agitation of the water molecules, a motion termed *Brownian movement.* Such a system is termed *dispersed.* Dispersion can be increased by adding materials that increase the repulsion forces without increasing the attraction. Dispersing agents like sodium silicate and sodium tetraphosphate are used in the sedimentation test for soil grain size to ensure that the individual particles do not stick together and give a false indication of their equivalent diameter.

When the particle spacing is small, as in a soil of a low water content, the attraction exceeds the repulsion and the particles hang together in a cohesive solid or semisolid, separated by their adsorbed layers. This effect can also be produced in a dilute suspension by reducing the repulsive forces. The addition of an electrolyte supplies ions to the soil particles, which partially neutralize the particle charges and thereby reduce their repulsion. The particles then attract each other even though widely spaced; they move together and stick in a heterogeneous loose arrangement termed a *floc.* Such flocs often contain hundreds of individual particles and are sometimes visible to the naked eye.

DISPERSED STRUCTURES / The structural arrangement formed from a dispersed soil is shown diagrammatically in Fig. 1.11a. The repulsion

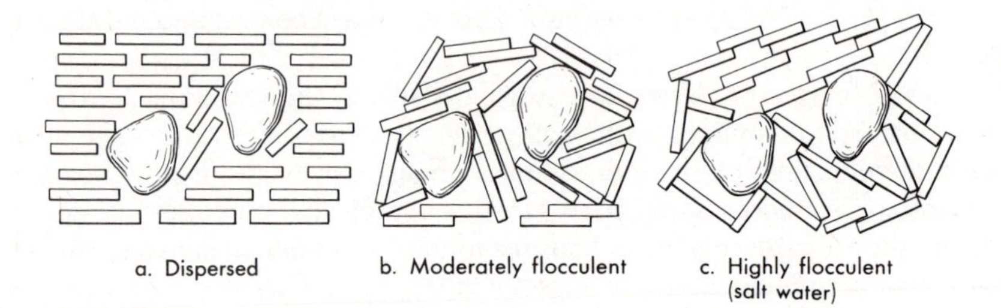

Figure 1.11 *Cohesive soil structures. (Adapted from T. W. Lambe.*[1:8, 1:9])

between the particles as they come close causes each one to position itself for the maximum grain-to-grain distance in a given volume soil. The resulting structure is very much like flat stones laid on top of one another to form a wall. The bulky grains are distributed throughout the mass and cause localized departures from the pattern. This arrangement is termed an *oriented,* or a *dispersed,* structure. It is typical of soils that are mixed or remolded, such as by glacial action (glacial till), of soils compacted under wet conditions in a man-made fill, or developed by sedimentation in the presence of a dispersing agent. Soils having a dispersed structure are likely to be dense and watertight. Typical void ratios are often as low as 0.5 but can be as high as 1 or 2, depending on the type of clay and the water content.

FLOCCULENT STRUCTURES / The arrangement in a flocculent structure is shown in Fig. 1.11b and c. It forms from a soil–water suspension which initially is dispersed, such as the suspended solids carried by a river. A sudden introduction of an electrolyte like saltwater brings about flocculation. The particles, suddenly with less repulsion, fall together in a haphazard arrangement. There may be considerable interparticle contact between the positively charged clay mineral edges and the negative faces, producing strong bonds that resist displacement. Considerable free water is trapped in the large voids between the particles, in addition to the adsorbed water already immobilized by the clay. Flocculent structures are typical of water-deposited clays. The degree of flocculation depends on the type and concentration of clay particles and on the electrolyte. Deposits formed in the sea, which is a strong electrolyte, are frequently highly flocculent, with void ratios as large as 2 to 4. Freshwater deposits, acted on by the weak electrolytes brought by rivers from different regions, are likely to be only partially flocculent or even dispersed. By way of contrast, organic acids from plant decay in shallow ponds and marshes may produce a high degree of flocculation.

Flocculent soils have low densities and are very compressible. However, they are relatively strong and insensitive to vibration because the particles are tightly bound by their edge-to-face attraction. A peculiar characteristic is their sensitivity to remolding. If the undisturbed soil is thoroughly mixed without the addition of water, it becomes soft and sticky as though water had been added to it. In fact, water has been added, for the bond between the particles has been destroyed so that the free water trapped between the grains has been released to add to the adsorbed layers at the former points of contact. This softening upon remolding is termed *sensitivity* and will be discussed at more length in Chapter 5. Construction operations in flocculent clays are difficult because the soils become softer as equipment works on top of them and may develop into a sea of mud even in dry weather.

COMPOSITE STRUCTURES / Composite or cemented structures (Fig. 1.12) consist of a framework of bulky grains arranged like the cohesionless bulky grain arrangements and held together by a binding agent. A wide variety of such structures can develop, depending on the relative amounts of the binder and the bulky grains, the type of binder, and the method of deposition.

A number of different types of binder are found. Clay that has been highly compressed or dried, so that it is stiff or hard, and calcium carbonate are the most widespread. They are usually strong but may be weakened by water. Various iron oxides and colloidal silica from rock weathering are also encountered as binders and are relatively insensitive to softening by water.

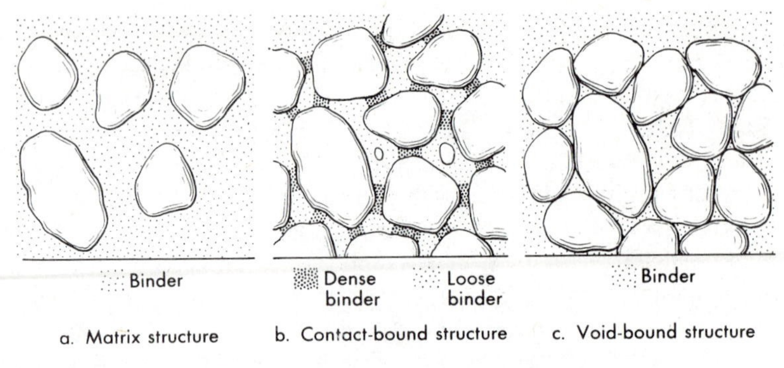

Figure 1.12 *Composite soil structures.*

In the *matrix* structure the volume of the bulky grains is less than about twice that of the binder so that the bulky grains float in a binder matrix, as shown in Fig. 1.12a. If the binder is clay, this is merely another form of a cohesive structure, and the physical properties are essentially those of a cohesive matrix. With other binders the matrix structure is a form of rock whose physical properties depend on either the binder or the bulky grain, whichever is the weaker.

When the volume of the bulky grains is more than about twice that of the binder, skeletal structures develop. These take two forms: *contact bound* and *void bound*, depending on the position of the binder between the grains. In the contact-bound structure (Fig. 1.12b) the binder is concentrated between the points of contact of the bulky grains, holding them apart similar to stones set in mortar. It can form in a number of ways. When bulky grains and clay settle simultaneously out of water, some of the clay is caught between the bulky grains and is compressed by the increasing weight of the sediment into a relatively rigid solid. A soft clay–water mixture occupies the voids between the grains but probably contributes little to the binding. The reweathering of a soil composed largely of quartz and with some feldspar, mica, or partially weathered clay can form a contact-bound structure in which the material in the voids is altered or leached away, leaving the material caught between the contact points of the grains largely unaltered. Contact-bound structures also form when large amounts of bulky grains and small amounts of clay are mixed and then consolidated or compacted. This occurs naturally by glacial action where the ice mass plows up and mixes the materials and the weight of the ice compacts the resulting *till* to a rocklike solid. It occurs artificially where clay–sand or clay–gravel mixtures are used in constructing highway or airfield subgrades.

Contact-bound structures are relatively rigid, incompressible, and resistant to shock and vibration as long as the binder remains strong. When the voids are large and open so that water can seep through them, the calcium carbonate and clay binders may soften. If the bulky framework is loose or honeycombed, the weakened soil will collapse like a loose cohesionless soil. If the bulky framework is dense, the softened binder will extrude into the voids, resulting in some settlement and weakening.

In the void-bound structure (Fig. 1.12c) the bulky grains touch each other and the binder occupies part or all the voids between them. This structure develops when the bulky grains are deposited first, and the binder subsequently is deposited between them. Water seeping through a bulky grained soil can precipitate calcium carbonate, iron oxides, or silica to form a cemented sand or gravel which is rigid, strong, and dense. This structure also forms by the weathering of a rock such as granite, which consists of a framework of interconnected quartz grains with feldspar and micas between. The decomposition of silicate minerals leaves a quartz framework supported by clay minerals. Clay and fine bulky grains washed into a coarse sand or gravel deposit also can act as binders but not to the same degree as the binders that are precipitated. The void ratio of void-bound skeletal structures may be as low as 0.2, but typical values are 0.3 to 0.5. The soil is rigid and incompressible and not likely to be softened by water.

CRYSTALLINE STRUCTURES / Crystalline structures or fabrics form by crystal growth, from cooling of plastic or molten rock, through recrystallization produced by heat or pressure or by precipitation out of water. The mass consists of crystals meshed together into a more or less continuous material. The crystal shapes may be badly distorted, because they are not free to grow equally in all directions. The particles are bonded together by interlocking, sometimes by intergrowth, and by molecular bonds at points of contact. Surprisingly, the areas of true grain contact may be small, with minute voids separating many of the particle faces. These voids may be interconnected, but most are independent.

Several types of crystalline fabric are of engineering significance. Oriented fabrics develop under heat, pressure, and shear, producing anisotropic behavior. Shear-oriented structures, with distorted broken crystals, are termed *mylonitic*. A *porphoritic* fabric consists of large crystals or *phenocrysts*, in a matrix of finer grains. Such materials may exhibit great variations in behavior because the phenocrysts are points of structural discontinuity. A *vesicular* fabric contains bubbles, from gases in the molten rock. It typically has a low density with both independent and some interconnected voids. *Vugs* are small cavities produced by solution or large gas voids.

MACROSTRUCTURE / The continuous structure or fabric of natural soils and rock is frequently altered by local conditions to develop *macrostructure* or secondary structure. The principal cause is the continuing advanced weathering of the materials near the ground surface. The effects are particularly noticeable when soil materials are deposited in a greatly different environment from that in which they were formed or when there has been a significant change in environment.

The most prominent feature is cracking, caused by the shrinkage and expansion produced by moisture and chemical changes. These divide the soil into blocks whose dimensions range from a fraction of an inch to a foot or two. These macrostructures are termed *prismatic*, *blocky*, or *columnar* depending on their size and shape. A second form is aggregation in which clumps of grains accumulate because of reflocculation from organic acids. The resulting structure is termed *crumby*. A third form is the localized cementing of portions of the soil and porous or fractured rock by concentrations of humic acids from plants, calcium carbonate, or iron oxides to form *nodular* or *concretionary* structures or *hardpan* layers. A fourth form is *slickensided* cracking from local shear. A fifth form includes the filling of cracks and fissures with soil particles washed down from the surface or with materials deposited from solution. Soils with pronounced macrostructures are characterized by zones of hardness and weakness, and by patterns of color and texture that reflect the discontinuities.

REFERENCES

1:1 "Glossary of Terms and Definitions in Soil Mechanics," *Journal of the Soil Mechanics and Foundations Division, Proceedings, ASCE*, **84**, SM4, October 1958.

1:2 D. Carroll, *Rock Weathering*, Plenum Publishing Corporation, New York, 1970, 202 pp.

1:3 C. E. Weaver, *Chemistry of Clay Minerals*, Elsevier Publishing Co., Amsterdam, 1973, 230 pp.

1:4 R. E. Grim, *Clay Mineralogy*, 2nd ed., McGraw-Hill Book Company, New York, 1968.

1:5 "Physico-Chemical Properties of Soil, A Symposium," *Journal of the Soil Mechanics and Foundations Division, Proceedings, ASCE*, **85**, SM2, April 1959.

1:6 *Annual Book of Standards*, American Society for Testing and Materials (issued yearly with some changes). D- designates soil, foundation rock, and paving tests, followed by a number that refers to a particular test; two numbers following the dash are the year of adoption or last revision; the letter "T" following the date indicates that it is a new standard and tentative; two numbers in ( ) indicate date of reapproval of old standard.

a. D-421-58(78): "Dry Preparation of Soil Samples for Particle Size Analysis and Determination of Soil Constants."

b. D-422-63(78): "Particle Size Analysis of Soils."

c. D-423-66(78): "Test for Liquid Limit of Soils."

d. D-424-59(78): "Test for Plastic Limit and Plasticity Index of Soils."

1:7 W. C. Krumbein and L. L. Sloss, *Stratigraphy and Sedimentation*, 2nd ed., W. H. Freeman and Company, Publishers, San Francisco, 1963, pp. 106–111.

1:8 T. W. Lambe, "The Structure of Inorganic Soil," *Proceedings, ASCE*, **79**, Separate 315, October 1953.

1:9 T. W. Lambe, "The Structure of Compacted Clay," *Journal of the Soil Mechanics and Foundations Division, Proceedings, ASCE*, **84**, SM2, May 1958.

SUGGESTIONS FOR FURTHER STUDY

J. Feld, "Early History and Bibliography of Soil Mechanics," *Proceedings, Second International Conference on Soil Mechanics and Foundation Engineering*, I, Rotterdam, 1949, pp. 1–7.

From Theory to Practice in Soil Mechanics (Selections from the life of Karl Terzaghi), John Wiley & Sons, Inc., New York, 1960, 425 pp.

J. E. Bowles, *Engineering Properties of Soils and Their Measurement*, 2nd ed., McGraw-Hill Book Company, New York, 1978, 213 pp.

Nyle C. Brady, *The Nature and Properties of Soils*, 8th ed., Macmillan Publishing Co., Inc., New York, 1974, 639 pp.

G. Millot, *Geology of Clays*, Springer-Verlag, New York, Berlin; Masson et Cie, Paris, 1970, 429 pp.

J. K. Mitchell, *Fundamentals of Soil Behavior*, John Wiley & Sons, Inc., New York, 1976, 422 pp.

R. N. Yong and B. P. Warkentin, *Soil Properties and Behavior*, Elsevier Scientific Publishing Co., Amsterdam, 1975, 449 pp.

U.S. Bureau of Reclamation, *Earth Manual*, U.S. Government Printing Office, Washington, D.C., 1974, 810 pp.

PROBLEMS

1:1 A cube of moist soil had a mass of 150 kg or a weight of 330 lb. Its volume was 84.9 liters or 3 ft^3. The water content was 27%, and its specific gravity of solids 2.72. Find e, n, s, ρ, and γ in English and SI units.

1:2 A 50-ml sample of moist soil has a mass of 95 g. Oven-dried, its mass is 75g. The specific gravity of soil averages 2.67. Find e, n, w, S, ρ, and γ of the moist soil.

1:3 A 558-ml volume of moist soil has a mass of 990 g. Dried the mass is 900 g. The average specific gravity of solids is 2.70. Find e, n, w, S, ρ, and γ of the moist soil.

1:4 A 75-ml sample of saturated soil from below the groundwater level has a mass of 120 g. Dried the mass is 74 g. Compute ρ, γ, w, e, n, and average G_s.

1:5 A 120-g sample of soil is 50% saturated. The average specific gravity of solids is 2.71, and the water content is 18%. Compute e, n, ρ, and γ.

1:6 A saturated soil has a water content of 38% and an average specific gravity of solids of 2.73. Find e, n, ρ, and γ.

1:7 A saturated soil has a water content of 40% and a unit weight of 17.9 kN/m^3 or 114 lb/ft^3. Find e, n, and average G_s.

1:8 A saturated soil has a water content of 47% and a void ratio of 1.3. Find ρ, γ, and average G_s.

1:9 A sand with an average specific gravity of solids of 2.66 and a void ratio of 0.6 is initially dry but becomes saturated by a rising water table.

a. Compute n.
b. Compute ρ and γ if the sand is dry.
c. Compute ρ and γ if the sand is 40% saturated.
d. Compute ρ and γ when the sand becomes saturated.

1:10 A soil has a unit weight of 110 lb/ft^3 and a water content of 6%. How much water should be added to each cubic yard of soil to raise the moisture content to 13% to make the soil easier to compact?

1:11 A soil has a unit weight of 15 kN/m^3, a water content of 10%, and a specific gravity of solids of 2.7 when dumped loosely from a scraper.

a. Find e, n, ρ, and γ.
b. For easiest compaction the water content (optimum) should be 15%. How much water should be added in liters/m^3 of soil to raise the water content to the optimum?
c. At optimum moisture, the soil is compacted until it is 95% saturated. Find the new e, n, ρ, and γ.

1:12 A soil has a unit weight of 20 kN/m^3 or 127.4 lb/ft^3, and a water content of 14%. Find its new water content if it dries to a unit weight of 18 kN/m^3 or 114.6 lb/ft^3 without a change of void ratio.

1:13 A peat (fibrous organic) sample of 1 kg had a volume of 0.85 liter. It is saturated. Its specific gravity of solids is 2.4.

a. Compute e, n, ρ, and γ.
b. Find ρ and γ if the peat dries without a change in void ratio.
c. What will happen if the dried soil is subject to a rapidly rising water table?

1:14 What is the difference in ρ and γ of a soil whose solids are pure quartz compared to a soil composed of 70% quartz, 20% biotite, and 10% iron oxide? Both soils are saturated and both have void ratios of 0.75.

1:15 Plot on 5-cycle semilog paper the following grain size curves. Compute D_{10} and C_u, where possible. Find the percentages of gravel, sand, silt sizes, and clay sizes by the ASTM grain size scale.

PERCENT FINER BY WEIGHT

Sieve No.	Lagoon Clay, Beaufort S.C.	Glacial Till Columbus, Ohio	Beach Sand, Daytona Beach, Fla.	River Sand-Gravel Teton, Dam, Idaho	Weathered Sand-stone Jasper, Ala.	Weathered Tuff, Central America	Core-Teton Dam
$\frac{1}{2}$ in.	—	94	—	52	—	98	—
No. 4	—	68	—	37	100	95	—
10	—	50	—	32	82	93	—
20	—	35	100	23	76	88	100
40	—	22	98	11	70	82	98
60	100	18	90	7	60	75	96
100	95	15	10	4	43	72	95
200	80	11	2	—	27	68	92
0.045 mm*	61	10	—	—	23	66	60
0.010 mm*	42	7	—	—	13	33	32
0.005 mm*	37	5	—	—	8	21	25
0.001 mm*	27	2	—	—	3	10	17

* From sedimentation test.

1:16 A soil has a liquid limit of 56 and a plastic limit of 25. The water content of the soil as it is excavated for use in a fill is 31%.

a. Compute the PI of the soil.
b. Is the soil likely to be stiff or soft when compacted at its existing moisture content in fill?
c. What would a light rain do to the consistency of this soil?

1:17 Compute the typical unit weight in kN/m^3 and lb/ft^3 of:

a. Dense, well-graded, subangular, dry sand.
b. Dense, well-graded, subangular, saturated sand.
c. Loose, uniform, rounded, dry sand.
d. Loose, uniform, rounded, saturated sand.
e. Honeycombed, saturated silt.

(Assume typical values for e and G_s.)

1:18 A sand with a minimum void ratio of 0.45 and a maximum of 0.97 has a relative density of 40%. The specific gravity of solids is 2.68.

a. Find the density and unit weight dry and saturated in the present state.

b. How much will a 3-m thick stratum of this sand settle if the sand is densified to a relative density of 65%?
c. What will be the new densities and unit weights, dry and saturated?

1:19 A sample of micaceous silt 10 cm in diameter and 2.5 cm thick is compressed to a 2-cm thickness without a change in diameter. Its initial void ratio is 1.35, and its specific gravity of solids is 2.70. Find the initial weight saturated, its void ratio after compression, and its unit weight, after compression, and the change in water content caused by compression. Assume that all of the compression is produced by a reduction in void ratio accompanied by a loss of water.

1:20 A soil has a specific gravity of solids of 2.72, a void ratio of 0.78, and a water content of 20%.
a. Compute its density unit weight and degree of saturation.
b. What will its new unit weight and void ratio be if it is compacted (a reduction in void ratio) without loss of water until it becomes saturated?

1:21 A sample of volcanic ash weighs 6.28 kN/m^3 dry and 8.95 kN/m^3 saturated. When crushed, the specific gravity of solids is 2.75. Find the total void ratio and the percentage of the voids that are not interconnected with the surface.

1:22 Two soils, clayey silty sands, have identical particle sizes from a sieve test; both have 20% finer than 0.074 mm. One dries out easily when exposed to air, the other does not. What explanation can you offer, and how would you investigate this?

1:23 A soil has a void ratio of 0.95, a degree of saturation of 37%, and a specific gravity of solids of 2.72.
a. Compute its water content, porosity, and unit weight.
b. How much water must be added to a cubic meter or cubic foot of soil to increase its saturation to 100%

1:24 A saturated soil weighs 20.4 kN/m^3 or 130 lb/ft^3. When oven-dried, it weighs 17.1 kN/m^3 or 109 lb/ft^3. Compute the void ratio, porosity, water content, and specific gravity of solids.

CHAPTER 2
Engineering Geology of Rocks and Soils

The construction specifications for an earth dam required that the contractor obtain most of the needed materials from sand and gravel from the flood plain upstream of the dam. The designer had made test pits into the deposits and had made grain size tests of the materials excavated by a small clamshell bucket. The specifications further required that the contractor separate the materials into coarse and fine components by a $\frac{1}{2}$-in. (12.7-mm) screen.

The specifications could not be followed. The composite bucket samples failed to disclose clay seams 25 to 80 mm (1 to 3 in.) thick. They were a small percentage of the total sample; however, they were sufficiently continuous to impede drainage of the excavation of borrow areas. Moreover, the clay clogged the $\frac{1}{2}$-in. screen, making simple screening impossible.

The contractor solved the problem by washing the gravel and sand during screening, and disposing of the clay in a sediment retention pond. The clean, coarse materials produced by this operation would not meet the specifications. The specifications required that the material retained on the $\frac{1}{2}$-in. screen could contain no more than 5% finer than $\frac{1}{2}$ in. after it had been compacted in the dam. However, the gravel contained numerous flat sandstone fragments, shale, and partially weathered rock that broke down during handling and compaction. Ultimately, it was never possible to meet the specifications using the flood-plain material specified.

A geologic study of the flood plain would have revealed the likelihood of clay seams lacing the sands and gravels in an unpredictable pattern. A geologic study of the tributaries of the river would have shown that the nearest sources of the sand and gravel deposits would contain weak sandstones, shales, and weathered igneous rocks. The lack of understanding of the geology of that flood plain caused loss of time, increased material handling costs, and increased haul distances for better materials. Because the source of the borrow and the procedure for handling it had been specified, the contractor submitted a claim to the owner for several million dollars for the extra costs entailed. Ultimately, he was paid for the extras.

This case history illustrates the problems that are generated by a lack of understanding of the geology of soil and rock as they occur in nature. Soil and rock formations comprise almost a limitless variety of structures that challenge the imagination. A rational solution to the engineering and construction problems involving these materials and their arrangement requires a thorough understanding of the formations and an accurate concise means of communicating that information to others.[2:1]

2:1 The Cycle of Soil and Rock Formation

The earth is dynamic. The billion years or so of documented geologic history have seen changes brought on by both evolution and revolution that leave their imprints in the soils and rocks that comprise the earth's crust (Fig. 2.1).

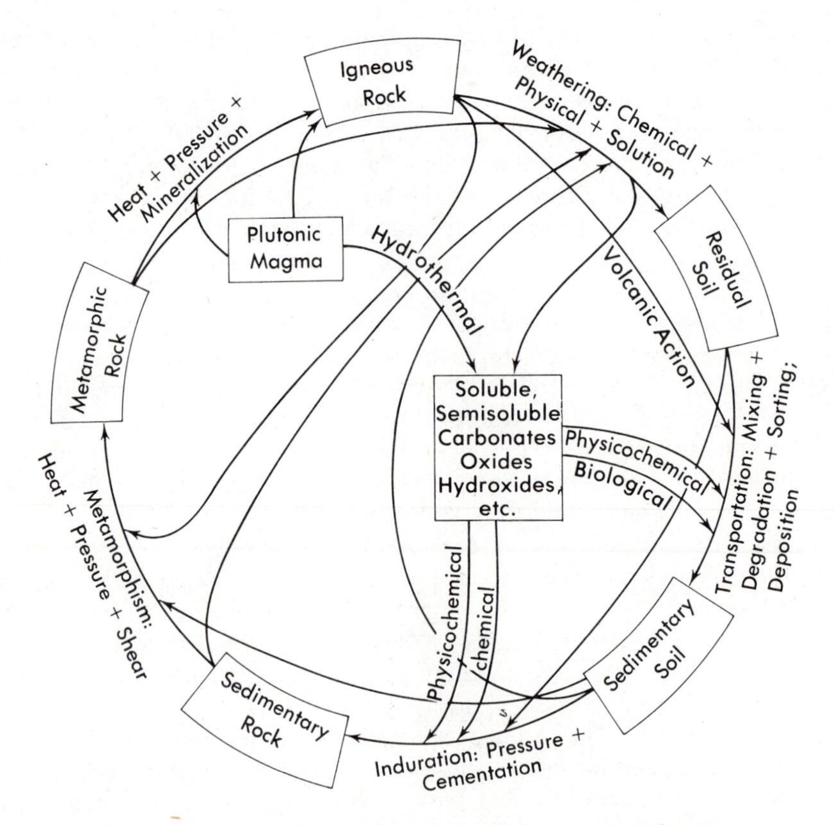

Figure 2.1 Cycles of change in soil and rock formation and alteration.

The entire earth's crust was probably at one stage a viscous liquid skin that slowly hardened into igneous rock. This crust is still being augmented by plutonic magma that occasionally flows or boils up from the depths.

The processes of weathering, aggravated by the wrinkling and cracking of the crust, attack the rock, creating *residual soils*: the in-place products of decomposition, solution, and local disintegration. Some of these materials are transported by gravity through creep and sliding to make *colluvial* deposits nearby. Other materials are transported by wind, water, and ice still greater distances. They are mixed together, sometimes sorted out by size, and then laid down in a new environment as *deposited soils.*

The deposited soils continue to weather, and some are retransported and redeposited in new formations. Others become indurated by consolidation and cementing into *sedimentary rocks.* Calcium and magnesium carbonates, sodium chloride, and calcium sulfate are precipitated from solution. Calcium carbonate and iron hydroxides biologically formed by marine organisms become sediments to add to the soils produced by weathering. They, too, can be hardened by cementing, consolidation, and recrystallization into sedimentary rocks. The sedimentary rocks are subject to the same distortion and fracturing produced by tectonic forces as the igneous rocks. Environmental changes subject them to in-place weathering that produces new residual soils, followed by the processes of erosion and transportation that eventually create new deposited rocks.

Instead of being exposed to weathering and erosion, the sedimentary rocks can be buried beneath accumulating sediments and subjected to increasing heat, pressure, and shear. The minerals are altered chemically and distorted or realigned physically to produce *metamorphic rocks.* The new rocks may resemble their ancestors but are usually more crystalline, denser, and harder. The metamorphic rocks are subject to weathering if exposed, forming residual soils that eventually are transported and mixed into new sedimentary deposits. Igneous rocks may also be metamorphosed by heat, pressure, and shear, but the changes are generally less drastic. Finally, the metamorphic rock can be transformed back to igneous rock by heat, pressure, and the addition of new minerals from molten masses below. The cycle is replete with short cuts and reversals; but it is continuous, with no definite beginning or endpoint. Engineering is merely another process added to the cycle, an insignificant one in the total pattern but locally often drastic. Therefore, all engineering works involving the earth must be evaluated in terms of their total impact on these dynamic processes.

2:2 Igneous Rocks

The ultimate source of igneous rock is still a cosmic mystery: Molten magma wells up from the deeper part of the mantle bringing the mineral ingredients of igneous rock—principally, quartz, orthoclase feldspar, plagioclase feldspar, muscovite and biotite mica, pyroxene and its variety augite, amphibole and its variety hornblende, and magnetite. These, in varying proportions, form the igneous rocks. Nearly all are silicates, from the simple quartz to the complex pyroxene, but they differ, particularly in the other metallic elements. Not all are present in any one rock: Certain of the ferromagnesian mineral combinations are not likely in the presence of quartz; similarly, large amounts of quartz and muscovite seldom occur when the rock is rich in plagioclase feldspar.

The grain size or texture depends on the rate of cooling: Rapid cooling means finer grains and slow cooling, coarse grains. Classification of igneous rocks (Section 2:8) is based on grain size and mineral content.

There are two primary forms of igneous structure: massive or *intrusive*, and *extrusive*. The intrusives include such regional masses as the Canadian shield of granite and more localized *domes* and *batholiths* (large knobs) such as Stone Mountain, Georgia, and Sugar Loaf, Rio de Janeiro. These are coarse-grained rocks because of the slow cooling of the large volume of plastic or molten magma. Most intrusives are granitelike, rich in quartz, but large masses of coarse-grained dark rocks like *gabbro* also occur. *Dikes* are wall-like intrusive bodies that fill fissures in older rock; and *sills* are similar intrusives between bedding planes. They frequently occur together, dividing the older rock by walls and sheets of the intrusion. Such smaller intrusions, which cool rapidly, are finer grained than the large bodies.

The extrusive rocks form at the earth's surface by flow or volcanic eruption. The flows may form uniform strata covering wide areas or irregular waves of porous glassy boulders.

The extrusive rocks are generally very fine grained because of their rapid cooling. Rocks rich in ferromagnesian minerals are more common extrusives, the *basalts* and *dolerites* (often collectively called *trap* rock), than those rich in quartz, the rhyolites. The *ash* and *bombs* formed by volcanic eruptions and explosions are another form of extrusive rock that falls to become sedimentary deposits. Often ash deposits and flows are interbedded; they later may be intruded with dikes and sills to form solid walls and floors that enclose the more porous ash.

Large intrusive bodies are rather homogeneous. The smaller intrusives and the extrusives and the boundaries of the large intrusives exhibit numerous discontinuities. Joint cracks from cooling and flexure divide the large mass into blocks or prisms of varying sizes, weakening it. Blocks of older rock are frequently engulfed in the surface of intrusive bodies. Exposed at the ground surface, the rocks spall or exfoliate in broad domelike shells several centimeters thick with small gaps beneath. Because of this, foundations on "bedrock" occasionally settle or move laterally. Joint cracks, shear cracks due to tectonic activity, and gas pockets in extrusives form seepage paths and accelerate weathering.

RESIDUAL SOILS FROM IGNEOUS ROCK[2:2] / Residual soils are found wherever the rate of weathering exceeds the rate at which the products of weathering are removed by gravity, erosion, and glacial action. In gently sloping terrain in the tropics they may be several hundred meters thick. In cool regions the weathering is slow and the residual blanket is thinner—only a few meters in Greenland. Moreover, in much of the Northern hemisphere, glacial action has plowed away the residual accumulations leaving ancient igneous rocks, such as parts of the Canadian Shield, naked except for local pockets of soil cover.

The soils reflect the mineralogy of the parent rock. Granites produce tan and yellow sandy silts and silty sands, with varying amounts of mica and clays of the kaolinite family. The rocks rich in ferromagnesian minerals, such as basalt, yield highly plastic montmorillonite clays with iron oxide coloration ranging from deep red to the dark browns of the rich "black cotton soils" of India. The degree of weathering varies with depth. At the surface the feldspars, micas, and ferromagnesian minerals are largely converted into clay minerals, while with increasing depth they are only partially altered and still retain some of their interparticle bonding. There is no well-defined boundary between soil and rock—only a transition. The weathering extends deepest and is most advanced along joints and in shear zones. In jointed rock, weathering from the joints inward creates plane-sided "unweathered" blocks with rounded edges that resemble boulders, which float in a matrix of the more completely weathered soil (Fig. 2.2).

The deeper residual soils retain the fabric of the original rock in concentrations of minerals and orientation of grains. Residual soils exhibiting this relic microstructure are termed *saprolites* (saprolites with strong relic structure are also derived from schists and gneisses).

Residual soils from granites vary in their foundation capabilities depending on weathering. They ordinarily are good construction materials. Those derived from the ferromagnesian minerals are less desirable because of their montmorillonite clays.

The transition between the saprolite and the unweathered rock consists of irregular seams of partially and more completely weathered materials, in alternating seams or in the form of the boulderlike bodies previously mentioned. This zone is very porous and is a cause of leakage of dam foundations.

2:3 Transported and Deposited Soils[2:3]

GRAVITY-TRANSPORTED SOILS / Most soils commence their movement by gravity alone. Residual soils in hilly areas tend to move slowly downslope—a process known as *creep*—but the general character of the soil deposit is usually not changed. Creep is important in that structures on

Figure 2.2 *Boulderlike blocks produced by weathering along joints in granite, north-central Georgia.*

shallow foundations may be moved out of position, and structures on deep, rigid foundations may be damaged by the pressure of the moving mass of soil. Landslides (see Chapter 12) are rapid downslope movements. The distorted landslide debris deposits are *colluvium.*

A *talus* is an accumulation of fallen rock and rock debris at the bases of steep rock slopes and faces. It is composed of irregular, gravel to boulder size particles and is very likely to be in an unstable condition. It is often a good source of broken rock and coarse-grained soil for construction.

Mud flows take place when loose, sandy, residual soils on relatively flat slopes become saturated. The soils flow like water and then come to rest in a more dense condition. The deposits are characterized by their heterogeneous composition and irregular surface topography.

RIVER DEPOSITS (ALLUVIUM) / Running water is one of the most active agents for soil transportation. As a transporting agent, water mixes soils from several different sources and then sorts and deposits them according to grain size. Small soil particles are lifted by the turbulence of the moving water and are carried downstream with little physical change, while the larger particles of sand, gravel, and even boulders are rolled along the stream bed and become ground down and rounded by abrasion.

The ability of running water to move solid particles is a function of the velocity and rate of flow. The total volume of particles that can be carried by a unit volume of water is proportional to the velocity squared. The volume of the largest particle that can be moved is proportional to the sixth power of the velocity. Therefore, during periods of high discharge, rivers carry tremendous volumes of coarse and fine particles; in periods of low flow, only small quantities of fine particles are transported. If the stream velocity increases, such as where steeper portions of its channel are reached or when rainfall swells the flow, the river erodes its channel until its ability to transport materials is satisfied. If the stream velocity decreases because of flatter slopes or decreased flow, some of the transported particles are deposited, with the largest particles dropped first.

Streams in arid regions are characterized by flash floods and prolonged periods of little or no flow. Tremendous quantities of small boulders, gravel, and sand are carried during periods of high water, but the volume of material transported in dry weather is negligible. The deposits formed in the steep portions of such a river fill the channel to great depths and also form narrow *terraces* of gravel and sand above the low-water channel. Both shift and change during every flood season. At the point the river enters flat country, its velocity is sharply checked, and some of its load is deposited in the form of a flat, triangular mass termed an *alluvial fan*. As the fan builds up, the river shifts its course to build a succession of these masses.

The fans (Fig. 2.3) join to form a sloping *piedmont* at the foot of the mountains, a thick, tilted, undulating surface underlain by thick erratic deposits of silty sand and gravel.

Figure 2.3 *Alluvial fan 1.5 km (1 mile) wide in a desert region, Death Valley, California.*

The valley below the piedmont becomes filled with irregular lenticular masses of silt, sand, and gravel brought by the rivers during flooding. The stream exhibits *braiding* because it chokes its channel with solids after every period of high discharge, forcing it to break through into a new course. These valley fills are usually very loose. They may develop some cementing because of continuing weathering, and precipitation of the soluble salts in hot desert climates. These cemented soils often collapse upon flooding, causing canals to slough and foundations to settle.

Both the alluvial fans and the braided stream deposits in the valley fills are good sources of sand and gravel for construction.

Streams in humid regions are characterized by floods and sustained dry weather flow. The particles carried by such streams are likely to be finer than those carried by streams in arid regions because the flood velocities tend to be smaller and because the greater degree of weathering in humid regions produces a much larger proportion of fines. The deposits in the steeper portions of streams in humid regions are similar to those formed by steep streams in arid regions but are smaller and less likely to shift during every period of high water. Where the rivers enter flat valleys they tend to form alluvial fans that are ordinarily broad and flat and composed largely of sands and fine gravels.

River deposits in flat valleys in humid regions are important because valleys are often the sites of highways, railroads, airfields, industrial plants, and large cities. During periods of low flow, the stream is confined in its channels and deposition is balanced by erosion. During flood periods, however, it overflows its banks and floods the valley to form immense lakes and broad, flat sheets of slowly moving water. The velocity in the overflow areas is so much smaller than in the channel that rapid deposition takes place along the banks of the channel, forming natural levees. The broad overflow areas act as settling basins in which the fine particles are deposited out of the slowly moving water. As the flood subsides, still finer particles are deposited until evaporation reduces the remaining puddles to dust. Floodplain deposits (Fig. 2.4) consist of broad, flat, thin strata of very fine sands and

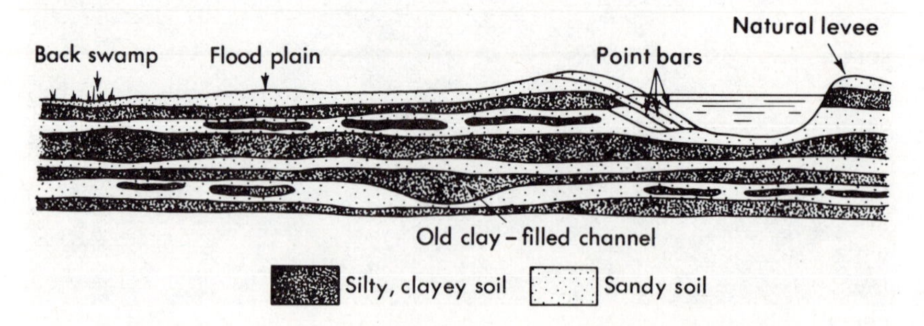

Figure 2.4 *Cross section of a flood plain of an old river in a humid region.*

clays with occasional elongated lenses of sand that formed in temporary channels or sloughs.

The lowest part of the flood plain is often farthest from the river and the last area to become dry after flooding. It is termed the *backswamp* because of the wet, soft soil and swampy organic matter that accumulates.

In old flat valleys the river meanders back and forth in sweeping S-curves. It erodes the outsides of the bends where the water velocity is highest and fills the insides with crescent-shaped beaches of sand known as *point bar* deposits. The river loops migrate downstream, cutting and building simultaneously. Sometimes the river cuts across a large bend to leave the old channel behind as an *oxbow lake*. This eventually fills with floodplain silts and organic accumulations. Finally, it becomes buried in the flood plain—an alien sinuous deposit of soft soils and organic matter trapped in the otherwise horizontal strata. These are hazards to heavily loaded foundations because they are weak discontinuities in otherwise regular formations.

The foundation capacity of floodplain deposits is often limited, depending on the relative thickness and compressibility of the clay strata, and construction is often complicated by high ground water. The old river bottoms and the insides of bends are good sources of sand and gravel, and the plains are sources of sand, silt, and clay for construction.

LAKE (LACUSTRINE) DEPOSITS / Geologically, lakes are temporary basins of water supplied by rivers, springs, and the outflow from glaciers. They act as giant desilting basins in which much of the suspended matter carried by the streams that feed them is deposited.

Streams in arid regions carry great quantities of suspended, coarse sands during periods of high discharge. These are deposited at the point the stream enters the lake and form a *delta*. Deltas are characterized by uniform grain size and by bedding at angles of about 30°. The finer suspended particles are carried out into deeper water where they settle out to form horizontal, thin strata of alternately coarse-grained, then fine-grained, particles. Lakes in arid regions soon fill with soil and become increasingly shallow. Many dry out in the hot summer sun. If they have no outlets, they become salty or alkaline, depending on the dissolved matter in the inflowing streams. The resulting deposits consist of thin strata of fine sands, silts, and sometimes clays that may be partially cemented with borax, gypsum, or calcium carbonate. At the edges of the deposits are thick, uniform beds of sand that represent the former deltas. Many are rimmed with salt.

Lakes in humid regions also accumulate deltas at the mouths of the inflowing streams, but the deposits are likely to be finer grained than those of arid regions. The silt and clay size particles are carried out into deeper water where the silts and the coarser clays are slowly deposited. During periods of low flow when there is little turbulence and when slight changes in the water chemistry produce flocculation, the colloidal clays are deposited. The result is alternate thin strata of silt and clay. As the lake fills up and becomes shallow, plant life around the edges increases. Rotting vegetable matter produces organic colloids that are deposited with the silts and clays to form organic soils. Microscopic organisms called *diatoms* contribute their silica skeletons, and other organisms add their calcium carbonate shells to the deposit. Finally, the lake chokes up with vegetation so thick and matted that only incomplete decomposition takes place. The result is a covering of fibrous organic matter known as *peat*, and at that stage the lake has become a *marsh* or *bog*. Lake deposits (Fig. 2.5) consist of alternate thin strata of silt and clay overlaid by organic silts and clays and finally topped with a stratum of peat. Thick beds of sand, the former deltas, are found at the edges of the deposit.

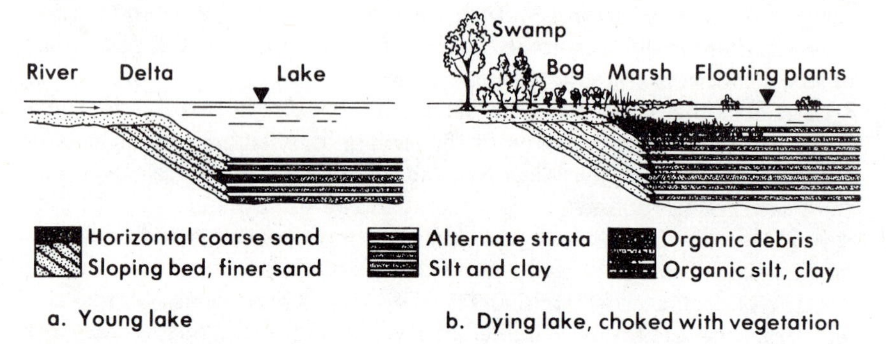

Figure 2.5 *Cross section of a lake deposit in a humid region.*

Lake deposits seldom make good foundations because the soils are likely to be weak and compressible. The deltas sometimes provide good structural support, and they are sources of uniform sand for construction.

MARINE DEPOSITS / Marine soils include two groups: offshore deposits and shore deposits. The offshore conditions are similar to those in lakes in that deposition takes place in relatively still water below the zone of wave action. The degree of flocculation is considerably greater because of the salt water, and calcium carbonate in the form of shells or microscopic particles may accumulate. Offshore deposits consist of horizontal strata of silt and clay that frequently have a highly flocculent structure. Occasional strata of shells or calcareous sands, silts, and clays termed *marls* may be formed that are partially cemented.

The shore deposits are highly complex, owing to the mixing and transporting activities of wave action and the many different shore currents. Materials brought to the sea by rivers and washed from the sea by wave action are swept along the shore by the shore currents, to be deposited in the form of *spits* or *bars* in areas where deep water or wide embayments reduce the current velocity. These materials are reworked by the waves to form the offshore bar at the line of breakers and the beach itself (Fig. 2.6). The deposits continually move parallel to the shore as *littoral drift*. The drift

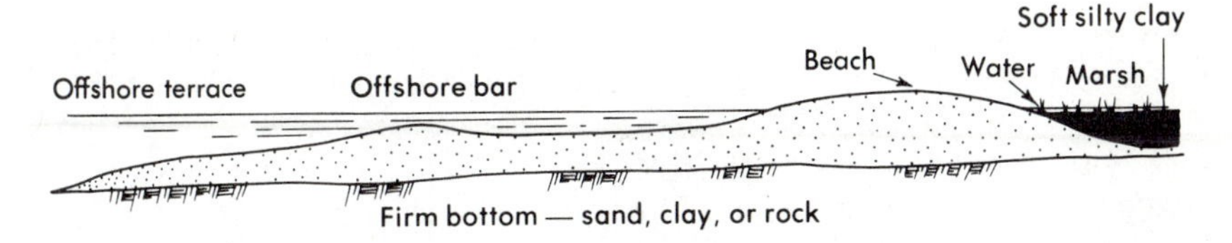

Figure 2.6 *Cross section of a shore deposit.*

accumulates behind structures that extend out from the shore, leaving the shore beyond starved for sand and subject to accelerated erosion. Spits, bars, and the beach are composed of coarser soil particles—sands, fine gravels, and shell fragments—which are uniform at any one point but which may vary considerably in size throughout the deposit. The coarser particles of sand and fine gravel are subrounded to rounded, but the finer sand grains are usually subangular. Irregular beds of broken shells are often included.

On many coastlines the spits or bars form barriers that eventually close off the beach from the sea and create shore or tidal lagoons. In some cases the lagoons are permanent lakes that rise and fall with the tide, but in others they are flat tidal marshes. Lagoon deposits are similar to those in shallow lakes. The clay deposits are likely to be thick and have a highly developed flocculent structure. They often contain lenses of sand or shells that are washed into them during large storms. Marine sands and gravels, shell beds, and the cemented strata provide excellent foundation support and are good sources of cohesionless materials for construction. The clays are ordinarily weak and highly compressive and capable of supporting only light loads. They are too wet for use in construction.

Coral reefs grow in warm, shallow salt water parallel to the shore. They sometimes grow on the offshore bar; where there is none, they form an analogous reef barrier. Dead, broken coral accumulates as sand both seaward and landward of the reef to form the picturesque shores of tropical islands such as in the West Indies and the South Pacific.

Along subtropical and tropical shores that are protected from wave action, mangrove swamps develop. The swamp is a wet jungle of the mangrove trees 5 to 10 m (15 to 30 ft) high and up to 80 mm (3 in.) in diameter. Their tangled roots extend as much as 1 m (3 ft) above the ground like spider legs. They trap sand and finer segments to form thick deposits of peat interlayered with inorganic soil. Decay of the mangrove peat produces hydrogen sulfide gas that has poisoned workers in excavations in those deposits. Water seeping through mangrove peat is extremely corrosive; at one construction site, stainless steel wells were destroyed in less than six months.

WIND DEPOSITS / Wind is a highly selective agent of particle transportation. Particles coarser than 0.05 mm, such as sand, are rolled along by the wind or may be lifted a few feet off the ground during violent windstorms only to be deposited a short distance away. Sand deposits formed by wind action are known as *dunes*. They form in desert regions where mechanical weathering produces an abundance of coarse particles and along lake- or seashores where the sands have been concentrated in beaches or bars by wave action. The most important characteristic of sand dunes is their continual migration in the direction of the prevailing wind—a migration that man is often powerless to halt. The moving sands cover highways, railroads, farmlands, and even towns, and efforts to stop them with sand "fences" or by attempting to cover them with protective coverings of vegetation have met with only sporadic success. Dunes take the form of irregular hills or ridges with flat slopes on their windward sides and steeper slopes equal to the angle of repose to the leeward. They are usually composed of relatively uniform, rounded to subrounded particles of sand sizes (Fig. 1.7b) and are a good source of such materials for construction.

Wind has the ability to lift and transport particles that are smaller than fine sand. Wind erosion is largely limited to dry silts of the arid regions, however, since cohesive or moist soils resist wind erosion.

Windblown silt may be carried for many miles before being deposited. Thick beds of windblown silt ordinarily accumulate in the semiarid grasslands that border the arid regions. The deposits build up slowly; therefore the grass growth keeps pace with the deposition. The result is high vertical porosity and vertical cleavage combined with an extremely loose structure. Such soils are termed *loess*. Most loess soils are hard because of deposits of calcium carbonate and iron oxide that line the former rootholes, but they become soft and mushy when saturated. Loess deposits are characterized by their uniform grain size, their yellow-brown color, and their pronounced vertical cleavage. Stream banks, gullies, and cuts in loess assume nearly vertical slopes because of the cleavage and because the high vertical permeability permits rapid saturation from rainfall and consequent sloughing of the soil on vertical planes (Fig. 12.11). Loess may be altered by weathering in humid regions, particularly if the soil grains consist of feldspars that were broken up by mechanical weathering alone. Such a soil is termed *loess loam* and it lacks the characteristic uniformity, high void ratio, and cleavage of true loess.

Loess provides good foundation support if it does not become saturated. It can be a source of fine-grained soil for construction if its structure can be broken down before use.

Volcanic ash may be grouped with wind-transported soils. It consists of small fragments of igneous rock blown out by the superheated steam and gases of a volcano. Fresh volcanic ash is a lightweight, porous-grained sand or sandy gravel. The deposits may be stratified or may be well-graded mixtures. Volcanic ash soaks up water readily and decomposes rapidly. When partially decomposed and then dried, it cements to form a soft rock known as *tuff*. Complete decomposition of the ash results in the formation of highly plastic clays with extremely high void ratios and high compressibilities. Although thick deposits of such clays are uncommon, the extreme settlement of structures built on them, such as Mexico City, makes them worthy of attention.

Cemented volcanic ash makes a good foundation. It is sometimes used as a construction material, but it tends to break down chemically and physically.

GLACIAL DEPOSITS / Ice, in the form of glaciers that plowed up great portions of North America and Europe, has been a very active agent of both weathering and transportation. The expanding ice sheets planed off hill tops, ground up rock, and mixed the materials together as they pushed their way southward. Some of the materials were directly deposited by the moving ice, while the remainder were transported by water flowing from the ice to be deposited in the lakes along the face of the ice sheets or transported in the rivers flowing away from the ice.

The direct deposits of the glacier are usually termed *moraines.* They are composed of *glacial till,* which is a term applied to the heterogeneous mixtures of particles, ranging from boulders to clay, that the ice accumulated in its travels. *End moraines* are irregular, low hills or ridges pushed up by the bulldozing action of the ice sheet. These mark the outermost limit of the glacier's travel, for they were left behind as the ice retreated. A *ground moraine* or *till plain* is the irregular veneer of till left on the areas once covered by the glacier. The upper surface of the gound moraine is undulating but rather level over broad areas; its thickness varies considerably, however, depending on the preglacial topography of the area. *Drumlins* are elongated low hills of till that point in the direction of the ice travel. They occur in areas of ground moraines.

The water-laid deposits of glaciers resemble those derived from mountain streams except that both the volume of water and the load of solids were considerably greater. *Eskers* are the remains of rivers that flowed in tunnels beneath the ice. When the ice retreated, the river bed materials formed sinuous ridges of coarse sands and gravels that resemble a crooked railroad embankment. *Kames* are terraces or irregular deposits of coarse sand and gravel formed along the margins of the ice sheets. Rivers flowing out of the edge of the ice sheet broke through the terminal moraine to deposit great quantities of sand and gravel in irregular, flat, braided beds termed *outwash plains.* In many areas the glacial streams flowed into large lakes that formed in depressions left by the retreating ice. The deposits in these lakes are similar to those formed in other lakes except that they are more extensive. Great deltas of sand formed at the mouths of the rivers, and thick beds of silt and clay formed in the still, deep waters beyond the shores. Occasional boulders and gravel found in the clays are believed to have been dropped by floating pieces of ice as they melted. The silts and clays often formed in thin alternate strata, which represent seasonal variations in the rate of ice melting and the resulting stream flow. The coarser particles were deposited in summer, during periods of high discharge, and the clays in winter. Such deposits are known as *varved clays* when the individual strata are more than 3 mm thick and as *laminated clays* when the strata are thinner.

Glacial sands, gravels, and till usually make good foundations. They are also good sources of construction materials. The glacial clays are only moderately strong and are often compressible. They often are problems in foundation design and usually are too wet to be used as construction materials.

2:4 Clastic Sedimentary Rocks

The clastic rocks are formed from soil deposits by a process of hardening of *induration.* The carbonate sediments also harden into sedimentary rock, but because of their special qualities they will be discussed separately in Section 2:5.

A number of different processes are involved in induration. Increasing weight from continued deposition or from glaciers as well as stresses induced by tectonic movements can consolidate the grains into a denser structure. The pressure may also produce added interparticle attraction in silts and clays. Cementing is the most important mechanism of induration. Silica, calcium carbonate, and iron oxides precipitated in the voids bind the solids together. Even clay, in a dry climate, can be a cementing agent. The degree depends on the amount and type of the cementing agent as well as the way in which it was precipitated.

STRUCTURE OF SEDIMENTARY ROCKS / The rock generally retains the structure of the original sediment. However, consolidation and tectonic forces can distort the original bedding, and solid particles can be broken and reoriented with their longest dimensions perpendicular to the direction of greatest normal stress.

The formations may be tilted and folded by crustal movements. Flexure of stratified deposits produces two sets of cracks, or joints, one parallel to the axis of folding and the other at right angles to it, and both perpendicular to the stratification. The former are *strike joints* and the latter *dip joints* because they point in the direction of the dip of the rock. The bedding planes may slide across one another during folding, so that the entire deposit becomes a mass of tightly packed, more or less rectangular blocks of hard rock that retains the sedimentary structure but which has lost its continuity completely.

Sedimentary rocks are sometimes invaded by igneous intrusions or blanketed by lava flows. The resulting structure can be a sandwich of rock forms, sliced apart by faulting and rejoined by intrusions.

TYPES OF SEDIMENTARY ROCKS / The classification of sedimentary rocks (Section 2:8) parallels the texture of soil deposits. *Siltstones* and *claystones* are hardened silts and clays. If the micas and clay minerals become reoriented so their surfaces are parallel, the claystone is *shale*. In engineering usage siltstones and claystones are often loosely described shales. A broader, all-encompassing term is *mudstone*. Indurated sand is *sandstone* and if the cementing is stronger than the sand particles it is an orthoquartzite. Indurated gravel is *conglomerate*. All can occur independently, but are frequently interbedded, as in the original sediment.

Angular broken rock produced by faulting is not properly a sediment. However, any angular fragmented masses, from faulting, volcanic action or accumulation that becomes indurated are *breccia*. Indurated volcanic ash is *tuff*, *tuffaceous sandstone* or *tuffaceous siltstone*, depending on its texture.

RESIDUAL SOILS FROM SEDIMENTARY ROCK / Sedimentary rocks weather back into soils if a changing environment reverses the induration processes. The minerals are often altered by the reweathering process. The new soil is somewhat different from the original sediment, although it retains much of the original structure. Generally, induration followed by weathering causes fracture of the harder particles, so that residual soils from sedimentary rocks are finer grained than their soil ancestors.

The depth of weathering of sedimentary rocks is generally less than for igneous rocks in the same environment because the mineral components had been weathered before induration. For example, in Georgia, the residual soil from porous sandstone may be 6 m thick, that from less pervious shale only 3 m thick but from granite 20 m or more, although all three have been exposed to similar environments for comparable periods.

SEDIMENTARY ROCKS AS ENGINEERING MATERIALS / Sedimentary rocks provide good foundations, depending on their induration. Some claystones and shales are likely to disintegrate rapidly upon stress relief or exposure to air; others expand because of chemical reactions between soluble salts in the clay and oxygen. Even biochemical disintegration has been observed in shales containing organic colloids. The loss of induration of sedimentary rocks is a serious problem in excavations and tunnels. It can be predicted by exposing small samples and noting their expansion and breakdown. Those that break down must be immediately protected from air and surface moisture by plastic or cement mortar spray coatings and kept under confining stress by rock bolts or other support to preserve their integrity.

The use of sedimentary rock as construction materials depends on their cementation; generally the sandstones and conglomerates make good fill. Shales, because of their tendency to swell and disintegrate after excavation and recompaction, should be considered suspect until tests show otherwise.

The residual soils are as variable in their structural qualities and their use in construction as the soil deposits from which they were derived.

Generally, if better strength or rigidity is needed in the virgin deposit, slightly increased depth can solve the problem.

2:5 Carbonate Sedimentary Rocks

Limestones and dolomites occur in a wide variety of forms and degrees of induration, depending on their mode of deposition and history. They are precipitated in warm, shallow water, often with tiny shells to form granular sediments. Sometimes these are soft and chalky but often indurated by reprecipitation into hard rock. A second form consists of shells, coral, and other calcareous debris that cements into porous *shellrock* or *coquina*. The third form appears to be the result of reprecipitation of the first two into a more crystalline rock, often with fossils embedded in the calcite or dolomite mass. A less common form is *travertine*, a soft limestone precipitated by hydrothermal action.

Limestones and dolomites frequently are deposited with other sediments, commonly clay. The combination may be mixed, as an argillaceous limestone, or interlayered as alternate seams of claystone (or shale) and limestone. Calcareous sandstones and interbedded sandstones and limestones also form, but are less common. Chemical alteration often takes place during or after induration with portions of the calcareous matter replaced by silica to form *chert* or *flint* nodules. In some old limestones the chert is a large part of the total volume.

STRUCTURE / Two forms of structural defects are prominent in limestones. The poorly indurated chalky, shell, and coral rocks are porous and are laced with interconnected voids of varying size, similar to soils termed *primary porosity*. These voids are frequently enlarged by solution into networks of small cavities. The second form, *secondary porosity*, consists of joints and bedding planes that divide the mass into more or less prismlike blocks. These cracks are also enlarged by solution.

WEATHERING[2:4, 2:5] / The limestones and dolomites weather by solution, leaving behind residual soils which comprised the insoluble parts of the original rock. The rate of solution depends on the acidity of the water, the carbon dioxide dissolved, the temperature, the water flow, the amount of previous saturation of the water by carbonate minerals, and the solubility of the rock. The surface area of the rock exposed to solution is a major factor. Solid, impervious limestone weathers from the surface down, creating an irregular, pitted rock surface. If the rock is jointed, solution proceeds deeper, enlarging the cracks and fissures into irregular slots that usually become narrower with increasing depth (Figs. 2.7 and 2.8). The slots may clog with the residual soil eroded down from above, redirecting the flow and the solution through smaller joints or across bedding planes. The rock mass is gradually changed into an irregular assortment of rock pinnacles and blocks with deep slots between and occasional horizontal caves and vertical

Figure 2.7 *Joints enlarged to slots filled with residual soil in limestone, exposed in highway cut 5 m (16 ft) deep in Tennessee.*

chimneylike holes. The porous limestones also dissolve, with the enlargement of the voids into tortuous channels of varying size and shape. Solution activity appears to be greatest near the water table; however, it occurs wherever there is groundwater movement. Moreover, limestones both above and far below the water table exhibit the results of solution produced during earlier geologic periods when the groundwater flow was different.

RESIDUAL SOIL / The residual soil consists of all the insoluble impurities of the rock: clay from kaolinite to montmorillonite, silica in the form of chert, from boulder sizes down, silica sand and silt, and iron oxides. The residual soil blanket is of varying thickness, depending on the age, the intensity of weathering, and the percentage of impurities. Some very cherty or clayey limestones accumulate more than 30 m of residual cover, while young or pure limestones in dry regions are bare. There is generally a sharp, although extremely irregular, line between soil and rock, in contrast to other forms of weathering. The soil immediately above the rock, and particularly in the slots and pits, is usually soft and pasty. That above is drier, stiffer, and sometimes partially indurated by cementing and desiccation. The residual soil mass is usually structureless, except for the soft zone above the rock. Very sandy or shaley limestones sometimes reflect the original rock structure in distorted bands of gravel, sand, or plastic clay.

CAVITIES, RAVELING, AND SUBSIDENCE[2:5] / The interplay between the solution slots and cavities in the rock, the sawtooth rock surface, and the blanket of residual soil above creates several serious engineering problems: (1) collapse of caves in the rock, (2) squeezing of the residual blanket at the points of support by pinnacles of rock, (3) raveling of the soil into the opening cavities and slots below, and (4) crushing of the weakened rock. These can occur independently or simultaneously and are responsible for many structural catastrophes. The most common and insidious problem is *raveling.* Downward seepage erodes part of the soil into the rock cavities leaving a dome-shaped cavity in the lower surface of the residual blanket. This enhances the seepage in that direction, aggravating the erosion. Pieces of the dome spall off when moisture weakens the dome surface, a process termed *raveling* or *roofing* in which the dome enlarges (Fig. 2.8). Depending on local weaknesses and on soil plasticity, the enlargement may be vertical to form a narrow chimneylike hole or lateral to

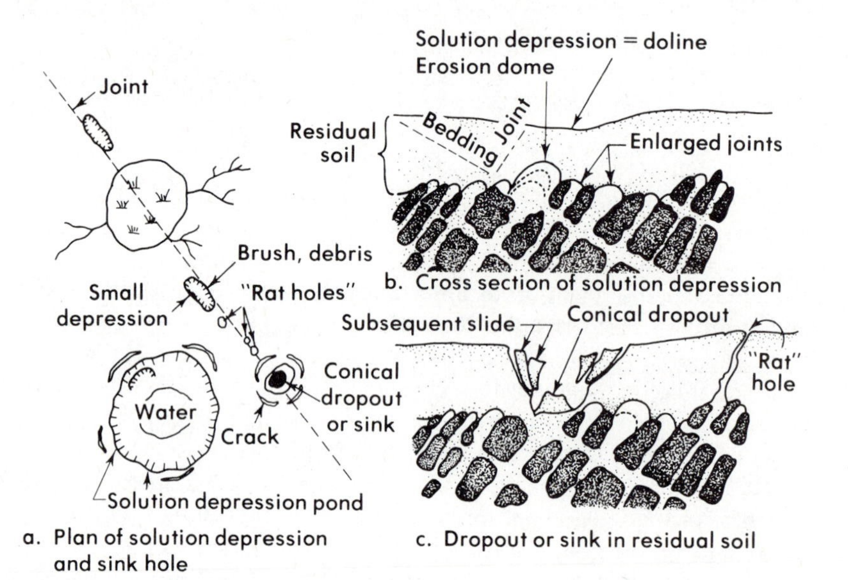

Figure 2.8 *Plan and cross section of a residual soil over limestone with solution channels, solution depressions, and sink holes in the residual soil.*

form a broad dome. Eventually, a dome becomes so large that the soil cannot span the opening, and a truncated cone drops downward creating a sink hole. Raveling is aggravated by changes in ground water, unusually dry weather, diversion of surface water into the ground, and sometimes through weakening the residual soil blanket by excavation.

ENGINEERING QUALITIES OF LIMESTONES / The harder, sound limestones and dolomites are among the best rocks for foundations, tunnels, and construction materials. The shell and coral rocks, shaley limestones, and the more earthy forms are variable; their qualities depend largely on the degree of cementation. Their crushing or disintegration must be evaluated from past experience or by tests before they can be used. The solution defects are major problems in foundations, reservoirs, dams, and tunnels. It is best to assume that they are present until adequate investigation (Chapter 7) proves otherwise.

The sandy, gravelly clay residual soils are usually good sources of clay for embankment construction, provided montmorillonite is scarce. They ordinarily are good foundations where not undermined by raveling. Subsidence must be suspected, until investigation proves otherwise.

2:6 Metamorphic Rocks

Metamorphic rocks are produced by heat, pressure, and shear sufficient to alter the minerals in the original materials, to reorient them, and to contort the rock mass as if it were a soft plastic. Generally, clay minerals are recombined to produce new claylike minerals such as chlorite. Other minerals are recrystallized and the crystals distorted by shear and pressure. Shales and mudstones with moderate metamorphism become *slates* and *phyllites*. Higher levels of heat and pressure produce more profound changes. Extreme orientation and segregation of platey, micaceous minerals is characteristic of *schists*, with their laminated or *foliated* structure, while *gneisses* exhibit segregation and orientation of minerals in their banding. Sandstones are altered into *quartzites*, which retain relics of the bedding, and which are among the toughest of rocks.

Limestones and dolomites are recrystallized into marble without significant changes in mineralogy. However, the impurities may be altered considerably.

STRUCTURE AND WEATHERING / Most metamorphic rocks exhibit joints and other structures similar to igneous rocks. In addition, their *schistosity* and *flow banding* are frequently surfaces of weakness that are twisted and contorted by the pressures which formed them (Fig. 2.9).

The marbles weather by solution similar to the nonporous limestones. The jointing is seldom as severe, and so the rate may be slower. The other metamorphic rocks weather in much the same way as the igneous rocks, with decreasing decomposition with increasing depth and no sharp boundary between residual soil and rock.

RESIDUAL SOILS / The soils range from sandy silts to silty sands with varying amounts of mica in those derived from gneiss and schist. The deposits are extremely variable in composition and extent. The minerals are arranged in the same laminations or bands as in the original rock. Those soils which retain the relic structure and fabric of the rock are also termed *saprolites*. Within the mass of unweathered rock there are often seams of

Figure 2.9 *Banding in a residual soil or saprolite derived from the decomposition of gneiss, exposed in a foundation excavation 0.6 m (2 ft) deep in North Carolina.*

partially weathered material, from some less resistant band, or along old joints or fault zones. Within the soil there are frequently pinnacles or sawtooth projections of hard, slightly weathered rock. At one site in the southeastern United States, the depth to sound rock varied from 2 to 20 m in a distance of 60 m. Seams of soft, highly kaolinized feldspar alternated with harder bands of quartz and thin seams of weak mica. Such a formation is often brightly banded, as can be seen in Fig. 2.9.

The residual soils from slates and phyllites are similar but more clayey, with slablike inclusions of less weathered rock alternating with clayey and silty strata.

Because of their variability, high mica and clay content, and mineral orientation, schists, gneisses, and similar rocks, as well as their residual soils, require careful study. The residual soils derived from quartzites more nearly resemble those from granites, and residuals of marbles resemble those of limestones in their engineering behavior.

2:7 Profile Development—Pedogenesis[2:6,2:7]

The continuing exposure of the uppermost soils develops a characteristic weathering *profile* from the ground surface down. A number of different mechanisms are involved: the accumulation and decay of organic matter, leaching, precipitation, oxidation or reduction, and additional weathering. The profile that ultimately develops depends on the *parent material* that comprises the initial deposit. Even more important are the environmental factors: temperature, amount and seasonal distribution of rainfall, ground surface slope, groundwater level, and vegetation. These factors are interdependent to some extent, and it is not always easy to identify the relative contribution of each. The degree of profile development depends on the environment and the period of time it has acted. In newly deposited materials the profile is shallow and poorly defined; in older deposits it may be as thick as 3 to 5 m and clearly defined.

The science of profile analysis is termed *pedology* or *soil science*. It is one of the basic sciences of agronomy, where tilth and fertility are directly related to the soil profile. It is also vital to the soil engineer who utilizes the uppermost portions of a soil deposit as a foundation or as a source of construction materials.

In the tropics where the secondary weathering occurs rapidly and in regions that have been geologically stable for long periods, the profile development is so deep that it comprises the greater part of the soil mass, obscuring the nature of the original deposit. These are sometimes termed *pedogenetic* soils or *geosols*.

PEDOLOGIC CLASSIFICATION / The soil profile and the environment that produced it (as reflected in the materials composing the profile) are the bases for classifying soils pedologically. Two different systems are in use. The first, which divides soils into *zonal orders* and great soil groups, is based on environment. Variations of this system are used worldwide, and were used in the United States before 1960. The second, termed the *Seventh Approximation*,[2:7] divides soils into *orders* and *suborders* based on the major differences in soil profile. It has been employed by the U.S. Department of Agriculture since 1960. Both systems include more detailed subdivisions ending with *series*, named for the locality in which the soil was first identified, and further divided into *phases* based on the texture or condition of the upper horizons.

COOL AND TEMPERATE HUMID REGION PROFILE / In cool and temperature regions with humid climates there is an abundant growth of vegetation and an accumulation of dead leaves, plants, and other organic debris. The slow decomposition of these materials and the secretion of plant roots make weak acids which accelerate the weathering. The prevailing soil–moisture movement is downward to the water table, and this creates a profile with three distinct layers or *horizons*: A, B, and C (Fig. 2.10).

The A-horizon is characterized by the chemical alteration of the soil materials in an acid, reducing environment. Clays of the kaolinite family, soluble carbonates, and semisoluble reduced iron minerals are likely to be produced. These are leached downward by the soil moisture, leaving the

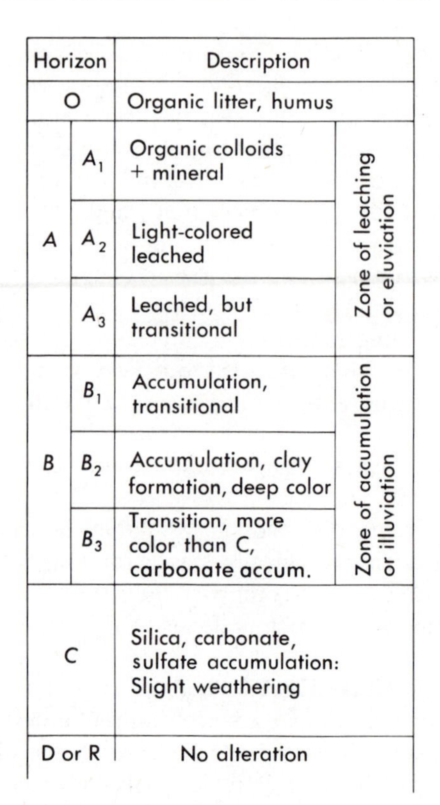

Figure 2.10 *Soil profile in a region of mild temperature and moderate rainfall.*

A-horizon deficient in them but rich in silica. As a result the lower part, designated the A_2-horizon, is usually very sandy and light colored. The upper part, designated A_1, is dark colored from its content of organic matter and has a spongy texture. The A_2-horizon is frequently a source of sandy soils in regions underlain by clays.

The leached materials accumulate in the B-horizon below. It is thicker than the A and contains a greater concentration of clay minerals, iron, and carbonates than does the original soil. The top part, B_1, is frequently partially cemented, and is deeply colored. The B_2 is rich in clay and soluble carbonates. The B_3 again suffers from downward leaching and is lighter colored. The B-horizon is the best source of clays in regions where they are scarce.

Below is the C-horizon, which is the slightly weathered parent material. When an unrelated stratum of different soil underlies the other materials, it is sometimes called the *R-horizon*. The D-horizon is the unaltered parent, or rock.

HOT, HUMID REGION PROFILES / In hot, humid regions the upper parts of the deposit are also subject to alternate wetting and drying and to downward leaching. However, the climate is favorable to the rapid decay of organic matter and of its consumption by such insects as termites. There is little or no organic acid produced, and because of the formation of soluble carbonates, the silicate weathering proceeds in a neutral environment. The soluble colloidal silica is leached downward. The aluminum and iron become highly oxidized, and are insoluble in the neutral environment. They remain to cement the quartz into a stiff rocklike solid. Advanced weathering and leaching causes the iron and aluminum to accumulate in nodules or *concretions*, giving the soil the texture of a loose, but cemented gravel. The color ranges from tan to bright red because of the highly oxidized iron, with mottling reflecting the local iron accumulations. The process is termed *laterization* and the well-indurated material is *laterite* or *ferricrete*. Well-developed laterites are strong and relatively incompressible, although often lightweight and porous. Some forms are sufficiently cemented to be used as gravel for road construction, while less well-developed laterites may soften upon wetting. Laterites are usually identified by their low ratio of silica to the aluminum and iron oxides.

DRY REGION PROFILE / In dry regions there is little or no organic matter. Any moisture movement is predominantly upward because of surface evaporation. This results in the accumulation of soluble materials such as carbonates near the surface and in the partial cementing of the soil.

Sometimes the carbonates are well distributed through the mass, while in other cases they are concentrated in lenses or concretions at the level of evaporation of capillary moisture in the deposit.

These soils are usually strong and incompressible when dry. Upon saturation they weaken and, in some cases, collapse with sudden loss of strength and rapid subsidence. They may be good construction materials, depending on the parent material.

In extremely arid regions, soluble salts are brought upward by capillarity following the brief periods of rainfall and are precipitated near and on the surface forming a saline or alkali topsoil and sometimes a white crust. Overirrigation in very dry regions can cause the same salt accumulation and eventually make the ground infertile.

HUMID—POORLY DRAINED / The wet, poorly drained soils form similar groups depending on the moisture. In a very wet environment, organic growth is rapid. Decay, however, is slow if the area is inundated and the water stagnant. The organic decay depletes the oxygen and the decay products inhibit further decay. The organic matter accumulates rapidly under such conditions. Higher rates of decay, associated with fluctuating water levels, produce nearly fiberless *mucks*, while slower decay, associated with continuing stagnant inundation, produces *fibrous peat*. Because the decay tends to be slowest in cool regions, the thickest peat deposits are frequently found in the subarctic. Thick peat deposits also form in the tropics in slowly submerging deltas or in old oxbow lakes.

2:8 Rock Classification and Description

The study of rocks, including their classification, is termed *petrography*. Because of the complexities of rock formation and their many different mineral constituents, it is a complex science. However, from the standpoint of engineering behavior, detailed classification is seldom necessary. Table 2:1 (pp. 70–71) is adequate for most engineering uses.

INDURATION-STRENGTH / The strength of the rock is of more importance in engineering than the texture or geologic classification. Tests of the intact rock (see Chapter 5) are necessary for a quantitative determination. A standard for describing induration in terms of unconfined compressive strength and simple filed tests for estimating are given in Table 2:2.

TABLE 2:2 / DESCRIBING ROCK INDURATION
(Adapted from Duncan)[2:8]

Description	Unconfined Compressive Strength	Field Test
Very hard	20,000 lb/in.^{2} (1400 kg/cm^2) or more	Difficult to break 10-cm piece with pick
Hard	8–20,000 lb/in.^{2} (560–1400 kg/cm^2)	10-cm piece broken with one hammer blow
Soft	2.5–8000 lb/in.^{2} (175–560 kg/cm^2)	Can be scraped, or dented slightly, with pick point
Very soft	1–2500 lb/in.^{2} (70–175 kg/cm^2)	Crumbles with pick, easily scraped with knife

STRUCTURE / The behavior of rock in engineering work is largely controlled by its mechanical structure. An infinite variety of arrangements occur; these can be grouped from the engineering point of view as described below.

Layering (Fig. 2.11a) is the segregation of like materials into more or less parallel sheets. It occurs in most sediments as deposit bedding and in metamorphic rocks due to pressure and flow. Many patterns are observed: thin layers of similar materials, hard layers over soft, soft over hard, alternating hard and soft, lying horizontally or dipping, straight or contorted, and even nonconformal. The character of the interface between layers is significant in establishing the behavior under load. Smooth surfaces slide more easily than rough ones in which the irregularities match or conform. Nonconforming irregularities contain points of stress concentration that can crush or tear under load. The orientation of the bedding with respect to the

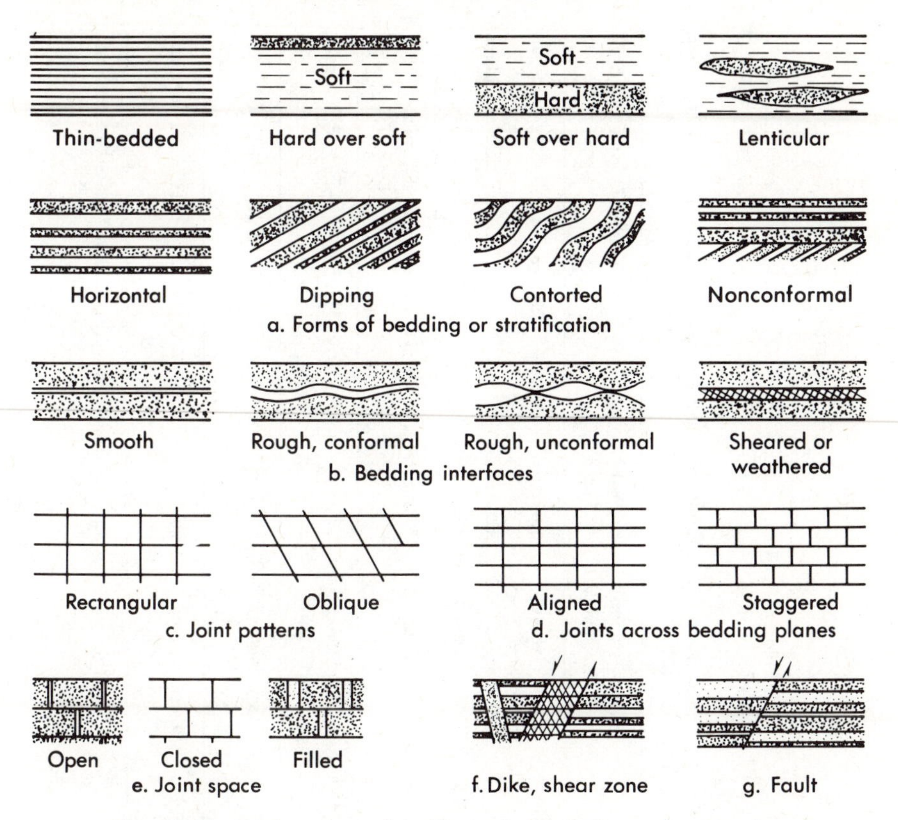

Figure 2.11 *Rock structure of significance in foundations and construction.*

applied loads determines the stresses and the tendency for movement to develop. The scale of the layering, expressed as the ratio of the engineering structure width, B, to the layer thickness, H, establishes the overall significance of the stratification and particularly of any weak interfaces.

The joints (Fig. 2.11c) are cracks more or less perpendicular to the bedding surfaces. They occur in groups or sets, with the joints of any one set approximately parallel and equally spaced. Typical joints lie parallel and at right angles to the axes of folds; these are *strike* and *dip* joints, respectively. They divide the rock into rectangular prisms. Other sets may occur at oblique angles that divide the rock into parallelepiped or wedges. Joints in one layer aligned with those in the next, are favorable to joint movement. If they are staggered, it is a more stable arrangement. The character of the joint itself is a major factor in its contribution to rock behavior. If closed, the mass behaves like a continuous solid. If it is open, considerable movement will take place before there is rock-to-rock transfer of load across the joint. If the joint is filled with a soft material, the behavior of the rock mass is controlled by the resistance of the filling. The roughness of the joint surface, similarly,

TABLE 2:1 / SIMPLE ROCK CLASSIFICATION

Class	Type	Family	General Features
Igneous	Intrusive	Granite	Light colored; 10–40% quartz, 40–60% O & M* feldspar, 2% mica, 0–10% FM‡
	(Coarse-grained)	Syenite	Light colored; 5–10% quartz, 25–50% O & M* feldspar, 0–20% P† feldspar, 5–20% FM‡
		Diorite	Light-dark "salt-pepper", 0–5% quartz, 0–25% O & M* feldspar, 20–50% P† feldspar, 20–30% FM‡
		Gabbro, ultra basics	Dark colored; 0% quartz, 0% O & M* feldspar, 40–50% P† feldspar, 30–50% FM‡
	Extrusive	Obsidian	Volcanic glass, often dark colored
	(Fine-grained)	Rhyolite	Light colored; 10–40% quartz, 40–50% O & M* feldspar, 2% mica, 0–10% FM‡
		Trachite	Light colored; 5–10% quartz, 25–50% O & M* feldspar, 0–20% P† feldspar, 5–20% FM‡
		Andesite	Dark colored; 0–5% quartz, 0–25% O & M* feldspar, 20–50% P† feldspar, 20–30% FM‡
		Basalt, Dolerite	Dark colored; 0% quartz, 0% O & M* feldspar, 40–50% P† feldspar, 30–50% FM‡
	Ejecta	Tuff (volcanic ash)	Cinderlike sand and silt-sized fragments, with occasional gravel-sized angular fragments
		Pumice (porous)	Frothy or very porous lava—usually light colored

Sedimentary	Carbonate	Limestone	Calcium carbonate	Range in texture: coarsely crystalline–fossiliferous–granular–earthy
		Dolomite	Calcium-magnesium carbonate	
	Silaceous	Siltstone, claystone mudstone	Nonoriented, indurated silt and clay	
		Shale	Oriented-laminated, indurated silt and clay	
		Sandstone	Cemented sand, arkosic if appreciable feldspar present	
		Conglomerate	Cemented rounded sand-gravel or gravel	
		Breccia	Cemented angular rock fragments	
Metamorphic	Foliated	Slate	Finely foliated or oriented; fine grained; thin smooth cleavage	
	(Oriented grains)	Schist	Pencil-line to paper-thin foliation	
		Gneiss	Bands of minerals 1 mm or more; rough cleavage	
	Nonfoliated	Quartzite	Dense sand gravel structure that fractures through the grains	
		Marble	Recrystallized limestone or dolomite	

* Orthoclase and microcline feldspar.
† Plagioclase feldspar.
‡ Ferromagnesian minerals.

influences potential movement through or across the crack. Orientation of the joints with respect to the loading and the ratio of structure width B to joint spacing, S, is of similar significance as in the layers.

Defects in the rock mass (Fig. 2.11e to g) include shear zones, slickensides, dikes, cavities, solution channels, and porous zones. Some discontinuities are zones of weakness; the more rigid ones concentrate stresses. Their spacing and orientation are major factors in their engineering effects, similar to the layers and joints.

In each situation all pertinent details of the structure must be established. In most cases the structure is the dominant factor, and the nature of the rock between joints or bedding planes is of minor importance.

Systems for describing the structure quantitatively are described in textbooks of geologic structure and treatises on rock mechanics. Useful terms for describing spacing are given in Table 2:3. Similar features are also found in soil formations, although, unfortunately, they are not always recognized.

TABLE 2:3 / SPACING OF STRUCTURE
(After Deere)[2:9]

Descriptive Term Bedding	Descriptive Term Joints	Beds or Joints per Meter or Yard	Spacing	
			Cm	In.
Very thin	Very close	>20	Less than 5	Less than 2
Thin	Close	3–20	5–30	2–12
Medium	Medium	3–1	30–90	12–36
Thick	Wide	1–1/3	90–300	36–120
Very thick	Very wide	<1/3	Over 300	Over 120

The orientation of the bedding foliation, joint, and fault surfaces is described by the strike and dip. The *strike* is the direction of the line of intersection between a level plane and the plane of the surface. Geologists generally express it as a bearing: N 35 E or N 40 W. Engineers ordinarily use azimuth—comparable values are 35 and 320°. Because the azimuth or bearing of a line can be expressed by two angles, depending on which end of a line is referenced, strike is usually referenced to the north half of the circle: 270 to 90°. Dip is the angle between the plane of the surface and the horizontal plane, measured perpendicular to the strike. The general direction of the down dip is also noted: dip 60° SW.

Neither the strike nor the dip of a single surface is constant; the surface is warped to varying degrees. Thus the orientation of systems of joints or bedding should be described statistically. Texts on field geology describe various representations that are in use.

2:9 Engineering Soil Classification

A soil classification system is an arrangement of different soils into groups having similar properties. The purpose is to make it possible to estimate soil properties or capabilities by association with soils of the same class whose properties are known and to provide the engineer with an accurate method of soil description. However, there are so many different soil properties in any natural soil deposit that any universal system of classification seems impractical. Instead the groups or classes are based on those properties which are most important in that particular phase of engineering for which the classification was developed. For example, the Public Roads Classification System groups soils according to their suitability for road construction. The same properties may be of little use in classifying soils for earth dams. The soils engineer should be familiar with the purposes and particularly the limitations of the important soil classification systems. He should be able to develop new systems to fit new problems rather than try to adapt the old systems to situations where they do not apply. But the engineer must not lose himself in this, for as A. Casagrande has said, "Those who really understand soils can, and often do, apply soil mechanics without a formally accepted classification."[2:10]

TEXTURAL CLASSIFICATIONS / Textural classifications group soils by their grain size characteristics. The gravel and larger sizes are disregarded and the particles finer than 2 mm in diameter are divided into three groups: sand sizes, silt sizes, and clay sizes. The soils are then grouped by the percentage of each of these three components.

Textural classification was developed by agricultural engineers, who found that grain size was an indication of the workability of topsoils. A number of different textural classification schemes have been employed in engineering work, but they have been superseded by the more complete engineering classification systems described below.

AASHTO OR PUBLIC ROADS CLASSIFICATION SYSTEM / The Public Roads Classification is one of the oldest systems of grouping soils for an engineering purpose. Since its introduction in 1929, it has undergone many revisions and modifications and is widely used for evaluating soils for highway subgrade and embankment construction. The modification proposed in 1945 is termed the *Revised Bureau of Public Roads, Highway Research Board*, or AASHTO (*American Association of State Highways and Transportation Officials*) System. This system divides all soils into three categories: granular, with 35% or less by weight passing a No. 200 sieve (finer than 0.074 mm); silt-clay, with more than 35% passing the No. 200 sieve; and organic soils. The first two categories are subdivided further, depending on their gradation and plasticity characteristics, as shown in Table 2:4. The symbols A–1 through A–8 are given to the classes which loosely indicate a decreasing quality for highway construction with

TABLE 2:4 / REVISED PUBLIC ROADS OR AASHTO CLASSIFICATION[2:11]

Group	Sub-group	Percent Passing U.S. Sieve No.			Character of Fraction Passing No. 40 Sieve		Group Index No.	Soil Description	Subgrade Rating
		10	40	200	Liquid Limit	Plasticity Index			
A–1			50 max	25 max		6 max	0	Well-graded gravel or sand; may include fines	
	A–1–a	50 max	30 max	15 max		6 max	0	Largely gravel but can include sand and fines	
	A–1–b		50 max	25 max		6 max	0	Gravelly sand or graded sand; may include fines	
A–2*				35 max			0 to 4	Sands and gravels with excessive fines	Excellent to good
	A–2–4			35 max	40 max	10 max	0	Sands, gravels with elastic silt fines	
	A–2–5			35 max	41 min	10 max	0	Sands, gravels with elastic silt fines	
	A–2–6			35 max	40 max	11 min	4 max	Sands, gravels with clay fines	
	A–2–7			35 max	41 min	11 min	4 max	Sands, gravels with highly plastic clay fines	

Group	Subgroup								
A–3			51 min	10 max		Nonplastic	0	Fine sands	
A–4				36 min	40 max	10 max	8 max	Low-compressibility silts	
A–5				36 min	41 min	10 max	12 max	High-compressibility silts, micaceous silts	
A–6				36 min	40 max	11 min	16 max	Low-to-medium-compressibility clays	Fair to poor
A–7				36 min	41 min	11 min	20 max	High-compressibility clays	
	A–7–5			36 min	41 min	11 min†	20 max	High-compressibility silty clays	
	A–7–6			36 min	41 min	11 min†	20 max	High-compressibility, high-volume-change clays	
A–8								Peat, highly organic soils	Unsatisfactory

* Group A–2 includes all soils having 35% or less passing a No. 200 sieve that cannot be classed as A–1 or A–3.

† Plasticity index of A–7–5 subgroup is equal to or less than LL–30. Plasticity index of A–7–6 subgroup is greater than LL–30.

increasing number. Some of the classes are subdivided to indicate differences in plasticity, but the subdivisions are not an essential part of the system. The classification is supplemented by the *Group Index*, or GI:

$$GI = 0.2a + 0.005ac + 0.01bd \qquad (2:1)$$

where a = percentage passing the No. 200 sieve greater than 35 and not exceeding 75; expressed as a whole number (0 to 40)
b = percentage passing No. 200 sieve greater than 15 and not exceeding 55; expressed as a whole number (0 to 40)
c = that portion of the liquid limit greater than 40 and not exceeding 60; expressed as a whole number (0 to 20)
d = that portion of the plasticity index greater than 10 and not exceeding 30; expressed as a whole number (0 to 20)

The value of the GI ranges from 0 to 20, with the low numbers indicating higher quality than the high numbers. The number is placed in parentheses following the class, such as A–2(0) or A–5(9).

Since the same basic symbols have been used for all the versions of the Public Roads System, the engineer should always note which he is using. The presence of the GI number, however, denotes the 1945 revision.

UNIFIED SOIL CLASSIFICATION SYSTEM / The Unified Soil Classification[2:12, 2:13] is an outgrowth of the Airfield Classification (AC) system developed by A. Casagrande[2:10] as a rapid method for identifying and grouping soils for military construction. The soils are first divided into coarse-grained and fine-grained classes. The coarse-grained soils have over 50% by weight coarser than 0.075 mm (No. 200 sieve). They are given the symbol G if more than half of the coarse particles by weight are coarser than 4.75 mm (No. 4 sieve) and S if more than half are finer. The G or S is followed by a second letter that describes the gradation: W, well graded with little or no fines; P, poorly graded, uniform, or gap-graded with little or no fines; M, containing silt or silt and sand; and C, containing clay or sand and clay. The fine-grained soils (over half finer than 0.075 mm) are divided into three groups: C, clays; M, silts and silty clays; and O, organic silts and clays. These symbols are followed by a second letter denoting the liquid limit or relative compressibility: L, a liquid limit less than 50; and H, a liquid limit exceeding 50.

The Casagrande plasticity chart (Fig. 2.12) is the basis for dividing the fine-grained soils. It also aids in comparing different soils. For example, clays having a similar geologic origin will usually plot in a narrow band parallel to the dividing area (often called the A-line) between the C and M-O soils. The different symbols, the soils they represent, and the classification criteria are given in Table 2.5 (pp. 78–79). For borderline soils a dual classification is sometimes given, such as GW-GC. The system involves more than the group symbol. It includes a description of grain shape, mineralogy, color, and, if

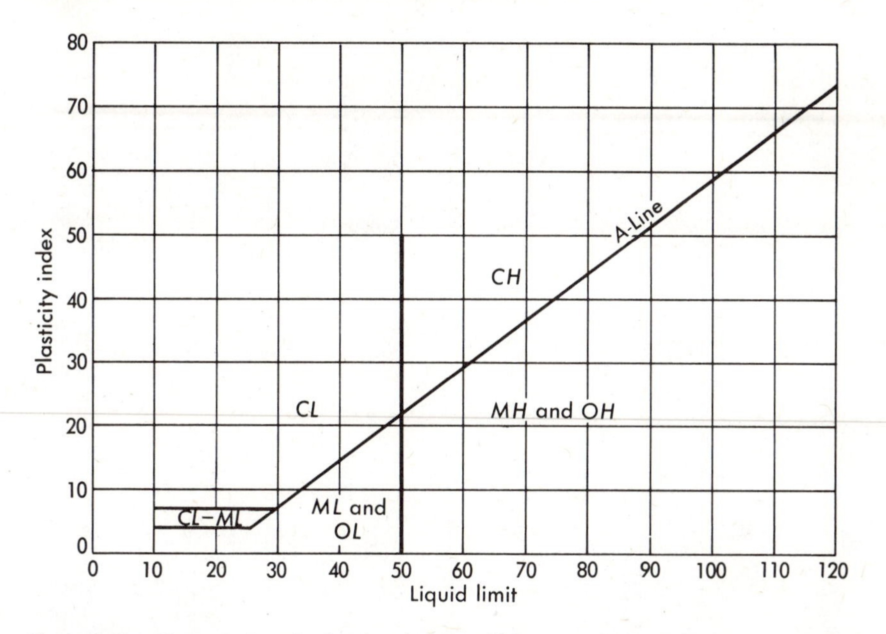

Figure 2.12 *Plasticity chart for the classification of fine-grained soils. Tests made on fraction finer than No. 40 sieve, 0.425 mm. (After A. Casagrande*[2:10] *and the U.S. Waterways Experiment Station.*[2:12])

undisturbed deposits are classified, the stratification of macrostructure, consistency, and density in the ground.

The original AC system differs from the Unified in the grouping of the coarse-grained soils: the division between coarse and fine is 0.1 mm; the symbol M is replaced by F; and all the coarse-grained subdivisions, W, C, P, and F, have slightly different meanings. For most practical purposes, however, the systems are the same.

The Unified System has proved very useful in classifying soils for many different purposes such as highway and airfield construction, earth dams, embankments, and even for foundations. It is frequently supplemented by tables showing the typical properties of each group, such as the drainage characteristics, and as such is a valuable guide for design and construction. The system is simple. Many soils can be grouped visually, and only tests for grain size and plasticity are necessary for accurate classification. It must be kept in mind, however, that no classification is a substitute for tests of the soil's properties and engineering analysis of the results.

2:10 Soil Identification and Description

The different soil classification schemes, while useful for grouping soils for a particular purpose, may be useless or misleading in other applications.

TABLE 2:5 / UNIFIED SOIL CLASSIFICATION
(After U.S. Waterways Experiment Station and ASTM D-2487-69)

Major Division		Group Symbol	Laboratory Classification Criteria		Soil Description
			Finer than 200 Sieve (%)	Supplementary Requirements	
Coarse-grained (over 50% by weight coarser than No. 200 sieve	Gravelly soils (over half of coarse fraction larger than No. 4)	GW	0–5*	D_{60}/D_{10} greater than 4 $D_{30}^2/(D_{60} \times D_{10})$ between 1 & 3	Well-graded gravels, sandy gravels
		GP	0–5*	Not meeting above gradation for GW	Gap-graded or uniform gravels, sandy gravels
		GM	12 or more*	PI less than 4 or below A-line	Silty gravels, silty sandy gravels
		GC	12 or more*	PI over 7 and above A-line	Clayey gravels, clayey sandy gravels
	Sandy soils (over half of coarse fraction finer than No. 4)	SW	0–5*	D_{60}/D_{10} greater than 4, $D_{30}^2/(D_{60} \times D_{10})$ between 1 & 3	Well-graded, gravelly sands
		SP	0–5*	Not meeting above gradation requirements	Gap-graded or uniform sands, gravelly sands
		SM	12 or more*	PI less than 4 or below A-line	Silty sands, silty gravelly sands
		SC	12 or more*	PI over 7 and above A-line	Clayey sands, clayey gravelly sands

Fine-grained (over 50% by weight finer than No. 200 sieve)	Low compressibility (liquid limit less than 50)	ML	Plasticity chart	Silts, very fine sands, silty or clayey fine sands, micaceous silts
		CL	Plasticity chart	Low plasticity clays, sandy or silty clays
		OL	Plasticity chart, organic odor or color	Organic silts and clays of low plasticity
	High compressibility (liquid limit more than 50)	MH	Plasticity chart	Micaceous silts, diatomaceous silts, volcanic ash
		CH	Plasticity chart	Highly plastic clays and sandy clays
		OH	Plasticity chart, organic odor or color	Organic silts and clays of high plasticity
Soils with fibrous organic matter		Pt	Fibrous organic matter; will char, burn, or glow	Peat, sandy peats, and clayey peat

* For soils having 5 to 12% passing the No. 200 sieve, use a dual symbol such as GW–GC.

In many fields, such as foundation engineering, there are so many significant soil properties that any scheme of soil classification is awkward. Instead, an accurate description of the significant soil properties can convey the necessary information without the restrictions of a definite classification scheme. The following soil properties are of significance in most soil problems and therefore form the basis of a complete soil description. They are also a required supplement to the Unified Classification.

1. Shear strength (cohesive soils).
2. Density (cohesionless soils).
3. Compressibility.
4. Permeability.
5. Color.
6. Composition (grain size, shape, plasticity, mineralogy).
7. Structure of soil.

For a precise description many of these properties must be determined by laboratory tests. An experienced soils engineer, however, can estimate most of these by careful field observation and examination of small samples of the soil.[2:14]

SOIL STRENGTH / Shear strength is a fundamental property of undisturbed cohesive soils, a knowledge of which is necessary in solving many problems. It is ordinarily defined in terms of unconfined compressive strength (Section 5:6) but may be estimated from the pressure required to squeeze an undisturbed sample between the fingers. If the soil is *brittle* (fails suddenly with little strain), *elastic* (rubbery), *friable* (crumbles easily) or *sensitive* (loses strength on remolding), these terms should be included in the description. (See Table 2:6.)

TABLE 2:6 / SOIL STRENGTH

Term	Unconfined Compressive Strength (After Terzaghi and Peck)[2:15]		Field Test (After Cooling, Skempton, and Glossop)[2:16]
	Kips/ft^2	kN/m^2	
Very soft	0–0.5	0–25	Squeezes between fingers when fist is closed
Soft	0.5–1	25–50	Easily molded by fingers
Firm	1–2	50–100	Molded by strong pressure of fingers
Stiff	2–3	100–150	Dented by strong pressure of fingers
Very stiff	3–4	150–200	Dented only slightly by finger pressure
Hard	4+	200+	Dented only slightly by pencil point

DENSITY / Relative density is as important for cohesionless soils as strength is for cohesive. It can be found by comparing the soil's actual void ratio with the range in void ratio from loose to dense for that soil. It may be estimated from the ease with which a reinforcing rod penetrates the soil, or from Tables 2:7 and 7:4.

TABLE 2:7 / SOIL RELATIVE DENSITY

Term	Relative Density (%)	Field Test
Loose	0–50	Easily penetrated with 12-mm or $\frac{1}{2}$-in. reinforcing rod pushed by hand
Firm	50–70	Easily penetrated with 12-mm or $\frac{1}{2}$-in. reinforcing rod driven with 2.3-kg or 5-lb hammer
Dense	70–90	Penetrated a foot with 12-mm or $\frac{1}{2}$-in. reinforcing rod driven with 2.3-kg or 5-lb hammer
Very dense	90–100	Penetrated only a few inches with 12-mm or $\frac{1}{2}$-in. reinforcing rod driven with a 2.3-kg or 5-lb hammer

COMPRESSIBILITY / Compressibility is determined by direct laboratory tests (Section 4:2) or is estimated from the liquid limit and void ratio. (See Table 2:8.)

TABLE 2:8 / COMPRESSIBILITY

Term	Compression Index	Liquid Limit (approx.)
Slight or low compressibility	0–0.19	0–30
Moderate or intermediate	0.20–0.39	31–50
High compressibility	0.40 and over	51 and over

PERMEABILITY / Permeability is determined by direct laboratory and field tests or may be estimated from Table 3:1.

COLOR / Color, although not an important physical property in itself, is an indication of more important properties. For example, yellow and red hues indicate that a soil has undergone severe weathering, for the colors are iron oxides. A dark greenish brown is often an indication of colloidal organic matter. A change in color encountered during excavation often means a different soil stratum with different properties has been uncovered. Color is usually the easiest property of a soil for persons untrained in soil mechanics to identify; therefore, a practical method of describing a certain soil to workers is by color. Soil color is described visually with the aid of the Munsell color charts.[2:17]

COMPOSITION / Composition includes the grain size, gradation, grain shape, mineralogy (or the coarser grains), and plasticity. Two groups of soils are recognized: predominantly coarse grained (over 0.075 mm) and predominantly fine grained (less than 0.075 mm), as in the Unified Classification. The coarse-grained soils are described primarily on the basis of the grain size, the fine-grained primarily on the basis of their plasticity. The amount of coarse or fine component required to predominate is not fixed, for it depends on the soil structure: If the coarse-grained particles can make contact with one another, the soil behaves essentially as a coarse-grained material; if they cannot touch but are separated by the fines, the fines predominate. The unified Classification arbitrarily defines predominant grain size at over 50% by weight.

However, for soils containing clay minerals, the volume of fines controls soil behavior, although they comprise considerably less than 50% by weight. For example, a well-graded, sand–silt–clay mixture in which the fines exhibit low plasticity may behave like a clayey or fine-grained soil with only 30% fines. If the fine fraction is highly plastic, 10 to 20% fines may be sufficient for the soil to behave as a fine-grained material. Therefore, no fixed percentage of fines can distinguish predominantly coarse-grained or fine-grained behavior, and the engineer must exercise judgment.

The sizes of particles as defined by ASTM–ASCE and their visual identification are given in Table 2:9.

TABLE 2:9 / GRAIN SIZE IDENTIFICATION

Name	Size Limits	Familiar Example
Boulder	12 in. (305 mm) or more	Larger than basketball
Cobbles	3 in. (76 mm)–12 in. (305 mm)	Grapefruit
Coarse gravel	$\frac{3}{4}$ in. (19 mm)–3 in. (76 mm)	Orange or lemon
Fine gravel	4.75 mm(No. 4 sieve)–$\frac{3}{4}$ in. (19 mm)	Grape or pea
Coarse sand	2 mm (No. 10 sieve)–4.75 mm (No. 4 sieve)	Rocksalt
Medium sand	0.42 mm (No. 40 sieve)–2 mm (No. 10 sieve)	Sugar, table salt
Fine sand*	0.075 mm (No. 200 sieve)–0.42 mm (No. 40 sieve)	Powdered sugar
Fines	Less than 0.075 mm (No. 200 sieve)	

* Particles finer than fine sand cannot be discerned with the naked eye at a distance of 8 in. (20 cm).

Gradation is estimated by the same criterion as for the Unified Classification (Table 2:5). A smooth grain size curve and a uniformity coefficient of more than 6 for sands or 4 for gravels denotes a *well-graded* soil. An irregular gradation denotes *gap-graded.* A uniformity coefficient less than the above limits indicates a *uniform* soil. The term *poorly graded* is sometimes applied to both uniform and gap-graded soils.

Grain shapes are indicated as angular to well rounded, as shown in Fig. 1.6. In addition, elongated or platey particles can be identified.[2:14]

The mineral composition of the grains can often be determined by a microscopic examination. The carbonates are easily identified by acid, which causes them to effervesce.

The fines are described on the basis of the Casagrande plasticity chart (Fig. 2:12). Soils above the A-line are clays and those below, silts. Soils that plot near the A-line are given a double designation: If the PI is less than 10% above the A-line, the soil is described as a silty clay; if it is less than 33% below the A-line, it is a clayey silt. A soil should not be termed a silty clay if its liquid limit is above 60, however.

Silts and clays can be identified in the field by the shaking test. A pat of wet soil (consistency of soft putty) is shaken in the hand. If it becomes soft and glossy with shaking or tapping the hand and then becomes hard, dull, and forms cracks when the pat is squeezed between the fingers, it has a *reaction to shaking*, or *dilatancy*. A rapid reaction indicates a nonplastic silt; a slow reaction means an organic silt, slightly clayey silt, or possibly a nonplastic silt with a very high liquid limit (over 100). No reaction indicates a clay or silty clay. (To be decisive, the test should be made at different water contents.)

The toughness of the thread that forms when the soil is rolled at the plastic limit also helps identify the fines. Inability to form a thread or a very weak thread indicates an inorganic silt of very low plasticity (ML). A weak spongy thread indicates an organic silt or an inorganic silt having a high liquid limit but low plasticity (MH). A firm thread indicates a low-plasticity clay (CL), while a tough, rigid thread indicates a highly plastic clay (CH).

Plasticity is defined by the plasticity index or can be estimated from the strength of an air-dried sample. The sample is prepared by first removing all particles coarser than a No. 40 sieve and then molding a cube at the consistency of stiff putty, adding water if necessary. The cube is dried in air or sunlight and then crushed between the fingers. (See Table 2:10.)

TABLE 2:10 / PLASTICITY

Term	PI	Dry Strength	Field Test
Nonplastic	0–3	Very low	Falls apart easily
Slightly plastic	3–15	Slight	Easily crushed with fingers
Medium plastic	15–30	Medium	Difficult to crush
Highly plastic	31 or more	High	Impossible to crush with fingers

Organic soils can be identified by their odor, which is intensified by heating, and their color, which is usually black, brown, dark green, or blue-black. However, some inorganic soils are black from certain iron, manganese, titanium, and ferromagnesian minerals. Organic matter can also be identified by oxidizing the soil with hydrogen peroxide and noting the loss in dry weight. It can also be identified by the loss of weight on ignition, provided no carbonate minerals and no adsorbed water or water of crystallization are present.

Soils containing fibrous organic matter are identified visually or by their loss of weight on ignition. A weight loss (organic content) of 80% or more denotes *peat.* If less fibrous organic matter is present, the soils are described as peaty sand or peaty clay, as the case may be.

In forming the description, a predominantly coarse-grained soil is termed either a *gravel* or a *sand*, depending on which component appears to be the more abundant. The less abundant component and the fines (either silt or clay) are used as modifiers, with the least important component first. For example, a soil with 30% fines (silt), 45% gravel, and 25% sand would be described as sandy, silty gravel. The grain shapes and sizes precede the component to which they apply. A predominantly fine-grained soil would be considered either silt or clay and the coarse components would be used as modifiers, with the least important first. For example, a soil with 70% fines (clay), 20% sand, and 10% gravel would be described as gravelly, sandy clay.

Modifiers and suffixes are used by some engineers to quantify grain size components: 0 to 15%, *trace*; 16 to 30%, *some*; 31 to 45%, *-y*; and 45 to 50%, *and.*

DESCRIBING SOIL STRUCTURES / The structure of the soil must be determined by careful observation. The following descriptive terms may be used:

Homogeneous (uniform properties).
Stratified (alternate layers of different soils), including strike and dip.
Laminated (repeating alternate layers less than $\frac{1}{8}$ in. thick), including strike and dip.
Banded (alternate layers in residual soils).

It is important to recognize defects in a soil structure. The following are often observed:

Slickensides (former failure planes), including strike and dip.
Rootholes.
Fissure (cracks, from shrinkage, frost), including strike and dip.
Weathering (irregular discoloration).
Jointed (regular, parallel cracks).

WRITING A SOIL DESCRIPTION / The soil should be described in essentially the order of the significant properties given previously. As many of the properties as are of interest should be included. Examples might be as follows:

1. Hard, moderately compressible, blue-gray, medium-plastic clay.
2. Dense, well-graded, clayey, sandy, well-rounded gravel.
3. Loose, brown, uniform, angular, fine sand.
4. Firm, black, slightly compressible silt and clay; laminated.

5. Loose, compressible, brown, micaceous, sandy silt; banded.
6. Loose tan subangular medium sand, trace of silt, and trace of clay (0–15% silt, 0–15% clay).

SHORTHAND FOR DESCRIPTION / Too often, engineers resort to soil classification alone or to inadequate descriptions because of the limitations of time required for their writing or the space necessary on small-scale drawings. A shorthand system (Table 2:11) gives the most useful abbreviations whose meanings are self-evident to engineers accustomed to soil and rock terminology.

TABLE 2:11 / SOIL AND ROCK SHORTHAND FOR DESCRIPTION

Consistency		Color		Soil Texture, Composition	
Ls	Loose	Gy	Gray	Bldr	Boulder
Fm	Firm	Bn	Brown	Gv	Gravel
Ds	Dense	Tn	Tan	Sa	Sand
VDs	Very dense	Yw	Yellow	Si	Silt
VSo	Very soft	Rd	Red	Cl	Clay
So	Soft	Bl	Black	Org	Organic
Stf	Stiff	Gn	Green	Pt	Peat
Hd	Hard	Wh	White	Cal	Calcareous
		Ora	Orange	She	Shells
				Lat	Lateric
				Cs	Coarse
				Fn	Fine
				Ang	Angular
				Rd	Rounded

Rock		Details	
SS	Sandstone	Sli	Slickensides
Shl	Shale	Sm	Seam
LS	Limestone	Lns	Lens
Gns	Gneiss	Por	Porous
Sch	Schist	WT	Water table
Q	Quartzite	Wth	Weathered
Gr	Granite	Jt	Jointed
Bas	Basalt	Dec	Decomposed
Sla	Slate	Cav	Cavity
Dio	Diorite	Rt	Roots
Tf	Tuff	Con	Concretions
Cg	Conglomerate	Nod	Nodular
Lv	Lava	Mot	Mottled
		Fr	Fractured
		Vrv	Varved
		Bnd	Banded

Graphic symbols for gravel, sand, silt, clay, peat, and colloidal organics are sometimes useful for depicting major formations on soil profiles and cross sections (see Chapter 7). They have not been standardized.

2:11 Soil and Rock Names

In addition to standardized terms used to describe soils, there are many names that have been given to soils, rocks, and deposits. Many of these names, such as *gumbo* and *buckshot*, have originated through association with other more familiar objects and have essentially a local meaning. Others, such as *silt* and *clay*, have been applied to soils having such wide ranges in characteristics that the names often have no definite meaning. In some cases a soil name that is associated with soils of certain characteristics in one region refers to soils of completely different characteristics in other regions. The soil names defined here are often encountered in engineering literature.

ADOBE refers to sandy clays of medium plasticity found in the semiarid regions of the southwestern United States. These soils have been used for centuries for making sun-dried brick. The name is also applied to some highly plastic clays of the West.

BENTONITE is a highly plastic clay, usually montmorillonite, resulting from the decomposition of volcanic ash. It may be hard when dry but swells considerably when wet.

BUCKSHOT is applied to clays of the southern and southwestern United States that cracks into small, hard, relatively uniform sized lumps on drying.

BULL'S LIVER is inorganic silt of very low plasticity. In a saturated condition it quakes like jelly from shock or vibration and often becomes quick and flows like a fluid.

CALICHE is a silt or sand of the semiarid areas of the southwestern United States that is cemented with calcium carbonate. The calcium carbonate is deposited by the evaporation of ground water brought to the ground surface by capillary action. The consistency of caliche varies from soft rock to firm soil.

CLAY is applied to any soil capable of remaining in the plastic state through a relatively wide range in water contents. As a soil name, the term *clay* has been badly abused. In some parts of the United States clay means a nonplastic silt; in other parts it is used to designate micaceous silts of low plasticity. Engineers should use the word clay in the restricted meaning given by Casagrande's plasticity chart (Fig. 2.12).

COQUINA is a soft, porous limestone made up largely of shells, coral, and fossils cemented together.

DIATOMACEOUS EARTHS are silts containing large amounts of diatoms—the silaceous skeletons of minute marine or freshwater organisms.

FILL is any man-made soil deposit. Fills may consist of soils that are free of organic matter and that are carefully compacted to form an extremely dense, incompressible mass, or they may be heterogeneous accumulations of rubbish and debris.

FULLER'S EARTHS are soils having the ability to absorb fats or dyes. They are usually highly plastic, sedimentary clays.

GRAVEL in United States practice denotes a material composed largely of particles between 4.75 and 76 mm (3 in.) in diameter. In some areas gravel implies rounded particles. The size ranges have not been standardized worldwide: the maximum size is 20 to 76 mm; the minimum size is 2 mm in most other nations.

GUMBO is a fine-grained, highly plastic clay of the Mississippi Valley. It has a sticky, greasy feel and forms large shrinkage cracks on drying.

HARDPAN is a term that should be avoided by the engineer. Originally, it was applied only to a soil horizon that had become rocklike because of the accumulation of cementing minerals. True hardpan is relatively impervious and does not soften upon exposure to air or water. Unfortunately, the term is also applied to any hard or highly consolidated soil stratum that is excessively difficult to excavate. Many lawsuits have centered about the meaning of hardpan because of its ambiguity. The name implies a condition rather than a type of soil.

KAOLIN is a white or pink clay of low plasticity. It is composed largely of minerals of the kaolinite family.

LATERITES are residual soils formed in tropical regions. The cementing action of iron oxides and hydrated aluminum oxides makes dry laterites extremely hard.

LOAM is a surface soil that may be described as a sandy silt of low plasticity or a silty sand that is well suited to tilling. It applies to soils within the uppermost horizons and should not be used to describe deep deposits of parent materials.

LOESS is a deposit of relatively uniform, windblown silt. It has a loose structure, with numerous rootholes that produce vertical cleavage and high vertical permeability. It consists of angular to subrounded quartz and feldspar particles cemented with calcium carbonate or iron oxide. Upon saturation it becomes soft and compressible because of the loss of cementing. Loess altered by weathering in a humid climate often becomes more dense and somewhat plastic. It is known as *loess loam*. *Swamp loess* is water-deposited loess. It does not have the loose structure or vertical cleavage of loess.

MARL is a water-deposited sand, silt, or clay containing calcium carbonate. Marls are often light to dark gray or greenish in color and sometimes contain colloidal organic matter. They are often indurated into soft rock.

MUCK OR MUD is extremely soft, slimy silt or organic silt found on river and lake bottoms. The terms indicate an extremely soft consistency rather than any particular type of soil. Muck implies organic matter.

MUSKEG is peat found in Northwest Canada and Alaska. The bogs in which the peat forms are often termed *muskegs.*

PEAT is fibrous, partially decomposed organic matter or a soil containing 80% or more fibrous organic matter. Peats are dark brown or black, loose (void ratio may be 5 to 10), and extremely compressible. When dried, they will float. Peat bogs often emit quantities of inflammable methane gas.

QUICKSAND is a condition and not a soil. Gravels, sands, and silts become "quick" when an upward flow of ground water takes place to such extent that the particles are lifted. (See Chapter 3.)

ROCK FLOUR is extremely fine-grained silt formed by the grinding action of glaciers.

SAND is a soil composed largely of particles from 0.075 to 4.75 mm in diameter. DIRTY SAND means a slightly silty or slightly clayey sand.

SILT is any fine-grained soil of low plasticity. Often it is applied to fine sands. In some parts of the United States the term *silt* is applied only to organic silts. Casagrande's plasticity chart (Fig. 2.12) may be used to differentiate between silt and clay.

STONE is sometimes used to designate angular gravel. It is more properly applied to gravel manufactured by crushing rock.

TILL is a mixture of sand, gravel, silt, and clay produced by the plowing action of glaciers. The name *boulder clay* is often given such soils, particularly in Canada and England.

TOPSOILS are surface soils that support plant life. They usually contain considerable organic matter.

TRAP includes all dark colored, fine-grained intrusive rocks, usually in the form of dikes or sills. The most common trap rock is basalt.

TUFF is the name applied to deposits of volcanic ash. In humid climates or in areas in which the ash falls into bodies of water, the tuff becomes cemented into a soft, porous rock.

TUNDRA is the mat of moss, peat, and shrubby vegetation that covers a gray, clayey subsoil in arctic regions. The deeper soil is permanently frozen, while the surface soil freezes and thaws seasonally. (See Section 3:13.)

VARVED CLAYS are sedimentary deposits consisting of alternate thin layers of silt and clay. Ordinarily, each pair of silt and clay layers is from $\frac{1}{8}$ to $\frac{1}{2}$ in. thick. They are the result of deposition in lakes during periods of alternately high and low water in the inflowing streams and are often formed in glacial lakes.

REFERENCES

2:1 R. F. Leggett, *Geology and Engineering*, 2nd ed., McGraw-Hill Book Company, New York, 1962.

2:2 D. Carroll, *Rock Weathering*, Plenum Publishing Corporation, New York, 1970, pp. 19–24.

2:3 R. W. Fairbridge, *Encyclopedia of Geomorphology*, Van Nostrand Reinhold Company, New York, 1968, 1295 pp.

2:4 G. F. Sowers, "Failures in Limestones of the Humid Subtropics," *Journal of the Geotechnical Engineering Division, Proceedings, ASCE*, **101**, GT8, August 1975, pp. 771–788.

2:5 G. F. Sowers, "Settlement in Terranes of Well-Indurated Limestone," in *Analysis and Design of Building Foundations*, H. Y. Fang, ed., Envo Publishing Co., Lehigh Valley, Pa., 1976, Chapter 24, pp. 701–727.

2:6 Nyle C. Brady, *The Nature and Properties of Soils*, 8th ed., Macmillan Publishing Co., Inc., 1974, 639 pp.

2:7 *Soil Classification, A Comprehensive System, The Seventh Approximation*, U.S. Department of Agriculture, Washington, D.C., 1960.

2:8 N. Duncan, "Rock Mechanics," a lecture delivered to the Southern Assoc., Institution of Civil Engineers (Great Britain), 1968.

2:9 D. U. Deere, "Technical Description of Rock Cores for Engineering Purposes," *Rock Mechanics and Engineering Geology*, **1**, 1, 1963.

2:10 A. Casagrande, "Classification and Identification of Soils," *Transactions, ASCE*, **1948**, 901.

2:11 "Classification of Highway Subgrade Materials," *Proceedings, Highway Research Board*, **1945**, Washington, D.C.

2:12 *Unified Soil Classification System*, Technical Memorandum 3-357, U.S. Waterways Experiment Station, Vicksburg, Miss., 1953.

2:13 ASTM D-2487-69(75): "Standard Method for Classification of Soils for Engineering Purposes," *Annual Book of Standards, ASTM*, Philadelphia, Pa.

2:14 ASTM D-2488-69(75): "Description of Soils: Visual Manual Procedure," *Annual Book of Standards, ASTM*, Philadelphia, Pa.

2:15 K. Terzaghi and R. B. Peck, *Soil Mechanics in Engineering Practice*, 2nd ed., John Wiley & Sons, Inc., New York, 1967, p. 30.

2:16 L. F. Cooling, A. W. Skempton, and R. Glossop, "Discussion," in Reference 2:10.

2:17 *Munsell Soil Color Charts*, Munsell Color Co., Baltimore, Md., 1954.

SUGGESTIONS FOR FURTHER STUDY

American Geological Institute, *Glossary of Geology*, Washington, D.C., 1974, 805 pp.

M. D. Gidigasu, *Laterite Soil Engineering*, American Elsevier Publishing Co., Inc., New York, 1976, 554 pp.

M. Herak and V. T. Stringfield, *Karst*, Elsevier Publishing Co., Inc., Amsterdam, 1972.

F. H. Lahee, *Field Geology*, 6th ed., McGraw-Hill Book Company, New York, 1961, 926 pp.

R. F. Legget, *Cities and Geology*, McGraw-Hill Book Company, New York, 1973.

PROBLEMS

2:1 Prepare a log of the soils you might expect when making an excavation in the following deposits:
1. Flood plain of a flat river in Mississippi.
2. Coastal marsh in North Carolina.
3. Edge of swampy lake in the glaciated portion of Michigan.
4. A dried-up lake bed in Nevada.
5. Sand dune in Indiana.
6. Prairie in North Dakota.
7. River valley in Connecticut.
8. Coastal plain of Texas.
9. A mountainside in Central America.
10. An area of low hills in Ontario, Canada.

2:2 a. Classify the following soils according to the Revised Public Roads or AASHTO System and find the GI number:

	Percent Passing		Characteristics of −40 Fraction	
Soil No.	No. 40	No. 200	LL	PI
1	95	57	37	18
2	72	48	31	4
3	100	97	73	45
4	18	0	—	—
5	63	8	—	—
6	97	65	50	6
7	45	18	14	3
8	70	30	17	5
9	55	20	—	—

b. Classify the same soils by the Unified System insofar as possible, showing a double symbol where more than one class is possible.

2:3 The following data were obtained by mechanical analysis and plasticity tests of soil samples. (Percentages finer than the given size noted.)

Size	Sample 1	Sample 2	Sample 3	Sample 4	Sample 5	Sample 6	Sample 7
No. 10	—	—	—	—	—	100	100
No. 20	86%	98%	93%	99%	98%	—	—
No. 40	72	85	79	94	95	86	—
No. 60	60	72	68	89	92	—	—
No. 100	45	56	56	82	86	—	85
No. 200	35	42	42	76	83	9	75
0.05 mm	33	41	41	74	82	—	68
0.01 mm	21	20	11	38	57	0	45
0.002 mm	10	8	4	23	36	—	29
LL	19	44	30	40	67	NP	70
PI	0	0	0	12	27	NP	42

Plot the grain size curve.

a. Classify each using the Unified System.

b. Classify each using Revised Public Roads or AASHTO System.

2:4 Classify the coarse-grained soils of Problem 1:15.

2:5 a. Describe visually a sample of soil you have secured from an excavation or highway cut.

b. Estimate the soil classification according to the Revised Public Roads System and the Unified System.

CHAPTER 3

Soil Water: Its Movement, Stress, and Effects

An excavation for a raw water pipe (Fig. 3.1) led from an intake structure on the river bank to a holding pond 200 m (650 ft) from the river. The soil was typical of many flood plains: alternate layers of clayey silt and silty clay 7 m thick underlain by a stratum of slightly sandy gravel 4 m thick. The bottom of the trench, excavated with slopes of about $1(H)$ to $1(V)$, was in the silty soils but only 4 m above the gravel.

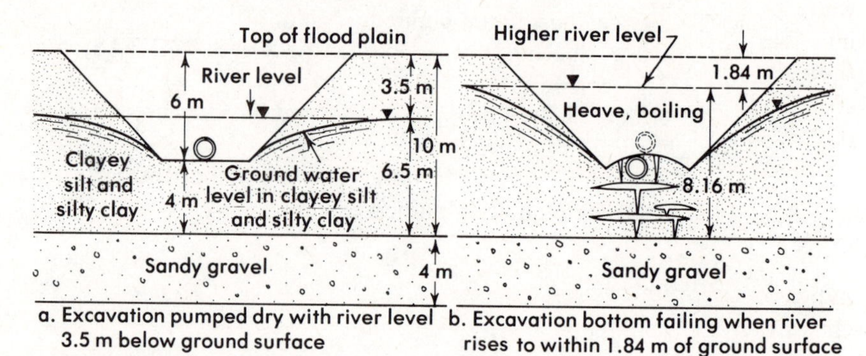

Figure 3.1 *Excavation failure from groundwater uplift and boiling (also Example 3:3).*

The contractor, working under the direction of the owner's engineer, had completed excavating the trench. Only a small pump was necessary to keep the excavation dry because the silty and clayey alluvium resisted seepage, despite the proximity of the sandy gravel below, which outcropped in the river. The contractor commenced laying the 1-m diameter concrete pipe in the trench when the river began to rise from heavy rainfall upstream. He expressed concern to the design engineer when the river level approached the floodplain level. The engineer directed that the work proceed without delay because he observed little seepage into the excavation despite the rising river.

Suddenly, the bottom of the trench heaved, lifting the pipe (Fig. 3.1b). Water began to spout upward in the heaving area, breaking up the soil. The pipe sank out of sight and the excavation filled with water. Completion of the work was delayed several months, awaiting installation of deep drainage in the gravel. The direct cost of reexcavation, relaying the pipe, and deep drainage was more than $250,000. The indirect cost of the delay in providing water to the new industrial plant was even more.

Failure could have been prevented by installing deep drains in the gravel as a protective standby before the river rose. Failure could also have been prevented by flooding the excavation after the river commenced to rise. Even the cost of building the pipe in the trench under water would have been less than that of reexcavating and rebuilding, with only a slight delay. The supervising engineer misjudged the condition because he observed little seepage despite the rising river.

This example illustrates the misconceptions of many engineers and contractors when they work below the groundwater level. They confuse the amount of seepage they can see with the potential effect of the seepage on the ground and on engineering structures. In some situations a large quantity of seepage can be a hazard; in others seepage is indicative of good groundwater control. Paradoxically, lack of seepage can be both a symptom of impending trouble or of successful drainage. The engineer must understand the apparent contradictions and paradoxes of soil moisture and groundwater stress and movement to control both the water and the engineering behavior of the soil and rock that it contacts.

3:1 Ground Water and Soil Moisture

Water is present in soils and rocks in five forms: ground water, capillary moisture, adsorbed water, water of hydration, and water vapor. All can change, all can interchange, but under different conditions and at different rates. All are major factors in the engineering behavior of the materials. Chapter 1 considered adsorbed and hydration water that are closely bound to the soil minerals. This chapter considers the other three.

GROUND WATER / Ground water is the continuous body of subsurface water that fills the soil, rock voids, and fissures and is free to move under the influence of gravity. It implies a degree of saturation approaching 100%. The *water table* or *phreatic surface* is the level of zero (atmospheric pressure) in a continuous body of ground water. It is the level of water in an open hole at least 10 mm in diameter that extends into the ground water. It is not a static level surface as the term *water table* implies. Instead, it is the sloping surface of a moving stream of water in the voids and fissures. It takes many shapes and changes with time, depending on the pattern and size of voids and fissures, on the sources of water, and the exits or points of discharge. During wet weather the water table rises; during dry weather or periods of depletion by water supply usage or drainage it falls. In formations consisting of alternate strata of more pervious and less pervious materials, the pattern is more complex.

CAPILLARY MOISTURE / Water in the voids that is stressed by *capillary tension* in addition to gravity is called capillary moisture. It requires an air–water interface within the soil or rock beyond which the voids are only partially saturated. The water is not free to move under the influence of gravity; instead, the capillary tension generally governs, usually opposing gravity flow.

VAPOR MOISTURE / Vapor moisture is present in the soil air. It can exist only if the degree of saturation is less than 100%. Deep within a formation the relative humidity of the soil air is virtually 100%; near the ground surface it is less, depending on temperature and the availability of water.

3:2 Permeability

The voids in a soil are not isolated cavities that hold water like storage reservoirs but are interconnected, small, irregular conduits (Fig. 3.2) through which water can flow in the same way as it flows through other conduits.

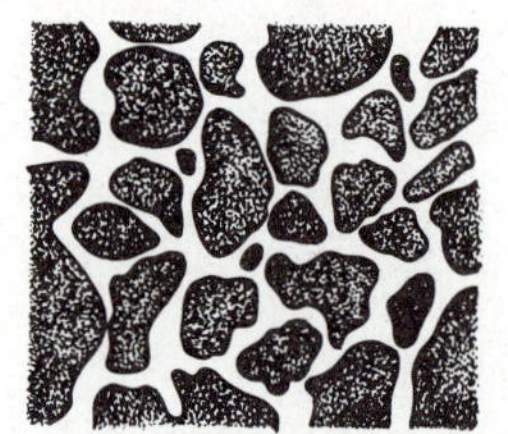

a. Labyrinth paths of conduits formed by voids, not in same plane

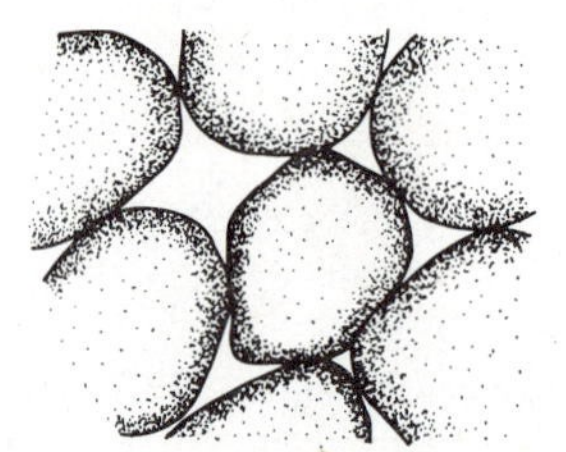

b. Cross sections of conduits formed by voids in a plane

Figure 3.2 *Conduits formed by soil pores.*

LAMINAR AND TURBULENT FLOW / Fluid mechanics defines two different flow regimes. *Turbulent flow* is characterized by chaotic, irregular movements of the fluid particles and by energy losses that are roughly proportional to the square of the velocity of flow. This takes place at relatively high velocities in large-diameter conduits, such as in pipes carrying air or water. Large open joints in rock-solution cavities and very porous formations, such as the scoria in lava flows and voids between boulders or clean, coarse gravel, exhibit turbulent flow. In *laminar flow* the water particles move in a smooth, orderly procession in the direction of flow, and the energy losses are directly proportional to the velocity. Laminar flow occurs at low velocities in small conduits. It is characteristic of soils except uniform coarse gravel, many porous rocks such as sandstone, and of narrow fissures and fracture zones in rock.

DARCY'S LAW / A French physicist, Henri Darcy, studied flow of water in soil by using apparatus similar to that shown in Fig. 3.3. He placed a

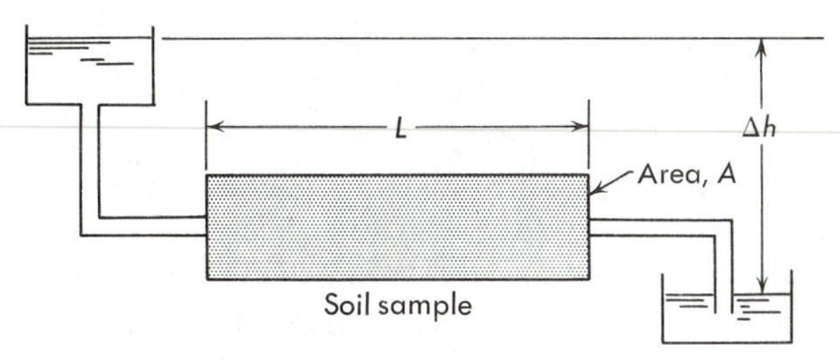

Figure 3.3 *Simple permeability test.*

sample of length L and cross-sectional area A on a tight-fitting tube with open ends. A reservoir of water was connected to each end of the tube. The level of the water in one reservoir was a distance of Δh above that in the other. (The term *head loss* is often applied to this difference in level, Δh.) He found by experiments that the flow of water, q, in ml per second was directly proportional to the area A and to the ratio $\Delta h/L$ (which is termed the *hydraulic gradient* and given the symbol i). This relation is expressed by

$$q = kiA \tag{3:1}$$

in which k is the constant of proportionality and is given the name *coefficient of permeability*, or simply *permeability*. This expression is valid as long as laminar flow exists. For wide fissures, or clean, coarse gravels with void diameters exceeding 10 mm, and for gradients exceeding 5, the expression does not apply.

COEFFICIENT OF PERMEABILITY / The coefficient of permeability is a constant (having the dimensions of a velocity) that expresses the ease with which water passes through a soil. Ordinarily, it is reported with the dimensions of millimeters per second or feet per minute, but for very impervious soils, meters per year is used.

The magnitude of the permeability coefficient depends on the viscosity and density of the liquid and on the size, shape, and area of the conduits through which the water flows. Viscosity is a function of the temperature: the higher the temperature, the lower the viscosity and the higher the permeability. Usually, the permeability is reported at 20°C or 68°F. At 0°C or 32°F it is 56%, and at 40°C or 104°F it is 150% of the value at 20°C or 68°F. The differences in density between water containing various dissolved salts are not sufficient to justify a correction. The influence of the factors that determine the size and shape of the conduits is less specific and no valid mathematical expression for their effect has been derived.

For clean cohesionless soils the permeability varies approximately as $(D_{10})^2$. Hazen's formula[3:1] for permeability of clean sands is

$$k = C(D_{10})^2 \tag{3:2}$$

in which k is given in mm/sec, D_{10} in mm, and C is a constant whose value ranges between 10 and 15. At the best, this formula gives only an indication of the order of magnitude of the permeability of clean sands. In soils having cohesion, the effect of grain size is even more pronounced, because part of the soil water around the clay particles is immobilized in the adsorbed layers. Void ratio is a factor in most soils with the permeability varying approximately proportional to e^2. Grain shape and graduation are also important, particularly in the coarser soils, but it is difficult to express their effects quantitatively. The degree of saturation is a major factor. The permeability of a partially saturated soil is significantly less than the same soil saturated because air in the voids reduces the flow cross section and blocks the smaller voids completely.

Clay minerals greatly influence permeability because some of the adsorbed water is so tightly bound to the clay surfaces that it cannot move from one particle to another without very high gradients. The permeability of a homogeneous clay, therefore, will usually be far lower than the grain size or void ratio alone suggests. Furthermore, the permeability is virtually zero at low gradients, but increases with increasing gradients.[3:2]

PERMEABILITY TESTS / Because of the numerous, complex factors that influence the permeability coefficient, only crude estimates of its magnitude can be made from a knowledge of the character of the soil. Therefore, tests must be performed to obtain the coefficient with any certainty. The simplest test is the *constant head* shown diagrammatically in Fig. 3.3. It is used primarily on sands and gravels.[3:3] For fine sands and silts the *falling head* test is used.[3:4] The upper reservoir of Fig. 3.3 is replaced with a vertical standpipe. During the test the level of water in the standpipe falls, and the volume of water that flows is equal to the volume difference in the standpipe. Extreme care is essential in testing fine-grained, cohesionless soils to avoid migration of the soil particles caused by hydraulic gradients greater than 5. For clay soils either the constant or falling head test is employed. The quantity of seepage is so small that great care is necessary to avoid leaks and evaporation, which could be many times greater than the flow through the soil.

3:3 Permeability of Soil and Rock Formations

The range in permeability of natural soils is even greater than the range in grain size. Table 3:1 can be used as a standard for describing permeability and as a guide for rough estimates.

TABLE 3:1 / RELATIVE VALUES OF PERMEABILITY (After Terzaghi and Peck)[3:5]

Relative Permeability	Values of k (mm/sec)*	Typical Formation
Very permeable	1	Coarse gravel, open-jointed rock
Medium permeability	1×10^{-2}	Sand, fine sand
Low permeability	$1\times10^{-2}-1\times10^{-4}$	Silty sand, dirty sand
Very low permeability	$1\times10^{-4}-1\times10^{-6}$	Silt, fine sandstone
Impervious	Less than 1×10^{-6}	Clay, mudstone without joints

*To convert to feet per minute, multiply above values by 0.2.

Permeability is sometimes expressed in other ways. The *Lugeon* unit is approximately 1×10^{-4} mm/sec. It is measured by pumping into a hole drilled in the pervious formation. The *transmissivity* of a confined aquifer of thickness H is kH. It is computed from well-pumping tests (Section 3.6 and Chapter 7) in which H the aquifer thickness, has not been measured.

VARIATION OF *k* IN A REAL FORMATION / In most soils and rocks the value of k depends on the direction in which the water is traveling. The k in the direction parallel to the stratification or planes of foliation is usually from 2 to 30 times that in the direction perpendicular to the bedding or foliation because of the layers with relatively low permeabilities. In soil deposits with erratic lenses of either coarse, pervious materials or fine impervious materials, the permeability varies greatly from point to point and is extremely difficult to determine.

Soils in which there is an orientation of flaky or slablike particles exhibit higher permeabilities parallel to the aligned faces than perpendicular to them. Similar *anisotropic permeability* is typical of some compacted soils.

In soils of low permeability and in most rocks, the permeability of the mass is governed by the cracks and fissures. The effective permeability will be far greater than that of the intact material between the cracks. On the other hand, cemented seams within a generally pervious formation will make the effective permeability across the seams very low. Because of anisotropy and nonhomogeneous defects, a large number of laboratory tests with flow in several directions is necessary for realistic values for the permeability coefficient.

Laboratory test results of fine-grained soils are often unrealistic because the gradients necessary to produce measurable flow are sometimes

10 to 100 times the gradients that would be present in the field. High gradients in clay produce proportionately more flow and higher k's than low gradients. High gradients in fine cohesionless soils can disrupt the grain structure, causing both flow blockage and internal erosion of the mass as described in Section 3.7. To avoid these problems, the laboratory tests should simulate field conditions as nearly as is practical. Field tests, described in Chapter 7, are usually more reliable because they integrate the effects of soil or rock discontinuities. They have the disadvantage of limited control of seepage direction. For solution of real problems, the choice of the appropriate k requires judgment. It is seldom possible to make an evaluation closer than the nearest decimal place; therefore, design decisions involving the coefficient of permeability should allow for such variations.

3:4 Aquifers, Water Tables, and Heads

Aquifers are relatively pervious soil and rock formations through which ground water flows, induced by gravity and resisted by fluid friction as expressed by the permeability. The most familiar is a stratum of coarse sand in which the groundwater surface rises and falls with changes in weather, discharge, and groundwater use. In such a *free aquifer*, the groundwater table is likely to slope in the same direction as the ground surface. The water table slope is generally flatter, and more uniform, without reversals.

An *aquiclude* is a stratum of relatively impervious soil or rock. Typically the ratio of permeabilities of an aquifer to an aquiclude exceeds 100 : 1. In formations that consist of alternate aquifer and aquicludes, the groundwater pattern or *regime* is more complex. A sagging aquiclude below an aquifier creates a basin that holds ground water perched above a general water table (Fig. 3.4). *Perched* water tables occur frequently in stratified

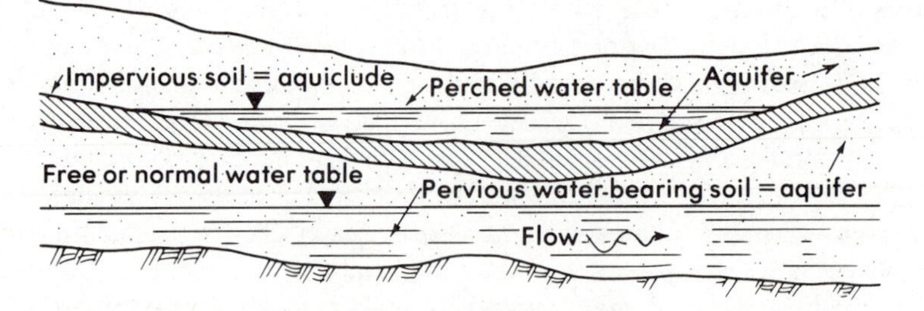

Figure 3.4 *Perched ground water.*

materials but are ordinarily of limited extent. They may be drained by drilling a hole through the impervious basin, allowing the water to seep downward. When an aquifer is confined between two aquicludes, it is capable of carrying water under pressure. When it does, the elevation of zero pressure is above the upper surface of the water, and the ground water is said to be under *artesian* pressure. (See Fig. 3.5.)

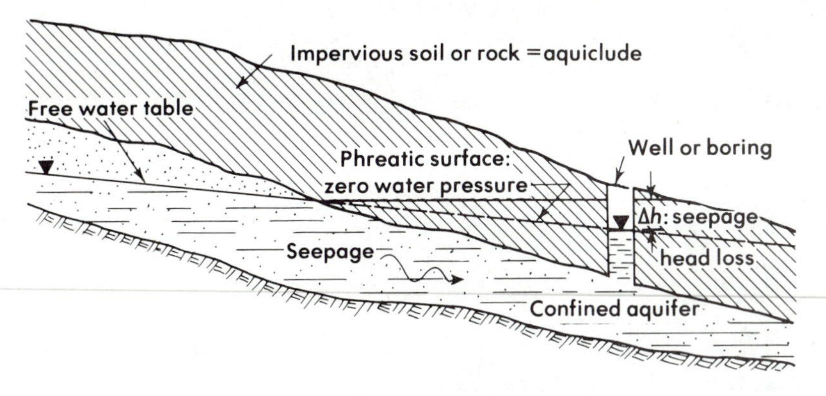

Figure 3.5 *Artesian ground water.*

Artesian pressures are usually developed by sloping aquifers where the point at which the water enters the confined previous stratum is higher than the point at which the pressure is measured. When a hole is drilled into an artesian aquifer, the water rises to the elevation of zero pressure. If this level is above the ground surface, a *flowing artesian well* results. Artesian aquifers may be local structures existing over an area of several hundred square meters, or they may be continuous over large areas like the vast artesian sandstones in North and South Dakota or the artesian limestones of south Florida. They are often troublesome to engineers because of the reduction of soil strength by water pressure as described in Chapter 5. Excavations that extend close to strata that are under artesian pressure may be damaged from "bottom blowouts." The water pressure, which formerly was balanced by the weight of the overlying soil, causes the remaining soil to burst upward into the excavation, as described in the introduction to this chapter, or if the soil is sand, it will create a "quicksand" condition discussed in Section 3:7.

SPRINGS AND SEEPS / When the groundwater table intersects the ground surface on a hillside (Fig. 3.6), a *spring* is formed, water trickles down the ground surface, and the soil may be softened by the added water and also by the seepage pressures. They may be corrected by intercepting the water with drains before it reaches the surface. The intersection of the water table and a level ground surface produces a marsh, bog, or swamp, depending on the vegetation. During wet weather the ground surface may be flooded, forming a shallow marshy or swampy lake; during dry weather it may be firm and dry. Upward seepage in such areas can produce loss of soil strength and a condition described as "quicksand," although the term is more accurately restricted to cohesionless soils as described in Section 3:7.

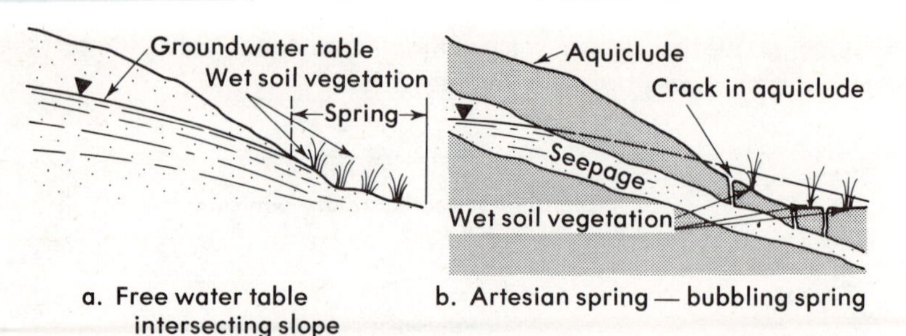

Figure 3.6 Spring or seep.

HEAD, WATER PRESSURE, AND HEAD LOSS / The energy in flowing water is in three forms: *potential energy*, owing to its height, h_z, above some arbitrary datum plane; *pressure energy*, h_p, owing to water pressure; and *kinetic energy*, h_k, owing to velocity. In laminar flow the velocities are so small that the kinetic energy is negligible; it is ignored in groundwater flow. Energy is usually expressed as head: the energy per unit of mass or weight. It is a linear dimension, meters, or feet: m-kg/kg or ft-lb/lb. Because energy is only relative, head must be expressed with respect to some arbitrary datum. The total head at any point in an aquifer is equal to the elevation of that point above the arbitrary datum plus the pressure head. If a tube is inserted in the aquifer to the point in question and the upper end of the tube is open to the atmosphere, the water pressure will force the water upward a distance equal to the pressure head, h_p (Fig. 3.7). If the water pressure is u and h_p is measured positive upward,

$$h_p = \frac{u}{\gamma_w} \tag{3:3a}$$

$$u = \gamma_w h_p \tag{3:3b}$$

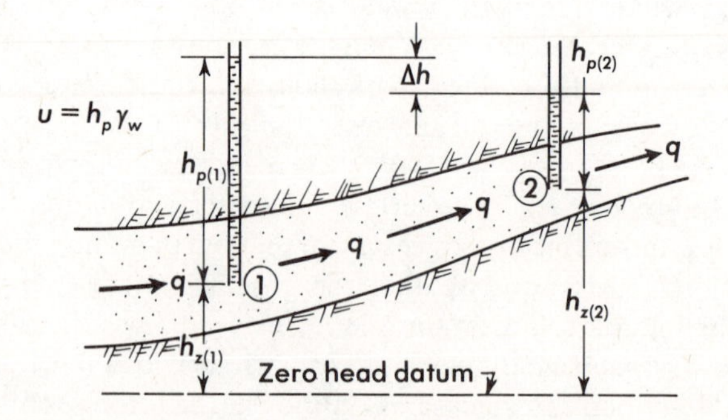

Figure 3.7 Head, water pressure, and head loss in an aquifer.

If the elevation of a point within an aquifer is h_z and the elevation of the total head at the same point is h, then

$$h_p = h - h_z$$

$$u = \gamma_w h_p = \gamma_w (h - h_z) \tag{3:4}$$

Example 3:1

In Fig. 3.7 the elevation of point (2) is 33 m or 108.2 ft above the datum. The total head, the elevation of water in piezometer 2 is 53 m or 173.8 ft above the datum. Find the water pressure at point (2).

$h_z(2) =$ 33 m $= 108.2$ ft

$h_{(2)} =$ 53 m $= 173.8$ ft

$h_{p(2)} = 53 - 33 = 20$ m $= 65.6$ ft

$u_{(2)} = 20 \times 9.81\ \text{kN/m}^3 = 196\ \text{kN/m}^2 = 65.6 \times 62.4\ \text{lb/ft}^3 = 4093\ \text{lb/ft}^2$

NOTE: In geotechnical engineering, pressure or compressive stress is positive; tension is negative.

Within an aquifer if the total heads, $h_{(1)}$ and $h_{(2)}$, at two points (1) and (2) are different, there will be a flow of water toward the one of lower energy, shown in Fig. 3.7. The head difference, Δh, is dissipated in the work of moving the water at a rate q, a net distance of L, as expressed by Darcy's law, Equation (3:1), rewritten as:

$$\Delta h = \frac{qL}{kA} \tag{3:5}$$

The movement of water within the soil voids and rock pores has a profound influence on the engineering behavior of the materials: (1) The amount of the liquid phase changes; (2) the water flow and its associated frictional drag can dissolve or move solid particles; and (3) the changing water pressure changes the distribution of stresses within the material. Most failures in geotechnical engineering are related to changing water and water pressure.

3:5 Saturated Flow: The Flow Net[3:6,3:7]

The flow of water through a saturated soil or porous rock can be represented pictorially by *flow lines* (Fig. 3.8a): the trajectories of water particles. Water tends to follow the shortest path from point to point but at the same time makes smooth curves when it changes direction. The flow lines are curved, and loosely parallel, often segments of ellipses or parabolas.

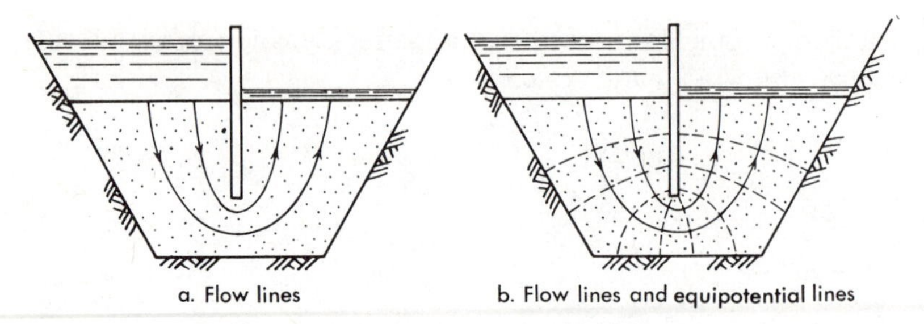

Figure 3.8 *Flow net of seepage beneath sheet piling.*

The different amounts of energy or head can be represented on the same picture by *equipotential lines* (Fig. 3.8b), which are the loci of all points having equal total heads. The equipotential lines can be thought of as contours of equal energy; the flow lines cross them at right angles, the water moves from higher energy levels to lower energy levels along paths of maximum energy gradient in the same way water flows down a hillside from higher levels to lower levels, following the steepest paths.

The pattern of flow and equipotential lines is termed the *flow net*, and it is a powerful tool for the solution of seepage problems.

DERIVATION OF THE FLOW NET / The mathematical expression for the flow net is derived on the basis that the soil is saturated, that the volume of water in the voids remains the same during seepage, and that the coefficient of permeability is the same at all points and in any direction at any point. The basic equation of seepage, Darcy's law [Equation (3.1)], is resolved into x and y components:

$$q_x = ki_x A_x$$

$$q_y = ki_y A_y$$

$$i = \frac{\Delta h}{\Delta L} = \frac{dh}{dL}$$

The seepage velocity v is the rate of seepage divided by the area of flow and so the equations can be rewritten:

$$v = \frac{q}{A}$$

$$v_x = k\frac{\partial h}{\partial x}$$

$$v_y = k\frac{\partial h}{\partial y}$$

The flow through a small element of soil having the dimensions of dx, dy, and 1 is shown in Fig. 3.9a and is expressed as follows:

In: $$v_x\,dy + v_y\,dx$$

Out: $$\left(v_x + \frac{\partial v_x}{\partial x}\,dx\right) dy + \left(v_y + \frac{\partial v_y}{\partial y}\,dy\right) dx$$

If the volume of water in the voids remains constant, then the flow *in* equals the flow *out*; so, equating the above expressions and collecting terms, we obtain

$$\frac{\partial v_x}{\partial x} + \frac{\partial v_y}{\partial y} = 0$$

By substituting the equations for velocity, the relation becomes

$$\frac{\partial^2 h}{\partial x^2} + \frac{\partial^2 h}{\partial y^2} = 0 \qquad (3{:}6)$$

This is the Laplace equation of mathematical physics, which describes the energy loss through a resistive medium. It represents two sets of lines, each set containing an infinite number of parallel curves and with each curve of one set intersecting each curve of the other at right angles, as shown in Fig. 3.9b. The *equipotential lines* comprise one set and the *flow lines* the other, and the entire pattern is the *flow net.*

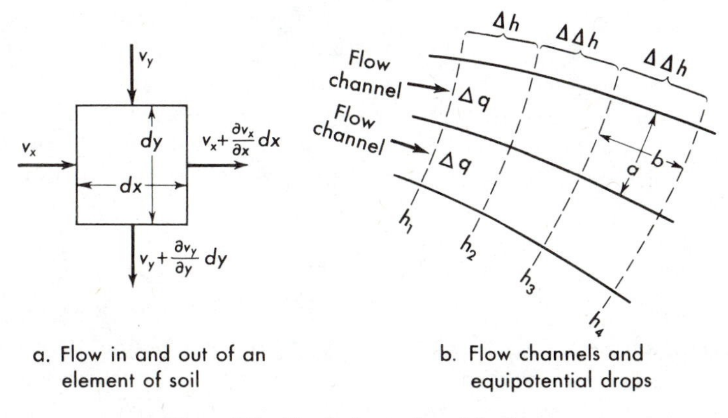

Figure 3.9 *Physical meaning of the flow net.*

FLOW NET CONSTRUCTION / The two-dimensional flow net derived above is a useful representation of the seepage patterns through earth dams, into large excavations, and below retaining walls and masonry structures. Unfortunately, the Laplace equation can be integrated mathematically for only a few, very simple conditions, and in practice the flow net must be obtained by successive approximations, or models.

The graphical procedure of Forcheimer is simple and is applicable to any problem of steady flow in two dimensions. The space between any pair of flow lines is a *flow channel.* If a certain number of flow channels, N_f, is selected so that the flow through each, Δq, is the same, then

$$\Delta q = \frac{q}{N_f}$$

The head loss between any pair of equipotential lines is the *equipotential drop* $\Delta\Delta h$. If a certain number of equipotential drops are selected, N_D, so that all are equal,

$$\Delta\Delta h = \frac{\Delta h}{N_D}$$

The width of any one element of such a flow net is a and the distance between the equipotential lines is b, as shown in Fig. 3.9b. (The third dimension is 1.) The gradient and discharge are given by

$$i = \frac{\Delta\Delta h}{b} = \frac{\Delta h / N_D}{b}$$

$$\Delta q = k\left(\frac{\Delta h / N_D}{b}\right) a$$

The total discharge for the net, whose third dimension is 1, is expressed by

$$q = \Delta q N_f = k\ \Delta h\left(\frac{a}{b}\right)\frac{N_f}{N_D} \tag{3:7a}$$

The ratio of (a/b) is fixed by the ratio of N_f/N_D and is the same throughout the net. If N_f and N_D are selected so that $a = b$, the equation for discharge (for a unit dimension perpendicular to the flow net) is

$$q = k\ \Delta h \frac{N_f}{N_D} \tag{3:7b}$$

This is termed a *square net* because all the intersections between the sides are at right angles and the average length and width are equal. However, it should be noted that the term *square* is used in a descriptive sense because the opposite sides of the figures are not necessarily equal, and they are seldom straight lines.

The first step in constructing a flow net is to make a scale drawing (Fig. 3.10a) showing the soil mass, the pervious boundaries through which water enters and leaves the soil, and the impervious boundaries that confine the flow. Second, two to four flow lines are sketched, entering and leaving at right angles to the pervious boundaries and approximately parallel to the impervious boundaries (Fig. 3.10b). Third, equipotential lines are drawn at right angles to the flow lines (Fig. 3.10c) so that the length and width of each

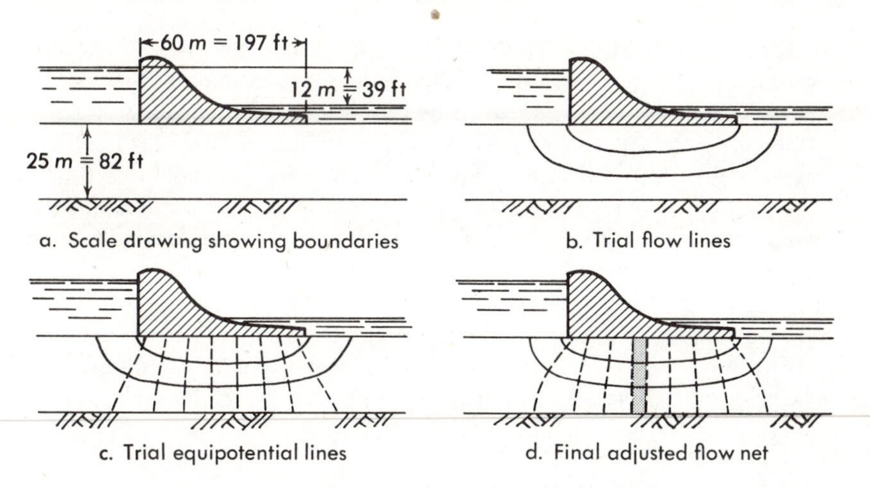

Figure 3.10 *Steps in constructing a flow net. (Shaded figures are rectangles.)*

figure will be equal. Of course, this will be impossible on the first attempt because the positions of the flow lines were only guessed, but the resulting net will guide the second attempt. Fourth, the flow lines and equipotential lines are readjusted so that all the intersections are at right angles and the length and width of each figure are equal (Fig. 3.10d). Between one pair of equipotential lines the figures may work out to be rectangles. However, each rectangle should have the same ratio of a/b. The resulting equipotential drop is a fraction of the others, equivalent to b/a. In Fig. 3.10d the shaded figures have $b/a = \frac{1}{2}$.

The quantity of seepage is computed by Equation (3:7b), using the values of N_f and N_D found by the graphical trial and revision. This is multiplied by the third dimension, perpendicular to the plane of the flow net, to get the total seepage.

Much practice is necessary to develop skill in drawing flow nets, and many cycles of trial and revision are required for an accurate solution.

Example 3:2

Compute the quantity of seepage under the dam in Fig. 3.10 if $k = 1.5 \times 10^{-3}$ mm/sec and the level of water upstream is 18 m above the base of the dam and downstream is 6 m above the base of the dam. The length of the dam (perpendicular to the direction of seepage) is 250 m.

1. From the flow net $N_f = 3$ and $N_D = 9.5$.
2. $k = 1.5 \times 10^{-6}$ m/sec, $\Delta h = 12$ m.
3. q per m $= 1.5 \times 10^{-6} \times (3/9.5) \times 12$
   q per m $= 5.7 \times 10^{-6}$ m^2/sec.
4. $q = 5.7 \times 10^{-6} \times 250 = 1.4 \times 10^{-3}$ m^3/sec $= 3$ ft^3/min.

FLOW NET WITH FREE SURFACE / In some cases, such as the flow of water through earth dams (Fig. 3.11a), one boundary flow line is a free-water surface that is not fixed by any solid, impervious mass. This is analogous to the free-water surface in open channel flow and is a more difficult problem to solve with the flow net. The upper boundary flow line is called the *line of seepage.* It is also the piezometric surface. It must satisfy all the requirements of any flow line, and in addition its intersections with the equipotential lines must be vertically spaced a distance equal to $\Delta\Delta h$ (Fig. 3.11b). Considerable juggling is necessary to construct such a net correctly, but in many practical problems even a rough net will be sufficiently accurate.

The line of seepage intersects the downstrean face of an embankment, (Figs. 3.11 and 3.12) tangent to the surface. Below that point, however, the

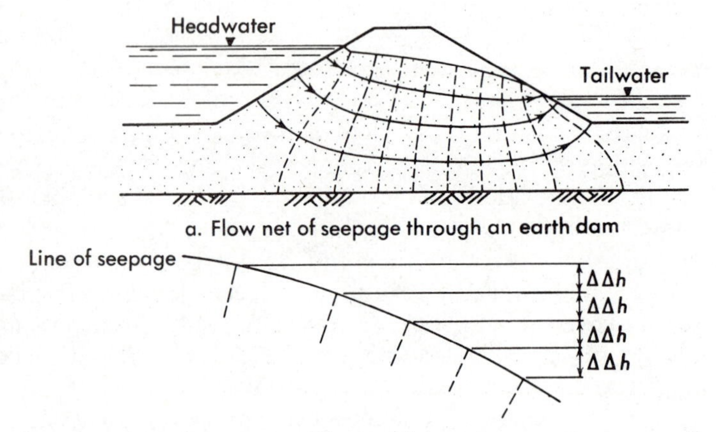

a. Flow net of seepage through an earth dam

b. Intersections of equipotential lines with line of seepage (uppermost flow line)

Figure 3.11 *Flow net with a free surface.*

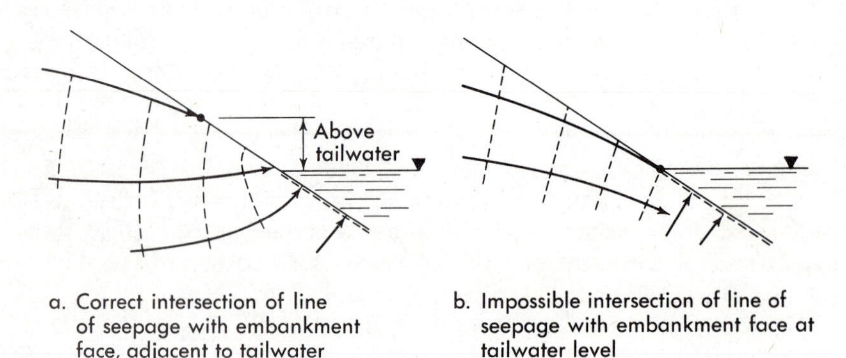

a. Correct intersection of line of seepage with embankment face, adjacent to tailwater

b. Impossible intersection of line of seepage with embankment face at tailwater level

Figure 3.12 *Intersection of line of seepage and flow net with a sloping embankment face adjacent to tailwater.*

embankment face is not a line of seepage. The face cuts across the "squares" of the flow net above the tailwater level, and each equipotential line intersects the face at the elevation equivalent to its piezometric level.

If tailwater is present, the intersection of the line of seepage with the embankment face must be above the tailwater level, as shown in Fig. 3.12a. If the line of seepage dropped to the tailwater level (Fig. 3.12b) the seepage conditions could not satisfy the requirements of the flow net. Below the free surface the embankment face is an equipotential line. If the line of seepage dropped to the tailwater level, the flow lines below tailwater would be at right angles to those above, an impossible condition, as can be seen in Fig. 3.12b. The correct intersection, found by trial, is shown in Fig. 3.12a.

A comprehensive paper describing methods of constructing flow nets with free surfaces and for constructing nets in soils whose permeability is not the same in all directions or at all points has been presented by A. Casagrande.[3:6]

OTHER METHODS OF ANALYSIS / Other methods are sometimes used to obtain the flow net. Seepage models can be constructed that are similar to hydraulic models. The soil or porous rock is modeled in sand that is coarse enough to minimize capillary rise but fine enough for laminar flow. There is no need to reproduce or model the permeability in the prototype because the flow net shape is independent of the value of k. Layers or zones of different permeability can be represented by the correct ratio of k's in the model, but a ratio of 20:1 or greater is impractical and rarely of value. The flow line can be traced in a glass-walled flume by injecting dye at points on the intake surface of the soil. Piezometric levels can be observed by miniature piezometer tubes forced into the soil.

Viscous fluids such as oil can be used to reduce velocities in models of unsteady flow. The most important uses of models are in studying complex three-dimensional seepage, unsteady flow, changing free surface, or non-homogeneous permeability within the soil deposit. Although accurate measurements are seldom possible, the results are ordinarily as good as the knowledge of the soil deposit and the boundary conditions justify.

Analogs are models employing phenomena that produce the same patterns of potential and flow as seepage. The flow of electricity through a semiconductor is also described by the Laplace equation. An electric potential applied to a graphite-coated paper produces the same potential distribution as water pressure applied to a soil cross section of the same shape. The equipotential lines plotted on the graphite paper from voltage measurements are identical in shape to those in the corresponding soil.

A network of electrical resistors similarly can represent the resistances of soil voids to seepage. Varying permeabilities can be simulated by different resistances. The potential can be measured at points throughout the network as with the semiconductor. The same thing can be done analytically using a digital computer with sufficient memory to describe the entire network. Such techniques are useful in solving problems of complex nonhomogeneity where the graphical analysis would be tedious. Their precision is limited only by the number of elements in the network. Mathematical networks are also used to solve the Laplace equation, where the digital computer can undertake the repetitive processes of trial and correction for the numerous elements of the net necessary to describe a flow system.

3:6 Well Drawdown

Flow into a well and the drop in the piezometric surface that results (termed *drawdown*) is a complex problem in three-dimensional seepage as well as unsteady flow. The three-dimensional flow can be analyzed by models or by mathematical approximations. Special cases of multiple wells closely spaced along a straight line can be approximated by a two-dimensional flow net drawn perpendicular to the line of wells. Although model studies and mathematical approximations are available for certain cases of unsteady flow, for many situations the simple approximation of steady flow is adequate.

The flow into an isolated well is depicted in Fig. 3.13. A pervious aquifer lies above an impervious, level stratum (Fig. 3.13a) or is confined between two impervious strata (an artesian aquifer) (Fig. 3.13b). The height of the piezometric surface above the impervious base at any distance from the well, r, is h. It is assumed that the average gradient i, at any radius, r, can be approximated by dh/dr, the slope of the water or piezometric surface. So long as the slope of the piezometric surface is no steeper than about 25°, this is reasonably correct. At any radius, r, the seepage in the homogeneous aquifer takes place normal to a cylinder whose area is $2\pi rh$. The equation for steady-state seepage can be derived as follows:

$$q = kiA$$

$$q = k\frac{dh}{dr}2\pi rh$$

$$\frac{q}{\pi k}\frac{dr}{r} = 2h\,dh$$

$$\frac{q}{\pi k}\log_e r = h^2 + C \tag{3:8a}$$

The constant of integration, C, is evaluated from the water levels, h_1 and h_2 measured at two points on the piezometric surface, r_1 and r_2.

$$\frac{q}{\pi k}\log_e\left(\frac{r_2}{r_1}\right) = h_2^2 - h_1^2 \tag{3:8b}$$

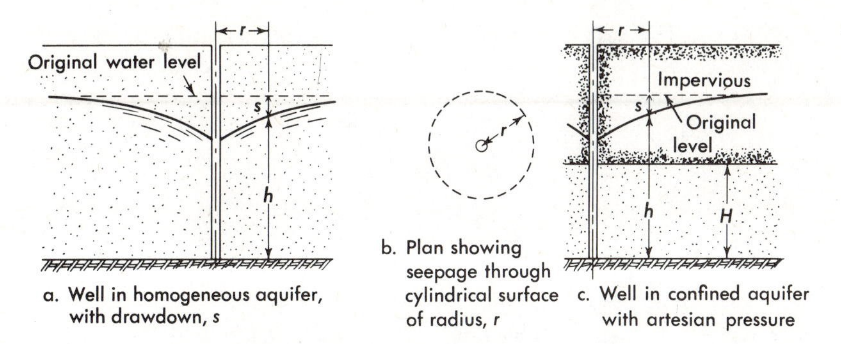

Figure 3.13 *Seepage into an isolated well.*

A similar analysis can be made for the confined aquifer whose thickness is H, provided the piezometric level does not fall below the top of the aquifer.

$$q = kH2\pi r\frac{dh}{dr}$$

$$\frac{q}{2\pi kH}\log_e\left(\frac{r_2}{r_1}\right) = h_2 - h_1 \tag{3:8c}$$

The analysis is frequently used with the well radius as r_1, and h_1 the level of the piezometric surface at the edge of the well. So long as the water surface slope is no steeper than 25° and the inflow into the well occurs throughout the height, h_1, the approximation is adequate. It must be kept in mind that the level of the water inside the well must be somewhat lower than h_1, however, to satisfy the contradiction shown in Fig. 3.12b.

At some distance from a well, the groundwater level remains virtually unchanged, despite the well pumping, due to recharge from a river or seepage from adjoining hills. In that case r_2 and h_2 are fixed by the site geology and the equations can be used to approximate the steady-state water levels for any given rate of pumping, q.

3:7 Neutral Stress, Uplift, and Seepage Erosion[3:7 3:8]

The effect of water on the soil or rock as well as on engineering structures is controlled by the pressure in the water, and its distribution within the soil or rock formation. The water is confined within the voids in soils and porous rocks and cracks, fissures, and larger openings in dense rock. The ability of the soil and rock to withstand water pressure is determined by the physical properties of the solid phase and particularly on the distribution of load (including soil or rock weight) between the water and solid phases.

TOTAL STRESS / The total load acting on an element of soil is the sum of all loads acting on the mass, including its own weight. The simplest case is a level mass of soil with a unit weight of γ, and a vertical load F (or F_z) at a depth z (measured positive down from the ground surface), acting over a horizontal area, A, as shown in Fig. 3.14.

$$F \text{ (or } F_z) = \gamma z A \qquad (3{:}9a)$$

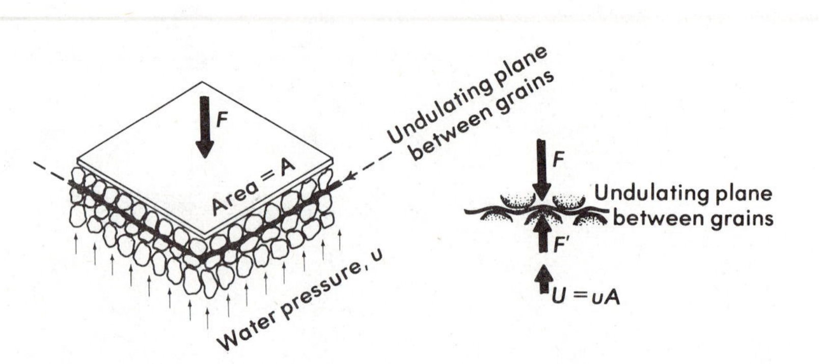

Figure 3.14 *Vertical load and effective stress in soil or rock.*

If both sides of the equation are divided by A, the expression becomes

$$\sigma_z = \gamma z \qquad (3{:}9b)$$

The symbol σ_z (or σ_v) denotes *total stress*: the total load divided by the gross area. This stress cannot be measured directly at any point. The total load can be measured or computed, and it is divided by the area over which it acts. The actual stress at any point within the area A will depend on the distribution of stresses between the liquid and solid phases; except for special cases the gaseous phase offers no load support, considering atmospheric pressure as zero.

NEUTRAL STRESS / The water pressure within the soil voids, u, is termed *neutral stress*. If the water is not moving, it can be computed from hydrostatics,

$$u = \gamma_w z_w \qquad (3{:}10)$$

where z_w is the vertical depth below the groundwater table or phreatic surface. If the water is moving, the pressure head can be found from a piezometer or computed by Equation (3:3b): $u = \gamma_w h_p$. It is uniform in all directions at one point but can change from one point to another.

UPLIFT / The neutral stress acting over an area A depends on the ratio of the void area A_v across the points of contact between the solid particles to the total area A, termed the *neutral stress coefficient**, N:

$$N = \frac{A_v}{A} \tag{3:11}$$

If the neutral stress acting on an area A is uniform, then the neutral force, U, is

$$U = uAN \tag{3:12}$$

This has been termed the *hydrostatic uplift* if acting upward or *hydrostatic force* acting in any other directions. At any point it is the same in all directions; over an area the pressure u varies hydrostatically or with seepage as described by the flow net. Values of the neutral stress ratio are given in Table 3:2.

TABLE 3:2. / NEUTRAL STRESS RATIO

Material	$N = Av/A$
Concrete	0.5 to 0.75
Fractured rock	0.5 to 1
Porous sandstone or limestone	0.75 to 1
Marble, granite	0.1 to 0.5
Soil	1

EFFECTIVE STRESS / Consider the undulating plane of area A at depth z in Fig. 3.14, passing through the points of contact between the grains. The force F is partially supported by the water force U. The remainder is supported by the grains. This is termed the *effective force* and is denoted F'.

$$F = U + F' \quad \text{or} \quad F = uNA + F' \tag{3:13}$$

If both sides of the equation are divided by the gross area A, it is expressed in terms of stresses

$$\frac{F}{A} = uN + \frac{F'}{A}$$

$$\sigma = uN + \sigma' \tag{3:14a}$$

$$\sigma = u + \sigma' \quad \text{(for soils, } N = 1) \tag{3:14b}$$

The term σ' is the *effective stress*. It is that part of the load that is transmitted by the particles divided by the gross area. It is not a stress that can be measured directly but a computed value. However, it is fundamental to

* Skempton has used a similar concept in terms of the fraction, a, of the total area that represents solid-to-solid contact. In this case $a = (1 - N)$.[3:8]

understanding the behavior of the solid phase under load, as will be described in Chapters 4 and 5.

A reduction in effective stress means reduced contact forces between the soil grains. In some cases this can be beneficial: Compression of the soil by structural load (Sections 4:1 to 4:4 and 10:3) can be offset by increasing the neutral stress and thereby reducing effective stress. On the other hand, a reduction in effective stress can also cause a loss in soil strength as well as rock strength (Section 5:5).

If the neutral stress should increase sufficiently, or the total stress decrease sufficiently, a condition can be reached where the neutral or water pressure force U = the total force F. At this point, the total load is supported by water pressure and interparticle contact ceases. The soil or rock mass is disrupted. Disruption is followed by failure if a sufficient volume of soil or rock is involved. Disruptive failure from seepage pressures is a major cause of both localized damage and catastrophic collapse of engineering works involving soil and rock. A margin of safety against such an occurrence is essential in design.

UPLIFT ON A STRUCTURE / When a structure rests on a soil, a part of the structure is in contact with the soil grains while the remainder bridges over the voids, as shown in Fig. 3.15.

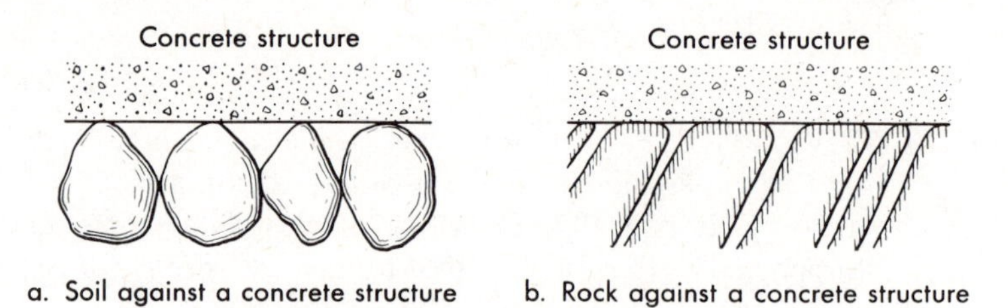

Figure 3.15 *Uplift and contact with a structure.*

As was described earlier, the actual contact areas of the soil grains with another solid are small (Fig. 3.15a). The neutral stress acts over a portion of the total area A, described by the neutral stress ratio N:

$$U = NAu = \gamma_w(h - h_z)NA \tag{3:15}$$

As given in Table 3:2, the value of N for soils is virtually 1; for jointed rock bonded to a concrete structure, the value of N can be as low as 0.5. Although lower values of N are theoretically possible for intact impervious rock such as granite and marble, it is unlikely that a perfect bond between structure and the rock can be realized. The minimum N used in design, therefore, should not be less than 0.75, the concrete in contact with rock.

When the uplift on the base of a structure exceeds the downward force of the weight of the structure and the loads it carries, the structure will rise or *heave*. In one case the empty concrete sedimentation basin for a sewage treatment plant under construction rose 3 m (10 ft) with the rising water table following a period of rain. Basement floors heave or even blow up as if they were blasted when subjected to excessive uplitt.

HEAVE AND BOILING IN A SOIL / Uplift develops within a soil mass similar to the way in which it occurs between soil and a structure.

Because the areas of contact between grains are very small, the neutral stress ratio, N, is virtually 1 and the uplift force, $U = uA$. If the upward force over an area A equals or exceeds the total load F of soil, water, and structure, a zone of instability and potential failure is created. At the point of failure:

$$F = U, \quad \sigma = u, \quad \text{and} \quad \sigma' = 0 \tag{3:16}$$

If the area is sufficiently great, any excess water pressure will force the overlying mass of soil and water to rise, a process called *heave*. The opening paragraph and Fig. 3.1 depict heave on a clay–silt stratum overlying a gravel layer. Similar heave can occur within a sand or silt stratum at any point where $\sigma = u$. The soil expands locally with an increase in void ratio. In some cases a blister of water forms within the soil mass (Fig. 3.16a). The soil particles from the roof of the blister fall to its bottom; by this process, termed *roofing*, the blister works its way to the surface. The soil surface bulges and then appears to explode as the blister reaches the top. Finally, the soil seethes and bubbles in a *boil* as if it were cooking. Heave can occur in any soil but particularly at the interface between an impervious soil underlain by a pervious stratum. Roofing and boiling are most pronounced in cohesionless materials.

Terzaghi has stated that heaving ordinarily will not take place unless the instability occurs over a width of $D/2$, where D is the depth of soil above the level of instability (Fig. 3.16b). The average neutral stress over different widths $D/2$ for different assumed depths D can be computed from a flow net. Where the average neutral stress equals or exceeds the stress produced by the weight of the overlying soil and water, there is a possibility of heave or boiling. It must be remembered that stability computations are approximate at best, and that a large safety factor should be used to be certain that boiling will not occur.

If the water pressure, u, is distributed so that a uniform upward gradient occurs, the heave can develop uniformly throughout the mass with many enlarged voids but without the formation of a big blister. In this condition, when neutral stress and total stress are in balance (or when u is slightly less than σ), the mass may look deceptively stable. However, a machine or structure on the soil surface will sink slowly if its unit weight exceeds that of the saturated soil.

Boiling is merely aggravated heave in which excess neutral stress causes concentrations of blisters. Small sand "volcanoes," Fig. 3.16c, often rise to

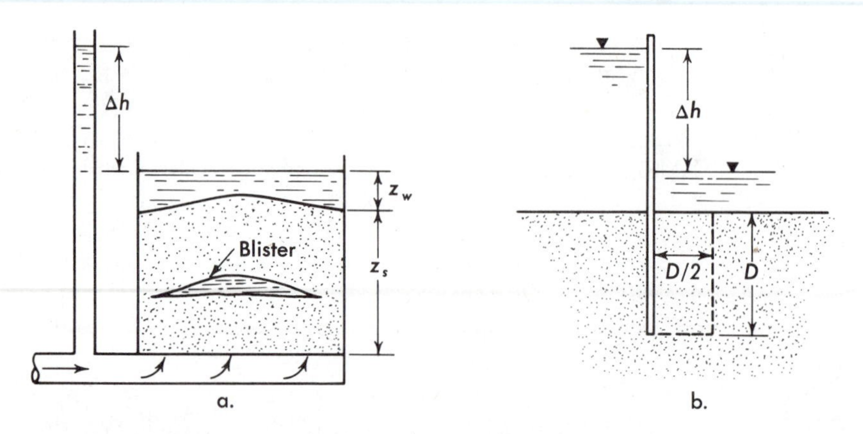

Figure 3.16 *Development of heave and boiling. (a) Formation of a blister from upward flow through soil; (b) width of critical zone for boiling (after Terzaghi); and (c) sand volcanoes from boiling.*

throttle the flow at one point and cause the boiling to shift to a weaker zone elsewhere. If the soil is cohesionless, the loss of grain-to-grain contact transforms the material into a fluid with a gross unit weight of γ but no static strength. This is *quicksand*, a condition, not a type of soil. Virtually any cohesionless soil can become quick if the effective stress is reduced to zero by an increase in neutral stress, decrease in total stress, or a combination of both.

CRITICAL GRADIENT / The hydraulic gradient associated with heave or boiling near an unrestrained soil surface is termed the *critical gradient*, i_c. For upward flow (Fig. 3.16a) the neutral and total stresses at the bottom of the sand are as follows:

$$u = \gamma_w(z_w + z_s + \Delta h)$$

$$\sigma = \gamma_w z_w + \gamma z_s \qquad (3:17)$$

At the instant heave when the quick condition develops, $\sigma = u$.

$$\gamma_w z_w + \gamma_w z_s + \gamma_w \Delta h = \gamma_w z_w + \gamma z_s$$

$$\gamma_w \Delta h = \gamma z_s - \gamma_w z_s = (\gamma - \gamma_w) z_s$$

$$i_c = \frac{\Delta h}{z_s} = \frac{\gamma - \gamma_w}{\gamma_w} \qquad (3:18)$$

For a typical saturated sand, $\gamma = 20\ \text{kN/m}^3$ or $127\ \text{lb/ft}^3$ and $\gamma_w = 9.81\ \text{kN/m}^3$ or $62.4\ \text{lb/ft}^3$ so that $i_c = 1$ approximately (in upward seepage). For seepage toward an unrestrained sloping surface the value of i_c is less: zero if the slope is barely stable without seepage.

Example 3:3

Consider the site conditions of Fig. 3.1. The soil profile consists of relatively impervious alluvial silts and clays 10 m thick underlain by very pervious gravel 4 m thick. A trench 6 m deep and 200 m long is excavated in the silt–clay alluvium and then pumped dry by a surface pump. The groundwater level in the silt–clay alluvium is drawn down by pumping from the excavation. The permeability of the gravel is 9 mm/sec and of the silt–clay is 7×10^{-5} mm/sec. The alluvium weighs $20\ \text{kN/m}^3$ or $127\ \text{lb/ft}^3$. At the time pumping commenced, the river is 6.5 m above the bottom of the silt–clay.

1. Compute the seepage upward through the trench bottom from the gravel for a 5-m-wide trench, assuming no head loss through the gravel from the river.
   a. $\Delta h = 6.5 - 4 = 2.5$ m; $L = 4m$ (vertically upward), $A = 200\ \text{m} \times 5\ \text{m}$.
   b. $q = kiA = 7 \times 10^{-5} \times 10^{-3} \times \frac{2.5}{4} \times 5 \times 200 = 4.4 \times 10^{-5}\ \text{m}^3/\text{sec}$ $= 1.6 \times 10^{-3}\ \text{ft}^3/\text{sec}$. Water inflow is negligible.
2. Compute the head loss in the first 25 m of gravel from the river assuming that all the flow computed above travels the 25 m from the river in a path averaging 6 m wide.
   c. $L = 25$ m $A = 6\ \text{m} \times 4$ m, $k = 9\ \text{mm/sec} = 9 \times 10^{-3}$ m/sec.
   d. $q = k(\Delta h/L)A$; $4.4 \times 10^{-5} = 9 \times 10^{-3} \times (\Delta h/25) \times 24$.
   e. $\Delta h = [4.4 \times 10^{-5} \times 25] + [24 \times 9 \times 10^{-3}] = 0.51 \times 10^{-2}\ \text{m} = 5\ \text{mm}$ $= 0.2$ in.

   This justifies the assumption that the head within the gravel at a distance from the river is virtually the same as in the river.

3. Compute the total, neutral (uplift) and effective stress at the interface between the gravel and the silt–clay below the trench bottom.
   f. $\sigma_v = 20 \times 4 = 80\ \text{kN/m}^2 = 1670\ \text{lb/ft}^2$.
   g. $u = 6.5 \times 9.81 = 64\ \text{kN/m}^2 = 1340\ \text{lb/ft}^2$.
   h. $\sigma'_v = 80 - 64 = 16\ \text{kN/m}^2 = 330\ \text{lb/ft}^2$.
4. At what river level would the bottom of the trench heave? What would be the seepage into the trench through the bottom from the gravel just before the heave?
   i. $\sigma'_v = 0$, $\sigma_v = 20 \times 4 = 80\ \text{kN/m}^2$, $u = \gamma_w h_p$.
   j. $h_p = 80/\gamma_w = 80/9.81 = 8.16$ m or 26.8 ft above the silt-gravel interface. The river is $10 - 8.16 = 1.84$ m or 6 ft below the top of the flood plain.

PIPING AND SEEPAGE EROSION / If the soil within the zone of boiling is washed away by the flowing water, an open pit will be created. This causes a concentration of flow into the pit and an increase in the hydraulic gradient because the seepage path is shortened. Consequently, the boiling is even more fierce, and the pit becomes deeper, working its way upstream at an increasing speed toward the source of the water, as shown in Fig. 3.17b. An opening or *pipe* is developed in the soil, and the process of continued backward erosion is called *piping.*

Piping also begins from very localized boiling or concentrations of seepage, as shown in Fig. 3.17a. When the upward hydraulic gradient approaches 1 at the soil surface, a small surface boil forms, and if the soil is washed away, a pit develops. This pit works its way upstream, becoming larger and moving faster as the seepage path is shortened.

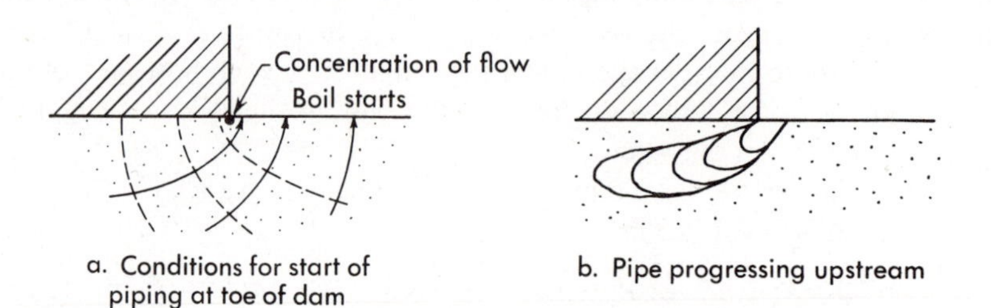

Figure 3.17 *Piping beneath a masonry dam on a sand foundation.*

If the seepage is horizontal toward the sloping face of an excavation or downstream face of an earth dam or downward into an improperly protected drain or leaking sewer, piping will develop at very small gradients. Extensive cavities have been created where cohesive strata support the remainder of the soil mass above the opening. In one situation erosion of fine sand through an 8-mm crack in a sea wall created a cavity 2 m deep beneath a concrete pavement. It was not discovered until a loaded truck broke through into the crater below.

When the pipe approaches the source of water, there is a sudden breakthrough and a rush of water into the pipe, which enlarges it. One such pipe a few inches in diameter through an earth dam was enlarged to 3 m in a few minutes after a breakthrough. Eventually, the enlarged hole collapsed from a lack of support, destroying part of the soil mass.

Cohesionless soils, particularly fine sands and silts, are most susceptible to piping failures. Most clays resist piping because the interparticle bonds help prevent the particles from washing away. Some clays, termed *dispersive clays* or *erodible clays*, either have limited bonding or the bonding is reduced drastically by ions in the water. The particles disperse into suspension and are eroded.[3.9]

Piping in erodible clays does not develop unless a small fissure is present, such as a root hole or shrinkage crack. The susceptibility is measured by a pinhole test in which a 1-mm diameter hole is made through a soil sample in a permeability test.[3:10] Clouding of the water or erosion of the hole when the soil is subjected to a gradient of 0.5 to 3, with water containing the same dissolved salts as in the field, is an indication of potentially erodible clay.

DISRUPTIVE GRADIENTS / Heaving, boiling, and most cases of piping involve insufficient restraint, F, to withstand the uplift U. In cases where extremely high gradients are involved, $i \gg 1$, piping can develop, although $F > U$. The frictional drag of the water moves some of the finer soil particles, concentrating them in lenses and leaving their source deficient in fines. Seepage then concentrates on the disrupted zone, aggravating the condition until true piping develops. Little information is available on disruptive gradients. Generally, they are well above 10 and increase with interparticle attractive forces and confining stress. For safety, gradients in restrained cohesionless soils are generally kept below 3 or 4. In clays the disruptive gradients sometimes exceed 50; they are seldom critical unless the clay is cracked, dispersive, or erodible.

3.8 Seepage Control

CONTROL OF SEEPAGE / Control of seepage involves reducing the flow, reducing the water pressure, or increasing the load that resists the water pressure. Excessive seepage is caused by high permeabilities, high heads, or short seepage paths. If the soil mass through which the seepage occurs is man-made, like a dam, the permeability can be reduced by the proper selection of materials. For example, mixing a small amount of clay with the sand used for constructing a levee can reduce the permeability greatly. A natural deposit is difficult to change. Small amounts of a dispersing agent such as sodium tetraphosphate mixed in the surface of a flocculent-structure clay, or injection of clay, chemicals, or cement into the voids of a coarse-grained soil, can reduce the permeability, but at considerable expense.

The seepage path can be lengthened, which will reduce the quantity of seepage and also reduce the water pressure at the downstream end of the flow. An impervious core in an earth dam (Fig. 3.18a) and an impervious cutoff trench in a pervious foundation for a dam can increase the path greatly. A complete cutoff (Fig. 3.18b) that extends to a deeper impervious stratum is much more effective than the partial cutoff (Fig. 3.18a). Cutoffs are constructed of an impervious soil or steel sheet piling, depending on availability of materials and ease of construction. An impervious blanket of clay upstream (Fig. 3.18c) is also useful but must not be used downstream because it will increase the uplift.

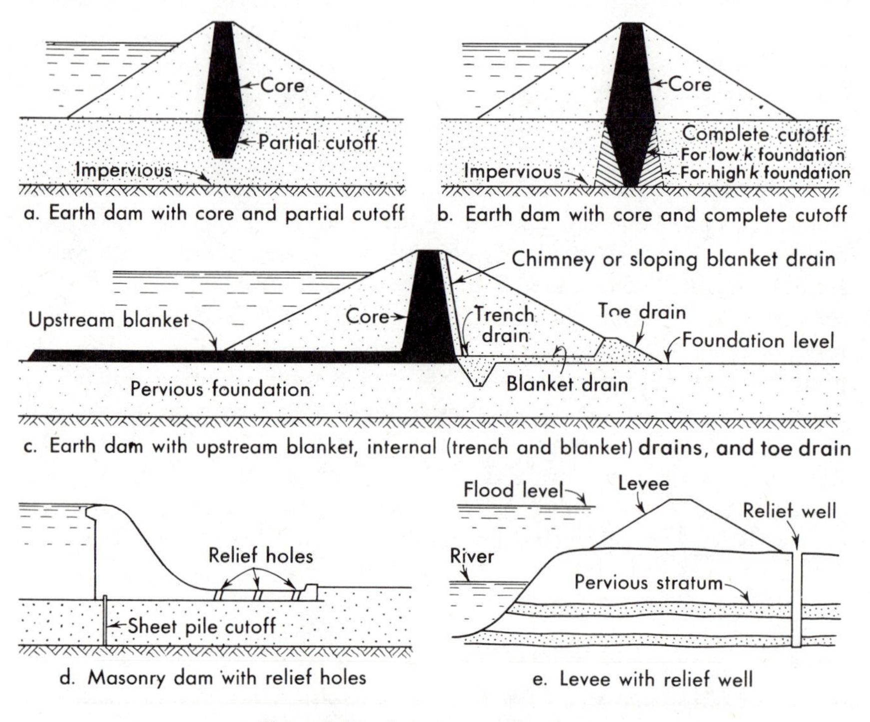

Figure 3.18 *Seepage control measures.*

A cutoff causes an increase in neutral stress upstream and a reduction downstream. A perfect, complete cutoff produces natural stresses upstream that are equivalent to the headwater level. A cutoff located too far downstream will reduce the seepage quantity and eliminate piping, only to create excessive uplift that destroys the structure by heave. An owner of a small dam attempted to correct piping that was slowly developing just downstream from the toe by driving sheet piling into the soil. He did this on the free advice of the piling salesman and against the warning of his engineer. The dam failed by shear in the downstream face that was already weakened by the increased neutral stress. The cutoff should be placed where the increased pore pressure is not harmful, at the center of the structure or upstream from the center and under the heaviest part of the structure, if possible. Incomplete cutoffs can increase seepage erosion by concentrating gradients.

Excessive water pressure can be controlled by drainage that short-circuits the flow and bleeds off the excess neutral stress at a point where it can do no harm. The *trench drain*, *blanket drain*, and *toe drain* (Fig. 3.18c) are used separately or in combination in earth dams to reduce neutral stresses in the downstream part of the embankment. Relief holes (Fig. 3.18d) reduce uplift on masonry dams. Relief wells (Fig. 3.18e) are used to reduce pressures in confined seams or pockets. Drainage has the disadvantage of shortening the seepage path and increasing the flow, but this can be corrected as previously described. It is essential that the drainage system be properly designed to avoid seepage erosion, as will be described in the section on filters.

FILTER DESIGN / A *filter* or *protective filter* is *any* porous material whose openings are small enough to prevent movement of the soil into a drain and which is sufficiently pervious to offer little resistance to seepage. Extensive experiments have shown that it is not necessary for a filter to screen out all the particles in the soil. Instead the filter openings need restrain only the coarsest 15%, or the D_{85}, of the soil. These coarser particles, D_{85} and larger, will collect over the filter opening as shown in Fig. 3.19a. Their voids will create smaller openings to trap even smaller particles of soil. Therefore, the diameter of the openings in the filter must be less than D_{85} of the soil. If the filter is a metal screen, filter fabric or holes in a perforated pipe, this limit fixes the finest soil that can be filtered by any

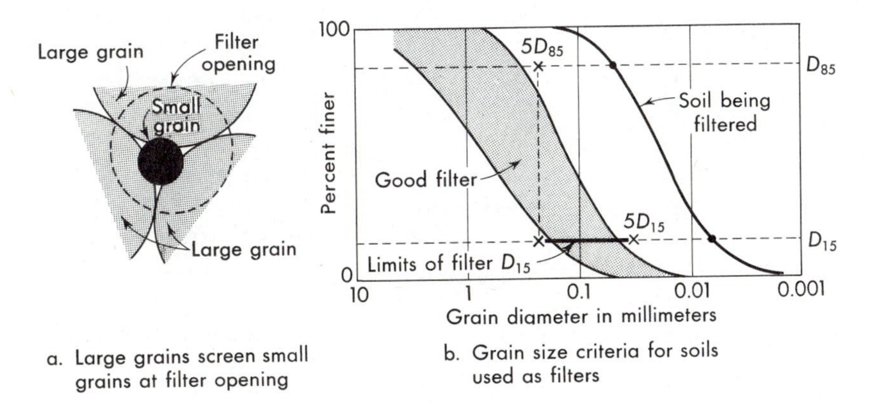

a. Large grains screen small grains at filter opening

b. Grain size criteria for soils used as filters

Figure 3.19 *Protective filters.*

given opening, or it establishes the largest opening that can be used with a given soil. Frequently, a soil is employed as a filter. This means that the effective diameter of its voids must be less than D_{85} of the soil being filtered. Since the effective pore diameter is about $\frac{1}{5}D_{15}$, then

$$D_0 \leq D_{85(\text{soil})} \tag{3:19a}$$

$$D_{15(\text{filter})} \leq 5D_{85(\text{soil})} \tag{3:19b}$$

If the filter is to provide free drainage, it must be much more pervious than the soil. Since the permeability coefficient varies as the square of the grain size, then a ratio of permeabilities of over 25 to 1 can be secured by

$$D_{15(\text{filter})} \geq 5D_{15(\text{soil})} \tag{3:19c}$$

These criteria (Fig. 3.19) are the basis for filter design.[3:11] In general, the filter soil should be well within these limits, and its grain size curve should be smooth and parallel to or flatter than the soil. If the soil being filtered is very fine grained, more than one filter layer will be required. The final filter layer is designed to fit between the openings in the conduit and the next finer filter. For many silty and clayey soils a well-graded concrete sand makes a satisfactory filter. A coarser pea-gravel second filter—usually described as ASTM No. 78 crushed stone— is then needed for the first.

If the soil being filtered is gap-graded, its grain size curve is redrawn, considering only the portion of the soil finer than the gap to be the total soil being filtered, and disregarding the part of the soil coarser than the gap. The filter is designed to fit the redrawn curve.

There have been many attempts to devise a "universal filter" that is small enough to filter the finest soil and yet having a D_{85} large enough so that it will not pass through the 80-mm or $\frac{5}{16}$-in. perforations of commercial drainpipe. However, such filter materials have such a wide range of sizes (high C_u) that the particles segregate during handling and construction. Therefore, they should not be used unless care is taken to maintain their gradation.

The thickness of a sand or gravel filter layer is controlled by the ability of the layer to undergo distortion without rupture and by construction ease. Around drainpipes, filter layers from 0.1 to 0.2 m (4 to 8 in.) are sufficient. In dams horizontal layers should be 0.2 to 1 m (1 to 3 ft) thick and trench or chimney filters 2 to 3 m (6 to 10 ft) wide.

Fabric filters of woven metal, woven plastic, and nonwoven fiber sheets are easy to install in restricted spaces. Their cost is offset by saving labor. Reliable data on their filtering performance must be obtained from tests utilizing the same soils and gradients anticipated in the installation. Under gradients of up to 10, the nonwoven filters appear to have an effective opening of about 0.25 mm. In woven fabric, the openings are somewhat larger; they can be measured with a microscope-comparator.

Both sand and gravel layers and fabric filters should not deteriorate physically, chemically, or biologically during their anticipated life. For example, soft sandstones and limestone fragments are questionable filter materials because of potential weathering. Fabrics should be resistant to oxidation, attack from dissolved minerals, and biologic attack. Sometimes filters clog with biologic slimes where there are sufficient nutrients and oxygen for their growth. Protection from air and from water containing organic materials minimizes clogging.

3.9 Capillary Tension and Capillary Rise

When the voids in soil or rock are only partially filled with water, the remainder of the void space is air (or possibly some other gas). The water forms small curved wedges between the grains (Fig. 3.20), with air separating adjacent water surfaces. The differences in density and intermolecular attraction between the air, water, and solid phases cause *surface tension*, a major factor in soil behavior and significant in fine-grained porous rocks. It causes water to rise above the groundwater table; it causes a wet clay to shrink upon drying; and it gives a damp, fine-grained, cohesionless soil the illusion of cohesion.

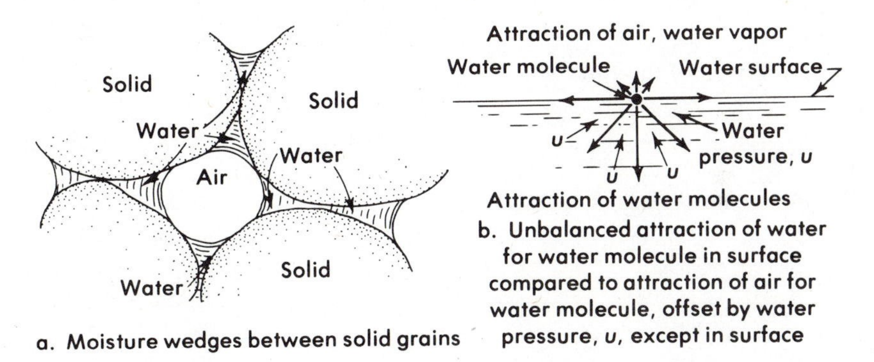

Figure 3.20 *Moisture wedges and unbalanced forces in surface tension.*

SURFACE TENSION AND CAPILLARY STRESS / The unbalanced forces on an element of water in the air–water interface are illustrated in Fig. 3.20b. The attraction of air for water is less than that of water for itself; thus the element of water is held in the surface by the attraction of the remainder of the water. The surface of the water, through force, behaves as if it were a tightly stretched membrane.

The surface tension force acts parallel to the water surface, with dimensions of a force per unit length of surface (see Table 3:3).

TABLE 3:3 / SURFACE TENSION

Temperature		Surface Tension, T_0	
1°C	34F	−0.076 N/m	−0.0052 lb/ft
20	68F	−0.073	−0.0050
40	104F	−0.070	−0.0048

NOTE: In soil mechanics, tension is negative.

Dissolved salts tend to increase surface tension of water in contact with air; however, some materials decrease it. The attraction of water for soil particles or clean glass is many times greater than surface tension with air; therefore, water in contact with such a solid spreads over the solid surface, extending the air–water interface. At the air–water–solid junction, the continuity of the water will be limited by the weaker of the water–air, water–solid tensions: T_0.

If the water is subject to a tensile stress u, confined in a uniform tube of diameter d (Fig. 3.21), the water surface will be stretched, forming a curved surface known as a *meniscus*. The tensile stress in water in equilibrium with the surface tension is capillary tension. If the angle of contact between the tube and the meniscus is α, the total unbalanced force caused by the meniscus along the perimeter of the tube is

$$F = \pi \, dT_0 \cos \alpha \tag{3:20a}$$

Since the area of the tube A is

$$A = \frac{\pi d^2}{4} \tag{3:20b}$$

the capillary stress, u, is found to be

$$u \frac{F}{A} = \frac{4T_0 \cos \alpha}{d} \tag{3:21}$$

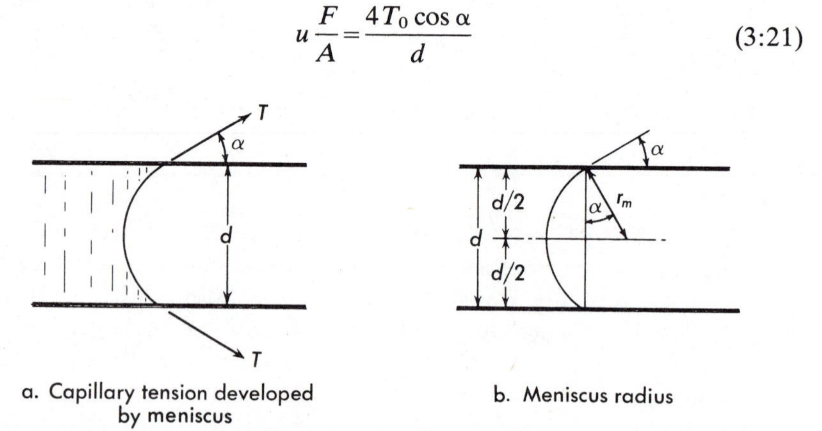

a. Capillary tension developed by meniscus

b. Meniscus radius

Figure 3.21 *Capillary tension and meniscus in a uniform tube of diameter d.*

(Note that u has a negative sign when the value of T_0 is substituted in the equation.)

MENISCUS RADIUS / The relation of capillary tension to the radius of the meniscus, r_m, can be found by considering the geometry of the meniscus (Fig. 3.21):

$$\frac{d}{2} = r_m \cos \alpha \tag{3:22}$$

If the expression for meniscus radius is substituted in the formula for capillary tension, then

$$u = \frac{2T_0}{r_m} \tag{3:23}$$

It can be seen that for water in contact with air, the capillary tension is dependent only on the meniscus radius, and it varies inversely with it.

MAXIMUM TENSION / The maximum tension will occur with a minimum meniscus radius. This will occur when the surface of the meniscus is tangent to the soil particles and $\alpha = 0$. In a cylindrical tube, therefore, the minimum r_m equals half the tube diameter. The maximum tension, therefore, will be

$$u_{\max} = \frac{4T_0}{d} \tag{3:24}$$

LIMITING TENSION / In large tubes water cannot sustain true tension; when the water pressure falls to the vapor pressure of water at that temperature, the water vaporizes and the tension is limited. The vapor pressure of water is a function of temperature as shown in Table 3:4. It decreases slightly with dissolved salts in the water.

TABLE 3:4 / VAPOR PRESSURE OF WATER
(Atmosphere = 101.3 kN/m^2 = 14.7 lb/in.^{2}, absolute)

Temperature	Vapor Pressure: Absolute		
0°C	0.45 kN/m^2	6.2 gf/cm^2	0.07 lb/in.^{2}
20	2.3	23.6	0.33
40	7.3	74.5	1.05

For equilibrium in a capillary tube, the limiting stress, u_L, will be the vapor pressure of water, u_v, added algebraically to the capillary tension, $u_{\max}$,

$$u_L = u_v + u_{\max} \tag{3:25}$$

In a large tube there is no capillary tension, $u_{\max} = 0$; the limiting water tension is equivalent to the vapor pressure of water. In capillary tubes, the limiting tension is equal to the capillary tension reduced by the vapor pressure of water. Thus, in very small tubes, the ultimate water tension exceeds the classic limit of vapor pressure—actual tensions equivalent to several atmospheres have been deduced from capillary rise.

CAPILLARY RISE / The interconnected pores or voids in soil and porous rock form irregular but definite capillary tubes (Fig. 3.2). The maximum tension that can develop will vary from point to point, depending on the pore diameter and the degree of saturation. In a saturated pore in a cohesionless soil, experiments show that the effective pore diameter for capillary tension is approximately $\frac{1}{5}D_{10}$.

If a soil mass is completely saturated and inundated, the air–water boundaries disappear and capillary tension becomes zero. When a saturated soil is exposed to open air, capillary tension develops as soon as evaporation creates meniscuses at the surface. Since the soil moisture in a saturated soil is continuous, the water–tension stress developed at the air–water boundary is felt throughout the mass in a way similar to the way pressure applied at one point in a continuous body of fluid is transmitted throughout the body. The water, however, also obeys the law of hydrostatics: from Equation (3:10)

$$\Delta u = \gamma_w \, \Delta z \tag{3.26}$$

where z is measured positive downward. Therefore, below the elevation of the meniscus, the tension decreases; or, in other words, the pressure increases.

The capillary rise of water in a soil above the groundwater table illustrates the combined effect of capillary tension and hydrostatic pressure. At the groundwater elevation, a free surface, the water pressure u is zero. Below the water table the pressure increases in accordance with Equation (3:10) or (3:26). Above the water table it decreases in the same way as shown in Fig. 3.22. The greatest negative stress in the water is the maximum capillary tension, u_{max}. The height, h_c, to which the water can rise above the surface of zero pressure, is found by equating the expression for hydrostatic pressure with that for maximum capillary tension:

$$h_c = -\Delta z \qquad u = u_{max} = \frac{4T_0}{d}, \quad h_c\gamma_w = \frac{-4T_0}{d}$$

$$h_c = \frac{-4T_0}{\gamma_w d} \tag{3:27}$$

In a partially saturated soil the moisture may be either *continuous* or *discontinuous*, depending on whether the moisture wedges (Fig. 3.20) are interconnected or discrete. If it is continuous, the variation of water stress

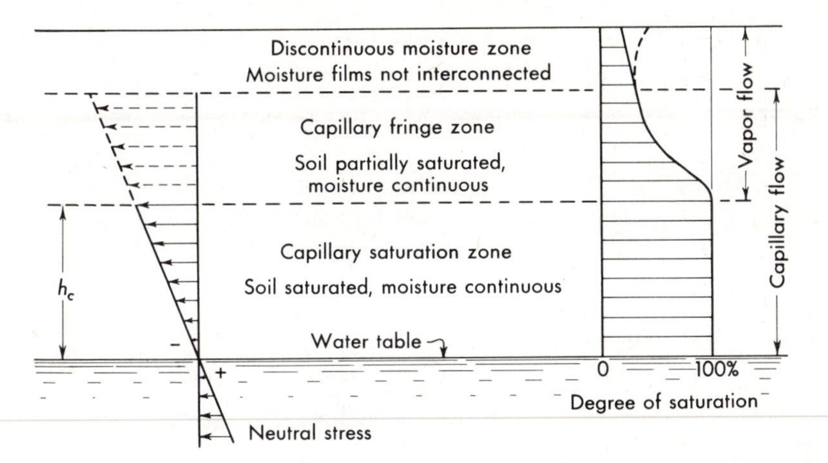

Figure 3.22 *Capillary equilibrium, neutral stress, and soil moisture distribution.*

with elevation follows Equation (3:10), and the meniscus radius in each wedge adjusts itself to conform to the water stress. If the moisture is discontinuous, each wedge is independent. The stress with the wedge is determined by its radius.

CAPILLARY EQUILIBRIUM / Moisture rises above the groundwater table or free-water surface as a result of capillary tension. When equilibrium is established, the soil moisture is distributed approximately as shown in Fig. 3.22. In the capillary zone the soil is saturated. The moisture is continuous and the neutral stress obeys the hydrostatic law. Above this zone is the *capillary fringe*. The degree of saturation falls off rapidly, but, although the moisture does not fill the voids, it is still continuous in interconnected wedges between the grains and in the smaller voids. The effective stress is no longer equal to the total stress minus the neutral stress as given by Equation (3:14b) because the neutral stress does not act over the entire void area. The degree of saturation becomes less with increasing height above the free surface until the moisture wedges are no longer interconnected. There is still neutral stress in the upper zone of *discontinuous moisture*, but it no longer follows the hydrostatic distribution. Each moisture wedge develops a different stress, depending on its radius, and, although the stress can be very great, it acts over only a small fraction of the void.

The height of capillary saturation varies from centimeters in sands to more than 30 m in some clays. The height of the zones of partial saturation and discontinuous moisture depend on the grain size of the soil, the water table level, and the climate, as will be described in Section 3:10. The maximum hypothetical capillary rise can be approximated from the effective pore diameter, as previously described. In most soils the real height is less because cracks and fissures, larger than the pores, control the maximum tension.

3:10 Capillary and Vapor Flow

Above the water table the soil moisture is acted on by both gravity and capillary tension. Head differences produced by the algebraic sum of both forces induces flow resisted by fluid friction as described by Darcy's law.

CAPILLARY FLOW / Capillary flow occurs in the zone of saturation and in the capillary fringe where the moisture is continuous. In the equilibrium condition (Fig. 3.22) the capillary tension, u, just balances the hydrostatic stress, $\gamma_w z$, and no movement occurs. If the capillary tension changes, flow will take place, depending on whether the tension is increased or decreased with respect to the hydrostatic gradient.

Evaporation of moisture in the fringe reduces the degree of saturation. As a result each meniscus radius at any given level in the fringe zone is reduced, increasing the tension. At the same time the level of capillary saturation is depressed to h_{c-1}, Fig. 3.23. At that level (the boundary between saturated and unsaturated voids), the capillary tension is unchanged. However, at the new lower level, the capillary tension now exceeds the hydrostatic stress $\gamma_w h_{c-1}$ and upward flow is generated by the gradient:

$$i = \frac{\Delta u/\gamma_w}{h_{c-1}} \tag{3:28}$$

as shown in Fig. 3.23a.

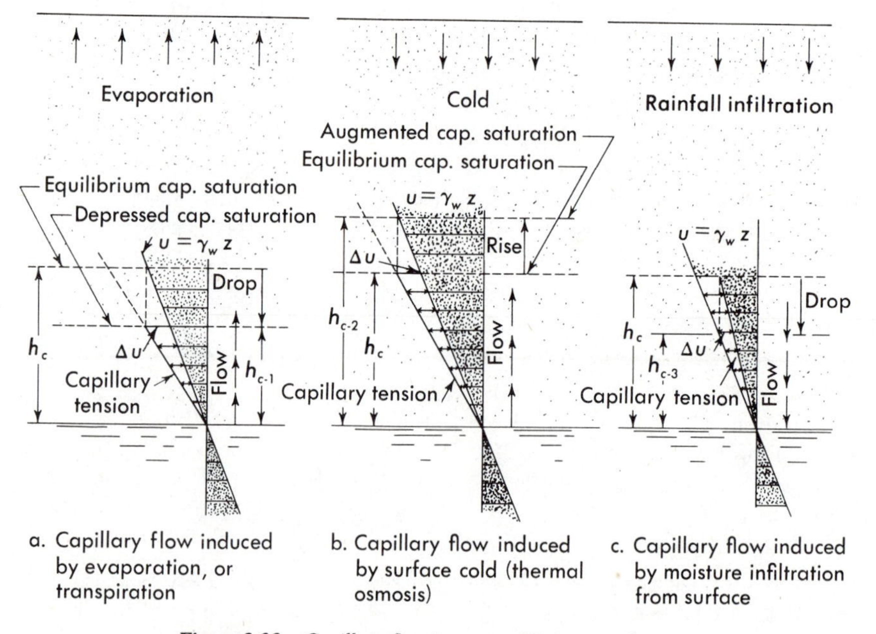

Figure 3.23 Capillary flow in nonequilibrium conditions.

In arid regions the continuous evaporation in the fringe zone maintains the state of nonequilibrium and upward capillary flow. Dissolved salts are brought up from the zone of saturation and precipitated in the fringe where the water is evaporated. This concentration of precipitated salts has two significant effects. First, the fringe zone becomes cemented and eventually made impervious by the salts, including calcium carbonate and calcium sulfate. A rocklike hardpan is created that presents serious problems for future excavation and influences drainage and seepage if the area is inundated by a reservoir or if it is irrigated for farming. The second, a more serious problem, is the effect of the salts on agriculture. They limit the crops to those that can survive in the saline environment and eventually make the land sterile. Vast areas of lands of ancient irrigation in India lie fallow because of the destruction of fertility by salt accumulation. The process can be reversed by sufficient irrigation and drainage to flush the salts downward.

In humid regions, upward flow can be induced by loss of water from the capillary zone by the transpiration of crops and other vegetation. Accelerated evaporation, produced by heat, can also increase capillary tension and upward moisture movement. If the soil is compressible, the increased capillary tension will produce shrinkage and settlement. Severe settlement from capillary desiccation is even caused by some species of trees.

In dry regions, the continuous flow of moisture upward to be evaporated is reduced by construction of a building or a pavement that prevents the evaporation loss. Static equilibrium is produced with a rise in the level of capillary saturation and an increase in the degree of saturation in the fringe zone. This can have two serious consequences. First, any cemented layers are weakened. Pavement rutting sometimes occurs due to a damp subgrade in a desert region. If the grain structure of the cemented soil is loose, there may be soil structure collapse, subsidence of the ground surface, and damage to buildings. Second, the moisture increase will cause expansion of any highly plastic clays. In arid regions this is severe because the clays are initially highly desiccated.

Upward flow is also induced by a drop in temperature at the ground surface that increases the capillary tension (Fig. 3.23b). The increased tension, Δu, produces the gradient that maintains upward flow until a new equilibrium is reached at the higher level $h_{c-2} = (u + \Delta u)/\gamma_w$. Temperature-induced movement is sometimes termed *thermal osmosis*. For example, construction work in the fall season when the air temperatures are dropping steadily is sometimes hampered by the increasing soil moisture in spite of no rainfall.

Downward flow is induced by an increase in the degree of saturation in the fringe accompanied by an increase in the radius of each meniscus in the fringe and on the boundary of capillary saturation. The tension is reduced (Fig. 3.23c), and the decrease, Δu, produces a gradient that causes downward seepage. In this way the ground water is replenished, although the water that reaches the water table may have come from the capillary zone. A similar downward movement can occur during periods of increasing ground surface temperature.

Capillary flow takes place horizontally as well as vertically if there are differences in capillary tension that induce a hydraulic gradient. The drying of soil in a deep excavation can induce capillary flow laterally to the exposed banks. Rain falling on these same banks can induce capillary flow into the soil mass.

The rate of capillary flow is proportional to three factors: the change in stress, Δu; the reciprocal of the distance through which flow occurs, $1/L$; and the permeability, k.

In sands, the stress changes, Δu, are small because the large voids limit the capillary tension. Therefore, although k is large, the rate of flow is small. In clays with minute voids, the Δu is likely to be large, but k is extremely low. In clays, therefore, the rate of capillary flow is also very small. In silty soils the optimum combination of small voids producing moderately large values of Δu and permeability coefficients that are not too small causes maximum rates of capillary flow.

SOIL MOISTURE ABOVE THE GROUNDWATER TABLE / Because of the environmental changes that occur daily, capillary equilibrium seldom exists for very long. Instead, the moisture in the capillary zone is constantly changing, and with the changes there are profound variations in the engineering properties of the soils. These changes are most significant in the fringe zones, but the effect of the capillary stress changes are felt below the line of saturation.

Only limited data are available regarding the possible moisture content variations that occur beneath structures. The major factors are climate, potential capillary rise in the soil, and the water table. When the water table is within 10 to 15 m of the ground surface in clays, 3 to 5 m in silts, and 1 to 2 m in sands, the equilibrium moisture is close to saturation. With a deeper water table, the equilibrium moisture near the ground surface will be somewhat less. Near-equilibrium moisture occurs beneath wide pavements or broad structures that shield the surface from both evaporation and rain.

VAPOR MOVEMENT / In the fringe and discontinuous zones, moisture flows in the vapor phase from points of high vapor pressure to points of low pressure. The saturated vapor pressure increases with increasing temperature. The actual pressure is equal to the saturated pressure multiplied by the relative humidity. Within the fringe zone the vapor pressure is close to saturation; above it is less because of evaporation. Although vapor permeability is far greater than water, the rate of moisture transfer is small because of the relatively small amounts of water in the vapor phase at soil temperatures.

Moisture in the vapor phase flowing to a point of lower vapor pressure increases the relative humidity until either saturation is reached or the increased relative humidity equalizes pressures. In the first case, the vapor water becomes liquid and flow continues. In the second case flow ceases. Temperature differences, rainfall infiltration, and surface evaporation are the major causes of vapor movement.

FIELD OBSERVATIONS / The complex interplay of capillary and vapor forces in the fringe and discontinuous moisture zone make it impossible to predict moisture analytically. Field observations in similar soils in similar environments are necessary to determine the range of moistures that is likely during the useful life of a structure.

3:11 Drainage[3.7]

Drainage means removal of water from the soil. It has three objectives: (1) Prevent seepage out of the soil, such as into an excavation where it would be a nuisance or a hazard; (2) improve the soil, such as increase the strength or reduce the compressibility; and (3) reduce water pressure. In fine-grained soils it can be effective in (2) and (3) even when little water is removed.

FORCES INVOLVED IN DRAINAGE / A number of forces control the ease with which water drains from the soil. First is the resistance to seepage, as indicated by the permeability coefficient. Second is the effect of the drainage on the soil. If the soil is relatively incompressible, the water lost is replaced by air in the voids. If the soil is compressible, the water loss is accompanied by a reduction in void ratio, and the soil may remain virtually saturated. Third are the forces that restrain water: capillarity and adsorption. Both the resistance to flow and capillary retention become greater with decreasing grain size. Coarse-grained soils, such as gravel and coarse sand and open rock joints drain rapidly and air replaces the water in the voids. Fine-grained soils that have low permeability and very high capillary retention drain very slowly and lose only as much water as the consolidation will permit, until evaporation commences.

In order to remove water from the soil, the force producing drainage must be greater than the retention and the resistance to flow. *Gravity* is the force most often employed: Water moves from the soil into the drain under the influence of its own weight. This is cheap and reliable but not strong enough in fine-grained soils. A *vacuum* can add atmospheric pressure to the head produced by gravity. With its aid, finer soils such as silty sands can be drained. A direct electric current will induce a flow of water in the soil toward a negative electrode. This *electroosmosis* can induce drainage of low-permeability soils such as silts.

Evaporation can be considered a drainage method, because it causes a loss of water. It is a slow but powerful force that can drain even clays. *Consolidation* described in Chapter 4 produced by a load on the soil surface is essentially a drainage process that is effective in compressible materials.

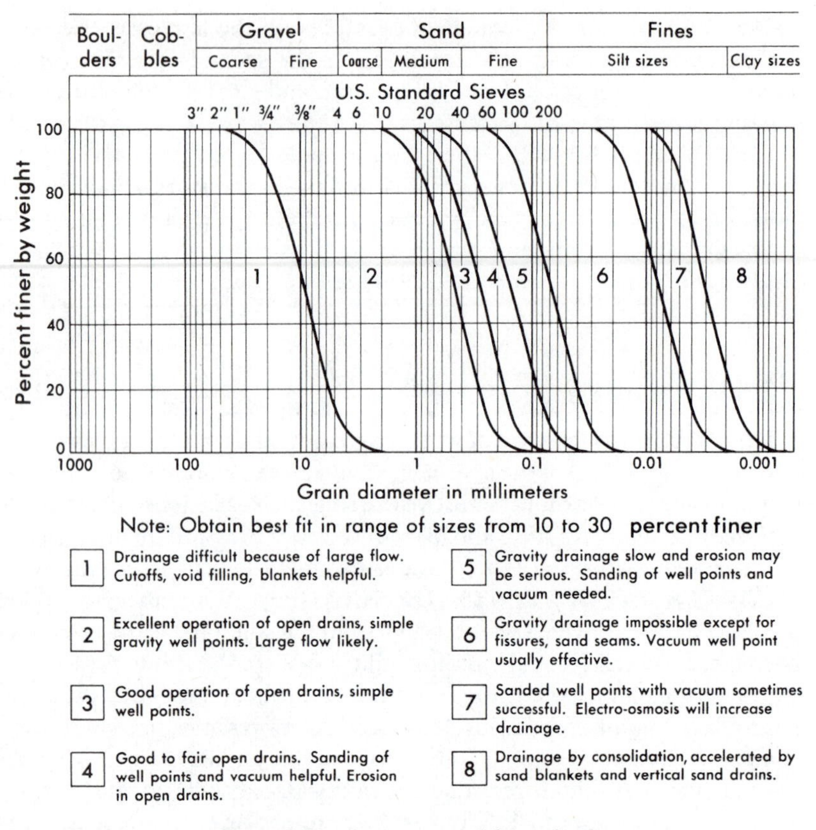

Figure 3.24 *Drainage capabilities of soils.*

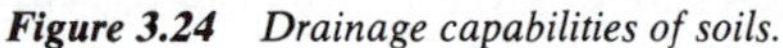

TABLE 3:5 / DRAINAGE POTENTIALITIES, UNIFIED SOIL CLASSIFICATION[2:12]

Soil Class	Drainage Characteristics	Soil Class	Drainage Characteristics
GW	Excellent	ML	Fair to poor
GP	Excellent	CL	Impervious
GM	Fair to impervious	OL	Poor
GC	Poor to impervious	MH	Fair to poor
SW	Excellent	CH	Impervious
SP	Excellent	OH	Impervious
SM	Fair to impervious	Pt	Fair to poor
SC	Poor to impervious		

DRAINAGE AND SOIL TYPE / The ease of draining a soil and the forces that are effective in producing drainage can be estimated from laboratory tests for permeability, consolidation, and shrinkage. The grain size distribution offers some indication of drainage properties, as shown in Fig. 3.24. Table 3:5 gives the drainage potential for the Unified Soil Classification.

The drainage of porous rocks such as earthy limestones and sandstones is similar to soils of equivalent pore size. Estimating the potential for other rocks is more difficult because, like permeability, drainage is controlled by cracks and localized zones of extremely high porosity, such as limestone caves and lava scoria. Local experience and full-scale field pumping tests are the only reliable means for evaluating drainage.

3:12 Drainage Systems

The design of a drainage system depends on the drainage characteristics of the soil, the length of time the system must operate, and the position of ground water. For temporary drains installed during construction, minimum interference with work and maximum effectiveness in a short time are essential. For permanent drains, long-term effectiveness and minimum maintenance are essential.

DRAINAGE LAYOUT / The position of the drainage system depends on the initial seepage pattern and the pattern that is to be established. If the drainage is installed in a dam, for example, the position of the water before and after drainage can be established by flow nets. If the drainage is for a building site, a highway, or an airfield, the initial groundwater conditions must be established by exploration, as described in Chapter 7. A contour map of groundwater elevation is prepared for the site and its surroundings. If the groundwater level fluctuates appreciably, more than one such map will be necessary, each representing a different condition.

Three locations are possible, as shown by Fig. 3.25: *intercepting*, *site*, and *downstream*. The intercepting drain removes the water before it reaches

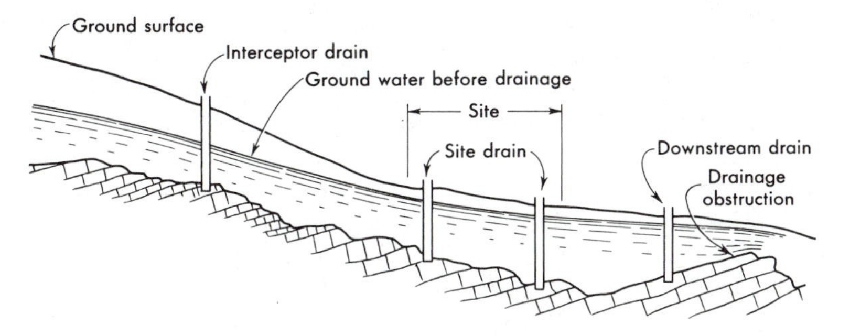

Figure 3.25 *Drainage layout.*

the site. It is particularly effective in artesian strata and where the groundwater surface slopes steeply. It does require use of land beyond the limits of the site, however. The site drain removes water directly from the area. In this way the quantity of water handled is less, and the drainage will be effective sooner than for the interceptor, but the drainage system may interfere with the work at the site. The downstream drain enables the water to leave the site more rapidly. It is most effective when an underground obstruction tends to dam up the ground water. In some cases one location is sufficient; in others, all three will be employed.

FILTERS, CONDUITS, AND DISPOSAL / A complete drain consists of three components: the filter, the conduit or collector, and the disposal system.

The filter is essential for continued efficiency of the drain and to prevent seepage erosion when the hydraulic gradients are high. The filter is pervious enough to permit the flow of water into the drain with little head loss and at the same time fine enough to prevent erosion of the soil into the drain. The design criteria were discussed in Section 3.8.

The drain conduit collects water from the filter and carries it away. Ordinarily, the conduit is 5 to 10 times larger than its hydraulics dictate to allow for variation in soil permeability and to accommodate some silting. Typical collecting perforations in commercial pipe are 8 to 9 mm (0.3 to 0.4 in.) in diameter and require a gravel filter with a maximum size of 12 to 15 mm. A *French conduit* or *French drain* is made of coarse gravel or crushed rock; where the amount of water is small, it can be cheap and effective. It is not, however, a substitute for a filter, and if employed as one, it will soon clog.

The disposal system removes the water from the area. Where possible, gravity is used because it is permanent and foolproof. However, the topography may make this impossible, particularly during wet weather and high water tables when the drain is needed most. Pumping will remove the water faster, but the cost of power over a long period will be appreciable, and maintenance is often uncertain.

OPEN DRAINS / The oldest method of draining excavations, roads, and similar projects is the open drain—either a *ditch* or a *sump.* A sump is merely a short ditch. Both are very effective in sands and gravels. Sumps and ditches are cheap. They can be constructed easily with unskilled labor and simple equipment, and ordinary construction pumps are suitable for pumping the water out of them. Boiling and piping sometimes commence in sumps and ditches, particularly if the soil is fine sand of low permeability; therefore, continuing surveillance is essential. Boils can be prevented by placing filter layers on the sides and bottoms of sumps or ditches below the phreatic surfaces but at increased cost.

CLOSED DRAINS / When seepage erosion or piping is troublesome or where a permanent drain is desired, perforated pipe can be laid in the ditches and the ditch backfilled with a filter material. It is important that the pipe be surrounded by one or two filter layers, as required, to prevent soil from clogging the openings (Fig. 3.26). The pipe should be laid in straight lines. Drains in silty soils should have an opening every 30 to 50 m (100 to 150 ft) through which a fire hose can be inserted to flush out the pipe occasionally. Manholes should be provided at changes in direction and at intervals of 100 to 150 m along straight sections.

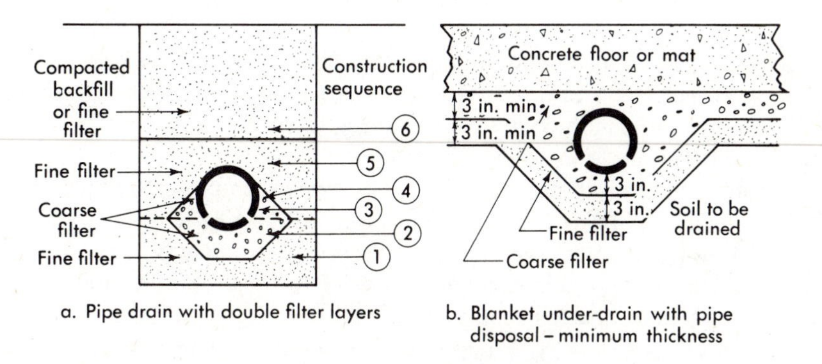

Figure 3.26 *Simple closed drains.*

BLANKET DRAINS / Continuous drainage blankets are sometimes provided beneath dams and basement floor slabs to reduce uplift pressures and beneath pavements to prevent capillary flow upward. The blanket consists of a filter layer in contact with the soil and a coarser collector layer, which also serves as a second filter (Fig. 3.26b). The latter is placed in contact with the underside of a masonry dam or basement floor, or is sandwiched between two filter layers in the base of an earth dam. Water is removed from the collector by conduits.

DEEP WELLS / Deep wells, such as are used for water supply, are occasionally employed in temporary drainage. Diameters of 0.3 to 0.6 m with spacings to 10 to 30 m and depths of more than 30 m or 100 ft have been used, depending on the size of the area to be unwatered and the amount the water table is to be lowered. They are also used in coarse soils and porous rock where the quantities of water drained are large.

The pump is in the bottom of the well, so the height the water is lifted is not limited to 8 m or 26 ft as with a suction well.

Like other drains, deep wells require filters. If the soil is coarse grained, or if a rock being drained resists erosion, a *well screen* alone is adequate. The screen is placed in the drilled hole in direct contact with the pervious stratum. It is generally attached to a pipe casing that supports the hole through any impervious strata above. If the soil is too fine grained to be filtered by the screen alone, a *pack* may be used. This is a gravel-sand filter placed around the well screen that filters the soil and in turn is filtered by the screen. The well is drilled 0.2 m or 8 in. in diameter larger than the screen to allow for the pack.

HORIZONTAL WELLS / Horizontal wells, about 50 to 80 mm (2 to 3 in.) in diameter and more than 60 m (200 ft) long, have been found useful in draining hillsides. The wells are installed by drilling into the hill at a slight upward angle to intercept water-bearing strata. The hole is then lined with a slotted or perforated pipe to keep it open and to discharge the water.

Combinations of vertical wells and horizontal holes have been used to drain stratified soil and jointed or pervious rock formations. The vertical well intercepts the strata, draining them to the horizontal well at the bottom. The latter acts as a drain, collector, and disposal pipe. Because of the difficulties in making the two wells intersect, the vertical well (the easiest to drill) is often more than 1 m in diameter.

Large horizontal drain tunnels have been employed to tap deep aquifers beneath hillsides. Smaller horizontal drains can be drilled from the tunnel to localized zones of excessive permeability, such as large joints or fault zones. The tunnel is principally a collector and conduit; however, it may directly drain the adjacent formations.

WELLPOINTS[3:12] / Wellpoints are small-diameter wells that are driven or jetted into the soil. Usually, they are placed in straight lines along the sides of the area to be drained and are connected at their upper ends to a horizontal suction pipe called the *header*, as shown in Fig. 3.27a. Depending on the type of soil to be drained, one or two wellpoints are usually installed for each 1 to 2 m (3 to 7 ft) of header. The header terminates in a self-priming pump specially designed for wellpoint work: one pump for each 50 to 100 points.

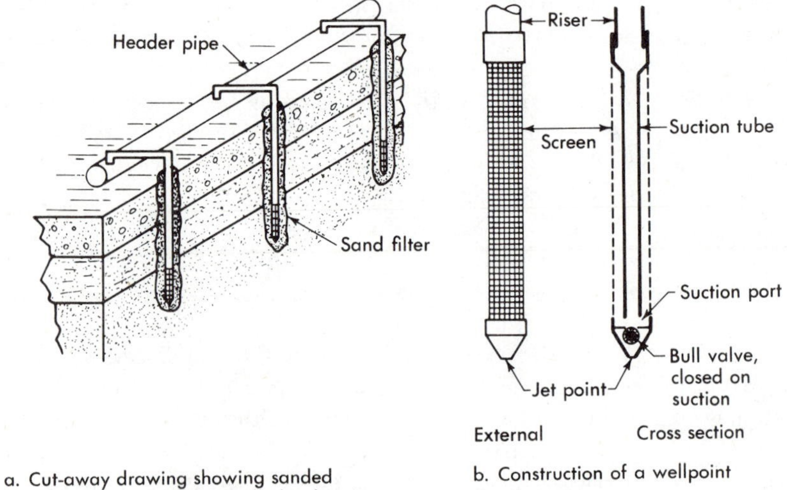

a. Cut-away drawing showing sanded wellpoints in ground attached to header

b. Construction of a wellpoint

Figure 3.27 Wellpoint installation and construction.

Many different types of points have been devised (Fig. 3.27b). The drivepoint consists of a length of heavy-gauge, 50-mm (2-in.) pipe. To its lower end is attached a perforated section 1 to 1.5 m (3 to 5 ft) long covered with a wire-gauze screen terminating in a conical steel tip. Points designed to be jetted into the soil are equipped with rubber ball valves at their lower ends. During jetting water is pumped into the wellpoint and is directed out of the tip by the valve. This washes a hole in the soil allowing the wellpoint to sink into position. When the wellpoint is connected to the suction header, the ball valve closes and the wellpoint takes in water through the gauze screen at its lower end.

The wellpoint screen (typically 0.3 to 0.6 mm opening, or 30 to 50 wires per in.) is an adequate filter for medium sands. In finer soils, a sand pack is placed around the wellpoint to increase the effective area of the well, minimize seepage velocities, and provide a better filter. The sand is installed by drilling or jetting a hole about 0.3 m (1 ft) in diameter. The wellpoint [7 cm (2.5 in.) OD] is centered in the hole, and a graded clean sand such as concrete sand is placed around the screen.

The effectiveness of the points in fine-grained soils is increased by sealing them into the uppermost soil strata with a plug of clay or mortar and maintaining a vacuum on the header at all times, even when little water is removed from the soil. The vacuum within the soil mobilizes atmospheric pressure to force the water out of coarser soils and to decrease the void ratio of finer soils. Pumping units designed for wellpoint work usually include vacuum pumps.

Since most wellpoints are operated by suction, the maximum vertical distance from the pump intake to the water level at the points is about 8 m (25 ft). If excavations extend deeper below the water table, two or more stages are used as shown in Fig. 3.28. The first stage consists of a row of points that are set in the ground and placed in operation as soon as the groundwater table is lowered by the first points. A second row of points is then placed when the excavation again reaches the groundwater table. Excavations as deep as 15 m (50 ft) below the original water table have been made by using from three to four lift stages of points.

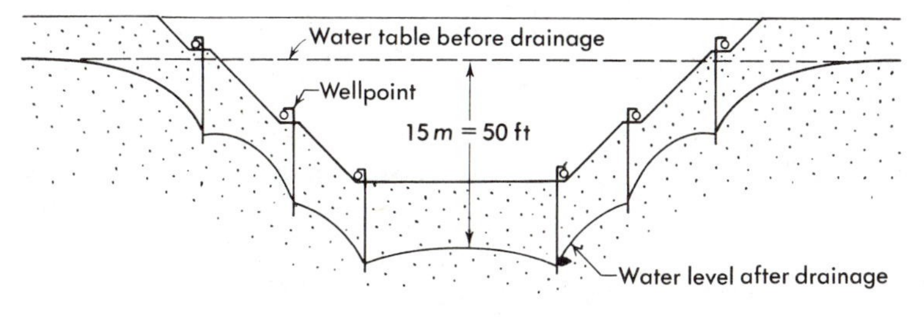

Figure 3.28 *Multiple-stage wellpoint system.*

Special wellpoints are available that employ small water-jet pumps at their bottoms. In this way the water is lifted out of the wellpoint rather than sucked out. Lifts of 15 m (50 ft) or more are practical with such wellpoints. There are two disadvantages. First, two headers are required: one to supply water to the jet, and the other for discharge. Second, the system is less efficient mechanically, and the flow capacity for each wellpoint is less than for the conventional points.

Wellpoints have proved very successful for draining soils of high and medium permeability, such as coarse sands and clean fine sands, and vacuum wellpoints for silty sands and sandy silts. Their success depends to a large extent on the experience and skill of the persons making the installation.

ADVERSE EFFECTS OF DRAINAGE / Drainage, in solving the problems of too much water or excessive water pressure, can create new problems. Although reducing neutral stress increases the effective stress and the strength of a soil, the increased effective stress sometimes causes consolidation settlement (Chapters 4 and 10) within the zone of drawdown.

Drainage of sands temporarily enhances their strength by producing capillary tension and increased effective stress. As a result small excavations sometimes can be made with steep slopes without bracing. If the sand dries, however, the capillary tension is lost and the steep excavation slopes will collapse with loss of property or life. (Such steep slopes often violate safety codes.)

Drainage without proper filters can produce piping, and cavities in the soil. The cavities soon collapse, causing the destruction of anything above. Seepage erosion (piping) into drains and leaky sewers causes occasional dropout of pavements and basement floors, particularly in sandy and silty soils.

RECHARGE / The adverse effect of drainage on adjacent structures can be minimized by *artificial recharge*. This is the pumping of water into the ground between the drainage system and the structure. Usually, this is done with a system of wellpoints, similar to that employed for drainage, but with a wider spacing. Water is pumped into the wellpoints at the rate that maintains a constant groundwater level under the endangered structure. This inflow rate is less than the drainage rate. Of course, the necessary drainage rate must be increased because of the locally high water table induced at the inflow wells. Continuing accurate control of inflow is necessary to avoid under- or overrecharging.

ELECTROOSMOSIS / If a direct current is passed through a soil of low permeability, the rate of drainage is increased greatly. Wellpoints serve as the negative electrodes, and steel rods driven into the soil midway between the wells form the positive electrodes. From 20 to 30 amp of electric current are used per well, at voltages from 40 to 180 d.c. The amount of energy required varies from 0.5 to 10 kWhr/m^3 of soil drained.

Electroosmosis requires expensive equipment and is relatively costly to operate; therefore, it is used only when cheaper methods cannot produce sufficient drainage.

DESICCATION / Drainage of a soil by evaporation is an extremely slow process and is ineffective if the soil mass, by capillarity, can replace the moisture evaporated. Ventilation galleries have been used to dry out clay strata in hillsides when the clay is subject to swelling and loss of strength during wet seasons.

DRAINAGE BY CONSOLIDATION / Soft, wet, cohesive soils are impossible to drain by gravity methods or even by vacuum or electroosmosis, yet they may require a reduction in their water contents before they have sufficient strength to support heavy, concentrated loads without undue settlement or failure. Consolidation—the removal of water by a reduction of the volume of the voids through compression—is an effective process in spite of its inherent lack of speed. It is discussed in Chapter 6.

3:13 Frost Action[3:13]

FROST ACTION AND ITS CONSEQUENCES / When the daily mean temperature remains below 0°C for a period longer than three or four days, the soil moisture at the ground surface freezes. The longer and the more intense the cold spell, the greater the depth to which the freezing extends. The result of the freezing is a rise of the ground surface, known as *frost heave*, that sometimes is as great as 0.3 m (12 in.) in the northern parts of the United States. If an excavation is made into frozen ground that has heaved, it will be seen that the soil has changed considerably. Its average water content has been greatly increased and much of the water is concentrated into ice layers or lenses that lie parallel to the ground surface. The amount of heave is rarely uniform, and the force exerted by the expanding soil lifts roads, walls, and buildings.

Frost heave is particularly damaging to highways and airfield pavements, as they are generally built directly on the surface of the ground. Unequal heave can crack concrete pavement slabs or tip the individual slabs at angles. Heave beneath flexible pavements causes bumps or waves in the surface. Small structures with shallow foundations, such as small bridges, culverts, walls, sewer inlets, and light buildings, often suffer if their foundations are above soils subject to frost heave. Cold storage warehouses are lifted and torn apart by unequal heave brought about by improperly insulated floors.

Frost action is not limited to the process of heave. When the weather becomes warm again, the frozen soil thaws from the top down, where the warmer air and sunshine are in contact. The uppermost portions of the soil become wet and soft as the ice layers melt. They remain wet until the excess water can drain downward through the deeper strata when the frost disappears.

Thawing beneath a highway or airfield pavement converts the soil into a liquid that supports the pavement. The weight of a truck or airplane under such conditions causes the liquid soil to spurt up through the expansion joints in concrete pavements (often called *pumping*) or to form holes known as *mud boils* in flexible (asphalt) pavements. Thawing beneath structural foundations can cause failure, since the soft, water-filled soil has little ability to support heavy loads.

Frost action, therefore, is a combination of two processes: first, freezing of the soil, and the formation of layers of ice that cause frost heave; and second, the thawing of the *ice lenses*, which provides an excess of free water in the soil and results in a lowering of the strength of the soil.

MECHANICS OF FROST ACTION[3:14] / When water goes from the liquid to solid form, it expands about 10%. In a saturated soil with a void ratio of 0.7, this means there would be a 4% increase in the volume of the soil, or about 40-mm heave in a soil frozen to a depth of 1 m. However, the observed heave in such a situation might be as much as 150 mm (6 in.). Furthermore, since the average water content and void ratio of the soil increase during the freezing process, it must be concluded that expansion of water in the voids is not the primary cause of heave. Examination of frozen soils indicates that the heave is about equal to the total thickness of all the ice layers formed during the process. Therefore, the formation of ice layers (and the increase in average soil water content) is the primary cause of heave.

The temperature deep in the ground remains nearly constant throughout the year, while the temperature just below the gound surface fluctuates with the air temperature. After a period of cold weather in which the air temperature is below freezing, a thermal gradient is established in which the 0°C point is below the ground surface, as shown in Fig. 3:29. This point defines the *frost line*. Neither the frost line nor the thermal gradient is fixed; both vary with the duration and intensity of the cold. The frost line is

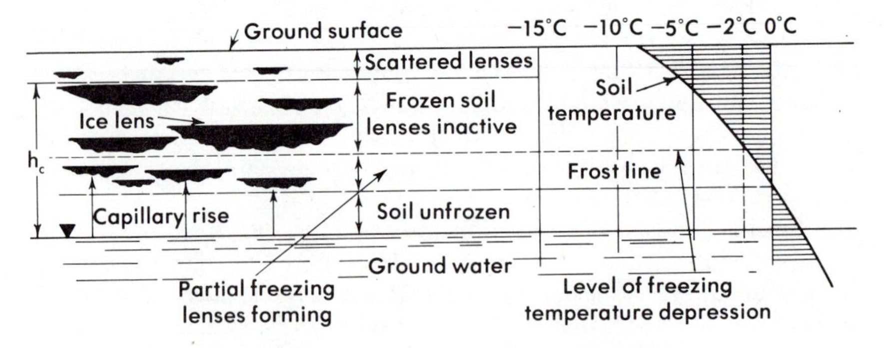

Figure 3.29 *Formation of ice lenses in zone of freezing.*

not found at a uniform depth, for it depends on the density, saturation, and composition of the soil.

Above the frost line the temperature is below the ordinary freezing point for water. However, in very small openings such as the voids of fine-grained soils, the freezing point may be depressed as low as −5 C or (23 F). Thus, just above the frost line, water will freeze in the larger voids but remain liquid in the adjacent smaller ones. When water freezes in a larger void, the amount of liquid water at that point is decreased. The moisture deficiency and the lower temperature in the freezing zone increase the capillary tension and induce flow toward the newly formed ice crystal. The adjacent small voids are still unfrozen and act as conduits to deliver the water to the ice. The ice crystal grows until an ice lens or layer forms. The capillary tension induced by the freezing and the low temperature sucks up water from the water table below or dehydrates and shrinks adjacent compressible strata, such as clays and micaceous silts, when the water table is beyond reach. The result is a great increase in the amount of water in the frost zone, and *segregation* of the water into ice lenses.

First, in order for the water to be drawn into the freezing zone by capillary forces, the soil must be saturated or approaching saturation. Partially saturated soils freeze, but the ice is scattered in tiny crystals and the heave is small. For continued heave, the freezing zone must be within the height of capillary rise above the water table so that water can be sucked up from below. If the freezing zone is saturated but above the height of capillary rise (for example, when the upper strata have been saturated by rain or leaking pipes), the ice lensing and heave will be limited by the amount of water that can be sucked from the adjoining soil.

Second, the soil must be fine grained. Segregation seldom occurs in coarse soils where the pores are so large that water freezes in them at 0°C. The heave, if any, is limited to a 10% expansion of the voids. On the other hand, very fine-grained soils are so impervious that the water migrates to the ice lenses very slowly. The most rapid segregation takes place in soils whose permeability is great enough to permit easy movement of the water.

Third, the temperature gradient in the soil must be favorable. When the rate of change in temperature with depth (temperature gradient) is very great, the zone of soil in which the pore water is unfrozen but below 0°C is narrow. The ice layers formed under such conditions are thin, and the amount of heave small. When the temperature gradient is small, the zone in which the pore water is unfrozen but below 0°C is wide, and the ice lenses tend to be thick and the amount of heave great.

A rapidly varying gradient with alternating freezing and thawing can aggravate frost heave in sands but will have little effect on silts and clays.

SOILS SUSCEPTIBLE TO FROST ACTION / The susceptibility of different soils to frost action has been studied by A. Casagrande.[3:15] He found that the coarse soils, sand, and gravel containing no fines are rarely subject to the formation of ice lenses and objectionable heave. On the other hand, fine sands and silts have the optimum combination of fine pores and relatively high permeability that results in maximum segregation and heave. Clays are usually considered to be frost-susceptible because cracks and fissures may permit rapid movement of the water through them. According to Casagrande, a uniform soil is susceptible to frost action if more than 10% of its particles by weight are finer than 0.02 mm, and a well-graded soil is susceptible if more than 3% of its grains are finer than 0.02 mm. Studies of frost action in Michigan, however, indicate that even sands may be susceptible to frost heave under some conditions. The potential frost action of the soil groups of the Unified Classification System are given in Table 3:6.

TABLE 3:6 / POTENTIAL FROST ACTION OF UNIFIED SOIL CLASSIFICATION[3:16]

Soil Class	Potential Frost Action	Soil Class	Potential Frost Action
GW	None to very slight	ML	Medium to very high
GP	None to very slight	CL	Medium to high
GM	Slight to medium	OL	Medium to high
GC	Slight to medium	MH	Medium to high
SW	None to very slight	CH	Medium
SP	None to very slight	OH	Medium
SM	Slight to high	Pt	Slight
SC	Slight to high		

DEPTH OF FROST PENETRATION / The depth below the ground surface to which a 0°C temperature extends is termed the *frost line.* Above the frost line, freezing occurs and ice lenses will form if the soil and water conditions are right. The depth of the frost line depends primarily on three factors: the air temperature, the length of time the air temperature is below 0°C, and the ability of the soil to conduct heat. The lower the air temperature and the longer it remains below 0°C, the greater the depth of the frost line; the higher the thermal conductivity of the soil, the greater the depth of frost penetration. The accompanying map (Fig. 3.30) shows the maximum depth of frost penetration in the United States. The map is only approximate: On mountaintops the depths will be much greater; in highly organic soils or coarse gravels above the water table, the depth will be smaller.

PREVENTING FROST DAMAGE / Frost heave and frost damage may be prevented by correcting one or more of the factors responsible for the segregation of water and the formation of ice lenses: frost-susceptible soil, capillary saturation by rise of water from groundwater table, and freezing temperatures in the soil.

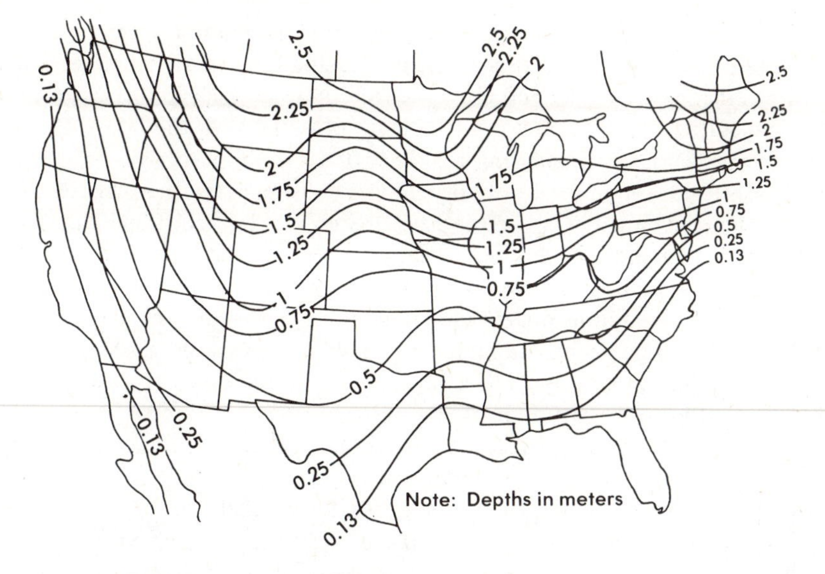

Figure 3.30 *Maximum depth of frost penetration in the United States.*

One of the most effective methods of preventing frost heave is to remove the frost-susceptible soil throughout the depth of frost penetration and replace it with a soil that is not affected. In regions where large quantities of clean sands and gravels are readily available, soil replacement is an economical and permanent cure for frost heave.

Capillary saturation caused by the rise of water from the water table may be controlled by lowering the water table below the height of capillary rise or by obstructing the upward movement of the water. When the groundwater table has a steep slope, or when it is perched on top of a saucer-shaped stratum of impervious soil, drainage may be very effective in preventing serious frost heave. In flat areas, and in areas in which excessive rainfall or snow melting are quickly followed by freezing, even extensive drains may not prevent a rise in the groundwater table and the subsequent capillary saturation. An impervious blanket, such as asphalt, plastic, or commercial bentonite, buried in the ground at the frost line may prevent movement of water upward into the zone of frost penetration. Such blankets are seldom used, for they are expensive, and they puncture and deteriorate. Blankets of coarse-grained soils such as clean coarse sand, gravel, or crushed rock (with fines removed) placed above the water are effective water stops, for they break the capillary tension. They should be thicker than the height of capillary rise through them and must be protected by filters so that the finer soils above or below do not penetrate their voids. Such blankets must be well drained because if they should fill with water, they would aggravate rather than prevent frost heave.

Insulating blankets between the ground surface and the frost-susceptible soil reduce the penetration of the freezing line. Ordinarily, these are well-drained coarse sand and gravel. Often these are placed directly on the ground surface and are employed beneath pavements or the floors of cold, storage warehouses. Insulating blankets of foam glass, plastic foam, and cork are used under very low-temperature storage facilities, where their high efficiency justifies their high cost. Similar blankets are being used beneath pavements on a limited scale.

Chemical additives show promise in preventing frost heave. Dispersing agents, such as sodium polyphosphates, mixed with the soil cause higher densities, lower permeabilities and less heave. Waterproofing materials and chemicals that change the adsorbed cations of the clay minerals reduce heave by altering the attraction for water.

PERMAFROST / In North America, north of about the Arctic Circle, the soil remains frozen to great depths throughout the year. The condition of the permanently frozen ground is termed *permafrost.* In some areas it is as deep as 300 m (1000 ft). The latitude of permafrost is not uniform, and in parts of Canada and Alaska it is found as far south as latitude N 62. Isolated islands of permafrost in high or sheltered areas are found south of the continuous zone, and very localized pockets occur in high mountains throughout the world.

The upper few feet of permafrost thaw in the summer. This is the *active zone,* and it frequently becomes a soft, soupy quagmire. Highways laid on permafrost fail and buildings settle in summer when the active zone softens. The heat of a building increases the depth of the active zone and aggravates the situation. Permafrost is often covered with an insulating blanket of moss and low, thickly matted vegetation called *tundra.* This minimizes the depth of the active zone, and removal of this natural insulation will greatly increase the active zone depth.

Foundations and subgrades must be placed below the active zone to avoid movement, or the soil in the active zone must be replaced with a non-frost-susceptible material like gravel or coarse crushed rock. Insulating blankets are sometimes used to minimize the thaw of the active zone. In extreme cases cold-air conduits through the soil and cooling systems are installed to offset the heat from buildings and boilers that would aggravate the thaw of the active zone.

3:14 Water: The Elusive Component

Unlike most solids, soils and rocks contain components that vary with time: air and water. These can change without a change in solid density or void ratio, by interchanging with each other. Their changes can alter the solid phase and its response to load. Many problems in soil and rock movement and rupture are caused by changes in water or water pressure as will be discussed in the remaining chapters. Therefore, the mechanics of change must be investigated, including the environmental and man-made factors responsible. Ultimately, geotechnical designs incorporate systems that either control water and water pressure or accommodate the changes. The handling of water is often a forgotten element of design; however, it is usually a major factor in success or, unfortunately, failure.

REFERENCES

3:1 A. Hazen, "Water Supply," in *American Civil Engineers Handbook*, John Wiley & Sons, Inc., New York, 1930.

3:2 D. Hillel, *Soil and Water*, Academic Press, Inc., New York, 1971, 288 pp.

3:3 ASTM D-2434-68(74), "Constant Head Permeability Test," *Annual Book of Standards, ASTM*, Philadelphia, Pa.

3:4 J. E. Bowles, *Engineering Properties of Soils and Their Measurement*, McGraw-Hill Book Company, New York, 1978, pp. 97–110.

3:5 K. Terzaghi and R. B. Peck, *Soil Mechanics in Engineering Practice*, 2nd ed., John Wiley & Sons, Inc., New York, 1968, p. 55.

3:6 A. Casagrande, "Seepage Through Dams," *Journal of the New England Water Works Association*, July 1937.

3:7 H. R. Cedergren, *Seepage, Drainage and Flow Nets*, 2nd ed., John Wiley & Sons, Inc., New York, 1977.

3:8 *Pore Pressure and Suction in Soils*, Proceedings of a Conference by British National Society on Soil Mechanics and Foundation Engineering, Butterworths, London, 1961.

3:9 J. L. Sherard, "Identification and Nature of Dispersive Clays," *Journal of the Geotechnical Engineering Division, Proceedings, ASCE*, **102**, GT4, pp. 287–301, April 1976.

3:10 J. L. Sherard, L. P. Dunningan, R. S. Decker, and E. F. Steele, "Pinhole Test for Identifying Dispersive Soils," *Journal of the Geotechnical Engineering Division, Proceedings, ASCE*, **102**, GT1, pp. 69–85, January 1976.

3:11 T. A. Middlebrooks, "Seepage Control for Large Earth Dams," *Third Congress of Large Dams*, **2**, Stockholm, 1948.

3:12 C. I. Mansur and R. I. Kaufman, "Dewatering," in *Foundation Engineering*, McGraw-Hill Book Company, New York, 1962, Chapter 3, pp. 241–350.

3:13 A. W. Johnson, "Frost Action in Roads and Airfields," *Special Report* 1, Highway Research Board, Washington, D.C., 1952.

3:14 G. Beskow, "Soil Freezing and Frost Heaving with Special Applications to Roads and Railroads," *Swedish Geotechnical Society*

26th Yearbook, Series C., No. 375, 1935 (Translated by J. O. Osterberg, Northwestern University, 1957).

3:15 A. Casagrande, "Discussion on Frost Heave," *Proceedings, Highway Research Board*, Washington, D.C., 1931, p. 168.

3:16 "The Unified Soil Classification System—Appendix B, Characteristics of Soil Groups Pertaining to Roads and Airfields," *Technical Memorandum 3–57*, U.S. Waterways Experiment Station, Vicksburg, Miss., March 1953.

SUGGESTIONS FOR FURTHER STUDY

H. Cedergren, "Drainage and Dewatering," in *Foundation Engineering Handbook*, H. F. Winterkorn and H. Y. Fang, eds., Van Nostrand Reinhold Company, New York, 1975, pp. 221–311.

M. E. Harr, *Groundwater and Seepage*, McGraw-Hill Book Company, New York, 1962.

A. Kedzi, *Handbook of Soil Mechanics*, Vol. 1, Elsevier Scientific Publishing Co., Amsterdam, 1974, 294, pp.

J. K. Mitchell, *Fundamentals of Soil Behavior*, John Wiley & Sons, Inc., New York, 1976, 422 pp.

Moisture Equilibria and Moisture Changes in Soils, A Symposium, Butterworths, Sydney, Australia, 1965.

A. W. Skempton, "Terzaghi's Discovery of Effective Stress," in *From Theory to Practice in Soil Mechanics*, John Wiley & Sons, Inc., New York, 1960 (A Tribute to K. Terzaghi's contributions), pp. 42–54.

G. F. Sowers, "Dewatering Rock for Construction," *Proceedings of Specialty Conference on Rock Engineering for Foundations and Slopes*, ASCE, **1**, pp. 201–216, 1976.

PROBLEMS

3:1 a. Compute the maximum capillary tension in kN/m^2 and lb/ft^2 in a tube 0.001 mm in diameter.

b. Compute the height of capillary rise in the tube in meters and feet.

3:2 Compute the capillary tension in kN/m^2 and theoretical height of capillary rise in meters and feet in a soil whose D_{10} is 0.002 mm if the effective pore diameter is about $\frac{1}{5}D_{10}$.

3:3 Compute the height of capillary rise in meters and feet in a sand whose D_{10} is 0.2 mm if the effective pore diameter is $\frac{1}{5}D_{10}$.

3:4 A sample of soil in a permeability test is 50 mm in diameter and 120 mm long. The head difference is 250 mm, and the flow is 1.5 ml in 5 min. Compute coefficient of permeability in mm/sec, ft/min, and m/year.

3:5 Given a block of soil 120 mm long and 360 mm^3 in cross section. The water level at one end of the block is 200 mm above a fixed plane and at the other end is 30 mm above the same plane. The flow rate is 2 ml in 1.5 min. Compute the flow rate is 2 ml in 1.5 min. Compute the soil permeability in mm/sec.

3:6 A canal and a river run parallel, an average of 60 m or 197 ft apart. The elevation of water in the canal is +200 m or +656 ft and in the river +193 m or 633.4 ft. A stratum of sand intersects both the river and the canal below their water levels. The sand is 1.5 m or 4.9 ft thick and is sandwiched between strata of impervious clay. Compute the seepage loss from the canal in m^3/sec^{-1} km^{-1} and ft^3 sec^{-1} $mile^{-1}$ if the sand's permeability is 6.5×10^{-1} mm/sec.

3:7 A wood crib filled with earth serves as a temporary cofferdam across a river to lower the water level in a construction site. The water level upstream is 6 m or 19.7 ft above the rock stream bed and downstream is 1.3 m or 4.3 ft above the stream bed. The cofferdam is 60 m or 197 ft long across the river and is 10 m or 32.8 ft wide upstream to downstream. It is filled with well-graded, slightly silty, sandy gravel having a coefficient of permeability of 5×10^{-2} mm/sec. Estimate the seepage through the cofferdam in m^3/sec and gal/hr (the unit in which construction pumps are rated). *Hint*: Assume that the average cross section of the water flowing through the cofferdam is the average of the intake area (6×60 m^2) and the outlet (1.3×60 m^2).

3:8 a. Draw a flow net for seepage under a vertical sheet pile wall penetrating 10 m into a uniform stratum of sand 20 m thick; $k = 3 \times 10^{-2}$ mm/sec.

b. If the water level on one side of the wall is 11 m above the sand and on the other side is 1.5 m above the sand, compute the quantity of seepage per unit width of wall.

3:9 Draw the flow net for a seepage under a concrete dam that is 50 m long (upstream to downstream) and rests on a 12-m-thick uniform stratum of silty sand. The bottom of the dam is 2 m below the upper surface of silty sand. Compute the quantity of seepage if the head on the dam is 20 m upstream and 5 m downstream and the permeability of the soil is 4×10^{-3} mm/sec.

3:10 Draw the flow net for an earth dam 25 m or 82 ft high, 5 m or 16.4 ft crest width a slope of 2.5 (horizontal) to 1 (vertical) upstream and 2 to 1 downstream. The dam rests on a 8-m- or

26.2-ft-thick stratum of soil with the same permeability. The headwater level is 22 m or 72.2 ft above the base of dam; the tailwater is 4 m or 13.1 ft above the base of the dam.

3.11 Compute the hydraulic gradient required to produce a "quick" condition at the surface of a level mass of sand through which water flows vertically upward. The void ratio is 0.63, and the specific gravity of the solids is 2.66.

3:12 Compute the safety against boiling in Problem 3:8 if the soil void ratio is 0.42 and the specific gravity of the solids is 2.67.

3:13 a. What would be the maximum head on the underside of the silty clay in Example 3:3 that would permit pumping the excavation down to the bottom, and would maintain a ratio of total stress to uplift of 1.3 (a safety factor of 1.3)?
b. What would the seepage rate be under this condition?
c. How could the seepage rate and uplift be controlled to maintain the safety factor described above?

3:14 A soil stratum overlies a bed of shale. The water table fluctuates greatly; the capillary rise is 2.5 m or 8.2 ft. The soil has a void ratio of 0.48 and a specific gravity of solids of 2.67; it is 10 m or 32.8 ft thick.
a. Draw graphs of the vertical stress as a function of depth for total neutral and effective stress, assuming the soil is dry.
b. Draw similar graphs, assuming that the groundwater table rises to the ground surface.
c. Draw similar graphs, assuming that the water table drops to 3 m or 9.8 ft above the shale and that the soil is 30% saturated above the height of capillary rise. (Above capillary saturation, effective stress and neutral stress are indefinite.)

3:15 A sheet pile cofferdam 12 m by 6 m or 39.4 ft by 19.7 ft is driven into a river bottom to permit construction of a bridge pier. The river is 5 m or 16.4 ft deep and the bottom is loose fine sand 2.5 m or 8.2 ft thick underlain by coarse gravel. Unknown to the contractor, the fine sand merely fills a depression in the gravel; most of the river bottom at some distance from the cofferdam is gravel. The sheet piles are driven until they reach the gravel, where they can penetrate no further. (The contractor's foreman thought that he had reached bedrock.) The fine sand weighs 19 kN/m^3 or 121 lb/ft^3 saturated and has a permeability coefficient of 4×10^{-3} mm/sec. After driving the entire cofferdam box, pumping is commenced to remove the water and to expose the sand bottom.
a. When the water level reaches x ft above the fine sand, the water turns muddy and commences to boil. What has happened?

b. Compute the level x and the rate of pumping in m^3/sec and gal/hr required to maintain equilibrium at this instant. Neglect any head loss in the gravel.

3:16 Specify the grain size distribution of soils that would serve as satisfactory filters for the soils given in Problem 1:15.

3:17 Sketch a wellpoint system for draining an excavation that is 17 m or 55.8 ft wide at the bottom and extends 15 m or 49.2 ft below the water table. Show stages of construction by separate diagrams.

3:18 a. Which of the soils listed in Problem 2:3 would you consider for drainage with wellpoints?

b. Which of the soils listed in Problem 1:15 could be drained with wellpoints based on the Unified System correlations or grain size curve matching?

3:19 A long excavation is 6 m or 19.7 m deep and 4 m or 13.1 ft wide at the bottom. The groundwater table is normally at a depth of 3 m or 9.8 ft below the ground surface. A single line of wellpoints extends 3 m below the bottom of the excavation and is 5 m or 16.4 ft from the center line of the excavation. The soil is sand, with $k = 1 \times 10^{-3}$ mm/sec and it is underlaid at a depth of 11 m or 36.1 ft by rock. The groundwater table remains at its original level 60 m or 197 ft from the excavation and is lowered 5 m or 16.4 ft below its normal elevation along the line of the points.

a. Draw the flow net for seepage into the wells, assuming the line of wells to be one continuous slot in the ground.

b. Compute the quantity of water pumped per well if the wells are 2 m apart.

c. At what level should the header be placed? Why?

3:20 Which of the soils in Problem 2:3 would you expect to be susceptible to frost action based on the percentages finer than 0.22 mm? Is this confirmed by Table 3:6?

3:21 From the grain size curve of the Teton Dam core (Problem 1:15) find the range in grain size and sketch the appropriate filter curve. Plot the curve for the Teton River sand-gravel, used as a filter in the dam. Is it satisfactory? (The dam failed by piping June 5, 1976.)

CHAPTER 4
Compression and Expansion with Changing Stress and Environment

A fill of of sand 2 to 4 m deep was placed on a low marshy island, to create a new industrial park, in a region where good building sites were scarce and expensive. After compaction, tests of the fill (Section 10:5) indicated sufficient bearing capacity for warehouses and one- or two-storey light manufacturing buildings. Roads, water lines, sewers, and other utilities were installed in anticipation of quick land sale. The developers even persuaded the city to invest nearly $1 million in a connecting highway and railroad bridge. The entire project appeared so lucrative that no time or money was wasted in engineering studies of the soils, other than the simple surface test of the fill required by the building code.

Within a year the roads resembled miniature roller coasters, and the ground surface contained a number of low, water-filled swales between undulating uplands. Some of the sewers reversed, flowing toward the area of deepest fill away from their designed outlets. Water mains cracked, and underground electric and telephone conduits were damaged. The developer went bankrupt and the city was left with a bridge to an abandoned site, interest on the bridge bonds, and no tax revenue to repay the costs. A subsequent soil investigation disclosed that the original ground surface was underlain by a thick stratum of organic clay and silt, with interbedded sand layers. The void ratios of the silts and clays reduced under load; the total volume decreased as the soil compressed. The greatest settlement was beneath the deepest fill; the irregularity of the surface settlement was caused by variations in the thickness and compressibility of the virgin soils. The fill was strong enough to support the simple surface test; the results were therefore misleading. The geotechnical engineer predicted that because of the great thickness of the compressible soils, the settlement would continue for a quarter century. This prediction was verified; the first constructive use of the site (other than as a waste dump) took place more than 20 years after the site was filled.

A shopping center was constructed on a compacted silty clay fill adjacent to a small creek. Because the developer knew that some garbage had been buried beneath the fill 15 years previously, he required the architect to support the buildings on pile foundations driven through the fill to rock. In order to save money, however, the floor slabs of the one-story buildings were supported directly on the ground surface. The architect reasoned that the garbage, after 15 years of supporting 2 m of fill, could support the additional weight of the floor. The architect's cost-saving decision initially appeared valid because for 10 years the shop and restaurants showed no signs of distress. However, following a two-year period of large variations in rainfall and alternating high and low water in the adjoining creek, the building floors and the parking areas outside the buildings began to settle. At some locations the settlement was as much as 0.15 m (6 in.). More alarming, there was an odor of gas coming out of cracks in the floor, particularly where gaps developed between the pile-supported building walls and the soil-supported floors. A climax was reached when the spark from a light switch ignited gas that had accumulated in a closet, and the explosion demolished a wall.

An investigation disclosed that the old garbage had not reached equilibrium after 25 years of fill load. Creek diversion raised the level of ground water and had wet the buried garbage, probably redistributing biochemical nutrients. Anaerobic bacteria were activated into consuming the organic materials and causing settlement. A by-product of decomposition was methane, which is explosive in air at a concentration of about 15%. The problem was eventually minimized by periodically pumping portland cement under the floors to compensate for the settlement and by ventilation trenches to the garbage to dissipate the methane.

A two-story office building for a new regional office of a nationwide engineering firm was supported by shallow spread footings resting on hard clay. Since the engineers' tests indicated high soil strength, they had decided on the shallow foundations to save money although similar neighboring buildings were supported on concrete shafts drilled 12 m into the hard clay. The new engineers' experience in their home region led them to believe that their local competitors were just overly conservative; the new building would serve as an advertisement for their superior out-of-town expertise. Within two years the building was badly cracked, and the demonstration of expertise had turned to embarrassment. The hard clay was highly plastic and dry. The new building shielded the clay from the hot summer sun. Capillary rise slowly saturated the soil, allowing it to expand. The center of the building rose about 0.1 m (4 in.); the outside walls rose somewhat more during wet weather and fell during dry weather, as if the building were flapping its wings in time with the seasons. There was no simple remedy except to underpin the building.

These three examples illustrate some of the more unusual properties of soils and soil-like materials compared to simple concrete or steel. Soils compress significantly under load. The settlements can continue for long times as has been demonstrated by 700 years of movement of the belfry (Leaning Tower) of the Cathedral at Pisa, Italy. Other mechanisms can be involved, such as biochemical decay. In some situations the soil or rock expands despite the increase in load of a building, a phenomenon that is impossible considering only elastic theory but which can be scientifically analyzed when the three-phase mechanics of soils are considered. The geotechnical engineer must expand his or her thinking to consider the changing three phases of composition, as well as the four dimensions of length, width, depth, and time, to solve elementary problems in soil behavior.

4:1 Effective Stress as Solid Phase Load[4:1]

Because of the different response of the solid and liquid phases to load, it is necessary to consider each phase independently. The gaseous phase supports neither compressive nor shear loads (except at very high stresses or unusual dynamic conditions) and is therefore not considered in most problems. The solid phase is relatively incompressible; moreover, it resists shear loads, although it distorts and at high stresses will rupture. The liquid phase is likewise incompressible; however, under differential compressive stress it flows because a liquid, by definition, is not capable of resisting static shear loads. Ultimately, the solid phase controls the resistance to compression and shear. Because of the very small grain-to-grain contact, the actual stress at any point in the solid phase is virtually impossible to compute or measure. The total load, Q, the total area over which it acts, A, and the water pressure, u, can be measured. The stress in the solid phase, therefore, can be expressed in terms of effective stress as described in Section 3:7.

$$\sigma = \sigma' + uN \quad \text{or} \quad \sigma' = \sigma - uN \tag{3:14a}$$

$$\sigma = \sigma' + u \quad \text{or} \quad \sigma' = \sigma - u \quad \text{(soils)} \tag{3:14b}$$

The actual stresses within the solid fabric vary greatly and at some points are probably several orders of magnitude greater. For a soil or rock formation not subject to major tectonic stresses, and with uniform density or unit weight, the vertical stress is

$$\sigma_z = \gamma z \tag{3:9b}$$

$$u = \gamma_w z_w \tag{3:10}$$

The computation of these stresses at any level in a soil deposit is illustrated in Example 4:1, which describes one location beneath the fill of the opening example in the chapter.

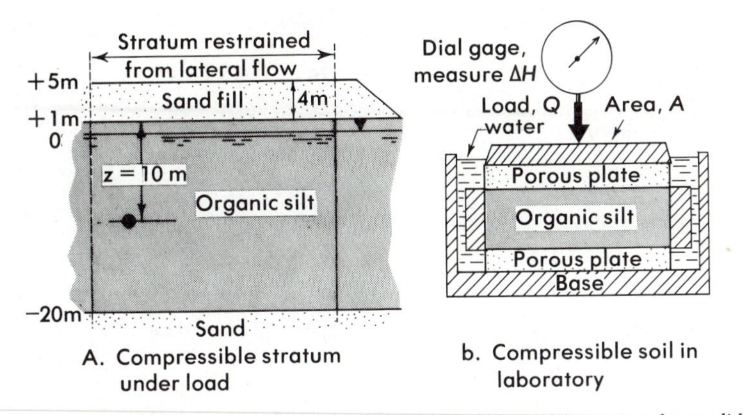

Figure 4.1 *Compressible stratum under load as modeled in a one-dimensional consolidation test.*

Example 1 (Fig. 4.1a)

A soil deposit consists of saturated organic silt with an average unit weight of 15.2 kN/m^3 (96.8 lb/ft^3). The groundwater level, sea level, is 1 m below the ground surface. The silt is 20 m thick.

1. Compute the initial vertical stresses at a depth of 10 m, ignoring the slight increase in density or unit weight of the organic silt with depth.
   a. Total Stress $= 10 \times 15.2 = 152$ kN/m^2; $32.8 \times 96.8 = 3175$ lb/ft^2.
   b. Neutral Stress $= (10-1) \times 9.81 = 88.2$ kN/m^2; $29.5 \times 62.4 =$ 1841 lb/ft^2.
   c. Effective Stress $= 152 - 88.2 = 63.8$ kN/m^2; $3175 - 1841 =$ 1334 lb/ft^3.
2. A fill consisting of saturated sand weighing 20 kN/m^3 (127.4 lb/ft^3) is spread 4 m (13.12 ft) deep over a large area. Compute the new vertical stresses and the increase in vertical stresses, considering that the water table will rise to the fill surface because the sand is placed hydraulically.
   a. Total Stress $= (10 \times 15.2) + (4 \times 20) = 152 + 80 = 232$ kN/m^2 $=$ $(32.82 \times 96.8) + (13.13 \times 127.4) = 3175 + 1673 = 4848$ lb/ft^2.
   b. Neutral Stress $= 14 \times 9.81 = 137.3$ kN/m^2; $45.95 \times 62.4 =$ 2867 lb/ft^2.
   c. Effective Stress $= 232 - 137.2 = 94.8$ kN/m^2; $4848 - 2867 =$ 1981 lb/ft^2.
   d. Effective Stress Increase $= 94.8 - 63.8 = 31$ kN/m^2; $1981 -$ $1334 = 647$ lb/ft^2.
3. The saturated sand drains until the water table again reaches sea level. The sand does not become dry; it remains 25% saturated with a unit weight of 17.27 kN/m^3 or 110 lb/ft^3. Find the vertical stresses in the organic silt after the sand drains.

a. Total Stress $= (10 \times 15.2) + (4 \times 17.27) = 221\ \text{kN/m}^2$
$= (32.82 \times 96.8) + (13.13 \times 110)$
$= 4619\ \text{lb/ft}^2$.

b. Neutral Stress $= 9 \times 9.8 = 88.2\ \text{kN/m}^2$; $29.5 \times 62.4 = 1841\ \text{lb/ft}^2$.

c. Effective Stress $= 221 - 88.2 = 132.8\ \text{kN/m}^2$; $4619 - 1841 = 2778\ \text{lb/ft}^2$.

d. Effective Stress Increase $= 132.8 - 94.8 = 38\ \text{kN/m}^2$ or $2778 - 1981 = 797\ \text{lb/ft}^2$.

Paradoxically, draining the sand fill reduces the total stress but increases the effective stress. This increase in effective stress produced by drainage has many benefits in inducing rapid soil loading, as will be brought out in Chapters 6 and 10.

4:2 Compressibility and Settlement

Settlement has plagued builders for centuries. The Leaning Tower of Pisa is continuing to tilt and subside after more than 700 years. It is famous because it has not yet fallen; other structures have been damaged or destroyed by settlement. The problem is not limited to loads imposed by structures. Some areas near Houston have settled as much as 5 m from increased effective stress induced by lowering the groundwater table by water supply wells. Long Beach, California, settled more than 5 m from the increased effective stresses produced by pumping from oil wells.

Until the early twentieth century most engineers vaguely blamed settlement on the extrusion of soft soils from beneath heavily loaded buildings. There are defects in this hypothesis. First, it does not explain regional subsidences due to groundwater depletion. Second, extrusion would necessarily produce an upward bulge of extruded soil around the settled area; instead, there is usually a saucer-shaped zone of settlement 2 to 3 times the building width with the building in its center.

Workers making excavations beneath old structures noted that the soils directly beneath were more dense than the same stratum nearby. Karl Terzaghi, an Austrian engineer of Czech birth, working in Istanbul, Turkey, in the early 1920s, followed this clue by sampling such soils.[4:1] He found that the amount of settlement corresponded to a reduction in void ratio of the soil under the building. He described the mechanics of the process in his *Erdbaumechanik*, published in 1925. It introduced the concept of effective stress and marked the beginning of modern soil mechanics and geotechnical engineering.

SOIL OR ROCK COMPRESSION[4:2–4:4] / Consider a soil stratum of thickness H that settles an amount ΔH, as shown in Fig. 4.2. The initial void ratio is e_0 and it undergoes a void ratio change of Δe. For a unit cross section of sample, the change in void ratio and sample height can be related to the

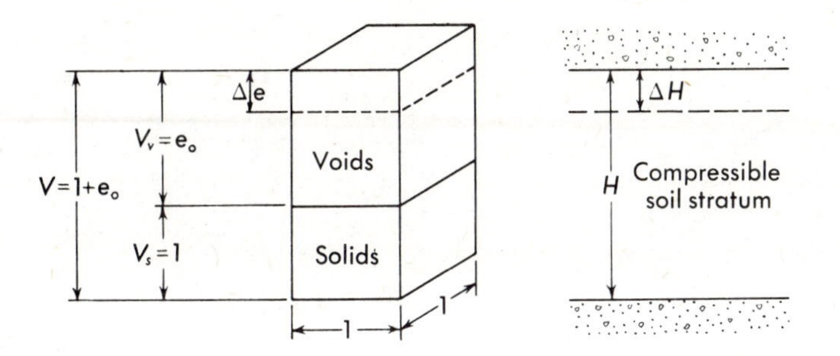

Figure 4.2 *Comparison of settlement of a soil or rock stratum of thickness H with change in height of a sample whose initial height is $1+e_0$.*

change in stratum thickness:

$$\frac{\Delta H}{H} = \frac{\Delta e}{1+e_0}$$

$$\Delta H = \frac{H\,\Delta e}{1+e_0} \tag{4:1}$$

There is some confusion in terminology. Geologists sometimes refer to the reduction in void ratio as *compaction*; engineers reserve that word for soil densification by rolling or tamping. The term *consolidation* is preferred by some engineers to denote strain related to a void ratio reduction. Others, including the author, reserve the term for the time rate of void ratio change controlled by the dissipation of neutral stress.

COMPRESSION TEST / The change in void ratio of a soil is measured in the laboratory in a *one-dimensional consolidation test* or *oedometer test* devised by Terzaghi. He reasoned that the lateral extrusion of a soft soil beneath a loaded area was resisted by the lateral resistance of that same soil surrounding the loaded area as well as confinement offered by more rigid cohesionless strata interbedded with the soft material (Fig. 4.1a). The test (Fig. 4.1b) involves a disc of soil of thickness H sandwiched between two rigid porous plates and confined laterally by a rigid ring. A vertical stress, σ'_z, is applied to the plates and the vertical displacement, ΔH, measured. The vertical strain, ε_z, is computed:

$$\varepsilon_z = \frac{\Delta H}{H} \tag{4:2}$$

Many different test procedures and methods for depicting the results have been devised. In the simplest form, increasing stresses $\sigma'_{z-1}, \sigma'_{z-2}, \ldots$ are imposed. Each is maintained constant until settlement virtually ceases and until there is no evidence of neutral stress remaining. The corresponding strains, $\varepsilon_1, \varepsilon_z, \ldots$ or void ratios are computed. The results are expressed graphically by strain or void ratio as a function of effective stress (Fig. 4.3).

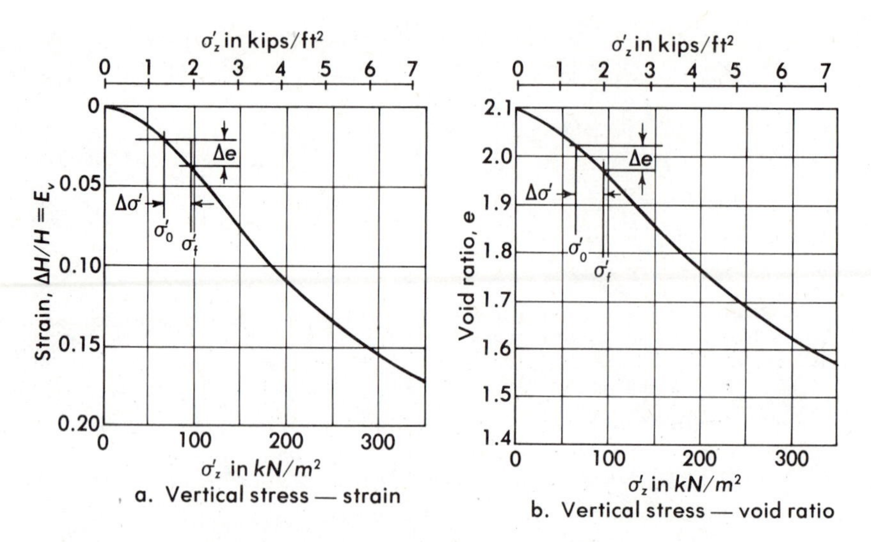

Figure 4.3 *Compression as a function of effective stress.*

The curves are similar in shape. The stress–strain form is more convenient for computations, while the void ratio form gives a better insight into the mechanics of the process.

STRESS–VOID RATIO RELATION / The shape of much of the curve is concave upward, indicating a decreasing rate of compression with increasing stress. If the stress is increased to a certain point, σ'_c, and then reduced, the soil will not swell to its original void ratio. Instead, it will increase in volume gradually along a flat, concave upward curve (Fig. 4.4) called the *decompression* curve. If stress is again applied, the recompression of the soil will follow a flat curve that is concave downward until the stress is nearly equal to σ'_c. At this point a more rapid decrease in void ratio takes place until the recompression curve practically joins the original curve. (The original curve is often termed the *virgin* curve.) Soil compression is not an elastic, reversible process; once compressed, a soil tends to remain so, even though the stress producing compression may be removed.

The slope of the curve is a measure of the soil compressibility or relative strain. In the stress–strain form it is termed the *coefficient of volume compressibility*, m_v, because if no lateral movement occurs, the vertical strain equals the volume strain (see Fig. 4.3a),

$$m_v = \frac{d\varepsilon}{d\sigma'} \qquad (4{:}3a)$$

This is the reciprocal of the modulus of deformation or modulus of elasticity commonly used to describe stress–strain relationships in solids.

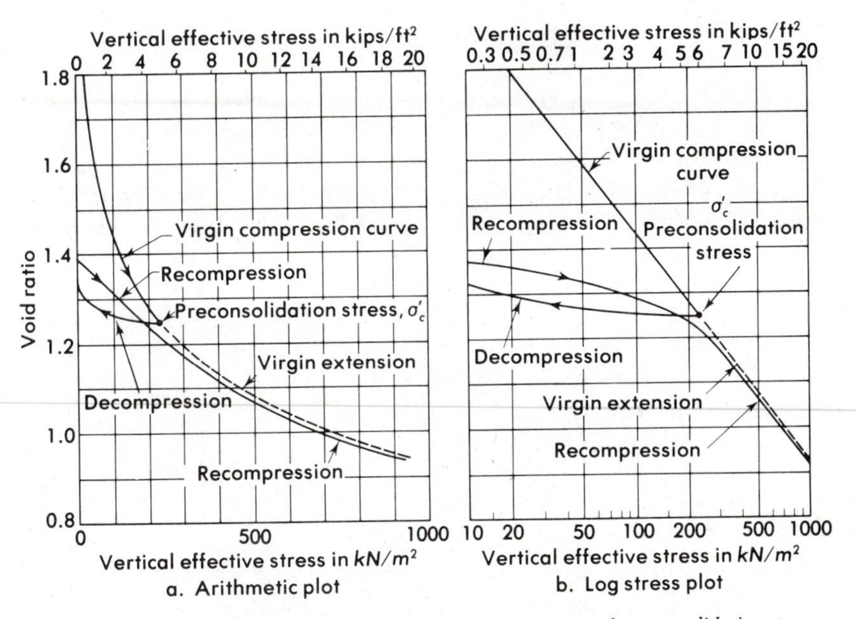

Figure 4.4 *Stress–void ratio: Compression, decompression, and preconsolidation stress.*

The second form is the *coefficient of compressibility*, a_v (see Fig. 4.3b),

$$a_v = \frac{de}{d\sigma'} = m_v(1+e) \tag{4:3b}$$

Neither m_v nor a_v are constants: Both decrease with increasing stress. Both are influenced by the past stress or stress history; the unloading and reloading curves are both different from the virgin curve, as shown in Fig. 4.4.

LOGARITHMIC REPRESENTATION / If the stress–void ratio relation is plotted with log stress as the abscissa and void ratio (or strain) as the ordinate, it will be seen that the virgin compression curve of many soils can be approximated by a straight line (Fig. 4.4b) in the range of stresses encountered in engineering work. At very high stresses and small void ratios, the curve flattens. At stresses less than preconsolidation, σ'_c, the curve is very flat. The logarithmic representation is widely used because it exaggerates the difference between behavior at stresses less than preconsolidation and stresses exceeding preconsolidation. Moreover, it provides a convenient parameter for expressing the relative compressibility of soils in the *compression index*, C_c, which is the slope of the straight-line approximation.

$$\Delta e = -C_c \log_{10}\left(\frac{\sigma'_0 + \Delta\sigma'}{\sigma'_0}\right) \tag{4:4}$$

(The negative sign denotes a reduction in void ratio for an increase in effective stress.)

COMPUTING STRESS–VOID RATIO CHANGE / For most problems in practice, the void ratio change, Δe, corresponding to an initial stress, σ'_0, is found directly from the arithmetic curve (Fig. 4.4a) or log curve (Fig. 4.4b). The appropriate virgin, recompression, or decompression curve is used, depending on the soil's stress history, as will be discussed in Section 4:3.

Example 4:2

Compute the settlement from parts 1 and 2 of Example 4:1 using the stress–void ratio curve of Fig. 4.3.

1. $\sigma'_0 = 63.8\ \text{kN/m}^2 = 1337\ \text{lb/ft}^2$
   $\Delta\sigma = 31\ \text{kN/m}^2 = 647\ \text{lb/ft}^2$.
   $\sigma'_f = 94.8\ \text{kN/m}^2 = 1984\ \text{lb/ft}^2$.
2. $e_0 = 2.02, \quad e_f = 1.97, \quad \Delta e = 0.05$
   $H = 20\ \text{m} = 65.6\ \text{ft}$.
3. $\Delta H = H\,\Delta e/(1+e_0)$
   $\Delta H = 20 \times 0.05/(1+2.02) = 0.33\ \text{m}$
   $\Delta H = 65.6 \times 0.05/(1+2.02) = 1.09\ \text{ft}$.
4. $\varepsilon_1 = 0.0215, \quad \varepsilon_2 = 0.0380, \quad \Delta\varepsilon = 0.0165$
   $\Delta H = H\varepsilon = 0.0165 \times 20 = 0.33\ \text{m}$.

For computing void ratio changes from stress changes with a digital computer, an analytical expression for stress–void ratio is expedient. The logarithmic representation is convenient because the entire curve can be approximated by a few straight lines, each with its different slope or *apparent compression index*, C_{c-1}, C_{c-2}, ... Of these, however, only the segment approximating the virgin curve should be designated *compression index*. Moreover, if the stresses involved in the engineering problem are in the vicinity of the preconsolidation load where the log slope changes rapidly, the straight-line approximations should closely fit the stresses anticipated.

THE MECHANISM OF COMPRESSION[4:4] / A number of different mechanisms are simultaneously responsible for compression. A major factor in rock and cohesionless, fibrous organic and micaceous soils (and probably to some extent in clays) is the bending and distortion of the solids. This is predominantly elastic and the soil compression that results from it is largely reversible. Fracture of the solids, particularly at their contact points, is probably a factor in all soils and porous rock. It is not elastic and not reversible. It partially accounts for decompression being less than virgin compression. Reorientation of the solid particles occurs in all soils to some degree, but is greater for soils of higher void ratio. In cemented soils or clays with strong bonds between particles, reorientation tends to develop suddenly when the stress exceeds preconsolidation. Reorientation is not reversible and is partially responsible for decompression–recompression being less than virgin compression. The electrical repulsion in the adsorbed water–cation field surrounding clay minerals is a major factor in clay soils. It is largely reversible. All these phenomena are related to the physical properties of the solid phase and the physicochemical behavior of the solid–water electrostatic double-layer systems. Their combined effects cannot be modeled analytically to the extent that the stress–void ratio relation can be deduced from data on mineralogy, water content, grain size, and void ratio. Therefore, consolidation tests are necessary to define the volume change of a soil or rock under stress.

4:3 Compressibility of Soils and Rocks

Although the compressibility cannot be predicted from other physical properties, similarities have been observed between some groups of soils and weathered rocks. Empirical relationships between compressibility and such qualities as void ratio and liquid limit are useful in interpreting test results and making estimates when test data are not available.

LOW TO MODERATE PLASTICITY, NORMALLY CONSOLIDATED / Soils of low to moderate plasticity include clays, silts, organic and micaceous silts with plasticity indexes up to 30, as well as sands, gravels, and porous rocks. When they have never been subjected to stresses greater than their present overburden load, they are termed *normally loaded* and exhibit the characteristic semilogarithmic stress–void ratio curve of Fig. 4.4b. The greater part of the curve is approximately straight, and the slope is expressed by the compression index, as was previously described. Terzaghi and Peck[4:5] have derived an expression for the compression index of clays from the work of Skempton:

$$C_c = 0.009(\mathrm{LL} - 10) \tag{4:5}$$

For soils of very low plasticity and porous rock, Sowers[4:6] has found the compression index to be related to the undisturbed void ratio

$$C_c = 0.75(e - a) \tag{4:6}$$

where a is a constant whose value is from 0.2 for porous rock to 0.8 for highly micaceous soils. Both relationships are approximate at the best, and considerable variations should be expected from the computed value of the compression index. Normally loaded clays can often be identified from their water content. Usually, it is near but less than the liquid limit.

The same soil remolded suffers a structural breakdown and a reduction in void ratio. It becomes somewhat less compressible, but exhibits a straight-line stress–void ratio curve on semilogarithmic coordinates, as shown in Fig. 4.4b.

A normally consolidated soil undergoes decompression during sampling. The stress can reduce to zero, or may be partially retained by capillary tension. In either case, the laboratory stress–void ratio curve represents recompression until the initial stress in the ground, σ'_0, is exceeded.

Most natural soils are somewhat disturbed by the sampling process. The stress–void ratio curve found in the laboratory, therefore, lies somewhere between that for the remolded soil and the true pressure–void ratio curve in nature.

The increments in stress employed in the laboratory test ($\Delta\sigma'_2 = 2\ \Delta\sigma'_1$ typically) are usually much larger than those produced by structural loads. In some soils these large increments cause disturbance and increased test settlement.

Because of these differences between compression of the undisturbed soil in place and the laboratory test, the settlement predictions based on the test results are usually excessive.

LOW TO MODERATE PLASTICITY—PRECONSOLIDATED / A *preconsolidated* soil is one that has been subjected to a stress, σ'_c, which exceeds the present overburden pressure. Most undisturbed soils are preconsolidated to some degree and exhibit the characteristic stress–void ratio curve of Fig. 4.5a. Preconsolidation is produced in a number of ways. Removal of overburden by erosion or excavation leaves the soil preconsolidated. This load can be estimated if the previous overburden thickness is known. Glacial ice produced preconsolidation in northern North America and northern Europe. The most important and widespread cause is capillary tension from desiccation or drying of the soil. Stresses greater than 500 kN/m^2 or 10,000 lb/ft^2 due to desiccation have been observed in warm and arid regions. The preconsolidation load caused by desiccation often decreases with increasing depth below the ground surface. The amount produced in this way cannot be predicted. A water table rising above a compressible stratum causes preconsolidation by reducing the effective stress. This is the reverse of the stress increase produced by lowering the water table in Example 4:1.

Chemical alteration produces the effect of preconsolidation by changing the physicochemical bonds between the clay particles or by introducing stresses by the expansion or contraction of the grains during the alteration process. Most residual soils, weathered rocks, and some partially indurated rocks exhibit *preconsolidation* from this source. Leaching that removes salt or high concentrations of cations may have the same effect in some clays, especially those deposited in salt water. All are pseudopreconsolidations.

The preconsolidation load σ'_c can be estimated from the stress–void ratio curve shown in Fig. 4.5a. Tangents are drawn to the initial, flat section and to the steep, straight-line portion of the curve. Their intersection is approximately the preconsolidation load.

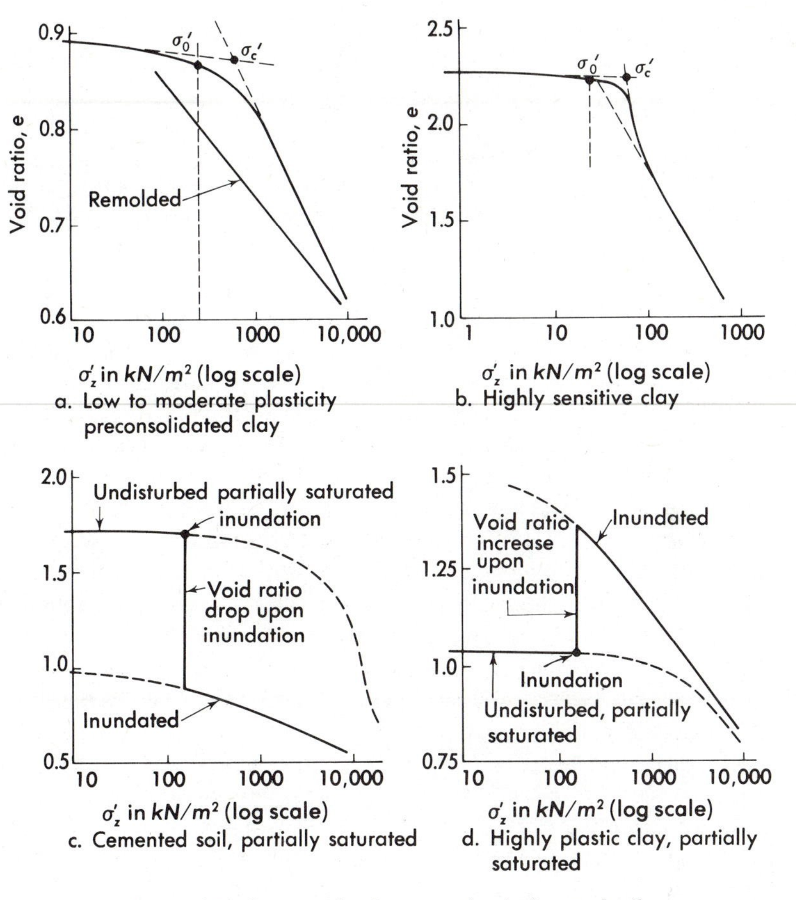

Figure 4.5 *Stress–void ratio curves of typical natural soils.*

Other more sophisticated but also empirical methods for estimating the preconsolidation (or pseudopreconsolidation load) have been proposed.[4:3] No one method appears consistent, based on geologic evidence.

Preconsolidated clays can be recognized from their water contents, which are usually much less than their liquid limits. In heavily preconsolidated soils (particularly if desiccation is the cause), the water contents are sometimes less than the plastic limits. Their compression indexes (above the preconsolidation load) can be found from the same relationships used for normally consolidated clays.

Preconsolidation is extremely important in foundation engineering. A soil that is inherently compressible usually will not settle appreciably until the stress imposed by the structure exceeds the preconsolidation load. If the natural preconsolidation load is not sufficient, it is sometimes possible to preconsolidate the soil by piling earth on the site until the soil compresses. This is often time-consuming, but the process can be accelerated by overloading or drainage. (See Section 6:6.)

SENSITIVE CLAYS / Sensitive clays and other soils with a flocculent or highly developed skeletal structure exhibit a third characteristic curve (Fig. 4.5b). Ordinarily, such soils are preconsolidated to some degree. The curve is flat up to the preconsolidation load; then it drops sharply and gradually flattens to approximate a straight line on semilogarithmic coordinates. It is suspected that the sharp drop in void ratio reflects a structural breakdown in which the bonds between the particles are broken and the grains rearrange themselves into a more dense orientation. The same soil badly disturbed or remolded has a straight or slightly curved stress–void ratio curve that is slightly flatter than the virgin curve at high stresses. The water content of most highly sensitive clays exceeds their liquid limit, which aids in their recognition.

CEMENTED SOILS / Cemented soils and similar soft rocks have stress–void ratio curves similar to preconsolidated soils. The appearance of preconsolidation probably reflects the bonding (and the interparticle stresses) between the grains. In some cases the bonding breaks abruptly with increasing load, producing a stress–void ratio curve similar to that for a sensitive clay (Fig. 4.5b). For such materials with water-sensitive bonding two pressure–void ratio curves can be developed: undisturbed (usually moist but not saturated) and inundated, as shown in Fig. 4.5c. The drop from the higher, undisturbed value will occur upon inundation under small stresses, such as produced by the weight of the soil itself. Serious sudden settlements occur in dry regions where cemented soils have been wet by constructing dams or by irrigation. Inundation produces near-saturation; complete saturation probably does not occur.

EXPANSIVE SOILS / Partially saturated highly plastic clays and some highly micaceous soils may be heavily preconsolidated by desiccation. If they have access to water at low stresses, they will adsorb water and expand (Fig. 4.5d). Two stresses–void ratio curves can be developed for such a soil: undisturbed but partially saturated, and inundated. The inundated soil approaches saturation; however, there is little evidence that 100% saturation occurs.

PEAT / Peats and highly organic soils are extremely compressible, depending on the void ratio. The pressure–void ratio curves resemble those for clays, with slight preconsolidation. The virgin curve on a semilog plot is seldom very straight. The compression index for peat can be estimated from its void ratio:

$$C_c = (0.5 \text{ to } 0.7)e \tag{4:7}$$

Because of their very high void ratios, approaching 20, peats are by far the most compressible soils.

WASTE FILL / The compressibility of waste fills, such as garbage and refuse, cannot be measured in the laboratory because of the non-homogeneity and the large particle size (imagine testing a mix of garbage, beer cans, rubber tires, and old autos). Overall, the compression of such mixes resembles that of peat. Much of the compression is crushing of hollow bodies and distortion of resilient materials. The compression index can be estimated from Equation (4:7); the coefficient of e ranges from 0.15 for rigid refuse to 0.7 for hollow bodies and garbage.

SANDS / Sands consolidate largely by grain reorientation and fracture, accompanied by some elastic distortion of the grains. The compression indexes are usually less than 0.1, and the curves resemble those for clays.

COMPRESSIBILITY OF ROCK / Well-indurated rocks with low void ratios (less than 0.2) are virtually incompressible. More porous rocks, including shale, tuff, and porous limestones consolidate similar to soils. The mechanism appears to be a combination of elastic distortion of the solid framework between the voids and a crushing of the rock where it is locally highly stressed. The pressure–void ratio curves for porous rocks resemble those for preconsolidated soils.

Continued leaching and weathering weakens rock and permits further consolidation. This has been observed in soft limestones and is suspected in tuff. The leaching appears to be aggravated by stress, and thus the settlement increases with increasing load as well as with the passage of time. This can only be evaluated by simulating the leaching under stress in the laboratory.

Broken rock, such as used in rock fills, consolidates by reorientation of the grains and by local crushing at the points of contact. The pressure–void ratio curves resemble those for virgin compression (Figs. 4.3 and 4.4).

SIMULATED FIELD COMPRESSION / The stress–void ratio curve from the conventional laboratory test frequently differs from that of the soil stratum in the ground. As mentioned previously, a number of factors are responsible: sample disturbance, large sudden test load increments, sudden inundation, and the limited time available. Schmertman[4:8] has proposed a technique for correcting the conventional test. The author suggests a test procedure that simulates the field loading and unloading sequence. This is illustrated in Fig. 4.6. The void ratio of the in-place soil was produced by loading in the sequence: (1) preconsolidation, (2) unloading by natural causes, (3) site excavating, and (4) reloading by the structure. Most of the stresses are less than preconsolidation and in the flat decompression–recompression range. The undisturbed sample, removed from the ground at overburden stress σ_0' is distorted by further unloading (5), although its void ratio may not change appreciably. Upon reloading (6), the recompression curve is steeper than for the undisturbed stratum (4) because of soil disturbance. If the laboratory specimen is first reloaded (7) above the overburden load (or preferably to the preconsolidation load), it is more or less restored to its earlier natural effective stress (although at a slightly

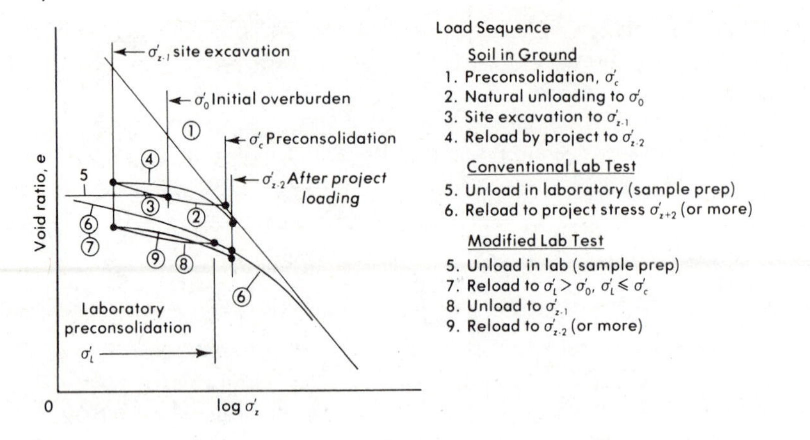

Figure 4.6 *Laboratory test sequence for simulating compression of undisturbed, preconsolidated, or overconsolidated soil.*

smaller void ratio). It is then unloaded (8) to the stress produced by site excavation σ'_{z-1} and then reloaded (9) to the stress eventually imposed by the structure, σ'_{z-2}. The stress–void ratio relation for this second sequence of unload–load (8–9) closely parallels that of the in-place stratum (2–3–4). In practice neither the effective stress produced by excavation nor structural load are known accurately at the time the laboratory tests are made, and estimates are necessary. The actual void ratio changes are found from the reload stress–void ratio curve, 9, interpolating and extrapolating as necessary.

Care must be taken to simulate field neutral stresses and saturation, particularly if the compressible soil stratum is above the groundwater table. The effect of possible groundwater changes can be evaluated by tests at natural moisture and after either inundation or drying. (Some engineers advocate testing all soils after complete saturation; this procedure can be misleading unless the soil naturally becomes fully saturated in the ground.)

4:4 Time Rate of Compression

The compression of a soil stratum does not occur suddenly. Instead, it often takes place so slowly that it is difficult to believe that any settlement is occurring. Buildings in Chicago continue to settle for 50 years, and the Leaning Tower of Pisa, commenced in 1174, is still moving.

The compression begins rapidly and becomes slower with increasing time. As shown in Fig. 4.7, it can be divided into three stages: (1) *initial*, (2) *primary or hydrodynamic*, and (3) *secondary*. The sum of the initial and primary stages is the compression computed from the laboratory stress–void ratio curve. The secondary stage is of importance principally in highly organic, highly micaceous, and highly sensitive soils.

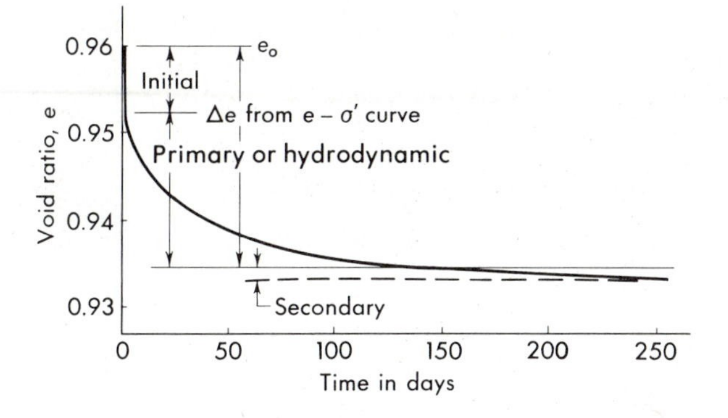

Figure 4.7 *Time-compression curve for an increment of load.*

INITIAL CONSOLIDATION / The initial stage occurs as soon as the load is applied. It occurs largely by compression and solution of the air in the soil voids. It also includes very small amounts of compression of the solid phase and the soil water.

The percentage of initial consolidation of soil compared to initial plus primary is largely controlled by the degree of saturation. It is virtually zero in a saturated soil but as great as 50% when the degree of saturation is 90%. It decreases with increasing load because of the decreasing void ratio (and thus higher saturation). Ordinarily, the initial compression is determined from the time-settlement data from the laboratory test and furnished as a part of the test results, as in Fig. 4.13a.

The total *settlement* of a thick soil stratum includes an additional component that develops simultaneously with initial consolidation: *distortion* or *immediate settlement.* This is the result of the vertical elastic deflection of the soil body beneath the structure, accompanied by lateral elastic bulging. This should not be confused with the sudden initial change in void ratio of a partially saturated soil. Both occur simultaneously but are caused by different mechanisms.

NEUTRAL STRESS[4:1] / The time rate of primary or hydrodynamic compression is controlled by the escape of water from the soil voids. The water is squeezed out by an external force applied to the soil mass. This force increases the water pressure beyond the hydrostatic, developing a hydraulic gradient that produces flow. The mechanics of this process can be demonstrated by the analogy of a spring, piston, and cylinder. A spring (Fig. 4.8) has attached to its top a piston whose cross-sectional area is 1 m^2 and whose weight is 4 kN. Under this weight the spring has a length of 1 m. When an added load of 2 kN is placed on the spring (Fig. 4.8b), it compresses to a length of 0.8 m. The compression is instantaneous as the load is applied. Suppose that, instead of in the open air, the piston and spring is placed in a tight-fitting cylinder (Fig. 4.8c), with the space below the piston filled with

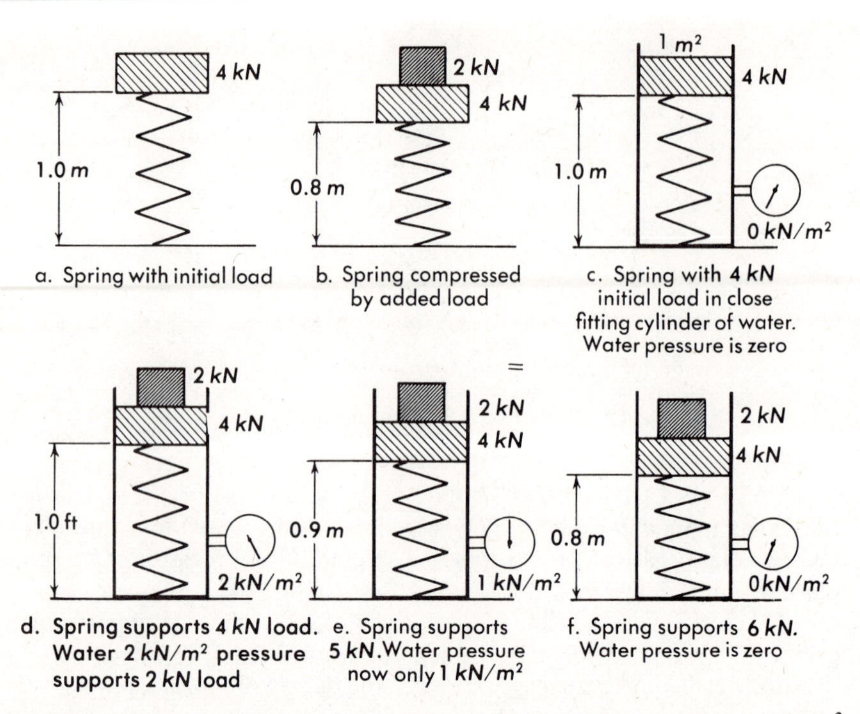

Figure 4.8 *Piston and spring analogy showing transfer of support of an added load of 2 kN/m^2 from water pressure or neutral stress to an elastic (spring) solid as effective stress.*

water. The spring initially is compressed by the weight of the piston, 4 kN, and the water is under no pressure at all.

If the 2 kN weight is now added on top of the piston (Fig. 4.8d), the spring does not compress because the water below the piston cannot escape. The spring still supports the 4 kN piston but offers no support to the added weight. The 2 kN are supported by water pressure on the piston of 2 kN/m^2. If the total load of 6 kN is denoted by σ, and the actual spring load by σ', then the following equation describes the way the total load is supported:

$$\sigma = \sigma' + u$$

$$6 = 4 + 2 \qquad (3{:}14b)$$

If 0.1 m^3 of water leaks out because of the pressure from beneath the piston, the spring will be compressed to a length of 0.9 m. The spring will now support 5 kN, and the neutral stress will be reduced to 1 kN/m^2, as shown in Fig. 4.8e,

$$6 = 5 + 1$$

When an additional 0.1 m^3 of water leaks away, the spring will carry 6 kN and the neutral stress will be zero, as in Fig. 4.8f. The spring compression is a process of transferring to it the 2 kN of added load that was supported initially by water pressure.

Consolidation of soil is similar to the above analogy. The resilient grain structure is represented by the spring, and the voids filled with water are represented by the cylinder. When a load is placed on the soil, the grain structure cannot immediately support it because compression cannot occur. Neutral stress, therefore, supports the load. As the water seeps out and the soil compresses, the grain structure assumes the load and the neutral stress becomes zero. The adjacent sequence (Fig. 4.9) illustrates the way in which the stress transfer occurs in a soil stratum that is bounded by pervious strata above and below. The initial stress in the soil is denoted by σ'_0, the added stress due to the weight of a structure by $\Delta\sigma$, neutral stress by u, and the stress in the grain structure by σ'.

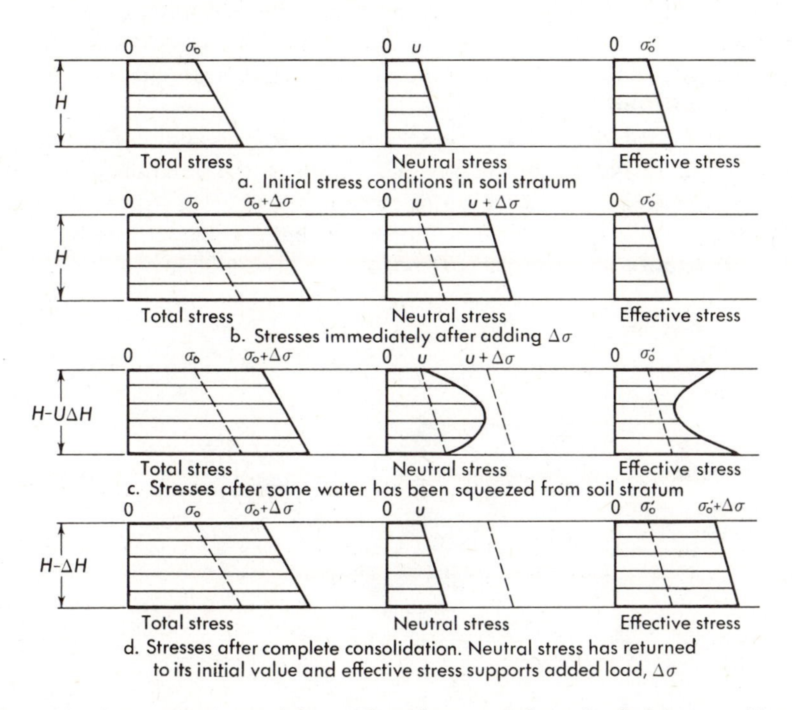

Figure 4.9 *Stresses during consolidation. The soil stratum is drained on both its top and bottom faces.*

The increment of stress $\Delta\sigma$ is initially supported by an increase in water pressure Δu (Fig. 4.9b) that is uniform throughout the stratum. Seepage is immediately induced by the water pressure difference with the highest gradients and highest rates of seepage at the boundaries. Seepage is upward vertically at the top boundary and downward vertically at the bottom. Within the stratum the gradients and seepage rates are smaller toward the center of the stratum. After some water has been squeezed from the stratum, the water pressure distribution will be as shown in Fig. 4.9c. At the boundaries it will have returned to its initial or ambient value. At the center of the stratum much of the increase in water pressure, Δu, will remain. The water pressure distribution will be somewhat parabolic. The effective stress increase will reflect the difference. At the boundaries it will be $\sigma_0' + \Delta\sigma$; at the center it will be less because of the remaining water pressure. Eventually, the excess pore water pressure is dissipated and the effective stress will be as shown in Fig. 4.9d. The hydrodynamic phase is now complete: The soil is said to be *fully consolidated* or 100% consolidated.

PERCENTAGE OF CONSOLIDATION[4:9] / The *percentage of consolidation*, *U*, is defined as the average percentage of the added stress $\Delta\sigma$ that is supported by increased effective stress. It represents the percentage of the total or ultimate compression that has already occurred in the stratum. For a saturated soil this is the percentage of the ultimate void ratio change, Δe (or volume change, ΔV) produced by an effective stress change, $\Delta\sigma'$, that has occurred after an elapsed time, *t*.

A simplified analysis of the seepage out of the soil, $\Delta\Delta q$, and the corresponding void ratio change, $\Delta\Delta e$, defines the relationship of the significant factors. Consider a saturated compressible soil stratum of thickness *H*, void ratio *e*, and permeability coefficient *k* (Fig. 4.10). Above and below are strata of much higher permeability and negligible compressibility.

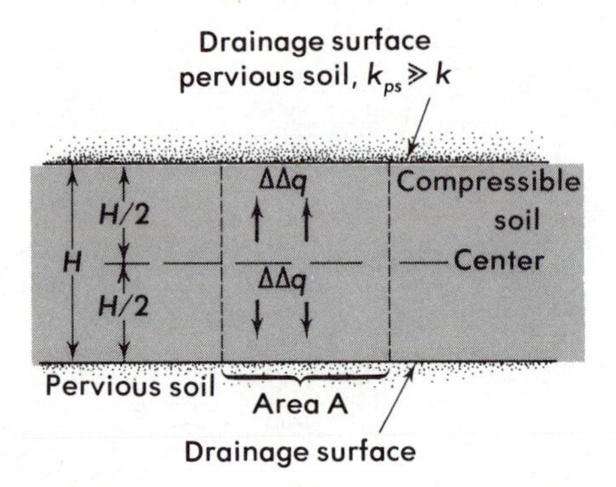

Figure 4.10 *Simplified model for time rate of consolidation.*

The number of horizontal pervious boundaries, *N*, is 2. The neutral stress in all the strata is hydrostatic. A uniform vertical stress increment of $\Delta\Delta\sigma$, a part of a larger increment $\Delta\sigma$, is immediately reflected in an effective stress increase of $\Delta\Delta u$ in the less pervious, compressible soil. The increased neutral stress, $\Delta\Delta u$, will create a gradient that forces water out of the soil similar to Fig. 4.9. Because the strata above and below are more pervious and the stress increase is uniform, seepage is largely vertical, with flow upward from the top half and downward from the bottom half, where

$\Delta u = \Delta\sigma$

$\Delta h = \Delta u/\gamma_w = \Delta\sigma/\gamma_w$

$L = H/2$ [or some fraction of $(H/2)$; if water can seep out of the stratum on only one side, $L = H$ (or some fraction)]

$$i = \Delta h/L = \frac{\Delta\sigma}{\gamma_w(H/2)}$$

For a unit area of soil in the stratum,

$$q = \frac{k\,\Delta\sigma}{\gamma_w(H/2)} \times 1 \times 1$$

In time t, the volume of flow, $\Delta\Delta V$ is qt,

$$\Delta\Delta V = \frac{k\ \Delta\sigma\ t}{\gamma_w(H/2)}$$

The total volume of water ΔV, squeezed out of the same unit area of soil by a void ratio change, Δe, is

$$\Delta V = \frac{\Delta e(H/2)}{1+e}$$

The proportion of the water that has been squeezed out in time t is expressed by the ratio $\Delta\Delta V/\Delta V$. This is the *percentage of consolidation*, $U\%$.

$$\frac{\Delta\Delta V}{\Delta V} = \frac{K\ \Delta\sigma t}{\gamma_w(H/2)}\ \frac{\Delta e(H/2)}{(1+e)}$$

$$\frac{\Delta e}{\Delta\sigma} = a_v$$

$$\frac{\Delta\Delta V}{\Delta V} = \frac{t(1+e)k}{\gamma_w(H/2)^2 a_v} \tag{4.8}$$

This simplified analysis shows that the proportion of water that has been squeezed out by consolidation in time t is a function of the dimensionless ratio, T, the *time factor*

$$T = \frac{t(1+e)k}{\gamma_w(H/N)^2 a_v} \tag{4:9}$$

A rigorous analysis that includes the change of Δu with time and position within the stratum leads to the expression

$$U\% = f\left(\frac{t(1+e)k}{(H/N)^2\gamma_w a_v}\right) \tag{4:10}$$

where the mathematical function is an infinite series involving the time factor, T, assumed to be a constant. The value of this function for most practical problems is shown in Fig. 4.11.

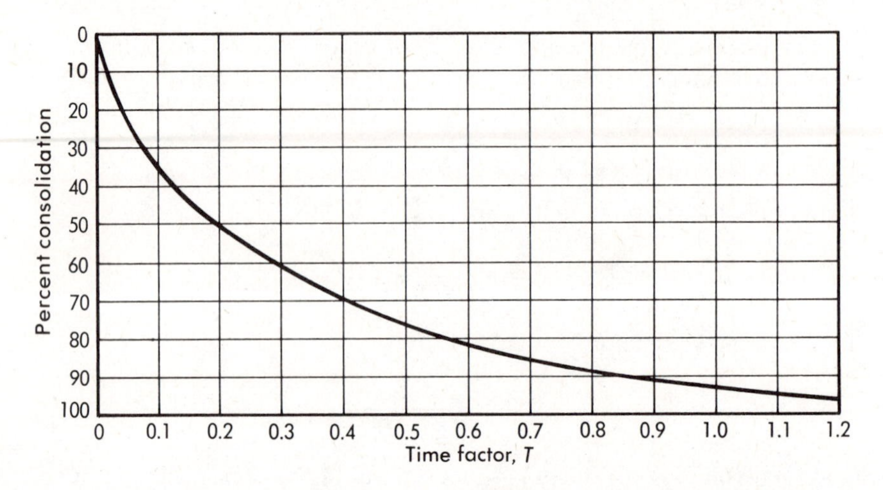

Figure 4.11 *Time rate of consolidation for a stratum drained on both surfaces and any distribution of stress or for a stratum drained on one surface and a uniform distribution of stress increase.*

The rate of consolidation is seen to be dependent on several factors:

1. *Permeability*: The greater the k, the greater the rate, a major factor.
2. *Void ratio*: The greater the void ratio, the greater the rate, although it is not a major factor.
3. *Compressibility*: The greater the a_v, the slower the rate, a significant factor.
4. *Stratum thickness*: The larger the H^2, the slower the rate, the most important factor.
5. *Previous surfaces*: The greater the number, N, (1 or 2), the greater the rate by N^2, a major factor.

The characteristics of the soil that control the percentage of consolidation are sometimes expressed by a *coefficient of consolidation*, c_v:

$$c_v = \frac{(1+e)k}{\gamma_w a_v} \tag{4:11}$$

It is assumed to be a constant in the rigorous derivation of the time factor function. For small stress changes, it is nearly so. For large changes in stress, it varies with the effective stress σ', as shown in Fig. 4.13b. It is small but often indeterminate at stresses less than the preconsolidation load, and increases at higher stresses, largely due to the increase in the resistance to compression as expressed by the coefficient of compressibility, a_v.

Example 4:3

Find the percentage of consolidation after one year for Examples 4:1 and 4:2 if k of the compressible stratum is 3×10^{-7} mm/sec.

1. From Examples 4:1 and 4:2, $H = 20$ m, $N = 2$, $e_0 = 2.02$, $\Delta e = 0.05$, $\Delta H = 0.33$ m, $\Delta\sigma = 31$ kN/m^2.
   Also: $\gamma_w = 9.81$ kN/m^3, $k = 3\times10^{-10}$ m/sec.
2. $T = \dfrac{t(1+e_0)k}{(H/N)\gamma_w a_v}$.
3. $H/N = 20/2 = 10$ m; $a_v = \Delta e/\Delta\sigma = 0.05/31 = 1.6\times10^{-3}$ m^2/kN;
   $t = 365\times24\times60\times60 = 31{,}536{,}000 = 3.154\times10^7$ sec.
4. $$T = \frac{3.154\times10^7 \text{ sec } (1+2.02)\times(3\times10^{-10}) \text{ m/sec}}{10^2 \text{ m}^2\times9.81 \text{ kN/m}^2\times1.6\times10^{-3} \text{ m}^2\text{/kN}}$$
   $$= \frac{(3.15\times3.02)\times(3\times10^{-3})}{1\times9.8\times(1.6\times10^{-1})}$$
   $T = 1.82\times10^{-2} = 0.0182.$
5. From Fig. 4.11 consolidation is 12% complete.
6. Settlement in one year $= 0.12\times0.33 = 0.0396$ m $= 0.13$ ft $= 1.6$ in. out of ultimate consolidation of 0.33 m $= 1.08$ ft $= 13$ in.

SECONDARY COMPRESSION / After the excess neutral stress has been dissipated, the compression does not cease. Instead it continues very slowly at an ever-decreasing rate indefinitely. This is *secondary compression.* This appears to be the result of a plastic readjustment of the soil grains to the new stress, of progressive fracture of the interparticle bonds, and progressive fracture of the particles themselves.[4:10, 4:11]

The secondary compression can be identified on a plot of settlement as a function of logarithm of time (Fig. 4.12). The secondary appears as a straight

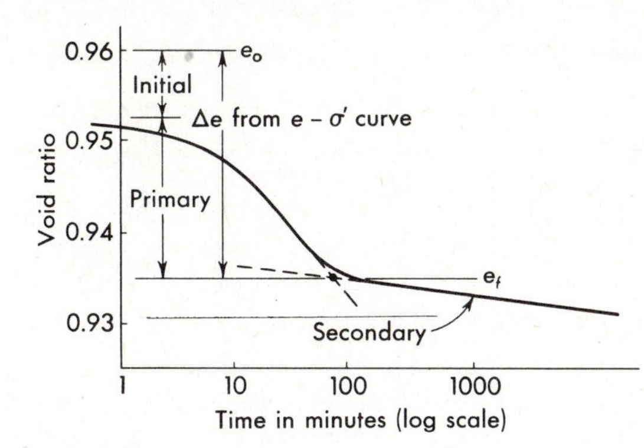

Figure 4.12 *Secondary compression; semilogarithmic time-settlement curve for an increment of stress.*

line sloping downward or, in some cases, as a straight line followed by a second straight line with a flatter slope. The void ratio e_f, corresponding to the effective end of primary consolidation, can be found from the intersection of the backward projection of the secondary line with a tangent drawn to the primary curve, as indicated in Fig. 4.12. The rate of secondary compression depends on the increment of stress increase, $\Delta\sigma'$, and on the characteristics of the soil. For inorganic soils of low to moderate compressibility, secondary compression is seldom important. It can be a major part of the compression of highly compressible clays, highly micaceous soils, fills of broken rock, and organic materials.

The equation for the rate of secondary compression can be approximated from the straight line on the log time plot

$$\Delta e = -\alpha(\log_{10} t_2 - \log_{10} t_1)$$

$$\Delta e = -\alpha \log_{10} \frac{t_2}{t_1} \tag{4:12a}$$

In this expression t_1 is the time required for hydrodynamic compression to be virtually complete and t_2 any later time. This ignores the secondary compression that occurs during the hydrodynamic phase, but the error is probably not serious. The value α is a coefficient expressing the rate of secondary compression. The coefficient of secondary compression C_α is another way of expressing the same thing in terms of percentage of settlement:

$$C_\alpha = \alpha/(1+e) \tag{4:12b}$$

$$\frac{\Delta H}{H} = -C_\alpha \log_{10} \frac{t_2}{t_1} \tag{4:12c}$$

Generally, α and C_α increase with increasing stress (Fig. 4.13).

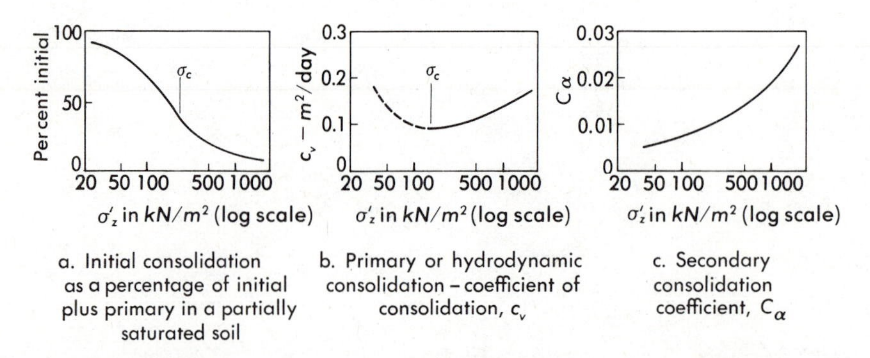

Figure 4.13 *Factors and coefficients describing the time rate of consolidation.*

The secondary compression is often irregular and only approximated by the straight line on the logarithm-of-time graph. Therefore, estimates of secondary compression are seldom accurate.

REPRESENTING TIME-RATE DATA / All three phases of the time rate are dependent on the stress level. Therefore, it is convenient to express all three as functions of the final stress (Fig. 4.13). The functions are not unique because they also depend on the stress increment $\Delta\sigma'$. However, if the test increment is similar to that in the prototype, the values will be sufficiently realistic for engineering estimates.

4:5 Shrinking, Swelling, and Slaking

Soils undergo volume changes that are not produced by external loads. Instead they are caused by changes in water content and in the effective stress produced by neutral stresses affected by the water.

SHRINKAGE / Shrinkage is caused by capillary tension. When a saturated soil dries, a meniscus develops in each void at the soil surface. This produces tension in the soil water and a corresponding compression in the soil structure. This can be expressed quantitatively from Equation (3:14b) where the external stress, σ, equals zero:

$$0 = \sigma' + u$$

$$\sigma' = -u$$

(Since u from capillary tension is negative, then $-(-u)$ is positive and the resulting effective stress σ' is positive.) This compressive stress is just as effective in producing soil compression as an external load, and pressures as great as 500 kN/m^2 or 10 kips/ft^2 can be produced in fine-grained soils.

SHRINKAGE LIMIT / During shrinkage the voids become smaller and the potential maximum capillary tension increases. This is shown graphically in Fig. 4.14. The resistance to compression, the stress–void ratio

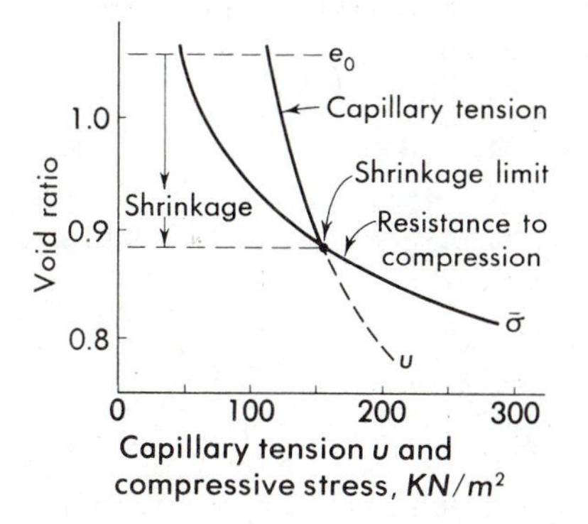

Figure 4.14 *Shrinkage limit as a function of capillary tension and resistance to compression.*

curve, is shown also. During drying and shrinking the void ratio decreases; both the maximum capillary tension and the resistance to compression simultaneously increase, but at different rates. The soil initially remains saturated, for the water loss causes an equivalent reduction in void ratio. A void ratio is reached, however, where the maximum tension equals the resistance. Further drying cannot produce a reduction in void ratio because the resistance will exceed the tension. At this point, known as the *shrinkage limit*, the reduction in the void ratio largely ceases. The meniscus in each void begins to retreat from the soil surface. The soil surface no longer has a damp appearance but now looks dry, and the soil ceases to be saturated.

The shrinkage limit is defined as the water content at the point that shrinkage ceases, and the soil is no longer saturated. It can be found by drying the soil slowly, visually observing the color change, and determining the moisture at this point. It can also be found by drying a saturated soil completely. If the weight and total volume at the beginning of shrinkage are W_1 and V_1 and at the end of shrinkage (oven-dried) are W_2 and V_2, then the following can be derived by assuming that there is no volume change after the shrinkage limit is reached and that the loss of weight by evaporation up to the shrinkage limit is accompanied by a corresponding loss of volume:

$$W_2 = W_s$$

$$(V_1 - V_2)\gamma_w = W_w \text{ (lost up to shrinkage limit)}$$

$$\mathrm{SL} = \left(\frac{W_1 - W_2 - (V_1 - V_2)\gamma_w}{W_2} \times 100\% \qquad (4{:}13)$$

(The % sign is usually omitted.)

Drier than the shrinkage limit, the capillary tension increases in the smaller voids; however, tension is released in some of the larger ones. Some soils, particularly those containing fibrous organic matter and mica, expand dry of the shrinkage limit; others continue to shrink slightly.

Soil shrinkage causes settlement of compressible soils. Because capillary tension is exerted in all directions, shrinkage occurs horizontally as well as vertically, causing vertical shrinkage cracks. Cracks 0.5 m wide and 5 m deep have been observed in highly compressible clays. Recurring shrinkage brought on by desiccation during dry weather will produce networks of shrinkage cracks in all directions and a blocky macrostructure in the soil.

SWELLING[4:7] / Some soils not only shrink on drying but also swell when the moisture is allowed to increase. The mechanism is more complex than shrinking and is caused by a number of different phenomena: the elastic rebound of the soil grains, the attraction of the clay minerals for water, the electrical repulsion of the clay particles and their adsorbed cations from one another, and the expansion of air trapped in the soil voids. In soils that have been precompressed by load or shrinkage, all these factors probably contribute. In soils that have never been precompressed, probably the attraction of the clay minerals for water and the electrical repulsion of the clay particles predominate.

High pressures can be developed if the soil has access to water but is prevented from swelling by confinement. If precompression is the cause, the swell pressure will approach the preconsolidation load. Where adsorption and repulsion predominate, as in clays of the montmorillonite family, pressures as great as 500 kN/m^2 have been measured.

Swell is produced in some soils and rocks by chemical changes. Strong bases, leaking from chemical processes, can cause expansion of layered minerals such as micas and clays. Oxidation of iron pyrite in shale, freshly exposed to air, and of some blast furnace slags will cause swelling that damages structures.

PREDICTING SWELLING AND SHRINKING / It is difficult to predict shrinking and swelling quantitatively, for they depend on the character of the soil and on the moisture changes. Shrinking can be measured by drying the soil and computing the relation between saturated water content and volume. In general, the lower the shrinkage limit, the greater the potential shrinking. Swelling can be estimated by tests resembling consolidation. The expansion (free swell) is found by flooding the soil when it is acted upon by a constant nominal pressure: 5 kN/m^2 or 10 kN/m^2 (100 to 200 lb/ft^2). The swell pressure is found by inundating the soil and measuring the pressure required to prevent its expansion. It has been found that the shrinkage limit and the plasticity index are some indication of the potential volume change, as given in Table 4:1.

TABLE 4:1 / VOLUME CHANGE POTENTIAL
(Adapted from Holtz and Gibbs)[4:12]

Volume Change	Shrinkage Limit	Plasticity Index
Probably low	12 or more	0–15
Probably moderate	10–12	15–30
Probably high	0–10	30 or more

The amount of swelling and shrinking (or the swell stress if the soil is confined) depends on the initial moisture in the soil.[4:13] If the soil is drier than the shrinkage limit, further drying will not produce significant additional shrinking. If it is wetter, then the maximum shrinking will be equivalent to the difference between the actual water content and the shrinkage limit.

Similarly, limited data indicate that little swell will occur after the soil moisture reaches the plastic limit, or slightly more, equivalent to a water–plasticity ratio of about 0.25. Smaller water contents produce higher swell amounts and swell pressures, as shown in Fig. 4.15.

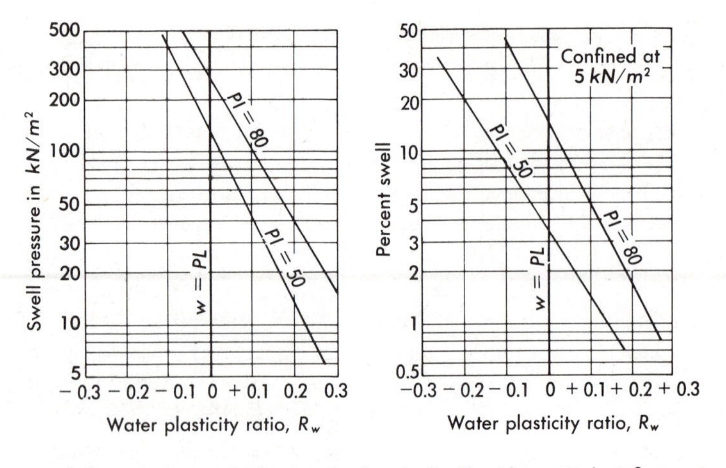

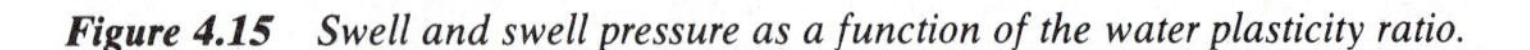

Figure 4.15 *Swell and swell pressure as a function of the water plasticity ratio.*

Few data are available to define the potential changes in soil moisture. Continuing field observations for several years are necessary to determine the moisture fluctuations that occur in any region, and these must include the effects of any structure or any changes in environment that are produced by man. These must be correlated with swelling and shrinking.[4:14]

SLAKING / If a cohesive soil that has dried well beyond the shrinkage limit is suddenly inundated or immersed in water, it will disintegrate into a soft wet mass, a process known as *slaking.*

Two factors are involved in slaking. First, the unequal expansion of the soil as the water penetrates from the surface causes pieces of soil to flake off the mass. Second, when the soil dries beyond the shrinkage limit, some of the voids fill with air. When the dried soil is immersed, water enters these air-filled voids on all sides. The air is trapped between the meniscuses of the entering water, and its pressure builds up as water fills the void. The result is an explosion of the void and disintegration of the soil. Both the flaking and the air bubbles can be seen by placing a lump of dried clay in a glass of water.

4:6 Evaluating Volume Change for Analysis and Design

One of the more unusual engineering properties of soils and rocks is their significant change in volume produced by both load and environment. All soil-like materials, including wastes and some rocks, compress similar to cork or sponge. But, unlike cork or sponge, many compressible soils are saturated and have low permeability coefficients, and therefore their compression rate is controlled by the rate of seepage of water from the voids.

In nonsaturated soils and porous rocks, the air in the voids can either compress rapidly or dissolve in the water permitting rapid initial consolidation. Their seepage or hydrodynamic consolidation also occurs rapidly and often cannot be identified. Some *secondary consolidation* continues indefinitely due to statistical readjustment of particles to load.

Some void ratio changes are not directly load-related, such as collapse of a loose cemented soil upon inundation or the biochemical decay of waste fill. Expansion occurs when prestressed micaceous soils or plastic clays gain water.

Consolidation by load can be quantified by the one-dimensional laboratory test and the three stages differentiated within limits: sometimes within 20% of the observed amount. The other phenomena can be studied in the laboratory; however, the environmental changes responsible are usually elusive.

Analysis for design considers all aspects of void ratio change, but the reliability of the analyses varies greatly with the quality and representativeness of the samples as well as the reliability of load and environmental change data. Typically, the total volume change is overestimated from laboratory tests, while the time rate is underestimated. The most reliable results are from consolidation tests of normally loaded medium plastic clay. Tests of very sensitive clays and overconsolidated clays produce results that can be 2 to 3 times as compressible as the virgin soil. The procedure described can reduce the discrepancies. The upper limits of biochemical compression and structural collapse can be measured, but the actual amounts and rates must be deduced from the laboratory data correlated with local experience. The same can be said for expansion of clays, micaceous soils, and some shales. Laboratory tests for compression of sands are seldom reliable because the soil structure of reconstituted samples is much more resilient than that of the virgin soil with a long-established structure. Field tests provide more realistic results (see Chapters 7 and 10).

REFERENCES

4:1 A. W. Skempton, "Significance of Terzaghi's Concept of Effective Stress," in *From Theory to Practice in Soil Mechanics*, John Wiley & Sons, Inc., New York, 1960, pp. 42–45.

4:2 T. W. Lambe, *Soil Testing for Engineers*, John Wiley & Sons, Inc., New York, 1951, p. 164.

4:3 J. E. Bowles, *Engineering Properties of Soils and Their Measurement*, 2nd ed., McGraw-Hill Book Company, New York, 1978, pp. 111–131.

4:4 J. K. Mitchell, *Fundamentals of Soil Behavior*, John Wiley & Sons, Inc., New York, Chapter 13, pp. 252–282.

4:5 K. Terzaghi and R. B. Peck, *Soil Mechanics in Engineering Practice*, 2nd ed., John Wiley & Sons, Inc., New York, 1967, p. 73.

4:6 G. F. Sowers, "Soil and Foundation Problems in the Southern Piedmont Region," *Proceedings, ASCE*, **80**, Separate 416, 1953.

4:7 F. H. Chen, *Foundations on Expansive Clays*, Elsevier Scientific Publishing Co., Amsterdam, 1975, pp. 16–56.

4:8 J. H. Schmertmann, "The Undisturbed Consolidation of Clay," *Transactions, ASCE*, **120**, 1201, 1955.

4:9 T. W. Lambe and R. V. Whitman, *Soil Mechanics*, John Wiley & Sons, Inc., New York, 1969, pp. 406–422.

4:10 K. Y. Lo, "The Secondary Compression of Clays," *Journal of the Soil Mechanics and Foundations Division, Proceedings, ASCE*, **87**, SM4, August 1961.

4:11 T. H. Wu, W. D. Resendiz, and R. J. Neukischner, "Consolidation by a Rate Process Theory," *Journal of the Soil Mechanics and Foundations Division, Proceedings, ASCE*, **92,** SM6, November 1966.

4:12 W. G. Holtz and H. J. Gibbs, "Engineering Properties of Expansive Clays," *Transactions, ASCE*, **120**, 1956.

4:13 G. F. Sowers and C. M. Kennedy, "High Volume Change Clays of the Southeastern Coastal Plain," *Proceedings of the Third Pan American Conference on Soil Mechanics and Foundation Engineering, II, Caracas, Venezuela*, 99, 1967.

4:14 R. L. Tucker and A. R. Poor, "Field Study of Moisture Effects on Slab Movements," *Journal of the Geotechnical Engineering Division, Proceedings ASCE*, **104**, GT9, 403–414, April 1978.

SUGGESTIONS FOR FURTHER STUDY

A. Kezdi, *Handbook of Soil Mechanics*, Vol. 1: *Soil Physics*, Elsevier Scientific Publishing Co., Amsterdam, Budapest, 1974, pp. 227–241.

G. A. Leonards, "Engineering Properties of Soils," in *Foundation Engineering*, G. A. Leonards, ed., McGraw-Hill Book Company, New York, 1962, pp. 139–176.

H. F. Winterkorn and H. Y. Fang, "Soil Technology and Engineering Properties of Soils." in *Foundation Engineering Handbook*, H. F. Winterkorn and H. Y. Fang, eds., Van Nostrand Reinhold Company, New York, 1975, Chapter 2, pp. 67–120.

R. N. Yong and B. P. Warkentin, *Soil Properties and Behavior*, Elsevier Scientific Publishing Co., Amsterdam, 1975, pp. 198–260.

PROBLEMS

4:1 The stress–void ratio curve for a saturated clay is shown in Fig. 4.4b. Compute compression index C_c. Find the change in void ratio from the curve if the stress increases from 50 to 200 kN/m^2 or 1044 to 4175 lb/ft^2. Find the change in void ratio from the curve if the stress changes from 300 to 1500 kN/m^2 or 6260 to 31,300 lb/ft^2. The change in void ratio in both cases, using Equation (4:4), and compare with the values found directly from the curve. Explain the differences.

4:2 A consolidation test had the following results:

kN/m^2	lb/ft^2	e	kN/m^2	lb/ft^2	e
12	250	0.755	400	8350	0.724
25	520	0.754	800	16,700	0.704
50	1040	0.753	1600	33,400	0.684
100	2090	0.750	400	8350	0.691
200	4180	0.740	12	250	0.710

a. Plot the stress–void ratio curves on semilog coordinates.
b. Compute the compression index.
c. If the initial soil stress is 67 kN/m^2 or 1400 psf and the soil stratum is 3 m or 9.8 ft thick, how high can the stress become before the ultimate settlement is 20 mm or $\frac{3}{4}$ in.?

4.3 Consolidation tests on samples of soil yield the following void ratios for 100% consolidation:

kN/m^2	lb/ft^2	e	kN/m^2	lb/ft^2	e
5	104	1.85	800	16,700	1.22
25	520	1.82	1600	33,400	1.05
50	1040	1.77	500	10,440	1.10
100	2090	1.68	100	2090	1.20
200	4180	1.56	25	521	1.28
400	8350	1.39	5	104	1.38

a. Plot the stress–void ratio curves on both arithmetic and semilog coordinates.
b. Compute the compression index, C_c.
c. Find the change in void ratio when the soil stress is raised from 79 to 129 kN/m^2 or 1650 psf to 2700 psf, using stress–void ratio curve.
d. If the soil stratum in (c) is initially 2.5 m or 8.2 ft thick, compute its settlement.
e. If the soil has a coefficient of consolidation of 1.5×10^{-3} m^2/day or 1.6×10^{-2} ft^2/day and the stratum in (d) is drained on both

sides, compute the time required for 25, 50, and 75% consolidation.

4:4 A soil has a compression index, C_c, of 0.31. Its void ratio at an effective stress of $125\,kN/m^2$ or 2600 psf is 1.04, and its permeability is 4×10^{-7} mm/sec.

a. Compute the change in void ratio if the soil stress is increased to $187\ kN/m^2$ or $3900\ lb/ft^2$.

b. Compute the settlement in (a) if the soil stratum is 5 m or 16.4 ft thick.

c. Find the time required for 25, 50, 75, and 90% of settlement in (b) to occur.

4:5 An organic silt loaded in a consolidation test had the following stress–void ratio curve:

Stress			Stress		
kN/m^2	lb/ft^2	Void Ratio	kN/m^2	lb/ft^2	Void Ratio
100*	2090*	1.68	150	3130	1.590
60	1250	1.71	200	4175	1.515
30	625	1.75	400	8350	1.335
60	1250	1.715	800	16,700	1.116
80	1670	1.690	200	4175	1.220
100	2090	1.660	60	4175	1.270

* Existing effective stress at sample level.

a. Plot stress–void ratio curve for this soil, on log stress coordinates.

b. Replot as a stress–strain curve using log stress coordinates.

c. The stratum is 5 m or 16.4 ft thick and is overlain by sand 7 m thick weighing $19\ kN/m^3$ or $121\ lb/ft^3$ saturated. The groundwater level is 1 m below the ground surface initially, but the sand above the groundwater level is saturated by capillary rise. The silt weighs $16\ kN/m^3$ or $102\ lb/ft^3$. Compute the initial effective stress and neutral stress at the top, middle, and bottom of the organic silt stratum.

d. A sand fill 8 m thick weighing $20\ kN/m^3$ or $127.4\ lb/ft^3$ moist and compacted is placed on the ground surface. Compute the stress increase at the middle of the clay stratum, assuming the fill is wide enough that beneath the fill center the fill weight approximately equals the stress increase in the clay.

e. Compute the settlement of the organic silt from the stress increase.

f. Compute the intial increase in piezometric level at the center of the silt stratum when consolidation commences from the fill weight.

g. Compute the time required for 25, 50, 75, and 95% of the consolidation to develop if $c_v = 1.2 \times 10^{-2}$ m^2/day or 1.3×10^{-1} ft^2/day.

h. If the coefficient of secondary compression, C_α, is 0.1, compute the continuing settlement after primary consolidation is virtually complete (use 90% consolidation as t_1) during the ensuing 10 and 50 years.

4:6 The organic silt of Problem 4:5 has an effective size, D_{10}, of 0.001 mm. The effective void diameter causing capillary tension is $D_{10}/5$. Compute the maximum capillary tension from Equation (3:20) and find the void ratio at the shrinkage limit from total stress.

a. If the organic silt should shrink to that amount, how much would the stratum settle?

b. What would its new water content be if $G_s = 2.7$?

CHAPTER 5
Rigidity and Strength with Changing Stress and Environment

An excavation 15 m deep with a slope of $1(H)$ to $2(V)$ or 63° was made in a gently sloping hillside. Strength tests indicated that the soil would be strong enough at this slope to permit building a concrete retaining wall for permanent support. Work was delayed by a strike for a month. The soil in the steep exposed cut dried somewhat. Shrinkage cracks 1 to 2 m deep developed in the freshly cut steep face. The foundation for the wall had just been poured when heavy rain occurred. The steep bank, which had remained stable at the $1(H)$ to $2(V)$ slope, suddenly failed, dumping several tons of soil on top of workmen who were placing reinforcing steel for the wall stem. Two men were skewered on the upright steel bars and bled to death; one other was crushed. Despite this tragedy, 2 hours after the failure the contractor backed a 20-ton concrete truck to the brink of the steep cut to pour concrete into a portion of the forms that were undamaged. A second section failed but this time without loss of life.

A deep trench 3 m wide at the bottom with $1(H)$ to $1(V)$ slopes was cut into alternate seams of stiff sand and clay residual soil to provide a foundation cutoff for a dam. Tests of the soil showed it to be strong enough to be stable on such a slope. However, a month after an initial stage of excavation 10 to 15 m deep and 2 to 3 m above rock, work was temporarily stopped. A month later it was found that the earth trench bottom had bowed upward 0.1 to 0.2 m or 4 to 8 in.; the bottom of the trench had narrowed by about 0.1 m; and the straight line slopes bulged inward. There was no cracking or other signs of soil failure. The soils had undergone *creep*: slow shear strain at stresses substantially less than those causing rupture. When the trench was filled with compacted impervious clay, instruments showed no further movement in the virgin soil.

A large excavation for a large office building was 16 m deep: 4 m of sand underlain by horizontally bedded limestone with vertical joints. The sand was restrained by a cylinder wall drilled into the rock. The rock was so strong that it was blasted in 2 to 6 m vertical lifts, without support. A few months after the excavation had been completed and one end of the basement constructed, the entire rock face moved outward like a door, remaining vertical with the hinge provided by the completed basement. The sand beneath the adjoining street dropped into the open space behind the doorlike rock wall. Ground water, seeping through the sand–rock interface slowly filled joints in the limestone parallel to the excavation face. The joints had probably opened up slightly from elastic strain relief from the unsupported rock face and possibly from blasting vibration. Water pressure increased in the joint until the horizontal force exceeded the friction between limestone beds at the bottom of the excavation. Motion stopped when the block movement was enough to drain the water from the joints. The designers and the contractor had ignored seepage through the rock joints and perched water in the sand during rainy weather. The problem was possibly aggravated by small strain relief movements that cracked water and sewer lines in the street, augmenting the available water. Further failure was prevented by horizontal drain holes drilled into the rock to remove water from the joints.

These three examples illustrate some of the real problems of strain and rupture induced by mass loading. In the first case, soil that was initially strong developed shrinkage cracks which destroyed tensile strength and permitted water pressure to enhance the stresses in the bank. Failure was delayed months after the excavation was made that commenced the process of soil change. In the second case, soil creep caused a costly reevaluation of the project, although there was no danger of failure. In the third case, a wall of strong dry rock deceived the engineer and contractor into overlooking small quantities of seepage and ignoring strain relief and its effects on the joints, water mains, and sewers.

Analysis and design must consider the geologic discontinuities of the formations, the possible effects of strain, and in all cases, the effects of water pressure. These concepts are discussed in the following sections.

5:1 Combined Stresses

Although the simple case of one-dimensional stress and compression describes many problems of settlement, many other problems of soil deformation and failure involve three-dimensional stresses. The simple cases of one direction of compression and tension, so useful in structural design of steel and concrete, are of little importance in the soil mass whose weight is a substantial part of the total load and where the structural loads are introduced in several directions and at different levels within the mass. Therefore, analysis of the effects of stress must begin with the entire stress field in three dimensions.[5:1]

A stress is defined as a force per unit of area. A stress applied to a plane surface of a solid can be resolved into two components: one perpendicular (normal) to the plane known as the *normal* stress, σ (sigma) and one acting in the surface of the plane known as the *shear* stress, τ (tau), as shown in Fig. 5.1a. When the stress acting on a plane consists only of a *normal* component and $\tau = 0$, that normal stress is termed a *principal* stress (Fig. 5.1b).

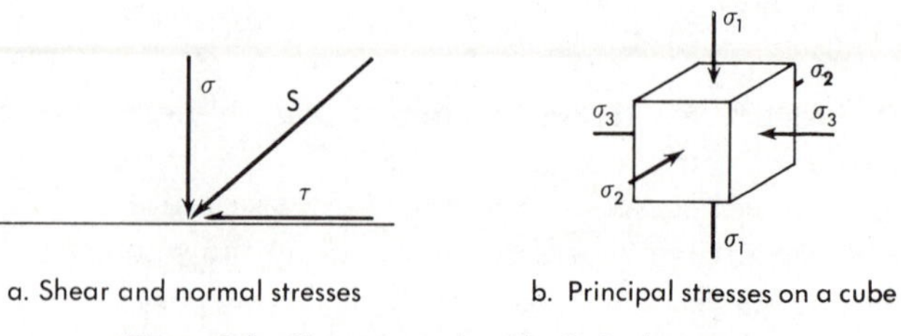

Figure 5.1 *Shear, normal, and principal stresses.*

When a cube of rock or mortar is to be tested to determine its strength, it is placed in a testing machine, and gradually increasing compressive forces are applied to its top and bottom faces. The compressive forces result in compressive stresses in the faces to which they are applied. These stresses are principal stresses, and the horizontal planes in which they occur are called *principal planes.* Although it is rarely done, it is possible to introduce compressive forces on the other two pairs of cube faces. These also produce principal stresses in the faces to which they were applied, and these faces would also be considered principal planes. Mechanics texts demonstrate that there are three independent, perpendicular, principal stresses acting on three perpendicular, principal planes. The largest of the three principal stresses is the *major principal* stress and is denoted σ_1. The smallest is the *minor principal* stress σ_3, and the third the *intermediate principal* stress σ_2.

In soil mechanics, tensile stresses are comparatively rare; therefore, to avoid many negative signs, compressive stresses are considered positive. In the case of the cube of mortar in an ordinary compression test, the compressive stress applied to the top and bottom faces is σ_1, and the other two principal stresses σ_2 and σ_3, are zero.

If an inclined plane cuts through the cube, it is possible to compute the shear and normal stresses on that plane from the three principal stresses and the laws of statics. The general case is complicated, for it involves the direction cosines of the plane from the principal planes. In many problems in soil mechanics, however, we are interested in stresses on planes perpendicular to the intermediate principal plane, which reduces the problem to two dimensions.

The direction of an inclined plane that is perpendicular to the intermediate principal plane is defined by θ, the angle the plane makes with the plane of the major principal stress, as shown in Fig. 5.2.

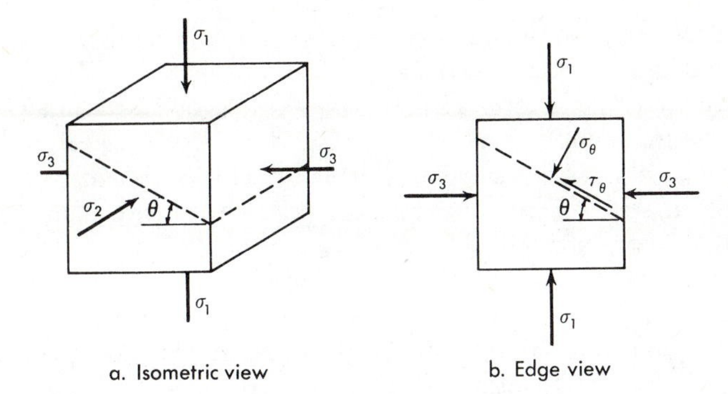

Figure 5.2 *Stresses in a cube cut by a plane that is perpendicular to the plane of σ_2 and makes an angle of θ with the plane of σ_1.*

Shear and normal stresses on the plane can be computed by the laws of statics from σ_1 and σ_3. If the cube is assumed to have dimensions $1 \times 1 \times 1$, then the forces acting on the plane in the directions of σ_1 and σ_3 are, respectively,

$$F_1 = \sigma_1 \times 1 \times 1, \qquad F_3 = \sigma_3 \times 1 \times 1 \tan \theta$$

The sum of the components of these forces normal to the plane is

$$F_n = F_1 \cos \theta + F_3 \sin \theta$$

$$F_n = \sigma_1 \cos \theta + \sigma_3 \tan \theta \sin \theta$$

The sum of the components parallel to the plane is

$$F_s = \sigma_1 \sin \theta - \sigma_3 \tan \theta \cos \theta$$

The area of the plane is $1/\cos \theta$; therefore, the normal stress on the plane of σ_θ, is

$$\sigma_\theta = \frac{\sigma_1 \cos \theta + \sigma_3 \tan \theta \sin \theta}{1/\cos \theta}$$

$$\sigma_\theta = \sigma_1 \cos^2 \theta + \sigma_3 \sin^2 \theta$$

$$\sigma_\theta = \frac{\sigma_1 + \sigma_3}{2} + \frac{\sigma_1 - \sigma_3}{2} \cos 2\theta \qquad (5:1a)$$

In the same way the shear stress in the plane, τ_θ, is

$$\tau_\theta = \frac{\sigma_1 - \sigma_3}{2} \sin 2\theta \qquad (5:1b)$$

By these expressions stresses on any inclined plane at an angle of θ can be computed, or if the stresses on any two planes are known, the principal stresses can be found. The expressions also lead to the following conclusions:

1. The maximum shear stress occurs when $\sin 2\theta = 1$ or $\theta = 45°$ or $135°$ and is equal to $(\sigma_1 - \sigma_3)/2$.
2. The maximum normal stress occurs when $\cos 2\theta = 1$ and $\theta = 0$.
3. The minimum normal stress occurs when $\cos 2\theta = -1$ and $\theta = 90°$ and the plane is parallel to the minor principal plane.
4. Shear stresses are equal in magnitude on any two planes perpendicular to each other.

MOHR CIRCLE / A German physicist, Otto Mohr, devised a graphical procedure for solving the equations for shear and normal stress on a plane perpendicular to one principal plane and making an angle θ with the larger of the two other principal planes. A system of coordinate axes is established (Fig. 5.3a), where the x distances represent normal stresses and the y distances represent shear stresses. Compressive (positive) normal stresses are plotted to the right; tensile, to the left. Shear stresses are plotted upward (+) if they produce a counterclockwise couple; downward (−) if clockwise. The coordinates of a point (σ, τ) represent the combination of shear and normal stress on a plane regardless of the plane's orientation.

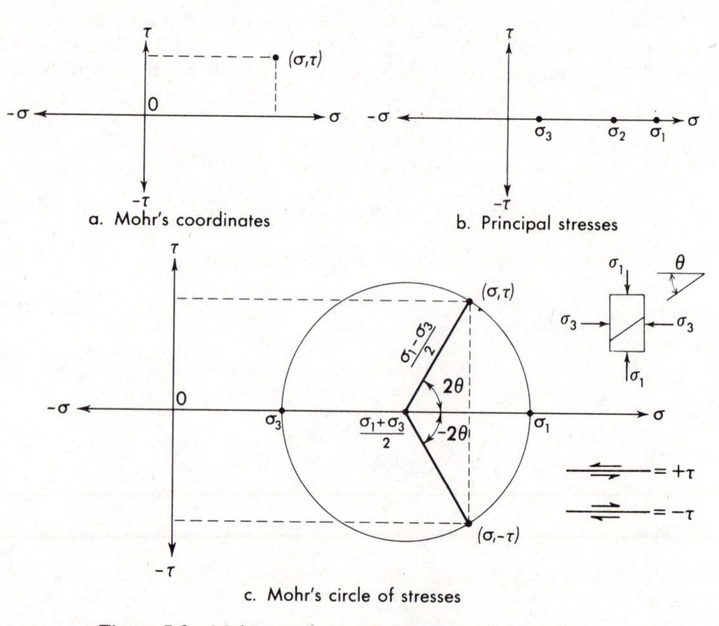

Figure 5.3 *Mohr coordinates and Mohr's circle of stresses.*

On this diagram are plotted the coordinates of σ_1 and σ_3 (Fig. 5.3b). Both lie on the σ axis because the shear stresses on the principal planes are zero. Through these points a circle is drawn whose center is also on the σ axis (Fig. 5.3c). The center of this circle is at the point $[(\sigma_1 + \sigma_3)/2, 0]$ and its radius is equal to $(\sigma_1 - \sigma_3)/2$. A radius is drawn at an angle of 2θ measured counterclockwise from the σ axis. The x coordinate of a point on the circle at the end of the radius is

$$\frac{\sigma_1+\sigma_3}{2}+\frac{\sigma_1-\sigma_3}{2}\cos 2\theta, \ =\sigma_\theta \tag{5:1a}$$

The y coordinate of the point is

$$\frac{\sigma_1-\sigma_3}{2}\sin 2\theta, \ =\tau_\theta \tag{5:1b}$$

Therefore, the circle represents all the possible stress combinations on any plane perpendicular to the intermediate principal plane. The stresses on a particular plane at angle θ can be found from the construction. The construction also shows that τ_{max} occurs on a plane with an angle of $2\theta=90°$ and is equal to $(\sigma_1-\sigma_3)/2$, or half the difference between the major and minor principal stresses. Also the shear stresses on two planes perpendicular to each other are equal but opposite sign.

The same construction can be applied to stresses on a plane that is perpendicular to the major principal plane, by using σ_2 and σ_3, or to a plane perpendicular to the minor principal plane, by using σ_1 and σ_2.

In three dimensions (Fig. 5.4) there are three circles, each representing the stresses on a plane perpendicular to one of the principal planes. The area between the circles represents the combined stresses on planes that are oblique to all three principal planes. These can be computed analytically, using the same reasoning as employed for Equations (5:1a) and (5:1b), as described in texts on the theory of elasticity and advanced strength of materials. The greatest shear stress is defined by the $\sigma_1-\sigma_3$ stress circle, which describes the stress combinations in any plane that is oblique to the major and minor principal planes and perpendicular to the intermediate. Therefore, the $\sigma_1-\sigma_3$ circle, and the planes it represents, are of greatest significance in most real problems of soil strength and failure.

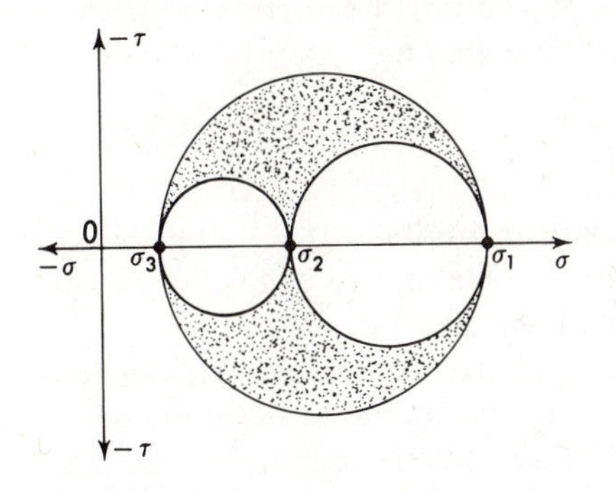

Figure 5.4 *Mohr circles for three-dimensional stress field defining all possible combinations of normal and shear stress within the shaded zone.*

ORIENTATION OF PLANES—THE POLE / A simple construction, involving a point on the circle, the *pole*, makes it possible to determine the orientation of the planes graphically. Figure 5.5a shows a Mohr circle and the stresses $(\sigma_\theta, \tau_\theta)$ on a plane, making an angle of $\theta = 35°$ with the major principal plane. The angle 2θ is plotted counterclockwise from the σ axis about the circle center. By geometry, the angle between any two lines intersecting on the circle perimeter and subtending the same arc as the central angle 2θ, will be equal to θ. This angle θ also is the angle between a line drawn from $(\sigma_\theta, \tau_\theta)$ to σ_3 and the σ axis.

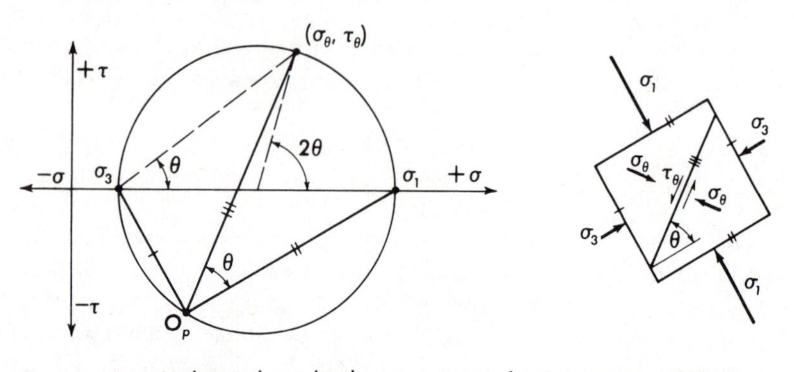

Figure 5.5 *Mohr circle with pole O_p to find the orientation of the planes.* (Note that the tic marks on the lines and planes denote those that are parallel.)

The orientation of the prism on which σ_1 and σ_3 act is shown in Fig. 5.5b. A line is drawn through σ_3 parallel to the plane of σ_3, intersecting the circle at point, O_p, the *pole*. A similar line is drawn from σ_1. It will be parallel to σ_1, making a 90° angle with the first line because the angle $\sigma_3 O_p \sigma_1$ subtends an 180° arc. A third line is then drawn from O_p to $(\sigma_\theta, \tau_\theta)$. This subtends the same arc as $(\sigma_\theta, \tau_\theta)\sigma_3\sigma_1$; therefore, the angle $(\sigma_\theta, \tau_\theta)O_p\ \sigma_1$ is also equal to θ. A plane cutting the prism of Fig. 5.5b parallel to $O_p(\sigma_\theta, \tau_\theta)$, also makes an angle of θ with the major principal plane.

Example 5:1

Given $\sigma_1 = 700\ \text{N/m}^2$ or $14{,}600\ \text{lb/ft}^2$, $\sigma_3 = -200\ \text{N/m}^2$ or $-4180\ \text{lb/ft}^2$. Find σ and τ on a plane making an angle of 50° with the major principal plane. Find the maximum shear stress.

1. Plot a Mohr circle through σ_1 and σ_3 (Fig. 5.6a) and a two-dimensional representation of the stress plane prism (Fig. 5.6b).
2. Plot $2\theta = 100°$ counterclockwise from the σ axis.
3. Find σ_θ and τ_θ graphically: $\sigma = 172\ \text{kN/m}^2$ and $\tau_\theta = 443\ \text{kN/m}^2$.
4. Alternatively, repeat step 1.

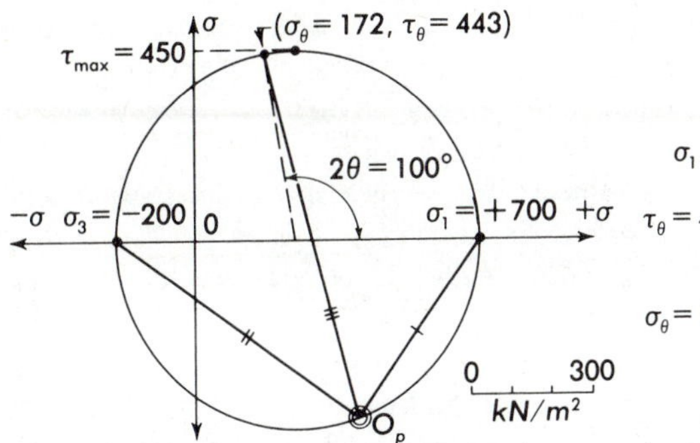

a. Example 5.1 Mohr circle and poles

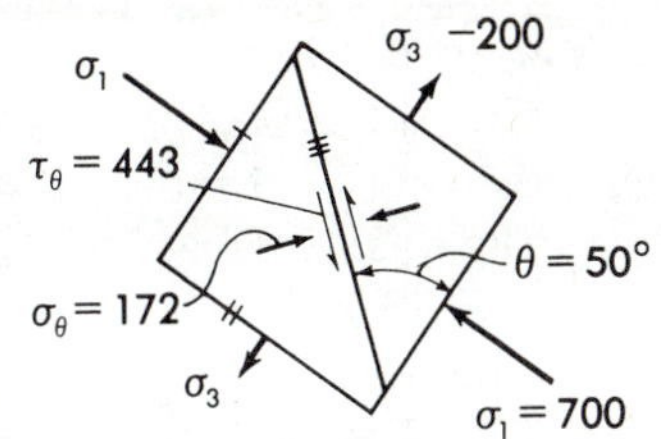

b. Example 5.1 stress orientation

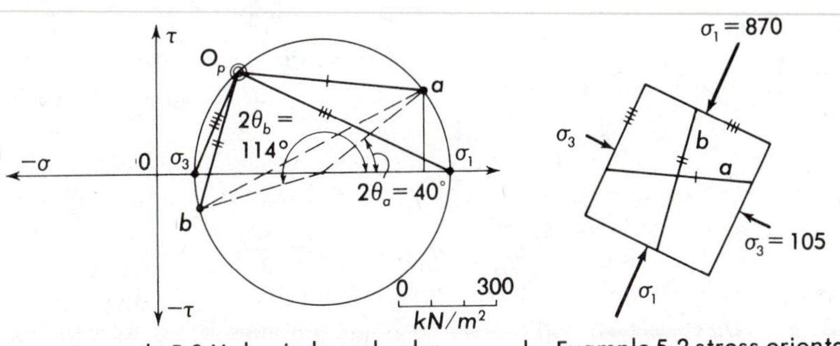

c. Example 5.2 Mohr circle and poles

b. Example 5.2 stress orientation

Figure 5.6 *Computing stresses and plane orientation with a Mohr circle and poles.*

5. Find pole O by drawing lines σ_1–O in Fig. 5.6a parallel to plane of σ_1 in Fig. 5.6b and σ_3–O in Fig. 5.6a parallel to plane θ in Fig. 5.6b.
6. Draw a line from the pole to the circle in Fig. 5.6a parallel to plane θ in Fig. 5.6b.
7. The intersection with the circle is $(\sigma_\theta, \tau_\theta)$: $\sigma_\theta = 172\ \text{kN/m}^2$ and $\tau_\theta = 443\ \text{kN/m}^2$, graphically.
8. Compute σ_θ and τ_θ by Equations (5:1a) and (5:1b).
   a. $2\theta = 100°$ $\sin 2\theta = 0.9848$ $\cos 2\theta = -0.1735$.
   b. $(\sigma_1 + \sigma_3)/2 = (700–200)/2 = 250\ \text{kN/m}^2$; $(\sigma_1–\sigma_3)/2 = (700-(-200))/2 = 450\ \text{kN/m}^2$.
   c. $\sigma_\theta = (\sigma_1 + \sigma_3)/2 + [(\sigma_1 - \sigma_3)/2] \cos 2\theta = 250 + 450(-0.1735) = 172\ \text{kN/m}^2$.
   d. $\tau_\theta = (\sigma_1 - \sigma_3)/2 \sin 2\theta = 450 \times 0.9848 = 443\ \text{kN/m}^2$.
9. Maximum shear stress $= (\sigma_1 - \sigma_3)/2 = 450\ \text{kN/m}^2$.

Example 5.2

Given plane a: $\sigma_a = 800\ \text{kN/m}^2$, $\tau_a = 235\ \text{kN/m}^2$.
Given plane b: $\sigma_b = 125\ \text{kN/m}^2$, $\tau_b = -100\ \text{kN/m}^2$.

Plane a is oriented as in Fig. 5.6d. Find σ_1, σ_3, θ_a, θ_b and the angle between the planes.

1. Plot $\sigma_a = 800$, $\tau_a = 235$, $\sigma_b = 125$ and $\tau_b = 100$ in Fig. 5.6c, adjacent to plane a and its orientation in Fig. 5.6d.
2. Find the center of the Mohr circle: the intersection of the perpendicular bisector of line a–b with the σ axis.
3. The intersection of the Mohr circle and the σ axis defines σ_1 and σ_3: $\sigma_1 = 870\ \text{kN/m}^2$, $\sigma_3 = 105\ \text{kN/m}^2$
4. Draw line a–O across the circle of Fig. 5.6c parallel to plane a in Fig. 5.6d. Pole O is the intersection of the line with the circle.
5. Draw lines O–b, O–σ_1, and O–σ_3 in Fig. 5.6c. Draw the planes of b, σ_1, and σ_3, respectively, in Fig. 5.6d parallel to the polar lines of Fig. 5.6c.
6. Determine the θ's (the angle between the plane and the major principal plane) directly in Fig. 5.6d or by measuring 2θ's in Fig. 5.6c. $\theta_a = 40°$; $\theta_b = 114°$.

CONJUGATE SHEAR / From the Mohr circle it can be seen that two planes with equal normal stresses and shear stresses that are equal in magnitude but opposite in sign exist in any given stress field. The angles of the planes with respect to the major principal plane are equal in magnitude but opposite in sign. These stresses are *conjugate stresses* and their planes. *conjugate planes*: one is in the upper half of the circle, the other in the bottom half. Each is a mirror image of the other. For solving many problems, only the top half of the circle is plotted; the conjugate stresses on the lower half are implied.

5:2 Strain and Failure

When stresses are applied to any material, including soil, it first undergoes deformation. The nature of the deformation depends on the resistance of the material and the combination of stresses. If the stresses are increased further, a point is reached at which the material fails to resist the increase. At this point, termed *failure*, different materials, including soils, react differently—some disintegrate, while others deform continuously with little or no stress increase. Soils with their three-phase composition exhibit a wider variety of deformation characteristics and more complex failure behavior than do the simpler engineering materials such as steel or concrete. Deformation is not always proportional to stress; moreover, it changes with time and environment. The resistance to failure also depends on the stress field, the environment, and time; further it is often difficult to define a point of failure. In spite of these differences between soil and rock and the simpler materials, the idealized concepts of applied mechanics, such as modulus of elasticity, Poisson's ratio, and Mohr's rupture theory, can be utilized in soil and rock engineering. These concepts must be viewed as approximations but if their limitations are understood they are useful in solving real problems.

ELASTICITY / If a normal stress increment, $\Delta\sigma_z$, is applied to a prism of soil (Fig. 5.7a), it will deform an amount ΔH in the direction of the stress increase. The direct strain increment, $\Delta\varepsilon_z$, is found from the deformation

$$\Delta\varepsilon_z = \frac{\Delta H}{H} \tag{5:2a}$$

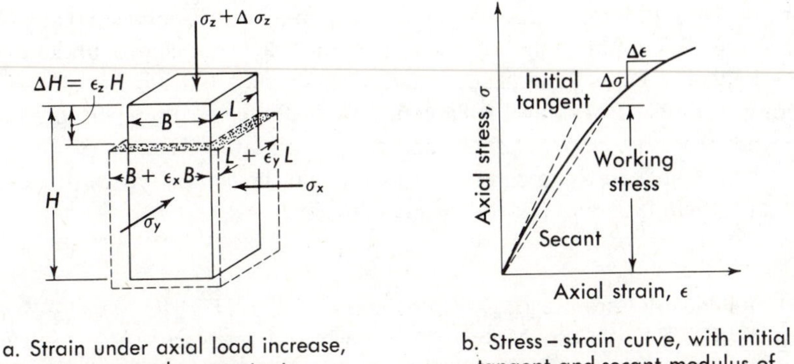

a. Strain under axial load increase, $\Delta\sigma_z$; σ_x and σ_y constant

b. Stress – strain curve, with initial tangent and secant modulus of elasticity as slopes of tangent and secant, respectively

Figure 5.7 *Vertical and lateral strain and stress–strain relationships.*

This is represented by a stress–strain curve (Fig. 5.7b). The ratio of the stress increment to the strain it produces is the *modulus of elasticity*, *E*. The general expression is

$$E = \frac{\Delta\sigma}{\Delta\varepsilon} \quad \text{or} \quad \frac{d\sigma}{d\varepsilon} \quad \left(\text{or} \quad E_z = \frac{\Delta\sigma_z}{\Delta\varepsilon_z}\right) \tag{5:2b}$$

Geometrically, it is the slope of the stress–strain curve. The modulus of elasticity of most soils and rocks is not a constant throughout the possible stress range and not quite the same for unloading as for loading. For this reason, the term modulus of deformation, *M*, defined in the same way as *E*, is used by some writers to call attention to the inelastic behavior of soils and rock. In this text *E* will be used with the understanding that it is neither a constant nor a unique function of the load applied. The range in *E* for soils and rock is almost limitless, from virtually zero for peats to greater than concrete for sound rock.

The value of *E* at very small strains, the *initial tangent*, E_T, is defined by a tangent to the stress–strain curve at its beginning, (Fig. 5.7b). The average value of *E* for a specified range of stress is the *secant modulus* E_s. The stress range is that for the particular problem (the working stress). Because E is a variable, it is always necessary to define the stress field and stress range for which a particular value applies.

POISSON'S RATIO / The stress increment $\Delta\sigma_z$ also produces a bulging in the lateral dimensions, ΔB and ΔL, and corresponding lateral strains ε_x and ε_y. The ratio of the lateral to the direct strain is Poisson's ratio, ν

$$\nu = \frac{-\varepsilon_x}{\varepsilon_z} = \frac{-\varepsilon_y}{\varepsilon_z} \tag{5:3}$$

The range in Poisson's ratio for ideal elastic materials is between 0 and 0.5. The value 0.5 implies a material whose volume does not change under load, like jelly. The value 0 implies no bulging under load, like a prism of cork or sponge rubber. Rocks and soils generally lie between these limits, but v exceeds 0.5 in some dense sands and preconsolidated clays.

The change in volume per unit of volume is the *volumetric strain*, ε_v. It is equal to the algebraic sum of the linear strains

$$\varepsilon_v = \varepsilon_x + \varepsilon_y + \varepsilon_z \tag{5:4}$$

For the consolidation test $\varepsilon_x = \varepsilon_y = 0$ and $\varepsilon_z = \Delta H/H$.

FAILURE / Otto Mohr also contributed to engineering science a theory of the failure of materials that represents more nearly the stress field involved than do the theories involving simple stresses only. It has been found to apply particularly well to soils, concrete and rock.

Mohr reasoned that yield or failure within a material was caused by critical combinations of shear and normal stresses. The failure occurs by shear, but the critical shear stress is governed by the normal stress acting on the potential failure surface.

The locus of the critical combinations of shear and normal stress, plotted on the σ, τ coordinates, define a pair of lines known as the *Mohr envelope of failure* (Fig. 5.8). Failure will occur if, for a given value of σ, the shear stress reaches that shown by the envelope.

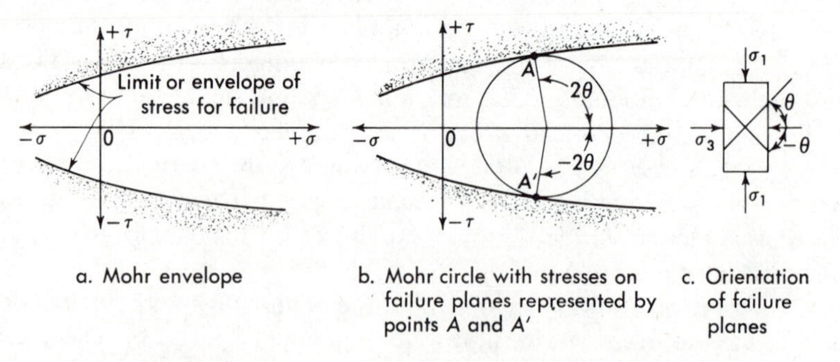

Figure 5.8 *Mohr envelope of failure and Mohr circle of stresses at failure.*

If the stresses on any two planes through a point are known, the stresses can be found on any other planes by means of the Mohr circle. Because the circle represents *all* possible combinations of shear and normal stress at that point, failure will occur on the plane represented by the point of intersection with the envelope (Fig. 5.8b). If the envelope is symmetrical about the σ axis, as it is for many materials, and if the material is homogeneous, failure will occur simultaneously on sets of conjugate planes all making angles of θ with the major principal plane, and no single surface of failure can be identified. If there is a small zone of weakness, failure will be localized in that plane passing through the zone; the multiple conjugate surfaces may not be evident.

Example 5:3

A cylindrical sample of soil cement is loaded on its ends until it fails in unconfined compression. Previous tests of similar samples found the Mohr envelope to be a pair of straight lines, making angles of 25° with the σ axis and intercepting the τ axis at 670 kN/m^2 or 96 lb/in.^{2}, as seen in Fig. 5.9a. Find (1) the axial stress (compressive strength) when failure occurred, (2) the normal and shear stresses on the failure planes, (3) the angle of the failure plane with respect to the major principal plane, and (4) the tensile strength.

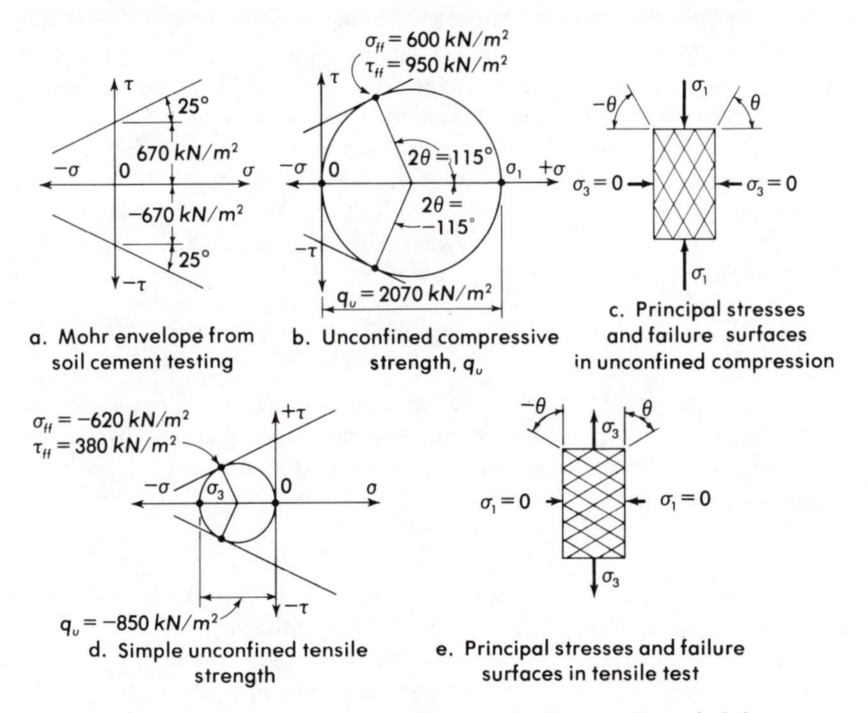

Figure 5.9 *Stresses during compression test of soil cement, Example 5:3.*

1. Draw the Mohr envelope, Fig. 5.9.
2. For unconfined compression or tension testing the forces acting on the sides of the specimen are zero.
3. At the beginning of the unconfined compression test the Mohr circle is a point where $\sigma_1 = \sigma_3 = 0$. During the test the axial load σ_1 increases, while σ_3 remains 0.
4. By trial find the largest circle that will pass through $\sigma_3 = 0$ and touch the envelope, Fig. 5.9b.
5. From the diagram, at failure, $\sigma_{1f} = q_u = 2070\ \mathrm{kN/m^2}$ or $300\ \mathrm{lb/in.^2}$. $\sigma_{ff}(\sigma_\theta$ at failure$) = 600\ \mathrm{kN/m^2} = 87\ \mathrm{lb/in.^2}$, and $\tau_{ff} = \pm 950\ \mathrm{kN/m^2}$ or $\pm 79\ \mathrm{lb/in.^2}$
6. From the diagram $\theta = \pm 57.5^0$, Fig. 5.9c.
7. By trial find the largest circle that can be drawn with $\sigma_1 = 0$ and σ_3 becoming increasingly negative or tensile and touching the same envelope, Fig. 5.9d.
8. From the diagram $\sigma_{3f} = 850\ \mathrm{kN/m^2} = -123\ \mathrm{lb/in.^2}$. The angle of the failure planes are the same, $\theta = \pm 57.5°$, but oriented differently with respect to the sample because the plane of σ_1, is now vertical.

NOTE: The normal and shear stresses on the *failure* plane are usually designated σ_{ff} and τ_{ff}, respectively. The difference between compressive principal stresses, $\sigma_1 - \sigma_3$, at failure is termed q_c, the *compressive strength* if $\sigma_3 > 0$. If $\sigma_3 = 0$, q_c is often denoted q_u, the *unconfined compressive strength.*

ADVANCED FAILURE CRITERIA / The Mohr theory of rupture implies that the intermediate principal stress has no influence on failure. High-pressure testing has shown that this is not always true. Although the effect is small, research requires a method of representing failure that includes the intermediate. One form is the principal stress plot (Fig. 5.10) in which all three principal stresses are shown. The stress field can be represented by normal and shear stresses on the surface of an octahedron. The octahedral normal stress, σ_{oct}, is defined by

$$\sigma_{oct} = \frac{\sigma_1 + \sigma_2 + \sigma_3}{3} \tag{5:5a}$$

This is represented by a vector equidistant from the three perpendicular axes, the *hydrostatic axis.* The octahedral normal stress is equal to the length of the vector, **ON**, multiplied by $1/\sqrt{3}$. The second component of stress is the *octahedral shear*, τ_{oct}:

$$\tau_0 = \tfrac{1}{3}\sqrt{(\sigma_1 - \sigma_2)^2 + (\sigma_2 - \sigma_3)^2 + (\sigma_1 - \sigma_3)^2} \tag{5:5b}$$

This can be represented by a vector perpendicular to the hydrostatic axis, *NN'*. The value $\tau_0 = NN'/\sqrt{3}$. The plane perpendicular to the hydrostatic axis is the *octahedral plane.* For any three-dimensional system, eight such planes are possible. These define an octahedron, from which the name is derived.

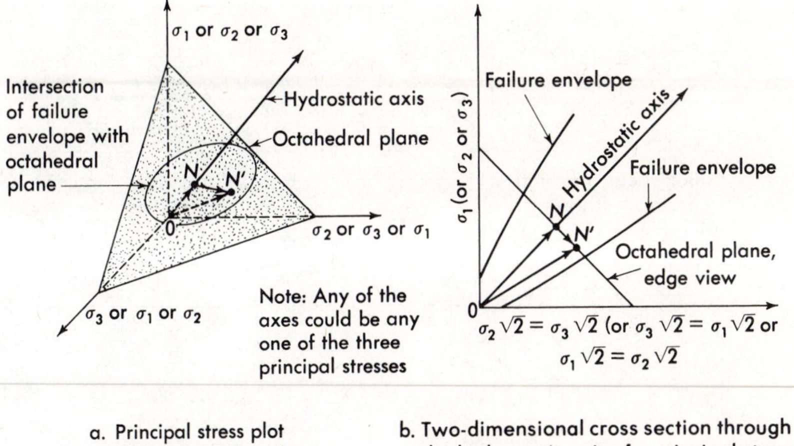

a. Principal stress plot ON′ represents a stress: $\overline{ON'}/\sqrt{3} = \sigma_o$; $\overline{NN'}/\sqrt{3} = \tau_o$

b. Two-dimensional cross section through the hydrostatic axis of a principal stress plot with $\sigma_2 = \sigma_3$.

Figure 5.10 *Principal stress axes and stress representations.*

For any given octahedral normal stress, failure occurs when the octahedral shear stress reaches a limiting value, determined by experiment. The limiting values define a three-dimensional envelope, centered about the hydrostatic axis, but not necessarily symmetrical to it.

Because of the problems of representing the three-dimensional plot on paper, two-dimensional "cross sections" are frequently employed (Fig. 5.10b). These are limited to the cases where two of the principal stresses are equal.

The Mohr criteria for failure and the Mohr envelope are sufficient for the solution of most problems in soil and rock engineering. The more sophisticated representations make it possible to evaluate the effects of the intermediate principal stress and to determine the possible error in the Mohr approximation.

5:3 Methods of Making Shear Tests[5:1, 5:2]

Because of the complex nature of the shearing resistance of soils, many methods of testing have been tried with varying success. The principal shear tests in use today are *direct*, *ring or double direct*, and *triaxial*. Of these, triaxial testing gives the most consistent and reliable results with varying soils.

DIRECT SHEAR TEST[5:2] / One of the earliest methods for testing soil strength, used extensively today, is direct shear. A sample of soil is placed in a rectangular box (Fig. 5.11a), the top half of which can slide over

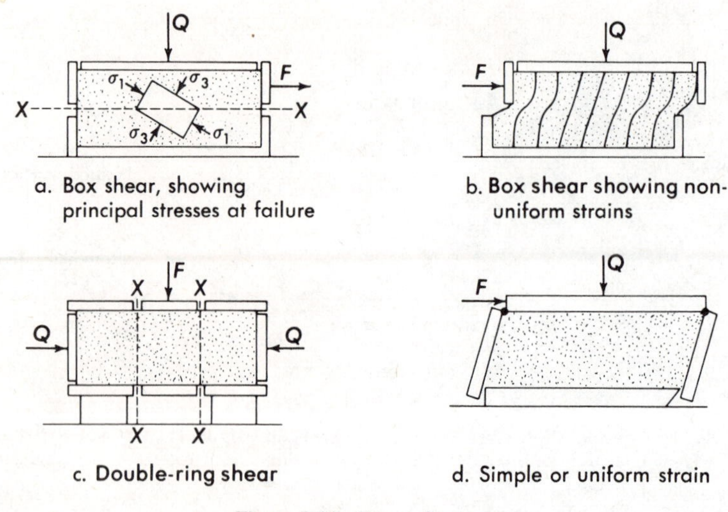

Figure 5.11 *Direct shear tests.*

the bottom half. The lid of the box is free to move vertically, and to it is applied the *normal load, Q. A shearing force, F,* is applied to the top half of the box, shearing the sample along plane *x—x*. In practice, the top and bottom of the box may be either porous plates to permit changes in the water content of the sample or projecting vanes to help develop a uniform distribution of stress on the failure surface. The test utilizes a relatively thin sample, which consolidates rapidly under load (when such consolidation is required). The sample preparation and test operation are simple for most soils, which makes the test attractive for routine work.

Inherent shortcomings limit the reliability of the test results. First, there is an unequal distribution of strains over the shear surface; the strain is more at the edges and less at the center, as shown in Fig. 5.11b. The result is progressive failure. In materials with highly developed structures, such as flocculent clays and cemented or very loose cohesionless soils, the strength indicated by the test will often be too low. Second, the soil is forced to shear on a predetermined plane, which is not necessarily the weakest one. The strength given by the test, therefore, may be too high. Finally, it is difficult to control drainage or changes in water content during the test, which limits its usefulness in wet soils.

RING SHEAR / The ring shear is a double-direct shear test. A cylindrical sample is supported laterally by a close-fitting metal tube (Fig. 5.11e). Normal pressures are applied to the sample by pistons on the ends. A section of the tube is forced downward, shearing the soil on two surfaces, *x—x*. This equipment makes it possible to control sample water content changes more closely than in the single direct shear, and, in addition, small-diameter samples can be used. It suffers from the same limitations of nonuniform stress distribution and a forced failure plane as does the direct shear test.

SIMPLE SHEAR / Several devices have been designed to produce more uniform shear strains in direct shear. One form (Fig. 5.11d) employs tilting sides on the shear box. Although it does develop uniform strain, it does not permit the localized increase in strain that accompanies failure. Further, it introduces stress concentrations within the mass. Another device employs a reinforced rubber tube for the walls of the shear box. This allows the irregular strains that develop in natural soils, but does not fully prevent lateral expansion. Both are research tools, and not well adapted to general shear testing.

PLANE STRAIN / The strains in the direct shear tests take place in two directions, vertically and parallel to the direction of shear. This condition is termed *plane strain*. It is similar to the strains developed in many real problems such as very long foundations or long walls, which can be represented by a two-dimensional cross section.

TRIAXIAL SHEAR TEST / The most reliable shear test for routine work is the triaxial direct stress (Fig. 5.12). A cylindrical sample is used with a diameter of 36 mm, 72 mm, or more and a length of at least twice the

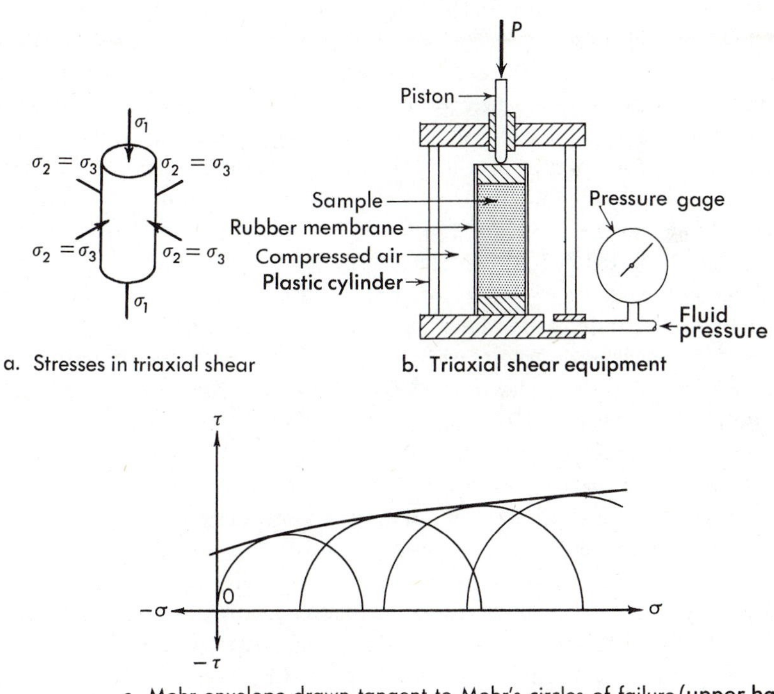

Figure 5.12 *Triaxial shear test.*

diameter. The sample is encased in a rubber membrane, with rigid caps or pistons on both ends. It is placed inside a closed chamber and subjected to a confining pressure σ_3 on all sides by air or water pressure. An axial stress σ_1 is applied to the end of the sample by a piston. Either the axial stress can be increased or the confining pressure decreased until the sample fails in shear along a diagonal plane or a number of planes. The Mohr circles of failure stresses for a series of such tests, using different values for σ_3, are plotted, and the Mohr envelope drawn tangent to them (Fig. 5.12c).

An alternate procedure is to hold the axial stress constant and increase the confining pressure until the sample bulges upward in the axial direction. In this form, the *triaxial extension* test, the confining pressure is $\sigma_1 = \sigma_2$ and the axial stress is σ_3. This is sometimes used to simulate the effect of a lateral thrust on a mass of soil. The Mohr envelope is similar to that for the compression test in a homogeneous isotropic soil; in stratified materials it is often different.

A special case of the triaxial shear is the *unconfined compression test*, in which $\sigma_3 = 0$.

The important advantages of the method are the relatively uniform stress distribution on the failure plane and the freedom of the soil to fail on the weakest surface. Furthermore, water can be drained from the soil or forced through the soil during the test to simulate actual conditions in the ground. Sample preparation is simple, and small-diameter cylindrical samples can be used. The chief disadvantage is the elaborate equipment required, including sample membranes, compressed air or water-pressure equipment, the triaxial cell itself, and auxiliary devices to measure the volume change of soil during testing. The conventional triaxial test utilizes rigid end caps. These restrain the shear and cause stress concentrations that change the conditions in failure. It is limited to values of $\sigma_2 = \sigma_3$ or $\sigma_2 = \sigma_1$ in compression and extension, respectively.

Special triaxial tests have been developed that utilize rectangular specimens with independently controlled stresses on all three principal planes.[5:2] One variant of such testing prevents deformation in the direction of the intermediate principal stress, a state of plane strain useful in simulating the loading conditions in two-dimensional problems. A triaxial test in which the sample is in the form of a hollow cylinder can induce a variable intermediate principal stress by maintaining the inside pressure different from the outside. Such hollow cylinders have also been subjected to torsion in order to measure the shear modulus of elasticity directly.[5:2]

The rectangular, plane strain, and hollow cylinder triaxial tests are primarily research tools for studying the mechanisms of soil behavior and minimizing some of the shortcomings of the conventional triaxial. In spite of the limitations, the triaxial tests utilizing $\sigma_2 = \sigma_3$ and axial compression continue to be the most useful methods for simulating soil and rock behavior for a wide range of engineering problems.

5:4 Strain and Strength of Dry Cohesionless Soils[5:3–5:5, 5:7, 5:8]

Cohesionless soils are composed of bulky grains; ranging in shape from angular to well rounded. A simplified representation of such a material subject to normal and shear forces, Q and F, is shown in Fig. 5.13. The particles are in contact at only a few points at which the stresses are extremely high, far greater than the average stresses on the soil $\sigma = Q/A$ and $\tau = F/A$.

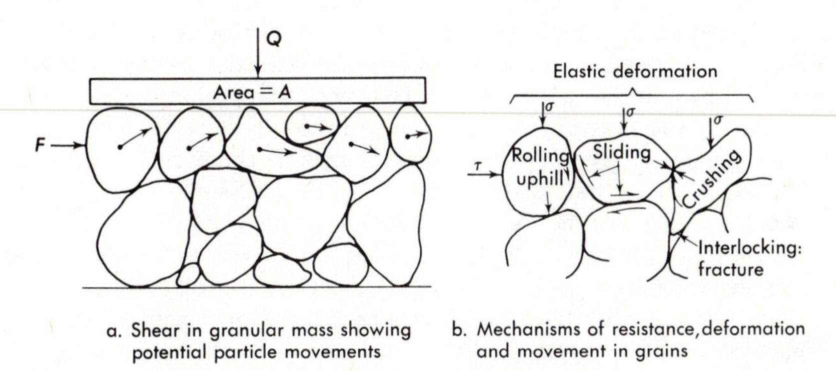

Figure 5.13 *Mechanisms for deformation and shear in a soil consisting of bulky grains.*

If the shear stress is increased, the soil particle system responds in several ways simultaneously (Fig. 5.13b). First, the particles deform more or less elastically. Although it might seem that a solid grain of quartz is exceedingly rigid, even small stress changes induce high localized stresses and strains in each particle. Second, there is local crushing at the most highly stressed points of contact. Third, both the elastic distortion and the crushing cause slight translation and rotation of the grains, increasing the size of some of the voids and decreasing others. The vector sum of all the small movements of each particle is the deformation of the mass, ordinarily described in terms of strain.

Both the previous confining stress and the stress level at the beginning of the stress increment influence the strain. The higher the degree of confinement, the greater the previous crushing and local adjustments, and therefore the less the additional strain produced by additional increment of shear stress.

If the shear stress is increased still further, two additional responses are evident. First, the particles tend to roll across one another (Fig. 5.13b). The resistance depends on their angle of contact and is proportional to the confining stress σ. The total resistance to rolling is the statistical sum of the behavior of all the particles: Some roll up, some roll down, but all do not move simultaneously. The second mechanism is the sliding of one grain across the other. The resistance to sliding is essentially friction, which is proportional to the confining stress. A third mechanism involves the interference and interlocking of the corners of the more angular irregular particles.

If the shear stress becomes sufficiently large, the statistical effect of the distortion, crushing, shifting, rolling, and sliding of the grains will be continuous movement and distortion of the soil body, or *shear failure*.

Moisture does not directly influence these mechanisms in most cohesionless soils because the intense stresses at the contact points between grains force the water molecules aside. In exceptional soils, such as porous volcanic ash or sands containing talc or chlorite, the particles may be weakened by moisture and thus their resistance to stress will be altered.

From the engineering point of view it is not necessary to identify nor evaluate quantitatively the contribution of each mechanism. Instead, their combined effect can be described in terms of the stress–strain relation and the Mohr envelope of the soil as a whole.

STRESS–STRAIN / Typical stress–strain curves for a cohesionless soil subject to an increasing shear stress with a constant confining stress, σ_3, are shown in Fig. 5.14. Both exhibit strains that are approximately proportional to stress at low stress levels, suggesting a large component of elastic distortion. If the stress is reduced, the unloading stress–strain curve is nearly

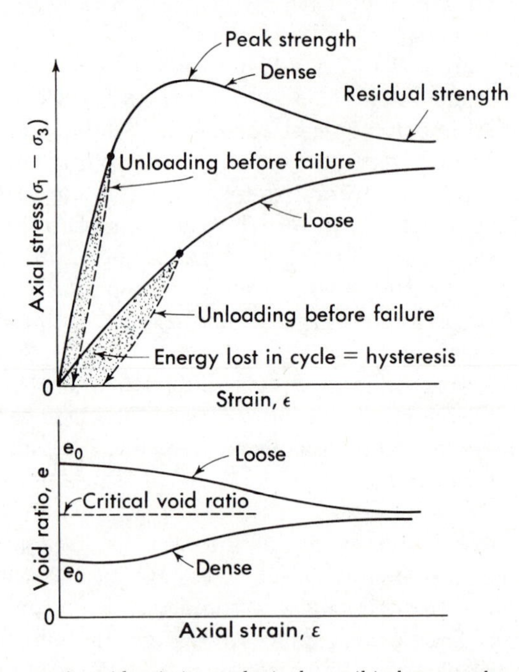

Figure 5.14 *Stress–strain void ratio in a cohesionless soil in loose sand and dense states at a constant confining stress.*

the same as the loading. Not all the strain is recovered upon unloading, indicating some particle reorientation and point crushing. The *hysteresis loss*, the area of the stress–strain loop, represents the energy lost in crushing and repositioning. The loose soil with larger voids and fewer points of contact exhibits greater strains and less recovery of strain upon unloading than the dense soil.

At higher shear stresses the strains are proportionally greater, indicating greater crushing and repositioning.[5:4]

A plot of void ratio as a function of axial strain and volumetric strain, ε_v [or void ratio change $(\Delta e/1+e)$] as a function of axial strain illustrates the effects of grain movement, as seen in Fig. 5.14b. At small axial strains both the dense and loose sand become denser. At larger strains, however, the dense sand must expand before its grains can turn and slide over one another. The loose sand, however, continues to densify but at a smaller rate. Eventually, at very large strains, both the loose and dense sand approach a similar void ratio.

ELASTICITY / As defined earlier, the modulus of elasticity is the slope of the stress–strain curve. From Fig. 5.14a it can be seen that for low shear stresses E is nearly constant, but with increasing shear it becomes smaller. The modulus of elasticity is also dependent on confining stress: the higher the stress level the greater the E for any given shear stress. An enlarged picture of E as a function of σ_3 and the shear stress (expressed in terms of σ_1/σ_3) for a typical sand is shown in Fig. 5.15. This can be approximated by the equation

$$E = C\sigma_3^n \tag{5:6a}$$

$$C, n = f\left(\frac{\sigma_1}{\sigma_3}, D_r\right) \tag{5:6b}$$

Typical values of n, the exponent are between 0.8 and 0.5 for the range in stresses involved in engineering problems. Because of the wide variety of soils and stress conditions, it is not possible to estimate the value of E from a description of the soil or from simple tests such as grain size. It should be determined directly by tests that simulate the soil-structure stress field, and the level of strains that will be involved in any real situation.

POISSON'S RATIO / Like E, Poisson's ratio is not a constant. It varies with D_r, σ_3, and σ_1/σ_3. It can exceed 0.5 when the volume and void ratio of a dense cohesionless soil increase as the peak strength is reached. Typical values for small axial strains, well below failure, are 0.25 to 0.4.

STRENGTH / As can be seen from Fig. 5.14 failure in the cohesionless soil is not easily defined.[5:6, 5:7] The dense cohesionless soil will sustain an increasing shear stress until a maximum is reached, usually at modest strains.

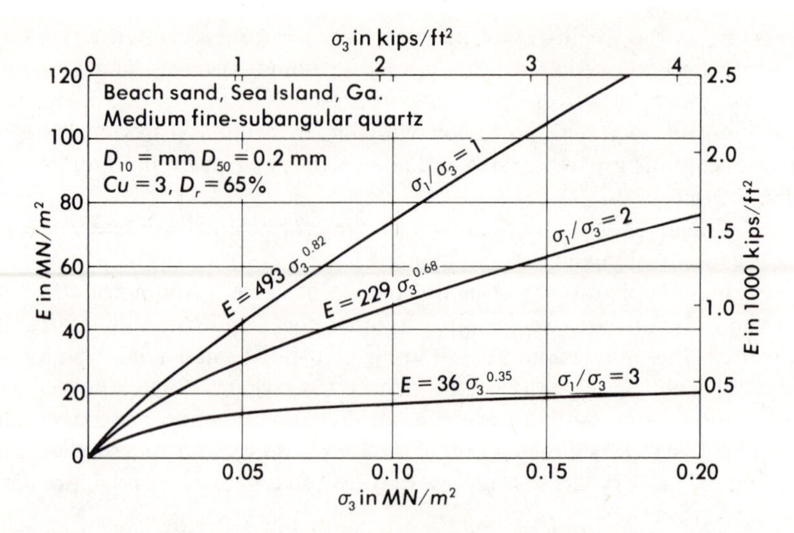

Figure 5.15 *Modulus of elasticity of a beach sand.*

This is the *peak strength*, τ_{pf}.* At larger strains the shear resistance decreases to a lower value, the *residual strength*, τ_{rf}.* It remains virtually constant for increasing strains. The decrease in strength is accompanied by a rapid increase in volume and void ratio. The loose cohesionless soil increases in shear resistances more slowly than the dense, and at a decreasing rate. Eventually, at significant shear strains, a maximum τ_{ff} is reached that approaches the residual strength for the denser soils. The gain in resistance is accompanied by a decrease in volume and void ratio until large axial strains are developed.

The results of many tests of cohesionless soils demonstrate that the maximum shear stress on the failure surface at failure, τ_{ff}, is nearly proportional to the effective normal stress on the failure plane at the moment of failure, σ'_{ff}:

$$\tau_{ff} = \sigma'_{ff} \tan \phi \tag{5:7}$$

(peak for dense and ultimate for loose unless otherwise designated). The angle ϕ is the *angle of internal friction*. The same expression also holds for residual strength of dense cohesionless soils, but the angle is ϕ_r. The actual Mohr envelope is slightly curved; it is approximated by a straight line through the origin in Fig. 5.16a. The angle of the failure plane θ can be found graphically from the Mohr circle at failure or from the straight-line equation and geometry:

$$\theta = 45 + \frac{\phi}{2} \tag{5:8}$$

* Peak shear stress on the failure plane; residual shear stress on the failure plane.

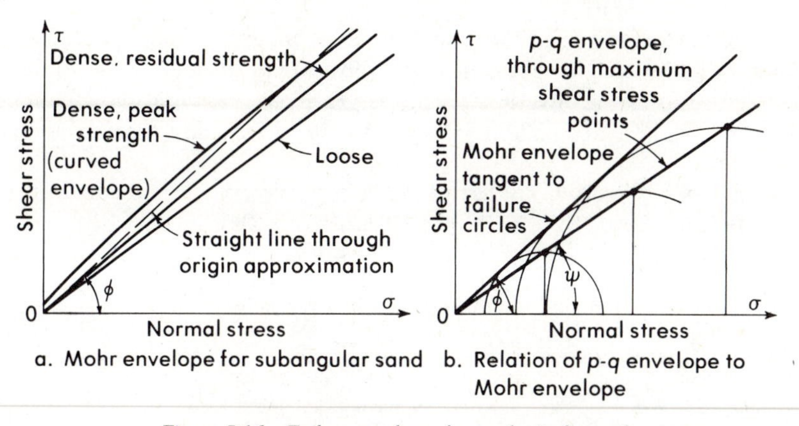

Figure 5.16 *Failure envelopes for a cohesionless soil.*

A second form of expressing the results is in the p–q diagram (Fig. 5.16b). This shows the maximum shear stress, τ_{mf} or q_f, as a function of the normal effective stress on the plane of maximum shear, σ'_{mf} or p'_f.

$$p'_f = \frac{\sigma'_1 + \sigma'_3}{2} \tag{5:9a}$$

$$q'_f = \frac{\sigma'_1 - \sigma'_3}{2} \tag{5:9b}$$

The advantage of this plot is that the p'_f and q'_f can be found directly from laboratory data, whereas σ_{ff} and τ_{ff} require that Mohr circles be plotted. The p–q data can be evaluated statistically in terms of median, standard deviation, and other parameters more readily than the circle, and the p–q envelope plotted directly by a computer. If the envelope is straight and the angle of the p–q envelope is ψ, then $\sin \phi = \tan \psi$ and

$$\phi = \arcsin(\tan \psi) \tag{5:10}$$

The angle of internal friction integrates all the factors of resistance to grain displacements: distortion, crushing, shifting, rolling, and sliding. These factors depend on the soil mineral, the particle angularity, roughness, and sphericity, gradation, and relative density. The real envelope is usually slightly concave downward. At high confining stresses (70 to 200 kN/m^2 or 500 to 1500 lb/in^2 in quartz sands), grain fracture predominates, and the Mohr envelope becomes much flatter. For weak grains, such as volcanic ash, fracture is significant at $\frac{1}{10}$ these stresses. Soils with high relative densities exhibit steep envelopes at small stresses and some strength at $\sigma_3 = 0$ as reflected in the envelope intercepting the axis above the origin. However, for most practical problems in the stress ranges involved in civil engineering structures, Equation (5:7), is a reasonable approximation.

Grain size (as opposed to gradation) does not greatly influence the angle of internal friction. However, the larger the particle, the greater the tendency to fracture; thus the envelope for dense gravel size materials is often more curved, flattening at smaller confining stresses than sands. (See Table 5:1.)

TABLE 5:1 / ANGLE OF INTERNAL FRICTION OF COHESIONLESS SOIL COMPOSED LARGELY OF QUARTZ OR SIMILAR HARD MINERALS

	Angle of Internal Friction	
Grain Characteristics	$D_r = 30\%$	$D_r = 65\%$
Rounded, uniform	29	35
Rounded, well graded	32	37
Angular, uniform	35	42
Angular, well graded	37	45+

EFFECT OF REPEATED LOADING[5:8 5:9] / If a cohesionless soil is subjected to stress that is significantly smaller than that producing failure, it will undergo strain, as illustrated by Fig. 5.14. The unloading stress–strain relation is steeper than the loading and does not recover the initial strain. There is an energy loss, *hysteresis*, in the process, from grain movement and fracture. The stress–strain curve for a second loading is significantly steeper than that for the initial loading, indicating increased E. The second unloading is slightly steeper than the first, and the second hysteresis loop is narrower. Less energy is lost because the soil became partially adjusted to the stresses during the initial loading. After about 5 cycles the loops are nearly closed and the unloading elastic modulus becomes nearly constant. During the process the soil becomes denser, with an increase in the amount of fine particles by grain fracture accompanying the void ratio decrease. The relative reduction in void ratio decreases with increasing shear stresses (below failure) and decreases with increasing relative density. When $D_r > 70$, there is little volume change during repeated loading.

5:5 Shear in Wet Cohesionless Soils

As stated in Chapter 3, the total stress applied to a soil is sustained by stress propagated through the grain structure, σ' (*effective stress*) and by *neutral* or *water stress*, u. At any time the stresses can be represented by the relation

$$\sigma = \sigma' + u \qquad (3:14b)$$

Since shearing resistance is a friction phenomenon, it depends on grain structure stress; therefore, at failure the equation for shearing strength of moist sand must be written:

$$\tau_{ff} = \sigma'_{ff} \tan \phi \qquad (5:7)$$

$$\tau_{ff} = (\sigma_{ff} - u) \tan \phi \qquad (5:11)$$

where σ_{ff} is the total normal stress on the failure plane at failure. The angle of friction of most bulky grains is not changed appreciably by water. Exceptions are shale fragments that slake slowly and such minerals as chlorite. For wet cohesionless soils, however, changes in effective stress are far more significant in soil behavior than small differences in ϕ. Therefore, engineering studies of potential failure must emphasize those factors involved in changing the effective stress: both total stress and neutral stress.

HYDROSTATIC NEUTRAL STRESS / The simplest case of neutral-total interchange occurs when neutral stress builds up hydrostatically: The piezometric level (or water table) rises from increased groundwater inflow or decreased drainage, and the strength falls correspondingly. If the soil strength of a potential failure surface should drop until the shear force resisting soil failure is equal to a shear force imposed on the soil, movement is imminent. A small increase in shear force or reduction in shear strength will cause failure. The interchange of forces is illustrated in the following example.

Example 5:4

A highway cut is excavated in level-bedded, jointed sandstone as shown in Fig. 5.17. The cut face is sloped 1(H) to 6(V) for appearance. An open vertical joint parallel to the cut face is 9 m from the cut toe. Just above the

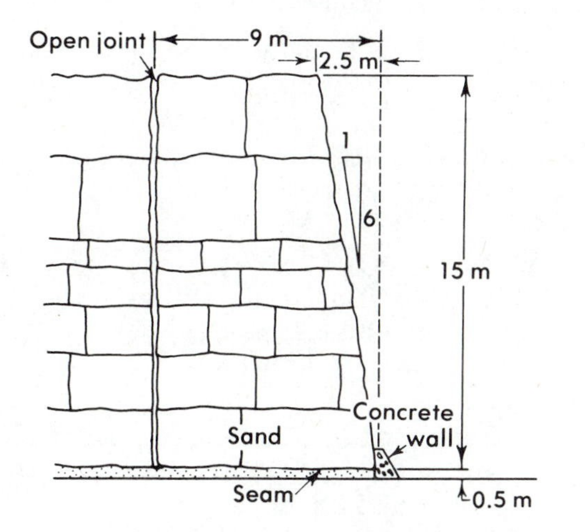

Figure 5.17 Effects of hydrostatic neutral stress in a highway cut made in jointed sandstone, Example 5:4.

base of the cut is a seam of slightly silty sand $\frac{1}{2}$ m thick from advanced sandstone weathering. A small concrete wall is poured at the toe of the cut to restrain the sand. The sandstone weighs 24 kN/m^3 or 152 lb/ft^3; the angle of internal friction of the weathered sandstone is 30°.

Case 1—No water

1. The weight of the sandstone on the weathered seam between the open joint and the cut for 1 m of cut length perpendicular to the cut face is $15\times[(9+6.5)]/2\times 25 = 2906$ kN/m^2.
2. $\sigma_v = (2906)/9 = 322.9$ kN/m^2 avg. (For simplicity use average σ_v, ignoring its distribution over the weathered seam.)
3. $\tau_{ff} = 322.9 \tan 30 = 186.4$ kN/m^2. It will resist a horizontal force of $186.4\times 9 = 1678$ k kN/m of cut.

Case 2—Water rises to 10 m above weathered seam joint because of blocked drainage of concrete wall

4. u is uniform in weathered seam because wall prevents seepage. $u = 10\times 9.81 = 98.1$ kN/m^2.
5. $\sigma' = 322.9 - 98.1 = 225$ kN/m^2, $\tau_{ff} = 225\times\tan 30 = 129.9$, a drop of $186.4 - 129.9 = 56.5$ kN/m^2 or 30%.
6. The weathered seam provides lateral resistance F_r of $129.9\times 9 = 1169$ kN/m^2.
7. The water pressure in the crack, 10 m deep, produces a lateral force.

$$F_w = \frac{\gamma h^2}{2} = \frac{9.81\times 10^2}{2} = 490.5 \text{ kN/m of cut}$$

8. Force provided by soil strength $F_r = 1169$ kN/m $> F_w = 490.5$ kN/m. The cut is stable.
9. The safety factor F_s, against sliding is

$$\frac{F_r}{F_w} = \frac{1169}{490.5} = 2.38$$

Case 3—Water rises in the weathered seam to 14 m

10. $u = 14\times 9.81 = 137.3$ kN/m^2.
11. $\sigma' = 322 - 137.3 = 185.6$ kN/m^2.
    $\tau_{ff} = 185.6\times\tan 30 = 107.2$ kN/m^2.
12. $F_r = 107.29 = 964$ kN/m lateral resistance per meter of cut.
13. $F_w = 14^2/2\times 9.81 = 961$ kN/m water force per meter length of cut.
14. F_s = safety factor = 964/961 = 1.003 Rock in face of cut is on the verge of sliding.

Case 4—Water at 10 m in joint: rock face cut back to top of joint

15. Weight of sandstone $= 9\times 15/2\times 25 = 1687.5$ kN/m of cut.
16. $\sigma = 1687.5 \div 9 = 187.5$ kN/m^2 (avg).
17. $u = 98.1$ kN/m^2 (Step 4); $\sigma' = 187.5 - 98.1 = 89.4$ kN/m^2, $\tau_{ff} = 89.4 \tan 30 = 51.6$ kN/m^2.

18. $F_r = 51.6 \times 9 = 465$ kN/m.
19. $F_w = 490.5$ kN/m (Step 7).
20. $F_r < F_w$. Safety factor, $F_s = 465/490.5 = 0.95$. The rock in the face will slide.

Example 5:4 illustrates a number of real problems. First, a rising water table reduces strength, although the total stress does not change (Cases 2 and 3 compared to Case 1). Second, a reduction in total stress such as by excavation reduces the effective stress if the neutral stress remains constant (Case 4 compared to Case 2). In Case 4 (simplified from a real situation), the design engineer sought to enhance the cut stability by flattening the slope—instead, he caused sliding. Third, structures (like the short concrete toe wall) designed to restrain soil must not block seepage.

NEUTRAL STRESS ACCOMPANYING SEEPAGE / As described in Chapter 3, the neutral stress in soil through which water is seeping cannot be computed from the elevation of the water in the soil. Instead, it must be computed from the piezometric level at that point: the height, h_p, to which water will rise in a static tube above the point in question, $u = h_p \gamma_w$. Alternatively, it can be found from the total head (expressed by the flow net Fig. 3.8) minus the elevation of the point [Equation (3:15)].

QUICKSAND / Chapter 3 describes the conditions leading to boiling in which $u = \sigma$, and the soil is lifted by water pressure. In a cohesionless soil, if $u = \sigma$, then the strength $\tau_{ff} = (\sigma_{ff} - u) \tan \phi$ will be zero. Thus *quicksand* is any cohesionless soil where strength is zero because of either hydrostatic or seepage pressures. It is a fluid whose viscosity is temporarily not much greater than that of water, but with a unit weight equal to the total weight of the soil—about twice the weight of water. There have been cases where buried tanks, filled with water, floated upward when seepage pressures produced a quick condition. The condition is almost always produced by a rising water table causing an increasing u or a diminishing total stress, σ, caused by excavation or even by pumping water out of an excavation.

Example 5:5

An excavation 6 m deep in an alluvial deposit 9 m thick is protected by steel sheet piling driven through silt underlain by sand to a pervious gravel and boulder stratum, Fig. 5.18. The boulders prevent driving the piling deeper. A river near the excavation intersects the gravel–boulder stratum and water seeps upward through the sand into the excavation so that the water levels are equal. The excavation is made by a clamshell bucket working through the water. After the final excavation level is reached, the contractor attempts to pump the excavation dry. When the water level has been lowered 3 m, the sand in the excavation bottom becomes quick. What is happening? The silt and sand weigh 19.8 kN/m^3 and the gravel is so pervious that head loss through it is negligible.

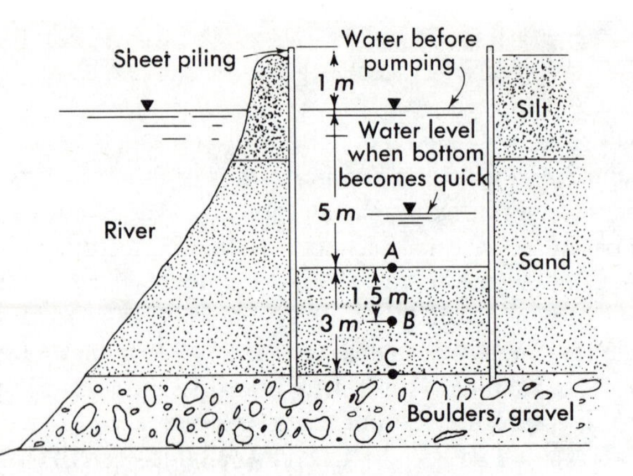

Figure 5.18 *Quicksand induced by pumping from an excavation, Example 5:5.*

1. After excavation, before pumping, the vertical stresses in the sand are listed in Table 5:2.

TABLE 5:2

Point	Total Stress, kN/m^2	Neutral Stress, kN/m^2	Effective Stress, kN/m^2
A	$5 \times 9.81 = 49$	$5 \times 9.81 = 49$	0
B	$5 \times 9.81 + 1.5 \times 19.8 = 78.7$	$6.5 \times 9.81 = 63.8$	15
C	$5 \times 9.81 + 3 \times 19.8 = 108.4$	$8 \times 9.81 = 78.4$	30

2. After pumping the water level down 3 m below the water surface, the stresses are changed, assuming a uniform gradient through the sand below the excavation bottom and no head loss in seepage from the river through the gravel to the sand immediately below the excavation. (See Table 5:3.) The entire sand stratum below the excavation bottom has negligible effective stress and thus negligible strength. It is virtually quick. It can no longer provide lateral support for the sheet piling; the piling is forced inward and upward at the

TABLE 5:3

Point	Total Stress, kN/m^2	Neutral Stress, kN/m^2	Effective Stress, kN/m^2
A	$2 \times 9.81 = 19.6$	$2 \times 9.81 = 19.6$	0
B	$2 \times 9.81 + 1.5 \times 19.8 = 49.3$	$5 \times 9.81 = 49.0$	0.3
C	$2 \times 9.81 + 3 \times 19.8 = 79$	$8 \times 9.81 = 78.5$	0.5

bottom of the excavation. Equipment resting on the bottom sinks downward. If pumping continues, the bottom will boil.

By way of contrast seepage can be beneficial. If drain holes were installed in the wall of Example 5:4, the head loss from seepage through the sand seam below the rock would reduce the neutral stress and increase the strength. Depending on the rate of water inflow into the joint, it might also reduce the water level in the joint and therefore reduce the water force, F_w, causing motion.

NEUTRAL STRESS THROUGH VOLUME CHANGE / When a cohesionless soil is sheared, its volume changes; if the soil is saturated, the volume change is accompanied by a change in the distribution of water in the voids. If shear and the change in volume occur so slowly that the movement of water requires negligible head, there will be only insignificant changes in neutral stress. Rapid shear, however, requires rapid changes in water content that develop large neutral stresses. This is particularly important in very fine-grained soils of low permeability.

In dense cohesionless soils, void expansion accompanies shear as shown in Figs. 5.14 and 5.19. Negative neutral stresses are developed in the pore water in the same manner as negative pressures are produced by pulling a cork from a tightly stoppered bottle. The limiting negative stress in a saturated soil is the maximum capillary tension as determined by the soil void diameter [Equation (3:24)]. The effect of the negative neutral stress is a temporary increase in the strength of the soil until seepage into the expanding voids restores neutral stress equilibrium.

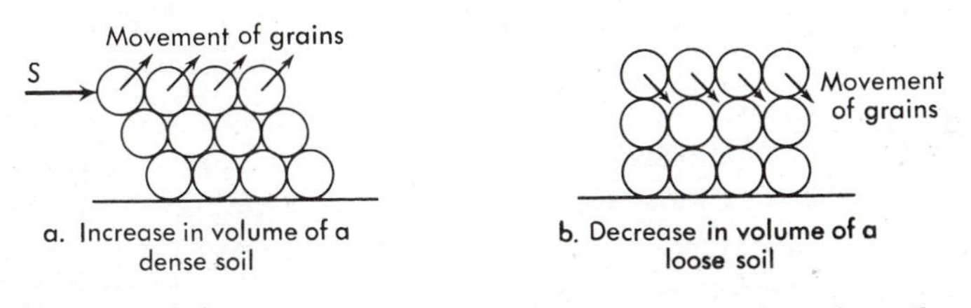

Figure 5.19 *Volume change that accompanies shear in cohesionless soils.*

In loose sands the opposite occurs. The voids decrease in volume, inducing a positive neutral stress. The maximum value of u is equal to the confining pressure on the soil, σ. The limiting effective stress $(\sigma - u)$ is zero, and the strength is zero. This is another case of quicksand—temporary but just as serious as that produced by hydrostatic conditions.

In either loose or dense soils, the pore pressure change induced by shear is initially limited to the zone of shear and the temporary gain or loss of strength is localized. Depending on the soil permeability and the geometry of potential seepage paths, the neutral stress can persist and spread through the mass until eventually it is dissipated by seepage. This is a serious problem in a fine loose sand or silt. A localized shear failure within the mass generates increased pore pressure. The pore pressure increase is transmitted through the continuous body of water in the voids. This weakens the soil beyond the zone of initial failure, and creates additional failure and more pore pressure. Thus failure propagates beyond the initial, isolated stress point of failure to engulf a larger body of loose, saturated soil. This is essentially a chain reaction or "snowballing" effect that can prove disastrous. Some of the most devastating landslides, such as the talus avalanche that demolished a large town in the Peruvian Andes following rapid snow melt, have commenced with local shearing in a loose, saturated cohesionless mass—sand, gravel, and even broken rock. On a smaller scale, a failure commenced in an insignificant sewer trench excavation but propagated into the sliding of an entire hillside and destruction of the construction site.

LIQUEFACTION[5:10] / Increases in neutral stress leading to failure can also be generated by repeated small loads. Each load repetition produces a small reduction in void ratio, which in turn, causes a small increase in neutral stress. When the soil is fine grained and the mass sufficiently large so that the neutral stress cannot be immediately dissipated, the accumulation of small increments of neutral stress adds up to significant pore pressures. The increased pore pressures weaken the soil, permitting even larger strains, increasingly larger volume changes, and greater pore pressures that eventually cause failure.

The cyclic loading produced by vibrating machinery, even though small, can induce pore pressure buildup and the eventual loss of soil strength. The repeated strains accompanying earthquakes have similarly produced increasing pore pressures in sand deposits and sufficient loss in strength so that large masses of soil have become quick or *liquefied.* In the large earthquake at Niigata, Japan, in 1963, the liquefaction of a sand deposit caused a large apartment building to drop suddenly about one story and to tilt more than 30°. The Turnagain Heights landslide accompanying the 1964 earthquake at Anchorage, Alaska, is another example of the rapid failure accompanying liquefaction.[5:11]

The number of cycles of load required to produce liquefaction in a cohesionless mass depends on a number of factors: the soil's initial relative density, the stress increment compared to failure stress, the permeability, and the geometry of the seepage paths. In general, the greater the stress increment and the looser the soil, the fewer the cycles of load required for liquefaction. If the initial relative density exceeds about 70%, the progressive, small void ratio decreases are followed by increasing void ratios at larger strains. Momentary liquefaction of dense soils ends with the generation of negative pore pressures when the strains increase. Limited data suggest that in such cases the liquefaction remains localized and does not propagate.

Blasting and the shock waves generated by the impact of excavating machinery also can cause liquefaction in very loose sands and cohesionless silts where only a few cycles of repeated stress are sufficient to build up large pore pressures. Such deposits must be considered as potential hazards to any construction operations. They must be corrected for permanent structural safety.

NEUTRAL STRESS FROM CAPILLARY TENSION / Capillary tension can be the cause of negative stress that increases soil shear strength. Moist sand owes its ability to pack and maintain a shape to capillary tension in thin water films between the grains. The small meniscus radii develop high tensile stresses in the moisture wedges that hold the grains in rigid contact (Fig. 5.20).

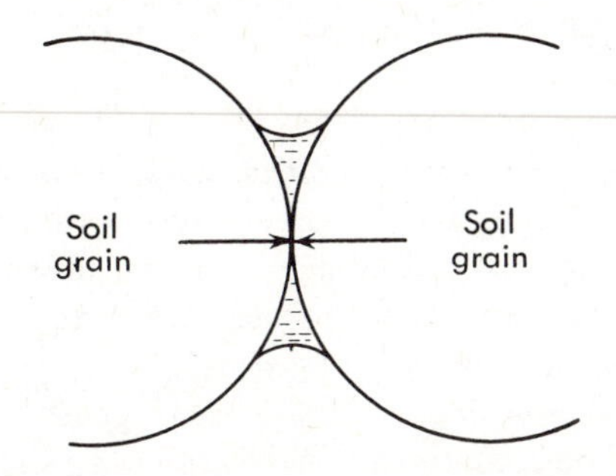

Figure 5.20 *Neutral stress from capillary tension producing compression between soil grains.*

Fine sand and silt above the groundwater table within the zone of capillary rise and fringe owe their strength to the capillary tension and the resulting effective stresses in the soil structure. Sometimes deep excavations can be made in such soils with very steep side slopes because of this strength. However, the strength is temporary; either drying or saturation will destroy the capillary tension. Serious excavation failures accompanied by loss of life can be traced to reliance on the strength temporarily induced by capillary tension.

A sample of saturated fine sand or silt will hold its shape when subjected to an unconfined compression test because capillary tension produces a positive effective σ_3. If loose, it will finally collapse when the load causes a reduction in the volume of the voids and a buildup of liquefaction. By way of contrast, dense deposits will expand and develop greater strength temporarily from capillary tension.

5:6 Strain and Strength of Saturated Cohesive Soils[5:12, 5:13]

Shear in a saturated cohesive soil (a clay) is more complex than in a sand or gravel. Like the cohesionless soil, the clay is made up of discrete solid particles which must slide, rotate, or fracture for shear to take place.

However, there are a number of significant differences. First, the soil is relatively compressible; therefore, when a load is applied to the saturated clay, it is initially supported by neutral stress and is not transmitted to the soil structure. Second, the permeability of the clay is so low that the neutral stresses produced by the load are dissipated very slowly. Therefore, it can be months or even decades before the soil structure feels the full stress increase. Third, there are significant forces developed between the particles of clay by their mutual attraction and repulsion.

RATE OF LOADING / Because of the slow changes in the neutral stress and the corresponding slow changes in effective stress, the strength of clays is defined in terms of neutral stress dissipation. Three limiting rates and their identifying symbols have been defined:

1. *Drained* (D) [Also termed *consolidated-drained* (CD) or *slow* (S)] *shear.* The confining and the shear stresses are applied so slowly that the neutral stress is not changed by the added loads; the applied stress produces an equal increase in effective stress; and the soil consolidates with the changing stress.
2. *Consolidated-undrained* (CU) [Also termed *consolidated-quick* (R)] *shear.* The confining stress is applied so slowly that the neutral stress is not changed and the soil consolidates fully. The shear stress, however, is applied so quickly that neutral stress carries all this change, and there is no further consolidation or change in water content.
3. *Undrained* (U) [Also termed *unconsolidated-undrained* (UU) or *quick* (Q)] *shear.* Both the confining and shearing stresses are applied so rapidly that the neutral stress supports all the added load, and there is no change in the water content.

DRAINED SHEAR—CONSOLIDATION AND STRESS–STRAIN / In drained shear there is no neutral stress change, and any increase in total stress produces a corresponding increase in effective stress. The soil consolidates reducing the void ratio and water content at stresses above the preconsolidation load.

The consolidation of the soil takes place in two stages: first, during the addition of the confining stress, and second, during the addition of the axial stress that produces shear. Although the mechanism is similar to the one-dimensional consolidation discussed in Section 4.2, the stress–void ratio curve is different because the stress fields are different. In the conventional consolidation test where $\varepsilon_2 = \varepsilon_3 = 0$, the lateral stress, $\sigma_2 = \sigma_3$ is a constant fraction, K, of the vertical stress σ_1. Successive Mohr circles for the increasing loads in such a consolidation test are given in Fig. 5.21. The successive positions of a point on the circles that represent a particular plane in the soil are termed the *stress path* for that plane.[5:12] The stress path is the

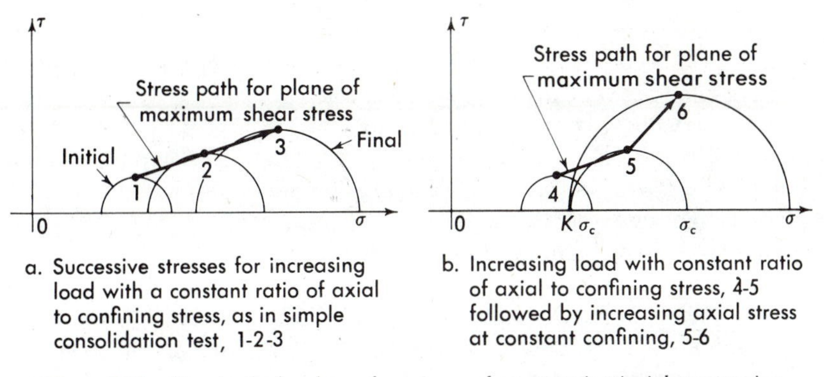

Figure 5.21 *Stress paths for plane of maximum shear stress in triaxial compression.*

locus of the combinations of shear and normal stresses resulting from a loading sequence on the soil. The particular plane of interest may be the plane of maximum shear (line 1—2—3) or the plane on which failure will eventually occur.

In the drained test the stress path is different. The initial consolidation is either hydrostatic, with $\sigma_1 = \sigma_2 = \sigma_3$, or anisotropic with $\sigma_1 > \sigma_2 = \sigma_3$. After consolidation the lateral stress is held constant, and the vertical increased (Fig. 5.21b) (or decreased). The stress paths for the plane of maximum shear stress (line 4—5—6) (or for the potential failure plane) are different from the path for simple one-dimensional consolidation.

The results of the three-dimensional consolidation can be expressed in several ways (Fig. 5.22): a stress settlement curve similar to Fig. 4.3, or a stress–strain curve similar to that for a cohesionless soil. The strain includes both the elastic deformation of the soil mass and the void ratio change.

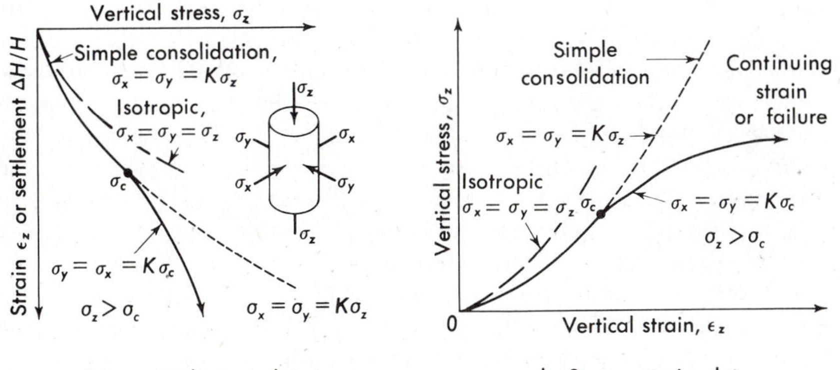

Figure 5.22 *Stress–strain in drained or slow shear: Consolidation followed by increasing axial stress.*

The simple one-dimensional consolidation test involves a special stress path (Fig. 5.22a) that describes the changes in relatively thin compressible strata confined between more rigid strata. The stress path for thick strata of compressible soils is more nearly like that of Fig. 5.21b, and the stress–settlement curve (Fig. 5.22a) is somewhat different. Therefore, if the laboratory test is to give an accurate indication of real settlement, the stress path of the laboratory test should be the same as for the field load. Following the real stress path in the laboratory is not always practical. The conventional single-dimension test is ordinarily used with due allowance for any errors involved.

STRENGTH IN DRAINED SHEAR / As a result of consolidation above the preconsolidation load, the water content and interparticle spacings are reduced and interparticle bonds are increased in proportion to the confining stress that overcomes their resistance to compression. The strength, therefore, increases in proportion to the effective confining stress increase. The Mohr envelope is a straight line through the origin (Fig. 5.23).

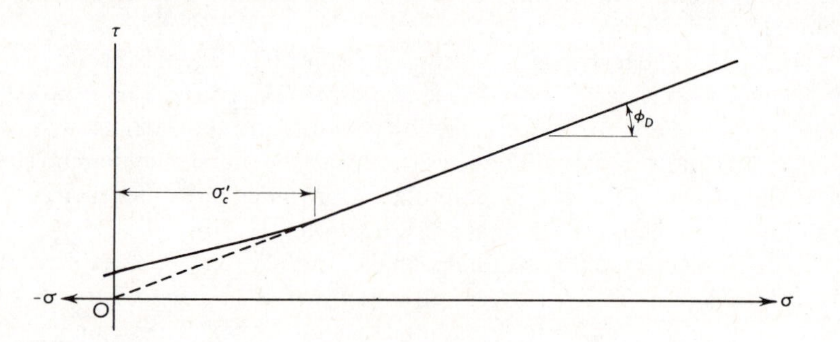

Figure 5.23 *Mohr envelope for saturated clay in drained,* D *(consolidated drained,* CD, *or slow,* S) *shear.*

The angle of the Mohr envelope is termed the *angle of shear resistance* or *effective internal friction* and is denoted ϕ_D, ϕ_{CD}, ϕ_S, sometimes ϕ'. Typical values lie between 12 and 30°. The higher angles are usually associated with clays having plasticity indexes of 5 to 10, while the lower values are for clays having plasticity indexes of from 50 to 100. This is a verification of the effect of particle repulsion and adsorbed water on the interparticle bonds, for the high PI indicates high adsorption and repulsion, large interparticle spacing, and correspondingly less interparticle attraction.

When a clay has been preconsolidated to a stress of σ_c' and then unloaded, the particles do not return to their original spacing and previously higher void ratio. As a result the interparticle attractive force is not reduced, and the strength at confining stresses less than the preconsolidation load is no longer proportional to the effective confining pressure but is somewhat higher (Fig. 5.23).

The strength above the preconsolidation load is given by the expression

$$\tau_{ff} = \sigma'_{ff} \tan \phi_D \tag{5:12a}$$

Below the preconsolidation load the strength must be obtained directly from the Mohr diagram. This curved portion of the envelope can be approximated by a straight line having the equation

$$\tau_{ff} = c_1 + \sigma'_{ff} \tan \phi_{D\text{-}1} \tag{5:12b}$$

In this expression c_1 is the intercept on the τ axis and $\phi_{D\text{-}1}$ is the angle that the straight line makes with the σ axis.

The drained shear condition represents the strength of the soil developed by a long-term stress change. It also is the strength that is related to the effective stress supported by the grain structure. It is probably the most reliable strength representation for analysis and design of long-term loading provided the neutral stress can be evaluated. If it cannot, the problem must be solved using total stresses and a form of test that simulates the rate of soil consolidation under changing loads. The drained shear test is time consuming; however, the Mohr envelope for this condition can usually be approximated from the consolidated-undrained test results, with pore pressure measurements.

CONSOLIDATED-UNDRAINED SHEAR / In consolidated-undrained shear, the soil consolidates completely under the confining stress σ_3, with a corresponding reduction in void ratio and water content. The axial load is then increased rapidly by an amount $\Delta\sigma_1$ without further changes in void ratio or water content until failure occurs. The total major principal stress at failure is $\sigma_1 = \sigma_3 + \Delta\sigma_1$, and the total minor principal stress is σ_3. Since no drainage or consolidation occurs from the added load $\Delta\sigma_1$, it is supported by neutral stress, or $\Delta u = \Delta\sigma_1$ (except as discussed under the following heading, *Pore Pressure Coefficient*). Water pressure at any point is the same in all directions, according to the laws of hydrostatics; therefore the neutral stress produced by $\Delta\sigma_1$ is exerted in the direction of both σ_1 and σ_3. The effective stresses at failure are, therefore,

$$\sigma'_1 = \sigma_1 - \Delta u = \sigma_3 + \Delta\sigma_1 - \Delta\sigma_1 = \sigma_3$$

$$\sigma'_3 = \sigma_3 - \Delta u = \sigma_3 - \Delta\sigma_1$$

A plot of these effective stresses on Mohr coordinates will produce a failure envelope that is almost identical with that for drained shear. The differences are most significant below the preconsolidation load, where the sudden changes in neutral stress in CU shear distort the soil structure slightly. If total stresses are plotted, however, a different envelope will be produced: all the failure circles are shifted to the right by $\Delta\sigma_1$ (Fig. 5.24). The total stress envelope will also be a straight line through the origin above the preconsolidation load, with an angle of shear resistance ϕ_{CU}, which is about half

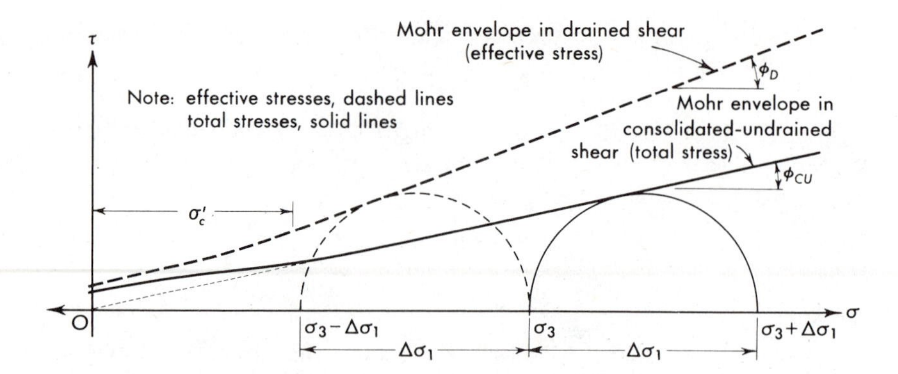

Figure 5.24 *Mohr envelope for saturated clay in consolidated, undrained* CU (*or* R) *shear.*

ϕ_D. The total stress equation is

$$\tau_{ff} = \sigma_{ff} \tan \phi_{CU} \tag{5:13}$$

Measurements of neutral stress can be made in the soil during the test to aid in plotting the effective envelope, but these require skill and sophisticated apparatus. The envelope for total stresses is sometimes used in solution of engineering problems where the neutral and effective stress produced by a long-term load can be determined, but where changes in neutral stress caused by sudden additions in load cannot be computed realistically. For example, the effective stresses imposed by an earth dam and its normal water load can be found from the total stresses and, with the aid of a flow net, the neutral stresses. However, the additional neutral stress generated by a flood surcharge can only be estimated. The CU total stress envelope could be employed for the load change produced by the flood.

The previous discussion of neutral stress equal to $\Delta\sigma_1$ implies that the soil volume will remain the same or decrease only slightly under the increased axial load. However, for overconsolidated clays and dense sand–clay matrix structures this is not correct. These soils dilate or expand with increasing axial load, similar to the dilation of a dense sand, causing a reduction in the neutral stress change and a greater proportion of the load supported by effective stress. A different mechanism with the same effect occurs in some porous rocks at high stresses. The compressibility of the soil water, compared to the rigidity of the rock fabric, causes the fabric to support a greater proportion of the load increment in effective stress and the neutral stress increment will be less than $\Delta\sigma_1$. By way of contrast, an increased axial load on a highly flocculent clay structure or a saturated cemented soil can cause a partial breakdown of the soil structure. The grain structure no longer supports the entire effective consolidating stress to which it previously had been loaded. Thus the increment in neutral stress exceeds $\Delta\sigma_1$.

PORE PRESSURE COEFFICIENT / The *pore pressure coefficient*, A, helps to interpret the pore pressure changes during CU loading:

$$\Delta u_{CU} = A(\Delta\sigma_1 - \Delta\sigma_3) \tag{5:14}$$

For most normally consolidated clays, A is approximately 1. For slightly overconsolidated clays, A lies between 0.3 and 0.8; for heavily overconsolidated clays, A can be slightly negative. For materials that break down structurally, A lies between 1 and 1.3. It is not a constant but varies with the stress field. It is determined by CU tests in which the pore pressures can be measured as the total stresses are changed.

UNDRAINED SHEAR / In undrained shear, both the confining and shear stresses are applied so rapidly that no consolidation takes place. The soil void ratio and water content remain unchanged, and neutral stress supports all the added loads. The soil initially supported an overburden pressure, σ_0' (or a preconsolidation load σ_c'), under which it consolidated to establish its void ratio, water content, and interparticle spacing. An increased confining pressure, $\Delta\sigma_3$, is supported by neutral stress, and the void ratio, interparticle spacing, and resulting soil strength remain unchanged. An increased axial load, $\Delta\sigma_1$, also is supported by neutral stress, and likewise produces no change in void ratio or water content. The stress conditions during loading are given in Table 5:4 assuming that the pore pressure coefficient, $A = 1$.

TABLE 5:4

Loading	Total Stress	Neutral Stress	Effective Stress
Overburden	$\sigma_1 = \sigma_0'$	$u = 0$	$\sigma_1' = \sigma_0'$
	$\sigma_3 = \sigma_0'^*$	$u = 0$	$\sigma_3' = \sigma_0'^*$
Adding confinement, $\Delta\sigma_3$	$\sigma_1 = \sigma_0' + \Delta\sigma_3$	$u = \Delta\sigma_3$	$\sigma_1' = \sigma_0'$
	$\sigma_3 = \sigma_0' + \Delta\sigma_3$	$u = \Delta\sigma_3$	$\sigma_3' = \sigma_0'$
Added axial load, $\Delta\sigma_1$	$\sigma_1 = \sigma_0' + \Delta\sigma_3 + \Delta\sigma_1$	$u = \Delta\sigma_1 + \Delta\sigma_3$	$\sigma_1' = \sigma_0'$
	$\sigma_3 = \sigma_0' + \Delta\sigma_3$	$u = \Delta\sigma_1 + \Delta\sigma_3$	$\sigma_3' = \sigma_0' - \Delta\sigma_1$

* In many cases the minor principal stress from the overburden load will be less than the major principal stress, but this does not alter the neutral stress effects described in the table.

The effective minor principal stress is independent of the added confining stress $\Delta\sigma_3$, and therefore the effective major principal stress at failure and the strength depend only on the original overburden stress σ_0 and the effective (drained shear) envelope. A plot of the total stresses, the solid lines in Fig. 5.25, shows a series of Mohr circles. All have the same diameter (since they are in reality the same circle), and the resulting envelope of total stresses is a horizontal straight line. As can be seen from the diagram, the intercept of the envelope on the τ axis is approximately equal to the shear strength of the soil on its original condition, consolidated by the overburden

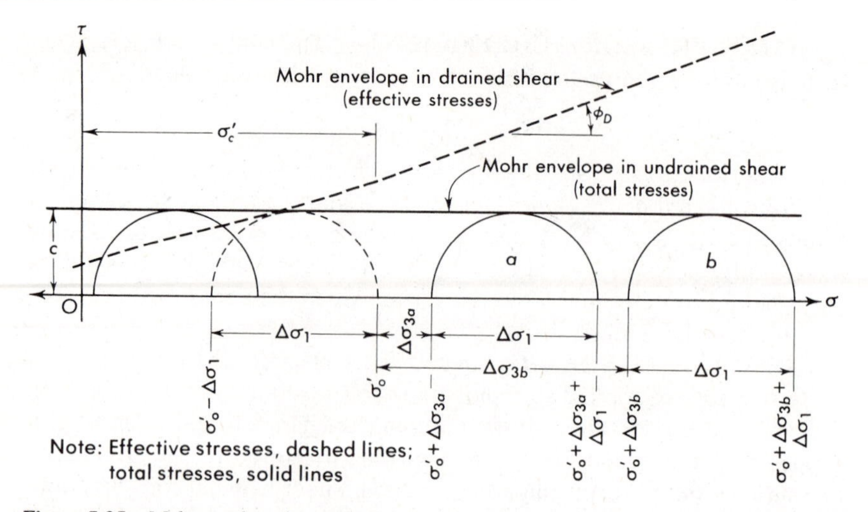

Figure 5.25 *Mohr envelope for saturated clay in undrained* U (*or* UU, *or* Q) *shear,* $\phi_u = 0$.

stress, σ_0. The intercept is denoted c and is called the *cohesion* of the soil. The strength of the soil under undrained conditions can be expressed by the equation

$$\tau_{ff} = c \tag{5:15}$$

The apparent angle of friction, ϕ_u, is zero. However, the angle of the failure plane is determined by the effective stresses; it is not 45° as might be presumed from the horizontal envelope ($2\theta = 90°$). The angle can be computed from the effective stress envelope (if the effective stresses are known).

The undrained strength represents the response of a soil consolidated under present overburden pressure and then loaded to failure without a change in water content. The total stress envelope and the simple expression of constant shear strength [Equation (5:15)] are expedient for those problems of design and construction where the load changes occur more rapidly than the water content can adjust. If load increases so slowly that some consolidation occurs, the drained envelope is more realistic, provided the neutral stress changes can be determined. Otherwise, the undrained, constant strength is used, recognizing that it will be conservative (too small) after the soil consolidates.

However, for situations in which the total stress is reduced, the strength drops as the soil decompresses. The undrained strength would be unsafe for analysis and design unless the time period is so small that no decompression occurs. For example, deep excavations in saturated clay may stand unsupported for a short period of time. The undrained strength describes the soils resistance adequately. However, the clay slowly expands and the strength gradually drops to the drained strength. Numerous cases of delayed failure have occurred in excavations in clay. The initial ability to stand without support (undrained strength) gives a false sense of security to those involved, until failure damages construction and kills people.

SENSITIVITY / If a sample of undisturbed saturated clay is completely remoulded without changing its water content, and then tested, it will be found that the undrained strength has been reduced. This is caused by a breakdown in the soil structure and a loss of the interparticle attractive forces and bonds. In clays with a dispersed structure, the loss is small, but in clays with a highly flocculent structure or soils with a well-developed skeletal structure, the loss in strength can be large. The ratio of the undisturbed to the remolded strength is defined as the *sensitivity*, S_t:

$$S_t = \frac{c\text{ (undisturbed)}}{c\text{ (remolded)}} = \frac{q_u\text{ undisturbed}}{q_u\text{ remolded}} \tag{5:16}$$

Clays with sensitivities exceeding 50 are sometimes termed *quick clays* because their strength after remolding resembles a viscous fluid or quicksand. (See Table 5:5.)

TABLE 5:5 / TYPICAL VALUES OF SENSITIVITY

Clays of medium plasticity, normally consolidated	2–8
Highly flocculent, marine clays	10–80
Clays of low to medium plasticity, overconsolidated	1–4
Fissured clays, clays with sand seams	0.5–2

The sensitive clay reaches a *peak strength*, similar to that of a dense sand, and then becomes weaker with increasing strain (Fig. 5.26). The strength remaining after large strains is the *residual strength* and is approximately equal to the remolded.

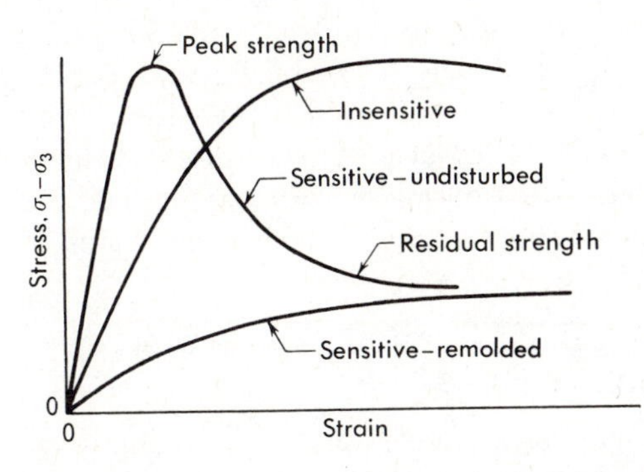

Figure 5.26 *Stress–strain in clays in undrained shear.*

STRESS–STRAIN IN UNDRAINED SHEAR / The relation of strain to stress in undrained shear involves little or no volume change, but only distortion of the mass (Poisson's ratio is nearly 0.5). The shape of the curve (Fig. 5.26) depends largely on the interparticle bonding imposed by preconsolidation and by the structure. For undisturbed clays the initial portion of the curve is straight, probably reflecting the distortion of the bonds. The curve flattens as increasing numbers of bonds are broken. The same soil remoulded without a water content change has a flatter stress–strain curve. The difference between the undisturbed and remolded soil is greatest in sensitive clays. Because of their flocculent structure and edge-to-face bonding they are relatively rigid and have much higher values of E for a given water content than an insensitive clay with a more oriented structure. When the structure is broken by remolding, the E is only a fraction of the undisturbed value.

CLAYS WITH FISSURES / Some clays in nature develop cracks or fissures from desiccation, high overburden stresses which produce local fracture, or physicochemical alteration and weathering. Often the fissures appear to be closed, but they still remain as planes of weakness and paths of seepage. The strength of such clays is dependent on the orientation of the cracks and fissures and on the effect of changing stress and water percolation on the clay along the fissures. Tests of the intact clay between the fissures are misleading; large enough samples must be used so that the fissures are included. Drained tests are most reliable because the fissures permit more rapid dissipation of neutral stress than in ordinary clays. If the cracks have a regular orientation, minimum and maximum Mohr envelopes are depicted similar to those for anisotropic materials Fig. 5.28.

THIXOTROPIC HARDENING / Some clays exhibit a gain in undrained strength after they have been strained or remolded to the degree their strength falls to the residual value. The gain is time-dependent; rapid at first but slower as time passes. The strength of some highly plastic clays may approach the undisturbed peak value. The strength of siltier but highly sensitive clays (whose virgin peak strength was produced by a highly flocculent structure) may be only partially regained. The mechanism of strength recovering is largely reestablishment of interparticle bonds between the clay minerals in their disturbed positions. This is analogous to *thixotropic hardening* of colloidal gels; the term is sometimes applied to clays that regain strength. Dilation and accompanying negative pore pressures sometimes produce strength increases with large strains.

5:7 Strength of Partially Saturated Cohesive Soils[5:13,5:14]

Shear of partially saturated cohesive soils involves the same forces as for saturated cohesive soils. However, the neutral stress in the soil pores is a complex combination of capillary tension and gas pressure which depends on the degree of saturation and the size of the voids.

Methods for analyzing neutral and effective stresses are presented by Gibbs, Hilf, Holtz, and Walker.[5:13]

Because of the uncertainties in measuring or evaluating neutral and effective stresses in partially saturated soils, the total stress envelope is ordinarily used for analysis and design. The test is conducted at moisture contents, rates of strain and stress paths that simulate the field conditions. The total stress Mohr envelope is ordinarily curved, with an intercept on the τ axis and with a decreasing slope at increasing normal stresses (Fig. 5.27).

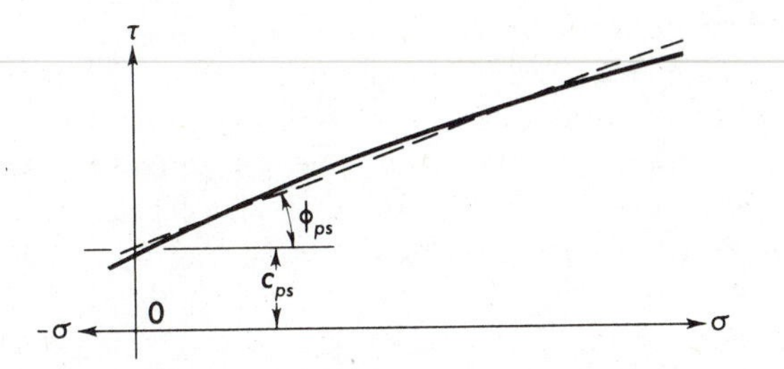

Figure 5.27 *Mohr envelope of total stresses for partially saturated clay.*

The intercept is probably the combined result of capillary tension in the voids and interparticle bonds from preconsolidation. The initially steep slope results from soil consolidation under the increasing confining pressure and is comparable to the drained shear of a saturated clay. As the soil consolidates under increasing pressure, however, the degree of saturation increases, capillary tension decreases, and positive pore pressures eventually develop. This is comparable to undrained shear and results in the Mohr envelope approaching a horizontal asymptote. The strength for any confining pressure is read directly from the envelope approximated by a straight line having the equation

$$\tau_{ff} = c_{ps} + \sigma_{ff} \tan \phi_{ps} \tag{5:17a}$$

or simply

$$\tau_{ff} = c + \sigma_{ff} \tan \phi \tag{5:17b}$$

where c_{ps} is the intercept on the τ axis and ϕ_{ps} is the *angle of shear resistance*. More than one straight line can be used to approximate any given curved envelope, depending on which part of the envelope is of most importance in that particular case. Therefore, c_{ps} and ϕ_{ps} should be considered to be empirical constants and not properties of the soil. The failure plane angle should be measured during testing. Because θ depends on the effective stress, the value computed from total stress parameters is likely to be incorrect.

Partially saturated clays often become saturated from high rainfall or a rising groundwater table. Therefore, the strength of a partially saturated clay should not be used in analyzing practical problems unless the soil remains in that condition. Frequently, partially saturated clays are first soaked in water and then tested as saturated clays to obtain data for design. Saturation is aided by *back pressure*, neutral stress added with increasing confinement.

5:8 Shear in Anisotropic Materials

Most rocks and many soils are *anisotropic* with less shear strength parallel to bedding, mineral orientation, or other weaknesses than perpendicular to it. The strength can be represented by two Mohr envelopes: one for shear along the weaknesses and the other for shear transverse to it as shown in Fig. 5.28a. Whether failure will occur at stress combinations represented by the weaker or stronger envelope depends on the orientation and magnitude of the principal stresses with respect to the surface of weakness, as can be shown by Mohr's circle.

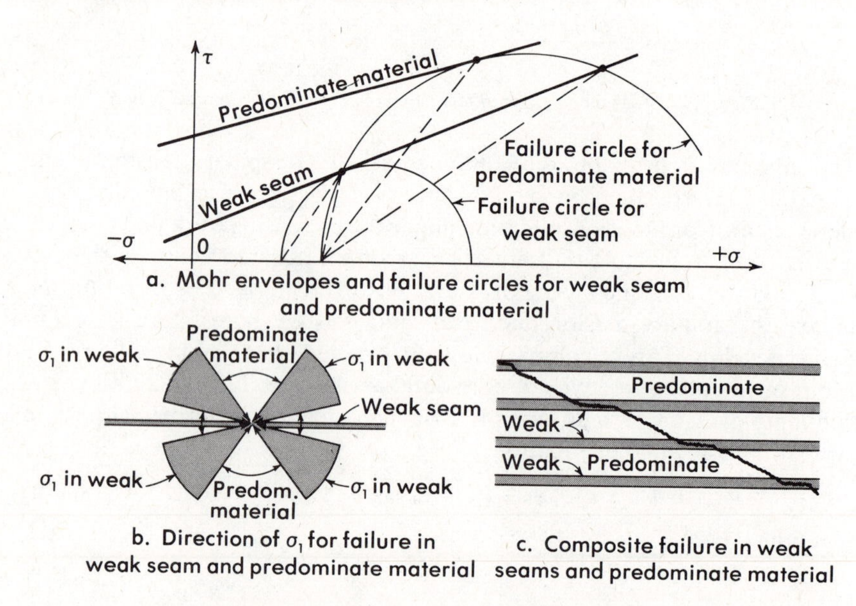

Figure 5.28 *Mohr envelopes for anisotropic shear.*

For anisotropic materials it is seen that the planes of weakness dominate failure. However, the deviator stress can exceed that required for the Mohr circle to touch the weaker envelope. Failure at the larger deviator requires the appropriate orientation of principal stresses to produce the critical stress combination on the weak plane. The angle of the failure plane is no longer $45° + \phi/2$. At sufficiently high stresses failure ideally will occur on both the weak planes and planes transverse to it as shown by Fig. 5.28. However, one plane will usually dominate (the one requiring the least strain at failure). Sometimes a zigzag failure surface develops that includes both the weak and the transverse planes. Testing an anisotropic material requires a number of orientations in order to define both envelopes. The consideration of anisotropic strength in design requires knowledge of the geometry of any surfaces of weakness.

5:9 Strength of Cemented Soil and Rock[5:15, 5:16]

Most rocks and cemented soils consist of relatively rigid mineral grains bonded directly by interparticle crystal bonds or by a cementing material between the grains. When load is applied, the bonds distort. The grains also distort but generally to a smaller degree.

In most cases there will be a small reduction in volume. The reduction will be greater in those materials having significant open voids than in the solid rocks such as marble in which the volume change occurs in the mineral crystals.

Increasing shear stress produces increasing strains with a high, nearly constant E for early stages in loading, reflecting the stretching of the crystal or cementing bonds. Failure generally occurs suddenly with breaking of the bonds.

STRENGTH OF INTACT ROCK / The Mohr envelope for most intact rocks and cemented soils (Fig. 5.29a) is similar to that for partially saturated soils, but it is initially steeper and more sharply curved with appreciable tensile strength and shear strength with no confining pressure. This envelope can be approximated by a straight line, with strength parameters c and ϕ as for the partially saturated clay. Sometimes the envelope is so curved that a reasonable approximation requires several straight lines and several sets of parameters, c_1, c_2, ϕ_1, ϕ_2, for specified ranges in normal stress.

EFFECT OF DEFECTS / Many rocks are anisotropic. Moreover, the stress–strain behavior and failure of most rock formations is dominated by the joints, bedding planes, and other defects or discontinuities.[5:15] Therefore, a realistic evaluation of both strength and elasticity must include these defects, and consider their orientation. Families of Mohr envelopes are defined as for fissured or anisotropic soils (Fig. 5.29b). Across the joints the tensile strength is zero, and the shear strength without confinement is small.

Shear along unweathered joints and cracks resembles that in cohesionless soils, a combination of sliding friction and resistance provided by the geometry of the surface in which the resistance is proportional to the normal face. The plane of weakness can be idealized as shown in Fig. 5.29. The plane of roughness forms *shelves* or *asperities* at an angle of ψ with respect to

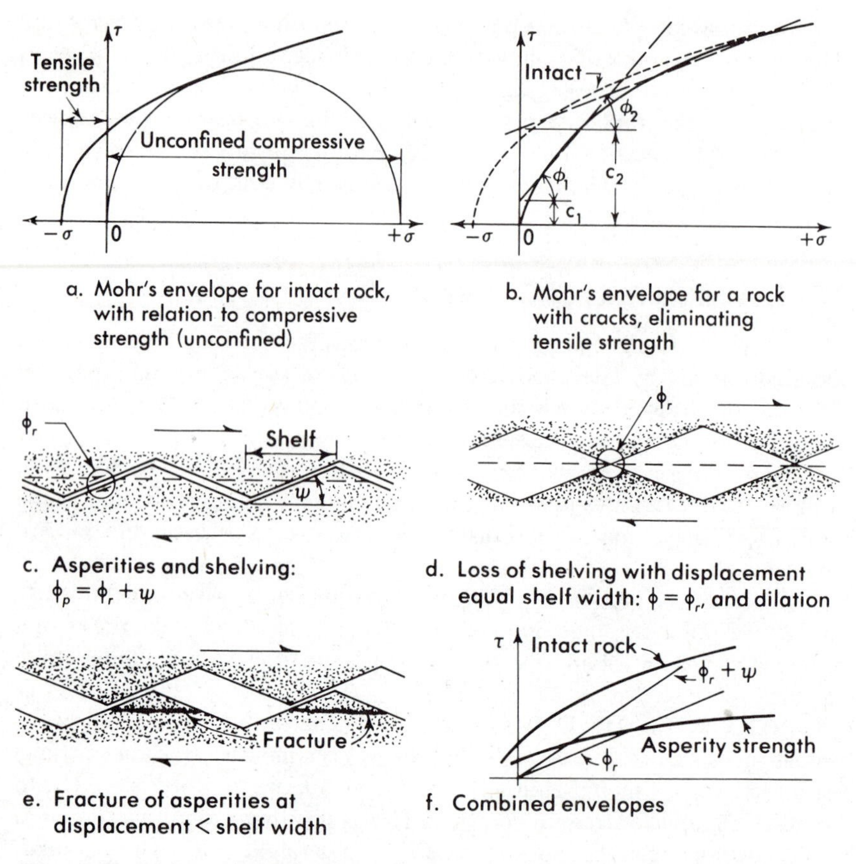

Figure 5.29 Mohr envelopes of total stress for rock and effects of asperities on frictional strength.

the "average plane of weakness" in the direction of potential shear. If the angle of friction of the rock-on-rock is ϕ_r, the effective angle of resistance, ϕ_p, for movement parallel to the plane of weakness for small displacement is

$$\phi_p = \phi_r + \psi \tag{5:18}$$

This defines a peak envelope. After there has been sufficient displacement that the shelves no longer mesh (Fig. 5.29d) the strength will be the residual defined by ϕ_r. If the shear resistance of the intact rock across the base of the shelf or asperity is less than that defined by ϕ_p, then a third envelope can be defined. It is equivalent to the product of the intact rock strength and the proportion of the failure plane area occupied by the asperity. This is illustrated by Fig. 5.29e. The three failure envelopes are shown in Fig. 5.29f. Failure can occur in three stages: (1) at peak strength with small displacements; (2) at the strength of the asperities, with some what larger displacements; and (3) at the residual strength for continuing strains.

A volume increase or dilation accompanies shear along irregular surfaces of weakness. This is analogous to the dilation of a dense cohesionless soil. Because dilation is concentrated in joints or cracks that enlarge, the negative pore pressure generated by capillary tension upon dilation is seldom significant. Moreover, dilation and crack enlargement can reduce hydrostatic pressure suddenly and also provide a larger drainage path for water. Thus pore water pressure effects are erratic. The most pessimistic view is taken for design unless field observations demonstrate otherwise. Dilation and pore/pressure decrease in rock sometimes occur before earthquakes.

The stress–strain behavior of rock may be largely the result of closing cracks. In such materials the value of E is erratic and low at small confining pressures. It suddenly increases when the displacement is sufficient to close the cracks.

EFFECT OF PORE PRESSURE / There is little information on the effect of pore pressure on rocks. The more porous rocks such as sandstones with void ratios exceeding 0.2 behave as soils. The neutral stress coefficients, N, range between 0.75 and 1. The pore pressure coefficient, A, is usually substantially less than 1 because of the rigidity of the grain structure.

Intact solid rocks such as marble and granite with void ratios of less than 0.03 show little effect of pore pressure, suggesting that the neutral stress coefficient, N is close to zero. For intermediate void ratios, there are probably intermediate values of N; however, there are insufficient data to define the transition. The problem is largely academic because of the cracks in most rocks. The rock formation, including cracks, behaves as if N is 1.

EFFECT OF WEATHERING / Weathering reduces the strength, increases compressibility, and reduces rigidity of intact rock. Weathering proceeds faster along cracks than in the intact rock (Fig. 2.2); therefore, three different levels of strength are exhibited in the same formation: (1)

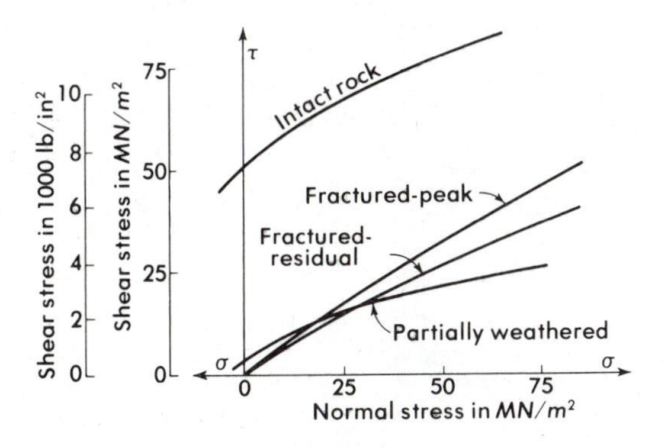

Figure 5.30 *Effect of weathering on rock strength.*

unweathered-intact, (2) unweathered-fractured, and (3) weathered. These are illustrated in Fig. 5.30. The intact envelope is high and curved as in Figure 5.29a. The closely fractured but unweathered rock resembles a cohesionless soil. The weathered rock resembles a cohesive soil and sometimes exhibits greater strength at very low confining stresses than the fractured rock.

5:10 Creep

Most of the data available on strain and strength of soils and rock is based on laboratory tests conducted for a relatively short period of time, generally a few minutes to a few hours but rarely more than a week. In real situations, however, the stress is maintained for years or centuries. Such long-term loads have been observed to produce continuing small strains, a phenomenon termed *creep*.[5:17]

Creep is observed in most materials under sustained high stress. For metals it is a critical factor in high-temperature design. For soils so many environmental factors are significant in producing strain that creep, although suspected, is difficult to identify. However, limited research has shown that continued strain at constant stress levels does occur in soils and rocks and must be considered in design.

A typical plot of strain as a function of time is shown in Fig. 5.31. For a low level of constant shear stress, the strain increases at a decreasing rate, and approaches a limit as shown by curve A. The strain measured in a conventional laboratory test is less than the ultimate strain. The modulus of elasticity for the conventional test, E, is therefore greater than the ultimate, E_u. Limited data suggest that, for low stresses, E_u/E is a constant.

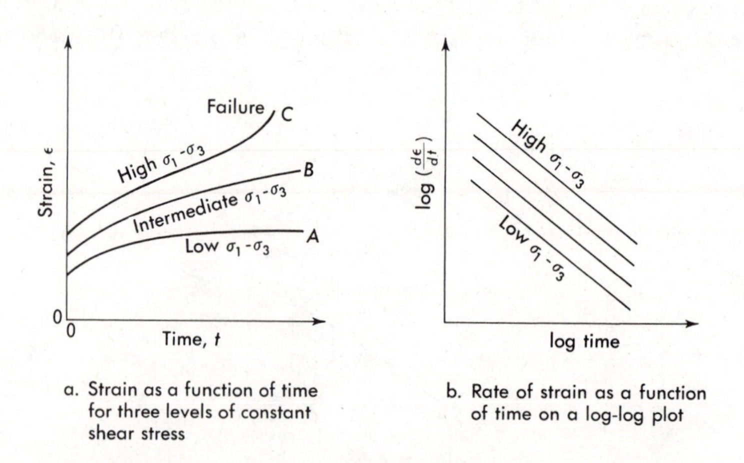

a. Strain as a function of time for three levels of constant shear stress

b. Rate of strain as a function of time on a log-log plot

Figure 5.31 *Creep in soils or rock. (Adapted from Singh and Mitchell.*[5:17]*)*

At higher stresses the strain continues indefinitely at a decreasing rate as shown by curve B. A plot of the logarithm of strain as a function of the logarithm of time is approximately a straight line for any given stress level (Fig. 5.31b).

At stresses close to failure as defined by a conventional test, the strain rate first decreases with increasing time and then increases sharply, ending in failure, as shown in curve C (Fig. 5.31a).

The stress range subject to creep, τ_c, appears to be approximately the same for many clays.

$$\tau_c > 0.3\tau_{ff} \tag{5:19}$$

In this expression τ_{ff} is the shear strength for a given confining stress as determined by conventional tests lasting a few hours at the most. The shear stress at which creep becomes significant, $\tau_c = 0.3\tau_{ff}$, is sometimes termed the *threshold stress*. Generally, the stresses employed in foundations are less than or approximately the threshold. Therefore, creep is seldom significant in properly designed foundations. The stress levels in earth retaining structures, open excavations, and embankments, however, are usually within the creep range. The creep rate is likely to be tolerable so long as $\tau < 0.7\tau_{ff}$, equivalent to a safety factor of 1.4 ($F_s = \tau_{ff}/\tau$ by one definition). Most designers consider that F_s, the minimum safety factor for earth structures under long-term loading, should be 1.5 or more.

5:11 Engineering Properties for Analysis and Design

The solution of engineering problems requires data on the enginering properties of those parts of the soil and rock formations that are influenced by the project: usually a volume soil whose dimensions are significantly greater than those of the immediate site. The portion of the formation that can be tested in the laboratory is an infinitesimal part of the total. Two questions are raised by this disparity.

First, are the samples tested representative? This can be answered only by testing enough samples to make a statistical evaluation. That is seldom feasible. Instead, the testing should concentrate on the poorer materials whose behavior will control the earth's performance. Design for safety is frequently based on the poorest 10% (*decile*) or 25% (*quartile*). This is meaningful only if the poorer materials are randomly distributed. If the poor materials occur in a well-defined zone, that zone is treated as an entity. When the probable performance of the soil is required, the median of the data is utilized. The possible deviations from the median are also required to complete the picture.

Second, what are the combined effects of the stratification, surfaces of weakness, and similar discontinuities? Large-scale field tests (Chapter 7) may be required to integrate the effects of such structural features. These large-scale tests are correlated with laboratory tests based on intuition and experience (judgment). The use of test data without interpretation can lead to disaster.

REFERENCES

5:1 T. W. Lambe and R. V. Whitman, *Soil Mechanics*, John Wiley & Sons, Inc., New York, 1969, pp. 105–121.

5:2 G. F. Sowers, "Strength Testing of Soils," *Laboratory Shear Testing of Soils*, Special Technical Publication 361, ASTM, 1963, p. 3.

5:3 H. K. Ko and R. F. Scott, "Deformation of Sand at Failure," *Journal of the Soil Mechanics and Foundations Division, Proceedings, ASCE*, **94**, SM4, July 1968.

5:4 Ref. 5:1, pp. 137–161.

5:5 J. E. Bowles, *Engineering Properties of Soil and Their Measurement*, 2nd ed., McGraw-Hill Book Company, New York, 1978, pp. 141–184.

5:6 J. K. Mitchell, *Fundamentals of Soil Behavior*, John Wiley & Sons, Inc., New York, 1976, pp. 283–338.

5:7 A. Schofield and P. Wroth, *Critical State Soil Mechanics*, McGraw-Hill Book Company, New York, 1968.

5:8 P. U. Lade and J. M. Duncan, "Elasto-Plastic Stress Strain Theory for Cohesionless Soil," *Journal of the Geotechnical Engineering Division, Proceedings, ASCE*, **101**, GT10, 1037–1059, October 1975.

5:9 V. Cuellar, Z. Bazant, R. J. Krizek, and M. L. Silver, "Densification and Hysteresis of Sand Under Cyclic Shear," *Journal of the Geotechnical Engineering Division, Proceedings, ASCE*, **103**, GT5, 399–415, May 1977.

5:10 H. B. Seed and K. L. Lee, "Cyclic Stress Conditions Causing Liquefaction, "*Journal of the Soil Mechanics and Foundations Division, Proceedings, ASCE*, **93**, SM1, January 1967.

5:11 H. B. Seed, "Landslides During Earthquakes due to Soil Liquefaction," *Journal of the Soil Mechanics and Foundations Division, Proceedings ASCE*, **94**, SM5, September 1968.

5:12 T. W. Lambe, "Stress Path Method," *Journal of the Soil Mechanics and Foundations Division, Proceedings, ASCE*, **93**, SM6, 304, November 1967.

5:13 H. J. Gibbs, J. W. Hilf, W. G. Holtz, and F. C. Walker, "Shear Strength of Cohesive Soils," *Proceedings of Research Conference on Shear Strength of Cohesive Soils, ASCE, New York, 1961*, pp. 33–162.

5:14 A. W. Bishop, I. Alpan, G. E. Blight, and I. B. Donald, "Factors Controlling the Strength of Partially Saturated Cohesive Soils," *Proceedings of Research Conference on Shear Strength of Cohesive Soils, ASCE, New York, 1961*, pp. 503–532.

5:15 F. D. Patton, "Multiple Modes of Shear Failure in Rock," *Proceedings of the First International Congress on Rock Mechanics, Lisbon, 1966*, **1**, 509.

5:16 *Testing Techniques for Rock Mechanics*, Special Technical Publication 402, ASTM, Philadelphia, Pa., 1966.

5:17 A. Singh and J. K. Mitchell, "General Stress–Strain–Time Function for Soils," *Journal of the Soil Mechanics and Foundations Division, Proceedings, ASCE*, **94**, SM1, 21, January 1968.

SUGGESTIONS FOR FURTHER STUDY

A. W. Bishop and D. J. Henkel, *The Measurement of Soil Properties in the Triaxial Shear Test*, 2nd ed., Edward Arnold & Co., London, 1962.

A. Kezdi, *Handbook of Soil Mechanics*, Vol. 1: *Soil Physics*, Elsevier Scientific Publishing Co., Amsterdam, Budapest, 1974, pp. 183–226.

C. C. Ladd, R. Fotl, K. Ishiharn, F. Schlosser, and H. C. Poulos, "Stress-Deformation and Strength Characteristics (of Soils): State of the Art," *Proceedings of the Ninth International Conference on Soil Mechanics and Foundation Engineering, Tokyo, 1977*, **2**, 421–494.

T. W. Lambe, *Soil Testing for Engineers*, John Wiley & Sons, Inc., New York, 1951.

R. E. Olson, "Shearing Strengths of Kaolinite, Illite and Montmorillonite," *Journal of the Geotechnical Engineering Division, Proceedings, ASCE*, **100**, GT11, 1215–1229, November 1974.

Proceedings of the Research Conference on Shear Strength of Cohesive Soils, ASCE, 1960.

M. L. Silver and T. Park, "Testing Procedure Effects on Dynamic Soil Behavior," *Journal of the Geotechnical Engineering Division, Proceedings, ASCE*, **101**, GT10, October 1975.

Stress–Strain Behavior of Soils, Proceedings, Roscoe Memorial Symposium, Cambridge, G. T. Foulis & Co., Ltd., Henley on Thames, 1971. (A conference collection of 36 papers.)

R. N. Yong and B. P. Warkentin, *Soil Properties and Behavior*, Elsevier Scientific Publishing Co., Amsterdam, 1975, 449 pp.

PROBLEMS

5:1 Given a major principal stress of 800 kN/m^2 and a minor principal stress of 100 kN/m^2, draw the Mohr circle. Find the maximum shear stress and the normal and shear stresses on a plane that makes an angle of 60° with the major principal plane.

5:2 Given a major principal stress of 575 kN/m^2 or 12,000 lb/ft^2 and a minor principal stress of 144 kN/m^2 or 3000 lb/ft^2, draw the Mohr circle. Find the maximum shear stress and the normal and shear stresses on a plane that makes an angle of 60° with the minor principal plane.

5:3 The normal stresses on two perpendicular planes are 1800 and 300 kN/m^2 or 18.3 and 3.06 kgf/cm^2 and the shear stresses are 600 kN/m^2 or 6.12 kgf/cm^2. Find the major and minor principal stresses graphically.

5:4 Given a major principal stress of 192 kN/m^2 or 4000 lb/ft^2 and a minor principal stress of 48 kN/m^2 or 1000 lb/ft^2, find the maximum shear stress and the angle of the plane on which it acts.

5:5 The shear and normal stresses on one plane are, respectively, 100 and 350 kN/m^2 or 2090 and 7300 lb/ft^2 and on a second plane, 190 and 140 kN/m^2 or 3970 and 2920 lb/ft^2.

a. Find the principal stresses.

b. Find the shear and normal stresses on a plane making an angle of 30° with the major principal plane.

5:6 Given the normal stresses on two perpendicular planes as 170 and 58 kN/m^2 or 3550 and 1210 lb/ft^2, the shear stresses on each as 110 kN/m^2 or 2300 lb/ft^2, draw Mohr's circle.

a. Can tension occur on any plane with this stress condition?

b. Find the principal stresses.

c. What are the shear and normal stresses on a plane making an angle of 74° with the direction of the major principal stress?

5:7 Given a major principal stress of 360 kN/m^2 or 7520 lb/ft^2, find the minimum value of the minor principal stress to limit shear stresses to 150 kN/m^2 or 3130 lb/ft^2.

5:8 A cylinder of concrete is tested in the ordinary manner and is found to have a "compressive" strength of 23,770 kN/m^2 or 3450 psi. The failure plane makes an angle of 63° with the major principal plane.

a. Draw Mohr's circle for the concrete at failure. (The minor principal stress is zero.)

b. Draw the Mohr rupture envelope, assuming it to be a straight line.

c. Find the "compressive" strength (difference between the principal stresses) if the minor principal stress is 7000 kN/m^2 or 1015 $lb/in.^2$.

5:9 Given ϕ of a sand, derive the algebraic relation between ϕ and θ.

5:10 Given ϕ of a sand, derive the algebraic expression for the ratio of the major principal stress to the minor principal stress when failure in the sand occurs.

5:11 A sample of sand subjected to a triaxial shear test failed when the minor principal stress was 153 kN/m^2 or 3200 psf and the major principal stress was 550 kN/m^2 or 11,500 psf. Draw Mohr's circle, and find ϕ and θ.

5.12 A sample of sand in a direct shear test fails when the normal stress is 300 kN/m^2 or 6260 lb/ft^2 and the shear stress is 200 kN/m^2 or 4175 lb/ft^2. Find the angle of internal friction and the principal stresses at failure.

5:13 Given the following stress conditions in a dense, angular, well-graded sand:

Plane A	Plane B
Shear stress 50 kN/m^2	-50 kN/m^2
Normal stress 170 kN/m^2	100 kN/m^2

Will failure occur?

5:14 A cylindrical sample of saturated rock flour composed of extremely fine-grained bulky particles is subjected to an unconfined compression test. The minor principal stress is developed by capillary tension in soil pores that have an effective diameter of 0.00075 mm. The angle the failure plane makes with the minor principal stress is 65°.

a. Compute capillary tension, u, and the effective minor principal stress.
b. Draw Mohr's circles of both total and effective stresses.
c. Find ϕ and the compressive stress necessary to produce failure.

5:15 A soil stratum 10 m or 32.8 ft thick overlies a bed of shale. The water table is 5 m or 16.4 ft above the surface of the shale and the height of capillary rise is 3 m or 9.85 ft. Above the zone of capillary saturation the soil is dry. The soil has a void ratio of 0.35 and a specific gravity of solids of 2.65. Draw diagrams showing the total, neutral, and effective vertical stresses in the deposit. (Remember that above the water table the neutral stress is negative, which denotes tension.)

5:16 A thin seam of sand lies inclined at an angle of 30° and intersects the base of a cliff. The drainage of the sand is stopped by accumulated talus. The sand is overlaid by clay 15 m or 49.2 ft thick and 1 m or 3.3 ft of topsoil. The ground surface slopes at 30°, and there is a deep vertical crack extending through the clay 20 m or 65.6 ft from the face of the cliff. The clay and topsoil weigh

18 kN/m^3 or 114.6 lb/ft^3 and the angle of internal friction of the sand is 40°. How high must the water rise in the sand before the block of clay slides upon the sand layer?

5:17 Derive the relationship between the angle of the p–q straight line envelope for sands and ϕ.

5:18 A saturated clay in consolidated-undrained shear had a ϕ_{CU} of 12°. Find the approximate value of ϕ_D graphically.

5:19 A saturated clay in drained shear was found to have a ϕ_D of 25°. Find the approximate value of ϕ_{CU} and the approximate unconfined compressive strength if the overburden pressure is 1200 psf.

5:20 Derive by means of Mohr's circle the relation between the major and minor principal stresses when given c from a quick shear test of a saturated clay.

5:21 Given the following data from an unconfined compression of saturated clay:

Stress kN/m^2	Stress (lb/ft^2)	Strain (dimensionless)
100	2090	0.0035
200	4180	0.0080
300	6260	0.0170
350	7310	0.0270
400	8350	0.0650

a. Plot the stress–strain curve.
b. Find the shear strength, c.
c. Find the average modulus of elasticity for 40% of the failure stress.

5:22 A sample of sandstone was tested in triaxial shear and found to have a cohesion, c, of 14,000 kN/m^2 or 2030 lb/in^2 and an angle of internal friction of 37°.

a. What will its compressive strength be in an unlined tunnel where the confining stress is zero and the water pressure is 1100 kN/m^2 or 160 lb/in^2?
b. If the water pressure should suddenly increase to 2100 kN/m^2 or 305 lb/in^2 because of the water hammer generated by a sudden shutdown of the turbines, what would the compressive strength be?

5:23 A sample of closely laminated schist has a strength defined by $c = 4200\ kN/m^2$ or 610 $lb/in.^2$ and $\phi = 42°$ for shear perpendicular to the laminations and a lower strength defined by $c = 3000\ kN/m^2$ or 435 lb/in^2 and $\phi = 33°$ for shear parallel to the laminations.

a. Plot the Mohr envelopes for this rock.
b. Find the range in the angle θ between the major principal plane and the plane of the laminations for which failure will occur parallel to the laminations when the minor principal stress is 7000 kN/m^2 or 1015 lb/in^2. (*Hint*: Draw failure circles for both conditions. The values of θ for the larger circles that intersect the lower envelope define the range where failure will occur with equal ease by both mechanisms. The polar diagrams will show relative orientation of planes.)

CHAPTER 6
Earth and Rock Construction

Rock and soil are among the oldest construction materials. Primitive man piled up stones for protection against weather, to shield the dead from scavengers or to worship his god. From this early random *rubble* he developed sophisticated coursed masonry and ornamental carvings that have survived longer than his bones. When stones were scarce, he molded soil into bricks, dried them in the sun and built his villages and cities. He excavated soil with the shoulder blade of a deer, hauled it in baskets on his head or back, and tamped it down to form both walls and mounds. Monks Mound of the Cahoika group near St. Louis, built from 900 to 1200 A.D. by American Indians, contain more than 1 million m^3—a major construction operation even with today's machinery.

Man's utilization of these simple materials was in the true spirit of engineering. He worked with materials that were locally available and readily worked with the tools he had, both of which were cheap. Moreover, the structures were durable, as is demonstrated by the number of ancient villages, mounds, and monuments that survive today.

The modern engineer uses earth and rock for the same reasons: They are locally available, readily handled by a variety of techniques, durable, and "dirt cheap."

The very ease with which earth can be handled and the potential for very low-cost construction mislead many developers and even some engineers. They presume that all earth and rock materials are alike and simple; neither effort nor skill are required to build earth structures properly, and they will endure forever without maintenance. Such thinking has caused needless failures and financial loss, as the following example illustrates.

In order to provide a level mall and parking area that would attract customers to a large shopping center in a hilly area, it was necessary to cut as deep as 12 m (40 ft) on the hill tops and fill 9 m (30 ft) in the low areas. The grading contract required placing the fill in 0.2-m (8-in.) layers and compacting the soil to a specified density as measured by field tests. In order to advance the opening date and take advantage of the pre-Christmas sales, the owner, over the protests of his engineer, waived the compaction requirements. The work was done during the hot dry weather of late summer. The fill was placed in layers as thick as 0.5 m (20 in.) and rolled with a heavy roller until the surface was hard. The owner was convinced of the quality of the fill when he found it impossible to drive a 12-mm ($\frac{1}{2}$-in.)-diameter rod more than 0.1 m with blows from a 4-kg sledge hammer.

December brought rain along with the crowds of people. Unpaved areas of fill soaked up water like a giant sponge. Settlements as great as 0.1 m (4 in.) developed in floor slabs and in building columns whose foundations were supported on fill. Portions of the parking area pavement settled 0.15 m (6 in.), causing water to pond and patrons to complain.

Tests of the fill in the area of settlement found very poor compaction. The hard appearance had been caused by lumps of dried clay that were rolled together into a loose, but rigid mass. The lumps melted upon wetting, however, and the fill settled into a wet spongy mass. In the parking areas it was necessary to replace the fill. In the building areas, foundations required underpinning, and floor slabs were releveled by pumping mortar beneath them. The ultimate cost and delay far exceeded the time and money saved by neglecting the project specification.

Modern technology has made it possible to build larger structures of earth and rock quickly and economically that will be even more durable than the monuments of the ancients if every step of the operation, from planning to construction, is properly engineered. The engineer must be intimately familiar with the materials, their ultimate behavior, and the construction operations that are necessary to obtain the required results if the product of construction is to perform as intended.

6:1 Soils and Rocks as Construction Materials

When soil or broken rock is used as raw material for construction, it undergoes so many changes that it ultimately has little resemblance to its undisturbed state. Excavation is the first step in the process of change: The structure is broken down by blasting or the action of a shovel or scraper; the different strata become mixed; and the water content increases or decreases, depending on the weather. In some instances the material is purposely modified to improve its characteristics: It can be mixed with other soils; it can be supplied with chemical admixtures that change its chemical and physical properties; or it can be bound together by a cementing agent. The final step of placing the soil or rock in the structure involves still more changes: Mixing produces more uniform composition, and compaction produces a controlled void ratio that is often considerably less than that of the original soil. Each step, from the undisturbed deposit to the finished structure, is a part of the manufacturing process; each requires engineering planning and quality control to insure a satisfactory product.

USES OF SOIL AND ROCK IN CONSTRUCTION / The most important use of soil and broken rock is in the construction of fills. A *fill* is a man-made deposit used to raise the existing ground surface or sometimes used to dispose of waste such as industrial by-products, garbage, or rubbish. The material from which it is constructed is termed *borrow*. Fills serve many purposes. Long narrow fills, termed *embankments*, carry railroads and highways across low areas or act as dams and levees to impound water. Fills are used in building construction to provide level sites in hilly country. If properly constructed, they can support heavy structures with safety and nominal settlement. Fills are placed behind retaining walls and bulkheads to secure the required ground surface contour and to bridge the gap between the wall and the original soil.

The foundation or supporting soil for a highway or airfield pavement is the *subgrade*. Subgrades can be the surface of the virgin soil or specially prepared, artificially compacted layers of soil or crushed rock.

Pavements themselves are sometimes constructed of soil or crushed rock. In such cases they often must be modified with admixtures and binders to give the pavements sufficient strength and resistance to deterioration.

In many countries earth is used for the walls of structures. Sun-dried bricks have been used for thousands of years in arid regions. In regions of more moisture, walls of earth are built by ramming the soil between temporary forms. Earth buildings are practical in areas where cheap construction labor and abundant supplies of clays and sands are available. They are durable when well constructed and protected from rain wash and flooding.

Another ancient technique, followed today, is to use soil as a plaster or filling over a loose framework of wood or reeds. This prototype of reinforced concrete is termed *wattle and daub*. It is used where wood is plentiful and where the environment (rainfall and earthquakes) are unfavorable to unreinforced earth.

CONSTRUCTION / Building structures of soil or rock requires a different approach from building structures of other materials. First, local materials must be used because it is expensive to transport large quantities a long distance.

Second, the design of the structure that is to be made of soil or rock must be very closely keyed to the construction. An engineer designs a concrete structure using assumed strengths, then writes specifications that will insure getting what was assumed. When the engineer plans an earth structure, however, he or she first must investigate the materials available, determine their suitability for construction, and then design the structure to fit the characteristics of available soils or rock. Often the design must be modified after construction has started to compensate for some unforeseen changes in the materials. The design of the structure is controlled by the engineering problem of manufacturing a suitable, economical material. Selection and processing are discussed in this chapter; the structural design of earth structures, such as embankments and dams, is discussed in Chapter 12.

Finally, the constructor must exercise engineering initiative in processing the materials, from excavation to compaction. He or she must guard against using methods that were found satisfactory on one job but which may not be suitable for handling different materials on a new project. For example, a contractor who has found that flooding helps to compact a damp sand too often wants to flood all fills with water. The result, if the soil is clay, is likely to be a soupy mass that will not support anything for years.

OBJECTIVES IN EARTH CONSTRUCTION / The product of earth construction, whether it be a fill for a highway, an embankment for a dam, the support for a building, or the subgrade for a pavement, must meet certain requirements:

1. It must have sufficient strength to support safely its own weight and that of the structure or wheel load on it.
2. It must not settle or deform under load so much that it damages the soil or the structure on it.
3. It must not swell or shrink excessively.
4. It must retain its strength and incompressibility for the life of the structure.
5. It must have the proper permeability or drainage characteristics for its function.

Strength is a major factor in the use of soil and rock in dams, high embankments, and subgrades. It depends on the nature of the material, the water content, and void ratio. In general, for any given earth material, the strength increases with a decrease in water content and with a decrease in void ratio (or increase in density). When the quality of the available material is poor, it is frequently possible to compensate for the deficiency by increased density.

Settlement from consolidation and elastic deformation is important in all applications of earth construction but particularly critical in subgrades and embankments that support pavements or structures. Settlement in the fill itself depends on the nature and density of the material of which it is composed. For most soil or broken rock, the elasticity increases and the compressibility decreases with increased density. By preconsolidation to a sufficiently high density through compaction, most soils and broken rock can be made to support reasonable loads without undue settlement.

Shrinkage of soil can be a factor in the settlement of pavements and structures on fill and is sometimes a hazard in the leakage of earth dams. The amount depends on the soil character, the density, and the loss in water content after construction; the greater the density and the less the water content change, the less the shrinkage. Generally, rock fills do not shrink.

Swelling is extremely hazardous because it disrupts the shape of the fill, damaging pavements and structures, and also because it is accompanied by a loss in strength. Swelling depends on the mineralogy of the rock or soil, the density, and the increase in moisture after construction. In general, the tendency to swell is increased with increased density. It can be controlled best by the proper choice of soils and by preventing (where possible) increasing water contents.

Loss of strength or increase in the compressibility are generally related to two mechanisms: the deterioration of the solid phase, and pore water pressure. Deterioration of the solids is a form of accelerated weathering induced by placing the material in a new environment. Moisture is a major factor in such physical and chemical deterioration: The clay minerals adsorb water, expand, and their bonding weakens; salts ionize to accelerate chemical reaction, and cemented bonds between particles soften.

Water pressure, as described in Chapter 5, is a direct factor in soil and rock strength. Seepage, induced by water pressure differences, can erode soil and some rocks and thus change them, as discussed in Chapter 3.

The losses in quality can be minimized by the selection of the material and by control of water as described in Chapter 3. Generally, the higher the density the slower the deterioration. Furthermore, a dense material, even in the deteriorated state, is generally better than the same material loose. However, if the material swells in the presence of moisture, too much density can promote deterioration.

Permeability is a factor in fills subject to temporary inundation, subgrades that must drain, and in dams. It is a function of the character of the soil and must be controlled by the proper selection of materials.

OBTAINING THE REQUIRED CHARACTERISTICS / In order to obtain the required properties in the end product of earth construction, the engineer controls the character of the material, the moisture, and the density. Moisture can be added or the material dried during construction. Afterward, however, the moisture is largely dependent on the environment and the use to which a structure is put, and usually is not subject to control no matter how desirable it may be. The material in some uses, such as the upstream face of a dam, is certain to become saturated. In other applications, such as a subgrade, the material can ordinarily be protected by drainage, but there is danger (however remote) that it could be saturated by abnormal rainfall. Proper water control by drainage is essential in the design of structures of earth in order to maintain the best properties under normal conditions. However, unless the engineer is certain that the drainage will always be effective, it is necessary to design earth structures on the basis that the moisture could increase to saturation.

Control over the character of the soil or rock is ordinarily limited by the materials available and the cost of their excavation and hauling to the job site. Too often, only a limited range of materials is present in any locality, and the engineer must select the best of what there is. It is sometimes possible to alter the soil or rock by processing, such as mixing two soils to improve the gradation of both, or by adding a material that alters their physical or chemical nature. Such a change to improve the material is termed *stabilization* and will be described in Sections 6:7 and 6:8.

The greatest control of the soil or rock properties is through densification. By densification it is usually possible to compensate for deficiencies in quality and for the deterioration in properties that results from increased moisture. The only property that is not improved by densification is the tendency to swell, and that must be controlled by proper selection.

6:2 Compaction Mechanics

From prehistoric times builders have recognized the value of compacting soil to produce a strong, settlement-free, water-resistant mass. Earth has been tamped by heavy logs, trampled by cattle, or compacted by rolling for more than 2000 years, but the cost of such crude work was often more than the value of the compaction. On the other hand, earth that was merely dumped in place without compaction sometimes failed under load and usually continued to settle for decades. Efficient compaction, the key to soil and rock construction, was promoted by R. R. Proctor, for whom the test for compaction is named.[6:1]

MECHANICS OF DENSIFICATION / Densification, or a reduction in the void ratio, occurs in a number of ways: reorientation of the particles; fracture of the grains or the bonds between them, followed by reorientation; and bending or distortion of the particles and their adsorbed layers. Energy consumed in this process is supplied by the *compactive effort* of the compaction device. The effectiveness of the energy depends on the type of particles of which the fill is composed and on the way in which the effort is applied. In a cohesive soil the densification is primarily accomplished by distortion and reorientation, both of which are resisted by the interparticle attractive forces of "cohesion." As the water content of the soil is increased, the cohesion is decreased, the resistance becomes less, and the effort becomes more effective. In a cohesionless soil or crushed rock the densification is primarily attained by reorientation of the grains, although fracture of the particles at their points of contact is an important secondary factor. The reorientation is resisted by the friction between the particles. Capillary tension in moisture films between the grains increases the contact pressures and increases the friction. As the moisture content increases, the capillary tension decreases and the effort becomes more effective.

In some sands and in broken rock, the local crushing of the points of contact between particles is a major mechanism of densification. Moisture accelerates the crushing and thus aids compaction. The acceleration of crushing reduces future settlement after construction.

If the moisture content is very high, however, the densification and accompanying decrease in void ratio leads to saturation. The buildup of neutral stress prevents further reduction in void ratio so that additional effort is wasted. Saturation, therefore, is the theoretical limit for compaction at any given water content.

In coarse cohesionless soils and broken rock, the permeability is so great that excess pore water can drain away. The limiting density is controlled by the particle geometry and the most favorable structural arrangement described as *packing*: the minimum void ratio.

MOISTURE–DENSITY RELATION / The importance of soil moisture in securing compaction is illustrated by the following experiments. A sample of soil is separated into six or eight portions. Each portion is compacted in a container with *exactly the same compactive effort.* Its water content and mass or weight of solids per unit volume of compacted soil, usually termed the *dry density,** ρ_d, or dry unit weight, γ_d, are determined:

$$\rho_d = \frac{M_s}{V}; \qquad \gamma_d = \frac{W_s}{V} \tag{6:1a}$$

$$\rho_d = \frac{\rho}{1+w}; \qquad \gamma_d = \frac{\gamma}{1+w} \tag{6:1b}$$

If a graph is plotted with water content as the abscissa and the dry density or dry unit weight as the ordinate, the resulting curve will be similar to Fig. 6.1. It will be seen that there is a particular water content, known as the *optimum moisture*, that results in the maximum dry density for the particular compaction method used. For a given soil, the greater the dry density, the smaller the void ratio, regardless of water content; so, maximum dry density is just another way of expressing the minimum void ratio or the minimum porosity.

For any particular water content, perfect compaction would expel all air from the soil and produce saturation. If the dry densities corresponding to saturation at different water contents are plotted on the graph, the result will be a curve that lies completely above the first. This is known as the *zero air voids* curve and represents the theoretical densities obtained by perfect compaction at different water contents. The theoretical maximum dry unit weight of the zero air voids curve γ_{zd} is computed from the specific gravity of

* In construction practice in the United States, the symbol γ_d is commonly termed *dry density*, and is expressed in lb/ft^3. In countries using the metric system both γ_d and ρ_d denote dry density in kg/m^3, g/cm^3, or $tons/m^3$.

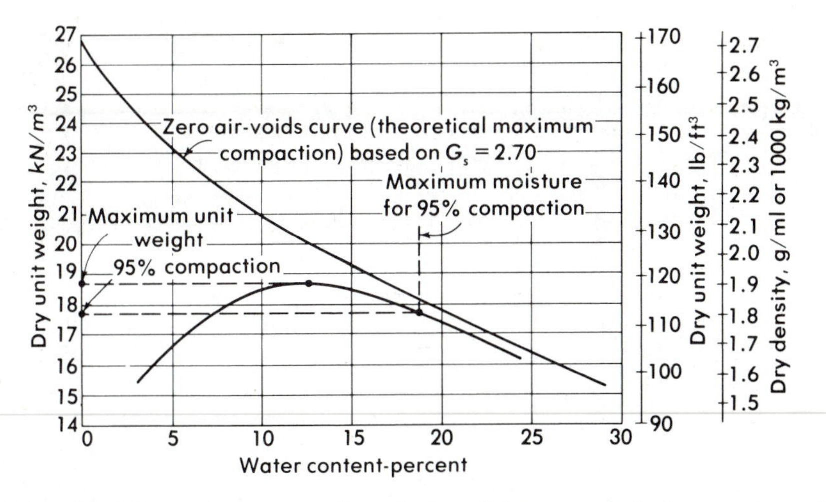

Figure 6.1 *Moisture-density curve of a cohesive soil for one method of compaction and maximum moisture for a specified degree of compaction.*

the soils for each given moisture:

$$\gamma_{zd} = \frac{\gamma_w}{w + (1/G_s)} \qquad \rho_{zd} = \frac{\rho_w}{w + (1/G_s)} \tag{6:2}$$

At high water contents the theoretical dry weight is low because much of the volume of the soil is occupied by water. At low water contents the theoretical weight increases until at a water content of zero it becomes equal to $\gamma_w G_s$, the weight of the soil particles.

Because of the geometry of the grains there is some limiting density (or minimum void ratio) beyond which no further change is possible without a major breakdown of the grains. This point can be fairly well defined for sands and is equivalent to the minimum void ratio (Section 1:11).

In materials having weak, porous grains, such as volcanic ash, coquina, and cinders, this limit cannot be defined and continued manipulation and compaction will produce increased dry weight until the material approaches a solid with $\gamma_d \rightarrow \gamma_w G_s$.

The compacted dry density increases with increasing moisture, as would be expected from the mechanics of the process previously described. The increase is limited by saturation—the zero air voids curve—where neutral stress prevents a further reduction in void ratio without a reduction in moisture. There is a small difference between the actual curve and the theoretical maximum caused by air trapped in the voids. The optimum moisture is a compromise where there is enough water to permit the grains to distort and reposition themselves by the compaction procedure employed but not so much water that the voids are filled.

In clays the optimum moisture for compaction by rollers is often close to or slightly below the plastic limit. In sands the moisture–density or unit weight curve drier than optimum is poorly defined (Fig. 6.2). In some cases it rises toward the maximum density at very low moistures because there is little capillary tension to resist repositioning of the grains.

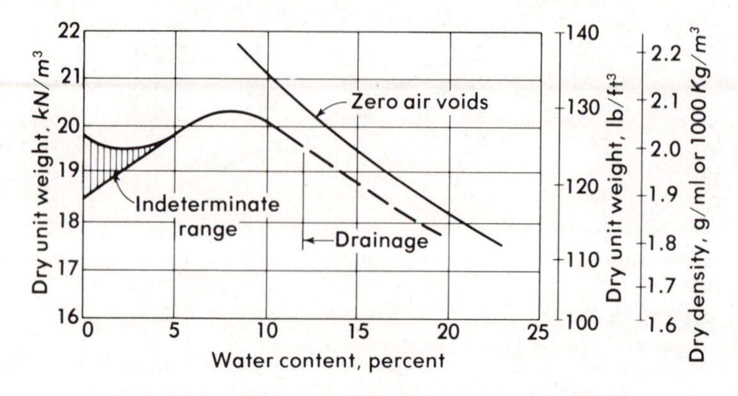

Figure 6.2 *Moisture density curve for a medium sand.*

COMPACTIVE EFFORT / If a second set of soil samples is prepared with different water contents, as previously described, and then compacted by a different effort, a similar moisture–dry density curve will be produced but with a different optimum moisture and maximum density. The greater the effort, the higher the maximum density and the lower the optimum moisture.

The relation between effort and maximum density is shown in Fig. 6.3. It is not linear, and a large increase in effort is required to produce a small increase in density. The way in which the effort is applied has a significant effect on the density. In cohesionless soils as well as crushed or broken rock,

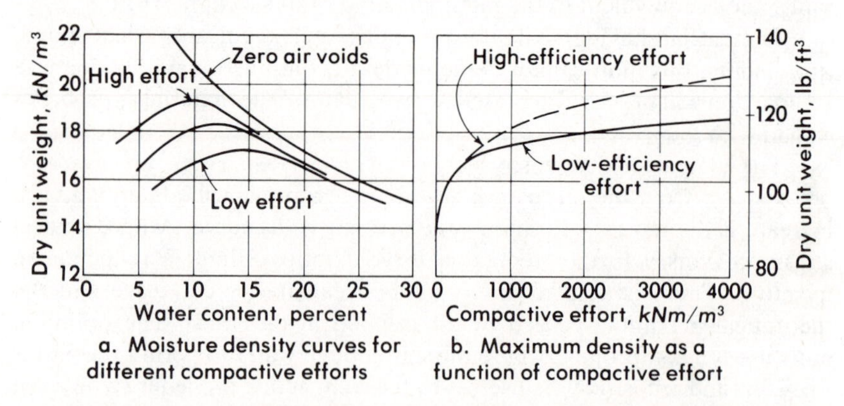

Figure 6.3 *Effect of compaction effort on moisture-density curves and maximum density.*

vibration, which reduces the friction between the gains, is particularly effective. In cohesive soils, pressure that bends and forces the grains into new positions is better. A large number of applications of small pressure is not so effective as the same total effort applied in a single application because a small force cannot overcome the cohesive resistance to grain movement, no matter how often it is applied. The duration of the effort is sometimes a factor in the density obtained. In coarse-grained soils the neutral stress that resists compaction at high moisture contents will not build up if the effort is applied so slowly that the water can drain away. In some clays a rapidly applied effort appears to mobilize viscous resistance in the water and is less effective than a slowly applied effort.

COMPACTION TESTS / A number of arbitrary standards for determining optimum moistures and maximum densities has been established to simulate different amounts of effort as applied by the full-sized equipment used in soil construction. The simplest and the most widely used are the "Proctor tests," named for R. R. Proctor, who first developed the optimum-moisture–maximum-density concept.[6:1]

Standard Proctor (ASTM D 698-78, AASHTO T-180-74)[6:2,6:3] Twenty-five blows of a 2.5-kg (5.5-lb) hammer falling 0.3 m (12 in.) on each of 3 equal layers in a 102-mm (4-in.)-diameter 0.943-liter ($\frac{1}{30}$-ft^3) cylinder.* The effort is 594 kNm/m^3 (12,400 ft-lb/ft^3), which is comparable to light rollers or very thorough tamping in thin layers.

Modified Proctor (ASTM D 1577-70, Modified AASHTO): T-180-74[6:3,6:4] Twenty-five blows of a 4.54-kg (10-lb) hammer falling 45 cm (18 in.) on each of 5 equal layers in a 102-mm (4-in.)-diameter 0.943-liter ($\frac{1}{30}$-ft^3) cylinder.* The effort is 2690 kNm/m^3 (56,200 ft-lb/ft^3), which is comparable to that obtained with the heaviest rollers under favorable working conditions.

A number of other procedures for obtaining optimum moistures and maximum densities have been developed, such as static pressures and kneading pressures, to simulate field conditions more closely.[6:5, 6:6] They are used primarily for research, however, because the results are not sufficiently different in many soils to justify the trouble.

These laboratory tests are limited to particles finer than 20 mm or $\frac{3}{4}$ in. Full-scale field pilot tests are necessary for coarser materials such as broken rock and gravels.

PERCENTAGE OF COMPACTION—RELATIVE DENSITY / It is frequently convenient to express the actual dry density or unit weight of a soil as a percentage of the maximum as defined by one of the two standards. This is the *percentage of compaction*. It can exceed 100%. The maximum dry density or unit weight from the laboratory compaction test is a function of

* If the soil contains many particles larger than a No. 4 sieve, a 152-mm (6-in.)-diameter cylinder of the same height is used and the blows increased to 55 per layer.

the method of compaction. The minimum void ratio (and its equivalent dry density) as defined in the discussion of relative density of cohesionless soils (Section 1:11) is the limit for all methods of compaction that do not break the particles. Therefore, there is no fixed relation between the minimum void ratio and maximum density as defined by any given level of compaction. Generally, the maximum densities for sands by the modified Proctor standard are equivalent to relative densities of between 95 and 100%.

6:3 Evaluation of Materials

The evaluation of the materials includes the determination of the quantity and quality of the materials available, the testing of the soils to find their physical properties when compacted, and the selection of the material and the degree of compaction to be used in construction.

FIELD SURVEY / The first step is a survey of all the soil deposits that can be used. The depth and extent of the different soil strata are determined by auger boring on a grid pattern, as described in Chapter 7. Samples weighing about 200 g ($\frac{1}{2}$ lb) are secured of each different material in each boring to the depth that it appears likely the soil will be removed.

PRELIMINARY EVALUATION OF SOIL / A preliminary evaluation of the soil samples is made on the basis of past experience. For this purpose, classification by the Unified System or the Public Roads System (Chapter 2) is helpful because both have been supplemented by performance ratings, such as Table 6:1. A number of state highway departments have developed systems that are applicable to their own peculiar soil problems. Since the ratings have been based on field behavior, classification is a cheap, rapid method for estimating the properties of the compacted soil and its suitability for construction.

COMPACTION STUDIES / Soils whose availability and estimated suitability for fill purposes have been found satisfactory by the field survey and classification are sampled again to secure enough material [20 to 40 kg (50 to 100 lb)] for more extensive testing. In order to avoid two sampling operations, representative large samples are often made in the initial survey, but the handling and transportation of many samples weighing 20 to 40 kg (50 to 100 lb) is a problem.

Tests are made of each sample to determine its natural or *field moisture* and to obtain its moisture density curve by one of the standard procedures. Samples are then prepared at different percentages of the maximum density, such as 92, 95, and 98%, usually at the highest moisture consistent with the percentage of compaction.

Swelling and shrinking tests can be made by soaking compacted samples in water and determining their volume increase and by drying samples to determine volume decrease. The sum of the volume increase and the decrease is called the *percentage volume change*. Fills made of soil having a volume change over 5% may require special provisions to prevent their moisture content from changing enough to cause damage by swelling and shrinking.

Strength and consolidation tests are made of the compacted samples, simulating the worst possible field conditions. Unless it is certain that the soil will never become saturated, the tests are made after soaking the compacted soil in water under the future confining load or saturating the soil under back

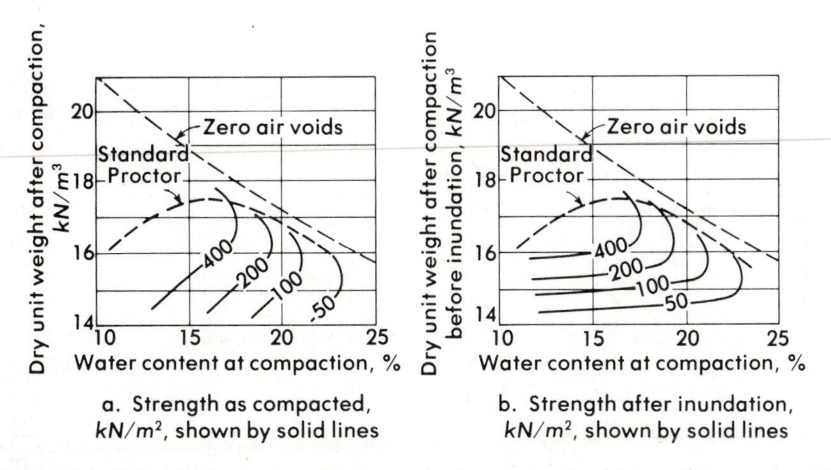

Figure 6.4 *Relation of undrained shear strength of a compacted cohesive soil to water content and dry density. (Adapted from Seed and Chan.[6:9])*

pressure. The results of the tests can be presented on moisture–density coordinates in terms of "contours" of settlement under a given load or contours of strength under a given confining pressure, as shown in Fig. 6.4, or as functions of density (Fig. 6.5).

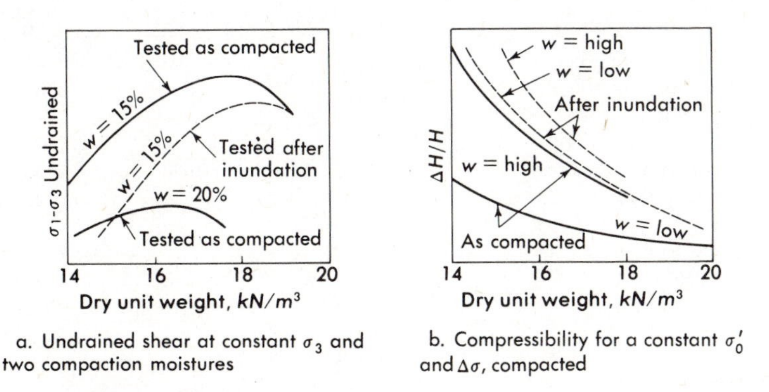

Figure 6.5 *Undrained shear strength and compressibility as functions of compacted dry density.*

TABLE 6:1 / CHARACTERISTICS AND RATINGS OF UNIFIED

Class	Compaction Characteristics	Maximum-Standard Proctor Dry Density (tons/m³)	Maximum-Standard Proctor Unit Weight (lb/ft³)	Compressibility and Expansion	Drainage and Permeability
GW	Good: tractor, rubber-tired, steel wheel, or vibratory roller	2.00–2.16	125–135	Almost none	Good drainage, pervious
GP	Good: tractor, rubber-tired, steel wheel, or vibratory roller	1.84–2.00	115–125	Almost none	Good drainage, pervious
GM	Good: rubber-tired or light sheepsfoot roller	1.92–2.16	120–135	Slight	Poor drainage, semipervious
GC	Good to fair: rubber-tired or sheepsfoot roller	1.84–2.08	115–130	Slight	Poor drainage, impervious
SW	Good: tractor, rubber-tired or vibratory roller	1.76–2.08	110–130	Almost none	Good drainage, pervious
SP	Good: tractor, rubber-tired or vibratory roller	1.60–1.92	100–120	Almost none	Good drainage, pervious
SM	Good: rubber-tired or sheepsfoot roller	1.76–2.00	110–125	Slight	Poor drainage, impervious
SC	Good to fair: rubber-tired or sheepsfoot roller	1.68–2.00	105–125	Slight to medium	Poor drainage, impervious
ML	Good to poor: rubber-tired or sheepsfoot roller	1.52–1.92	95–120	Slight to medium	Poor drainage, impervious
CL	Good to fair: sheepsfoot or rubber-tired roller	1.52–1.92	95–120	Medium	No drainage, impervious
OL	Fair to poor: sheepsfoot or rubber-tired roller	1.28–1.60	80–100	Medium to high	Poor drainage, impervious
MH	Fair to poor: sheepsfoot or rubber-tired roller	1.20–1.60	75–100	High	Poor drainage, impervious
CH	Fair to poor: sheepsfoot roller	1.28–1.68	80–105	Very high	No drainage, impervious
OH	Fair to poor: sheepsfoot roller	1.12–1.60	70–100	High	No drainage, impervious
Pt	Not suitable			Very high	Fair to poor drainage

* Adapted from Reference 6:7.

† Not suitable if subject to frost.

SOIL SYSTEM CLASSES FOR SOIL CONSTRUCTION*

Value as an Embankment Material	Value as Subgrade when Not Subject to Frost	Value as Base Course for Pavement	Value as Temporary Pavement: With Dust Palliative	Value as Temporary Pavement: With Bituminous Treatment
Very stable	Excellent	Good	Fair to poor	Excellent
Reasonably stable	Excellent to good	Poor to fair	Poor	Fair
Reasonably stable	Excellent to good	Fair to poor	Poor	Poor to fair
Reasonably stable	Good	Good to fair†	Excellent	Excellent
Very stable	Good	Fair to poor	Fair to poor	Good
Reasonably stable when dense	Good to fair	Poor	Poor	Poor to fair
Reasonably stable when dense	Good to fair	Poor	Poor	Poor to fair
Reasonably stable	Good to fair	Fair to poor†	Excellent	Excellent
Fair stability, good compaction required	Fair to poor	Not suitable	Poor	Poor
Good stability	Fair to poor	Not suitable	Poor	Poor
Unstable, should not be used	Poor, not suitable	Not suitable	Not suitable	
Fair to poor stability, good compaction required	Poor	Not suitable	Very poor	Very poor
Fair stability, expands, weakens; shrinks, cracks	Poor to very poor	Not suitable	Very poor	Not suitable
Unstable, should not be used	Very poor	Not suitable	Not suitable	Not suitable
Should not be used	Not suitable	Not suitable	Not suitable	Not suitable

STRENGTH AND COMPRESSIBILITY OF COMPACTED COHESIVE SOILS[6:8–6:11] / The physical properties of a compacted soil depend largely on the soil material, moisture, and density. In addition, the structure and the conditions of compaction that produced it are important in cohesive soils. When the cohesive soil is compacted at moisture contents less than optimum, an aggregated structure is formed; when the soil is compacted at high moisture contents, a dispersed structure is formed, with the flaky particles aligned in parallel.

The typical undrained strength of a cohesive soil, as compacted, at a constant confining pressure is shown in Fig. 6.4a. At constant density the strength decreases with increasing moisture; at constant moisture the strength increases with increasing density. An exception to the latter occurs as the zero air voids curve or saturation is approached. The increasing density brings about increased pore pressure and thereby decreased undrained strength. This sometimes occurs when a soil is compacted at a moisture content well above the optimum and can result in shear failure during construction. This is termed *overcompaction* and is prevented by proper moisture control.

The undrained strength of a compacted cohesive soil after inundation (under a constant confining pressure) is shown in Fig. 6.4b. Although the moisture content after inundation is nearly the same for equal densities, the strength decreases slightly with increasing initial (compaction) moisture, probably because the soil structure changes from the stronger aggregated to the weaker dispersed form. As before, close to the zero air voids curve the strength drops rapidly with increasing density because of the greater buildup of neutral stress in the dispersed structure.[6:9, 6:11]

The strength as a function of density for various water contents is shown in Fig. 6.5a. Generally, the undrained strength increases with increasing density, although it falls off as saturation is approached.

The compressibility of a typical cohesive soil as a function of compaction is shown in Fig. 6.5b. In general, compressibility decreases with increasing density (or decreasing void ratio) and decreasing compaction moisture. If the compacted soil is inundated before being subjected to consolidation, the compressibility is increased greatly. However, the soil that was compacted dry is less compressible than that which was compacted wet of optimum because compaction in a dry state produces more strong edge to face bonding than wet compaction.

The potential swell of clays increases with density and decreases with the compaction moisture. Generally, the swell is far greater for a soil compacted drier than the plastic limit than for one compacted wet. Swell can be minimized by limiting the degree of compaction and increasing the moisture content to the plastic limit or slightly above (somewhat more than the optimum moisture by the Standard AASHTO method).

SELECTION OF SOIL AND DEGREE OF COMPACTION / The final selection of the soil depends on its availability and compacted characteristics, and the cost of excavation, hauling, and compaction. The density specified is the minimum percentage of the maximum dry density that will provide the necessary strength and incompressibility under the worst possible, future moisture conditions. Table 6:2 is a guide for preliminary estimates when soil test data are not yet available.

TABLE 6:2 / TENTATIVE REQUIREMENTS FOR COMPACTION, UNIFIED SOIL SYSTEM CLASSES

	Required Compaction—Percentage of Standard Proctor Maximum		
Soil Class	Class 1	Class 2	Class 3
GW	97	94	90
GP	97	94	90
GM	98	94	90
GC	98	94	90
SW	97	95	91
SP	98	95	91
SM	98	95	91
SC	99	96	92
ML	100	96	92
CL	100	96	92
OL	—	96	93
MH	—	97	93
CH	—	—	93
OH	—	97	93

Class 1 Upper 3 m (9 ft) of fills supporting 1- or 2-story buildings
Upper 1 m (3 ft) of subgrade under pavements
Upper 0.3 m (1 ft) of subgrade under floors

Class 2 Deeper parts of fills under buildings
Deeper parts [to 10 m (30 ft)] of fills under pavements, floors
Earth dams

Class 3 All other fills requiring some degree of strength or incompressibility

EVALUATION OF ROCK FILL / The evaluation of rock and some industrial wastes for use in fills is less certain because the experiences with them have been difficult to evaluate quantitatively. Two major factors are involved: (1) the physical changes, particularly the fragmentation and crushing during excavation, handling, and compaction; and (2) the deterioration and physico-chemical changes after placement. These

TABLE 6:3 / ROCKS AS CONSTRUCTION MATERIALS

Rock	Excavation method	Fragmentation	Deterioration
Granite, diorite	Explosives	Irregular fragments, depending on blasting pattern.	Likely to be resistant.
Basalt	Explosives	Irregular fragments depending on joints.	Likely to be resistant.
Tuff	Machines* to explosives	Irregular, often with excessive fines.	Many forms deteriorate rapidly.
Sandstone, conglomerate	Explosives; machines if weathered, thin-bedded	Slabby to irregular, depending on bedding. Excess fines, depending on cementing.	Depends on character of cementing. Some deteriorate into silty sand.
Mudstone, siltstone, shale	Machines*	Small blocks to thin slabs and chips.	Many slake or break down rapidly into clay, accompanied by consolidation and loss of strength: should be considered suspect unless tests show otherwise.
Limestone: massive	Explosives	Irregular fragments, sometimes slabby.	Shaley seams deteriorate, otherwise resistant except to acids.
Coquina, chalk	Machines*	Porous fragments, excess fines common.	Some porous forms soften on wetting, others become partially cemented with alternate wetting and drying.
Quartzite	Explosives	Irregular, very angular.	Likely to be resistant.
Slate and schist	Explosives	Irregular, slabby to flaky, depending on laminations.	Some deteriorate on alternate wetting and drying.
Gneiss	Explosives	Irregular fragments, sometimes elongated or slabby.	Likely to be resistant.
Mine waste and industrial waste	Machines	Depends on the material, in most cases irregular.	Most forms (except igneous rock mine waste) should be considered susceptible until experience or tests show otherwise.

* Although excavation machinery alone can be used, loosening by explosives before excavation can be cheaper.

same factors apply to soils; however, they are more critical in rocks and industrial wastes. Most soils have already undergone major physical and chemical changes during weathering and are therefore at least partially adjusted to their environment. Freshly excavated rock and industrial wastes have not been subject to weathering; in their new environment of exposure to air, water, and physical punishment, they often change drastically.

A preliminary evaluation of rock for construction purposes is given in Table 6:3. So many factors are involved, however, that such a classification can only indicate what problems are likely to be encountered.

A more complete evaluation requires both field and laboratory tests. The field test is essentially a model of the proposed excavation and construction operation on a sufficiently large scale that it is realistic. The excavation, manipulation, and compaction utilize a variety of techniques that simulate full-scale construction. Tests of the behavior of the rock at all stages of the operation will show the degree of breakdown as well as the engineering properties.[6:12]

Deterioration is more difficult to evaluate. The best method is to examine old fills, spoil areas, and waste heaps of the material, as well as exposures in open cuts that are several years old. Changes in texture, color, and shear strength with increasing depth below the surface are indications of deterioration.

Deterioration can be tested qualitatively by exposing identical samples of compacted rock to different numbers of cycles of wetting and drying or to different periods of exposure to the environment. After exposure the samples are tested for volume, gradation, and strength. A change in any of these properties with time is an indication of deterioration. An interpretation of such test data must be based on a correlation with the performance of fills made of the same material.

6:4 Excavation, Placement, and Compaction

Excavation of the raw materials, processing, hauling, placement, and compaction are important from the standpoint of cost and the time required for construction. All, and particularly compaction, are also vital in determining the quality of the completed soil or rock structure.

EXCAVATION METHOD / The excavation procedure is usually selected by the constructor on the basis of the type of material and the layout of the borrow pit. However, because the excavation method affects the breaking of the material into small pieces and the way the material is mixed, it is sometimes necessary for the engineer to specify the methods to be used, to ensure the desired result.

Hand excavation has been used throughout history and is still important in areas of cheap labor. Stratified materials can be either mixed or segregated by layers with equal ease, and small pockets of objectional materials can be removed with little added expense. The soil is well broken up and usually requires little additional pulverization.

The *power shovel* and large *front loader* are adapted to a wide variety of materials, from soft soil to boulders and layered or soft rock. They are also suitable for hard rock that has been broken by explosives. Stratified materials are easily mixed because the buckets remove a nearly vertical slice; but segregation by layers is more difficult and expensive. At their most efficient output, they are likely to excavate soil in large chunks, but can be made to break the soil thoroughly, with some sacrifice in efficiency. They are most effective in deep borrow pits above the water table.

The *dragline* is adapted to most soils except tough clays and hard or cemented materials. Stratified materials can be either mixed or segregated if the layers are not thin. The soil is well broken up and mixed by the boiling action in the bucket. The dragline is most efficient when it excavates below its own base level, and for that reason it is effective below water. Of course, the materials are waterlogged in that case.

The *scraper* or *pan* is adapted to most soils except very soft or sticky clays. It can excavate broken rock. Stratified materials are easily segregated because the scraper can excavate layers as thin as 0.15 m (6 in.). Some mixing of layers is possible by making a deep sloping cut across several strata. The soil is broken up by the cutting blade and by the boiling action in the scraper bowl. The scraper is best in long borrow pits because it must travel in a straight line during excavation.

The *elevating grader* is similar to the scraper in the types of materials handled, but it tends to provide more pulversizing. A long level pit is required for best operation.

Continuous excavating machines originally developed for mining thin layers of coal have been adapted to large-scale excavation of soil and soft rock. These consist of a wheel 3 to 5 m (10 to 16 ft) in diameter, with teeth and shallow buckets on the perimeter. The wheel is operated in a vertical plane, at the end of a long boom that is pivoted so as to sweep across a pie-shaped sector of ground, at an ever increasing radius. The rotating wheel cuts the material into small fragments and drops them on to a belt conveyor mounted in the boom. The belt transfers the material to the operations center where it is transferred to the hauling equipment.

LOOSENING / Rock and the harder soils cannot be excavated without loosening. Thinly stratified rock and some hard soils can be loosened with *rippers* or *rooters*. Such devices consist of hook-shaped blades that are towed by a tractor or mounted on a road grader. The blade lifts rock layers, breaking them into slab-shaped pieces, and tears hard soils into fragments that can be excavated by loaders, shovels, or scrapers. They are very effective in thinly bedded sandstones, shales, and schists, partially weathered rock, and hard, brittle soils.

Very hard soils and most rocks require blasting. This is a highly developed art that must be planned and executed by experienced personnel. However, the engineer must maintain control over this work because the engineering properties of the fragmented material are largely controlled by the blasting operations. The size and gradation of the particles depend on (1) the joint and bedding plane spacing; (2) the hardness of the rock; (3) the spacing of the blast holes; (4) the depth blasted at one time (termed the *bench height*); (5) the spacing of the explosive throughout the drill hole; and (6) the amount and type of explosive. To a great degree an experienced *powder man* can compensate for variations in the rock by his control over the blast hole spacing and the *powder factor* (weight of explosive per cubic yard of material). In most cases full-scale experimenting is necessary to establish the best combination of drill pattern and explosive and powder factor to produce the size and shape of particles needed for the subsequent construction operations.

The soil–rock boundary presents many problems in excavation, both technical and contractual. The rock surface is usually irregular and often hard strata are underlain by soft seams. The technical problem arises as soon as the earth-moving equipment strikes the hard knobs or pinnacles in the rock surface: Earth moving is either severely hampered or practically impossible. It may be necessary to resort to drilling and blasting the knobs the same as in solid rock, although the total percentage of rock may be small. On the other hand, continuous rock consisting of alternate hard and soft seams sometimes can be loosened with rippers making it possible to excavate with equipment designed for soil. The contractural problem arises because of the change in procedure that is necessary and the greatly increased cost of work below the uppermost point of rock. It is essential that the contract documents describe the soil–rock interface as accurately as possible, and that they clearly define the limits of payment for each class of material. In some instances it is so difficult to do this that the contract lumps all excavation in one category, *unclassified*, that includes everything from muck to rock. Although this supposedly minimizes controversies, it makes it necessary for the contractor to include a large contingency for the uncertain quantities of cheap soil and expensive rock work.

BORROW PIT CONTROL / Moisture control is often necessary in the borrow pit to obtain efficient operation of the excavation equipment and to condition the soil for future compaction. Predrainage is required in low, wet pits. Plowing and exposure to air and sunshine often help dry the soil. Moisture addition makes it easier to excavate hard, dry soils. If the soil is drier than optimum, water addition is necessary for compaction; and when the moisture is added at the borrow pit, the moisture is mixed by the subsequent handling and is likely to be uniformly distributed.

Some gradation control may be necessary in the borrow pit or rock quarry. A *rake* attached to a tractor can remove boulders or oversized rock fragments. A bulldozer or front end loader can push aside an occasional large lump of hard soil or boulder that will be objectionable in the fill. In extreme cases, a coarse screen is fabricated of structural shapes, and all the material dumped through it to *scalp* the unwanted sizes. Such a *grizzley* impedes the normal excavation process, and is used only when other methods of oversize control fail.

Supervision of the borrow pit is necessary to ensure the quality of the materials. This requires experienced technicians who can recognize the specified soils, and sometimes even a field laboratory for checking moisture gradation, and plasticity.

HAULING / Transport of the soil and rock is primarily the concern of the contractor; although the operation is intimately involved in the cost of construction, it generally has no direct influence on the qualities of the soil or rock. However, if materials from several borrow pits are to be combined, or if different materials from a single borrow pit are to be segregated, such as sand for drainage layers and clay for impervious zones, then the hauling system must be compatible with these requirements. For example, a belt conveyor is not well adapted to transporting several different materials at the same time. Different materials can be easily segregated in single truck or scraper loads; if different materials from different sources are to be mixed, the proportioning can be controlled by the proportion of truck loads.

Some wetting or drying may occur during transportation; this is seldom significant, except for long belt conveyors. Addition of moisture to the trucks and scraper loads is sometimes practiced. Better mixing is usually possible when moisture is added before excavation or after spreading on the fill.

PLACEMENT AND PROCESSING / The placement of the fill materials depends on the method of hauling, the processing that is necessary before compaction, and the size of the area to be filled. The materials excavated by shovels, draglines, and elevating graders are hauled by trucks and wagons and *end-dumped* from the rear of the hauling unit in uniformly spaced piles, or *side-dumped*, or *bottom-dumped* from special units while moving to form long, narrow windrows. The scraper spreads the materials in ribbonlike layers the width of the blade.

The piles and windrows are spread out into uniform layers with a bulldozer or road grader. Leveling of the layers spread by the scrapers is sometimes needed. At the same time objectionable materials such as roots, clumps of grass, and large stones are removed.

If the materials are too wet, they are cut and turned with a disk plow so that they will aerate and dry in the sun. Cutting is also necessary if the soil is in lumps that are too large to compact. If the soil is too dry, the correct amount of water (plus an allowance for evaporation) is added by sprinkling irrigation and mixed into the soil by plowing.

In some cases it is desirable to mix different materials, such as a wet and a dry soil, to obtain optimum moisture, or a sand and a clay to secure a sandy clay. If the materials are hauled by trucks or *bottom-dumpwagons* they are placed in alternate piles or windrows and then mixed by blading them across each other. Materials hauled by scrapers are placed in thin layers, one on top of the other, and are mixed vertically by plowing. Different types of traveling mixers are available to pick up the soils, mix and pulversize them, and place them back, ready for compaction.

COMPACTION CHARACTERISTICS OF SOILS / Compaction occurs by the reorientation, fracture, or the distortion of the particles and their adsorbed layers. In a cohesionless soil, compaction is largely by fracture and reorientation of the grains. Static pressure is not very effective in this process, for the grains wedge against each other and resist movement. If the grains can be momentarily freed, then even light pressure is effective in forcing them into a more dense arrangement. Both vibration and shock are helpful in reducing the wedging and aiding compaction. Flowing water will also reduce the friction and permit easier compaction. However, water in the voids also prevents the particles from assuming a more dense arrangement. For this reason flowing water can be used to aid compaction only when the soil is so coarse grained that the water can drain quickly.

In cohesive soils compaction occurs by both reorientation and distortion of the grains and their adsorbed layers. This is achieved by a force great enough to overcome the cohesive resistance or interparticle forces. Vibration and shock are of little help, for, although they provide a dynamic force in addition to the static, this is largely offset by the increased cohesive resistance that accompanies dynamic loading.

For greatest efficiency the compaction force must be high enough to distort the particles and shift the individual grains but not great enough to shear the mass. In a cohesionless soil the strength is dependent on confinement. This can be provided by a wide area of load application. In cohesive soils the strength is dependent on void ratio and moisture and largely independent of the confinement.

In cohesionless soils efficient compaction requires a moderate force applied to a wide area, or shock and vibration. In cohesive soils efficient compaction requires higher pressure for dry soil than for wet, but the size of the loaded area is not critical. Increasing the pressure during compaction, as the density and strength increases, promotes efficiency.

COMPACTION METHODS / Many different compaction methods are in use, each with its adaptations and limitations that must be understood by the engineer. In many jobs the constructor claims that the specified degree of compaction is impossible, whereas it is only the wrong equipment or its improper use that is responsible for low density. Too often the performance data on compaction equipment do not define the job conditions or the character of the soil sufficiently for the engineer to decide if the equipment will work in the new application. Therefore, the engineer must study the mechanical behavior of the equipment and, if uncertain, conduct performance tests.[6:13]

Tamping is the oldest method of compaction. It provides momentary pressure at the instant of impact and some vibration. Because of this dual action it is effective in both cohesive and cohesionless soils. The hand tamper, a block of iron or stone with a mass of 3 to 4 kg or weighing 6 to 9 lb is the simplest, but the compactive effort is so small that the soil must be tamped in 25- to 50-mm (1- to 2-in.) layers at a moisture 2 to 4% above the Standard Proctor optimum. Furthermore, it is slow. Pneumatic tampers are faster, but because the moving part is light, they can seldom compact thicker layers. Greater weight and energy are supplied by a wide variety of tampers powered by internal combustion engines. They tamp rapidly, but the effort per tamp is seldom effective in layers thicker than 0.1 m (4 in.), Fig. 6.6a.

Weights as great as 20,000 kgf (44,000 lb), with end areas of 2 to 4 m^2 (22 to 43 ft^2), dropped 15 m (50 ft) by a crane produce both high impact forces and vibration that are transmitted deep within the mass. A pattern of 2 to 5 tamps every 10 m^2 (100 ft^2), termed *dynamic consolidation* by Menard, can densify loose sands and broken rock to a depth of 5 to 10 m (16 to 32 ft). The procedure is also effective in damp (but not saturated) cohesive soils and even loose debris and rubbish fills. The spacing, number of tamps, and effective depth are found by full-scale testing. The tamping loosens the uppermost 1 m (3 ft); the surface is compacted by rolling or vibration.

Rolling produces pressure that is applied for a relatively short time, depending on the roller speed. The *sheepsfoot roller* (Fig. 6.6b) consists of a steel drum with projecting lugs or feet. It applies a high static pressure to a small area with an equivalent diameter of 80 to 130 mm (3 to 5 in.). The pressure exerted depends on the number of feet in contact with the ground at any one time and on the roller weight (which can be varied by changing the water or wet sand ballast in the drum). Although pressures as low as 700 kN/m^2 (7 kgf/cm^2 or 100 $lb/in.^2$) and as high as 8500 kN/m^2 (85 kgf/cm^2 or 1250 $lb/in.^2$) are available, most equipment now in use falls into two categories: light, in the lower third of that range and medium, in the middle of the range.

Many different foot shapes are used. There is little experimental evidence that favors any one of these.

Because of the small width of the loaded area, the sheepsfoot roller is adapted best to cohesive soils such as clays. The medium rollers are capable of producing densities greater than the Standard Proctor maximum in layers 0.15 to 0.3 m (6 to 12 in.) (after compaction) and at moistures slightly below the optimum in six to eight passes over the surface. The light roller can produce 95% of the Standard Proctor maximum at optimum moisture with 0.1- to 0.15-m (4- to 6-in.) layers.

A modified sheepsfoot roller with feet 0.2 m (8 in.) wide is far better for silty soils of low cohesion because the increased width of the loaded area produces greater confinement. This device, termed the *elephant's foot roller* by the author, is made by removing some of the feet from the ordinary sheepsfoot and welding flat plates on those remaining. Segmented drum rollers have the same effect.

Tamping rollers are often operated in tandem. The trailing roller should have a different foot arrangement than the lead roller to prevent the trailing roller feet from following in the exact footsteps of the lead roller. The principal of increasing pressure can be realized by using larger feet and lower pressures on the lead roller.

Sometimes a sheepsfoot roller is employed to break up slabby fragments of soft rock, to pulverize a hard dry soil, or to aid in mixing or in the addition of water. Any compaction is incidental, and probably not of benefit in this stage of the operation. High pressures and narrow or worn feet are most effective.

The pneumatic tire has proved to be an excellent compactor for cohesionless and low-cohesion soils, including gravels, sands, clayey sands, silty sands, and even sandy clays. It applies a moderate pressure to a relatively wide area so that enough bearing capacity is developed to support the pressure without failure. The *light rubber-tired roller*, also known as the *traffic roller*, employs from 7 to 13 wheels mounted in two rows and spaced so that the wheels of the rear row track in the spaces between those of the front row. The wheels are mounted in pairs on oscillating axles so that they can follow the ground irregularities. The tires are similar in size to those used on small trucks. Each exerts a load of about 1500 kgf (1.5 tons) on the soil, depending on the ballast in the box mounted above the wheels. With a tire contact pressure of 350 kN/m^2 (3.5 kg/cm^2 or 50 lb/in.^{2}), the load is applied to an area with an equivalent diameter of about 0.2 m (9 in.).

The *heavy rubber-tired roller* (Fig. 6.6c) consists of four large tires mounted side by side on a suspension system that permits them to follow the ground irregularities. The load is provided by a ballast box, which is filled with earth, water, or even steel billets. A number of sizes are available, from a 35-ton maximum load to a 200-ton maximum load, with tire pressures of from 500 to 1000 kN/m^2 (5 to 10 kg/cm^2 or 75 to 150 lb/in.^{2}). The widely used 50-ton roller applies its 12,500-kgf wheel load to an equivalent circle of 0.4 m (16 in.).

The light rollers are capable of compacting soils in 0.1-m (4-in.)-thick layers to densities approaching the Standard Proctor maximum at optimum moisture in three or four passes. The heavy rollers can obtain densities far greater than the Standard Proctor maximum (as high as 105% of the Modified Proctor maximum in one case) in layers up to 0.5 m (20 in.) thick with four to six passes, at moistures comparable to the Modified Proctor optimum.

(*a*)

Figure 6.6 *Soil compaction equipment*: (*a*) *Mechanical tamper* (Courtesy of Wacker Mfg. Co.); (*b*) *sheepsfoot roller* (Courtesy of Wabco, Inc.); (*c*) *50-ton rubber-tired roller* (Courtesy of Bros, Inc.); *and* (*d*) *15-ton vibratory roller.*

(*b*)

(*c*)

(*d*)

The smooth-drum paving roller is sometimes used for compacting cohesionless soils in thin, level layers, but it tends to bridge over low spots. A drum-type roller with a surface of circular segments with space between, provides low pressure over a wide area and is useful in compacting sand, gravel, and crushed rock. The grid roller employs a heavy steel grid in place of the smooth steel drum. It develops a relatively high pressure over small areas and has been found useful in compacting cohesive soils in thin layers.

Vibrators in a number of forms have been developed for compacting cohesionless soils. The vibratory tampers consist of a curved metal plate or shoe on which is mounted the vibrator. Typical plate units have a weight of 40 to 60 kgf (90 to 130 lb), and impulse frequencies of 1000 to 2500 Hz (1000 to 2500 cps). They are capable of compacting cohesionless soils to densities as high as the Standard Proctor maximum in layers 0.15 to 0.3 m (6 to 12 in.) thick. They can also be used in silts and clays of low cohesion to obtain 95% of Standard Proctor maximum in layers 80 mm or 3 in. thick.

Vibrating rollers have been developed to provide both greater weight and greater intensity of vibration. One form consists of a steel drum 1 to 1.5 m (3.5 to 5 ft) in diameter and weighing several tons (Fig. 6.6d). An engine driven vibrating unit is mounted on the roller and delivers the impulse to the drum. They are capable of compacting cohesionless soils in layers from 0.3 to 1 m (1 to 3 ft) thick to the Standard Proctor maximum in two or three passes. The largest drum vibrators, weighing up to 15 tons are very effective in compacting broken rock up to 0.5 m (20 in.) in diameter in layers as thick as 1 m (3 ft). Significant density increases in sands and gravels have been measured 3 m (10 ft) below the surface after 3 to 6 passes of these machines.

The *vibroflotation* process employs a giant cylindrical vibrator that is suspended from a crane. Water jetted from within the vibrator loosens the soil and permits the vibrator to penetrate as deep as 15 m (50 ft). The vibrator and water help the particles reorient themselves, and produce cylindrical column 2 to 3 m (8 to 10 ft) in diameter. The device is slowly withdrawn while sand is placed in the annular space between the vibrator and the soil, filling the hole left behind. The water opens the hole for the vibrator and, in some cases, helps the particles reorient themselves. The method is suitable only for free-draining soils where the water in the voids will not hamper compaction. It can be used below the water table.

The *terra probe* consists of a 0.75-m (30-in.)-diameter pipe up to 15 m (50 ft) long, connected to a vibrator. The vibrator imparts a vertical impulse normally at 15 Hz (15 cps) (but variable between 12 and 18 Hz). The vibration enables the vertical pipe probes to penetrate a loose sand. Water jetting is sometimes used to aid penetration but is stopped during withdrawal. The vertical vibration will densify a cylinder of soil about 1 m (3 ft) in diameter, about 1 m deeper than the probe insertion.

Pile driving is very effective in compacting loose, cohesionless soils throughout great depths. The hammer vibration coupled with the displacement force of the pile is ideal, but the method is slow and expensive. The method ordinarily used is the same as for driving uncased concrete piles, but instead of concrete, sand can be used. The resulting piles are called *sand piles.*

The treads of a crawler tractor are efficient compactors of cohesionless soils placed in layers 10 cm (4 in.) thick. The action produced is light, static pressure combined with vibration.

Explosives can be used to densify very loose cohesionless soils, utilizing both the vibration and the transient pressure produced by the detonation.[6:14] The typical spacing of charges is 3 to 8 m (10 to 25 ft) on a grid pattern, with the explosive at a depth below the surface equal to or slightly less than the spacing. From 1.5 to 2 kg (3 to 5 lb) of dynamite are used per charge. The method is most effective in either dry or saturated coarse sand. Capillary tension in damp sand resists densification. Although dramatic increases of density are possible in some loose sands, the results are seldom uniform or as reliable as vibratory compaction.

Jetting and flooding have been used with some success in very permeable cohesionless soils. Flooding destroys the capillary tension, which prevents the grains from moving into more compact arrangements. Jetting with water under pressure in some cases provides flooding and a little vibration that compacts medium sands, but the results are extremely erratic. With modern, controlled compacting methods available, jetting and flooding should not be used.

FILL OPERATION[6:15, 6:16] / The selection of the proper compaction method, layer thickness, contact pressure, and moisture is the joint responsibility of the engineer and the constructor, for these affect the quality and uniformity of the soil structure and the speed and cost of construction.

The most important factor is the pressure in the layer being compacted. As shown in Fig. 6.7, the pressure beneath the compaction device decreases with depth (see Chapter 10). A high pressure applied over a small area, such as by the sheepsfoot roller, decreases rapidly, while a moderate pressure over a large area produces a more uniform pressure throughout the layer. The average pressure in a layer can be increased by decreasing the layer thickness or increasing the surface load. Usually, the best compaction efficiency is obtained with the maximum possible pressure that will not produce bearing-capacity failure. This is found by full-scale experiment. Bearing failure is indicated by rutting of the surface by rubber tires or failure of the sheepsfoot roller to rise out of the ground or "walk out" with each successive pass.

A loose, uncompacted soil has a low bearing capacity even when the compacted soil has high capacity. Heavy equipment on the loose soil is likely to create bearing failure and accompanying loss of compaction efficiency

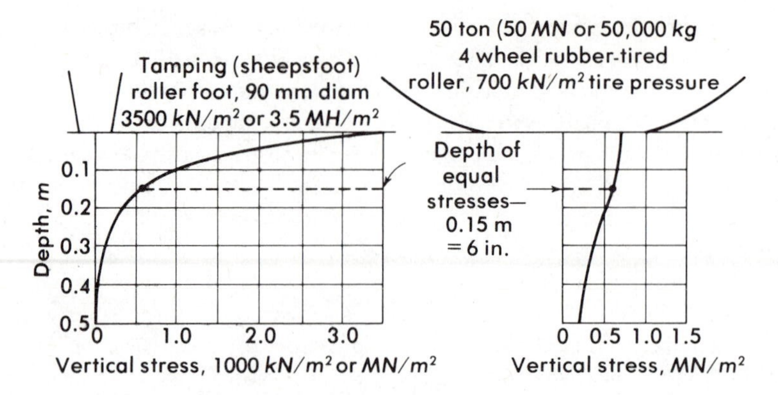

Figure 6.7 *Comparison of vertical stresses below a sheepsfoot and 50-ton rubber-tired roller.*

until the soil is densified enough to support the loading. If the soil is first partially compacted by a light roller, the bearing can be increased enough to support the heavy roller. Such *stage compaction* can be very effective where high densities are required.

The best moisture for compaction is the optimum for that particular method. The laboratory optimum has no exact counterpart in the field, but it serves as a guide. The Standard Proctor optimum is indicative of the needs of light rollers; the Modified Proctor optimum (a few percent less than the Standard) is indicative of the needs of the heaviest equipment.

On many projects a test area is constructed to try different combinations of equipment, pressure, layer thickness, and moisture to determine the best for each different soil. Such testing ultimately saves much time and money.

ROCK FILLS / Both coarse gravel and broken rock make excellent fills because they are free draining and frost-free. However, they require compaction to obtain their best strength and to minimize settlement. If the largest particles are smaller than about 0.3 m (1 ft), the fill is compacted by 50- to 100-ton rubber-tired rollers.

The 15-ton heavier vibratory rollers are even more effective. As mentioned previously, layers as thick as 1 m (3 ft) and particles as large as 0.6 m (2 ft) can be compacted.

Wetting promotes the compaction of most rock and minimizes future settlement. Water jets wash the fines from between the contact points of the larger particles allowing them to wedge more tightly. Water also helps break down the highly stressed projections of the rock, and allows them to crush and produce closer contact. The amount depends on the rock permeability; as little as $\frac{1}{5}$ volume of water to 1 of rock is required for fine rock while as much as 1 to 1 may be necessary for very coarse rock.

Very large rock, up to 20 tons, can be compacted by dumping and sluicing. The rock is dumped from a hillside or from one end of a completed embankment so that it slides and rolls down the slope. The momentum wedges the pieces together. At the same time water jets [of 75-mm (3 in.) nozzles at 700 kN/m^2 (100 psi) or more] wash the fines inward so that the coarser particles make intimate contact. Dumped rock fills are strong but tend to settle from $\frac{1}{4}$ to 1% of their height during the first 5 to 10 years.

DENSITY CONTROL / Continuous testing of the moisture and density of the compacted fill is essential to ensure that the finished product meets the requirements.

The density test procedure depends on the character of the soil. If little gravel is present, a thin-walled tube 0.1 m (4 in.) in diameter and 0.13 m (5 in.) long with a sharp cutting edge (Fig. 6.8) is used to secure a sample of known volume that is weighed and then tested for moisture to determine

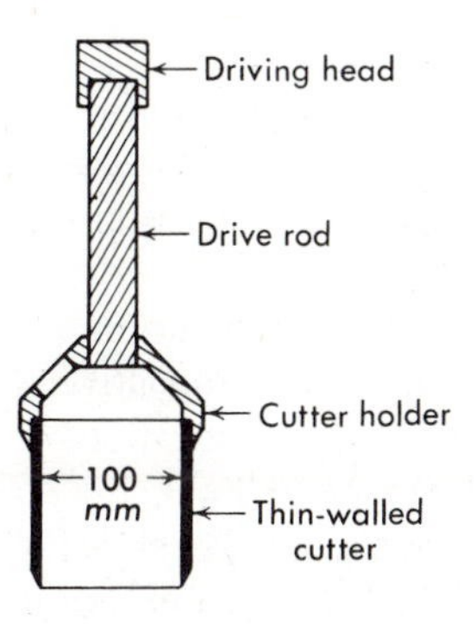

Figure 6.8 *Field density sampler (thin-walled drive type). Volume: $\frac{1}{30}$ ft^3 or 1 liter.*

the dry density. For gravelly soils the *sand cone method* is often used. A hole 0.1 to 0.15 m (4 to 6 in.) in diameter of 0.15 m (6 in.) deep is dug, and the soil from it is weighed and tested for moisture. The volume of the hole is measured by filling it with calibrated loose, dry sand that falls from a fixed height through a cone-shaped stand. An alternate method of measuring the hole volume is to line it with a thin rubber membrane and fill it with water. A number of similar devices, called *balloon density* meters, incorporate a retracting rubber membrane and self-indicating water supply to expedite measuring the hole volume. The compacted density is not uniform within the layer. It is often lower than average in the upper 25 mm (1 in.) because of surface disturbance, higher in the center of the layer, and falls off toward the bottom. The test gives the average for the depth sampled; that average is implied in writing compaction specifications. If the layer is thicker than the test depth, a second test should be made in the bottom of the layer.

The most reliable moisture test is conventional oven drying at 105°C. The drying time can be accelerated by drying the soil in wide shallow pans. A number of accelerated moisture tests are used to minimize construction delay in awaiting results: (1) drying at 200 to 250°C, and (2) alcohol absorption followed by burning. Direct moisture readings can be provided by electrical resistance of an absorbing medium and by gas pressure produced in a chemical reaction with soil moisture.

Moisture and density determinations can be made by measuring the effects of nuclear radiation through the compacted soil. Both surface tests of the uppermost 0.3 m (1 ft) of fill and deep tests throughout the fill depth are possible.

The accelerated tests are most reliable in cohesionless soils. For soils containing clays and mica, they can be seriously in error. The accelerated test results should always be checked against oven drying at 105°C for each different soil. If the error is constant, a correction factor may provide a reasonable approximation; if the error is inconsistent, the test must not be used.

QUALITY EVALUATION[6:16, 6:17] / The requirements for compaction are ordinarily specified in terms of the percentage of compaction. For cohesionless soils, the requirements alternatively can be defined by relative density. These requirements are established either by experience or by tests in order to produce a fill with the required strength or incompressibility. Within the fill there will be variations in quality due to differences in the soil composition, moisture, and compactive effort. Any evaluation must concern itself with the effect of these variations on the behavior of the total.

Sufficient tests must be made to define the range of quality. In large fills with uniform compaction, one test for each lift of fill and for every 1000 to 2000 m^2 (10,000 to 20,000 ft^2) of each different soil is adequate. In small fills, where the uniform operation of compaction equipment is impossible, 2 to 3 tests per lift in each area, however small, are necessary.

The results are best depicted on a frequency distribution graph (Fig. 6.9). The average and *median* quality as well as the variations can be easily determined.

Ordinarily, the necessary quality is assured by requiring that all test results exceed a specified minimum. The median quality, therefore, is greater than the specified minimum. Statistically, however, this approach can be misleading. Areas found to be less than the specified density are recompacted until they meet the requirements. However, assuming a random distribution of tests and a random variability of quality, there are probably other points, not tested, at which the compaction is less, and which will not be recompacted.

Requiring a minimum compaction level generally assures adequate fill performance because the median is well above the minimum. Furthermore, if a sufficient number of tests have been made, areas of unseen poorer compaction are likely to be scattered and have little detrimental effect on the whole.

For large fills a statistical evaluation of compaction is more realistic. A median density level is specified to assure the all-over behavior of the fill.

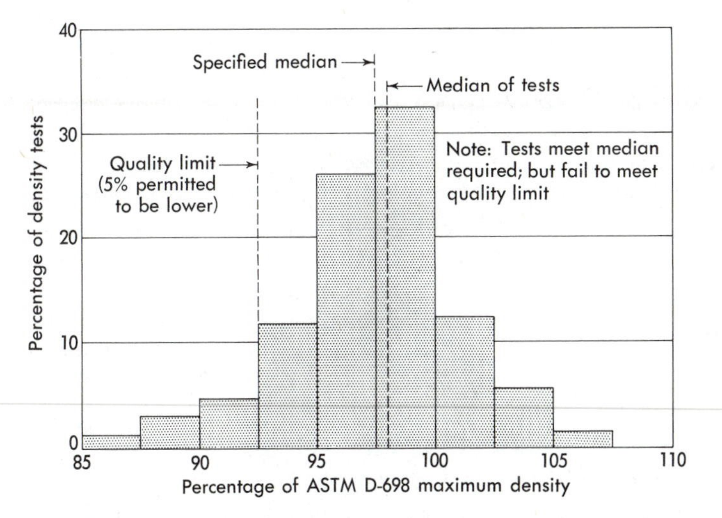

Figure 6.9 *Frequency distribution of percentage of compaction.*

A limit of variation or *quality limit* is permitted, usually 3 to 5% of the maximum density less than the median, with no more than 5% of the test values below that limit provided that the low densities occur at random. If the low densities are concentrated in one layer or area, that area must be recompacted until the median meets the requirements.

6:5 Hydraulic Filling

When large volumes of soil must be excavated and transported, hydraulic methods may be economical. This is particularly true along rivers or near sea- or lakeshores where sufficient quantities of water are available.

HYDRAULIC EXCAVATION AND TRANSPORTATION / Hydraulic excavation is most effective in cohesionless or slightly cohesive soils. Jets of water, forced by pressures as high as 1500 kN/m^2 (15 kg/cm^2 or 200 lb/in.^{2}) through 50- to 100-mm (2- to 4-in.) nozzles, wash the soil from the borrow pit into sluices. The mixture of soil and water can then be pumped and transported by pipe for miles.

Suction dredging lifts solids below water through a *suction head* connected to the intake of a large centrifugal pump. A rotary cutter on the suction head loosens sand, cuts clay into balls, and even breaks soft rocks, such as coquina, shale, and siltstone. The solids (grains, balls, or fragments) are carried by turbulence through the pump and into the discharge pipes in a suspension that is typically 90% water by weight. Too much solid matter increases the head loss and abrasion; too little means inefficient excavation. The maximum particle (or ball) size is about one third the suction diameter.

The suspension can be pumped several kilometers with only the dredge pump; greater distances are possible with booster pumps.

FILL CONSTRUCTION / If a hydraulic pipe line discharges directly on the ground, a fan-shaped mound is formed, with the pipe at its focus. The coarsest particles settle out close to the outlet, while the finer ones are carried away until the water velocity is low enough so that they will settle. Such a *run-out fill* has surface slopes between $5(H)$ to $1(V)$ in coarse sand, gravel, or clay balls, near the discharge, and slopes between $20(H)$ to $1(V)$ and $40(H)$ to $1(V)$ in the fine sands in the broader more distant parts of the fan. The discharge pipe is moved periodically to form a relatively flat surface, but the fill edges feather out on flat slopes, as described.

To confine the fill with steeper slopes, diking is employed (Fig. 6.10a). A starter dike, typically 2 m (6 ft) high, is placed around the area to be filled. The hydraulic pipe rests on the dike with discharge valves or sluices at regular intervals. The suspension fills the dike area forming a pond; the solids settle with the coarsest close to the dike, and the fines in the pond center. Unwanted silt and clay in suspension are discharged with the water over small spillways spaced around the perimeter dikes.

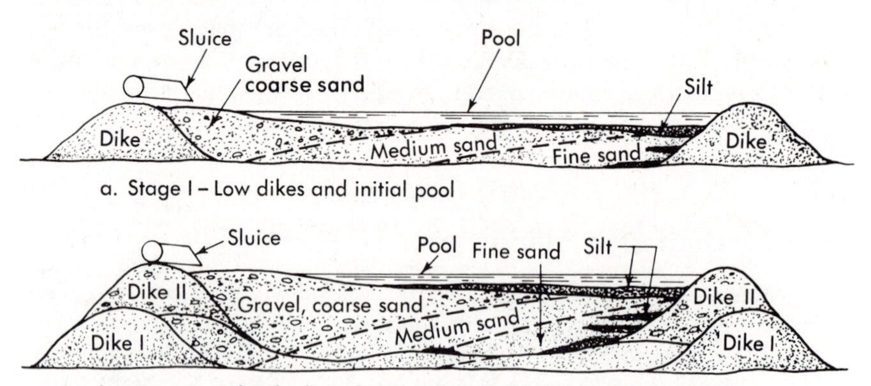

Figure 6.10 *Hydraulic filling with dikes.*

To reduce water pollution, the unwanted fines are recovered in an auxiliary diked area. The proportion of fines that settle can be controlled by the proportion of water to solids, and the spillway levels: The less the sedimentation time and the greater the water velocity, the coarser the fill will be. When the pond is full, the dike is raised by excavating the coarser material with a dragline. The process can be repeated to 2- to 3-m (7- to 10-ft) lifts. Heights greater than 30 m (100 ft) are practical with good control. In constructing dams by the hydraulic fill method, it is possible to make use of these different rates of sedimentation to create an impervious core of fines with outside shells of coarse, pervious soil. The outlets are placed at the outside of the fill where the coarse materials settle, and the fines for the core settle in the pool in the middle.

COMPACTING HYDRAULIC FILLS / Hydraulic fills are inherently loose throughout their depth. If they must support loads, they require densification or stabilization (Section 6:7). The first step is to drain sandier zones with ditches or well points. Capillary action produces temporary cohesion in sands and effective stresses that help consolidate silts and clays.

In-depth compaction by surface and deep vibration or heavy tamping described in Section 6:4 are effective in sands. However, it must be done in small increments with time for drainage in between.

Hydraulic fill construction is very cheap and has been used for many years in marine and dam construction. Unfortunately, the quality of the resulting fill is often poor, and expensive methods of compaction may be required to make the fill suitable for its purpose.

6:6 Densification by Preloading

Compressible soils, organic silts, clays, peats, and even waste fill can be densified and stabilized in place by consolidation. The increased stress is usually produced by a load at the ground surface, such as a fill. If the soil is inundated, lowering the water table by drainage can increase the effective stress (reduce buoyancy). Drainage can have a secondary effect by increasing capillary tension above the water table (negative pore pressure produces positive effective stress). This can be enhanced near the ground surface by evaporation and desiccation in hot, dry climates. The benefit of capillary tension and desiccation on effective stress increase are difficult to predict, however.

CONSOLIDATION MECHANICS / The technique is most effective in normally consolidated soils of high compressibility (Fig. 6.11a). The original, effective vertical stress is σ'_0 and the corresponding void ratio, e_0. The drained shear strength on a horizontal plane can be found from the

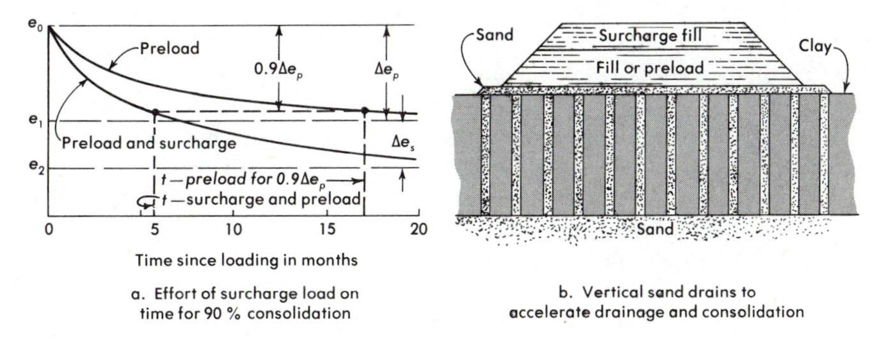

Figure 6.11 Stabilization by consolidation.

drained Mohr envelope [Equation (5:12a)], $\tau_{f0} = \sigma'_0 \tan \phi_D$. An increased stress from a structure, $\Delta\sigma$, to a final vertical stress of σ'_1 causes a void ratio decrease of Δe_1 and corresponding settlements. If a preload, $\Delta\sigma_p$, equal to $\Delta\sigma$ is placed on the soil and allowed to remain sufficiently long for the hydrodynamic consolidation to be virtually complete, the same void ratio reduction as will be produced by the structure will occur (Fig. 6.11b). The surcharge is then removed, followed by decompression, and a void ratio increase, Δe_2, and a higher void ratio, e_2. When the structural load equal to the preload is added, $\Delta\sigma$, the soil again consolidates. The void ratio decrease, Δe_3, is somewhat greater than the expansion, Δe_2, but much less than Δe_1.

The new strength on a horozintal plane can be computed from the drained shear parameter $\tau_{ff\text{-}1} = (\sigma'_0 + \Delta\sigma_p) \tan \phi_D$. The strength will decrease slightly when the surcharge is removed, corresponding to the decompression, Δe_2. The strength after the soil is reloaded will be equal to or slightly more than $\tau_{ff\text{-}1}$.

ACCELERATING PRECONSOLIDATION / The time rate of void ratio decrease is controlled by the rate of soil drainage through consolidation, described in Section 4:4. Whether the process is effective depends on the time required for consolidation compared to the time available. Because many, very compressible soils are also thick-bedded and impervious, preconsolidation is not feasible unless the process can be accelerated or unless low shear strength or future settlement can be tolerated.

By adding a second load, $\Delta\sigma_s$, which exceeds that required for eventual full consolidation, $\Delta\sigma_p$, the rate of consolidation is increased. The coefficient of consolidation, c_v, is nearly the same for both preconsolidation load, $\Delta\sigma_p$, and the load with surcharge, $\Delta\sigma_p + \Delta\sigma_s$. The percent consolidation, U, at any time, t, will be the same; however, the amount of void ratio reduction in time t with the larger load will be greater because the ultimate void ratio change will be greater, as shown in Fig. 6.11. Thus 90% of the ultimate settlement of Δe_p can be achieved by the surcharge load $(\Delta\sigma_p + \Delta\sigma_s)$ in a fraction of the time for $\Delta\sigma_p$ alone. There are two drawbacks. First, the stress increase produced by the surcharge must not cause shear failure of the soil. This will be investigated in Chapters 10 and 12. In some cases the loading can be added in increments, corresponding to the strength gain. This offsets some of the time benefits of surcharging, however. Second, although the average void ratio decrease will be nearly equal to that eventually produced by the surcharge alone, the center of the compressible stratum will have a higher void ratio and the top and bottom surfaces less, as illustrated in Fig. 4.9. When the surcharge is removed, the center will continue to consolidate while the surface decompresses. The net effect will be additional but small amounts of settlement. This added settlement and nonuniform strength must be within the tolerance of the project.

VERTICAL DRAINS / Consolidation can also be accelerated by shortening the seepage path with vertical drains (Fig. 6.11d). Two types are used. One consists of vertical holes 0.3 to 0.4 m (12 to 16 in.) in diameter filled with sand. The second consists of flat porous plastic or cardboard tubes that are punched vertically through the soil. The upper ends of both drain into a horizontal sand filter-blanket beneath the fill or surcharge. Typical spacings of the sand drains are 2 to 3 m (6 to 10 ft), and 1.5 to 2 m (5 to 6 ft) for the flat tubes. The horizontal, radial drainage[6:18] is computed in the same way as vertical consolidation, based on radial time factor curves and radial consolidation test data. Vertical drains are effective in thick homogeneous strata of clays when consolidation otherwise would be very slow. They are ineffective in alternate strata of sand and clay where the vertical seepage path is already short. In peat the permeability is high enough so that primary consolidation is so rapid that vertical drains are unnecessary. Moreover, there will be significant secondary compression that is not accelerated by drainage. The secondary compression, however, is accelerated by surcharge.

6:7 Soil Stabilization[6:19, 6:20]

Frequently, the soils available for construction cannot meet the strength and incompressibility requirements imposed by their use in embankments or subgrades. The process of improving the soil so that it does meet the requirements is *stabilization.* In its broadest meaning, stabilization includes compaction, drainage, preconsolidation, and protection of the surface from erosion and moisture infiltration. However, the term *stabilization* is often restricted to one aspect of soil improvement: the alteration of the soil material itself.

REQUIREMENTS OF STABILIZATION / The mode of alteration and the degree of alteration necessary depend on the character of the soil and on its deficiencies. In most cases additional strength is required. If the soil is cohesionless, this can be provided by confinement or by adding cohesion with a cementing or binding agent. If it is cohesive, the strength can be increased by drying, making the soil moisture-resistant, altering the clay-adsorbed water system, increasing cohesion with a cementing agent, and adding internal friction. Reduced compressibility can be obtained by consolidation, by filling the voids, cementing the grains with a rigid material, or by altering the clay-mineral adsorbed water forces. Freedom from swelling and shrinking can be provided by cementing, altering the clay-mineral water-adsorbing ability, and by preventing moisture changes. Permeability can be reduced by filling the voids with an impervious material, or by altering the clay-mineral adsorbed-water structure to prevent flocculation or internal erosion. It can be increased by removing fines or creating an aggregated grain structure.

Many different methods for stabilization have been proposed. From the standpoint of their function or effect on the soil they can be classified as follows:

1. Moisture-holding: retaining moisture in the soil.
2. Moisture-resisting: preventing moisture from entering the soil or from affecting the clay materials.
3. Cementing: binding the particles together without their alteration.
4. Void-filling: plugging the voids.
5. Mechanical stabilization: improving the soil gradation.
6. Physicochemical alteration: changing the clay mineral or the clay-mineral adsorbed-water system.
7. Consolidation.

A satisfactory process must provide the required qualities and in addition must satisfy the following criteria: (1) be compatible with the soil material, (2) be permanent, (3) be easily handled and processed, and (4) cheap and safe. Many methods have been employed but few with success. No one method meets all the requirements and most are deficient in the last criterion—cost. The principal methods and materials and their typical applications are described in the subsequent sections.

MOISTURE-HOLDING ADMIXTURES / Moisture in the soil provides some cohesion in sands and silts by capillary tension and prevents dust in all materials. It prevents shrinkage and cracking of cohesive soils and thereby reduces their surface disintegration in the first rain following a dry spell. Ordinary salt is excellent moisture-holding material in hot, relatively humid regions; it is applied at a rate of about 15 kg/m^3 or 25 lb/yd^3. Calcium chloride at 9 kg/m^3 (15 lb/yd^3) is very effective, particularly in dry areas, because it is deliquescent (capable of taking moisture from the air).

MOISTURE-RESISTING ADMIXTURES / Moisture-resisting or waterproofiing materials help keep water away from the soil particles and prevent softening or swelling. This is done in two ways: coating the grains or by preferential adsorption. Bituminous materials, such as RC-3 or 5 or MC-3 or 5 cutback asphalts, are the most widely used waterproofing agents. The amounts required vary from 2 to 7%, increasing with the percentage of fines. They are most successful in soils of low plasticity but in many soils tend to lose their effectiveness with time. Resinous waterproofing agents have been used but are expensive and likely to deteriorate badly.

Hydrophobic agents such as silicones and stearates, which are adsorbed on the particles in preference to water, appear to be relatively permanent in their effect. Processing is difficult, for each clay particle must be treated. The method is promising but requires development.

CEMENTING / A wide variety of cements or binding agents are employed in cementing, the most widely used and most successful method for stabilization. Although the most pronounced benefit is increased strength by cohesion, there is also a reduction in permeability of most soils by filling the voids with the cementing agent. When the cementing agent is relatively rigid, the modulus of elasticity of the soil is increased and its compressibility decreased.

SOIL CEMENT / *Soil-cement stabilization* employs portland cement to form a mixed-in-place concrete in which the soil is the aggregate. It has proved very successful in making low-cost pavements for light traffic and rigid base courses for the heaviest traffic.

The proper mix is determined by a trial procedure to obtain the required durability and strength. Samples of soil are prepared with different amounts of cement and compacted at optimum moisture by the Standard Proctor method. Higher densities are occasionally used; however, there is little evidence to show there is a significant improvement in the soil-cement properties.

The soil cement is cured under moist but not saturated conditions, because field curing is usually limited to retaining the compaction moisture by an impervious membrane or asphaltic seal coat. The typical curing period is 7 days, for Type I cement, because the greatest strength gain is in that period.

Evaluation of the compacted soil-cement depends on its use. For base courses or other exposed applications, durability controls the minimum cement content. Test samples are subjected to 12 cycles of freezing and thawing or 12 cycles of wetting and drying. The maximum volume change (swell plus shrink) permitted is 2%. The maximum loss of weight permitted (after brushing) (a measure of durability) ranges from 7% for A-6 and A-7 soils to 14% for A-1, A-2, and A-3 soils. In some cases a minimum, unconfined compressive strength is also specified. The required cement content is the least percentage that will satisfy these criteria. Typical amounts are 6% for sandy soils to 15% by weight for clayey soils.

Typical unconfined compressive strengths for soil-cement contents that provide the standard durability are from 1400 kN/m^2 (200 lb/in.^{2}) for clayey soils to 5500 kN/m^2 (800 lb/in.^{2}) for sandy soils. Higher strengths can be obtained by increasing the cement content, more or less in accordance with the water-cement ratio rules used in concrete mix design. Higher cement contents, however, increase the optimum moisture.

The key to successful soil-cement stabilization (as well as other stabilization) is thorough mixing. Most problems with soil cement arise from poor mixing of cohesive soils, where clay lumps containing no cement are locked in a matrix of soil containing excess cement. Proper compaction is also essential. Ordinarily, this is done in the same way as the soil alone would be compacted. The surface is finished by rubber-tired rolling. Soil cements are not highly abrasion-resistant. For pavements and areas exposed to percolation, surface sealing with a thin bituminous coating is necessary. This is usually applied after compacting the soil cement so that it also serves as the curing membrane.

Some organic colloids in the soil inhibit hardening. Even slightly organic soils should be considered suspect until tests prove otherwise. Sodium hydroxide treatment is often effective in correcting the organic effect before the cement is added. High sulfate contents in the soil are also damaging, and even sulfate-resistant cements may not prevent trouble. By far the widest application is for highway base courses. Tests indicate that a 6-in. soil-cement base course is as effective in spreading wheel loads as 0.3 to 0.4 m (12 to 15 in.) of crushed stone or bituminous concrete. Soil-cement linings for irrigation and drainage ditches are proving economical and desirable. Soil-cement erosion protection on dam faces has replaced riprap in areas where good stone is not available.

Soil-cement modification uses about one fifth the usual amount of cement to improve the strength and rigidity of soils that do not require complete stabilization. This is particularly useful when proper soil compaction is prevented by excessive soil moisture, and when sufficient time is not available to wait for dry weather.

Lime-fly ash stabilization resembles soil cement in that a pozzolanic cement is created by the reaction of lime on the silica of the fly ash. The proportions of lime to fly ash and the amount of cementing material are found by trial. Typical requirements are 10 to 15% of a mix consisting of two parts of ash to one part of lime mixed with the soil and compacted in the same way as portland cement.

BITUMINOUS CEMENTING / *Bituminous binders,* usually asphaltic cutbacks such as RC-1, RC-3, MC-1, and MC-3, have been used for both subgrades and low-cost pavements. Emulsified asphalts are also used, but they require a long period of warm, dry weather to permit the mix to cure properly. The amount of bitumen is determined by trial or from past experience, and usually is from 4 to 7% by weight. Bituminous stabilization finds its widest use in sandy soils having little or no clay, such as SW, SP, and SM classes.

CHEMICAL CEMENTING / Chemical cementing consists of bonding the soil particles with a cementing agent that is produced by a chemical reaction within the soil. The reaction does not necessarily include the soil particles, although the bonding does involve intermolecular forces of the soil. (Soil cement properly is classified as chemical cementing but is generally given a separate category.)

The first chemical cements were the soluble silicates, such as sodium silicate. In the presence of a weak acid or metallic salts the silicate breaks down into silica gel or an insoluble silicate. A hypothetical reaction is as follows:

$$Na_2SiO_3 \cdot 2H_2O + 2HCl = 2NaCl + SiO_2 \cdot 2H_2O$$

$$Na_2SiO_3 \cdot 2H_2O + CaCl_2 = 2NaCl + SiO_2 \cdot H_2O + Ca(OH)_2$$

$$Na_2SiO_3 \cdot 2H_2O + Ca(OH)_2 = CaSiO_3 + 2H_2O$$

The silica gel, $SiO_2 \cdot H_2O$, is a viscous, jellylike mass that solidifies into silica, with the release of water. The calcium silicate is an insoluble precipitate, reached after rapidly passing through the silica gel phase. The silica gel–soil mixture can be manipulated and rolled to form a membrane that becomes hard and impervious.

The calcium silicate precipitate fills the soil voids with an impervious binder in a flash reaction that does not permit manipulation. This form is utilized for injection into large voids to block the flow of water.

A delayed reaction is possible with the use of an organic reagent, such as *formamide*, that slowly breaks down to form the acid that produces the colloidal silica gel. The time of gelling can be controlled so that a period of from several minutes to several hours is available in which the soluble silicate and the reagent remain in their initial condition. This process is used in injection stabilization, where the lower viscosity of the ungelled silicate permits greater penetration of low-permeability soils.

The silicates have had their widest use in injection stabilization of sands and open-jointed rock to provide both strength and reduction in water flow. Although they shrink on drying and become brittle, they appear to be relatively permanent in a moist state.

Organic monomers include a wide range of complex chemicals that are either liquids, colloids or water-soluble solids that can be mixed with the soil or (in the most common applications) injected into the voids. A second liquid or solution, variously termed an *activator* or *catalyst*, causes the organic molecules to link together, processes that can be described as *polymerization*. The linked molecules form a lattice that binds soil grains, traps water molecules, and prevents free movement of water through voids. The polymer is an elastic solid whose strength and rigidity are controlled by its chemistry and its concentration. Although thousands of such reactions are known, those that can be used for stabilization must be capable of bonding with the soil or rock solids and remain stable indefinitely, be compatible with water, and have low enough viscosity to penetrate or mix readily. A controlled delay or induction time is also desirable to permit time for penetration of the mix into the soil.

The first of these to be extensively used in soil stabilization had the trade name AM-9. It is a mixture of acrylamide–methylene–bisacrylamide used in a 10% water solution with a viscosity only 1.5 times that of water. It polymerizes into a rubbery gel similar to very stiff gelatin. The rate of reaction can be controlled from a few minutes to 10 hours by the choice and proportion of activators. Its use has declined despite its versatility because the nonpolymerized chemical is neurotoxic.

A second system is *chrome-lignin*, which utilizes the waste lignin black liquor from a sulfite paper manufacturer. Potassium or sodium dichromate reacts with the lignin to form an organic monomer, chrome-lignin, which slowly polymerizes into a brown gel. The typical concentrations are from 10 to 20% by weight. The rate of gel formation is controlled by temperature and concentration; typical times are 15 minutes to 1 or 2 hours.

A phenolic liquid, bisphenol A epichlorohydrin, catalyzed by an amine with polyamide modifiers, polymerizes to a strong *epoxy* resin some of which have shear and tensile strengths exceeding that of concrete. However, the base liquids are so viscous that they must be diluted with organic solvents for most stabilization uses, which reduces the strength. Even with dilution they are among the strongest (and most-expensive) stabilizers.

Other synthetic resins include urea-formaldehyde and resorcinal-formaldehyde. Polyisocyate mixed with triethylene glycol, castor oil, or pentaneidol, polymerizes into *polyurethane*, a rigid resin, in the presence of water. The polymerized liquid hardens immediately when it reacts with soil moisture or with water in rock cavities or joints. By treating the unpolymerized liquid with diacetone alcohol, a liquid foam is produced which hardens in contact with water. This is particularly effective for stopping flowing water in large voids.

Chemical reaction with the soil particles to form a cementing agent shows some promise. Phosphoric acid with a wetting agent acts on the clay minerals to form insoluble aluminum phosphates.

FREEZING / *Freezing* the soil is, in effect, cementing with ice. Coaxial pipes are forced into the soil, and a refrigerant at −20°C or −4°F is forced down the center pipe and up the outside one to cool the soil. Typical tube spacings are 0.6 to 1.5 m (2 to 5 ft) around the perimeter of the area to be stabilized. They are slightly longer than the required solidification depth. To speed freezing in large areas, additional pipes within the area are helpful.

The water in the voids expands about 10% upon freezing, producing an increase in soil volume. In addition, continued cold below the water table can produce frost lenses (Chapter 3) and a corresponding stress in the soil on the outside of the frozen zone. The more rapid the freezing and the shorter the total time of freezing, the less the ice formation and increase in pressure adjacent to the zone.

The frozen soil becomes a temporary sandstone or mudstone. Its strength increases with decreasing temperature. Typical values adapted from Tsytovich[6:21] are given in Table 6:4.

TABLE 6:4 / STRENGTH OF FROZEN SOIL

Temperature	Clean Sand	Sandy Silt	Clay
−2°C +28 °F	6 MN/m^2 125 kip/ft^2	4 MN/m^2 80 kip/ft^2	1.5 MN/m^2 30 kip/ft^2
−12°C +10 °F	14 MN/m^2 290 kip/ft^2	13 MN/m^2 270 kip/ft^2	6 MN/m^2 125 kip/ft^2

The frozen soil creeps significantly, especially at higher temperatures. Upon thawing the soil becomes weaker than before freezing and decreases in volume because of the void expansion produced by freezing. The strength, creep, and volume changes should be measured by laboratory tests. They should be included in analyses for design and construction.

Freezing has been used to solidify soil beneath foundations that are failing: to stop mud flows (Chapter 12); to provide a rigid wall in soft, saturated soil; and to make it possible to tunnel through loose silt, fine sand, and fractured rock below the water table. Excavation is aided by steam-air jets that locally thaw the frozen soil. Both the initial installation of the refrigeration plant and the heat loss during freezing are significant in the cost of freezing.

SLURRY TRENCH / A trench filled with an artificial mixture of sand-gravel and semifluid plastic clay has proved to be an effective deep impervious membrane in pervious formations. The trench is excavated 2 to 4 m (6 to 13 ft) wide by a dragline in sands and gravels and 1 to 2 m (3 to 6 ft) wide by a clamshell bucket or rotary paddle. In most formations, the excavation walls are supported by keeping the hole filled with a viscous mixture of highly plastic clay and water, termed *slurry*. The slurry is usually made from commercial bentonite, similar to the drilling mud employed in oil wells, with the consistency of softened ice cream. The permanent membrane is formed by a mixture of the same slurry with sand and gravel that displaces the lighter slurry to form a void-bound structure. The cohesionless particles contact one another to form a reasonably rigid mass with enough slurry in the voids to provide imperviousness. The slurry material either stiffens thixotropically or hardens with the aid of a cementing agent. Membranes deeper than 60 m (200 ft) have been constructed by this process.

VOID FILLING / The filling of the voids in cohesionless soils reduces the permeability and at the same time maintains the soil strength by reducing water penetration. A number of materials are used for this purpose, including portland cement, soluble silicate gels, and the organic monomer-polymers that are also used in other applications as cementing agents.

Fine-grained materials such as silt, fly ash, and clay can also be used in void filling. They may gradually wash out under high gradients, however. Swelling materials with fine particles that can lodge in the voids and later expand to fill the voids have been developed. An example is a highly plastic clay that is treated so as to delay its water absorbency. Emulsified asphalt or latex can be treated so that the emulsion breaks in the soil voids and creates an impervious mass. Most of these materials are placed by injection to seal pervious, natural soil formations where excessive seepage is hazardous or expensive.

Bentonite, a dry, pulverized processed clay, largely montmorillonite, is an effective permeability-reducing admixture for lining reservoirs and waste treatment lagoons. From 3% for clayey soils to as much as 10% for sands and sandy gravel (dry bentonite to dry soil solids) can reduce the permeability to 1×10^{-6} mm/sec or less. Damp soil and the bentonite are thoroughly mixed, compacted in 3- to 6-in. layers at the optimum moisture for the mix to 85 to 90 percent of the Standard Proctor maximum. Immediately after compaction the layer is flooded to a depth of 0.3 m (1 ft) to allow the dry bentonite to hydrate or swell. It must remain inundated permanently or protected from drying to avoid shrinkage cracks that destroy the impermeability. If the liquid impounded contains dissolved salts, chemically treated bentonite is often necessary to minimize increasing permeability by leaching and base exchange.

MECHANICAL STABILIZATION / *Mechanical stabilization* is the improvement of the soil by changing the gradation. It usually consists of mixing two or more natural soils to secure a composite material that is superior to any of its components, but it also includes adding crushed rock or slag, or involves the screening of the soil to remove certain particle sizes.

The soil is considered to be made up of two components. The *aggregrate* includes all particles coarser than some arbitrary size, such as a No. 40 sieve (0.42 mm) or a No. 200 sieve (0.075 mm), and consists of predominantly bulky grains. The *binder* is the finer fraction and includes fine bulky grains and clay minerals. The aggregate provides internal friction and incompressibility and ideally consists of well-graded, strong, angular particles. The binder provides cohesion and imperviousness. It should have sufficient plasticity to develop high cohesion but not so much that it tends to swell. Based on experience, the best binders (finer than the No. 40 sieve) are CL soils with liquid limits less than 40 and plasticity indexes between 5 and 15.

The relative amounts of aggregate and binder determine the physical properties of the compacted stabilized soil. With no binder, the soil has high internal friction and is relatively incompressible because the loads are carried by grain-to-grain contact of rigid bulky particles, but the cohesion is negligible. With small amounts of binder, some is trapped between the bulky grains and is highly compressed by compaction, while the remainder partially fills the voids. The result is a sharp increase in cohesion, a slight decrease in the angle of internal friction, slightly greater compressibility, and relatively high permeability (and potential softening from water circulating in the voids). The optimum amount of binder is reached when the compacted binder fills the voids without destroying all the grain-to-grain contact of bulky particles. Increasing the binder beyond this point results in a sharp drop in the internal friction, a small increase in cohesion, and greater compressibility.

The design of a mechanically stabilized mixture is the determination of the proportions that will provide the optimum binder. A number of standard gradation specifications have been developed for this purpose, based on past experience, but most do not consider the effect of the grain shape and the volume of water adsorbed in the clay. A rational procedure separates each soil into aggregate and binder. The aggregates and the binders are compacted separately to determine the volume of voids in the compacted aggregate and the density of the compacted binder. The mix is proportioned so that the total binder (from all the ingredients) is from 75 to 90% of that required to fill the voids. Typical binder requirements for maximum strength are 20 to 27% solids by weight and are somewhat less than the amounts that result in the maximum compacted density. Mechanical stabilization is primarily used for pavement subgrades and for low-cost pavements, where some improvement in the soil is needed but great expense is not justified.

PHYSICOCHEMICAL ALTERATION / Physicochemical alteration, including *chemical stabilization*, consists of changing the properties of the soil grains, principally the clay minerals, and their adsorbed water. Ion exchange or *base exchange* is the changing of the cations in the adsorbed water films. The plasticity of the clay tends to decrease with increasing valence of the cations. By the addition of sufficient concentrations of chemicals with high valence cations, the soil is forced to exchange with a resulting lowering of its plasticity. Lime and calcium chloride provide calcium ions with a valence of 2, which bring about a marked improvement of high-plasticity clays having sodium or potassium cations. Aluminum sulfate and certain organic chemicals have also been used for this purpose experimentally. The amount of the chemical required is small, as low as 0.1% in some cases.

Lime stabilization includes additional mechanisms. The pH is raised, which makes iron and aluminum oxides less soluble and creates some cementing. Pozzolanic reaction with silica forms calcium silicate gels, similar to those of portland cement. Finally, the hydroxide can react with carbon dioxide to form insoluble calcium carbonate, a weak cement. Because of the complex mechanisms, the benefit of lime stabilization must be determined by tests at different lime concentrations. A typical requirement is 5%. Soils of high plasticity are improved most; cohesionless soils usually are not changed appreciably.

Electrochemical stabilization induces base exchange by an electric current. Aluminum cations leave a positive electrode of aluminum in the soil and migrate toward the negative, and in the course of their movement, bring about base exchange. At the same time electroosmotic drainage toward the negative electrode in the form of a well helps harden the soil.

Dispersing agents, such as sodium silicate and sodium polyphosphate, increase the repulsion in the clay-adsorbed water layers and cause the soil to develop a dispersed or oriented structure. The liquid limit, plasticity index, and permeability are reduced, and the maximum compacted density is increased. The procedure is inexpensive, for it requires only from 0.1 to 0.2% of cheap chemicals. The method has been employed successfully for sealing leaky ponds. The process may be reversible in chemical waste treatment lagoons.

Coagulating or aggregating chemicals provide the opposite effect to dispersion. The particles link themselves together in chains with large voids between. The permeability and plasticity are increased and the maximum compacted density is decreased. They improve cultivation and drainage and are used to help establish grass on steep slopes.

Surface-active agents and enzymes (also used in laundry detergents) increase the ability of water to wet a material. If the soil is too dry for compaction, a small amount of the agent, generally less than 0.1%, mixed with the added water will aid in uniform diffusion of the moisture. The maximum density may be increased slightly by their use and the strength and incompressibility improved somewhat.

Thermal alteration is the application of intense heat to desiccate the soil and even produce limited fusion and some vitrification. The bricklike mass, 2 to 3 m (7 to 10 ft) in diameter, that forms is permanently stabilized.[6:22]

A hole 0.1 to 0.2 m (4 to 8 in.) in diameter is drilled vertically through the depth requiring stabilization. A burner utilizing fuel oil or gas with compressed air is introduced near the bottom to create a column of burning gases that heats the walls of the hole to temperatures exceeding 1100°C or 2000°F. In porous dry soils such as loess, it is claimed that the burning gases penetrate into the mass. The walls of the hole rapidly vitrify into a glassy cylinder, and thereafter heat is diffused into the soil only by conduction. Within a few feet of the hole the soil becomes partially virtrified into a bricklike mass that does not deteriorate. Beyond, the soil is stabilized by desiccation. The cost of thermal stabilization depends on the soil moisture, the heat conductivity, and the availability of fuel. In dry soils in areas of abundant cheap fuel, it is apparently economical.

STABILIZATION PROCESSING / Processing is the most critical part of stabilization because the effectiveness of any method depends on what proportion of the soil particles are treated. The ease of processing depends on the cohesion: Cohesionless soils break up readily and are easily processed, whereas cohesive soils tend to form impenetrable lumps that defy treatment. Dry materials such as cement are spread mechanically on each soil layer, while soluble materials and liquids such as asphalt are added by sprinkling.

Proportioning of dry materials is usually by *layering*. A layer of soil 0.15 to 0.2 m (6 to 8 in.) thick is placed loosely but as uniformly as possible. The weight of soil per unit of area is established by testing selected samples. The dry stabilizer is spread across the surface at the weight per unit of area required by the soil. The batch method applies a measured weight to each unit of area defined by a grid laid out on the soil layer. The spreader method utilizes a mechanical device that spreads a thin uniform layer over the soil surface. This may be calibrated for a particular stabilizer, such as cement, but must be readjusted by trial for variations in stabilizer texture. Liquid admixtures are applied to the soil layer by spray, utilizing metering pumps. In either case variations in the proportioning of ±10% are normal, and ±20% are not unusual. If the soil is placed in windrows, the same approach is utilized. If the amount of stabilizer is large, it is placed in a parallel adjacent windrow. If the quantity is small, the stabilizer is placed on the windrow directly.

Mixing requires either extensive manipulation of the layer or the use of a traveling mixer. Manipulation includes cutting the stabilizer into the layer with a disc plow or harrow that penetrates the entire thickness. Many passes are required in alternating and different directions so as to obtain mixing in three dimensions. Finally, the layer is turned over back and forth with a road grader to ensure that there are no zones that are under- or overstabilized. Traveling mechanical mixers are more effective and can produce better blending of the materials. These consist of high-speed rotating blades that accurately cut through and lift the entire layer, throw it back and forth between blades or paddles within the machines, and then deposit the soil-stabilizer mix in a uniform layer. Similar devices are adapted to mixing windrows. In some cases portable premixing plants similar to those used for asphalt paving are used.

The mix is compacted as for soils. Compaction must follow stabilization before any permanent reaction develops; otherwise the compaction will break down any hardening that has occurred. Cement and emulsified asphalt stabilization require curing. A plastic membrane or a bituminous seal coat is applied to soil cement immediately after compaction to hold the mixing moisture, which is adequate for curing. Curing, ranging from a week or two for soil cement and lime to a few days for the RC bitumen, is necessary for stabilized base courses.

6:8 Grouting—Injection Stabilization[6:23–6:27]

Injecting the stabilizing agent into the soil, loosely termed *grouting*, makes it possible to improve the qualities of natural soil and rock formations as well as existing fills without excavation, processing, and recompaction. Grouting ordinarily has two objectives: to improve the structural properties or to reduce permeability. This is done by filling cracks, fissures, and cavities in rock and the voids in soil with a stabilizer that initially is in a liquid state or in suspension, and which subsequently solidifies or precipitates.

GROUTING MATERIALS / Many of the stabilizing agents previously discussed, and particularly the cements, are utilized in grouting. Grouting mixtures must meet a number of requirements:

1. Sufficiently liquid to be pumped.
2. Viscosity and particle size compatible with the size of opening to be filled.

3. Reaction or hardening time compatible with both the pumping requirement and the diffusion through the soil or dilution by ground water.

The properties of the grout must fit the soil or rock formation being injected. The dimensions of the pores or fissures determine the size of grout particle that can penetrate. Generally, the D_{85} of the grout must be smaller than $\frac{1}{3}$ the crack width B, or diameter of the smallest pores. If a soil is being grouted, the effective pore diameter is about $\frac{1}{5}D_{15}$.

$$D_{85}(\text{grout}) < \tfrac{1}{3}B(\text{fissure}) \qquad (6:3a)$$

$$D_{85}(\text{grout}) < \tfrac{1}{15}D_{15}(\text{soil}) \qquad (6:3b)$$

The ratio of the D_{15} of the soil to the D_{85} of the grout is sometimes termed the *groutability ratio*, R_g.

$$\frac{D_{15S}}{D_{85G}} = R_g \qquad (6:4)$$

Although the ratio should exceed 15, penetration is occasionally successful with ratios of 5 to 10.

The viscosity largely determines the rate of grout penetration under a given gradient produced by the grout pressure. Viscosities as low as possible are necessary in fine-grained soils or thin fissures. In large voids or cavities a high viscosity is desirable to restrict grout flow to the area where stabilization is required. The rate of hardening also controls the penetration. Rapid hardening restricts flow to large voids, while slow hardening permits maximum penetration through small voids.

The grout must not be unduly diluted or washed away by ground water. Insoluble or rapid-setting grouts are used in large water-bearing voids or cavities to minimize the grout loss by dilution and to stop flow.

The typical range of use of grouts is shown in Fig. 6.12.

INJECTION METHOD / The grouting plant includes the material handling system, mixers, pumps, and delivery pipes or hoses. Slow-setting grouts such as portland cement utilize a single mixing and pumping system. Rapid-setting or controlled time-of-set grouts require two or more systems, one for each component. The mixing of components is done by a proportioning valve or pump at the point of injection.

The injection point for soil consists of a perforated pipe with a conical point that is driven to the level at which the grout is to be injected. In rock or hard soil a hole is drilled to the grouting level. A grout pipe is sealed into the hole by an expanding gasket, termed a *packer*, just above the level to be grouted. If the rock is too broken or soft to hold a packer, the grout pipe is sealed into the formation by a quick-setting cement.

The injection pattern depends on the purpose. If an area is to be stabilized to increase its bearing capacity or reduce its compressibility, or

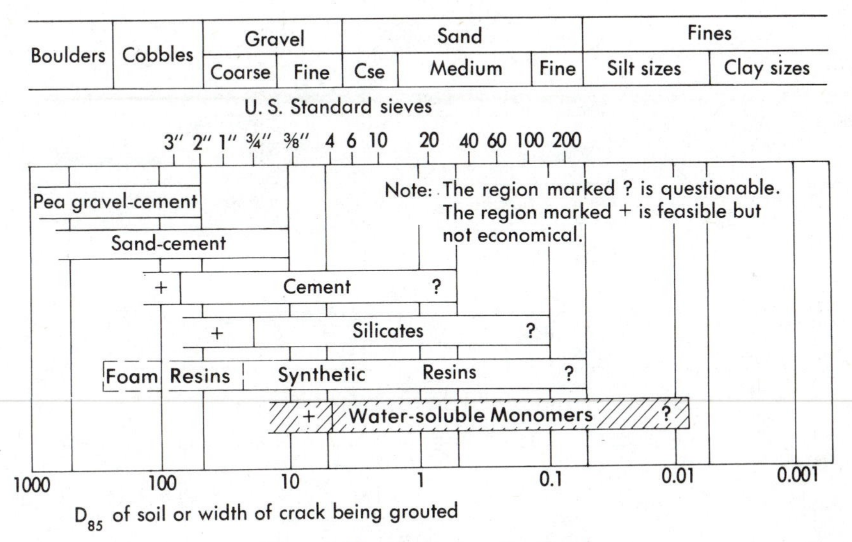

Figure 6.12 *Grouts for soil and rock.*

minimize hydraulic gradients, a grid pattern known as *compaction grouting* is used. Typical spacings are 6 to 15 m (20 to 50 ft) initially. If an impervious barrier is to be constructed, one to three parallel lines of injection are used. The line spacing may be 6 to 12 m (20 to 40 ft) and the hole spacings along the lines 3 to 6 m (10 to 20 ft) initially.

Each hole is grouted in increments of depth of from 3 to 15 m (10 to 50 ft), using sufficient pressure to force the grout into the voids and fissures, but not high enough to damage the formation. Generally, the pressure measured at the ground surface is initially limited to 25 $kN/m^2 \cdot m$ (0.25 $kgf/m^2 \cdot m$ or 1 $lb/in^2 \cdot ft$) of depth below the surface. Higher pressures may be used if no ground heave occurs. Usually, each level is grouted until the volume of grout is sufficient to fill the voids in a hypothetical cylinder of soil (Fig. 6.13), the diameter of which equals the hole spacing or until the resistance builds up to the pressure limit.

After the first pattern is complete, its effectiveness is tested by drilling *split spacing* or intermediate holes halfway between the initial grid holes or halfway between the holes on each grout line. If the intermediate holes take no grout, that portion of the grout grid or line is considered complete. If grout can be pumped, each new hole is grouted as before. A second split spacing is employed between all holes taking grout on the first split. Sometimes four or five splits are required to effectively grout an area.

The hole orientation is designed to fit the porosity of the formation. In porous or horizontally stratified soils, vertical holes are used. In jointed rock, the holes are inclined to intersect as many fissures as possible (Fig. 6.13b). In cavernous limestone holes 0.3 to 1 m (1 to 3 ft) in diameter may be

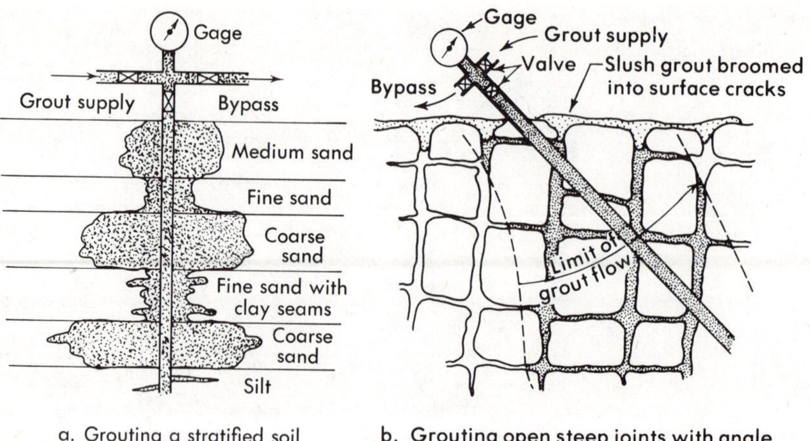

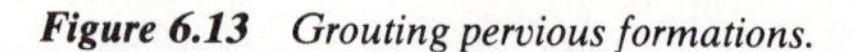
Figure 6.13 *Grouting pervious formations.*

drilled to permit direct access and pumping concrete through a hose into the larger cavities.

GROUTING FINE-GRAINED SOILS / Because of the low permeability, it is impossible to secure much penetration of the most fluid grouts into the voids of silts and clays. Instead, the grout forms irregular fingers and sheets that penetrate weaker seams and force them apart. This fingering or *hydraulic fracturing* produces some consolidation of partially saturated soils. Their strength will be increased and compressibility reduced. This is termed *compaction grouting.* The grouts used are very viscous portland cement–sand–bentonite–water mixtures, injected at pressures as high as 4000 kN/m^2 (600 $lb/in.^2$). Grouting of poorly compacted clay fills has arrested building settlements and has reduced lateral soil movement. The results are usually difficult to predict; therefore, grouting silts and clays always should be considered experimental until the results show otherwise.

PLANNING AND EXECUTING GROUTING / Grouting is a highly specialized art requiring an intimate familiarity with the structure of the soil or rock formation and a high degree of experience with the materials, equipment, and procedures that might be used. Any grouting program must be considered tentative and is revised as the work progresses. Therefore, the work must be supervised by experienced engineers who can make the required decisions without delay.

GROUTING CAVITIES / Enormous quantities of grout are required to fill cavities and old mine chambers, both because of the size of the opening and because the grout flows easily, far beyond the area where it is needed. Running water aggravates the grout loss. Hot asphalt grout solidifies in contact with the water, forming an elastic membrane that confines the liquid, yet allows the mass of grout to expand and plug the opening. Cloth bags attached to a grout pipe and filled with cement grout also can plug openings with running water. Pea gravel washed into the cavity through a 15-cm (6-in.)-diameter hole will partially plug the cavity and can later be grouted with sand–cement to form a solid mass. Cavity fill grout is generally placed in batches of 3 to 10 m^3 (100 to 1000 ft^3) and then allowed to harden before a second batch is introduced. In this way the loss beyond the area needing grout can be limited.

6:9 Old Technologies—New Applications

Although humans have been building earth and rock fills for more years than they have recorded their thoughts in writing, the objectives and the techniques have not changed greatly. Soil improvement is somewhat younger, although mixtures of soil and dung or straw have a recorded history of more than 3000 years.

In many respects, the earth construction industry appears stagnant. There have been few significant developments in the past 30 years, only modest improvements, with tomorrow's problems just yesterday's repeated. There is room for major innovation as the better soil borrow areas are depleted, and at the same time more complex structures require higher-quality embankments.

There have been many dramatic developments in stabilization during the same 30 years. Some have been witches' brews promoted with the same enthusiasm and respect for facts as the snake-oil cure-all of the legendary medicine man. Others have been successful (such as AM-9) but proved too hazardous. Still others, such as slurries, are almost magically effective in most applications but occasionally failures. Systematic important research is needed to raise stabilization from the level of alchemy to that of science-based engineering. Until then, experience and intuition still play a major role in the quality of the product.

REFERENCES

6:1 R. R. Proctor, "Fundamental Principles of Soil Compaction," *Engineering News-Record*, August 31, September 9, September 21, September 28, 1933.

6:2 ASTM D-698-70, "Moisture Density Relations of Soils Using 5.5 lb (2.5 kg) Hammer and 12 in. (304.8 mm) Drop," *Annual Book of Standards, ASTM*, Philadelphia, Pa.

6:3 J. E. Bowles, *Engineering Properties of Soils*, 2nd ed., McGraw Hill Book Company, New York, 1968, pp. 79–96.

6:4 ASTM D-1557-70, "Moisture Density Relations of Soils Using 10 lb (4.5 kg) Hammer and 18 in. (457 mm) Drop," *Annual Book of Standards, ASTM*, Philadelphia, Pa.

6:5 S. D. Wilson, "Small Compaction Apparatus Duplicates Field Results Closely," *Engineering News-Record*, pp. 34–36, November 2, 1950.

6:6 "Factors Influencing Compaction Test Results," *Bulletin 319*, Highway Research Board, Washington, D.C., 1962.

6:7 "The Unified Classification, Appendix A—Characteristics of Soil Groups Pertaining to Roads and Air-fields" and "Appendix B—Characteristics of Soil Groups Pertaining to Embankments and Foundations," *Technical Memorandum 357*, U.S. Waterways Experiment Station, Vicksburg, Miss., 1953.

6:8 T. W. Lambe, "The Engineering Behavior of Compacted Clay," *Journal of the Soil Mechanics and Foundations Division, Proceedings, ASCE*, **84**, SM2, May 1958.

6:9 H. B. Seed and C. K. Chan, "Structure and Strength Characteristics of Compacted Clays," *Journal of the Soil Mechanics and Foundations Division, Proceedings, ASCE*, **85**, SM5, October 1959.

6:10 H. B. Seed and C. K. Chan, "Undrained Strength of Compacted Clays After Soaking," *Journal of the Soil Mechanics and Foundations Division, Proceedings, ASCE*, **85**, SM6, December 1959.

6:11 Jack Hilf, "Compacted Fill," in *Foundation Engineering Handbook*, H. F. Winterkorn and H. Y. Fang, eds., Van Nostrand Reinhold Company, New York, 1975, Chapter 7, pp. 244–311.

6:12 R. J. Marsal, "Large-Scale Testing of Rockfill Materials," *Journal of the Soil Mechanics and Foundations Division, Proceedings, ASCE*, **93**, SM2, 27–43, January 1967.

6:13 "Factors That Influence Field Compaction of Soils," *Bulletin 272*, Highway Research Board, Washington, D.C., 1960.

6:14 B. J. Prugh, "Densification of Soils by Explosive Vibrations," *Journal of the Construction Division, Proceedings, ASCE*, **89**, CO1, March, 1963.

6:15 F. C. Walker and W. G. Holtz, "Control of Embankment Materials by Laboratory Testing," *Transactions, ASCE*, **118**, 4, 1953.

6:16 W. J. Turnbull, J. R. Compton, and R. G. Ahlvin, "Quality Control of Compacted Earthwork," *Journal of the Soil Mechanics and Foundations Division, Proceedings, ASCE*, **92**, SM1, January 1966.

6:17 J. L. Beaton, "Statistical Quality Control in Highway Construction," *Journal of the Construction Division, Proceedings, ASCE*, **94**, CO1, January 1968.

6:18 F. E. Richart, "Review of the Theories for Sand Drains," *Transactions, ASCE*, **124**, 709, 1957.

6:19 O. G. Ingles and J. B. Metcalf, *Soil Stabilization*, John Wiley & Sons, Inc., New York, 1973, 374 pp.

6:20 T. W. Lambe, "Soil Stabilization," in *Foundation Engineering*, McGraw-Hill Book Company, New York, 1962, Chapter 4.

6:21 N. A. Tsytovich, *The Mechanics of Frozen Ground*, Scripta Book Co., Washington D.C., McGraw-Hill Book Company, New York, 1975 (published in USSR, 1973), pp. 138–180.

6:22 I. M. Litvinov, "Discussion on Thermal Consolidation," *Proceedings of the Fourth International Conference on Soil Mechanics and Foundation Engineering, 3, London, 1957*, p. 169, 1957.

6:23 T. B. Kennedy and W. F. Swiger, "Symposium on Grouting," *Transactions, ASCE*, **127**, 1337, 1962.

6:24 W. E. Perrott, "British Practice on Grouting Granular Soils," *Journal of the Soil Mechanics and Foundations Division, Proceedings, ASCE*, **91**, SM6, November 1965.

6:25 H. B. Erickson, "Strengthening Rock by Injection of Chemical Grout," *Journal of the Soil Mechanics and Foundations Division, Proceedings, ASCE*, **94**, SM1, 159, January 1968.

6:26 R. H. Karol, "Chemical Grouting Technology," *Journal of the Soil Mechanics and Foundations Division, Proceedings, ASCE*, **94**, SM1, 175, January 1968.

6:27 "Guide Specifications for Chemical Grouts," *Journal of the Soil Mechanics and Foundations Division, Proceedings, ASCE*, **94**, SM2, 345, March 1968.

SUGGESTIONS FOR FURTHER STUDY

C. Caron, P. Cattin, and T. F. Herbst, "Injections," in *Foundation Engineering Handbook*, H. F. Winterkorn and H. Y. Fang, eds., Van Nostrand Reinhold Company, New York, 1975, Chapter 9, pp. 337–353.

J. W. Hilf, "Compacted Fill," in *Foundation Engineering Handbook*, H. F. Winterkorn and H. Y. Fang, eds., Van Nostrand Reinhold Company, New York, 1975, Chapter 7, pp. 244–311.

O. G. Ingles and J. B. Metcalf, *Soil Stabilization*, John Wiley & Sons, Inc., New York, 1973.

P. A. Lenzini and B. Bruss, *Review of Grouting and Freezing Techniques for Underground Openings*, U.S. Department of Commerce, NTIS PB-253-12, 1975.

N. A. Tsytovich, *The Mechanics of Frozen Ground*, Scripta Book Co., Washington, D.C., McGraw-Hill Book Company, New York, 1975 (published in USSR, 1973).

U. S. Bureau of Reclamation, *Earth Manual*, 2nd ed., Denver, Colo., 1974, 810 pp.

H. F. Winterkorn, "Soil Stabilization," in *Foundation Engineering Handbook*, H. F. Winterkorn and H. Y. Fang, eds., Van Nostrand Reinhold Company, New York, 1975, Chapter 8, pp. 312–336.

PROBLEMS

6:1 The following data were found from a moisture–density (compaction) test for a soil with an average specific gravity of solids of 2.71.

Water Content (%)	Total Density (g/ml)	Total Unit Weight (lb/ft^3)	Water Content (%)	Total Density (g/ml)	Total Unit Weight (lb/ft^3)
10	1.57	98	20	2.07	129
13	1.70	106	22	2.05	128
16	1.91	119	25	1.97	123
18	2.00	125			

a. Plot the moisture–dry density or dry unit weight curve. Find the maximum density and optimum moisture.
b. Plot the zero air voids curve.
c. If the contractor is required to secure 90% compaction, what is the range in water contents that would be advisable?

6:2 Compaction tests were made on the same soil, using first the Standard Proctor method and second the Modified Proctor method. The following results were obtained:

STANDARD PROCTOR			MODIFIED PROCTOR		
Water Content (%)	Dry Density (g/ml)	Dry Unit Weight (lb/ft^3)	Water Content (%)	Dry Density (g/ml)	Dry Unit Weight (lb/ft^3)
6	1.63	102	6	1.71	107
9	1.70	106	9	1.81	113
12	1.73	108	12	1.89	118
14	1.75	109	13	1.89	118
16	1.73	108	14	1.88	117
19	1.68	105	16	1.80	112
22	1.60	100	18	1.73	108

a. Plot both curves on the same graph and determine maximum density and optimum moisture for each.
b. Plot the zero air voids curve if the specific gravity of solids is 2.67.
c. How much increase in maximum density results from the Modified compaction? What decrease in optimum moisture occurs when using modified compaction?
d. The soil is classed CL by the Unified System. What densities would be required for a highway fill 12 m or 39 ft high? What would the range in permissible moisture contents be if the field compaction methods were comparable to the effort of (1) the Standard Proctor test, and (2) the Modified Proctor test?

6:3 A Standard Proctor test on a ML soil having a specific gravity of solids of 2.68 was as shown in the following tabulation:

Water Content (%)	Dry Density (g/ml)	Dry Unit Weight (lb/ft^3)	Water Content (%)	Dry Density (g/ml)	Dry Unit Weight (lb/ft^3)
12	1.38	86	24	1.46	91
15	1.43	89	27	1.41	88
18	1.46	91	30	1.35	84
21	1.49	93			

a. Plot the moisture density and zero air voids curve. What is the maximum degree of saturation of the soil?
b. The soil is to be used in a fill 5 m or 16 ft high that supports a one-story building. What densities should be specified, based on Table 6:2?
c. The soil moisture is 25%. The constructor is able to obtain a dry density of 1.38 g/ml or unit weight of 86 lb/ft^3, using a sheepsfoot roller developing 4800 kN/m^2 or 700 psi. The roller fails to walk out. What should he do to obtain the required density (1) in the deeper part of the fill, and (2) in the upper part?

6:4 Make an estimate of the suitability of each of the soils listed in Problem 2:3 for
a. Fill for a highway.
b. Subgrade for an airfield pavement in Illinois.
c. Core of an earth dam.
d. Shell (structural supporting part) of an earth dam.

6:5 List in the order of their importance the properties necessary for a soil to be used in the following ways:
a. Highway fill.
b. Railroad embankment.

c. Earth dam.
d. Subgrade for major airport.
e. Surface for a secondary road.

6:6 Prepare a report on available compaction equipment, showing width compacted, pressure, coverage for
a. Sheepsfoot rollers.
b. Heavy rubber-tired rollers.

6:7 Prepare a report, based on an article appearing in an engineering or construction journal, that describes the construction of a large fill. Include the following points:
a. Soil description.
b. Method of excavation.
c. Method of compaction.
d. Control of compaction.

6:8 Prepare an outline of the procedure required for testing soil–cement mixtures. (Secure information from bulletins of the Portland Cement Assocation.)

6:9 From Fig. 6.9, compute the median and average compacted densities.
a. How do the median and average compare?
b. What percentage of tests fell below the quality limit?
c. If the soil moistures are approximately the optimum, and the soil is a silty sand, SM, what might be done to make the compaction meet the requirements?
d. Would these steps change the median? Why?

6:10 A contractor is compacting a soil classified as MH using a sheepsfoot roller with a 5170 kN/m^2 or 750 $lb/in.^2$ foot pressure, with compacted layers 0.6 m or 2 ft thick. The optimum moisture is 22% but the soil moisture is 30%. The median density obtained is only 93% of the maximum (ASTM D698), although a 98% minimum is specified.
a. List all the things that are wrong with this situation.
b. What are the alternatives to obtain a strong, incompressible material if the layer is to be a subgrade for a pavement?

CHAPTER 7
Soil and Rock Investigations

The plans for a railroad cut 50 m (165 ft) deep required slopes of 1(H) to 6(V) loosened by blasting in rock and 2.5(H) to 1(V) excavated by scrapers in the badly weathered rock and soil overburden. The resident engineer for the designer ruled that the boundary between rock and weathered rock-soil slopes would be determined by the level of rock outcrops in the adjoining hillside and confirmed by the level at which the contractor commenced ripping to loosen the weathered rock. The upper 12 m (40 ft) was cut on a 2.5(H) to 1(V) slope; the remainder of the cut was the nearly vertical rock slope.

Within six months the exposed sandstone and shale in the steep "rock slope" had commenced to slide. The resident engineer ruled that the failing areas had been damaged by excessive explosives and that they would have to be laid back to the soil slope of 2.5(H) to 1(V). By the time this corrective excavation had been completed, more slides had occurred. Eventually, the entire slope was laid back to the flatter slope.

The work was delayed because additional property had to be secured: The cut was now 87 m (285 ft) wide on each side. The excavation was tedious and expensive, requiring loosening the shale and sandstone on dangerous narrow ledges above the sliding areas, bulldozing the spoil downslope and excavating it at the bottom. The cost per unit volume was 5 times that of the original excavation. Moreover, the blasting of the shale and sandstone required by the plans and specifications opened fractures along joints and bedding planes, contributing to the sliding.

The fiasco was the result of inadequate information on the nature of the soil and rock. In order to save money, the designer had eliminated 95% of the soil and rock borings. Instead he relied on nearby rock outcrops to define the level of sound rock. The contractor, looking at the same outcrops, reached the same conclusion. However, the outcrops were misleading. Long exposure had hardened the sandstone layers and they protected the fractured, easily weathered shale beneath. What had appeared to be sound, massive rock was badly fractured, thinly bedded, soft rock that weathered rapidly on exposure. The blasting procedure specified had aggravated the fracturing and probably contributed to the deterioration of the rock. If the contractor had been permitted to rip the sandstone shale as he requested, the sliding might not have been so serious. The additional excavation cost the contractor several million dollars. The lack of quantitative data on the nature of the rock generated litigation that encumbered more engineering effort than the original design and which required years for resolution.

This was an expensive lesson for all concerned; the owner, who was eventually liable for the contractor's claim; the designer, who thought he could save money by eliminating borings but who had to admit in court that his design was not based on facts; the resident engineer, who embarrassed his employer by field decisions based on expediency instead of knowledge; and the contractor, who assumed that the rock was as sound as the outcrops appeared. The cost of the borings saved was about 0.1% of the contractor's claim for additional compensation.

The economic feasibility of a housing project in a lowland region was based on using 90% of the land area for streets, buildings, parking lawns, and recreation areas and 10% for ponds. The developer had engaged an architect-engineer to survey the site, investigate soil conditions, and prepare preliminary plans for development before he purchased the land. The plans were reviewed by the bank before it committed a loan to finance land purchase, final design, and construction. The plans required excavating soil from low areas and filling to provide good surface drainage; the excavations were to become small recreation lakes.

A month after construction commenced the contractor reported peat and muck in the low areas. The project engineer, who had also supervised the field surveys, assured the bank and the developer that a small amount of muck had been anticipated, "based on studies of color infrared air photographs," but that there would only be enough to improve the topsoil. A month later, however, all of the excavation shown on the plans had been completed. Most of the soil excavated was peat, which could not be used for fills beneath roads and buildings, and there were many alligator and snake-filled marshy ponds in which a person would sink to his armpits where the plans showed buildings and roads.

The owner engaged another engineer to investigate the site. A study of air photographs, obtained from the U.S. Department of Agriculture, showed that two thirds of the property was covered with circular depressions filled with lush vegetation: some with shrubs and trees, others with a ring of trees surrounding a reed-filled marsh. Borings made in these circular areas found peat 1 to 3 m (3 to 10 ft) deep in the areas with shrubs and trees and peat interlayered with organic clay 3 to 6 m (10 to 20 ft) deep beneath the reed-filled areas. With these conditions, the project was not economically feasible. The bank withdrew its loan, although it had already advanced nearly a million dollars for construction; the owner went bankrupt, defaulting on the loan. The developer sued the engineer, on the basis that the faulty site investigation had misled him to purchase the site and to undertake the project.

The court testimony disclosed that air photographs were the basis for site topography. These photographs revealed the low swampy areas, but apparently the topographers did not realize their significance. The testimony raised doubts about the ground control survey because some of the property corners fell in areas with peat so deep that only marsh buggies could reach them. A few borings had been made for design, but these were in the higher ground between the low areas in order to reduce the cost of moving the drill rigs. Significantly, the architect-engineer never produced the color infrared photographs in court, from which he had initially claimed he could measure peat depth. The court found the architect-engineer negligent in not realizing the significance of the low circular areas and in confining borings to the higher ground between the low spots.

This disaster illustrates the need for careful visual observation of site conditions and good liaison between the architect-engineer-planner and field personnel who make ground surveys and soil borings. A brief study of air photographs, available in the local county farm agent's file, would have revealed the marsh and swamp areas and the likelihood of peat to even a layman familiar with the region. The out-of-state topographer, using air photographs for mapping, did not appreciate their significance. Instead of allowing the driller to select boring locations for ease of access, a boring plan in each of the different features revealed by the photographs would have disclosed the seriousness of the peat problem, and at less cost than the borings made. Unfortunately, the value of the lesson has been partly lost; the architect-engineer-planner is no longer in business in that area.

The designer of a steel structure cannot proceed without knowing the physical properties of the steel; the designer of reinforced concrete must know the physical characteristics of both steel and concrete; yet both often complete their superstructure designs without quantitative data on the material that supports these structures—the earth. The soil and rock formations are just as much a part of the structural system as the concrete and steel above. However, whereas designers can control the character of the man-made materials, they have little control over the soil and rock. Therefore, either the design must be adapted to the site conditions or else the site conditions must be improved. In either case it is imperative that these conditions be evaluated accurately.

The designer is not the only one concerned with underground conditions, as the two case histories show. Even before a site is purchased, the prospective owner should determine if the property is suitable for the purpose. For example, expensive pile foundations were required for a low-cost building site in a marsh. A more expensive site nearby with a stable sand foundation proved cheaper when the total development cost was evaluated.

The builder must understand the site conditions in order to plan earth moving and foundation construction. Superstructure construction is often controlled by the time and operation sequence required for foundations. It is not surprising, therefore, that the largest part of the contractor's allowance for the unknown (the bid *contingency*) is often in the "below ground" work. When the soil and rock are significantly different from what was reasonably anticipated, either from the underground investigation or from past experience in the area, this may constitute a *changed condition*. Many contracts include provisions for added time and payment for changed conditions; in other cases these unforeseen differences have generated tedious, expensive lawsuits.

The need for data on the underground conditions at a site is generally recognized by the engineer and the contractor and to varying degrees by the prospective owner. The differences between evaluating site conditions and obtaining data on other engineering materials are not always understood, however.

When a structural material, such as steel, is purchased, the specifications define the minimum quality. The manufacturer is responsible for meeting the specification; quality control measures limit the variability of the material, and independent laboratory tests establish that at least the minimum quality is furnished.

The soil and rock, however, did not form under rigid quality control.[7:1] Defects are frequently hidden from view with blankets of topsoil and thick vegetation. The Creator cannot be called upon to meet a specification, to certify a minimum quality, or to answer to demands for damages. Therefore, evaluating the quality of the underground conditions at a site is far more difficult and leaves a much greater margin for uncertainty than establishing the properties of the other materials of construction.

The responsibility for meeting this uncertainty lies with all three parties to construction: the *owner*, the *engineer-architect*, and the *constructor*. The constructor must plan the operations to allow for the unknown, and the engineer and architect must design the structures with enough margin for safety to cover the variable conditions; both must be prepared to revise the design when unforeseen site conditions are encountered. The owner, whose peculiar site includes the unknowns, is responsible for the extra costs resulting from these conditions, as well as for the cost of investigating the site.

Because nature provides no assurances of site quality, the engineer-architect must plan a program of site investigation that will identify the significant undergound conditions and define the variability as far as is practical. *There will always be some risk of unknown conditions*; it can be minimized by a more intensive investigation but never eliminated. This risk is inherent in all human endeavor: No project of any kind is undertaken with a trouble-free guarantee. The degree of success reflects the skill and imagination of those involved, but it also depends on circumstances beyond their control.

7:1 Planning Investigations

INFORMATION REQUIRED / A complete investigation of underground conditions includes the following points:

1. Nature of the deposits (geology, recent history of filling, excavation, and flooding; possibilities of mineral exploitation).
2. Depth, thickness, lateral extent, elevations, and composition of each soil and rock stratum, and of geologic discontinuities.
3. Groundwater elevations, their differences across the site, and their changes with time and environmental change.
4. The engineering properties of the soil and rock strata that affect the performance of the structure.

In many cases all this information is not necessary, and in others estimates suffice. The best investigation is the one that provides adequate data at the time it is needed for planning and design and at a cost consistent with the value of the information.

VALUE AND COST / The value of an investigation is measured by how much money might be spent for the project if no investigation were made. A designer confronted with inadequate data compensates for the lack by overdesign; A contractor furnished with incomplete information increases the cost estimates to allow for possible trouble. In most cases the cost of inadequate data is considerably more than the cost of the investigation. When unforeseen soil conditions cause a change in design or construction, the cost of the work increases greatly. If the structure fails, the entire project becomes a loss. In such cases, the cost of an adequate investigation would be but a small fraction of the money lost.

The cost of an adequate investigation (including laboratory testing and geotechnical engineering) has been found to be from 0.05 to 0.2% of the total cost of the entire structure. For critical structures or unusual site conditions, the percentage is somewhat higher, from 0.5 to 1%.

PROCEDURE FOR INVESTIGATION[7:1,7:2] / A complete investigation consists of three steps:

1. *Reconnaissance*, to determine the geology of the formations and to estimate the soil, rock, and water conditions.
2. *Exploratory investigation*, to determine the depth, thickness, and composition of the soils and rocks, the levels of ground water, and to estimate the engineering properties of the materials.

3. *Intensive investigation*, to secure quantitative data on critical strata from which design computations can be made.

For project feasibility, reconnaissance supplemented by limited exploratory boring can often disclose site conditions such as swamps that would make the site economically unattractive. By way of contrast, for a skyscraper whose location is dictated by available land, reconnaissance is limited. Exploratory boring and detailed investigation proceed simultaneously to provide the data needed for early design decisions, such as a deep basement and its risks to surrounding structures and impact on construction time and cost.

Planning and conducting a soil investigation is among the most intricate of engineering problems. Careful coordination is necessary between the engineer, the laboratory, and the men in the field in order to secure the best information in the least time and at the lowest cost. If the field men send soil samples to the laboratory immediately, time-consuming tests can be started before the field work is finished. If the geotechnical engineers are promptly furnished with the test data, they can make changes in the field and laboratory procedures without expensive delays and without having to repeat some operations.

7:2 Reconnaissance

GEOLOGIC STUDY / A geologic study is essential in planning and interpreting a complete soil investigation.[7:1] The primary purpose is to establish the nature of the deposits underlying the site. The types of soil and rock likely to be encountered can be determined, and the best methods of underground exploration can be selected before boring, sampling, or field testing is commenced. The geologic history may reveal changes such as faulting, flooding, or erosion that have taken place, which have altered the original character of the soil or rock. The possibility of defects in the rock, such as cracks, fissures, dikes, sills, sinkholes, and caves, may be indicated. Geologic information is essential in interpreting sampling and field tests. The presence of minerals having economic value can be assessed. If there is a likelihood of future mining or oil- or water-well drilling on the site, this must be considered both in the design of the structure and in planning for the use of the site. Legal and engineering problems have arisen where structures settle because of the collapse of mine workings beneath them or where valuable minerals are discovered below expensive buildings.

SEISMIC POTENTIAL / Potential seismic activity is a major factor in structural design in many parts of the world. Even in regions that have rarely experienced an earthquake, seismicity must be considered in the design of structures such as dams and nuclear reactors whose damage or failure could cause widespread injury or loss of life, or life-support structures, such as hospitals, that must remain functioning in times of disaster.

Earthquakes are generally the result of deep-seated accumulated strains in the crust, which are climaxed by their release in cracking or faulting. Earthquake damage to foundations occurs in two forms: the direct tearing of structures that lie on the fault and the acceleration of structures within the zone of more intense motion. Preliminary estimates of the seismicity experienced in the United States can be had from a map of earthquake zoning (Fig. 7.1).[7:3] The potential damage is not directly related to the zone. Instead, the zones reflect the level of engineering consideration for good design. In Zone 0, experience suggests no influence on design (except for the most critical structure, such as a nuclear power plant). Zone 1 suggests nominal effects from distant earthquakes or very small local events that should be considered in design of critical structures. Zone 2 implies moderate intensities equivalent to accelerations as great as $0.15g$. Their effect can be included in design of all but critical and life support structures by semiempirical methods.

Zone 3 encompasses all larger earthquakes whose effect should be evaluated by dynamic analyses. The design earthquake is developed by a team of geologists, seismologists, and engineers who consider the earthquake, the dynamic response of the site, and the site–structure interaction. The same approach applies to critical and life support structures of Zone 2 and critical structures of Zone 1.

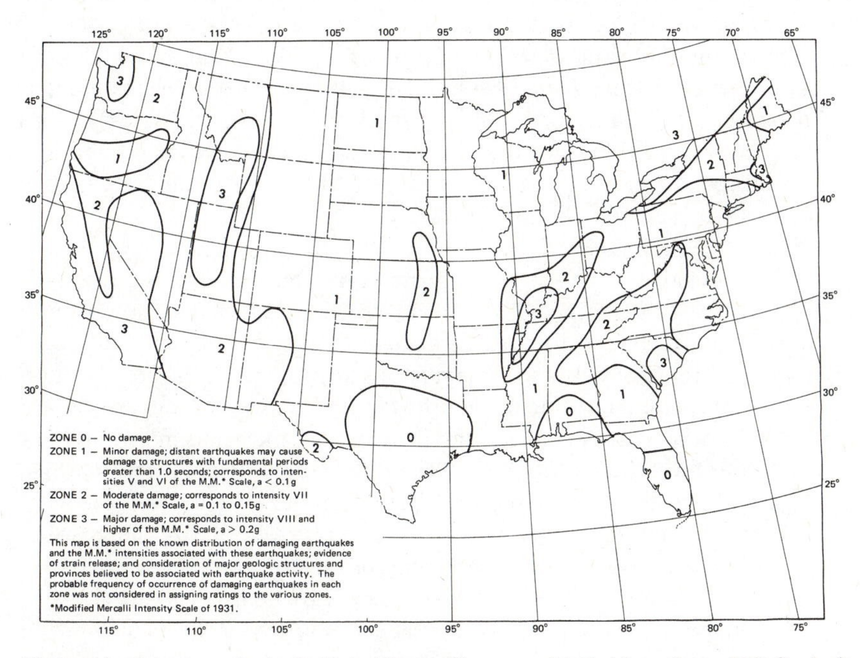

Figure 7.1 Seismic zoning in the United States. (Courtesy of S. T. Algermissen, U.S. Geological Survey.)

SOURCES OF GEOLOGIC INFORMATION / Geologic studies have been made of most parts of the earth by state and national geologic surveys, oil and mining interests, and industrial concerns. Water-well or oil-well records are frequently available that will show the depth of soil, its approximate nature, ground water, and important rock changes. In many cases, soil and rock profiles along highway or railroad cuts can be studied. Geologic maps often show ancient shorelines and river and lake locations, with their terraces, deltas, and fills that are now soil strata of gravel, sand, and clay. The "deltas" of Mississippi and Louisiana and the shorelines of glacial Lake Maumee in Ohio are examples of such deposits.

SOIL SURVEY DATA / The U.S. Department of Agriculture soil survey bulletins provide maps and descriptions of the soil horizons in the United States, generally by counties.[7:4] As described in Chapter 2, the information applies to the A, B, and C horizons, the uppermost 2 to 3 m (6 to 10 ft). However, they usually provide geologic background and some basic engineering data such as texture and compacted densities. They are useful for studies of residential developments and rural roads where the effect of structural loads or excavation depths will be less than 2 m (6 ft).

The United States Geological Survey (USGS), State Geological and Mining Departments, and State Transportation Departments have collected and published maps and bulletins on geologic conditions. Special geologic maps, such as areas of highly plastic clay subject to swelling and shrinking, or areas in which landslides occur, are being developed by the USGS. Many of these excellent publications are out of print or not easily found in catalogs of local libraries. The state geological surveys maintain libraries and catalogs of published geologic data for their area (and often adjoining areas). It has been the author's experience that the state geological surveys and the state USGS offices are the best sources of maps and bulletins, purchased when available or copied when out of print.

SITE INSPECTION / An examination of the site and the adjacent areas will reveal much valuable information. The topography, drainage pattern, erosion pattern, vegetation, and land use reflect the underground conditions, particularly the structure and texture of the soil and rock. Highway and railroad cuts and stream banks often disclose the cross section of the formations and indicate the depth of rock. Outcrops of rock or areas of gravel and boulders may indicate the presence of dikes and more resistant strata. Groundwater conditions are often reflected in the presence of seeps, springs, and the type of vegetation. For example, marsh grass on what appears to be a dry hillside shows that the area is wet during the growing season. The water levels in wells and ponds often indicate ground water, but these also can be influenced by intensive use or by nearby irrigation.

The shape of gullies and ravines reflects soil texture. Gullies in sand tend to be V-shaped, with uniform straight slopes. Those in silty soils often have a U-shaped cross section. Small gullies in clay often are U-shaped, while deeper ones are broadly rounded at the tops of the slopes.[7:5]

Special features such as sinkholes, sand dunes, old beach ridges, and tidal flats are often obvious to the layman. More obscure forms require intensive geological training for their recognition.

Valuable information about the presence of fills and knowledge of any difficulties encountered during the building of other nearby structures may be secured by talking to old residents of the area adjacent to the site. Settlement cracks in nearby buildings often indicate that poor foundation conditions will be encountered. However, it must be remembered that good soil conditions at one site do not necessarily mean good conditions at an adjacent site.

AERIAL RECONNAISSANCE / Examination of the site from the air can reveal the broad patterns of topography and land form, drainage, and erosion even more effectively than site inspection on the surface. Features that are obscured because they cover too large an area or because of poor access are easily observed from the air. Large areas can be inspected in a short time, especially if the site is in rugged country.

A study of aerial photographs permits reconnaissance in any weather and without exciting local residents. Mosaics covering large areas are good base maps for extensive surface reconnaissance. Low-altitude photographs or enlarged high-altitude photographs are better base maps for site inspection than the usual property maps. Surface features can be studied in detail at leisure, and if there is sufficient overlap between adjacent photographs, a three-dimensional examination is possible. More intensive photo study, termed *air photo interpretation* is discussed in Section 7:3.

A personal inspection from a small aircraft or helicopter often permits examination of outcrops as close as 30 to 60 m (100 to 200 ft), as well as observation of the site as a whole from altitudes as high as 2500 m (8000 ft). A permanent record of the inspection made with a good camera in color, and color infrared, is extremely useful in later detailed studies of the soil and rock conditions.

THE VALUE OF RECONNAISSANCE / A reconnaissance investigation provides an estimate of site conditions. Obvious problems such as sinkholes or coastal marsh can be identified and their impact on site suitability evaluated before much more has been spent for land purchase or detailed engineering studies. If the site appears to be suitable, the reconnaissance aids in planning an effective boring and sampling program.

7:3 Remote Sensing[7:5–7:8]

Remote sensing is the estimation of underground conditions from their reflection in surface features. It consists of three steps:

1. Identifying natural and man-made features on both a broad regional scale and local detail.
2. Associating the features with the geologic deposits or structures with which they usually occur.

3. Estimating the three-dimensional distribution of soil and rock materials from the geologic associations. In some cases even the soil texture and some engineering properties can be estimated.

REMOTE SENSING TOOLS / The era of space exploration and the associated development of sophisticated instruments to sense different forms of energy and ground surface topography has provided many different techniques for observing the ground.[7:8] All involve sensing or measuring some form of energy radiation from the ground. The wavelength (or frequency) of the energy level and the patterns are all clues that are evaluated. The possible range of radiation ranges from cosmic rays and X rays to long-wave radio, as shown in Fig. 7.2. The ranges useful in remote sensing are limited by the transmission of the earth's atmosphere and the instruments that can measure them.

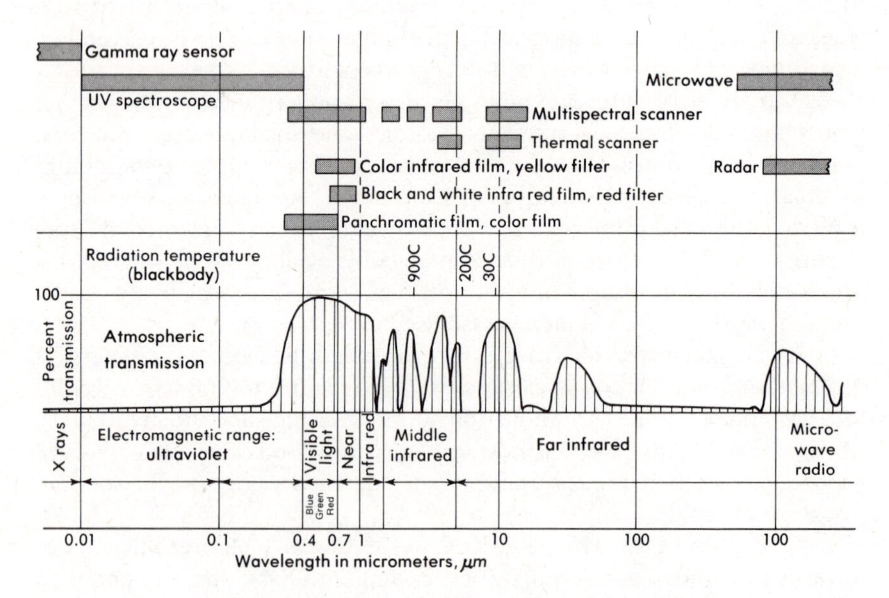

Figure 7.2 *Radiation spectrum adapted to remote sensing.*

No single instrument is capable of recording wavelength and radiation intensity throughout the range. Moreover, the natural radiation emitted by the earth, either directly or by reflection, termed *passive radiation*, varies greatly. Therefore, some systems employ sources of radiation such as a light beam, laser, or very short-wave radio (radar) that is reflected back to the instrument. Two forms of observation systems are employed. The *viewer* records a portion of the earth's surface in a fraction of a second; a photographic camera is the best example. Several views of the same area from different positions permit stereoscopic examination. The *scanner* successively examines small portions of a narrow scan strip (Fig. 7.3a).

Repeated scanning at right angles to the instrument's path of travel provides a more or less continuous strip or swath of readings. The data can be recombined electrically to produce a photographlike strip or pseudophotograph similar to a moving television image (Fig. 7.3b).

IMAGERY AND ITS USES / Black and white photography is adapted to visual light, 0.4 to 0.7 μm wavelength. Selected wavelength bands can be isolated by color filters. Photography in several narrow-band wavelengths simultaneously with separate cameras is *multispectral* photography. The radiation intensity, the passive reflection of sunlight, is quantitatively indicated by gray tones: light-tone positive prints indicate high radiation. Color photography differentiates wavelength by color and intensity by lightness or darkness. Color is helpful for identifying such features as vegetation and soil color.

Color infrared is technically similar to ordinary color photography. There are two differences: (1) The wavelength recorded has been extended beyond red into the long-wave or *near-infrared* range, to nearly 1 μm; (2) the colors displayed have been shifted from the color sensed—green photographs as blue, yellow photographs as greenish, red photographs as yellow-orange, and the invisible near-infrared as red. Blue light is blocked out by a special yellow filter. The major source of the near-infrared radiation is reflectance and fluorescence of the chlorophyl of living vegetation. Thus the high near-infrared radiation of growing green vegetation overwhelms its green reflectance, making it photograph as red. Green paint, however, photographs blue. Vegetation vigor as well as vegetation type is indicated by the brilliance of the red. Thus dry grass and trees in late summer appear dull red or greenish while grass that maintains vigor from springs will be bright red. The red tone, therefore, is an indicator of relative soil moisture. Water photographs blue because of some green reflectance, whether it appears blue, greenish, or even brown from sediments. Thus even small bodies of water stand out as blue in contrast to the red vegetation surrounding them, making it easier to identify small streams and lakes. During periods that vegetation is dormant, such as winter, color infrared is of limited value. It does not photograph heat despite claims otherwise.

Photographic contact print scales range from 1:3000 (1 in. = 250 ft) to 1:130,000 (1 in. = 2 mi). They are obtained from aircraft at altitudes of 1500 to 20,000 m. They discern objects, *resolution*, of 0.3 m or 1 ft at 1:10,000 and 3 m or 10 ft at 1:100,000. Even higher altitude photographs have been obtained from various manned spacecraft; the proportion of the earth's surface that has been so photographed is limited, however.

Infrared can be sensed through the atmosphere at wavelengths of 2 to 5 μm and 8 to 14 μm. The shorter wavelength displays vegetation reflectance and fluorescence. The longer wavelength, 8 to 14 μm, is in the range of heat radiation between 15 and 40°C (60 to 104°F). Thus the radiation intensity is proportional to the temperature and the radiation characteristics of the ground. Figure 7.3b shows a thermal infrared scan pseudophotograph of a swampy area made in early morning in late spring. The cool creek is the black sinuous line. Seeps in the swamp are shown by the patterns of converging dark lines. The scan strip width is 1.2 to 1.5 times the height of the scanner above ground. The image resolution is about 2 m/1000 m altitude for objects directly beneath the scanner, but is wider (and more distorted) toward the scan strip edge. Ground and water temperature differences as small as 1 or 2°C can be measured. The equipment is complex, and the ability to resolve small areas is limited.

Radar employs pulses of short-wave radio, focused into a narrow beam that is reflected back to the sensory instrument. Because it employs its own radiation source, it can be used day or night. The radar beam is projected at a flat angle (Side Looking Airborne Radar = SLAR) that is reflected differently from different surface slopes. The resulting pseudophotographic image (Fig. 7.3c) looks like a relief map of the area. The long waves of radar can penetrate light clouds and, at some wavelengths, through open vegetation to reflect the ground surface. Thus it is very useful in rainy tropical jungles that are impossible to photograph, and for wide-area mapping. The scan swath or strip is typically 20 to 30 km (12 to 19 mi.) wide, commencing 10 km (6 mi.) from an aircraft flying at 10 km (6 mi.) height with a resolution of 3 to 10 m (10 to 35 ft). Because the radar pulse also gives object distance, the geometry of the imagery can be corrected accurately to provide a strip pseudophotograph with geometric distortion less than 1%.

ERTS/LANDSAT / The ERTS-1 Earth Resources Satellite launched in 1972 and its later companions 2 and 3, have provided continuous multispectral scanning of the earth surface at a height of about 900 km

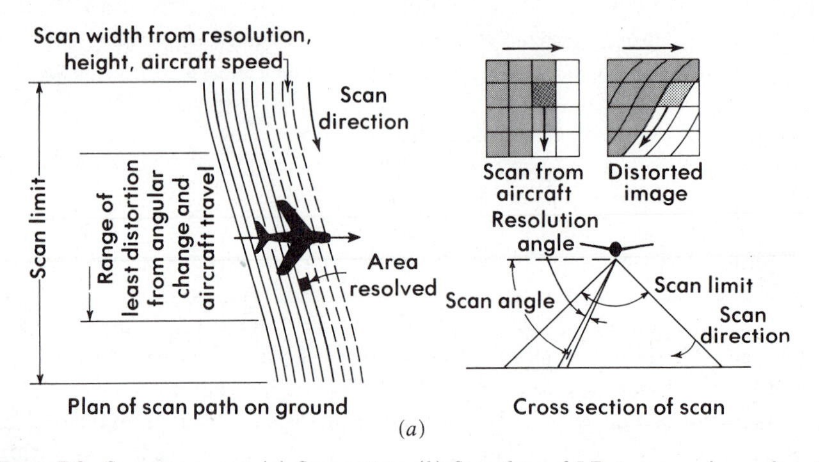

(*a*)

Figure 7.3 *Scanning sensor: (a) Scan pattern. (b) Scan thermal I.R. imagery of a creek and freshwater marsh. Sinuous black lines are cool creeks, black areas cool marsh, and light areas warm grass (8–14 mm). (c) Scan imagery (SLAR K-band Radar) in ridges of sandstone-shale crossed by a river in northeast Alabama, 16×24 km (10×15 miles).*

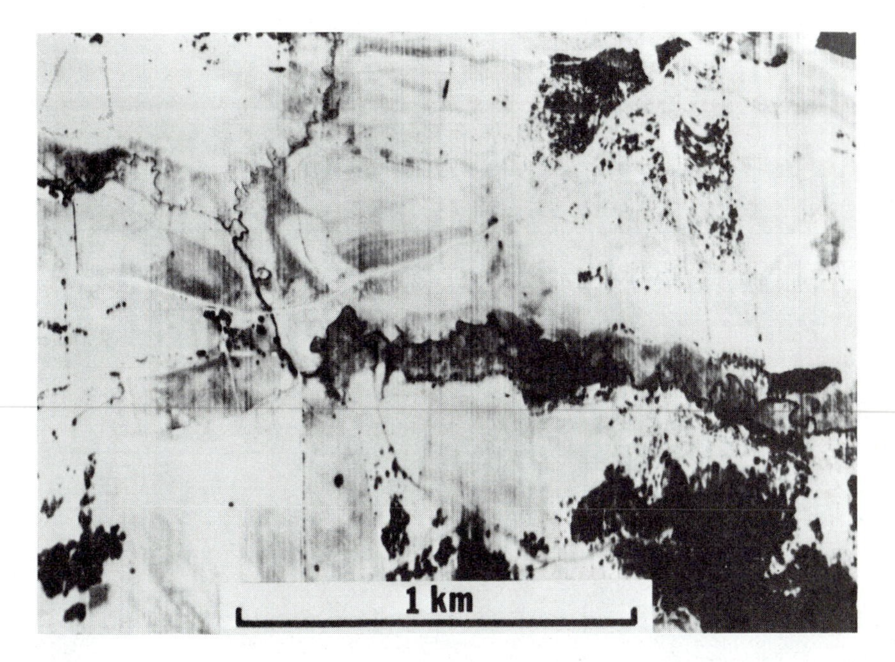

(*b*)

(*c*)

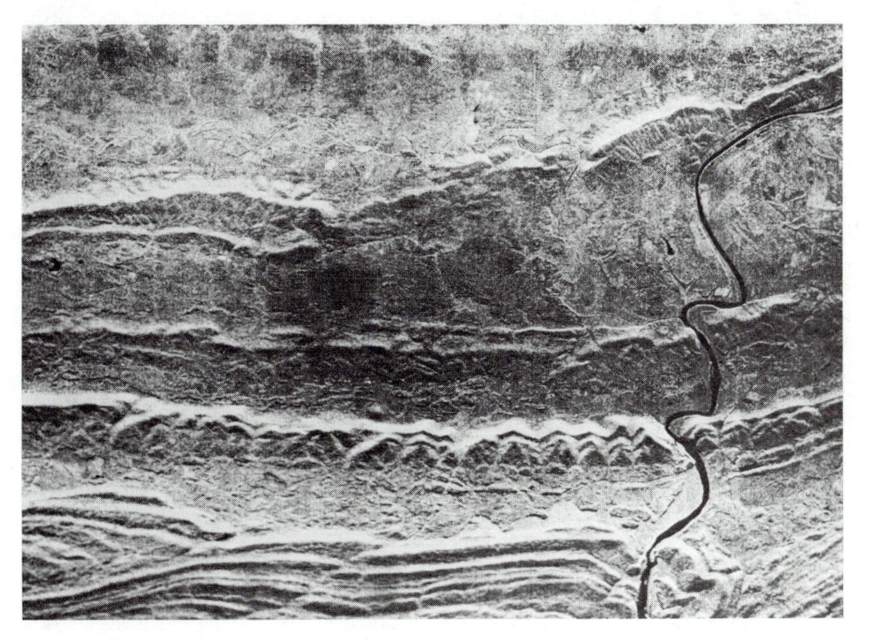

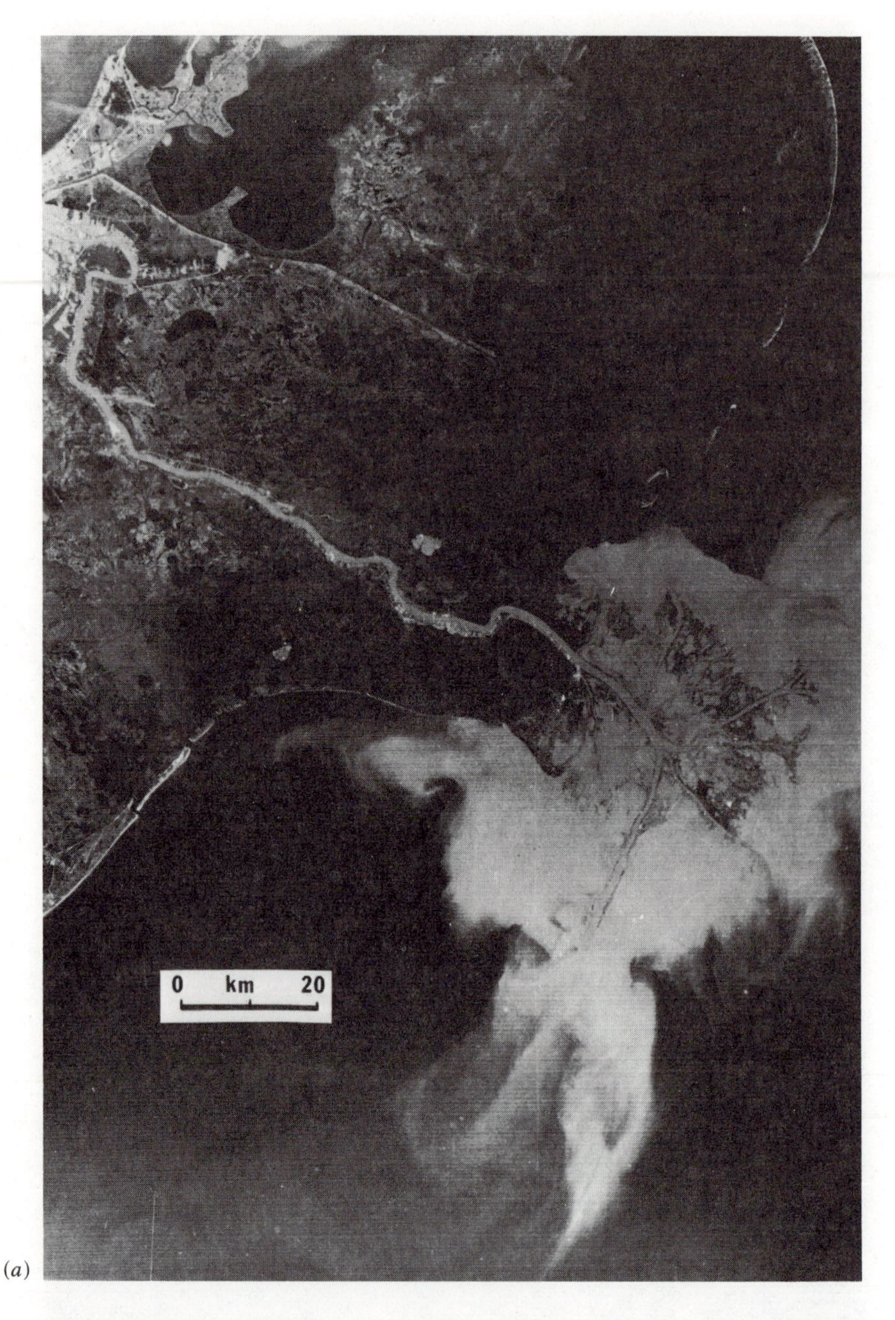

(*a*)

Figure 7.4 *ERTS-LANDSAT images of the Mississippi River delta (left) and Gulf of Mexico (right). (a) Band 5, orange, 0.6 to 0.7 μm; river and shore, light gray; farm lands and buildings, light gray; silt plumes in the Gulf, light; and roads, light. (b) Band 7, near infrared, 0.8 to 1.1 μm; river, ponds, and the Gulf, black; green vegetation, light, and soil, dark.*

(b)

(570 mi) since 1972. The satellites orbit the earth daily in nearly N–S paths, completely covering the surface in 18 days. The imagery is in the form of parallel strips about 185 km (115 mi) wide. Four wavelengths are obtained: Band 4, 0.5 to 0.6 μm (green); Band 5, 0.6 to 0.7 μm (orange-red); Band 6, 0.7 to 0.8 μm (red-near-infrared); and Band 7, 0.8 to 0.9 μm (near-infrared). The resolution is 56 × 80 m (185 × 260 ft), about the size of a football field. Bands 4 and 5 show vegetation, land use, water siltation, and geologic features, Fig. 7.4a; Band 7 emphasizes water and green vegetation (Fig. 7.4b). The imagery is available in 180 × 185 km (115 × 115 mi) parallelograms in each band and in a color composite of Bands 4, 5, and 7 that resembles color infrared photographs. The imagery is unique in that it is repeated at 18-day intervals by each satellite, showing seasonal and long-term changes. Only haze and cloud cover limit worldwide coverage.

PRINCIPLES OF INTERPRETATION[7:5, 7:6] / Interpretation begins with analysis of all the natural and man-made features. Air photographs on a scale of 1 : 10,000 to 1 : 40,000 are most useful with overlapping coverage that permits stereoscopic viewing of topography.

1. Topography.
2. Stream patterns.
3. Erosion and gully details.
4. Gray tones or color.
5. Vegetation.
6. Land use–man-made boundaries.
7. Natural or man-made boundaries.
8. Topographic details.

The topographic study defines the shape of the ground surface, such as hills, valleys, terraces, and similar features, both on a large scale involving several kilometers and a small scale of a hundred meters. The shape, size, slope, and the sequence or interrelationship of adjacent shapes are identified.

The drainage or stream pattern is a fundamental unit of topographic shape. The largest or *primary* streams, easily seen on ERTS images, frequently are clues to geologic age, while the *secondary* streams or tributaries and the *tertiary* streams (creeks and smallest permanent water courses sometimes only seen in color infrared photographs) often reflect structure and the sequence of geologic events. For example, parallel streams usually reflect a gradual tilting of the ground surface. Streams that flow parallel, with right-angle tributaries, and which join together at right angles in a trellis pattern, reflect long parallel folds. Sharp bends with straight reaches between them in primary and secondary streams sometimes indicate the major joint sets in the rocks below. Lazy meandering loops with swampy zones and sand deposits on the insides of the bends are typical of flood plains of an old river (Fig. 7.5); sharp changes in direction of adjacent streams may indicate a fault or shear zone if all the changes are in the same direction and along the same straight or curved line, or *lineation.*

Figure 7.5 *Air photo of a flood plain and the curved fine sand-silt deposits on the inside of a river bend.*

Gully profiles and erosion details reflect the permeability and the strength of the surface materials. In clean sand the few gullies are short with uniformly sloping sides, and stream banks slope at approximately the angle of internal friction. In silts, weak sandstones, and clayey sands of low cohesive strength, the gullies are long and deep. The gully and stream banks drop vertically, from tension cracking, while the gully bottoms are rounded from accumulating sloughed soil. The U-shaped cross section is often interrupted by isolated pinnacles of the original bank that either became detached and slid partially downward or that resisted erosion. In clays and shales the gullies are long but shallow with the tops of the banks rounded by the progressive surface softening of the materials.

The color or gray tones reflect the color of the formations where exposed: The reds and blacks appear dark, the tans and yellows, light. Damp materials appear darker than dry materials. Differences in vegetation are reflected in color differences: Pine trees appear darker green than deciduous trees and blacker in the black and white photographs. Growing crops are darker than dryer vegetation.

Vegetation differences frequently reflect both drainage and soil character. For example, pine trees require drainage, while cypress trees flourish in swamps. Some vegetation prospers on particular soils: The eastern cedar is often associated with residual soils derived from limestone, and particularly phosphatic limestone. Differences in vegetation are easily observed and sometimes denote geologic boundaries; the implications of particular types require the aid of agriculturists and botanists.

Micro details include limited features such as sinkholes, rock outcrops, and boulder accumulations. These usually require on-foot examination to define their significance.

Differences in land use and man-made boundaries may be the result of arbitrary decisions, but they also can reflect differences in the character of the underlying strata. For example, an irregular-shaped pastureland surrounded by rectangular cultivated fields may be the result of too rough a surface or too shallow soil for cultivation. An abrupt jog in a fence line could indicate a rock outcrop, a swampy area, or a sinkhole. Natural boundaries between river patterns, vegetation, or color tone usually reflect major differences in geology.

LANDFORM / The landform is the basic unit of association of the various features, particularly topography and stream pattern. It represents the total effect of environment and geologic history on the underlying soil and rock formations. The study of the evolutionary processes that produce a given landform is termed *geomorphology*. Once the landform is identified, the geologic associations are defined. From this the geologic structure can be established and the probable soil and rock stratification estimated. Groundwater levels and even soil texture can be deduced from tone, drainage density, and erosion patterns.

Remote sensing requires a thorough grounding in geology and geomorphology as well as an understanding of related fields such as agriculture and hydrology. The results indicate which areas are favorable for development, the locations where trouble is likely to occur, and the best places to look for materials of construction. The technique is a valuable preview and supplement to site reconnaissance, particularly in the planning stages where large areas for potential development must be compared without the time for extensive field work.

7:4 Exploratory Investigations[7:2, 7:9]

PLANNING EXPLORATORY WORK / The purpose of the exploratory investigation is to secure accurate information about the actual soil and rock conditions at the site. The depth, thickness, extent, and composition of each stratum; the depth of the rock; and the depth of ground water are the primary objectives. In addition, approximate data on the strength and compressibility of the strata are secured in order to make preliminary estimates of the safety and settlement of the structure.

A carefully planned program of boring and sampling is the best method of obtaining specific information at a site and is the heart of an exploratory investigation. Many different methods have been developed for doing this work, and many organizations offer such services. Too frequently exploratory work is so poorly planned, carelessly performed, incompletely reported, and incorrectly interpreted that the results are inadequate or misleading. Soil or rock boring and sampling to obtain information that will give an accurate picture of underground conditions is an engineering problem requiring resourceful, intelligent personnel trained in the principles of geotechnical engineering.

SPACING OF BORINGS / It is impossible to determine the spacing of borings before an investigation begins because the spacing depends not only on the type of structure but also on the uniformity of the formations. Ordinarily, a preliminary estimate of the spacing is made based on experience and site geology; this is decreased if additional data are necessary or is increased if the thickness and depth of the different strata appear to be about the same in all the borings. Spacing should be smaller in areas that will be subjected to heavy loads; greater, in less critical areas. The spacings given in Table 7:1 are often used in planning boring work. For uniform, regular soil conditions the spacings given in the table are often doubled; for irregular conditions they are halved.

TABLE 7:1 / SPACING OF BORINGS

Structure or Project	Spacing of Borings (m)	(ft)
Highway (subgrade survey)	60–600	200–2000
Earth dam, dikes	15–60	50–200
Borrow pits	30–120	100–400
Multistory buildings	15–45	50–150
One-story manufacturing plants	30–90	100–300

DEPTH OF BORINGS / In order to furnish adequate information for settlement predictions, the borings should penetrate all strata that could shear or consolidate materially under the load of the structure. For very important heavy structures, such as large bridges or tall buildings, this means that the borings should extend to rock. For smaller structures, however, the boring depth is estimated from geologic evidence, the results of previous investigations in the same vicinity, and by considering the extent and weight of the structure.

Experience indicates that damaging settlement is unlikely when the added stress in the soil due to the weight of the structure, $\Delta\sigma_z$, is less than 10% of the initial stress in the soil due to its own weight. For typical steel or reinforced concrete buildings of S stories, a stress increase $\Delta\sigma_z$ of $0.1\sigma_0'$ will occur in a wide range of soil deposits within the following boring depths, z_b:

Condition	Meters	Feet	
Narrow, light building	$z_b = 3S^{0.7}$	$z_b = 10S^{0.7}$	(7:1a)
Wide, heavy building	$z_b = 6S^{0.7}$	$z_b = 20S^{0.7}$	(7:1b)

The old rule of boring depth equal to building width is ridiculous, especially for wide one- or two-story buildings. A simple guide for small buildings is 2 to 3 m (7 to 10 ft) per story (or equal load).

Example 7:1

A manufacturing plant and warehouse will be a one-story building with a floor load of 25 kN/m^2 or 520 psf constructed on fill 1.5 m or 4.5 ft thick. The building plan will be 75 m × 250 m. What will be a reasonable boring plan and depth?

1. The 1.5 m fill will weigh 30 kN/m^2 or 625 lb/ft^2. This is equivalent to 3 stories of concrete building at 10 kN/m^2 per floor.
2. The floor load is 25 kN/m^2 equivalent to 2.5 stories. The roof will be equivalent to 0.5 stories.
3. The total load is equivalent to a 6-story building. $z_b = (3 \text{ to } 6) \times 6^{0.7} = (3 \text{ to } 6) \times 3.5 = 10 \text{ to } 20$ m.
4. The boring plan depends on site geology:
   a. For a uniform flood plain, a total of 8 borings: two rows each with 3 borings 65 m apart, 5 m inside the 2 longer sides with a single row of 2 midway between producing a staggered pattern will be enough. The first 6 borings will be 20 m deep; the remainder 10 to 20 m depending on the soils encountered.
   b. For nonuniform site conditions, 2 rows of 4 borings each and one of 3 are suggested.

7:5 Boring and Sampling

Many different exploratory techniques are in use. Some are suited to a wide range of soil and rock conditions; others are limited to special conditions. A summary of the principal methods is given in Table 7:2. They are discussed in the following sections.[7:10]

AUGER BORING / The soil auger is a simple tool for making a hole and for secondary samples in a mixed, disturbed condition. Hand-powered augers to 0.1 m (4 in.) in diameter suitable for holes as deep as 6 m (20 ft). Power-driven augers up to 0.25 m (10 in.) can drill a wide range of soils and some soft rocks as deep as 30 m (100 ft) in less than an hour.

The power auger with continuous flutes will eventually raise the soil cuttings to the ground surface. However, there is an indeterminate time lag between drilling and the emergence of the pulverized sample. Therefore, it is essential to drill in increments of 1 to 1.5 m (3 to 5 ft) and then remove the auger. The strata depths can be determined by the amount of soil retained in the auger helix compared to the increment of depth drilled.

Soil augers have the advantage of obtaining a dry hole until the water table is reached and of providing easy visual recognition of changes in soil composition. On the other hand, they are difficult to use in very soft clays and coarse gravel and impossible to use in some soils below the water table. Hand augers are seldom economical when boring deeper than 6 m (20 ft).

The auger sample is a well-disturbed mixture of the materials penetrated. It is useful for determining the average water content, grain size, and plasticity characteristics and is sufficient for most borrow pit exploration. It gives little information on the character of the undisturbed soil. In soft rock and some hard soils, the particles will be broken by the drilling, making grain-size tests misleading; the heat of drilling can reduce the water content.

DRILL CUTTINGS / Most drilling techniques that do not include sampling produce fragments of the soil or rock that are returned to the ground surface by the drilling fluid. The solid cuttings can be recovered from the fluid-circulation settling tank for examination. Although changes in the cuttings may indicate changes in the soil or rock, estimating the nature of the formation from cuttings is like identifying a cow from the hamburgers.

TEST BORING / Test boring (ASTM D-1586)[7:11] is the most widely used method of soil exploration. It consists of two steps: *drilling*, to open a hole in the ground; and *dry sampling* to secure an intact sample that is suitable for visual examination and tests for water content, classification, and even unconfined compression.

Drilling is done by augering, wash boring, or rotary drilling with a high-speed revolving cutter and circulating liquid or air to remove the cuttings. In firm soils the hole remains open by arching. In soft clays and in sands below the water table, it is kept open by inserting steel tubing (casing) or preferably by keeping the hole filled with a viscous liquid known as *drilling mud.* Drilling mud, usually a mixture of bentonite clay and water, has the advantage of supporting both the walls and the bottom of the hole. The mud also serves as the circulating liquid in wash and rotary boring and maintains a cleaner hole by washing out coarse sand and gravel, which accumulate in the bottom.

The *sampler* (Fig. 7.6), also called a split spoon, consists of a thick-walled steel tube split lengthwise. To the lower end is attached a cutting shoe; to the upper end, a check valve and connector to the drill rods. The standard size is 35 to 38 mm (1.38 to 1.5 in.) ID and 51 mm (2 in.) OD, but similar samplers with 51 mm (2 in.) ID × 64 mm (2.5 in.) OD and 64 mm (2.5 in.) ID × 76 mm (3 in.) OD are occasionally used.

The hole is drilled as previously described until a change in the soil is detected. The drill tools are removed and the sampler lowered to the bottom

TABLE 7:2 / SUBSURFACE EXPLORATION—EXPLORATORY BORING METHODS

Method	Procedure	Use	Limitations
Auger boring (ASTM D-1452)	Hand or power auger with removal of material at regular short intervals.	Identify changes in soil texture above water table. Locate ground water.	Grinds soft particles—stopped by rock, etc.
Test boring (ASTM D-1586)	Drill hole, sample at intervals with 35 mm (1.4 in.) ID, 50 mm (2 in.) OD, split barrel sampler driven 0.45 m (18 in.) in three 15-cm (6-in.) intervals by 63.6 kg (140–lb) hammer falling 0.76 m (30 in.). Below water, maintain hydrostatic balance with fluid.	Identify texture and structure; estimate density or consistency in soil or soft rock.	Gravel, hard seams.
Continuous core: soil (ASTM D-2113)	Force or drill tube into soil until resistance prevents further movement. Remove cuttings with air or water.	Identify soil texture and structure continuously in cohesive soils.	Gravel, hard seams, sands. Misleading squeeze in some clays.
Borehole camera, TV	View inside of bore hole.	Examine stratification in place.	Textural changes indistinct. Sometimes obscure below water table.
Continuous core: rock (ASTM D-2113)	Rotate tube with diamond-studded bit to cut annular hole. Cuttings removed by circulating water. Core retained in tube by cylindrical wedge. Best with stationary inner tube to protect core.	Identify rock strata and structural defects continuously.	No data on soft seams, etc.

Dynamic sounding	Drive enlarged point on end of rod with weight falling fixed distance, in increments of 0.15 to 0.3 m (6 in. to 1 ft).	Identify significant changes in density or consistency of materials.	Misleading in gravel, or cemented seams.
Static penetration (ASTM D-3441)	Force enlarged cone (Dutch cone: 3.6-mm (1.4-in.)-diameter, 60° angle) on end of rod into soil, measuring resistance of point at regular intervals. Other sizes can be adapted to local condition.	Identify subtle changes in density or consistency. Possibly identify soil by ratio of point load to skin friction.	Stopped by hard strata, misleading in gravel, hard seams.
Pits, trenches	Excavate pit or trench, by hand, large auger, and by excavator.	Visual examination of structure and stratification above water table.	Caving of walls, ground water.
The following drilling methods are frequently used to advance the hole in test boring and core drilling; for quick but crude exploration, they are occasionally used independently.			
Wash boring; rotary wet drilling; rotary air drilling	Chop with chisel bit or rotate toothed cutter. Cuttings washed to surface by circulating water or drilling mud through bit.	Identify coarser fraction from cuttings, hardness from drilling rate.	Misleading if appreciable fines present.
Churn or cable drilling	Pound and churn soil boulders and rock to slurry by dropping heavy chisel bit in wet hole. Bail water and cuttings at intervals.	Drill and identify broken rock, etc., from cuttings.	Strata difficult to define. Quick condition formed in sands.
Percussion drilling	Impact—drill with jack hammer; remove cuttings with compressed air.	Identify rock from cuttings, hardness from rate.	Plugged by wet soil.

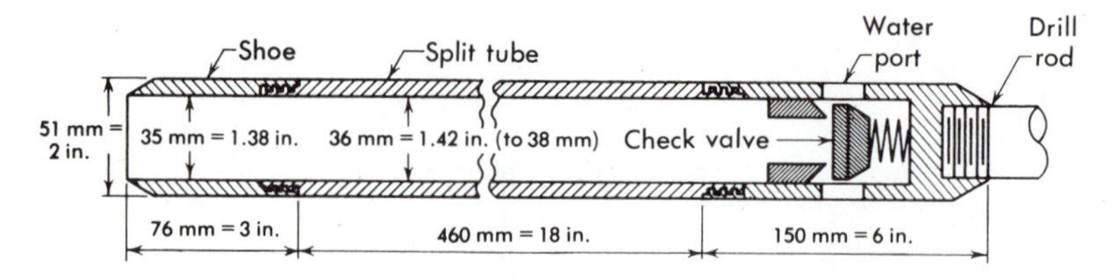

Figure 7.6 *Standard split-tube sampler.*

of the hole attached to the drill rods. It is first driven 0.15 m (6 in.) into the soil to ensure that the cutting edge is seated in virgin material.

It is then driven 0.3 m (12 in.) in 0.15 m (6 in.) increments with a 63.6-kg (140-lb) hammer falling 0.76 m or (30 in.). The number of blows for each 15 cm (6 in.) of penetration are recorded. The *standard penetration resistance*, *N*, is the sum of the blows for the second and third increments.[7:11] Standard penetration sampling is shown in Fig. 7.7.

The sample is examined and classified by a field technician in charge of boring and then sealed in a glass or plastic container for shipment to the laboratory. The sample maintains the average water content, composition, and stratification of the soil, although there is appreciable distortion of the structure. Good samples can often be used for preliminary unconfined compression tests but are not of sufficient quality for accurate strength or consolidation testing.

The penetration resistance is an indication of the density of cohesionless soils and of the strength of cohesive soils. In effect, it is an in-place dynamic shear test. Tables 7:3 and 7:4 have been proposed to describe density and strength from the standard penetration test results.[7:9,7:12,7:13]

The resistances measured with a 51 mm ID × 64 mm OD (2 in ID × 2.5 in. OD) sampler driven with a 136-kg (300-lb) hammer falling 0.46 m (18 in.), as specified by some building codes, are roughly equivalent to those measured by the standard test.

Test boring is the most widely used method for securing data on the depth, thickness, and composition of the soil strata and approximate information on the soil strengths.[7:14] It is economical, rapid, and adapted to most soils (except coarse gravel) and even to soft rock.

CORE DRILLING / When a soil boring encounters a material so hard that its standard penetration resistance, *N*, exceeds 100 blows, further progress with soil-boring equipment is difficult and often impossible. This resistance is termed *refusal.* This indicates a very hard soil, a boulder, a cemented seam, an obstruction, or rock.

Core drilling is used to penetrate such hard materials in order to determine whether refusal represents continuous hard formations, such as rock, or simply a boulder or other obstruction underlain by softer materials. Holes 0.6 to 1.4 m (24 to 54 in.) in diameter permit an engineer or geologist

(*a*) (*b*) (*c*)

Figure 7.7 *Auger boring and standard penetration sampling:* (*a*) *Power-driven auger prior to sampling;* (*b*) *driving the split-tube sampler with a 63.6-kg* (*140-lb*) *hammer;* and (*c*) *split-tube sampler disassembled showing soil in one-half of the tube in the foreground, soil in the cutting head to the right, and the check valve head and drill rod connector at the right rear.* (Courtesy of Law Engineering Testing Co.)

to examine the strata in place, but the cost of drilling is great. Small-diameter cores that are brought to the surface reveal the composition, soundness, and defects of the rock for great depths at a moderate cost.

Diamond drilling is the most common method for obtaining small-diameter cores. Although the detailed procedures must be adapted to the

TABLE 7:3 / RELATIVE DENSITY OF SAND—STANDARD PENETRATION TEST (After Terzaghi and Peck[7:12])

Blows	Relative Density
0–4	Very loose
5–10	Loose
11–20	Firm
21–30	Very firm
31–50	Dense
51+	Very dense

TABLE 7:4 / CONSISTENCY OF COHESIVE SOILS—STANDARD PENETRATION TEST (After Terzaghi and Peck[7:12])

Blows	Consistency
0–1	Very soft
2–4	Soft
5–8	Firm
9–15	Stiff
16–30	Very stiff
31+	Hard

rock and its fracture patterns, the ASTM standard D-2113 is suited to a wide range of conditions.

The sampler, or *core barrel*, is a piece of hardened steel tubing from 0.6 to 6 m (2 to 20 ft) long with a *bit* attached to the lower end. The bit (Fig. 7.8)

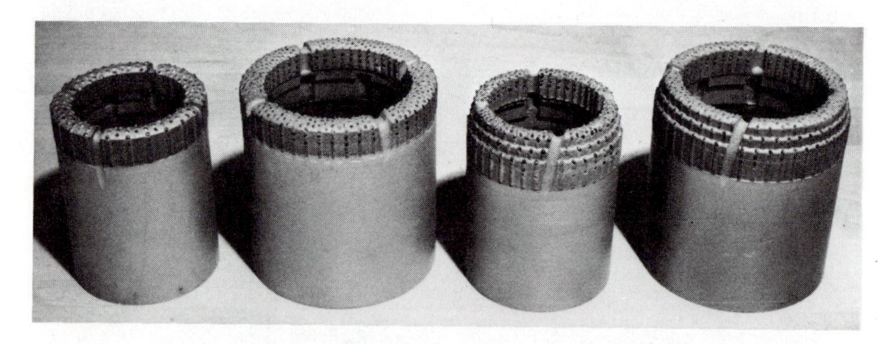

Figure 7.8 *Diamond-studded rock-coring bits.* (Courtesy of Law Engineering Testing Co.)

is ordinarily set with borts (black diamonds), although tungsten carbide or other very hard, tough materials can be used for drilling soft rocks. The six most widely used sizes in the United States are given in Table 7:5).

In drilling, the core barrel and bit rotate while water or thin drilling mud under high pressure is forced down the barrel and into the bit. The cuttings, ground to a powder, are carried up the hole with the water or mud. The rock core extends upward into the barrel and is retained by a circular wedge *core catcher*. The ratio of the length of core obtained to the distance drilled is known as the *core recovery* and is expressed as a percentage. It is an indication of the quality of the drilling and of the soundness of the rock; in homogeneous, sound rock a recovery of 100% may be expected; in rocks with seams a recovery of about 50% is typical; however, in decomposed or seamy rock the recovery may be little or nothing.

TABLE 7:5 / SIZES OF DIAMOND BITS[7:11]

	Outside Diameter		Core Diameter	
	(mm)	(in.)	(mm)	(in.)
EX	38	$1\frac{1}{2}$	21	$\frac{13}{16}$
AX	49	$1\frac{15}{16}$	30	$1\frac{3}{16}$
BX	60	$2\frac{3}{8}$	41	$1\frac{5}{8}$
NX	76	3	54	$2\frac{1}{8}$
$2\frac{3}{4} \times 3\frac{7}{8}$ in.	98	$3\frac{7}{8}$	68	$2\frac{11}{16}$
$4 \times 5\frac{1}{2}$ in.	140	$5\frac{1}{2}$	100	$3\frac{15}{16}$

Note: To obtain good cores in soft or fractured rock, BX or larger is desirable.

Deere proposed a modified recovery, RQD:[7:15] the ratio of the length of intact rock in NX core sections longer than 100 mm (4 in.) to the distance drilled.* A ratio of 90% or more denotes excellent rock; 75 to 90%, good rock; 50 to 75%, fair rock; and 25 to 50%, poor rock.

In fractured or soft rock a double-tube core barrel obtains better core recovery. This employs a thin steel tube that fits snugly around the core and remains stationary while the outer tube rotates. It protects the core from vibration and from erosion of the drilling wash water. In highly fractured or weathered rock a modified technique, *wire line drilling*, permits better recoveries. The drill rods are as large as the core barrel. After drilling as far as practical, the core is lifted out of the core barrel through the drill rods to the surface by a retriever attached to a wire rope from which the name is derived.

The nonrotating inner tube can be fitted with three diamonds that inscribe longitudinal, unequally spaced grooves in the core. The tube also contains a compass whose direction is recorded with respect to the grooves. By such *oriented cores*, it is possible to measure the directions of the dips of bedding and joints.

In very fragile rocks, core recovery is low; moreover, the weakest material that governs design is the part that is lost. The *integral core* technique of Rocha[7:16] minimizes the loss. A small-diameter hole is first drilled into the weak formation to be cored. The hole is filled with quick-setting cement and a steel reinforcing rod placed in the hole. After the cement sets, the rock is cored with a 100-mm (4-in.) ID core barrel or larger, keeping the reinforcing rod in the center of the core. The reinforced core retains its integrity; the weaker strata may be partially destroyed, but enough remains attached to the reinforcing rod to determine their nature.

A soft plastic sleeve, inserted in the core hole and expanded by hydraulic pressure, will retain an imprint of the cracks and fissures in the walls.

* Core sections that exhibit fresh breaks that were obviously produced during drilling are included in the intact lengths of core.

This technique, the *bore hole copier,*[7:17] can obtain wall prints as long as 2 m (7 ft) with the orientation indicated by a compass incorporated in the expansion cylinder.

VISUAL EXPLORATION / Sequence cameras and TV cameras have been developed to fit inside bore holes and record the strata exposed in the hole sides. These require special lenses to scan the entire perimeter or rotating lenses to picture a segment at a time. Although the records (including the TV data on magnetic tape) require experience for interpretation, they make it possible to examine the strata below the practical limit for pit construction and sometimes below ground water. Bore hole periscopes are also available for limited depths. Large test holes or *pits* are described in Section 7:10.

7:6 Penetration Tests[7:10, 7:14, 7:18]

Changes in underground conditions can be identified by differences in the resistance of the strata to being pierced by a *penetrometer.* Ancient builders, who drove poles into soft marsh mud to locate a firm sand seam, practiced this technique. Although the equipment is more sophisticated today, the principle is the same.

Most penetrometers consist of a conical point attached to a drive rod of smaller diameter. Penetration of the cone forces the soil aside, creating a complex shear failure, resembling the point penetration of a foundation pile. The test, therefore, is an indirect measure of the in-place shear strength of the soil.

Two forms of penetration are used: *quasi-static* and *dynamic.* In the *static* test the point is forced ahead at a controlled rate and the force required for movement is measured. In the *dynamic,* the penetrometer is driven a specified distance by hammer blows of equal energy. The number of blows or the total energy required for the specified distance is the measure of resistance. The static test is very sensitive to small differences in soil consistency. The test operation probably does not seriously change the structure of loose sands or sensitive clays. The dynamic test is adapted to a much wider range of soil consistencies and can penetrate gravels and soft rock that would stop a static device.

STATIC / The *Dutch cone* (Fig. 7.9a) is the most widely used static test. The cone has a 60° point angle, a diameter of 36 mm (1.4 in.) and a projected area of 1000 mm^2 (0.01 ft^2) approximately. A number of variants are in use, differing in cone angle and diameter. In the form illustrated (Fig. 7.9a), an independent sleeve is attached behind the cone. The force developed by friction between the sleeve and the soil can be measured independently of the cone resistance. The ratio of sleeve resistance to cone resistance is higher in cohesive soils than in cohesionless. This ratio helps to estimate the type of soil. The mechanical systems for measuring the resis-

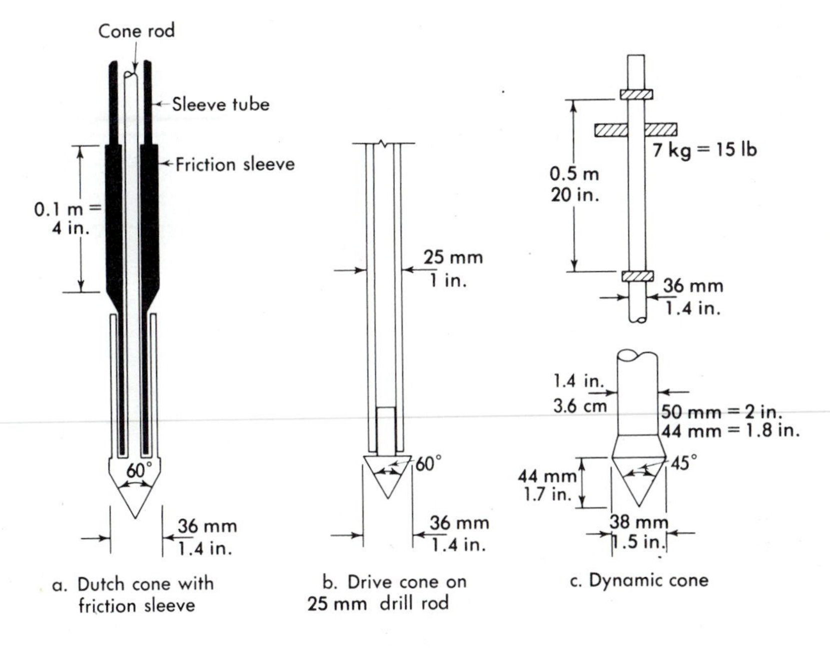

Figure 7.9 *Penetrometers.*

tances vary with the manufacturer. They range from simple rack and pinion drives with spring balance weighing devices to automatic hydraulic driven machines with continuous load indicators and recorders. All are limited in the penetration force that can be developed: from half a ton in simple equipment to several tons for large machines that are anchored to the ground.

DYNAMIC / The dynamic test is utilized in many forms. The Standard Penetration Test has a dual function, penetration testing and sampling, that makes it possible to identify changes in the soil by two independent methods. For this reason it is such a useful tool in exploration. Cones and points of various size and shape for dynamic resistance measurement alone are also used because of their simplicity and adaptability to a wide range of conditions. In one form (Fig. 7.9b) an expendable cone point or spherical point, 36 mm (1.4 in.) in diameter, is placed on a 25-mm (1-in.) OD drill rod and driven with a 63.5-kg (140-lb) hammer falling 0.76 m (30 in.). The number of blows required to drive the cone a foot is comparable to the standard penetration resistance, *N*. The portable cone penetrometer (Fig. 7.9c) employs a 7-kg (15-lb) hammer falling 51 mm (20 in.). The number of hammer blows required to drive the sampler 44 mm ($1\frac{3}{4}$ in.) is approximately the standard penetration resistance, *N*. Although the dynamic tests disturb some soils by shock and vibration, they are simple and adapted to both very soft and very hard materials.

PROJECTILE TESTS / A projectile penetrometer is dropped onto the soil surface. The depth of penetration, related to the kinetic energy and geometry of the device, is an indication of the soil strength. One simple form is a long pointed rod dropped through water to locate the boundary between soft recent silt accumulations and harder strata below. Similar sounding devices are employed by dredging contractors to estimate the character of the river and harbor bottoms they must excavate. The results of these simple tests are qualitative, and the value depends largely on the experience of those who interpret them.

More advanced projectile devices have been designed to be dropped from aircraft. Their rate of deceleration through the strata can be recorded automatically and even transmitted by radio to a remote recorder. The deceleration rate and its changes can indicate the consistencies of the strata and their boundaries.

INTERPRETATION / A penetrometer is similar to a miniature pile foundation that forces the soil aside in a complex pattern of shear. Although the force required to advance the point Q_0 is related to shear strength, many other factors are involved, similar to those of importance in pile bearing capacity (Chapter 11). The most significant factors are

1. c and ϕ of soil in undrained shear.
2. Effective overburden stress, σ'_z.
3. Neutral stress, u.
4. Geometry of the penetrometer.
5. Method of driving, including the equipment.
6. Effect of driving on ϕ, c, and u.

For a clay in which the undrained strength can be approximated by a single parameter c that is independent of changing confining stress and changing hydrostatic neutral stress, u, the resistance Q_0 can be approximated by

$$Q_0 = cN_pA \tag{7:2a}$$

$$c = \frac{Q_0}{N_pA} \tag{7:2b}$$

In this expression the dimensionless penetrometer factor, N_p, embodies the shape of the device and the mode of driving, and A is the projected area of the cone or sampler walls in the direction driven. For the Dutch cone, values of N_p range between 5 for very sensitive soils to 15 for medium-plasticity clays of low sensitivity.

For dynamic penetrometers, the energy of the hammer of weight W, falling a distance h with a total mechanical efficiency of m is mWh. If the distance penetrated by N hammer blows is S, then the static and dynamic resistance are related:

$$WhmN = Q_0S$$

$$Q_0 = cN_pA \tag{7:2a}$$

$$c = \frac{WhmN}{SN_pA} \tag{7:2c}$$

For the standard penetration test, limited data indicate that the value of m ranges between 0.1 and 0.5 with an average of about 0.25.

Numerous empirical expressions have been developed[7:9, 7:14] relating penetration resistance to engineering properties.[7:19–7:21] These are useful correlations for preliminary estimates. There is considerable scatter of data, reflecting the many variable factors that are not included in each relation. Therefore, in any particular association of soils, it is best to collect sufficient data to verify the relationship that is used, or to adjust it to fit local conditions. For cohesionless soils, the angle of internal friction (Fig. 7.10a) is related to N. In cohesive soils, the in-place undrained shear strength, c, at the test level is related to N (Fig. 7.10b).

The modulus of compressibility in simple consolidation of sands (Fig. 7.10c) and the relative density above the groundwater table (Fig. 7.10d) are also approximately related to the value of N. Below the groundwater table both relations are reasonably valid for coarse sands if effective stresses are used. In fine sands, the pore pressure effects in both loose and dense states make the results less reliable.

All penetration tests are only indirect indications of soil behavior. Therefore, penetration testing should always be considered a supplement to direct methods of soil exploration such as boring and sampling. Once the general pattern of soil conditions has been established, penetrometers can be helpful in providing detail between borings. Direct uses in estimating bearing capacity are discussed in Chapters 10 and 11.

7:7 Ground Water

Determining the level of ground water is an essential part of every exploratory investigation. In most cases it is measured in the exploratory borings; however, it is frequently necessary to make borings expressly for ground water measurement when perched or artesian water is expected or if a drilling technique (such as the use of drilling mud) obscures the water.

The first hint of ground water may be wet samples or moisture trickling into the boring. Such observations are possible only with dry drilling methods. However, while these conditions must be recorded, they may represent only capillary saturation or a perched water table. Caving of uncased auger holes in sand is also indicative of ground water, but not conclusive. A more valid indication of the general groundwater elevation is found by allowing the water in the boring to reach an equilibrium level. In sandy soils a few hours will be enough, but in clays a week or more is required.

It is more difficult to identify perched water tables and artesian pressures, particularly if the aquifers have low permeabilities. A series of

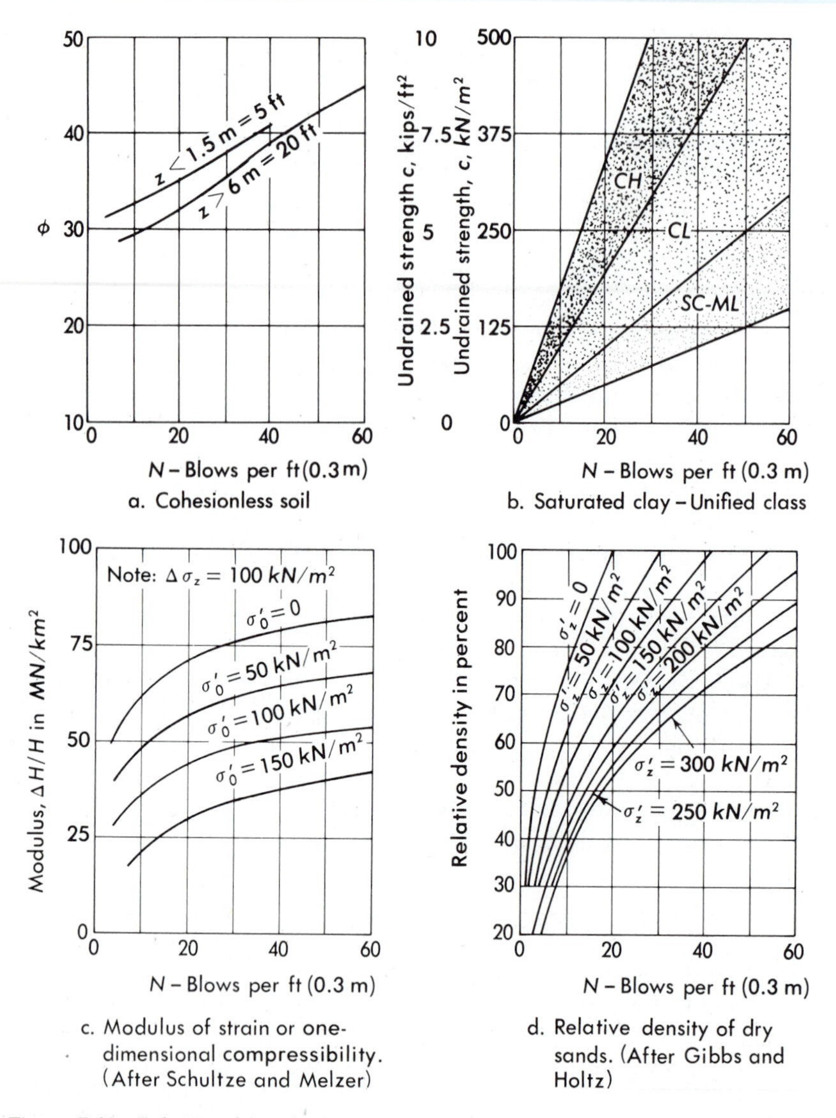

Figure 7.10 *Relation of standard penetration resistance to soil properties for preliminary estimates. Note that all the relationships are approximations and should be verified or modified for each different soil formation and drilling-sampling team.*

borings, each terminating in a different suspected pervious aquifer will generally find the perched water tables. However, very localized perched water can be drained by the deeper borings themselves. In such cases sealed piezometers, discussed below, are necessary. Artesian water is probable when the water level in the boring suddenly increases, when drilling mud thins or increases in volume, or if deep borings find a higher water level after stabilization than nearby shallow borings. A sealed piezometer is essential in measuring the artesian pressure. These should be installed in all pervious strata in which abnormal pressures are suspected.

OBSERVATION WELLS / Observations for a year or more are required to show seasonal fluctuations of the ground water. In most cases it is necessary to provide casing to maintain an open hole and to assure that the water level in the hole does not lag behind the changing water levels in the ground. In sandy soils a simple well can be made from plastic pipe with a 25-mm (1-in.) or larger inside diameter. The lower end of the pipe is slotted with fine saw cuts throughout the level of the aquifer. The well is sealed with mortar at the top to keep out surface water, and fitted with a vented cap. In fine soils the construction is more elaborate. Perforated or slotted plastic pipe is wrapped with plastic screen or filter cloth. The annular space between the soil and the screen is filled with clean concrete sand throughout the aquifer thickness up to the maximum probable groundwater level. The top of the well is sealed as previously described, Fig. 7.11a.

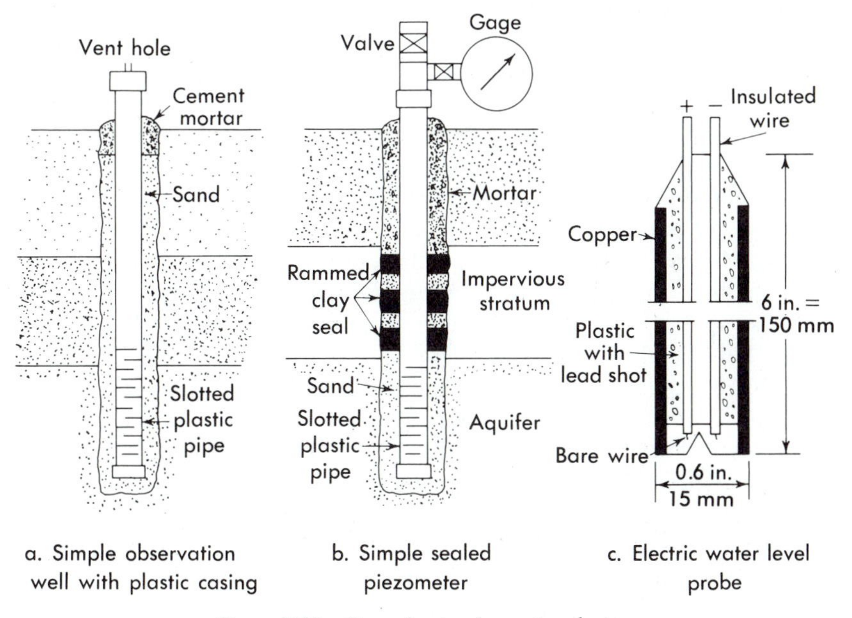

Figure 7.11 *Groundwater observation devices.*

PIEZOMETER / In order to measure artesian pressure, the well must be sealed into the impervious stratum immediately above the aquifer, to form a piezometer. It is extremely difficult to form a good seal, but without it the piezometer is of little value. Balls of clay or damp bentonite are dropped onto the annular sand fill in the aquifer. They are rammed into a continuous plug by a cylindrical drop weight that fits around the plastic pipe and inside the bore hole. Several layers of rammed clay balls alternating with cement mortar are required.

A quick-setting sand mortar, placed in layers of 0.2 to 0.5 m (6 to 18 in.), accompanied by gentle vibration of the piezometer tube, also makes a good seal. Each increment is allowed to set before placing the next; three or four layers are usually adequate. To minimize the time required for a piezometer to respond in soils of low permeability, the riser tube should be as small as practical compared to the surface area of the sanded hole. Some commercial piezometer kits are available with 12 mm ID riser tubes connected to a porous ceramic tube. In most applications semirigid slotted plastic pipe 20 mm ($\frac{3}{4}$ in.) ID (larger in pervious aquifers) is more foolproof and as effective.

WATER-LEVEL PROBE / In order to detect the groundwater level accurately, a slender electric probe (Fig. 7.11c) is necessary. It consists of two insulated wires embedded in a weighted sleeve that will fit inside the piezometer tube. The wire ends, uninsulated, extend 1 to 2 mm below the notch in the sleeve. When the wires touch water, there is sufficient conductivity so that the current can be indicated by a milliammeter. Recording probes of many forms are available for continuous readings.

7:8 Geophysical Exploration

In geophysical exploration the structure is inferred from distortions of physical force patterns, either those inherent in the earth or imposed on the earth by the exploration work. In a theoretically homogeneous, isotropic mass, the shape of the pattern can be defined mathematically. Any deviation from the theoretical pattern, termed an *anomaly*, is the result of a nonhomogeneity, such as stratification. In many cases it is possible to interpret the anomalies in terms of the depth and thickness of the different strata and to measure or estimate some of the engineering properties of the materials.

Many different force systems have been investigated, and most find some use in the study of geologic structure and the exploration for minerals. In most cases the anomalies are so large geographically that they are of limited use in civil engineering. However, a number of techniques have been found valuable in site investigations, as summarized in Table 7:6.

REFRACTION SEISMIC / This method is based on the physical principle that an elastic compression wave in a homogeneous elastic material having a density ρ and a modulus of elasticity E travels at a velocity V_p, expressed by

$$V_p = \sqrt{\frac{Eg(1-\nu)}{\gamma(1-2\nu)(1+\nu)}} = \sqrt{\frac{E(1-\nu)}{\rho(1-2\nu)(1+\nu)}} \qquad (7:3a)$$

For shear wave velocities, V_s, where G is the shear modulus of elasticity

$$V_s = \sqrt{\frac{Gg}{\gamma}} = \sqrt{\frac{G}{\rho}} = \sqrt{\frac{Eg}{2\gamma(1+\nu)}} \tag{7:3b}$$

Although the densities of soil and rock vary within narrow limits, the values of E, G, and ν vary greatly depending on strain level and the structural qualities of the material. Typical compression wave velocities are given in Table 7:7. Shear wave velocities are lower: about $\frac{1}{3}$ the compression wave for $\nu = 0.45$ to $\frac{3}{5}$ the compression wave for $\nu = 0.25$.

A small explosive charge is placed at or below the ground surface. Detectors, called *geophones*, are placed on line at increasing distances, $d_1, d_2, \ldots$, from the charge. The explosive is detonated, and the time required for the elastic wave to reach each detector is automatically recorded by a *seismograph*. The time required for the first shock to reach each geophone is plotted as a function of the distance from the charge, as shown in Fig. 7.12. A simple interpretation is possible if each stratum is uniform in thickness, $H_1, H_2, \ldots$, and that each successively deeper stratum has a higher velocity of transmission: $V_2 > V_1, \ldots$. The wave to the first few geophones travels directly through the upper stratum. Therefore, the slope of the time distance graph is inversely equivalent to the velocity.

$$V_1 = \frac{d_2 - d_1}{t_2 - t_1} \tag{7:4}$$

At the same time a shock wave is traveling down into stratum 2, where it is refracted to travel through stratum 2 and eventually return to the surface to be recorded by the geophones. Close to the explosive charge the time of wave travel is less by the more direct surface route. Eventually, if $V_2 > V_1$ a distance is reached where the time of travel by the longer route is less than by the surface. The time–distance graph in this range is flatter than for the first, and V_2 can be computed similar to V_1, as shown in Fig. 7.12. The two lines intersect at a point equivalent to the distance, d', from the shot. The thickness of the stratum, H_1, is given by

$$H_1 = \frac{d'}{2}\sqrt{\frac{V_2 - V_1}{V_2 + V_1}} \tag{7:5}$$

The velocity and thickness of each successive stratum can be computed, provided its velocity is higher than that of the one above. Typical velocities are in Table 7:7. The method is best adapted to horizontal or gently sloping strata with well-defined contrasts in velocity, such as soil overlying rock, or loose, dry sand overlying a sand saturated by the groundwater table. Under ideal conditions it can define the depth of boundaries several hundred feet deep with an accuracy of 1 to 2%.

TABLE 7:6 / GEOPHYSICAL METHODS FOR GEOTECHNICAL ENGINEERING*

Method	Principle	Use	Limitations
Refraction seismic	Shock wave by hammer impact or small explosive near ground surface. Time of travel to geophones at different distances measured. Shock may travel to distant phone faster through deeper hard strata than by shortest path.	Depth to ground water; depth to successively harder strata; possible estimate of rigidity and location of sink holes. Measure modulus of elasticity in dynamic loading.	Interpretation questionable with irregular or poorly defined boundaries; will not identify soft strata under more rigid strata.
Cross-hole seismic	Compression or shear shock wave transmitted from source in one drill hole to geophone in adjoining hole.	Compression or shear wave velocity horizontally through a stratum. Modulus of compression or shear elasticity.	Hole spacing must be narrow compared to thickness of soft stratum.
Down-hole, up-hole seismic	Compression or shear wave transmitted upward from source in hole to geophones surface or from sources on surface to geophone in hole.	Compression or shear wave velocity of thin weak strata below more rigid strata.	Interpretation difficult in multilayered materials.

Geophysical logging	Probe lowered into drill hole measures geophysical parameters as function of depth.	Correlation of stratification from hole to hole. Determine subtle changes in soil properties.	Uniform hole, indirect measurements.
Electrical resistivity	Electric current passed between electrodes at varied spacings. Potential drop between intermediate electrodes and current define apparent resistivity. Depth and resistivity of strata determined by resistivity–electrode spacing relations.	Depth to strata of different resistivities and ground water. Location of masses of dry sands and gravels or hard rock.	Interpretation questionable with irregular or poorly defined boundaries.
Gravity	Measure earth's gravity by sensitive torsion balance.	Locate major structural anomalies: faults, domes, large cavities.	Will not identify structures unless major differences in density involved, and anomaly large compared to its depth.
High-frequency sound	Time of travel of sound or supersonic wave through water and soft silt and reflected upward by stratum changes.	Depth of water and soft silt above hard bottom.	Of little or no use in homogeneous formations.

* Other techniques, such as reflection seismic and magnetic, which are useful in mineral exploration, are sometimes helpful in exploring regional geologic structure.

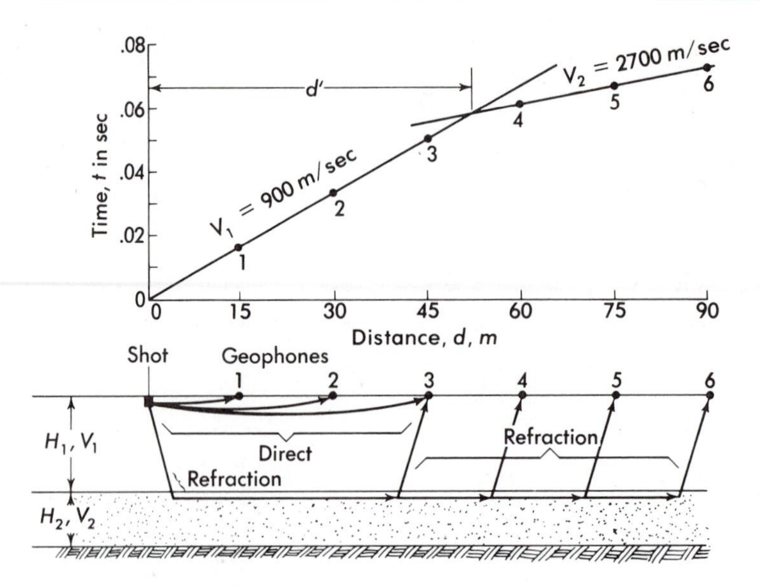

Figure 7.12 *Refraction seismic exploration.*

TABLE 7:7 / SEISMIC COMPRESSION WAVE VELOCITIES

Material	Velocity (m/sec)	Velocity (ft/sec)
Loose, dry sand	150–450	500–1500
Hard clay, partially saturated	600–1200	2000–4000
Water; loose, saturated soil	1600	5200
Saturated soil; weathered rock	1200–3000	4000–10,000
Sound rock	3000–6000	10,000–20,000

CROSS-HOLE / Seismic velocities can also be measured between two drill holes, with the energy source in one hole and a geophone opposite in the other. This makes it possible to measure the velocity in a stratum of low velocity beneath one of higher velocity so long as the drill hole spacing is small compared to the distance to a layer of higher velocity. Both compression and shear wave velocities can be measured cross-hole. The shear impulses are generated by dropping a weight in the hole, creating up–down motion in the hole wall; they are detected by special geophones that are less sensitive to the faster compression waves that mask the slower shear waves. The corresponding dynamic moduli E and G can be computed from the velocities V_p and V_s.

UP-HOLE—DOWN-HOLE / Compression and shear wave velocity measurements can be made by a shock source in a drill hole and geophones at the ground surface at different distances from the hole, or by a geophone in the hole and shock sources at varying distances from the hole. Although several layers are present, it is sometimes possible to compute the shear (or compression) wave velocity of each by this technique.

ENGINEERING PROPERTIES / Although the greatest use of seismic exploration has been to determine the depth of different strata, the velocities have direct engineering significance. The modulus of elasticity, E, and the shear modulus, G, can be computed from Equations (7:5) and (7:6). These values correspond to the very small strain levels and the relatively high random frequencies produced by the shock source. The E and G values for larger strains and lower frequencies associated with vibration problems, such as induced by machinery, are significantly lower. The seismic values probably represent their upper limits. Poisson's ratio, ν, can be computed from G and E. It also corresponds to small strains and high frequencies.

It should be noted that beneath the level of soil or rock saturation the compression wave velocity can be no smaller than that of water. Thus the effective E for loose saturated soils may be meaningless.

ELECTRICAL RESISTIVITY / The electrical resistivity method is based on the fact that the conductivity of soil and rock varies with the ionized salts present. Dense rock with few voids, little moisture, and little ionization will have high resistance, while saturated clay will have low resistance. Although a number of procedures are available, the Wenner method with four equally spaced electrodes is simple and widely used in site investigations. The four electrodes are placed in a straight line at equal distances, d, as shown in Fig. 7.13. An electrical current, typically 50 to 100 milliamperes, is passed between the outer electrodes, and is precisely measured. The voltage drop in a segment of the mass is measured between the two inner electrodes by a null-point circuit that requires no current flow at the instant of measurement. Either alternating current with its less sensitive measuring systems or direct current with nonpolarizing potential electrodes is required

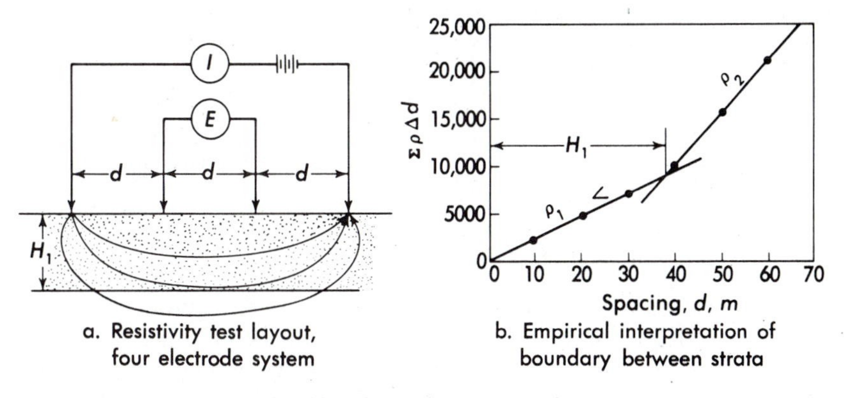

Figure 7.13 *Electrical resistivity exploration.*

to avoid polarization (the accumulation of hydrogen ions at the negative electrode) and an error in potential.

In a semi-infinite, homogeneous isotropic material the electrical resistivity, ρ, is given by the expression

$$\rho = \frac{2\pi\, dE}{I} \tag{7:6}$$

where I is the current in amperes, E, the potential difference between the center two electrodes and d, the spacing between electrodes. If the soil mass consists of strata of different resistivities, then the apparent resistivity as computed by this expression will be changed. The pattern of apparent resistivity as a function of electrode spacing or test location as the basis for interpretation.

The variable spacing technique is used to locate the depths of strata of different resistivity. A series of tests are made centered about one point, with increasing spacings, such as 3, 6, 9, . . . , m. A plot of apparent resistivity as a function of spacing can be interpreted in terms of the depth of the boundary between strata, using theoretical standard curves. An empirical interpretation used in site exploration work plots the sum of the apparent resistivity values as a function of spacing, as shown in Fig. 7:13. The resulting curve consists of relative straight segments if the strata are horizontal and uniformly thick. Tangents are drawn to the curve. The spacings equivalent to the intersections of the tangents are the depths of strata boundaries. The slope of the curve is proportional to resistivity—a steep curve indicates dry soil or rock and a flat curve wet soils or other materials of low resistivity. Typical resistivity values are given in Table 7:8.

TABLE 7:8 / ELECTRICAL RESISTIVITIES OF SOILS AND ROCKS

Material	Resistivity in Ohm-meters
Saturated organic clay or silt	5–20
Saturated inorganic clay or silt	10–50
Hard, partially saturated clays and silts; saturated sands and gravels	50–150
Shales, dry clays, and silts	100–500
Sandstones, dry sands, and gravels	200–2000
Crystalline rocks, sound	1,000–10,000

A different method is used in locating areas of shallow rock or high groundwater level. A constant electrode spacing is used, approximately equal to the estimated depth of the material. Measurements are then made at widely spaced locations in a grid pattern. A contour map of apparent resistivity shows areas of highs and lows. Bedrock or dry sand and gravel are most likely to be found in the high areas, and shallow ground water or clay in the low areas.

BOREHOLE LOGGING / A number of geophysical methods have been used to measure rock properties from inside drill holes for deep oil wells where direct sampling is extremely expensive. These techniques are finding increasing use in geotechnical engineering. The instrument is sealed into a watertight heat-resistant probe that is lowered into the drill hole. It measures the average properties of a spherical or ellipsoidal volume of soil or rock adjacent to the hole, typically 0.02 m^3 (0.7 ft^3). The results are plotted graphically (usually automatically and continuously) as a function of depth. A number of properties found useful in oil well interpretation are applicable to site exploration (Table 7:9).

TABLE 7:9 / BOREHOLE GEOPHYSICAL LOGGING

Property	Typical Use
Gamma radiation	Identify certain strata that are radioactive.
Density	Determine γ_d or ρ_d.
Neutron scatter	Determine water content; correlate with density to compute void ratio and estimate permeability.
Self-potential	Identify mineral changes.
Resistivity	Identify soil and rock type.
Sonic velocity	Estimate *E*.
Temperature	Determine hydrothermal activity, groundwater flow patterns.
Caliper (hole diameter)	Identify cohesionless, fractured material.

The graphs of each property in each hole are compared to find similar patterns. These are correlated with boring and sampling data to identify strata and define their continuity from hole to hole. Thus low-cost drill holes without sampling can help establish stratagraphic continuity. In addition, thin strata that have been overlooked in sampling can be identified and then sampled in new borings.

ADVANTAGES AND LIMITATIONS / The geophysical methods have two important advantages. First, they permit a rapid coverage of large areas at a relatively small cost. Second, they are not hampered by boulders or coarse gravel that obstruct borings. These qualities make them useful for both reconnaissance and exploration.

Their lack of unique interpretation is a distinct disadvantage. This is particularly serious when the strata are neither uniform in thickness nor horizontal. Irregular or transitional contacts often are not identified, and strata of similar geophysical properties sometimes have greatly different engineering properties. If the contact is very irregular, the boundary defined by resistivity is the average depth for a distance approximately equal to the depth. The same boundary defined by seismic refraction is the depth to the more or less continuous deeper stratum. For these reasons geophysical methods always must be used as a supplement to the direct methods, and the results verified by boring before definite conclusions can be reached.

7:9 Analyzing the Results of an Exploratory Investigation

LABORATORY TESTS / Although a visual examination of the soil samples obtained from exploratory borings may provide the engineer with a preliminary picture of the soil conditions, a study of the results of laboratory tests clarifies the picture and makes it possible to analyze the soil conditions on the basis of factual data.

The samples are ordinarily described in the field by the engineer in charge of the boring and sampling work, but they should be reexamined in the laboratory, and the field identifications should be verified. Tests can then be made on the samples to confirm their identification and to determine their physical properties. Table 7:10 summarizes the tests most useful in exploratory work.

TABLE 7:10 / LABORATORY TESTS FOR EXPLORATORY INVESTIGATIONS

Test	Types of Soils	Size of Sample (gram)	Type of Sample	Use of Data
Specific gravity of solids	All	100	Auger or split barrel	Void ratio, minerals.
Grain size	Cohesionless (sands, gravels)	200	Auger or split barrel	Classification. Estimate permeability, shear strength, frost action, and compaction.
Grain shape	Cohesionless (sands, gravels)	200	Auger or split tube	Classification. Estimate shear strength.
Liquid plastic limits	Cohesive (silts, clays)	200	Auger or split tube	Classification. Estimate compressibility and compaction.
Water content	Cohesive	100	Auger or split tube	Correlate with strength, compressibility, and compaction.
Void ratio	Cohesive	200	Split tube*	Estimate compressibility and strength.
Unconfined compression	Cohesive	200	Split tube*	Estimate shear strength.

* Sample must be relatively undisturbed.

Other tests, such as the loss of weight by ignition which identifies organic materials, or treatment with hydrochloric acid which indicates the presence of soluble carbonates, may be useful in identifying some soils. A microscopic examination of coarse soils and of the particles coarser than 0.075 mm in fine-grained soils is very useful in correlating similar strata in different borings.

PLOTTING BORING RECORDS / The first step in analyzing the data obtained by exploratory investigation is to plot the boring records graphically as profiles, as shown in Fig. 7.14. Each boring is represented by a vertical bar graph, with the different soils indicated by appropriate symbols or abbreviations. All should be plotted to the same scale with elevation (above the site datum) as the vertical ordinate. If possible, borings that are adjacent on the site should be plotted adjacent to one another, but a space of 5 to 8 cm (2 or 3 in.) should be left between the plot of each boring to provide room for the laboratory data.

Although the soil descriptions can be shown by graphic symbols, a simple shorthand system is preferable, as described in Chapter 2.

The soil penetration resistances are plotted as a broken-line graph next to the boring plot. This makes it possible to correlate the resistances of the different soils encountered. On the same graph can also be plotted the unconfined compressive strength data from the laboratory tests.

A second graph, also plotted adjacent to the boring record, shows water content and the liquid and plastic limits. The water content may be plotted as a broken-line graph and the limits as isolated points. The characteristics of cohesionless soils such as grain size and shape cannot be represented so conveniently on such a graph but may be indicated by notes or symbols.

PREPARING CROSS SECTIONS / Soil profiles or geologic cross sections (Fig. 7.15) for critical parts of the site are prepared by correlating the soils encountered in each of the borings. For example, a hard clay layer found in each of three adjacent borings at about the same elevation is probably the same continuous stratum, especially if the liquid limits and plasticity indexes are the same. By interpolating between borings, a reasonable soil profile may usually be established. In some very erratic soils, such as glacial moraines, interpolation is dangerous, since apparently continuous strata may be discontinuous lenses. Silts and organic soils often occur in limited lens-shaped deposits and should be viewed with suspicion. Bore hole logging and geophysical exploration between borings can aid in correlation.

THREE-DIMENSIONAL REPRESENTATION / In most formations the soils vary not only with depth, but also with location. A geologic cross section or profile represents the changes between borings in one direction only. Therefore, a number of cross sections parallel and at right angles to one another may be necessary to depict the three-dimensional variations. Cross sections are usually oriented with the building, such as parallel to column lines, and thus may not always display the most critical

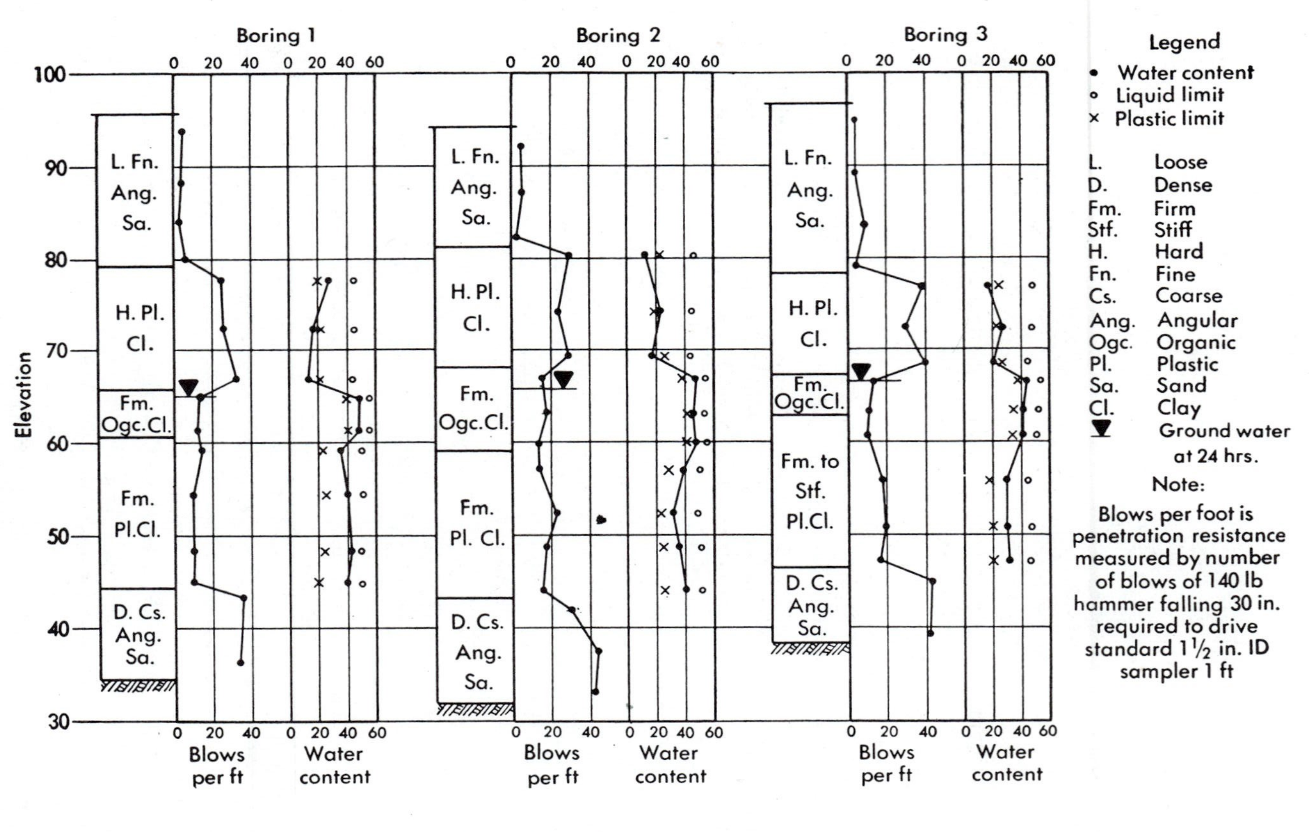

Figure 7.14 *Profiles of exploratory boring data.*

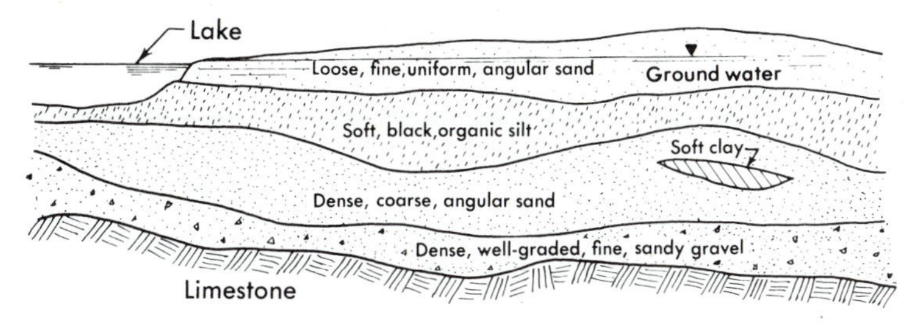

Figure 7.15 *Geologic cross section of a site.*

orientation of the soils. For example, if the strata dip steeply, a cross section parallel to the strike would depict horizontal boundaries, and might lead to a false sense of security. Cross sections, therefore, should be oriented to show the most critical soil and rock variations.

Where the variations in soil and rock are irregular, such as where faulting and folding are significant, three-dimensional representations are helpful in visualizing the underground structure. Combined cross sections plotted in a form of isometric projection, termed *fence diagrams* (Fig. 7.16), convey a three-dimensional impression. Similarly, block diagrams can give a limited three-dimensional view.

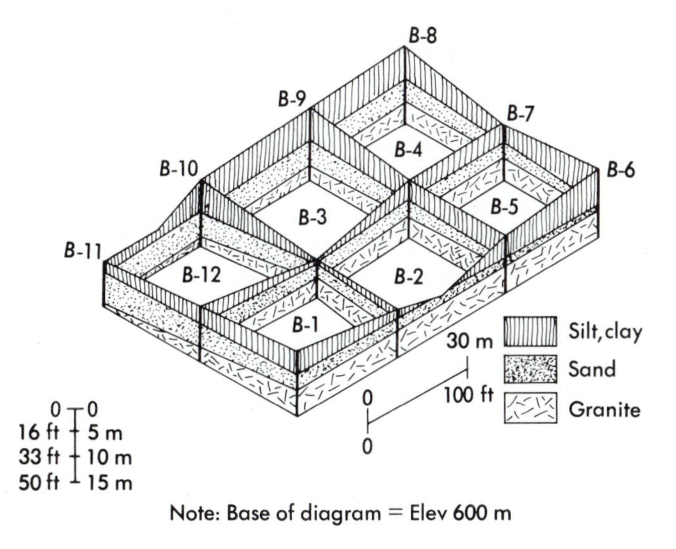

Figure 7.16 *Fence diagram showing site conditions in three dimensions.*

Three-dimensional models permit a quantitative representation. The simplest is a *peg board.* This consists of a base board that represents the site plan at a reference level well below the zone of interest. Rods are set in this board at the boring locations, and the significant soil and rock features are marked at the appropriate levels on the rod above the base. The ground surface is shown by the upper end of the rod. The level of the building, such as a foundation mat, can be indicated by a transparent plastic sheet pierced to fit over the rods at the proper depth. Boundaries between strata can be constructed with solid plastic foam or strings between the borings. If the horizontal and vertical scales are equal, complex problems in orientation can be solved quantitatively with the model, and the results can be meaningfully presented to a layman. Unfortunately, models are bulky and expensive and not easily preserved in files or in reports.

In attempting to correlate the boring records and determine the soil profiles, the engineer often finds that additional data would be very helpful. If the records are plotted at the same time the boring work is progressing, then the number or spacing of the borings can be changed to produce a clearer picture of the soil strata. In many cases low-cost auger borings can be used effectively to determine the extent of strata between the more expensive test borings, and in some cases geophysical methods or penetration tests prove very useful for the same purpose.

PRELIMINARY COMPUTATIONS / The unconfined compressive strength, void ratio, and compressibility of clays, and the unit weight and angle of internal friction of sands and gravels are necessary for most studies involving the safety and settlement of earth masses and the structures they support. The average values of these for each stratum in the soil profiles may be estimated from the laboratory data and the penetration resistance.

Preliminary computations for safety and settlement may be made by utilizing the soil profiles and the estimated soil properties. The results of these computations can be placed in four categories:

1. The structure is so safe from failure and excessive settlement that further study is unnecessary, and the estimated soil properties can be used as a basis of design without the sacrifice of economy.
2. The structure is safe and free from excessive settlement, but additional detailed soil studies may lead to a more economical design.
3. The structure appears to be unsafe or will probably settle too much; therefore, additional detailed soil studies will be necessary before a satisfactory design can be developed.
4. The structure is so unsafe or will settle so much that further soil studies would be useless.

On the basis of these computations, the designer decides whether to go ahead with the plans without further study, to secure additional, more accurate data, or to abandon the project as originally planned.

7:10 Intensive Investigation

The intensive investigation provides the engineering data on the soil and rock strata that are necessary for a quantitative design. Permeability, for projects involving seepage or drainage, strain and strength under changing loads, and volume changes produced by both stress and environment are as essential for the success of a project as the strength of the steel and concrete used in the superstructure. The more complex the soil conditions and the heavier the structure, the greater the cost savings possible with adequate data. The savings embrace both the construction operations and the ultimate structure. Moreover, when the soil conditions are marginal, estimates of soil behavior based only on exploratory data can lead to either ridiculously safe designs or to serious risks of future trouble.

The detailed investigation ordinarily focuses on the critical strata pinpointed by the exploratory work. For example, if a clay stratum is found that has a low penetration resistance, it is likely that its shear strength and compressibility will be significant factors in final design. Two approaches are possible in obtaining the needed data.

1. Secure representative samples of sufficient quality for laboratory testing.

2. Test the soil in place.

Testing the soil in the laboratory has the advantage that the environment, including stress, can be changed at will to simulate the changes produced by the construction and the future structure. Further, laboratory tests are available to measure nearly all the required soil qualities. The results are limited, however, by the soil samples: their alteration or disturbance produced by sampling, and their degree of representation of the total stratum. Testing the soil in the ground evaluates its behavior in its present environment. The test may integrate the effects of many variables that are difficult, if not impossible, to simulate in the laboratory, and the soil disturbance is limited to that produced by the test instrumentation. Because of time and cost the number of soil properties that can be evaluated and the changes in stress and other environmental factors that can be induced are severely limited. For most projects, undisturbed sampling with laboratory testing is sufficient; for complex soil conditions a program of both is essential. The field test results verify the laboratory work and the laboratory tests expand the range of conditions possible in the field.

In both cases the intensive work can be done either by interrupting the exploratory boring and investigating each critical stratum at the time it is identified, or by a separate phase of work conducted after the exploration has been completed. Combining the exploratory and intensive phases of the investigation saves time and on small projects is more economical. In large projects a better integration of work is possible if the detailed program is planned from the correlated results of the exploratory phase.

DATA REQUIRED / Analysis of the safety of a structure and of limiting earth pressures requires the soil shear strength. The strength of cohesive soils can usually be measured adequately in the laboratory. Very soft or sensitive clays that cannot be sampled or that are easily disturbed are tested better in the field. The strength of cohesionless soils can be determined either by undisturbed samples or on disturbed samples reconstructed at the same relative density. Because of the difficulties in securing undisturbed cohesionless samples, and their tendency to change state during transit, it is often expedient to test the samples for density in a field laboratory. However, it is impossible to reproduce the field grain structure that evolved from countless natural mechanisms; the reconstituted laboratory specimen does not fully reflect field performance. Alternatively, in-place testing may give more realistic results.

When settlement is critical, modulus of elasticity and compressibility data are essential. Laboratory consolidation and stress–strain tests are conducted on undisturbed cohesive samples and undisturbed or reconstructed cohesionless samples. Field tests for soil strain from changing stress can also be obtained from small- or large-scale field loading tests as well as from bore hole expansion. Permeability testing can be done on laboratory specimens; however, very large numbers of tests are necessary to define the range in values. Permeability is so profoundly influenced by seemingly insignificant changes in soil stratification that field testing is advisable when possible. Field testing is limited by the present groundwater conditions, and largely to horizontal seepage.

SAMPLES REQUIRED / In most investigations the critical strata prove to be composed of cohesive soils—clays, organic silts, and organic clays—which require undisturbed samples of sufficient size for laboratory tests. Table 7:11 lists the typical sizes of samples for testing.

TABLE 7:11 / SIZES OF SAMPLES FOR TESTING

Test	No. of Samples for One Test	Sizes of Samples Tested	
Unconfined compression	2	3.5 cm × 7.5 cm	1.4 in. diam × 3 in. long
	2	7 cm × 15 cm	2.8 in. diam × 6 in. long
Triaxial shear	3–6	3.5 cm × 7.5 cm	1.4 in. diam × 3 in. long
	3–6	7 cm × 15 cm	2.8 in. diam × 6 in. long
Direct shear	3–6	2.5 cm × 7.5 cm	1 in. × 3 in. diam
Consolidation	1	6 cm × 2.5 cm	2.5 in. diam × 1 in. thick
	1	11 cm × 2.5 cm	4.25 in. diam × 1 in. thick

The number of samples to be made depends on the uniformity of the stratum. A perfectly homogeneous soil would require only one sample large enough for the necessary tests; unfortunately, most actual soil deposits are far from uniform. The range in variation in soil properties can be determined from the results of the exploratory investigation. Typical points and extreme points within the stratum are selected from the boring logs and the plots of penetration resistance, water content, Atterberg limits, and unconfined compressive strength. The undisturbed samples are secured as close to these points as it is practical to do so. In many instances, however, it is necessary to secure an unbroken or continuous series of undisturbed samples throughout the depth of the critical stratum. One series of undisturbed samples is made beneath each important structure or beneath each different part of large structures, but more or less may be necessary, depending on the uniformity of the soil. Methods for deep sampling are summarized in Table 7:12.

7:11 Undisturbed Sampling[7:10]

The most important step in the detailed investigation is securing a sample with as little disturbance as possible. Unfortunately, it is impossible to secure a completely undisturbed sample. The removal of a portion of soil from the ground produces changes in the soil stresses which change the soil structure to some extent. The best "undisturbed samples" are those in which the soil water content and composition remain unchanged and the void ratio and structure are changed as little as possible.

CHUNK SAMPLING / A *chunk sample* excavated carefully by hand is usually the best undisturbed sample obtainable. Figure 7.17 illustrates the steps for securing a chunk sample. A pit or shaft is excavated in the proper location to the depth from which the sample is desired. The soil is carefully removed from around the sample, leaving it projecting in the side or bottom of the excavation like a small stump. If the sample is strong and rigid, it can be cut free with a flat shovel, then wrapped with plastic film to preserve its water content, placed in a substantial box for support, and transported to the laboratory by automobile. If the sample is weak, or if it must be transported by railroad or truck, additional protection is required. A heavy wooden box, with lid and bottom removed, is placed around the sample so as to leave a space of 25 mm (1 in.) on all sides. This is filled with melted paraffin. The sample and the box are removed from the excavation, paraffin is poured on the top and bottom of the sample, and the top and bottom lids replaced. Another method is to slide a cylindrical container over the sample, which has been carefully trimmed to make a snug fit. Beer or larger cans with their tops removed, sections of stove pipe, and large thin water pipes have been used. Wood or metal caps protect the open ends of the sample from damage. A layer of paraffin poured over the open ends prevents evaporation of the moisture. These containers should be placed in a substantial wooden box and surrounded with sawdust or shavings in order to protect them for shipment.

A test pit or trench not only makes it possible to secure undisturbed samples, but also provides a "window" from which to observe the soil structure in place. Pits as large as 10,000 m^3, 20 m deep, have been excavated. A visual examination of the soil strata that are uncovered by excavation of the pit will disclose the arrangement, uniformity, and the

TABLE 7:12 / METHODS FOR DEEP SAMPLING

Method	Equipment and Procedure	Type of Sample and Use	Limitations
Auger (ASTM D 1452)	Retain cuttings from short increments of auger boring.	Disturbed for soil identification, water content above water table.	Structure destroyed. Soil mixed with water below water table.
Split barrel (ASTM D-1586)	Split barrel sampler, driven 0.46 m (18 in.) into stratum, 36 mm ID×51 mm OD (1.4 in. ID×2 in. OD); 51 mm ID×64 mm OD (2 in. ID×2.5 in. OD), without or with liners for sample protection.	Intact but disturbed. Soil identification, structure, water content; density of very wide range of soils.	Sample distorted—disturbance too great for strength, consolidation tests.
Thin-walled tube (ASTM D-1587)	76 mm OD to 127 mm OD (3 in. OD to 5 in. OD) thin-walled tube with sharp edge forced into soil, 10 to 20 diameters.	Relatively undisturbed sample for shear density, consolidation, etc., of most soils.	Sample lost in very soft clay or loose sand below water
	Same—driven with hammer.	Stiff clays only.	Slight disturbance.
Thin-walled tube, fixed piston	76 mm OD to 127 mm OD (3 in. OD to 5 in. OD) thin-walled tube with sharp edge. Piston keeps cuttings out, remains stationary while tube advances and fixed after driving to help hold sample. Rod-activated, internal hydraulic piston (Osterberg type), ratchet control (Hong).	Relatively undisturbed of very soft silts, clays; loose sands if hole filled with heavy drilling fluid.	Sample sometimes lost in soft clay, loose sand.
Swedish foil	Thin strips of metal foil stored above cutting edge surround sample to prevent contact with sample tube and help hold sample.	Relatively undisturbed continuous [to 12 m (40 ft) long] sample of soft clay, for shear, consolidation, etc.	Gravel, coarse sand, or hard strata strain damage sampler.
Rotary core: soil	Outer tube with teeth rotates; stationary inner tube protects and holds soil. Cuttings removed by drill fluid. (Denison type with fixed cutter; Pitcher type with automatic variation of cutter position.)	Relatively undisturbed of firm to stiff cohesive soils, soft rock; continuous.	Torsion failure in soft soil and sometimes in sands.
Rotary core: rock	Tube with diamond bit on end rotates. Core protected with stationary inner tube in double-tube core barrels. Cuttings removed by drill fluid.	Continuous core in hard rock; nearly continuous in soft or fractured rock with M-type double tube.	Fractured or very soft rock not recovered.

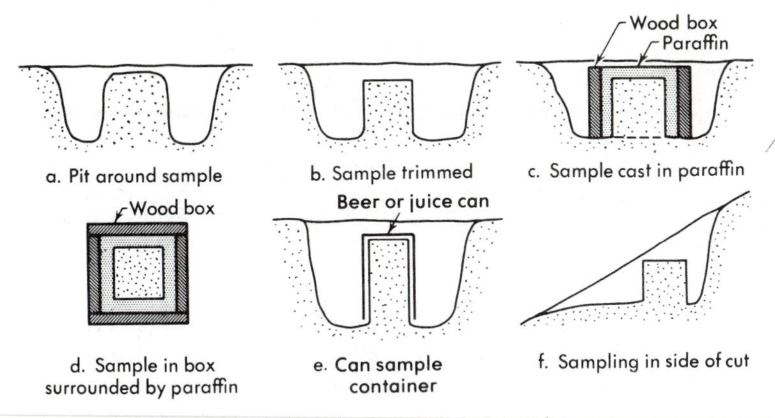

Figure 7.17 *Chunk undisturbed sampling.*

inclination or dip of the strata—information that is not easily secured from borings. Color photographs made of the pit walls provide a permanent record.

DEEP UNDISTURBED SAMPLING / Many types of equipment have been designed to secure undisturbed samples from deep bore holes, but not all types are successful in every soil. The undisturbed quality of the sample has been found to be dependent on the following factors:

1. Displacement of the soil by the sampler.
2. Method of forcing sampler into ground.
3. Friction on inside of sample tube.
4. Squeezing of soil, owing to pressure of overburden.
5. Handling and storing of samples until tested.

The displacement of the soil by the walls of the sampler is probably the most important source of disturbance. The soil is forced aside and upward, which severely distorts it and changes its structure. This can be minimized by excavating around the sample, by using a long, thin cutting edge on the sampler, and by keeping the cross-sectional area of the walls of the sampler as small as possible. The relative displacement of the sampler is expressed by the area ratio A_r:

$$A_r = \frac{D_0^2 - D_s^2}{D_s^2} \times 100\% \qquad (7:7)$$

where D_0 is the outside diameter of the sampler and D_s the diameter of the sample. According to Hvorslev,[7:10] displacement disturbance is minimized by keeping the area ratio less than 10 to 15%.

A thin sharp cutting edge also minimizes displacement. This is easily damaged by gravel or hard seams, and therefore can be used only in continuously soft strata.

The method of driving the sampler into the ground is important in loose sands and very sensitive clays. In such soils continuous hammering and the accompanying shock and vibration are harmful, although a single blow of a very heavy hammer does not appear detrimental. The best method is to force the sampler into the ground with a steady movement such as is provided by hydraulic-feed drilling machines.

Friction between the sample and the walls of the sample tube produces disturbance in the edges of the sample that is readily visible in stratified or laminated soils but is present in all cases. This friction can be reduced by drawing in the cutting edge of the tube about $\frac{1}{2}$ mm ($\frac{1}{50}$ in.). The diameter of the sample is then about 1 mm ($\frac{1}{25}$ in.) less than the inside diameter of the tube. If too much draw-in is used, the sample expands, changing its structure; if too little is used, the sample will be distorted. Friction is also minimized by limiting the sample length and by excavating around the sample.

Removal of soil from the boring reduces the downward stress at the hole bottom without changing the lateral and upward stresses produced by the weight of the overburden. As a result the soil squeezes inward and upward, distorting itself in the process. This condition is very serious in soft clays but exists to some extent in all materials. In cohesionless soils any attempt to lower the water level in the hole below the groundwater table by pumping or bailing will have an added effect because the unbalanced water pressure in the hole bottom creates a quick condition. The unbalanced stress can be overcome by keeping the hole filled with liquid at all times. Water is adequate in many soils, but in loose sands and soft clays, drilling mud (described in Section 7:5) is better. In extremely soft clays it is sometimes necessary to match the unit weight of the drilling fluid to that of the soil by adding baroids or iron filings to the mixture.

Suction due to the removal of the sampler distorts the bottom of the sample or even pulls the soil out of the tube. This is difficult to prevent. A piston or check valve above the sample can help by developing comparable suction above, as the sample tends to slide out. A small tube or duct fastened to the outside wall of the sampler and open at the cutting edge can allow the drilling fluid from above to fill the gap produced by removing the sampler. As an alternative air can be pumped through the tube from the ground surface. A simple suction release device is a 6 mm ($\frac{1}{4}$ in.) diameter rod or bar welded to the side of the sampler. The sampler is rotated before extraction, creating a temporary duct in the soil. In any case the duct will increase the sampler displacement slightly.

Improper handling and storing cause shock, distortion, and drying. By sealing the sample immediately, packing it in a cushioned container, protecting it from temperature extremes, and exercising due care in transportation, these causes of disturbance are minimized.

Many types of equipment have been developed to minimize disturbance; however, the relative importance of the factors differs with the soil, the depth, and the ground water. Therefore, sampler design is a compromise. The principal types are listed in Table 7:12 and the more versatile are described in the following paragraphs.

THIN-WALLED SAMPLER / The simplest and most widely used, deep undisturbed sampler is the *thin-walled* or *Shelby tube* (Fig. 7.18a). It is made of cold drawn steel tubing (sometimes known as *Shelby* tubing) from 50 to 130 mm (2 to 5 in.) OD, with walls 1 mm (0.04 in.) thick for the 50-mm (2-in.) tube to 3 mm (0.125 in.) thick for the 130-mm (5-in.) tube. The lower end is beveled to form a tapered cutting edge, and it can be drawn in to reduce the wall friction. The upper end is fastened to a check valve that helps hold the sample in the tube when it is being withdrawn from the ground.

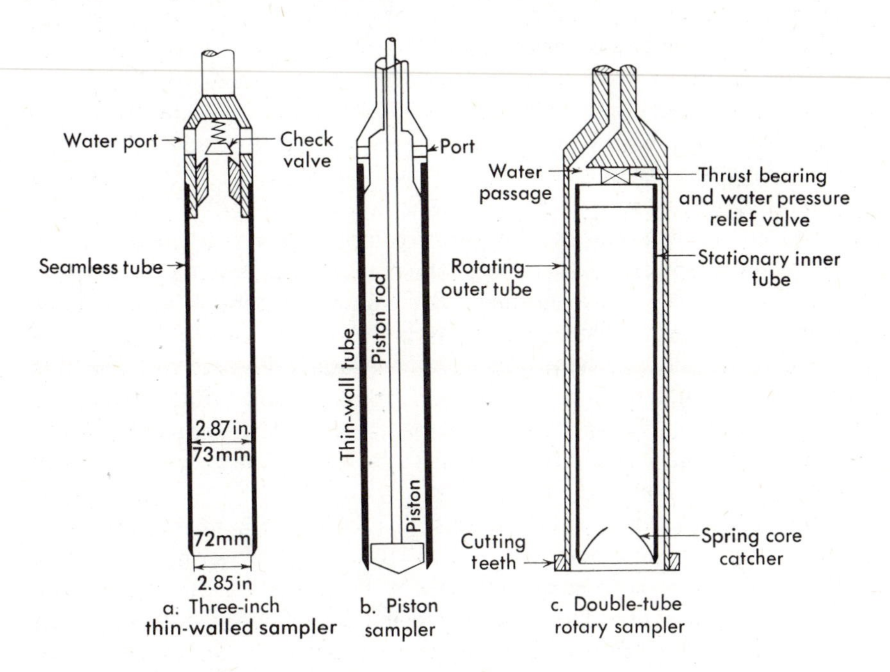

Figure 7.18 *Undisturbed samplers.*

The sampler is introduced in the bore hole and forced into the soil a distance of 10 to 20 diameters so as to minimize friction between the sample and the walls of the tube. The sample is sealed in the tube with melted wax and shipped intact to the laboratory.

The thin-walled sampler minimizes the most serious sources of disturbance: displacement and friction. When used in a bore hole that is stabilized with drilling mud, excellent samples can be made in a wide variety of soils.

PISTON SAMPLER / In extremely soft soils, even the small displacement of the thin-walled tube tends to cause the soil to squeeze into the tube faster than the sampler is advanced, causing distortion. The distortion due to friction also can be serious, but if friction is reduced too much by draw-in and by limiting the sample length, the sample will slide out of the tube when it is being withdrawn from the ground.

These difficulties are reduced by placing a piston in the thin-walled sampler (Fig. 7.18b). At the start of sampling, the piston is at the tube bottom and in contact with the soil surface. The piston is fixed in this position by its actuating rods, which extend to the ground surface and which are locked to a rigid support. The sample tube is forced ahead of the piston into the soil below. The fixed piston prevents the soil from squeezing upward. If the sample tends to slip out of the tube, a vacuum is created between the piston and the soil, which helps hold it. Good samples can be secured in very soft soils by this method but at a considerable increase in cost.

The Osterberg fixed-position sampler incorporates a hydraulic cylinder in the head of the sampling device to actuate the advance of the thin-walled tube. Hydraulic pressure for operation is supplied through the hollow drill rods, eliminating the need for cumbersome, double concentric drill rods, one to hold the piston and one to force the sampler.

A number of variations of the piston sampler are in use. The free piston is locked in the lower end of the tube while the sampler is being lowered into the hole. This keeps loose cuttings from choking the tube. When sampling begins, the piston is unlocked and floats up the tube on top of the soil. When the tube is withdrawn, the piston again locks and helps hold the sample in place. It does not, however, prevent squeezing of the soil during sampling. In another form the piston is torpedo-shaped, which allows the sampler to be advanced through very soft soil without a bore hole. None of these variations is so effective as the fixed piston.

FOIL SAMPLER / The Swedish foil sampler minimizes the friction between the sample tube walls and the soil by feeding thin strips of metal foil into the tube. These form a moving liner of thin metal that prevents the soil from touching the tube. The foil is pulled ahead by a piston at the same rate the sampler is advanced, so that there is no tendency to distort the soil by squeezing or wall friction. By this method continuous undisturbed samples as long as 15 m (50 ft) have been obtained in soft to firm clay or silts. It is of little use in sands and gravels, however.

The foil lining is made up of 16 strips, each strip 10 mm ($\frac{1}{2}$ in.) wide so as to nearly cover the surface of the 70 mm (2.7 in.) diameter sample. The foil is supplied from rolls that fit in a retainer 0.3 m (12 in.) above the cutting edge so that the displacement of the retainer will not cause undue soil disturbance. The strips feed to the sample 127 mm (5 in.) above the edge.

ROTARY SAMPLERS / The rotary samplers combine drilling and sampling, which minimizes disturbance due to sampler displacement. An example is the *Denison sampler* (Fig. 7.18c), which consists of two concentric tubes. The inner tube is actually the sampler and is 75 to 130 mm (3 to

5 in.) OD, 0.6 m (2 ft) long, with a thin steel liner and a heavy cutting shoe at the lower end. The outer tube rotates, cutting the soil, and the cuttings are washed up the bore hole by water that is pumped down the drill pipe and flows between the two tubes. The inner tube remains stationary and protects the sample from the wash water. At the same time, both tubes are forced downward into the soil. The samples secured by this device are excellent, particularly in hard clays and slightly cohesive sands that are difficult to sample in any other way.

A number of variants of the double tube rotary sampler are adapted to different conditions. The *M-series* of double-tube diamond core drills was originally designed for sampling coal. The stationary inner tube extends almost to the cutting edge, and the cutting edge is altered. With it, good cores, 43 to 53 mm (1.7 to 2.1 in.) in diameter, can be made in very soft, easily eroded rock and hard soils. The *Pitcher sampler* employs a spring to control the advance of the inner tube. In hard formations, the tube retracts so that the rotating cutting edge leads. In soft formations, the inner tube, with a sharp edge, extends ahead of the rotating cutter. In this way the sampler resembles a thin-walled tube with a rotary cutter to minimize displacement slightly above the tube entrance.

DEEP SAMPLING IN SAND / It is difficult to secure undisturbed samples of cohesionless sand and gravel from bore holes, since the sampling operation may displace the grains and the samples often flow out of the sampler. One method has been to freeze the soil and then drill through it with a rotary drill, but freezing increases the void ratio.

Undisturbed samples of sand can be secured below the groundwater table if a very heavy drilling mud is employed. The mud forms a coating over the lower end of the sample, which prevents the sand from running out. The sample is drained of excess water before sealing so that capillary tension will help to maintain its structure during shipment. It is also helpful to measure the sample so that any change in density during shipment can be detected, and the original void ratio can be computed from the weight as received in the laboratory.

The Bishop sand sampler utilizes a larger cylinder that is forced over the thin-walled sample tube after it penetrates the sand. Air is pumped into the larger cylinder so that the sampler with its loose sand is surrounded by air. If the replacement of water by air around the sample tube is sufficiently slow so that the sand can drain partially, it will be held intact in the tube by capillary tension.

LINER TUBE SAMPLERS / An early type of undisturbed sampler that is still in use is similar to the split-barrel sampler except that it is equipped with a seamless liner tube that fits snugly in the barrel. The sample is shipped and stored in the liner to minimize handling disturbance. The large area ratio often causes severe displacement distortion, and for this reason such samplers are not widely used.

7:12 Field Tests[7:18, 7:22–7:25]

Testing the soil in place has the theoretical advantage of minimizing the disturbance caused by stress changes and similar sampling distortion as well as eliminating the shock and vibration of transport and subsequent handling. Moreover, the effect of associated features of the formation are included in the field test, so that it probably is the most realistic measure of the physical properties in the existing environment. The test itself, however, may provide some disturbance and it is usually impossible to evaluate the effects of drastic environmental changes. The methods are summarized in Table 7:13.

The various penetrometers described in Section 7:6 are simple field tests. The static cone is a direct measure of *end bearing* of the formation. Extrapolation of the results of full-sized foundations is discussed in Chapter 10, but as in all models the *scale effect* must be evaluated. The dynamic tests similarly can be related to bearing capacity. Because bearing capacity is a function of the soil's engineering properties (see Chapter 10) a relationship should exist between these properties and either static or dynamic penetration resistance. Typical empirical relationships involving the in-place strength of clays and the relative density and angle of internal friction of cohesionless soils are given in Section 7:6. However, because other factors are involved, these relationships are not well defined and should be used only as indications of the desired property. For any given site and stratum, the scatter of data is limited and well-defined relations can be derived for that one situation.

VANE SHEAR / The vane shear is adapted to testing soils below the bottom of a bore hole at great depths and with a minimum disturbance. Field vanes (Fig. 7:19) employ two crossed blades attached to a vertical rod. The typical vane diameters are 50, 65, and 75 mm or 2, 2.5, and 3 in. wide with lengths of 3 to 5 diameters. The vane is forced into the soil so its top is 2 diameters below the bottom to the bore hole. Rotation of the vane shears the soil on a cylindrical surface. The torque required to initiate shear is measured, and often the increasing torsional strain is indicated as a function of torque. After failure, a second torque is made after several revolutions, to measure the soils remolded strength. The shear strength, τ_{ff}, for a long vane of diameter D, and length L, is found from the torque T, as follows

$$T = \frac{D}{2}(\pi DL)\tau_{ff}$$

$$\tau_{ff} = \frac{2T}{(\pi D^2 L)} \qquad (7:8)$$

Corrections can be made for the end shear; if L/D exceeds 3, the correction is negligible. The cylindrical shear surface does not resemble soil failure in real problems nor in other laboratory tests. Neither do the strains that

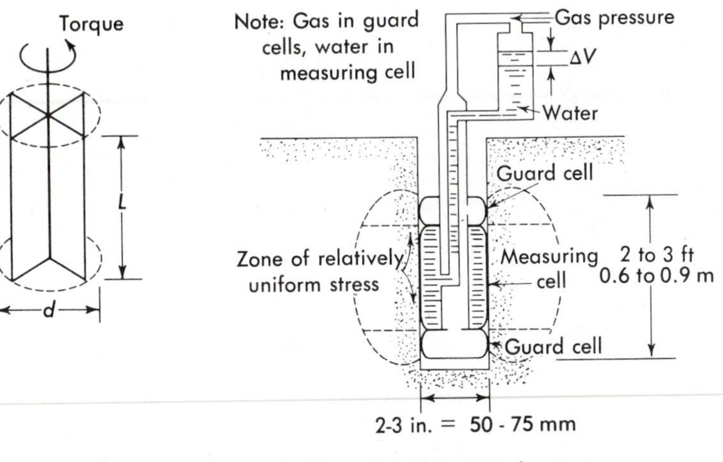

a. Vane shear

b. Menard pressiometer for strain, pseudo-strength. (Note: Not to scale)

Figure 7.19 *In-place strength tests.*

develop before failure. Therefore, the results of vane tests do not always agree with other shear tests. Although empirical corrections have been proposed, the vane strengths should be used with caution in highly plastic or very sensitive soils, where experience shows that vane strengths exceed other test results.

INTERNAL OR BOREHOLE EXPANSION / The stress–strain character of soil or rock can be evaluated by the resistance of a bore hole to expansion from internal pressure. The simplest device is a rubber sleeve with rigid end caps that fit snugly in a portion of the hole. Pressure in the sleeve is transmitted to the soil or rock. The radial strain can be measured by internal probes or indirectly by the volume change of liquid in the sleeve. The Menard pressuremeter (Fig. 7.19b) employs triple sleeves, all three with equal pressure. The stresses and strains around the two end or *guard cells* are three-dimensional; they force two-dimensional strain on the soil adjacent to the central cell. The radial strains corresponding to increasing internal pressures are measured by the volume increase in the central or measuring cell. From this a radial modulus of deformation is obtained corresponding to the initial and increased state of stress in the soil at that level. The modulus can be related empirically to the elastic modulus, E, and to soil compressibility, depending on the rate of test.

Liquid pressure, confined in a short section of a hole in impervious rock or soil, can be increased sufficiently to overcome the existing lateral stresses in the ground and even produce tensile stresses and rupture. This is termed *hydraulic fracturing.* The rock or soil splits in the direction of least tensile strength, as determined by the Mohr envelope and the stresses. If the tensile

TABLE 7:13 / FIELD TESTS

a. Direct Tests

Method	Procedure	Properties Measured	Limitations
Vane shear	Rotate vane in undisturbed soil. Measure maximum and post maximum torque.	Undrained strength of soils difficult to sample (soft clay).	Progressive failure in sensitive soil—variable effect of gravel, questionable in sand.
Load test (ASTM D-1194, ASTM D-1143)	Load large model or full-size footing or pile in increments; measure settlement.	Ultimate bearing; short-term deflection.	Interpretation in terms of prototype difficult.
Well test (Field permeability)	Pump water into or out of bore hole, measure draw-down in adjacent holes or rate of level change in well.	Effective horizontal permeability of mass.	Questionable above water table; not effective for vertical k.
Field density (D-2167, D-1556, D-1587)	Excavate hole; weigh soil; measure hole volume by volume of standard sand or volume of water-filled rubber membrane introduced into hole. (Measure, weigh undisturbed Sample—D-1587.)	Unit weight of virgin soil or fill.	Limited by gravel, water.
Test fill	Place fill equal in weight to structure on surface.	Settlement, pore pressure, ultimate bearing.	Time, cost.
Test pit	Excavate pit; measure hole volume, soil weight; obtain undisturbed samples and photographs.	Density of coarse gravel, broken rock; samples for detailed testing; in place dip, stratification, cracking.	Cost, ground water.
Auger plate	Screw disc auger into soil to test level. Load in increments, either compression or uplift.	Ultimate bearing, short-term deflection.	Interpretation difficult in terms of prototype.

b. Indirect Tests

Dynamic penetration	Drive standard sampler or cone point by hammer of known weight. Measure number of free-fall blows to drive a fixed distance (ASTM D-1586 for "Standard Penetration Test".)	Estimate strength and/or density. Estimate allowable bearing capacity.	Changes density of loose sands—sensitive to procedure changes.
Static penetration	Force conical point into soil, measuring force to penetrate at fixed rate; rotate cone point with helical blade, measure force to advance cone ahead of auger.	Estimate strength and/or density. Estimate allowable bearing capacity.	Interpretation varies with soil; identification of soil questionable. Sensitive to procedure changes.
Borehole expansion	Expand elastic tube in bore hole. Measure expansion volume.	Estimate strength elasticity and compressibility.	Interpretation controversial.
Hydraulic fracturing	Pressurize fluid in bore hole until walls rupture.	Estimate in-place lateral pressure.	Full stress picture requires holes in three different directions.
Overcoring	Core around measured drill hole. Measure external stress that returns hole to initial diameter.	Estimate in-place lateral pressure.	Full stress picture requires holes in three different directions.
Electric-resistivity	Pass current between two electrodes; measure potential between intermediate electrodes; compute apparent resistivity.	Estimate soil identity (clay or sand). Estimate corrosion.	Interpretation of data controversial—rough estimates at best.
Seismic	Measure time of travel of shock wave produced by impact or explosive. Compute velocity.	Estimate rigidity and material identification.	
Vibration response	Induce vibration by counter-rotating eccentric weights. Measure response by seismometers at varying distances.	Estimate dynamic response.	
Nuclear	Introduce nuclear radiation source into soil.	Density, water content.	Variable reliability; some soils, rocks.

strength of the rock or soil is determined by laboratory tests, the initial horizontal stresses can be computed. The direction of the fracture is found from a rubberlike plastic that is forced into the crack and then retrieved without turning. With drill holes at different angles, it is sometimes possible to compute all three initial principal stresses at the depth of testing.

Another form of measuring in-place test in rock involves *overcoring.* A hole is drilled in rock and its internal diameter is measured. A larger diameter rock core with the hole centered in it is removed from the ground. The hollow cylinder of rock, no longer confined by the existing stress in the ground, expands. The cylinder is then compressed externally by hydraulic pressure until the hole returns to its original diameter. If the rock is elastic, that external compression is comparable to the initial stress in the ground.

The initial state of stress in soil or rock can be investigated by one-dimensional expansion. Reference points are placed in the surface of the mass. A slot is cut between them, and if there is sufficient internal stress, the slot will close partially. A thin disk-shaped hydraulic jack is placed in the slot. (Very thin, wide jacks for this purpose are termed *flat jacks.*) The jack pressure is increased until the distance between the reference points returns to its initial value. If the response of the mass is elastic, the pressure required to force the mass back is equivalent to the initial stress in that direction.

FIELD PERMEABILITY / The permeability of the mass as a whole can be measured by test wells, in which water is introduced into or pumped out of the ground. The simple expressions for steady-state flow, derived in Chapter 3, can be used to evaluate either inflow or outflow wells. Better control of the operation is possible with pumping into the ground and a small cheap well is adequate. If the groundwater level is initially high or if water to pump in is not available, pumping out is satisfactory. The test well, 80 to 150 mm (3 to 6 in.) in diameter penetrates (as far as practical) the aquifer thickness. Two and preferably three sets of observation wells are placed around the test well (Fig. 7.20). Typical distances of the sets from the test well are 3, 10, and 30 m (10, 35, and 100 ft). Pumping is continued in the test well at a constant rate until the levels in the observation wells stabilize. The average coefficient of permeability can be computed from the average of the drawdowns of any two sets. The differences in the permeabilities computed between different sets indicate variations in k with distance or insufficient time to establish equilibrium. If significant differences are noted, pumping should be continued until the results become more consistent.

The rate of flow into a bore hole can be used as an approximate measure of the permeability of a pervious stratum. The hole is cased into the pervious layer, as shown in Fig. 7.20c, so that the end of the casing is no closer than 5 diameters from the stratum boundaries. The original groundwater level is measured. Water is then added to the boring to maintain it a constant level in the casing. When a constant rate of inflow, q, is established, the permeability

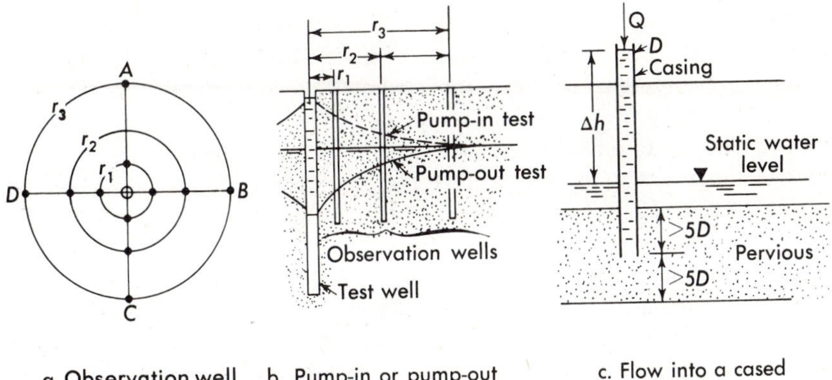

Figure 7.20 *Field tests for permeability.*

is computed by the relation,

$$k = \frac{q}{2.75d\,\Delta h} \tag{7:9}$$

LOAD TESTS / A load applied to a limited area of a soil or rock formation is a model test of the effect of a full-scale foundation. A near-surface plate load test is described in Chapter 10. It measures elastic deflection, possibly some consolidation, and, with sufficiently great loads, soil shear. However, the load–deflection–failure data are difficult to interpret in terms of the fundamental soil parameters of E, C_c, c_v, and τ_{ff}.

Load tests can be performed at depths by a disk or plate auger that is screwed into the ground to the test level. The plate area of the Norwegian equipment is 0.02 m^2 (0.2 ft^2) with a diameter of about 0.16 m (6.3 in.). It can be forced downward or pulled upward, with the load measured by an internal hydraulic cell. Like the surface load test, interpretation in terms of fundamental parameters is difficult.

LARGE-SCALE TESTS / Large-scale tests of soil or rock can integrate the effects of stratification and structural defects and will provide a more realistic evaluation of the behavior of the total mass than can the smaller conventional tests. Because of time and expense, large tests are conducted only when the peculiarities of soil and rock structure prevent reliable conventional testing. The equipment arrangement and procedure is designed to fit the mass structure. Ingenuity and imagination must be exercised in devising each test program.

Shear tests of the mass can define the total significance of defects such as old shear zones, cracks, and seams of weakness. Direct shear is probably the simplest approach. Trenches are excavated in the mass to isolate a block which is representative of the whole. This may be several cubic meters (or cubic yards) in volume. The orientation of the trenches is such that thrust can be applied parallel to the weakness to be evaluated. A normal stress is applied by direct weighting if the potential shear surface is horizontal; otherwise the normal stress is developed by hydraulic jacks reacting against a beam anchored to the soil or rock beyond the test mass (Fig. 7.21). Shear loads are applied by jacking against the trench walls, and strains measured by dial gauges. The results are presented in the same form as for other direct shear tests. Such tests have been used in hard soils in jointed rock in dam foundations and tunnels, and to evaluate the sliding resistance of concrete (poured in a block) on a rock foundation.

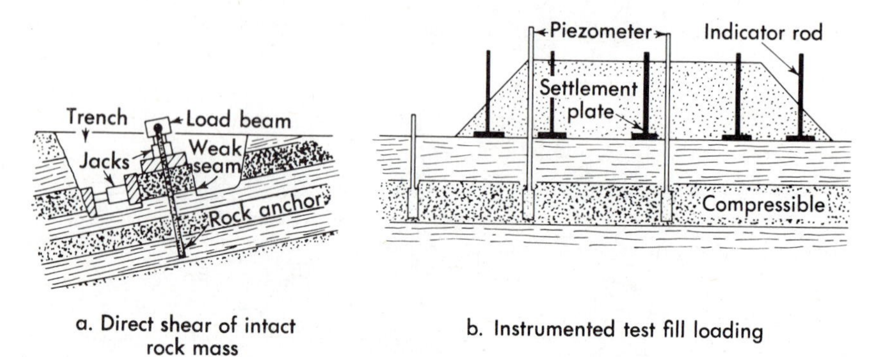

Figure 7.21 *Large-scale in-place tests for strength and settlement.*

Large-scale bearing capacity and consolidation tests are conducted by test fills, or other weights, such as water-filled tanks. Settlement points consisting of steel plates 1 m (3 ft) square, with threaded indicator rods extending upward are placed on the ground. The plates are arranged in regular patterns both under and adjacent to the test area. Elevations are secured on the indicator rods, and fill is added in increments. Settlement measurements are obtained on the rods at regular intervals, as in a consolidation test. If the required load is great, extensions are added to the rods so that they are always above the fill surface. Fluid actuated remote reading elevation sensors are also used. Piezometers are frequently installed in the critical strata to evaluate pore pressure increases, and samples of the critical strata are secured before and after loading to determine any differences in soil void ratio, water content, or strength. Such tests are expensive and time-consuming, but if properly executed they offer the most reliable test possible of the total soil or rock mass (short of the behavior of the structure itself).

7:13 Marine Investigations

The design of marine structures requires data on the materials below open water. The field work is complicated by working through water (where the soil surface is not visible) and by wind, waves, and water currents that interfere with work.

Three approaches are used: (1) conventional investigative techniques from the water surface; (2) cable or projectile devices, suspended from floating platforms through the water into the bottom; and (3) bottom-operated devices. All are highly specialized and will be only briefly described.

WATER SURFACE / Conventional boring and sampling can be done from the water surface in depths as great as several hundred feet by modifications of the drilling equipment. Where wave and tide action are limited, floating platforms are used. Stiff casing must be employed to withstand the unsupported length and the platforms require multiple anchors for stability. Various forms of telescoping drill rods are used that permit rotary motion while absorbing vertical movement of waves and tides. Alternatively, a fixed platform on spuds or pile foundations above the water surface offers a steadier work surface at a much greater cost.

CABLE OR PROJECTILE DEVICES / A second category is devices suspended from a barge or boat that is either loosely anchored or slowly moving. Projectile samplers are long thin-walled tubes, equipped with fins to guide them through the water, and sufficient weight to provide velocity and momentum to penetrate the bottom. One such device consists of a 100 mm (4-in.) ID stainless steel tube 5 to 10 m (17 to 35 ft) long, with variable weights up to 500 kg (1100 lb) at the upper end. The weight holder incorporates four vertical fins 0.5 to 1 m ($1\frac{1}{2}$ to $3\frac{1}{2}$ ft) long and 0.5 m ($1\frac{1}{2}$ ft) wide to guide the device through the water. The sampler is suspended vertically on a trigger release attached to a cable from the ship above. A bottom-sensing weight hangs 5 to 10 m (17 to 35 ft) below the sampler opening and is attached to the trigger. The device is lowered slowly to the bottom. The trigger releases the sampler 5 to 10 m (17 to 35 ft) above the bottom. The tube is propelled downward by the weight and penetrates into loose sand or soft clay several meters. It is retrieved by a slack cable on the trigger mechanism. The sample quality approaches that of thin-walled tubes in conventional sampling. The depth is limited, however.

A second similar device employs an explosive charge to achieve greater penetration. Although harder soils can be penetrated, the depths seldom exceed 10 m (35 ft).

GEOPHYSICAL EXPLORATION / Compression waves from an ultrasonic generator, explosives, or electric spark traveling down through water are reflected from the bottom and from boundaries between strata of different rigidities and densities. The water depth can be computed with great accuracy; the depths of the other boundaries are computed from estimated or computed wave velocities. The resolution (width of the compression wave) varies with the equipment; however, with irregular surfaces the height of the irregularities is subdued in the record. A continuous depth graph can be obtained from a moving boat. High-resolution side scanning ultrasonic beams have been used to explore marine cliffs and the underwater faces of dams.

Some seismic exploration devices are designed for bottom placement, operating similar to surface refraction seismic exploration. Others are designed for towing behind a moving boat.

BOTTOM DEVICES / Increasing numbers of bottom-actuated devices are being developed that are essentially remote-controlled submersible boring and sampling rigs. The actuating device rests on the bottom with any required mast supported in the water by a buoy. A drill or sampler is advanced into the bottom, powered and controlled electrically from a barge or boat anchored above. One device, the *Vibracorer*, forces a 56-mm (2.2-in.) ID tube about 5 m (17 ft) into the gound by a 22-Hz (cycles per second) vibrator. Water, jetted around the outside of the tube, helps it to advance in sands. Foil liners can be employed in clays similar to the foil strip undisturbed sampler. Depths of 80 m (270 ft) have been reached in clays.

Vane shear and cone penetrometers have also been designed for bottom actuation. One such design utilizes a vibrating cone, an electrical transducer to measure cone resistance, and an electrical pressure gauge to record depth below the water surface. Hydraulically actuated piston samplers, reacting against weights on the bottom, are similar to the Osterberg sampler but remotely operated.

With the growing importance of utilizing the resources of the sea, there should be continuing improvement in marine boring and sampling.

7:14 Laboratory Testing and Evaluation

The objective of the detailed investigation, including the field testing, is to provide the quantitative data necessary for the engineering analyses and design of the project. The total structure of the soil and rock system is defined by the exploratory phase (Section 7:9); the final step is to evaluate the appropriate engineering properties of the critical strata that were previously identified, utilizing the results of laboratory tests of undisturbed samples and the field test data, correlated stratum by stratum with the identification tests, the geophysical parameters, and the penetration resistances.

SAMPLE EXAMINATION / Undisturbed samples removed from bore holes are encased in the thin-walled tubes or liner tubes. Sometimes these are ejected in the field by a hydraulic piston. However, it is difficult to examine the soil in the field and then reseal it for transfer to the laboratory without drying or disturbance. Instead, most are left in their original tubes for shipment; only the sample ends can be examined in the field. In the laboratory the entire sample must be extruded to select the sections for testing. As an expedient alternative, the sample is X-rayed to identify soils of different texture. The selection of the portion to be tested can be made quickly without removing all the soil from the tube. However, X rays are not a substitute for a continuous examination of thick strata.

SAMPLE PREPARATION / Undisturbed samples delivered to a laboratory are of two types: continuous and intermittent. The continuous samples are cut into sections representing 0.15 to 0.3 m (0.5 to 1 ft) of depth. The samples are weighed and then small portions may be removed for visual classification and water content determinations. The average unit weight and void ratio for each section can be computed. The average unit weight and void ratio for each intermittent sample can also be determined. The test results are plotted on the same work sheet as the exploratory boring data.

A thin, vertical slice from each section or sample may be set aside to dry. This is examined at intervals to determine the presence of laminations or thin strata which ordinarily become more visible after partial drying. The disturbing effects of sampling can often be seen from the distortion of the strata. Finally, the thickness of each distinctive soil seam is measured.

TESTS OF COHESIVE SOILS / A study of the classification test and void-ratio results will show that the soil characteristics vary within any one stratum. The points where the soil characteristics of a stratum appear to be both typical and extreme can be determined from the plots of water content, void ratio, plasiticity, and penetration resistance. The undisturbed samples to be tested for shear or consolidation are usually selected from the typical and near extreme points in order to determine the variation of soil properties. Only selected samples are tested, for the cost of testing every one is seldom cost-effective. The test results can be correlated with the other soil properties to determine the average strength and range or compressibility of the stratum.

TESTS OF COHESIONLESS SOILS / Strength tests of cohesionless soils usually are made at different void ratios with disturbed samples. The angle of internal friction, correlated with the relative density of the soil, can be used to estimate the strength of the actual soil deposit.

CORRELATING TEST RESULTS / By correlating specific test data with the total information available for each stratum, a more realistic appraisal can be made than from limited shear, consolidation, and permeability tests alone. The factors suitable for correlation depend on the soil and rock and thus vary from project to project. Table 7:14 gives relationships that are frequently suitable.

PARAMETERS FOR ANALYSIS / In any stratum there is ordinarily a range of values for any given quality such as c, ϕ, C_c, c_v, and k. Too often average values or median values are utilized. If the range in test results is narrow, this approach is adequate. In most strata the deviations from the average are so great that unsafe conclusions will be reached from averages. For design, therefore, the lower values are given emphasis. Although some designers argue that the design should be based on the

TABLE 7:14 / CORRELATION OF TEST DATA

Simple Test	Possible Correlation
Water content	Shear strength of clay. Compression index of clay.
Grain size (D_{10}, D_{15}, Cu)	Permeability, strength, and drainability of cohesionless soils.
Liquid limit, LL	Compressibility.
Plastic index	Swell-shrink, ϕ_D, of clays.
Water plasticity ratio, R_w	Potential swell-shrink; preconsolidation load.
Void ratio, e, unit weight, γ	Compressibility, shear strength.
Relative density, D_r	Strength, compressibility of cohesionless soil.
Seismic velocity, V	Modulus of elasticity; strength of soil, rock.
Electrical resistivity, ρ	Water, clay, organic, and salt content.
Penetration resistance, static and dynamic	Shear strength, relative density, modulus of compressibility.

poorest conditions observed in each stratum, this is overly conservative because localized bad spots seldom control the behavior of the entire stratum. If the weak areas occur at random, a reasonable basis for design is the *lowest quartile*—the value for which 25% of the data are poorer and 75% better. If the range is very wide, the *lowest* 10 *or* 20% is better. Moreover, the final design is analyzed using the poorest single value and the design possibly revised if the safety factor is significantly less than 1.

The results of all the testing, extended by correlations, are indexed on the underground cross sections prepared from the explanatory phase of the investigation. Superimposed on the same sections are the outlines of the proposed structure, so that these sections become the starting point for the engineering analyses of stability, bearing capacity, and settlement.

7:15 Total Investigation

The underground investigation does not stop with the completion of design or the commencement of construction. As was described in the introductory case histories, unforeseen conditions are encountered or new evidence of underground conditions uncovered that could have a profound influence on both design and construction. Careful records of all soil, ground water, and rock conditions should be kept, documented by photographs and by measurements. If these differ from what the earlier investigations indicated, the designer is informed so that these deviations can be evaluated. If the conditions are the same, the records will do much to minimize unwarranted claims for changed conditions.

The very safety of the construction work, particularly of deep excavations and high embankments and dams, is dependent on continuing observation of groundwater elevations, piezometric levels in confined strata, and movements of the soil or rock, both lateral and vertical. Instrumentation for the regular observation of these features is routinely incorporated in earth dams and deep excavations. In even small projects simple instrumentation, such as groundwater observation wells and strategically placed bench marks, are useful. The most important instrument, however, is an observant engineer who regularly inspects the site, notes all changes, and then interprets these in terms of the construction operations and their potential effect on the project.

Underground investigation is a continuing geotechnical endeavor that requires a multidisciplinary approach as well as ingenuity and imagination. It requires constant liaison between field work and design interpretation and flexibility to change procedures as the changing information warrants. The work should not be undertaken as a competitive bid contract—a necessary evil preceding the more challenging design, but the first step in the design process. The final engineering can be no better than the data on which it is based.

REFERENCES

7:1 R. F. Leggett, *Geology and Engineering*, 2nd ed., McGraw-Hill Book Company, New York, 1962.

7:2 "Subsurface Investigation for Design and Construction of Foundations of Buildings," *ASCE Manual of Engineering Practice 79*, ASCE, New York, 1976, p. 61.

7:3 S. T. Algermissen, "Seismic Risk Studies in the United States," *Proceedings of the Fourth World Conference on Earthquake Engineering, Santiago, Chile, 1969.*

7:4 *List of Published Soil Surveys*, Soil Conservation Service, U.S. Dept. of Agriculture.

7:5 D. R. Lueder, *Aerial Photographic Interpretation*, McGraw-Hill Book Company, New York, 1959.

7:6 O. S. Way, *Terrain Analysis*, 2nd ed., Dowden, Hutchison and Ross, Inc., Stroudsburg, Pa., 1978.

7:7 R. G. Reeves, Editor-in-Chief, *Manual of Remote Sensing*, American Society of Photogrammetry, Falls Church, Va., 1975, 2144 pp.

7:8 P. F. Krumpe, *The World Remote Sensing Index*, Tensor Industries, Fairfax, Va., 1976, 619 pp.

7:9 G. F. Sowers, "Modern Procedures for Underground Exploration," *Proceedings, ASCE*, **80**, Separate 435, May 1954.

7:10 M. J. Hvorslev, *Subsurface Exploration and Sampling of Soils for Civil Engineering Purposes*, U.S. Waterways Experiment Station,

Vicksburg, Miss., 1949. (Reprinted by Engineering Foundation, New York.)

7:11 ASTM Standards:
D-420-69(75), "Surveying and Sampling Soils for Highway Subgrades."
D-1452-65(72), "Soil Investigation and Sampling by Auger Borings."
D-1586-67(74), "Penetration Test and Split Tube Sampling of Soils."
D-1587-74, "Thin-Walled Sampling of Soils."
D-2113-70(76), "Diamond Core Drilling for Site Investigation."
D-2573-72(78), "Field Vane Shear Test of Cohesive Soil."
Annual Book of Standards, American Society for Testing and Materials, Philadelphia, Pa. (issued yearly).

7:12 K. Terzaghi and R. B. Peck, *Soil Mechanics in Engineering Practice*, 2nd ed., John Wiley & Sons, Inc., New York, 1968, pp. 341 and 347.

7:13 G. Fletcher, "Standard Penetration Test, Its Use and Abuses," *Journal of the Soil Mechanics and Foundations Division, Proceedings*, ASCE, **91**, SM4, July 1965.

7:14 G. Sanglerat, *The Penetrometer and Soil Exploration*, Elsevier Publishing Co., Amsterdam, 1972, 464 pp.

7:15 D. U. Deere, "Technical Description of Rock Cores for Engineering Purposes," *Rock Mechanics and Engineering Geology*, **1** (1), 17, 1964.

7:16 M. Rocha, "A Method of Integral Sampling of Rock Masses," *Rock Mechanics*, **3** (1), 1–12, 1971.

7:17 "Bore Hole Copier," Mori Geotechnique, Inc., Tokyo, 1977.

7:18 J. Schmertmann, "Measurement of In-Situ Strengths," *Specialty Conference on Measurement of Soil Properties*, Vol. 2, Geotechnical Engineering Division, ASCE, 1976, pp. 57–138.

7:19 G. G. Meyerhof, "Penetration Tests and Bearing Capacity of Cohesionless Soils," *Journal of the Soil Mechanics and Foundations Division, Proceedings, ASCE*, **82**, SM1, January 1956.

7:20 E. Schultze, "Determination of Density and Modulus of Compressibility of Non-Cohesive Soil by Soundings," *Proceedings of the Sixth International Conference on Soil Mechanics and Foundation Engineering, 1, Montreal*, 1965, p. 354.

7:21 H. J. Gibbs and W. G. Holtz, "Research on Determining the Density of Sands by Spoon Penetration Testing," *Proceedings of the Fourth International Conference on Soil Mechanics and Foundation Engineering, London*, 1957.

7:22 *Proceedings, Conference on In-Situ Investigations in Soils and Rocks*, British Geotechnical Society, London, 1970, p. 323.

7:23 J. K. Mitchell, "In-Situ Measurement of Volume Change Characteristics," *Specialty Conference on Measurement of Soil Properties*, Geotechnical Engineering Division, ASCE, 1976, 2, pp. 279–345.

7:24 G. P. Wroth, "In-Situ Measurement of Initial Stresses and Deformation Characteristics," *Specialty Conference on Measurement of Soil Properties*, Geotechnical Engineering Division, ASCE, 1976, 2, pp. 181–230.

7:25 V. Milligan, "In-Situ Measurement of Permeability in Soil and Rock," *Specialty Conference on Measurement of Soil Properties*, Geotechnical Engineering Division, ASCE, 1976, 2, pp. 3–36.

SUGGESTIONS FOR FURTHER STUDY

M. J. Hvorslev, *Subsurface Exploration and Sampling of Soils for Civil Engineering Purposes*, U.S. Waterways Experiment Station, Vicksburg, Miss., 1949. (Reprinted by Engineering Foundation, New York.)

R. F. Leggett, *Geology and Engineering*, 2nd ed., McGraw-Hill Book Company, New York, 1962.

Proceedings, Conference on In-Situ Investigations in Soils and Rocks, British Geotechnical Society, London, 1970.

Proceedings, Specialty Conference on In-Situ Measurement of Soil Properties, ASCE, New York, 1975, 2 vols.

G. Sanglerat, *The Penetrometer and Soil Exploration*, Elsevier Publishing Co., Amsterdam, 1972, 464 pp.

Soil Sampling, Proceedings, Specialty Session 2, *Ninth International Conference on Soil Mechanics and Foundation Engineering, Tokyo*, 1977. (Papers by 23 authors.)

U.S. Bureau of Reclamation, *Earth Manual*, U.S. Government Printing Office, Washington, D.C., 1974.

PROBLEMS

7:1 Prepare a typical soil profile for your locality. Secure data from contractors, city and county engineers, engineers specializing in underground exploration work, and from geologic and engineering reports.

7:2 A single test boring made for a water tank led to the following data:

Soil Data			Sampling Data				
Depth m	Depth ft	Soil Stratum	Depth m	Depth ft	Penetration Data (N)	*w*	*LL*
0–1.2	0–4	Fill: cinders, brick	0.6	2	3	—	—
1.2–2.1	4–7	Hard, slightly plastic clay	1.5	5	32	21	44
2.1–7.6	7–25	Firm, uniform,	2.4	8	15	—	—
		coarse, subrounded	4.0	13	19	—	—
		quartz sand	5.5	18	27	—	—
			7.0	23	20	—	—
7.6–9.8	25–32	Firm, highly	7.9	26	7	55	62
		plastic clay	9.4	31	6	57	64
9.8–15.5	32–51	Dense, uniform,	10.0	33	35	—	—
		fine, angular	11.6	38	37	—	—
		sand	13.1	43	40	—	—
			14.6	48	46	—	—
15.5–22.8	51–75	Dense to very dense	15.8	52	48	—	—
		graded, angular,	17.4	57	55	—	—
		coarse, sandy,	18.9	62	72	—	—
		subrounded fine	20.4	67	68	—	—
		gravel. Coarser with depth	21.9	72	63	—	—
6.7	22	Ground water					

a. Plot the boring log.
b. Determine which strata, if any, require more detailed study. List tests necessary.
c. Should more or deeper test borings be made? Tank weighs 4000 kN (900 kips) and rests on four columns arranged in a square 9 m (29.5 ft) apart.

7:3 Sketch and list the essential equipment for test boring, using both hand auger and wash drilling, and spoon sampling.

7:4 A 10-story office building with a basement floor 2 m (6.6 ft) below the ground surface is 30 m (100 ft) wide and 75 m (246 ft) long. The dead load per floor is 5.75 kN/m^2 (120 psf) and the live load 2.4 kN/m^2 (50 psf). The soil weighs 18 kN/m^3 (114.5 lb/ft^2).

a. Show layout of borings for average conditions.
b. Determine depth of boring, using the rule of 10% of increase in effective stresses.

7:5 A one-story manufacturing plant is 40 m × 200 m (131 ft × 656 ft). It rests on new fill 1 m (3.3 ft) thick. The geology of the area indicates that it is underlain by horizontally stratified coastal lagoon deposits covered with an old man-made fill.

a. Show the recommended boring layout.
b. Estimate the required boring depth. Under what conditions encountered during boring should this depth be changed? When should undisturbed samples be made?
c. Estimate the cost of making a soil investigation at this site, assuming that soil testing and engineering costs will be 50% of the cost of boring and sampling.

7:6 A 20-story medical office building 30 m×45 m (98.5 ft×147.7 ft) in plan is centered in a 90 m×90 m (295 ft×295 ft) 2-story shop and parking plaza. Both structures include a one-story basement. The site is located in an urban renewal area, once occupied by old houses and shops with basements, but now filled in for parking space. The city is located in a flood plain of a broad meandering river, with limestone cliffs overlooking the valley.

a. Describe the steps advisable in a reconnaissance or preliminary evaluation of the site.
b. Prepare a boring plan, and outline or guide to the procedures required for exploratory phase of the investigation.
c. Estimate the cost of this investigation.

7:7 The exploratory borings for Problem 7:6 found the following typical soil profile, relatively uniform across the site.

0–6 m	0–20 ft	Rubble
6–15 m	20–49 ft	Alternate seams of soft clay and clayey sand; N from 5 to 15 blows per 0.3 m or ft.
15–21 m	49–70 ft	Limestone with clay seams: average core recovery 15%
21–28 m	70–92 ft	Hard limestone, average core recovery 95%, bedding dipping 60° with horizontal: joint spacing 1 m (3.2 ft)
6 m	20 ft	Groundwater level—in September

a. Draw a hypothetical underground cross section of this site.
b. Prepare an outline for the required final, detailed investigation, including boring locations and undisturbed sampling methods.
c. List tests required for samples obtained in this phase.
d. Estimate the cost of this final phase of the investigation, excluding engineering work.

CHAPTER 8 Mass Response to Load

A deep open excavation was made close to an old building supported by shallow spread footings. The engineer required that the excavation slopes be flatter than 2(H) to 1(V) to insure against shear failure of the banks and destruction of the building. The bank did not fail; however, the building cracked badly. The soil mass deformed elastically with both outward and downward movement of the building wall closest to the excavation. This possibility was never considered by the engineer, yet it led to the condemnation of the building.

A new retaining wall 10 m (33 ft) high formed one side of a large industrial plant under construction. The construction sequence required (1) building the wall, (2) placing the backfill behind it, and (3) constructing the remainder of the plant. The backfilling of the wall had just been completed when a crack was observed in the backfill surface about 6 m (20 ft) from the wall. The engineer examined the wall and found that its top had tilted more than 50 mm (2 in.) away from the backfill. He blamed the trouble on faulty construction and ordered the wall and backfill reconstructed. The contractor refused and completed the building. In 1978, 45 years afterward, the wall and the building stood intact; there had beeen no further movement of the wall nor additional cracking. The soil had sheared during backfilling and the wall had deflected, but both the normal, essential responses to the mass load that did not endanger the ultimate performance of the structure.

A deep excavation for a building was constructed without bracing for the banks, because a trial excavation demonstrated that the soil could stand on a vertical face without support. Five weeks later, a large part of one bank fell outward, killing two workmen and injuring five. The weather had been good; there had been no construction work in the vicinity of the bank that failed and no apparent cause for the sudden soil shearing. Instead, it appeared to the layman that the soil had become tired of its continued load. In effect this was true; continued, but unnoticed, creep at high stress led to a delayed failure.

These three examples illustrate the extremes of soil mass response to stress changes. In the first case the soil deformed more or less elastically; although the soil was safe against failure, the structure was damaged beyond repair. In the second case the soil failed, but, although the engineer was not aware of this, the soil failure was essential to the ultimate safe performance of the retaining wall. The earth pressure theory used in the wall design assumed shear failure in the soil; without the shear the pressure on the wall would have been twice as great. The third example illustrates the time-dependent nature of soil behavior, even when the environment remains unchanged. If the weather is unusually wet, or extremely hot and dry, or if freezing and thawing occur, the effect of time is more easily explained; the changes that occur with creep at high stresses can be just as serious, however.

Soil and rock mechanics deal with the response of the earth to the loads imposed. The principles of engineering mechanics apply to the solution of problems of soil and rock as they do to other engineering materials. The total problems, however, are usually more complex, because of three factors.

First, most problems include both the soil or rock mass and the engineering structure it supports. The two components of the system, the soil and the structure, obey the same physical laws. Although the structure is generally made of a stronger, more rigid material than the earth, the two components respond or move together. Therefore, their interaction must be considered. The case of the soil slope that was safe, although the structure failed, illustrates the importance of interaction.

Second, the weight of the soil or rock mass is usually far more significant than the loads imposed by the engineering structure. For example, for a highway bridge across a river valley, the live load will be 20 to 40% of the dead load of the structure. For a deep fill that supports the same highway across the same valley, the live load will be less than 1% of the dead load of the fill. Because total weight of the soil mass is usually far greater than that of a structure it supports, or of a live load, it must be given proportionally more consideration.

Finally, the response of the soil to load depends on a wide variety of factors, including the load itself, as described in Chapters 4 and 5. Although the properties of other engineering materials change, the changes in soil and rock are generally far greater and have much greater effects on the response of the soil structure system.

The general approach to mass response to loads will be considered in this chapter. Specific cases will be covered in Chapters 9 through 12.

8:1 The Mass Load—Body Forces

The forces within a body generated by mass are termed *body forces.* Temperature stresses are an example of one form of important body force in steel or concrete; the most important body force in soil or rock mass is that of gravity. Certain simple cases will be considered here.

VERTICAL STRESS IN A LEVEL MASS / The simplest case of a gravity body force is the vertical stress in an infinitely wide mass of level soil or rock. If the unit weight of a stratum of thickness H_1 is γ_1 (Fig. 8.1a), then the total vertical stress σ_v (σ_z in a three-dimensional coordinate system) at any depth z (positive downward) below the surface is

$$\sigma_z \text{ (or } \sigma_v) = \gamma_1 H_1 + \gamma_2 H_2 \ldots) \qquad (8:1a)$$

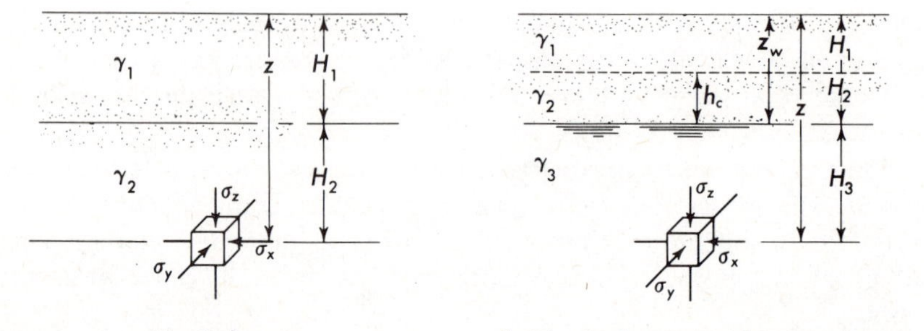

a. Vertical stress b. Vertical stress with ground water

Figure 8.1 *Vertical stresses due to soil or rock weight in a level stratified mass not subject to tectonic loading.*

If the soil is saturated, with a groundwater table, and a capillary rise zone of h_c (Fig. 8.1b) the vertical stress of any level can be separated into effective and neutral stress components. It should be kept in mind that the neutral stress or water pressure at any point is the same in all directions.

$$\sigma_z' = (\gamma_1 H_1 + \gamma_2 H_2 \ldots) - (z - z_w)\gamma_w \qquad (8:1b)$$

CHANGES IN STATIC VERTICAL STRESS / Changes in the static vertical stress occur in several ways: added surface load, such as a fill over a wide area, changes in the unit weight of the soil, and changes in neutral stress from water level fluctuations or capillary tension. These can cause significant decreases (or increases) in soil strength, breakdown of structure, and expansion or consolidation of compressible strata. The added stress produced by a uniform surface load is computed by Equation (8:1a) if the dimensions of the loaded area are much greater than the depth, z. If the area loaded is small, the increase in stress is computed as shown in Chapter 10.

The effects of neutral stress changes are less obvious, but of great significance. If the groundwater level falls by an amount Δz_w, the total stress at any point below the capillary fringe level will decrease slightly due to the reduced weight of water in the soil voids. However, the neutral stress will be reduced by $\Delta u = -\gamma_w \Delta z_w$. The net effect will be an increase in effective stress (weight of water lost from the voids is seldom significant). Also, z is positive down.

Example 8:1

Compute the change in vertical effective stress at the surface of the clay stratum in Fig. 8.2 if the water table falls 2 m or 0.6 ft. The sand has a void

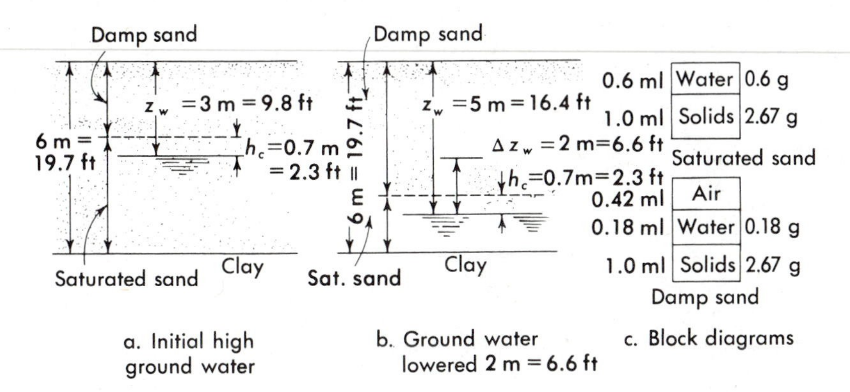

Figure 8.2 *Changes in vertical stresses of Example 8:1 on a clay stratum due to groundwater changes.*

ratio of 0.60 and a specific gravity of solids of 2.67. The height of capillary saturation is 0.7 m (2 ft). Above the capillary line the sand is 30% saturated.

1. The saturated density of the sand is $(2.67+0.60)\div(1+0.60) = 2.045$ g/ml $= 2045$ kg/m^3. The unit weight is 127.5 lb/ft^3 = 20.0 kN/m^3.
2. The drained sand density above the capillary saturation line is $(2.67+0.3\times0.6)\div(1+0.60) = 1.78$ g/ml $= 1780$ kg/m^3. The unit weight is 111 lb/ft^3 = 17.43 kN/m^3.
3. Initially, the vertical stresses at the clay surface are
   Total: $2.3\times17.43+3.7\times20.0 = 114.1$ kN/m = 2381 lb/ft^2.
   Neutral: $3\times9.81 = 29.4$ kN/m^2 = 614 lb/ft^2.
   Effective: $114.1-29.4 = 84.7$ kN/m^2 = 1767 lb/ft^2.
4. After the water drops the vertical stresses at the clay surface are
   Total: $4.3\times17.43+1.7\times20 = 108.9$ kN/m^2 = 2275 lb/ft^2 (a loss of 5.2 kN/m^2 = 106 lb/ft^2).
   Neutral: $1\times9.81 = 9.8$ kN/m^2 = 205 lb/ft^2.
   Effective: $108.9-9.8 = 99.1$ kN/m^2 = 205 lb/ft^2 (a gain of 14.4 kN/m^2 = 303 lb/ft^2).

The change in effective vertical stress caused by water level changes of z_w can be expressed as

$$\Delta\sigma'_z = \Delta z_w[\gamma_w - (\gamma_{sat} - \gamma_{drained})] \tag{8:2}$$

The increase in effective stress is $14.4/2 = 7.2$ kN/m^2 for a 1-m drop in water level or $303 \div (2 \times 3.28) = 46$ lb/ft^2 for a 1-ft drop in water level. A 3-m drop in water level would produce an effective stress increase equivalent to slightly more than 1 m of fill; a 3-ft drop in water level is equivalent to slightly more than 1 ft of fill.

EARTHQUAKE STRESS IN A LEVEL MASS / The dynamic effect of an earthquake can be approximated by the inertia stresses developed by the acceleration. Consider a column of soil of depth z and length and width of L and B (Fig. 8.3). The weight of the column will be $LBz\gamma$, and the vertical stress at the bottom will be γz. In order to produce vertical acceleration of $\pm a_v$, it will be necessary to increase the vertical force by ΔF:

$$\Delta F = \pm M a_v$$

$$\Delta F = \pm \frac{LB\gamma z a_v}{g}$$

The vertical stress change due to earthquake load, $\Delta\sigma_{ve}$, is

$$\Delta\sigma_{ve} = \pm \frac{LB\gamma z}{LB}\frac{a}{g} = \pm \gamma z \frac{a_v}{g} \tag{8:3a}$$

The shear stress change, $\Delta\tau_e$, on the base of the column can be computed from a horizontal acceleration, a_h,

$$\Delta F_h = \pm \frac{LB\gamma z a_h}{g}$$

$$\Delta\tau_e = \pm \frac{\gamma z a_h}{g} \tag{8:3b}$$

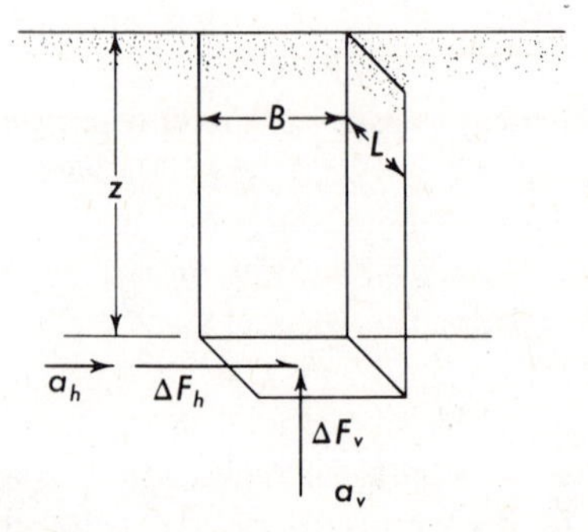

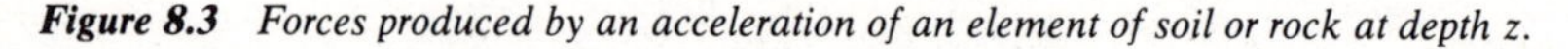
Figure 8.3 *Forces produced by an acceleration of an element of soil or rock at depth z.*

The ratio a/g is often expressed as a percentage and in that form is termed the earthquake acceleration factor. This approximation assumes a steady-state acceleration, whereas the accelerations in an earthquake change in recurring cycles of random frequency and intensity. For preliminary estimates at low acceleration ($a < 0.1$) such analyses are helpful. For larger accelerations dynamic analyses utilizing earthquake time histories are necessary.

RESIDUAL STRESSES / Other sources of body stress include the effects of previous soil or rock loading that are retained because the mass is confined. Expansion and contraction due to long-term temperature change or chemical alteration cause stresses in many rock masses and some soils. Tectonic movements and, on a smaller scale, landslides, produce lateral stresses that exceed the vertical stress due to soil weight. Previous loads, now removed, are maintained by mineral bonds or by capillary tension. The magnitude of these residual stresses seldom can be computed theoretically. Field measurements as described in Chapter 7 are necessary to determine their magnitude and large-scale tests are required to evaluate their possible changes.

8:2 Elastic Equilibrium

As was discussed in Chapter 5, changes in stress acting on soil produce strains. When the stress is well below that required for failure, there is a definite relation between a stress increment and the strain it produces. If the stresses acting on a mass of soil are in equilibrium, and if a definite relation between stress and strain is present throughout the mass, and if the deformations are compatible with the strains, the mass is in a state of *elastic equilibrium*. When the relationship between stress and strain can be defined mathematically, the effects of a system of stress can be described by a *theory of elasticity*.[8:1,8:2] A number of such theories have been developed based on simple relationships between stress and strain. If the stress level in the soil is sufficiently low that one of the simple expressions for stress–strain is approximately correct, then that theory can be applied to the solution of soil engineering problems.

STRAINS / A portion of an element of soil is shown in Fig. 8.4. The angle of ABC, initially a right angle, is displaced a small amount to a new position $A''B'C''$, as the result of a stress change. It can be seen that in the new position the sides of the angle have first moved by translation to $A'B'C'$ and then the ends have further rotated to $A''B'C''$. The x-displacement of B is u, and the y displacement is ν (the z displacement is w). The x displacement of A to A'' because of its distance, dx, from point B is $u + (\partial u/\partial x)\, dx$; the y displacement of A to A'' is similarly $v + (\partial v/\partial x)\, dx$.

The strain in the x direction is equal to the elongation of the line AB divided by the initial length, dx. Similar expressions can be derived for all

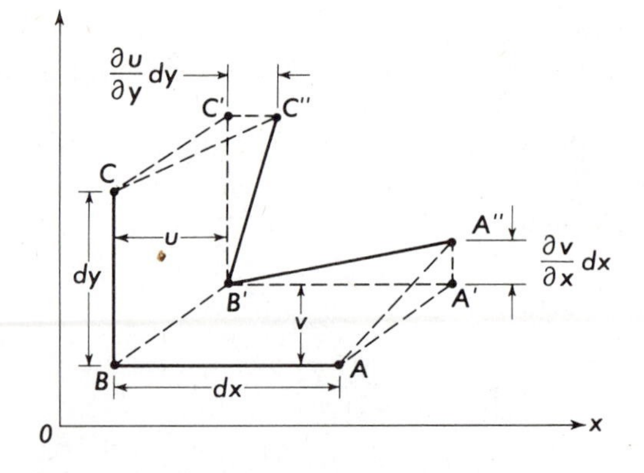

Figure 8.4 *Linear and angular displacements in the x-y plane.*

three directions of linear strain.

$$\frac{A''B' - AB}{dx} = \frac{dx + (\partial u/\partial x)\,dx - dx}{dx} = \varepsilon_x$$

$$\varepsilon_x = \frac{\partial u}{\partial x}; \qquad \varepsilon_y = \frac{\partial u}{\partial y}; \qquad \varepsilon_z = \frac{\partial u}{\partial z} \tag{8:4a}$$

The angular change of line AB, angle $A''B'A'$, can be found approximately from the y displacement of its end from A' to A'', $(\partial v/\partial x)\,dx$. The angular change in radians is

$$A'B'A'' = \frac{(\partial v/\partial x)\,dx}{dx} \qquad C'B'C'' = \frac{(\partial u/\partial y)\,dy}{dy}$$

The angular strain, γ_{xy}, is the sum of the two angles:

$$\gamma_{xy} = \frac{\partial v}{\partial x} + \frac{\partial u}{\partial y}; \qquad \gamma_{xz} = \frac{\partial w}{\partial x} + \frac{\partial u}{\partial z}; \qquad \gamma_{yz} = \frac{\partial v}{\partial z} + \frac{\partial w}{\partial y} \tag{8:4b}$$

The complete picture of strains is described by the six equations (similar to 8:4a and b).

COMPATIBILITY / If the body remains intact, then the angular strains must be compatible with the linear strains. This condition is illustrated by Fig. 8.5 in the distortion of the set of squares (a) into (b). If the strains are not compatible, the body becomes discontinuous (Fig. 8.5c). This is assured mathematically by taking the second derivatives of the linear strain functions and equating them to the second derivative of the angular strain function:

$$\frac{\partial^2 \varepsilon_x}{\partial y^2} \; \frac{\partial^2 \varepsilon_y}{\partial x^2} = \frac{\partial^2 \gamma_{xy}}{\partial x\,\partial y} \tag{8:5a}$$

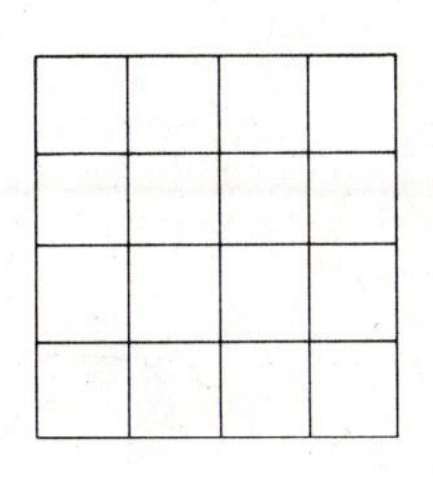

a. Solid body before deforming

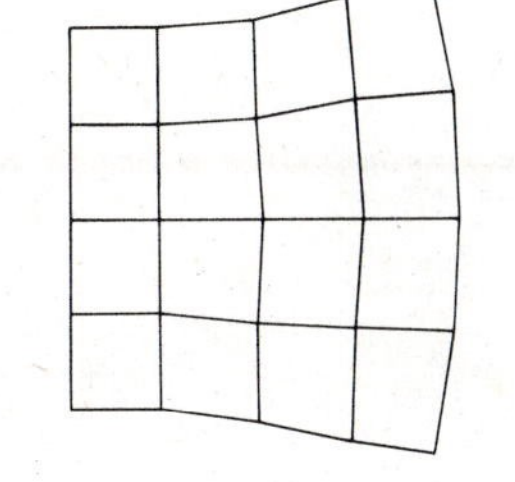

b. Compatible strains in deformed body

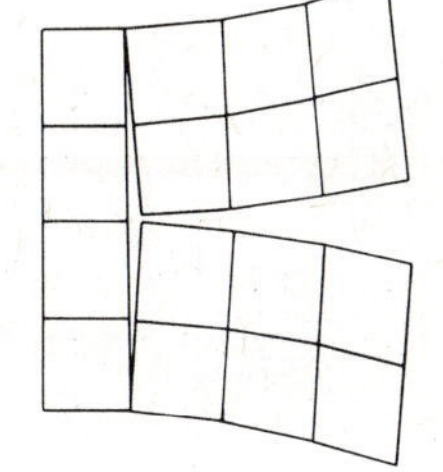

c. Incompatible strains in deformed body

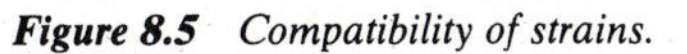

Figure 8.5 *Compatibility of strains.*

$$\frac{\partial^2 \varepsilon_y}{\partial z^2} + \frac{\partial^2 \varepsilon_z}{\partial x^2} = \frac{\partial^2 \gamma_{yz}}{\partial y\, \partial z} \tag{8:5b}$$

$$\frac{\partial^2 \varepsilon_z}{\partial x^2} + \frac{\partial^2 \varepsilon_x}{\partial z^2} = \frac{\partial^2 \gamma_{xz}}{\partial z\, \partial x} \tag{8:5c}$$

These are the *equations of compatibility* and must be satisfied in an intact body.

ELASTICITY / An elastic theory assumes that the relationship between stress and strain can be defined mathematically. The only limitation imposed is that the relation be simple enough so that it can be utilized in analysis and that it is valid for both increases and decreases in stress (hysteresis is negligible). The simplest form is when strain is proportional to stress, a linear function described by the modulus of elasticity, E (Fig. 8.6a).

$$E = \frac{\Delta\sigma}{\Delta\varepsilon} \tag{5:2b}$$

The modulus at one point may be the same in all directions, an *isotropic* material, or it may be different in which case E must be identified with a specific direction:

$$E_x = \frac{\partial\sigma_x}{\partial\varepsilon_x}; \qquad E_y = \frac{\partial\sigma_y}{\partial\varepsilon_y}; \qquad E_z = \frac{\partial\sigma_z}{\partial\varepsilon_z} \tag{8:6}$$

This state is analogous to a spring that is infinitely long, and where deformation is proportional to stress, whether increasing or decreasing.

The lateral strains produced by an axial stress, $\Delta\sigma$, are defined by Poisson's ratio, ν,

$$\nu = \frac{-\varepsilon_L}{\varepsilon_D} \tag{5:3}$$

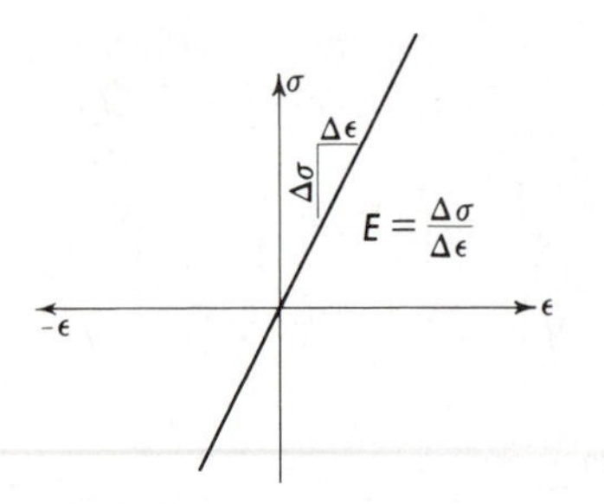

a. Ideal elastic stress – strain

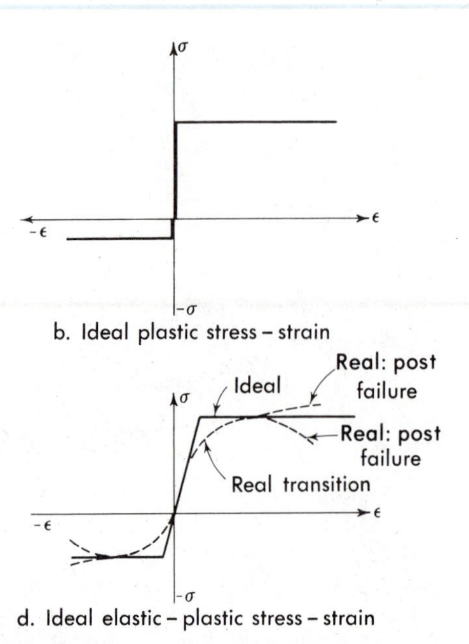

b. Ideal plastic stress – strain

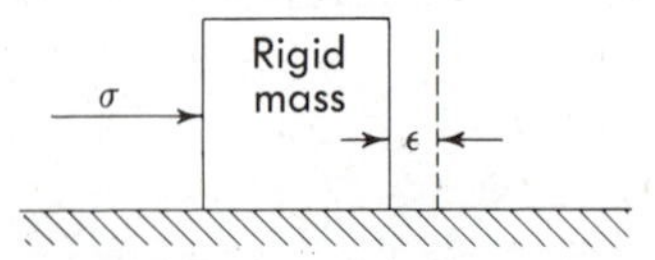

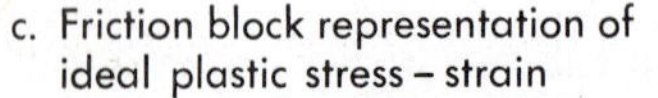

c. Friction block representation of ideal plastic stress – strain

d. Ideal elastic – plastic stress – strain

Figure 8.6 *Elastic and plastic stress–strain relationships.*

In a three-dimensional stress field with x, y, z coordinates:

$$\Delta\varepsilon_z = \frac{\Delta\sigma_z}{E_z} - \frac{\nu\,\Delta\sigma_x}{E_x} - \frac{\nu\,\Delta\sigma_y}{E_y} \tag{8:7}$$

The solution of elastic equilibrium problems requires combining the equations of stress–strain, (8:6) and (8:7) with the equations of compatibility. Equilibrium requires that the sums of the forces and moments acting on each portion of the mass be zero. The equations are integrated with the constants of integration determined by the boundary conditions. The results express the stresses, strains, and displacements in terms of the loads or stresses imposed on the mass.

ELASTIC THEORIES / Direct mathematical expressions of stress and strain have been developed only for certain special simple cases. The derivations can be found in textbooks on the theory of elasticity. The results of some of these are given in Chapter 10 without proof. However, it is important that engineers understand the basis for these equations even if they do not undertake their derivation.

The theories differ in the extent of the solid body and in the variations of E with position and direction. Although variations of E with stress and variations in ν could also be evaluated, the complexities of integration have limited most analyses to a constant E and homogeneous, isotropic ν.

The simplest theory, developed by Boussinesq in 1888, describes the stresses and strains produced in a semi-infinite homogeneous isotropic elastic solid. The semiinfinite mass, comparable to an extensive, deep deposit of soil or rock has a level plane surface and extends infinitely far in all directions below it. The homogeneous character means that E and ν are the same at all points, and isotropic means that they are the same in all directions. Although many soil and rock deposits do not meet these criteria, the theory is a useful approximation for many real situations, as will be discussed in Chapters 9–12.

Other theories have been developed for layers having different E values, materials in which E varies with depth or with direction, and for a material in which lateral strain is prevented by thin layers of more rigid materials. The latter, developed by Westergaard, is an approximation of a soil mass consisting of elastic clay strata alternating with more rigid sand seams. It is useful in foundation analysis (Chapter 10). A more complete discussion of these theories and their equations for stress and strain can be found in the references.

LATERAL STRESS WITH NO STRAIN / A simple example of the application of elasticity to the computation of stresses in soil is the case of the uniform level mass described in Section 8:1. The vertical pressure at any depth, z, is equal to the overburden stress, $\Sigma\gamma H$. The lateral pressure produced by that vertical stress depends on the strain.

In the level mass of infinite extent, each little element of soil tries to bulge laterally under the vertical stress, σ_z, in accordance with Poisson's ratio. One element cannot bulge without the neighboring element contracting. If the soil is homogeneous at that level and all elements are loaded equally, all must respond identically; therefore, there can be no lateral expansion or contraction and all lateral stresses, σ_x and σ_y, are equal. Rewriting Equation (8:7) and then simplifying, we obtain

$$\Delta\varepsilon_x = \frac{\Delta\sigma_x}{E_x} - \frac{\nu\,\Delta\sigma_y}{E_y} - \frac{\nu\,\Delta\sigma_z}{E_z}$$

and if

$$E_x = E_y \qquad \text{and} \qquad 0 = \frac{\Delta\sigma_x}{E_x} - \frac{\nu\,\Delta\sigma_y}{E_x} - \frac{\nu\,\Delta\sigma_z}{E_z}$$

and if

$$\sigma_x = \sigma_y \qquad \text{and} \qquad \frac{\Delta\sigma_x}{E_x}(1-\nu) = \frac{\nu\,\Delta\sigma_z}{E_z}$$

then

$$\Delta\sigma_x = \left(\frac{\nu}{1-\nu}\right)\left(\frac{E_x}{E_z}\right)\Delta\sigma_z \qquad (8:8a)$$

The relationship between the lateral stress $\sigma_x = \sigma_y$ and the vertical stress σ_z when there is no lateral strain depends on Poisson's ratio and the ratio of the lateral to vertical E. If the soil is isotropic, so $E_x = E_y = E_z$,

$$\Delta\sigma_x = \left(\frac{\nu}{1-\nu}\right) \Delta\sigma_z = K_0 \, \Delta\sigma_z \tag{8:8b}$$

$$K_0 = \frac{\nu}{1-\nu} \tag{8:8c}$$

The lateral stress, $\Delta\sigma_x$, with no strain is essentially the earth pressure developed against a very rigid wall, as will be described in Chapter 9. The ratio K_0 is the *coefficient of earth pressure at rest.*

APPLICATIONS AND LIMITATIONS OF ELASTIC THEORIES / The elastic theories have proved extremely useful in computing stresses and strains in a wide range of materials, including soil and rock masses. The validity of the theories is limited by how accurately they represent the real material behavior. As was discussed in Chapter 5, the stress–strain relations for soils and rock are not linear nor the same in unloading as loading. Furthermore, the degree of linearity depends on the stress level and on the stress history. Small changes in stress compared with the total stress level, and repeated stresses are more nearly elastic than large single loadings that approach failure. In spite of their theoretical limitations, elastic analyses of stress and strain in soil and rock provide reasonable approximations when the stress levels are low and when the theory selected fits the geometry and the variations in E of the soil or rock mass.[8:2]

8:3 Plastic Equilibrium

Plastic equilibrium is essentially a state of impending failure. The forces acting on each element of the mass are in equilibrium. The stress within the zone of plasticity are those which produce failure. The strains, instead of being related to the stress, are indefinite.[8:3]

The stress–strain curve for an ideally plastic material is given in Fig. 8.6b. The failure stress is reached without appreciable strain, after which the strain increases without a stress increase. This is analogous to a weight resting on a rough surface, and acted upon by an increasing load, σ as in Fig. 8.6c. There is no movement until σ overcomes friction, thereafter, the block moves without an increase in σ.

STRESSES IN A PLASTIC STATE / If the stresses required to produce failure can be defined mathematically, then the state of stress in plastic equilibrium can be described by the equations developed in Chapter 5. For many materials the Mohr envelope is a reasonably accurate criterion for the stresses on the surface of failure, and it can be approximated by the simple expression originally proposed by the French engineer-physicist Coulomb:

$$\tau_{ff} = c + \sigma_{ff} \tan \phi \tag{5:17b}$$

$$\theta = 45 + \frac{\phi}{2} \tag{5:8}$$

The relationship between the principal stresses in the plastic state defined by the Coulomb equation can be found from the geometry of the Mohr circle (Fig. 8.7a) either graphically or by analysis:

$$\sigma_1 = \sigma_3 \tan^2\left(45 + \frac{\phi}{2}\right) + 2c \tan\left(45 + \frac{\phi}{2}\right) \tag{8:9a}$$

Two sets of shear surfaces develop, at angles of $\pm\theta$ between their normals and the direction of the major principal stress, as shown in Fig. 8.7b.

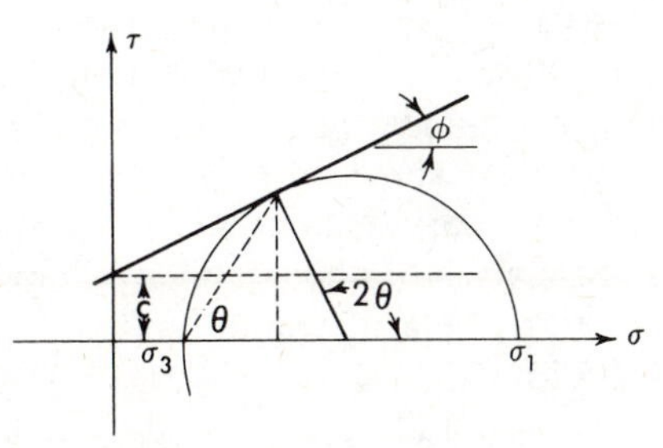

a. Mohr's circle solution of Coulomb's equation

b. Shear surfaces in plastic equilibrium

Figure 8.7 *Plastic equilibrium and the Mohr circle.*

The earth pressure theory of Rankine (Section 9:1) is a direct application of this definition of plasticity, utilizing the simple case of plane rupture surfaces and the principal stresses at the same orientation throughout the mass.

Example 8:2

Compute major principal stress, which acts vertically in a mass of soil that is in a state of plastic equilibrium if the horizontal (minor) principal stress is 10 kN/m^2. The soil is clay with an undrained strength of $c =$ 50 kN/m^2. From the Mohr circle (Fig. 8.8),

$$\sigma_1 = \sigma_3 + 2c \tag{8:9b}$$

$$\sigma_1 = 10 + 100 = 110 \text{ kN/m}^2$$

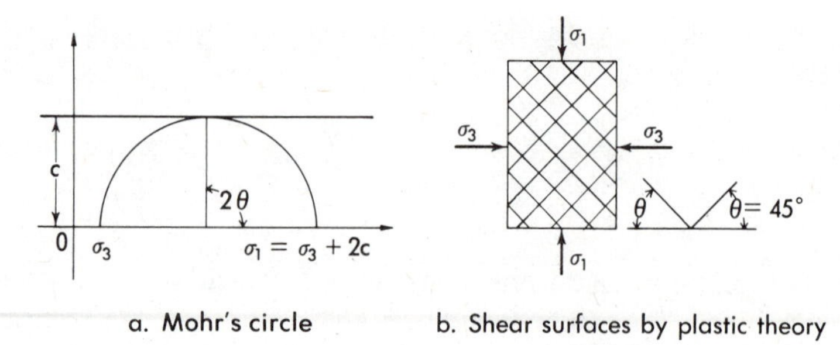

Figure 8.8 *Plastic equilibrium in a hypothetical soil with $\theta = 0$ (a representation of saturated clay in undrained shear).*

The state of plasticity may involve a large part of the mass or only a limited zone. The boundary conditions, the load, and the trajectories of soil movement all must be compatible with the failure.

PLASTICITY APPROXIMATIONS / In many practical problems it is possible to approximate the state of plastic failure by considering the failure stresses along the boundary of the zone of plasticity. The geometry of the boundary that satisfies the plastic state of stress as well as the equilibrium of the portion of the mass defined by the boundary is not necessarily known but can be found by trial. The circular arc analysis of slope stability (Chapter 12) is an example of such an approximation of plastic equilibrium.

APPLICATIONS AND LIMITATIONS OF PLASTIC THEORIES[8:4] / The plastic theories define a state of failure or impending failure in a part of a mass. If the stress conditions for failure can be defined simply, as by the Coulomb equation (5:17b) the analysis may be a valid approximation of the conditions of failure in a mass. What occurs before the instant of failure, and what the consequences of failure might be, such as amount or rate of movement, cannot be computed.

As with most approximations, the plastic theories neglect certain real conditions that influence the results of the analysis. First, the state of plasticity commences after the mass has been distorted by strains (Fig. 8.6d) rather than without strain as the definition of simple plasticity suggests (Fig. 8.6b). The distortion to varying degrees influences the geometry of the real failure surfaces. A few theories of elastic–plastic action have been proposed but are of limited use because of their complexities. Moreover, most materials exhibit a transition in which neither ideal elasticity nor plasticity occur (Fig. 8.6d).

A second shortcoming is in representing the state of stress at failure by a simple expression such as the Coulomb equation. Part of the loads applied to the mass are supported by neutral stress, the remainder by effective stress, and no simple expression of plasticity suffices to represent the total stresses at failure. For example, it was shown in Chapter 5 that for a saturated clay in undrained shear, where $\phi_u = 0$, θ should be 45°. However, the actual failure plane angle is determined by effective stress, and $\theta > 45°$.

A third shortcoming is the effect of the plastic strains on the state of stress. A loose sand becomes denser and stronger, a sensitive clay becomes weaker. The real state of stress at failure is strain-dependent, whereas simple plasticity presumes the strain and stress are independent.

In spite of these shortcomings, plastic theory is a valuable tool for solving real problems.

8:4 Rheology

In many materials including most soils and rocks the response of the mass to load is time-dependent. The time rate of consolidation of soil, Chapter 4 is an example. The analytical treatment of time-dependent behavior is termed *rheology.*

BASIC ASSUMPTIONS OF RHEOLOGY / Theoretical rheology is an analytical representation of time-dependent response to load in terms of simple mechanical models or analogs (Fig. 8.9). The immediate elastic response that is proportional to load is represented by a spring termed the *Hookean substance.* A rate of movement that is proportional to stress is analogous to viscous flow of a liquid and is represented by a liquid-filled cylinder with a small orifice through which laminar flow can occur. This is often termed a *dashpot, dashpot analog,* or *Newtonian substance.* The failure that occurs at a certain stress level is represented by a *stress link* or *fuse,* termed the *St. Vennant substance.* The characteristics of all of these elements in terms of stress response can be described mathematically.

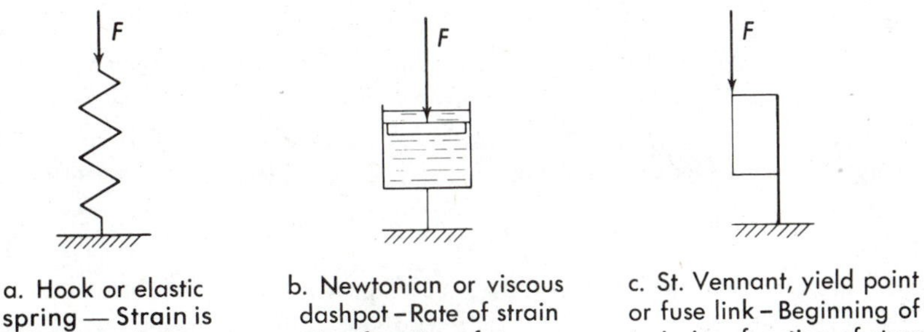

Figure 8.9 *Rheologic analogs.*

A *rheologic theory* is constructed by combining these simple elements in such a way that they simulate the real behavior of the mass. This requires a clear mathematical representation of observed behavior under simple load condition. The combination of elements to synthesize the real response is found by trial.

SIMPLE RHEOLOGIC MODELS / A number of rheologic models are useful in soil and rock mechanics. The compression of air in soil voids, followed by its flow from the soil can be represented by a spring and dashpot in series (Fig. 8.10a) the *Maxwell model.* The Terzaghi time rate of consolidation theory is a rheologic model consisting of a series of springs and dashpots in parallel (Fig. 8.10b). This is a *Kelvin model.*

A combination of the Kelvin and Maxwell models (Fig. 8.10c) is termed the *Burger model.* This might represent the time rate of consolidation of a partially saturated soil with initial plus primary consolidation.

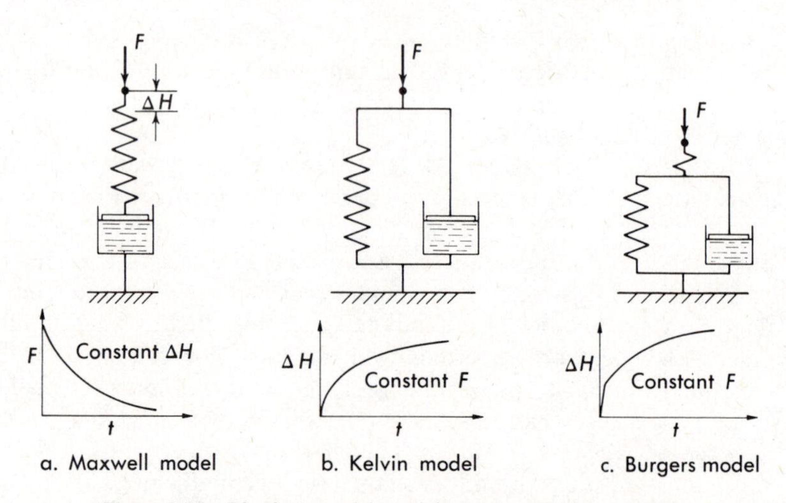

Figure 8.10 *Rheologic models useful in soil and rock mechanics.*

The rheologic models are helpful in approximating the response of soil to complex load variations, where observations of real soil behavior are difficult to measure. The model combination is usually based on simple loading. Therefore, if a more complex load pattern alters the soil, the model may not be a valid representation of the behavior under complex load changes.

There is a real danger in utilizing rheologic models in assuming that the analog which represents the time-dependent behavior is a representation of the mechanical process that determines that behavior. Although the analog is mechanically similar in time rate of consolidation of soils, others are not necessarily valid.

8:5 Solution of Problems in Mass Behavior

CHARACTER OF REAL PROBLEMS / The solution of real engineering problems concerning the response of soil and rock bodies to stresses requires an understanding of the character of the materials and the nature of the loads.

Rocks and soils have engineering properties that may be approximated by simple equations; however, the inherent errors in these approximations (such as the difference between the true θ and $45+\phi/2$ in undrained shear of clay) must be evaluated before using the approximation in analysis.

The bodies of soil and rock are three-dimensional, nonhomogeneous, and seldom isotropic. Although they may be approximated by simple forms, such as the semiinfinite homogeneous isotropic elastic body, the effects of the simplification must be evaluated. The localized discontinuities, such as cracks, cavities, and unusually rigid or weak zones also have a profound effect on the response of the mass.[8:5]

The loads are known with varying degrees of accuracy. The dead load due to soil weight and the weight of an engineering structure can be determined within a few percent. The live loads due to wind, occupancy, vehicles, and earthquakes can often be only estimated.

The environment, including the effect of excavation and nearby construction profoundly influences soils or rocks, changing their behavior. Hopefully, these changes can be predicted by laboratory tests; in some cases, however, there are surprises. For example, pore pressures in joints in the rock abutments of the Malpasset Dam in France reduced the strength sufficiently so that the mass sheared, taking a portion of the dam with it. Over 300 people were killed when the valley below the dam was scoured by the deluge.

There are three types of problems the engineer solves. The foremost is design. Here the objective is to predict the behavior of the soil or rock. On the basis of that prediction a design is developed either to control behavior or to adapt to it. Because the design generally influences the soil and rock behavior, the process is iterative; design and analysis alternate until a satisfactory solution is reached. What constitutes a satisfactory design depends only partly on the accuracy of the prediction of soil or rock behavior. For example, an analysis of foundation-bearing capacity shows that a wall footing 0.3 m (1 ft) wide is adequate. However, the cheapest way to build the footing will be to use a back hoe that digs a trench 0.8 m (2.6 ft) wide. The design soil-bearing capacity, therefore, is only 38% of the computed permissible value so that refined analyses are not justified.

By way of contrast, a refined analysis of the stability of the shell of a large earth dam made it possible to reduce the slope from 2.7(H) to 1(V) to 2.5(H) to 1(V) and save 1,000,000 m^3 (1,307,000 yd^3) of fill.

A second objective of analysis is to determine corrective measures for some observed shortcoming in the original design. The analysis starts from a known condition and works back to the cause of trouble, which is much easier than prediction. A new analysis must then predict the effectiveness of the correction or of a redesign.

The third objective is to develop more realistic analyses for future design, or to check the validity of an old method by comparing the prediction with measured performance. The accuracy is limited only by the data available and the time and resources of the investigator. The value can only be measured in the use of the information in new designs or in improving old ones.

THE GENERAL APPROACH / The approach to any problem is controlled by four factors:

1. The existing theories are not always adequate. Moreover, there are no theories available for some situations.
2. Past experience does not include some aspects of a new problem.
3. The soil and rock parameters needed for either analysis or a logical extrapolation of past experience cannot be measured reliably.
4. Time, facilities, and money to evaluate the uncertainties fully may not be available.

The engineer, therefore, must find a satisfactory answer despite these controls. What is satisfactory depends on the price of the solution compared to its value, considering the structure's useful life and the cost of its possible premature end. The design analysis for a temporary construction cofferdam that could be flooded without loss of life in case of an excessive river flow would thus be far less demanding than that for a permanent dam that must supply enough power to recover its cost in 50 years and whose failure could cause loss of life and extensive property damage.

A number of tools are available to the engineer. Specific ones are found in Chapters 9 through 12; the general procedure is outlined below:

1. The problem is identified and then simplified by dividing it into segments and delineating the significant factors.
2. Appropriate theories are utilized with allowances for their limitations.
3. Physical models and mathematical models are utilized where theories prove inadequate.
4. The material parameters to be used in analysis or models are evaluated from site data, appropriate tests, and related experience.
5. The results of the theoretical and model studies are interpreted in the light of experience.
6. The gaps in knowledge are filled intuitively.
7. The answers are reevaluated and revised when observations of actual performance show their inadequacies.

Of these steps, the second and third are capable of mathematical evaluation; the remainder require understanding, initiative, and—courage.

TWO-DIMENSIONAL REPRESENTATION / A common simplification is to reduce the problem in two dimensions. If the body or the loading is uniform and of great extent in one dimension, the two-dimensional representation or cross section may depict all the relevant factors. For example, the shear that develops behind a long retaining wall involves no displacements parallel to the face of the wall, a condition termed *plane strain*. A two-dimensional analysis, including shear strengths determined by plane strain testing, is a realistic evaluation of this problem except at the ends. An analysis of foundation-bearing capacity by plastic theory is feasible in two dimensions, but has eluded solution in three dimensions. However, empirical correction factors have been developed from model tests to extend the results of the two-dimensional analyses to three-dimensional problems.

Other simplifications include representing complex structural arrangements of the body by simple equivalents. For example, in seepage analysis, alternate strata of pervious sand and low permeability silt can be represented by a single stratum with a high permeability parallel to the bedding and a lower value perpendicular to it.

In any case it is essential that the effect of the simplification on the results be investigated by comparing the results of the analysis with observations of performance.

NUMERICAL APPROXIMATIONS, MATHEMATICAL MODELS / The high-speed digital computer with large storage or memory capacity has made it possible to solve many problems by successive approximations or by very large numbers of simultaneous equations. Without the computer the labor and time required would be prohibitive.

Many approximations involve dividing the body into segments whose response to stress can be simply represented. The response of the body is the sum of the responses of the segments.

In the *lumped mass* approximation, each segment is visualized as consisting of a discrete mass, connected with the adjoining segment by springs that represent the elasticity (Fig. 8.11a). Anisotropy and variations in elasticity from point to point can be represented by differences in the spring constants, and masses.

In the *finite element approximation*[8:6] (Fig. 8.11c) the body is divided into discrete elements each of whose response to load is equivalent to that of the portion of soil it represents. The elements, rectangular or triangular solids, are connected together by frictionless pins at their corners or nodal points. The forces act on each element through the nodal points, and are in static equilibrium. The corresponding strains of each element are compatible with those of the adjoining element because they are connected at the nodal points. The total response of the body is reduced to a system of simultaneous equations that satisfies equilibrium and compatibility for each element. The response of each element need not be isotropic or linear so long as the computer can handle the variables.[8:7]

The finite element approach has made it possible to obtain realistic solutions to problems that previously could only be approximated crudely. One example is the stress patterns in embankments resting on an elastic foundation, including the effect of building the embankment in layers.[8:8] A

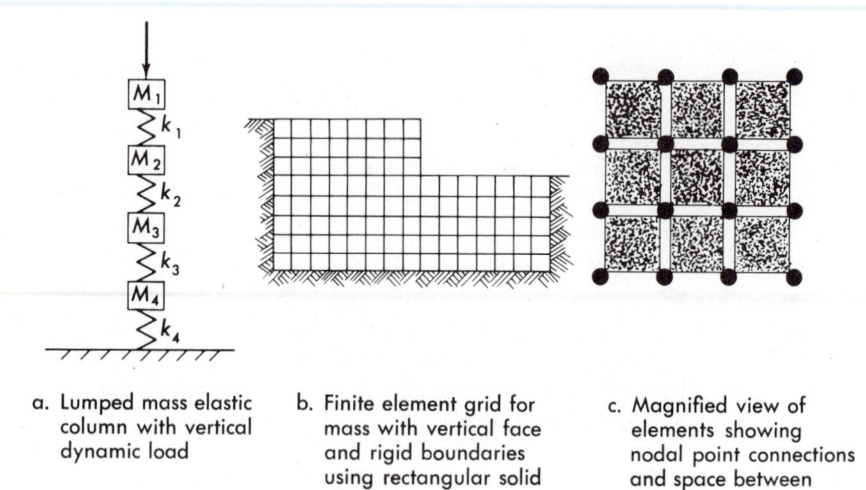

a. Lumped mass elastic column with vertical dynamic load

b. Finite element grid for mass with vertical face and rigid boundaries using rectangular solid elements

c. Magnified view of elements showing nodal point connections and space between elements

Figure 8.11 *Approximations of elastic masses for computer analysis.*

second is the response of an earth dam or slope to earthquake accelerations.[8:9] It is also possible to express the results for certain general cases in the form of graphs or charts suitable for routine calculations.

The numerical approximations are capable of solving complex two-dimensional problems in stress and strain as well as three-dimensional problems that are symmetrical about a vertical axis. General solutions to three-dimensional problems are limited by the capacity of the computers available.

HYPERBOLIC STRESS–STRAIN / The finite element analysis and others involving nonlinear stress–strain can be extended into the transition between elastic and plastic behavior by approximating the real stress–strain relation of the soil with a hyperbola.[8:10] The relation between strain, ε, and the corresponding maximum shear stress, $\sigma_1 - \sigma_3$, are expressed in terms of an initial modulus of elasticity, E_1, and the maximum shear–stress of failure, $\sigma_{1f} - \sigma_{3f}$.

$$\sigma_1 - \sigma_3 = \frac{\varepsilon}{1/E_1 + \varepsilon/(\sigma_{1f} - \sigma_{3f})} \tag{8:10}$$

The real stress–strain curve (Fig. 8.12a) is transformed by ordinates of $\varepsilon/(\sigma_1 - \sigma_3)$ (Fig. 8.12b). A straight-line approximation is made that fits the real data of two points that best define the range of interest. The y intercept is $1/E_1$, the slope of the line is $1/(\sigma_{1f} - \sigma_{3f})$. Typically, the 0.7 and 0.95 values of $(\sigma_1 - \sigma_3)_f$ are selected for the straight-line fit in problems of large strains; however, this approximation introduces errors at both smaller and larger strains. It still cannot describe fully plastic conditions where strain is independent of stress nor the effect of loss of strength with continuing strain.

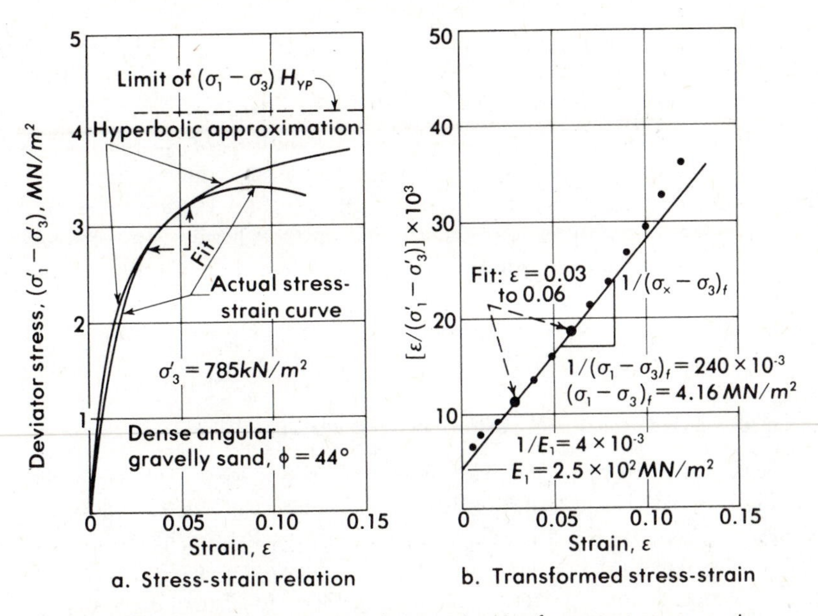

Figure 8.12 *Hyperbolic stress–strain approximation for computer programning.*

Various other numerical approximations, such as the method of finite differences, are used in solving the differential equations of elasticity for which no direct solution is possible. The details of all these methods of numerical analysis are beyond the scope of this textbook. They are powerful tools for the solution of complex problems. Both their applications and limitations must be clearly understood before the results of such methods can be used.

MODELS / Physical models have been widely used in engineering to solve problems where analysis is impractical or to verify and modify the results of analyses based on approximate theories. In addition, they can indicate the areas of high and low stress or the shapes of zones of plastic flow that can lead to more realistic theories in the future.

The problem in any model is scale, unless the model has the same dimensions as the prototype. Dimensional analysis is necessary to establish the characteristics of the model that will properly reflect the behavior of the prototype. For models that are geometrically similar to the prototype it can be shown that the dimensionless ratios (known as π terms) must be the same in model and prototype. For simple problems in static plastic and elastic equilibrium,

$$\pi_{(\text{strength})} = \frac{\tau_{ff}}{\gamma L} \tag{8:11a}$$

$$\pi_{(\text{rigidity})} = \frac{E}{\gamma L} \tag{8:11b}$$

In these expressions L is a linear dimension, γ is the unit weight of soil or rock, and τ_{ff} and E are the shear strength and modulus of elasticity. Both π terms indicate that for similarity a model with length dimension L_m of $\frac{1}{10}$ that of the prototype, L_p would require a model strength τ_m or elasticity E_m of $\frac{1}{10}$ that of the prototype for similarity if the soil and model unit weights were the same. Alternatively, the model unit weight could be 10 times that of the prototype if the strengths were equal. Such models have been constructed with the unit weight increased by a large centrifuge. For complex problems several π terms are required which are not necessarily compatible. The model, therefore, does not fully represent the behavior of the prototype, and only a full-scale model will be realistic.

8:6 Analysis—A Tool, Not a Product

Analysis of the response of a soil or rock mass to load is an essential tool in design, corrective action, and for finding new solutions to old and new problems. Like any tool, analysis is limited to the quality of the material used with it as well as the understanding and ingenuity of the user. Although the tool has its limitations, the first two are more likely to control the value of the results. Ultimately, the tool helps shape the final product: a safe economical design. Without it the product is crude, but the tool is useless without the skill and creativity that put it to work.

REFERENCES

8:1 S. P. Timoshenko and J. N. Goodier, *Theory of Elasticity*, 2nd ed., McGraw-Hill Book Company, New York, 1951.

8:2 H. G. Poulos and E. H. Davis, *Elastic Solutions for Soil and Rock Mechanics*, John Wiley & Sons, Inc., New York, 1974, 411 pp.

8:3 A. Nadi, *Theory of Flow and Fracture in Solids*, McGraw-Hill Book Company, New York, 1950.

8:4 W. D. L. Finn, "Application of Limit Plasticity in Soil Mechanics," *Journal of the Soil Mechanics and Foundations Division, Proceedings, ASCE*, **93**, SM5, 101, September 1967.

8:5 R. E. Goodman, R. L. Taylor, and T. L. Brekke, "A Model for the Mechanics of Rock," *Journal of the Soil Mechanics and Foundations Division, Proceedings, ASCE*, **94**, SM3, 637, May 1968.

8:6 O. C. Zienkiewicz, *The Finite Element Method in Structural and Continuum Mechanics*, McGraw-Hill Publishing Co., Ltd., London, 1967.

8:7 C. V. Girijavallabhan and L. C. Reese, "Finite Element Method for Problems in Soil Mechanics," *Journal of the Soil Mechanics and Foundations Division, Proceedings, ASCE*, **94**, SM2, 473, March 1968.

8:8 R. W. Clough and R. J. Woodward, "Analysis of Embankment Stresses and Deformations," *Journal of the Soil Mechanics and Foundations Division, Proceedings, ASCE*, **93**, SM4, 529, July 1967.

8:9 H. B. Seed, I. M. Idriss, K. L. Lee, and F. I. Makdisi, "Dynamic Analysis of the Slide in the Lower San Fernando Dam During the Earthquake of February 9, 1971, *Journal of the Geotechnical Engineering Division, Proceedings, ASCE*, **101**, GT9, 889–911, September 1975.

8:10 J. M. Duncan and C. Y. Chang, "Non-Linear Analysis of Stress and Strain," *Journal of the Soil Mechanics and Foundations Division, Proceedings, ASCE*, **96**, SM5, 1629–1654, September 1970.

SUGGESTION FOR FURTHER STUDY

J. M. Duncan, "Finite Element Analysis of Stresses in Dams, Embankments, and Slopes," in *Symposium on Applications of Finite Element Method in Geotechnical Engineering*, U.S. Waterways Experiment Station, Vicksburg, Miss., 1972.

PROBLEMS

8:1 Compute the settlement of the clay stratum in Example 8:1 if the initial void ratio is 1.6, the compression index is 0.43, and the clay stratum is 6 m (20 ft) thick.

8:2 After the groundwater lowering of Example 8:1, the ground water rose until it was 2 m (7 ft) from the ground surface.

a. Compute the new effective stress on the surface of the clay stratum.
b. What has been the total change in effective stress produced by the changing water table?
c. If the sand has an angle of internal friction of 33°, what has been the percentage change in sand shear strength just below the clay due to both lowering of the water table and raising of the water table?

8:3 A mass of clean fine sand 6 m (20 ft) thick overlying a shale rock has a void ratio of 0.45 and a specific gravity of solids of 2.68. The height of capillary rise is 2 m (7 ft). Within the capillary zone the soil is saturated; above it is virtually dry.

a. Compute the vertical effective stress in the sand at several levels, and plot the stress as a functional depth.
b. The groundwater level rises to the ground surface. Compute total, neutral, and effective stresses and plot as functions of depth.

c. The water level falls to a depth of 4 m (13 ft). Plot the vertical total, neutral, and effective stresses as functions of depth.

d. Compare the sand strengths for all three conditions at the following levels: (1) ground surface; (2) 2 m deep; (3) 4 m deep; (4) 6 m deep.

8:4 Compute the approximate shear stress in the upper surface of the clay stratum produced by an earthquake whose horizontal acceleration is 0.08 g, in Example 8:2.

8:5 A thick clay stratum with a $c = 30\ \mathrm{kN/m^2}$ ($626\ \mathrm{lb/ft^2}$) and a unit weight of $16\ \mathrm{kN/m^3}$ ($102\ \mathrm{lb/ft^3}$) is excavated with a vertical bank. Consider that a small cube of clay in the face of the bank is unrestrained (the lateral stress is zero). How deep could the excavation be made before a state of plastic equilibrium is reached in a clay cube at the bottom of the face?

CHAPTER 9
Earth Pressure and Structures to Support Earth

The bracing required to support the earth, adjoining streets, and office buildings for an underground subway station was so critical that the design earth pressure and soil strengths were made a part of the contract drawings. However, to provide innovative competitive prices, the contractor was required to provide the bracing design to restrain the design earth pressure. His design consisted of vertical H-piles 2.5 m (8 ft) apart driven through firm silty sands and sandy silts to rock below. The H-piles were restrained by horizontal struts where the excavation was narrow but by earth anchors where the excavation was wide and came within 3 m (10 ft) of the pile foundations of an 8-story office building.

Excavation to a depth of 6 m (20 ft) produced no problems and excavation to the final depth of 11 m (36 ft) in the narrow strutted area was uneventful. However, in the wider excavation, close to the office building, the bracing moved. The outer wall of the building shifted 50 to 75 mm (2 to 3 in.) toward the excavation; the ground surface and the floor slabs and walkways supported on it tilted toward the excavation 25 to 50 mm (1 to 2 in.). The building and the floor slabs cracked.

A review of the bracing design disclosed that it was marginally safe against failure. The stresses were so high, however, that there was significant elastic distortion of the soil that carried the ground floor slabs and which provided lateral support for the piles. The wall of soil behind the bracing system stretched and sagged; the side of the building closest to the excavation shifted with the soil, cracking the structure. The money saved on the contractor's cheap design was considerably less than the cost of building repair.

A steel sheet pile anchored bulkhead system formed a continuous wharf in a small port. The sheet piles were 18 m (59 ft) long, providing 4 m (13 ft) of water depth and 3 m (10 ft) of freeboard. They were driven into virgin clay soils, tied together with a system of horizontal steel beams, and anchored

3 m (10 ft) below their tops with steel rods extending back 20 m (66 ft) to a second line of shorter steel sheet piles with horizontal wale beams. Both lines of sheet piles were driven, the horizontal wales installed, and all were tied together with anchor rods. Wet sand backfill was then bulldozed into the area between the piles to raise the ground surface 3 m (10 ft) above the original clay surface (at approximately sea level).

Suddenly, the wall moved outward but tilted slightly backward toward the shore. The top of the outer wall dropped 1 m ($3\frac{1}{2}$ ft) and then split along the interlocks. The sand fill between the walls flowed outward, propelled by the high water table generated by compacting the saturated sand.

An evaluation of the design disclosed a number of defects. The piles were supported in soft clay that was not strong enough to restrain the lateral loads on the deep piles. There was no provision for wall drainage; the ground water level between the two walls was at the new ground surface, 3 m (10 ft) above sea level. The rigid structural connections between the sheet piles and the anchors were not able to absorb rotation, and tore, allowing the outer wall to collapse and lose the backfill.

In his own defense the designer argued that a similar design nearby was stable; however, at the other site the clay was stronger and underlain by sand. He claimed that drains would only allow the fine sand backfill to wash away, as it did during the failure. Moreover, he argued that, so close to the sea, the groundwater level in the backfill should be little higher than sea level. Finally, his stress analyses showed that the structural connections were both rigid and strong enough to resist the load in the piling, including the high ground water in the backfill.

These examples illustrate the apparent paradoxes in conventional structural analyses as well as the shortcomings of simple design concepts for earth retaining structures. First, conventional safety of earth support structures with respect to strength of the components is not enough. Deflection must also be considered. Second, although the support structure is structurally safe, the soil it restrains can deflect excessively or even partially fail. Third, groundwater pressure and its control by drainage is a vital part of design that is often overlooked by structural engineers. Finally, rigid connections, however strong, can fail.

Real problems in lateral earth pressure involve more than the simple loads produced by soil against a retaining wall. Earth pressure is not a unique property of the soil or rock. Instead, it is a function of the material that the retaining structure must support, of the loads that the soil behind the structure must carry, the groundwater conditions, and the amount of deflection the retaining structure undergoes.

Engineering structures such as retaining walls, trench and excavation bracing, bulkheads, and cofferdams have a common function—to provide lateral support for a mass of soil. The pressure exerted by the soil on these structures is known as *earth pressure* and must be determined before a satisfactory design can be made.

Some of the earliest theories of soil mechanics dealt with earth pressure on retaining walls. Unfortunately, the engineers using these theories have not always realized the significance of the assumptions made in their development. The result has been many failures and a discrediting of soil mechanics by engineers who have had to deal with soils in construction work—an attitude that persists today.

9:1 Theory of Limiting Earth Pressure

The general theory of earth pressure can be developed from the stresses in an extremely large, level mass of normally consolidated soil. The total vertical stress in homogeneous soil at a depth of z is equal to the weight of the soil above. When ground water is present, the vertical stress can be separated into two components: neutral stress and effective stress, as previously described in Chapters 3, 4, and 8:

$$\sigma_z = \gamma z \tag{3:9b, 8:1a}$$

$$\sigma'_z = \gamma z - u \tag{8.1b}$$

EARTH PRESSURE AT REST / The stress conditions of an element of soil at depth z in a level mass are shown in Fig. 9.1a. The element

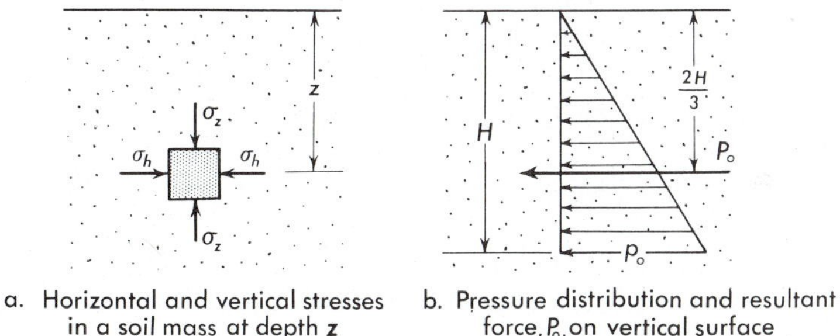

Figure 9.1 *Earth pressure at rest.*

can deform vertically under load, but it cannot expand laterally because it is confined by the same soil under the same loading conditions. This is equivalent to the soil being placed against an immovable frictionless wall that maintains the same lateral dimension in the soil regardless of the vertical load. The soil is in a state of *elastic equilibrium*, and the stresses in the lateral direction can be computed from the stress–strain relationships of the soil. The relation between lateral and vertical strain is described by Poisson's ratio (Section 8.2), and for the condition of zero lateral strain the principal stresses above the water table are related by

$$\Delta\sigma_x = \left(\frac{\nu}{1-\nu}\right)\Delta\sigma_z \tag{8:8}$$

The lateral pressure exerted in the at-rest state is given the symbol p_0 and can be computed from the vertical stress σ_z in a dry soil by

$$\sigma_x = \sigma_h = \sigma_z \frac{\nu}{1-\nu} = p_0$$

$$p_0 = \gamma z\left(\frac{\nu}{1-\nu}\right) = K_0 \gamma z \qquad \text{(dry soil)} \tag{9:1a}$$

Below the water table the pressure is found from its effective and neutral components:

$$p'_0 = (\gamma z - u)K'_0 \qquad \text{(wet soil, effective)} \tag{9:1b}$$

$$p_0 = (\gamma z - u)K'_0 + u \qquad \text{(wet soil, total)} \tag{9:1c}$$

where K_0 is the *coefficient of earth pressure at rest* and is found from Poisson's ratio. The effective coefficient of earth pressure at rest has been found related to ϕ by both approximate theories and empirical data.[9:1]

$$K'_0 = 1 - \sin\phi \tag{9:1d}$$

The value of K_0 for saturated clays in undrained loading or quick loading is sometimes also expressed in total stresses that include the neutral stress, and for which Equation (9:1a) should be used. (See Table 9:1.)

TABLE 9:1 / VALUES OF K_0

Soil	K'_0, Effective, Drained	K_{0u} Total, Undrained
Soft clay	0.6	1.0
Hard clay	0.5	0.8
Loose sand, gravel	0.6	
Dense sand, gravel	0.4	
Overconsolidated clay	0.6 to 1+	
Compacted, partially saturated clay	0.4 to 0.7	

The resultant force per unit of length of the wall, P_0, acting on a wall of height H, can be found by integrating Equation (9:1a) or from the pressure diagram (Fig. 9.1b). For a dry soil (or a saturated clay in undrained loading), the diagram is triangular and the resultant force is

$$P_0 = \frac{K_0 \gamma H^2}{2} \tag{9:2a}$$

and the location of the resultant is at a depth of

$$z_p = \frac{2H}{3} \tag{9:2b}$$

With ground water the effective and neutral pressure diagrams must be computed separately and the magnitude and location of the resultant found by the methods of mechanics.

The stress conditions in the soil mass at rest are far from failure, as can be seen in Fig. 9.2. Stresses in an oblique direction can be computed by the Mohr circle.

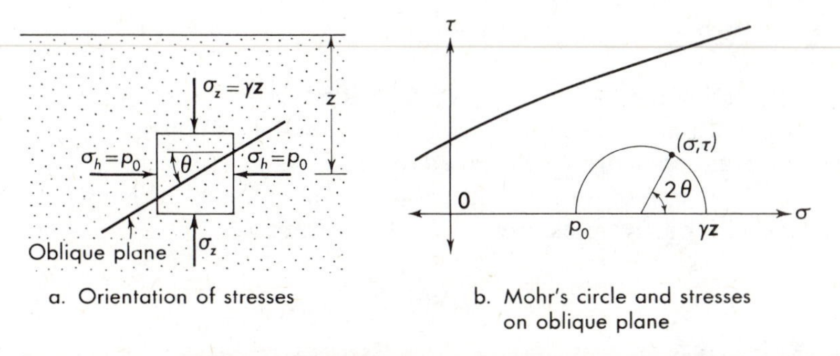

Figure 9.2 *Mohr circle for earth pressure at rest.*

ACTIVE EARTH PRESSURE / If the unyielding frictionless vertical wall of the at-rest condition is allowed to move away from the soil, each element of soil adjacent to the wall can expand laterally. The vertical stress remains constant, but the lateral stress or earth pressure is reduced in the same way the stress in a compressed spring becomes less as the spring is allowed to expand. Initially, the stress reduction is elastic and proportional to the deformation; but as the difference between the major and minor principal stresses increases with the reduction in lateral stress, the diameter of Mohr's circle grows until the circle touches the rupture envelope. The lateral pressure has reached a *minimum* at this point; the state of stress is no longer elastic; the soil mass behind the wall is in a state of shear failure or plastic equilibrium, and further movement of the wall will just continue the failure with little change in pressure.

The minimum horizontal pressure p_A at any depth z for dry sands and gravels can be found from the Mohr diagram at failure (Fig. 9.3a) and is

$$p_A = \frac{\gamma z}{\tan^2 [45 + (\phi/2)]} \tag{9:3a}$$

$$p_A = \gamma z \tan^2 \left(45 - \frac{\phi}{2}\right) \tag{9:3b}$$

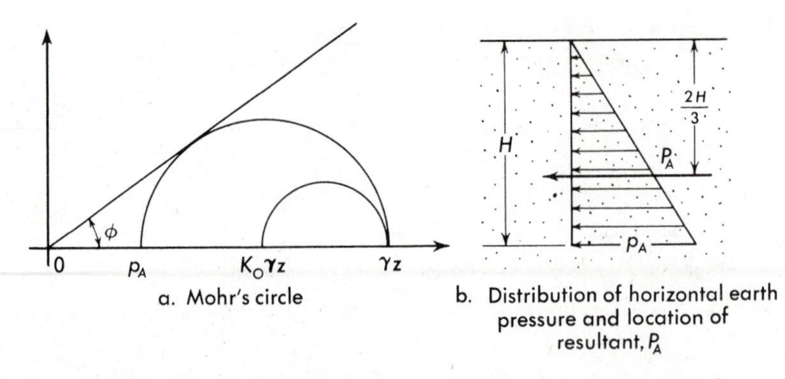

Figure 9.3 *Active earth pressure in cohesionless soils, sands, and gravels.*

The expression $\tan^2[45-(\phi/2)]$ is often called the *coefficient of active earth pressure* and is given the symbol K_A. The state of shear failure accompanying the minimum earth pressure is called the *active state*. The resultant force P_A per unit of length of wall for the dry sand can be found by integrating the expression for active pressure or from the area of the pressure diagram:

$$P_A = \frac{\gamma H^2 K_A}{2} \tag{9:4}$$

The action line is through the centroid at a depth of $2H/3$ (Fig. 9.3b).

If the soil is below water, neutral stress must again be considered. The effective active pressure is computed from the effective vertical pressure and K_A. The total is the sum of the effective and the neutral stress:

$$p'_A = (\gamma z - u)K_A \tag{9:5a}$$

$$p_A = (\gamma z - u)K_A + u \tag{9:5b}$$

When a dry cohesionless soil is inundated by a rising water table, the effective pressure is reduced to about half its original value. The total pressure, however, is approximately tripled. The location and magnitude of the resultant for a cohesionless soil below water is found by combining the effective and neutral stress diagrams.

Example 9:1

Compute the active earth pressure at a depth of 6 m or 19.7 ft in a clean sand whose angle of internal friction is 34° and which weighs 15.4 kN/m^3 or 98 lb/ft^3 drained and 20 kN/m^3 or 127.4 lb/ft^3 saturated.

1. The sand is dry throughout

$$p_A = \gamma z \tan^2 = [45-(\phi/2)] \tag{9:3a}$$

$p_A = 15.4 \times 6 \times (0.532)^2 = 26.15$ kN/m^2

$p_A = 98 \times 19.7 \times (0.532)^2 = 546$ lb/ft^2.

2. The groundwater table is at a depth of 2 m or 6.57 ft
   $p'_A = [\gamma_1 H_1 + \gamma_2 H_2 - (z - z_w)\gamma_w] \tan^2 (45 - \phi/2)$
   $p'_A = (15.4 \times 2 + 20 \times 4 - [(6-2)9.81)] \times (0.532)^2$
   $(30.8 + 80 - 39.2) \times 0.532^2 = 20.3 \text{ kN/m}^2$
   $p'_A = [98 \times 6.57 + 127.4 \times 13.13) - (19.7 - 6.57) \times 62.4] \times 0.532^2$
   $(644 + 1673 - 819) \times 0.532^2 = 424 \text{ lb/ft}^2$.
3. $p_A = p'\hat{A} + u$
   $20.3 + 39.2 = 59.5 \text{ kN/m}^2$
   $424 + 819 = 1243 \text{ lb/ft}^2$.
4. $20.3/26.15 = 0.78$; $59.5/26.15 = 2.28$.
5. The water table rising $\frac{2}{3}$ of the thickness of the sand reduced the effective pressure to 78% of the drained, but increased the total to 228% of the drained.

A similar analysis for saturated clay in undrained loading, using a Mohr circle (Fig. 9.4a), gives the formula for the active pressure as

$$p_A = \gamma z - 2c \tag{9:6}$$

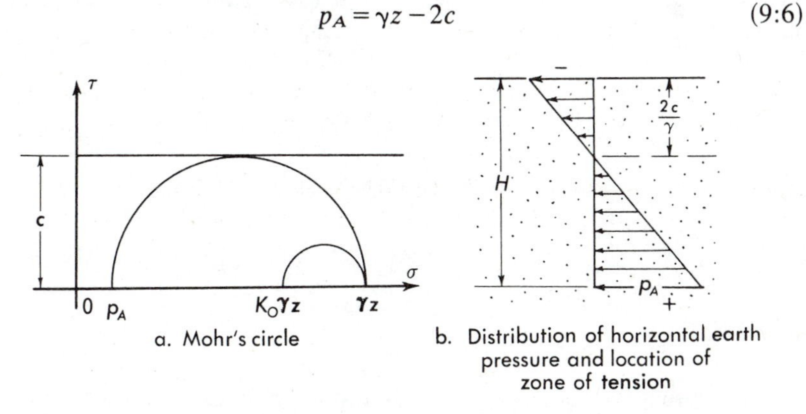

Figure 9.4 *Active earth pressure in saturated clay in undrained shear.*

The total force per foot length of wall, P_A, is given by the expression

$$P_A = \frac{\gamma H^2}{2} - 2cH \tag{9:7}$$

According to these expressions the total earth pressure is zero at a depth $z = 2c/\gamma$, as shown in Fig. 9.4b. Above that level the soil is in tension; below it is in compression. If the total force $P_A = 0$, expression (9:7) can be rewritten

$$2cH = \frac{\gamma H^2}{2} \qquad H_0 = \frac{4c}{\gamma} \tag{9:8}$$

This implies that the resultant force against a wall would be zero if the wall height, H_0, were $4c/\gamma$. This simple expression explains why a vertical cut can be excavated in saturated soil without immediate failure. It also suggests that there would be no force against a wall whose height was only $4c/\gamma$. These implications, however, are misleading and dangerous. First clay cannot support tensile stresses for very long; a vertical crack will soon develop in the tension zone, destroying much (or all) of the tension force above the theoretical level of zero pressure, $2c/\gamma$. The soil now requires support below that level. Thus the safe vertical bank in a saturated clay in the undrained state will be less than $4c/\gamma$ but more than $2c/\gamma$. A wall built against such a vertical bank will not experience the tension force above the depth $2c/\gamma$; the clay will pull away from the wall, but press against the wall below the level $2c/\gamma$. Therefore, a wall that supports saturated clay in undrained conditions should be designed on the assumption of a tensile crack to a depth of $2c/\gamma$ and pressure below increasing at the rate of $\Delta p_A = \gamma\,\Delta z$ (Fig. 9.5b). A more insidious complication is surface water accumulation in the crack. It adds a positive pressure to a wall or within the soil mass that changes the total force system drastically, as shown in Fig. 9.5c. This accounts for many failures of unsupported clay banks and walls supporting clay during rainy weather.

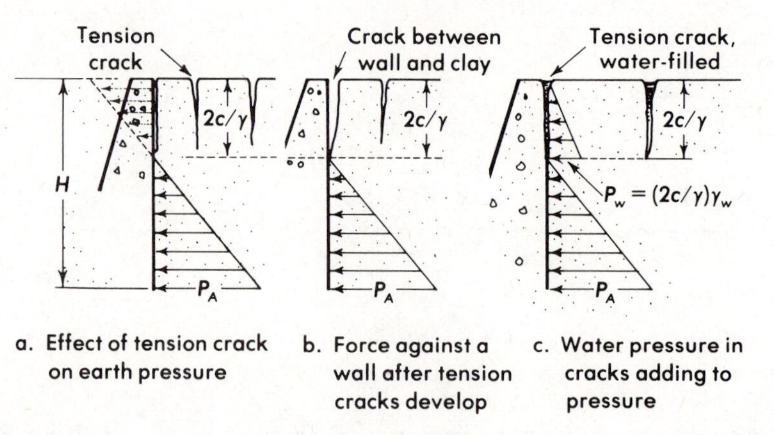

Figure 9.5 *Effect of tensile cracks on active earth pressure of saturated clay in undrained shear.*

A final complication is that total stresses in undrained shear approximate the failure conditions only so long as the water content and void ratio of the clay remain unchanged. In the new stress environment there is a redistribution of pore water pressures. Once the pore water pressures stabilize and can be evaluated, the ultimate earth pressure for clay is computed by effective and neutral stresses using Equations (9:5a) and (9:5b) and the drained shear strength parameters of the clay. The active earth pressure of clay, therefore, changes with time and the environment. Its calculation involves uncertainties that must be considered in design.

PASSIVE STATE / If, instead of moving away from the soil, the wall moves toward the soil, the pressure against the wall increases. The stress circles increase to the right of the vertical stress γz, which now becomes the minor principal stress. The maximum pressure against the wall is reached as shear failure again occurs in the soil behind the wall.

For dry cohesionless soils the pressure at any depth can be found by a Mohr diagram (Fig. 9.6) and is

$$p_p = \gamma z \tan^2\left(45+\frac{\phi}{2}\right) \tag{9:9}$$

where p_p is the maximum or *passive earth pressure*. The expression $\tan^2[45+(\phi/2)]$ is often called the *coefficient of passive earth pressure* and is given the symbol K_p.

The total force per unit length of wall of height H is found from the pressure diagram (Fig. 9.6b):

$$P_p = \frac{\gamma H^2}{2} K_p \tag{9:10}$$

The action line is horizontal and is at a depth of $2/3H$.

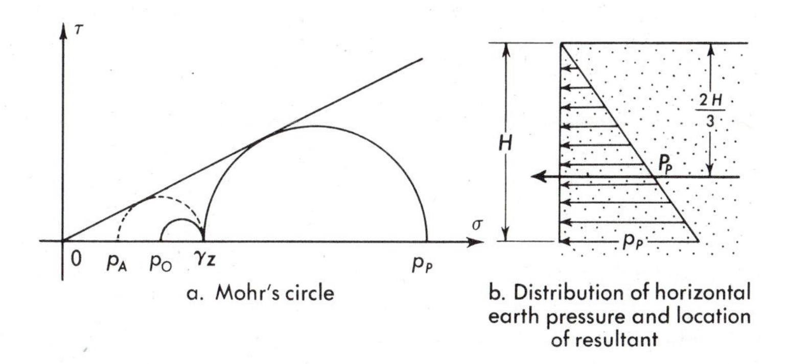

Figure 9.6 *Passive earth pressure in cohesionless soils: Sands and gravels.*

Below the water table the effect of neutral stress is handled in the same way as for the active state.

For saturated clays in undrained loading, the passive pressure is found by a Mohr circle (Fig. 9.7a) to be

$$p_p = \gamma z + 2c \tag{9:11}$$

The total force for a unit length of wall is found from the pressure diagram (Fig. 9.7b) and is

$$P_p = \frac{\gamma H^2}{2} + 2cH \tag{9:12}$$

For soils such as partially saturated clays whose shearing resistance is given by the formula $\tau_{ff} = c + \sigma_{ff} \tan \phi$, the following are derived with the aid

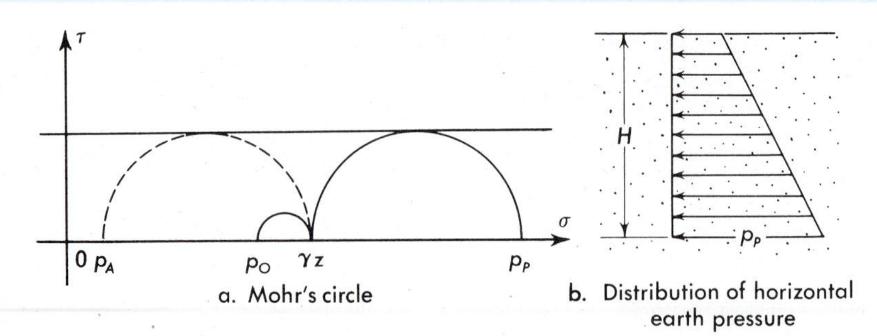

Figure 9.7 *Passive earth pressure in saturated clay in undrained shear.*

of Mohr diagrams. For the active state:

$$p_A = \gamma z \tan^2\left(45 + \frac{\phi}{2}\right) - 2c \tan\left(45 - \frac{\phi}{2}\right) \tag{9:13}$$

$$P_A = \frac{\gamma H^2}{2} \tan^2\left(45 - \frac{\phi}{2}\right) - 2cH \tan\left(45 - \frac{\phi}{2}\right) \tag{9:14}$$

For the passive state:

$$p_p = \gamma z \tan^2\left(45 + \frac{\phi}{2}\right) + 2c \tan\left(45 + \frac{\phi}{2}\right) \tag{9:15}$$

$$P_p = \frac{\gamma H^2}{2} \tan^2\left(45 + \frac{\phi}{2}\right) + 2cH \tan\left(45 + \frac{\phi}{2}\right) \tag{9:16}$$

The pressure diagrams for these conditions are similar to those for saturated clays.

This analytical approach to earth pressure is termed the *Rankine method* after that famous Scottish engineer who first applied such reasoning to soil masses.[9:2]

9:2 Deformation and Boundary Conditions

In both the active and passive states the zones of soil adjacent to the frictionless wall that are in a state of shear failure or plastic equilibrium form plane wedges (Fig. 9.8). Because the angle between a failure plane and the major principal plane is $\theta = 45 + (\phi/2)$, the wedge is bounded in the active state by a plane making an angle of θ with the horizontal, and in the passive state by a plane making an angle of θ with the vertical. Within the wedges in both cases are an infinite number of failure planes making angles of θ with the major principal plane.

The amount of horizontal movement of any point on the wall necessary to produce either the active or passive state is proportional to the width of

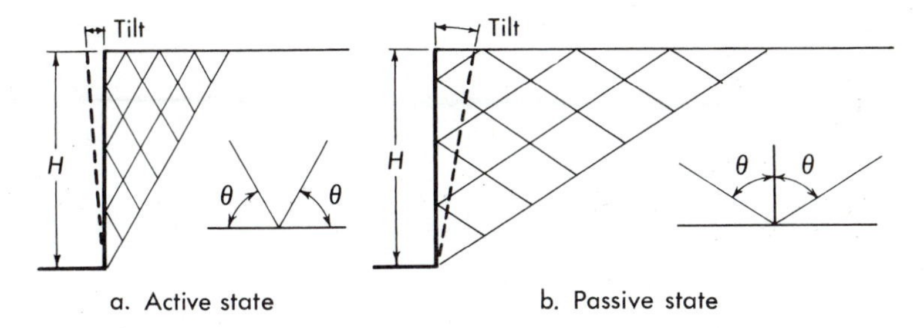

Figure 9.8 *Failure planes and shear wedges in soil behind a vertical frictionless wall in the Rankine states of plastic failure.*

the shear zone adjacent to that point. As can be seen from Fig. 9.8, the minimum movement consists of tilting about the base of the wall.[9:3] The amount of tilt is small and depends on the soil rigidity and wall height H, as given in Table 9:2.

TABLE 9:2 / TYPICAL MINIMUM TILT NECESSARY FOR ACTIVE AND PASSIVE STATES

Soil	Active State	Passive State
Dense cohesionless	$0.0005H$	$0.005H$
Loose cohesionless	$0.002H$	$0.01H$
Stiff cohesive	$0.01H$	$0.02H$
Soft cohesive	$0.02H$	$0.04H$

Soft cohesive soils do not remain in either the passive or active condition for long. Slow yield of the soil (ofter termed *creep*) tends to return the soil mass to the "at-rest" state. In the case of walls supporting a soft clay backfill, this means that there will be a continual slow, outward movement of the wall if the wall is designed to support only active pressure. Over a period of a few months, however, the change in pressure due to creep is usually negligible.

INCOMPLETE DEFORMATION / The relation between wall movement and earth pressure is shown qualitatively in Fig. 9.9. The minimum and maximum limits are defined by soil shear, the Rankine active and passive states. The pressure at no deformation is defined by Poisson's ratio in the elastic state. If the wall movement is less than that required for shear, the earth pressure is between the limiting values, and is defined by the relation of soil deformation to stress.

The finite element analysis can be used to estimate pressures in both the elastic range and in the transition to the limiting shear of active and passive pressures. The results are only as valid as the stress–strain parameters used in the analysis. Sometimes the computed movements are 2 to 3 times those

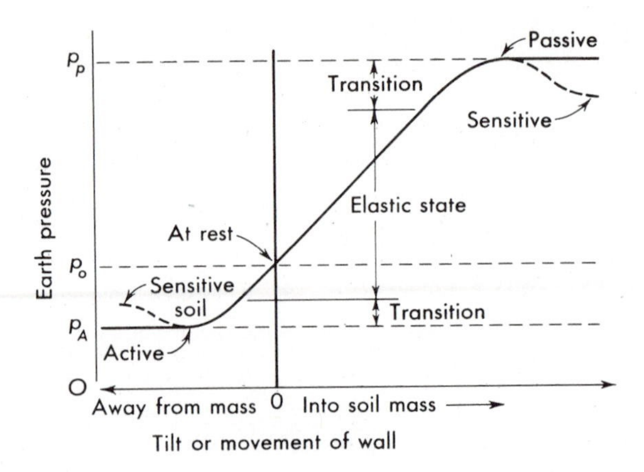

Figure 9.9 *Effect of wall movement or tilt on magnitude of resultant of earth pressure.*

observed on real structures, suggesting that the soil elasticity constants (determined in the laboratory) are too low. Limited data suggest that in-place parameters, such as determined by bore hole expansion, are more realistic. Despite the observed discrepancies, the finite element analysis is valuable for *parametric studies*, where the relative effect of changing soil properties, wall movement or of wall geometry can be evaluated.

Alternatively, the earth pressures in the in-between range can be estimated from the limiting at-rest, active, and passive pressures, correlated with measured pressures and movements on full-sized structures.

At very large movements, more than those needed for active or passive conditions, the pressures can shift toward the at-rest in sensitive soils whose strength is reduced by continuing shear, or decrease slightly in soils that gain strength with shear.

IRREGULAR DEFORMATION: ARCHING[9:2, 9:3] / When the movement of the wall is different from the tilting required to establish the active or passive states, both the magnitude and the distribution of the earth pressure are changed. If a section of the wall deflects outward more than the neighboring sections, the soil adjacent to it will tend to follow, as shown in Fig. 9.10a and b. Horizontal shear develops along the boundaries of this section of soil, and this restrains it, transferring part of the lateral load it carried to the adjoining soil. The result is a *redistribution of pressure by shear*, sometimes called *arching*, and an irregular pressure distribution. Examples of the effect of arching on earth pressure against typical structures are shown in Fig. 9.10b and c. The magnitude of the redistribution is estimated from observations of pressures and deflections of actual structures, or from finite element analyses.

EFFECT OF WALL FRICTION / The Rankine analysis considers an extensive zone in plastic equilibrium, with the shear pattern undistorted by the wall. This is equivalent to assuming that no shear can develop

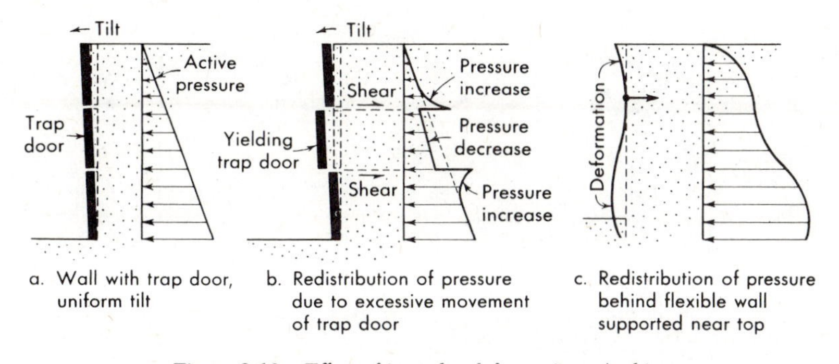

Figure 9.10 *Effect of irregular deformation: Arching.*

between the wall and the soil (the *wall is frictionless*). In reality the movement between the wall and the backfill does develop shear or friction, which distorts the shear pattern and the magnitude and direction of the resultant force that acts on the wall. The error involved is not serious for small walls with smooth faces, but it can be significant for structures higher than 10 m (33 ft) and with rough faces that induce appreciable shear. Methods for solving such problems are discussed in Section 9:3.

Simple problems of active and passive earth pressure can be adequately solved by the Rankine analysis. For more complex problems either modifications or corrections are applied to the Rankine method, or analyses compatible with the real boundary conditions are used. Some of these are discussed in this chapter. More extensive treatments can be found in the references.

9:3 Computing Earth Pressure

APPROXIMATE ANALYSES OF SLOPING WALLS / For walls less than 10 m (33 ft) high a number of useful approximations based on the Rankine analysis have been developed. When the back of the wall is inclined or *battered*, the force is assumed to act on a vertical plane through the heel of the wall, as shown in Fig. 9.11a. In such cases the weight of the wedge of soil between the vertical plane and the wall is added vectorially to the resultant P_A.

EFFECT OF SURCHARGE LOADINGS / If a uniform surcharge load of q_q acts on the soil behind the wall, as shown in Fig. 9.11b, it produces additional pressure on the wall. In the active state the resultant of this pressure P_q in force per unit length of wall, is

$$P_q = q_q H \tan^2\left(\frac{45-\phi}{2}\right) \tag{9:17}$$

It acts midway between the top and bottom of the wall.

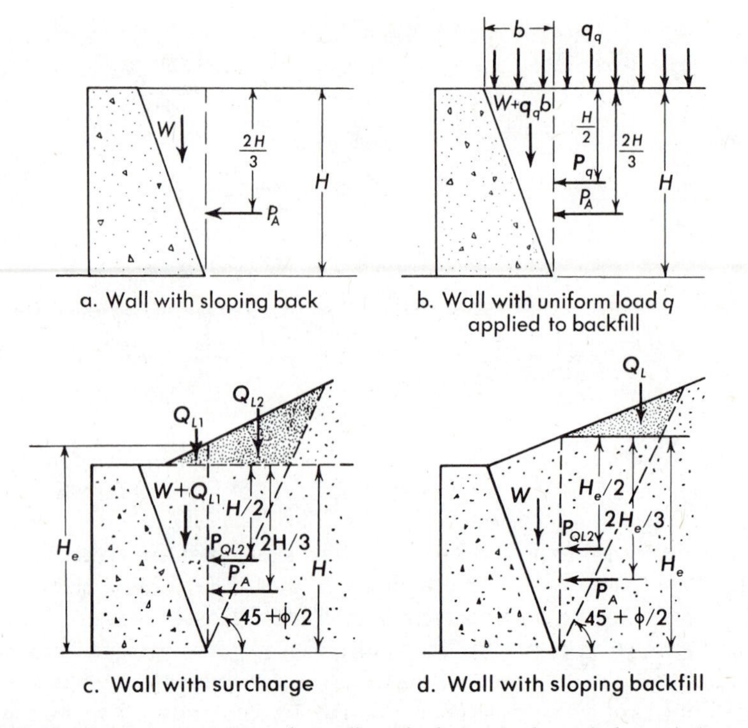

Figure 9.11 *Approximate pressure for walls with sloping backs and sloping cohesionless backfills.*

Sloping surcharges such as piles of materials on a level backfill can be approximated by a uniform surcharge equivalent to 0.8 the material weight, Q_L, within the shear zone.

APPROXIMATION FOR SLOPING BACKFILL / A backfill sloping at an angle of β with the horizontal is shown in Fig. 9.11c and d. An equivalent wall of height H_e is measured vertically through the wall toe. The earth pressure p_A is computed at the face of that equivalent wall. The backfill above the equivalent wall height is considered to be a surcharge force per unit of wall length. Q_{L2} The resultant force per unit of wall length, P_{QL2}, is $0.8Q_{L2}K_A$ (the reduction of 0.8 compensates for ignoring shear in the soil above H_e).

EFFECT OF CONCENTRATED SURCHARGES / In many real situations loads of limited extent are placed on the backfill. For example, a building foundation, a highway pavement, or railroad is often supported on the backfill close enough to the wall that additional earth pressure is produced. This creates a local distortion of the soil mass and usually a change in the shear pattern. The distortion is, in effect, a state of elastic equilibrium superimposed on a state of plastic equilibrium. While there are theoretical objections to this concept, large-scale model studies show it to be realistic.

According to the Boussinesq analysis of elastic equilibrium, a load of Q at the surface produces a stress increase $\Delta\sigma_{hq}$ at a depth of z and at a horizontal distance of x from a wall and a length of y along the wall (Fig. 9.12a) of:

$$\Delta\sigma_{hq} = \frac{Q}{2\pi}\frac{3x^2z}{R^5} = \frac{0.48Qx^2z}{R^5} \tag{9:18a}$$

$$R = \sqrt{x^2 + y^2 + z^2}. \tag{9:18b}$$

For a continuous load of Q_L per unit of length parallel to the wall and at a distance of x from it (Fig. 9.12b) the increase in stress, $\Delta\sigma_{hq}$ uniformly along the wall length is given by

$$\Delta\sigma_{hq} = \frac{0.63Q_Lx^2z}{R^4} \tag{9:18c}$$

$$R = \sqrt{x^2 + z^2} \tag{9:18d}$$

In either case R is the direct distance from the load to the point on the wall the stress increment acts.

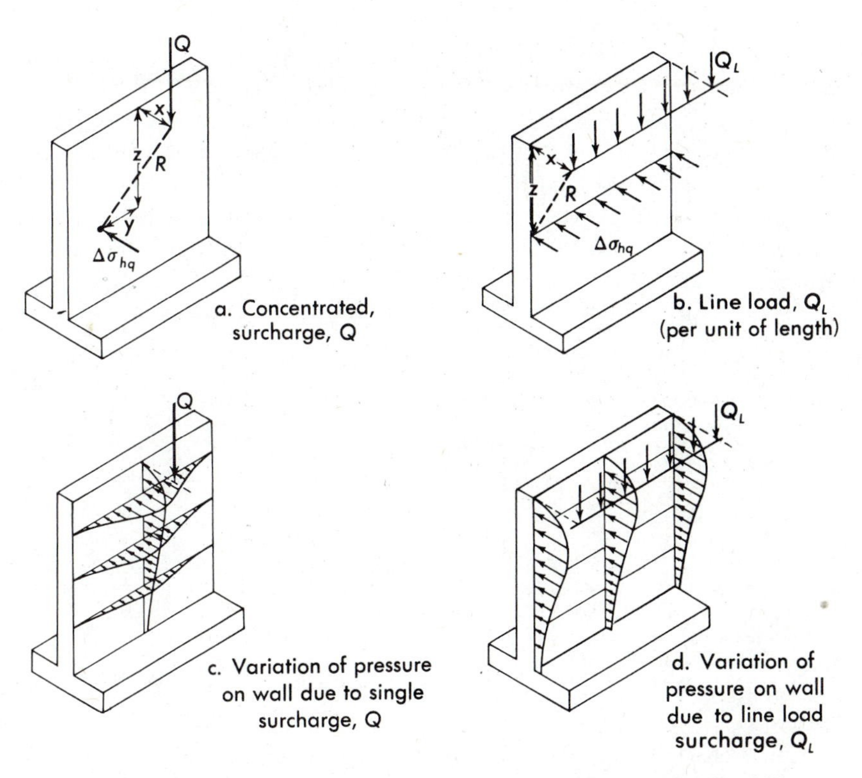

Figure 9.12 *Pressures produced by surcharge concentrations on a level backfill.*

The stress increments above do not consider any restraining effect of the wall on the elastic equilibrium. Tests by Spangler[9:4] indicate that for the active state these stresses should be increased by a factor of about 1.5. For the at-rest condition they should be increased by a factor of 2.

The pressure increase is not uniform along the wall surface. For the concentrated load it varies in both the y and z directions (Fig. 9.12c). The greatest intensity of increase is opposite the load at a depth below the top of the wall of $z = x/2$ approximately. For a continuous line load of Q_L the pressure varies vertically as shown in Fig. 9.12d. The greatest increase is similarly at a depth of $z = x/2$ approximately.

If the surcharge is of great magnitude, the total shear pattern is changed. This effect can be approximated by adding the surcharge load to the backfill weight, utilizing the approximate analysis of plastic equilibrium described in the next section. This approach is also theoretically weak. The real effect is probably between that computed by the Boussinesq or elastic analysis and that found by assuming the surcharge to add to the soil weight in the zone of plastic equilibrium.

COULOMB ANALYSIS / In 1776 the French scientist Coulomb published a theory of earth pressure that includes the effect of wall friction and which can be applied regardless of the slope of the wall or its backfill. He discovered through many observations with dry sand that a retaining wall tilts outward until earth pressure becomes a minimum—the active state. In this condition, the backfill is in a state of shear failure along a series of inclined, parallel, slightly curved surfaces (Fig. 9.13a). The wedge-shaped section of backfill, bounded by the shear surfaces, slides downward and outward as the wall moves outward. Coulomb approximated the shape of the failure wedge by assuming it to be sliding on a plane surface, and derived the active earth pressure from the forces producing equilibrium in the wedge as it commences to move, as shown in Fig. 9.13b.

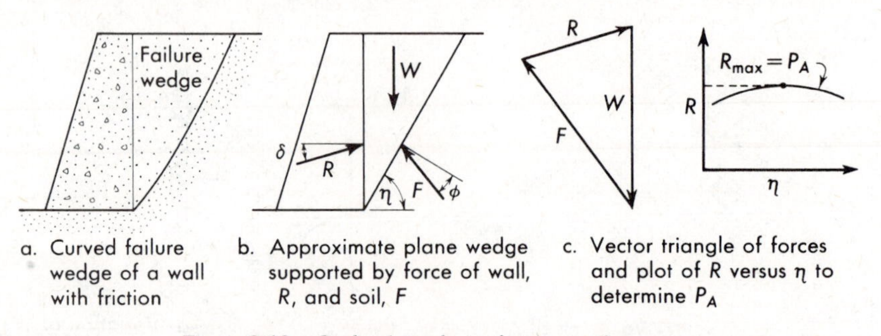

a. Curved failure wedge of a wall with friction

b. Approximate plane wedge supported by force of wall, R, and soil, F

c. Vector triangle of forces and plot of R versus η to determine P_A

Figure 9.13 *Coulomb analysis of active earth pressure.*

The weight W of the wedge of earth is computed by the assumed angle of the failure plane η, the soil weight, and the wall and backfill dimensions.

Its direction is vertical. The resultant force F of the wedge on the soil is inclined at an angle ϕ with the normal to the shear plane, but its magnitude is unknown. The force of the wall R on the wedge is inclined at the angle of wall friction δ to the normal to the wall. Its magnitude is also unknown.

The three forces form a vector triangle (Fig. 9.13c) from which the magnitude of R (and P_A) is obtained graphically. Of course, the correct angle of the failure wedge η is unknown. It is found by computing values of R for several assumed values of η and plotting the results graphically. The peak of the curve represents the critical failure plane, and the maximum R is equal to, but opposite in direction from, P_A. If the soil has cohesion, the cohesive force c along the failure plane is added to the vector polygon. Because of tension cracks, some engineers conservatively omit the shear force resulting from c to a depth of $2c/\gamma$. Others utilize only $\frac{1}{2}c$ in undrained or partially saturated clay analyses. For drained analyses, the drained ϕ_D and the c' multiplied by the length of the failure surface below $2c'/\gamma$ are appropriate for analysis.

A number of graphical shortcuts have been developed for the Coulomb analysis. All reduce to the same essentials: The maximum R, for all values of the failure plane η, is equal and opposite to P_A. An inherent error is the assumption of plane failure; this does not satisfy the equilibrium of moments. The error in active earth pressure is not significant.

The value of the angle of wall friction δ can be found by laboratory friction tests. For smooth concrete it is often $\frac{1}{2}\phi$ to $\frac{2}{3}\phi$, and for rough stone it is equal to ϕ.

The Coulomb analysis gives an active earth pressure equal to the Rankine for a frictionless vertical wall and a level backfill. When ϕ exceeds 0, the resultant pressure computed by the Coulomb method is as much as 10% lower. The location of the resultant is the same.

The analysis can be applied to the case of the sloping backfill or an inclined wall face as easily as for the simple case of the level backfill and vertical wall; only the geometry of the wedge is changed. The effect of a uniform surcharge or a large concentrated surcharge can be estimated by adding the load to that of the soil wedge, but if a concentrated load is beyond the wedge it does not contribute. The plot of R as a function of the wedge angle, η, will include a sharp drop at the angle where the surcharge is no longer within the wedge.

For passive pressure, the curvature of the failure surface increases greatly with increasing values of δ (Fig. 9.14a). Furthermore, a thrust load against the wall in the upward direction could develop a reversed or negative value of δ, as shown in Fig. 9.14c, and a completely different shear surface. The Coulomb analysis with its assumed straight line shear surface can be seriously in error in computing passive pressure. For the more common case of passive pressure with a downward component of friction between the wall and the soil (Fig. 9.14a), the magnitude of the real resultant of passive

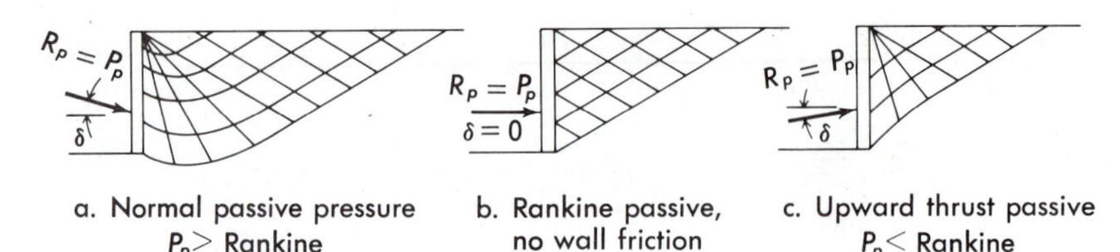

Figure 9.14 *Shear patterns in passive earth pressure on walls with friction.*

pressure lies between the Rankine and Coulomb values. More rigorous solutions involving curved failure surfaces should be used for computing passive pressure when δ exceeds about $\frac{1}{4}\phi$.[9:1,9:2]

LATERAL PRESSURE IN ROCK / Because of the high permanent cohesion, the lateral pressure developed by an intact mass of rock is negligible. According to Equations (9:6) and (9:7) tension would be required to produce shear failure, and the net force will be zero for faces of great height in intact rock of even moderate strength.

Intact rock, however, seldom exists. The pressure exerted by rock is largely controlled by the joints, bedding planes, and water pressure. The lateral pressure exerted by a rock mass can be found by a simplified Coulomb's wedge (Fig. 9.15). The potential shear surface in three dimensions is defined by joints or bedding planes whose angles can be measured. The shear characteristic along such surfaces is established by laboratory tests made on large samples that include the potential surface of weakness oriented in the most favorable angle for failure. Large-scale field tests for shear on the suspected plane of weakness are more realistic than the laboratory tests and should be made for important structures whose cost justifies the expense.

Groundwater pressure must be included in the analysis if there is seepage through the joints and bedding planes. Even horizontally bedded rock can develop high lateral pressure from water in joints (Fig. 9.15c).

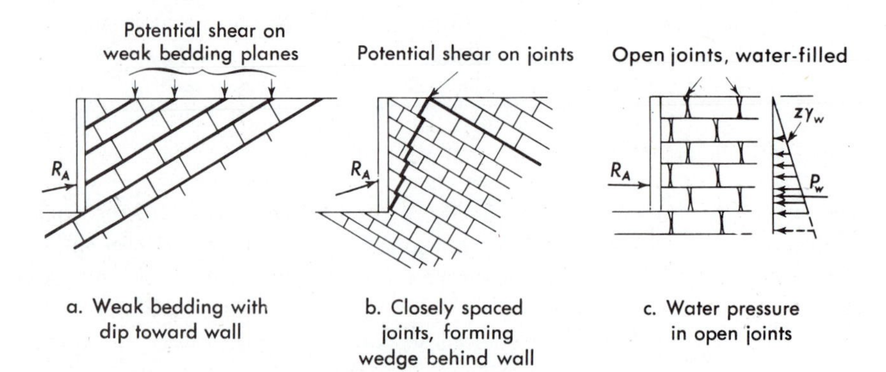

Figure 9.15 *Pressures developed in rock masses.*

Because of the uncertain water pressures and the variability of strengths along joints and bedding planes, pressure computations for rock masses are approximate at best. Moreover, three-dimensional solutions are required for intersecting, sloping joints.

9:4 Retaining Walls

A retaining wall is a permanent, relatively rigid structure of cribbing, masonry, or concrete that supports a mass of soil. It substitutes the steep face of the wall for the gentle natural slope of the earth to provide usable space in highway and railroad cuts, in and around buildings, and in structures below ground level.

The wide use of retaining walls is accompanied by many failures and partial failures because of designs based on rules and formulas that fit only limited conditions. For example, the design of walls backfilled with soft clay is often based on analyses that apply only to sand, and the design of walls that support structures that will be cracked by foundation movement is often based on the active earth pressure that requires the wall to tilt outward. The only margin between success and failure in many cases has been an overgenerous safety factor.

REQUIREMENTS FOR DESIGN / A satisfactory retaining wall must meet the following requirements:

1. The wall is structurally capable of withstanding the earth pressure applied to it.
2. The foundation of the wall is capable of supporting both the weight of the wall and the force resulting from the earth pressure acting upon it without:
   a. Overturning or soil failure.
   b. Sliding of the wall and foundation.
   c. Undue settlement.

The earth pressure against a retaining wall depends on the deformation conditions or tilt of the wall, the properties of the soil, and the water conditions. For greatest economy, retaining walls are ordinarily designed for active pressure as developed by a dry cohesionless backfill; but, if necessary, a design can be developed for any conditions of yield, soil, and water.

Organizations such as state highway departments and railroads, who design many retaining walls, have developed charts and tables that give earth pressures with a minimum of computation. Nearly all such charts and tables are based on the Rankine formula with assumed values for ϕ and γ of the backfill soil. Some designers use the hydrostatic pressure developed by an imaginary fluid whose unit weight γ_f is termed the *equivalent unit weight.*[9:5] This is a modification of the Rankine formula, where $\gamma_f = K_A\gamma$. These simple approaches are no more reliable than the assumed soil parameters on which they are based: They mislead designers into believing they have made a refined design. Moreover, some designers become so intrigued with manipulation of such charts that they fail to consider the most important aspects of design: allowance for tilt, drainage, and adequate foundations.

RETAINING WALL TILT / The earth pressure must be compatible with the wall tilt, which is limited by the rigidity of the wall, the foundation, and any connections with adjoining structures. The tilt required for active pressure is given in Table 9:2. The designer generally allows 10 mm of tilt per meter of wall height ($0.01H$) unless he or she is certain of the backfill quality and its installation. Ordinarily, tilting stops within a month after the backfill is placed, although small movements sometimes continue in loose sands subject to vibration and in clays. For isolated straight walls on a soil foundation such tilting is no problem. For curved walls, long walls, or walls of widely varying height, it is necessary to provide joints that will permit movement. These can be sealed after the tilt stops or provided with flexible joints.

Walls on piles are somewhat restrained, and the earth pressure is somewhat greater than the active pressure. For walls smaller than 10 m (33 ft) high, the difference is usually within the accuracy of the earth pressure computations. For larger walls supported on batter piles, the effect of restraint should be investigated or the pressure increased about 20% above the active.

The foundations of walls resting on bedrock cannot deflect. If the wall itself is flexible, like a thin reinforced cantilever, it will probably deflect enough to establish active or near-active conditions. A massive wall, however, cannot tilt and must be designed for at-rest pressure. A cushion of sand between the wall foundation and the rock permits some movement and can result in a lower pressure on the wall.

The tilt is often limited by adjoining structures. If construction of such structures is delayed until after the wall is backfilled, little trouble should develop. Building walls supported on retaining walls often suffer from cracking, owing to retaining wall tilt. In one case, the anchor bolts for columns resting on a 9 m (30 ft) high wall were 50 mm (2 in.) out of line after the wall was backfilled; in another case, a brick wall on top of a retaining wall was split apart by the tilting.

DRAINAGE PROVISIONS / The greatest proportion of failures of walls higher than 3 m (10 ft) is caused by water pressure in a backfill that was assumed by the designer to remain dry. The most important single consideration in wall design is insuring good drainage. There are two approaches:

1. Remove water from backfill.
2. Keep water out of backfill.

In all cases the first method should be used, and in some cases both are incorporated in design.

The simplest water control measure is a weep hole (Fig. 9.16a). They are 1.5 to 3 m (4 to 10 ft) apart horizontally and vertically and 75 to 100 mm (3 to 4 in.) in diameter to permit easy cleaning. Unless the backfill is coarse gravel, a filter system is essential to prevent erosion of the backfill and clogging the weep hole. Filter fabrics or a few shovels of peagravel over the hole inlet (Fig. 9.16b) will be adequate if the backfill is a well-graded sand. Weep holes have the advantage of easy maintenance, demonstrating their function during wet weather. They have the disadvantage of discharging the water at the base of the wall where the foundation pressures are greatest.

More elaborate drains consist of 0.15 to 0.2 m (6 to 8 in.) perforated pipe laid in a filter trench at the wall base (Fig. 9.16c). A blanket drain against the wall or an inclined drain between the backfill and the original ground (Fig. 9.16d) are used with backfill soils that drain slowly.

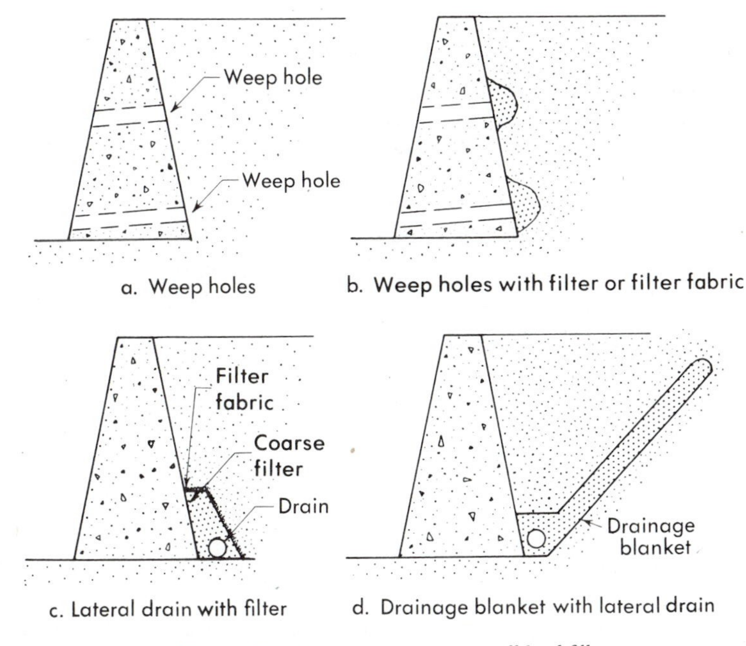

Figure 9.16 *Drains for retaining wall backfills.*

If it is necessary to use highly plastic clay (that could swell) or soils that are difficult to drain, water must be kept from the backfill. In one extreme case, a plastic clay was encapsulated in a water- and vapor-tight membrane. Surface drains should be provided to keep water from the inevitable crack between the wall and the backfill, and surface paving can minimize infiltration.

PREVENTING FROST ACTION / In northern climates frost action has caused many retaining walls to move so far that they have become useless. Because stone and concrete are relatively good conductors of heat, the temperature along the back side of the wall is close to that of the air. When freezing temperatures prevail and the backfill soil is susceptible to frost action, ice lenses form parallel to the wall. If the drains freeze (as often occurs with weep holes), a plentiful source of water will help generate dramatic frost lensing and large continuing movements, up to 1 m (3 ft) in a season. When the lenses thaw, the released water can exert enough additional pressure on the wall to cause it to overturn.

Frost action can be prevented by substituting a thick blanket of relatively coarse, cohesionless soil, such as sand or gravel, for the portion of the backfill adjacent to the wall. This blanket should be as thick as the depth of frost penetration in the region. It can be constructed by dumping small loads of sand or gravel against the wall as the backfill is being placed.

Drainage is the key to preventing catastrophic frost damage. It is incorporated into any sand-gravel backfill. Preferably, the drain should be protected so that it will not be blocked by freezing; deep lateral drain pipes are preferred to weep holes.

BACKFILL MATERIAL / The best backfill is rigid, free-draining, and with a high angle of internal friction, so as to develop minimum earth pressure with the least movement. Table 9:3 rates the soils of the Unified Classification for selection.

TABLE 9:3 / RETAINING WALL BACKFILL

GW, SW, GP, SP	Excellent, free-draining backfill.
GM, GC, SM, SC	Good if kept dry but require good drainage. May be subject to some frost action.
ML	Satisfactory if kept dry but requires good drainage. Subject to frost. Neglect any cohesion in design.
CL, MH, OL	Poor. Must be kept dry. Subject to frost. Wall deflection likely to be large and progressive at-rest pressure is used.
CH, OH	Should not be used for backfill because of swelling.
Pt	Should not be used.

Lightweight artificial materials such as expanded shale and crushed slag often make good backfill. All the cohesionless backfills are best when well compacted because the higher internal friction angle and the resistance to vibration offset the higher weight.

A wedge-shaped backfill of sand, gravel, or slag at least 50% wider than the failure wedge makes it possible to design the wall for the low pressure of a cohesionless soil, even though the remainder of the backfill is clay.

RETAINING WALL DESIGN / The design of a retaining wall is based on the materials available, appearance, the space required, the forces acting, and finally, cost. The materials for wall construction are stone masonry, plain concrete, reinforced concrete, steel and earth or broken stone. The choice is based partly on appearance. Walls used in conjunction with stone-faced buildings or in residential areas and parks are often of brick or stone masonry. Walls in industrial areas or adjacent to bridges and dams are usually concrete. Cost and availability of materials and labor are important factors in the choice of wall materials. Stone masonry is expensive in the United States and requires skilled workmen; plain concrete is relatively easy to form and requires no steel but may use excessively large quantities of concrete; reinforced concrete is economical for large structures but requires accurate fabrication of the steel and forms and controlled-quality concrete.

Space is an important factor in wall design, since the function of retaining walls is to make more usable, level space than a natural slope will provide. Walls should not be designed with vertical faces because the inevitable slight tipping will cause them to lean outward and appear unstable, even though they actually may be quite safe. To give the appearance of stability, it is better to provide the face of the wall with an inward *batter* or slope of at least $1(H)$ to $10(V)$.

Gravity walls (Figs. 9.11 and 9.16) resist earth pressure by their weight. They are constructed of stone and concrete that resists compression and shear but no appreciable tension; therefore, structural design is mainly concerned with preventing tension. Tentative dimensions are selected: a top width of 0.3 to 0.6 m (1 to 2 ft) and a bottom width of about 40% of the height are typical. Sections are taken through the wall at the base and at one or two intermediate levels. The resultant of all the forces acting above the section, including the resultant of earth pressure, the weight of the wall, and any load acting on the top of the wall, pass through the middle third of the section to avoid tension.

Two types of monolithic reinforced concrete walls are used: *cantilever* (Fig. 9.17a) for heights typically to 12 m (40 ft) and *buttress* or *counterfort* (Fig. 9.17b) for heights greater than 6 m (20 ft). The cantilever wall is a vertical cantilever beam loaded with pressures that increase from the upper or outer end to the restrained or bottom end. Variable vertical reinforcement provides resistance to bending; nominal horizontal reinforcement prevents cracking. The bending of the wall plus deflection of a pile or soil foundation is usually sufficient to establish active earth pressure. With a rock foundation, wall bending alone may not establish the active state. The deflection should be computed and if it does not approach that given in Table 9:2, a larger earth pressure or provisions for foundation movement should be incorporated in design.

The counterfort or buttress wall consists of a vertical slab supported on two sides by *counterforts* in tension on the backfill side or *buttresses* in compression on the outside face. The structural design usually ignores the support of the wall base by the foundation. There is seldom enough wall bending to establish active earth pressure; therefore, active pressure design

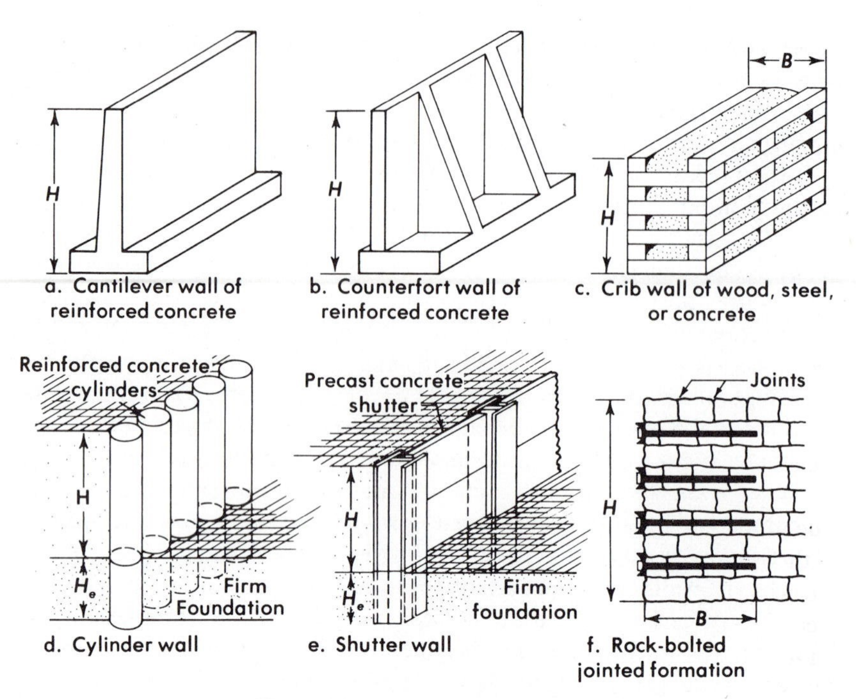

Figure 9.17 *Types of retaining walls.*

must incorporate provisions for foundation deflection that permit movements compatible with Table 9:2.

The crib retaining wall (Fig. 9.17c) consists of a hollow rectangular cribwork of logs, timbers, reinforced concrete beams or steel beams filled with soil or rock. The cribwork can be vertical or tilted toward the backfill for greater stability. Crib walls are relatively cheap and are usually flexible enough to be used where settlement is a serious problem. Structurally, the crib is a gravity wall and is designed so that its width is sufficient to keep the resultant forces within the middle third. In addition the shear at any cross section must not exceed the shear strength of the soil. The deflection is sufficient for active design pressure.

Baskets or containers formed of wire mesh and filled with gravel or crushed stone, termed *gabions*, are a form of gravity wall. These depend on the shear of the fill for internal stability, and their mass to resist the backfill loads.

Vertical reinforced concrete cylinders 0.4 to 1 m (16 to 40 in.) in diameter in line, touching one another, can form a simple cantilever wall (Fig. 9.17d). They are formed by augering holes in the ground, placing reinforcement, and then pressure grouting or concreting the cylinder in place. The backfill is the existing ground; the open face is constructed by excavation. They are best suited to free-draining soils underlain by a firm foundation that resists overturning and vertical loads by the embedment, H_e. Although drainpipes cannot be placed in such a system, the cracks between cylinders act as weep holes. The system has the disadvantage that misalignment of cylinders can lead to soil erosion or raveling through the wall.

Shutter walls (Fig. 9.17e) are formed by vertical cantilever beams, such as driven H-piles or H-shaped precast concrete beams embedded to a depth of H_e in auger holes. Horizontal slabs (shutters) of precast concrete are placed between the vertical members, supported in the slots formed between the H-sections. The wall is then backfilled as any other retaining wall. The cantilever bending is ordinarily sufficient to establish active earth pressure. Curved shutters, arched against the fill, provide a pleasing appearance. Terraced shutter walls, set back 1 m ($3\frac{1}{2}$ ft) for each 3 to 4 m (10 to 14 ft) of height have been used to minimize a high, sheer face in park areas.

ROCK SUPPORT / Retaining walls for intact rock masses are not true retaining structures. Instead they are facings to prevent the weathering of the rock, to prevent the local spalling of loose fragments, or to minimize erosion of a soft rock below a dam spillway or hydro plant tailrace. The facing is anchored to the rock by rock bolts (see Section 9.7) or by reinforcing bars grouted into the rock.

Walls to support rock pressures developed in masses with joints and bedding planes can be designed in the same way as walls that support soil, once the resultant force has been computed.

Sometimes the rock mass can be made to support itself as a gravity-retaining wall. The rock blocks formed by the joints are tied together by reinforcing rods or rock bolts through the mass and by bearing plates, or beams along the rock face, as shown in Fig. 9.17f. This creates an intact body similar to a crib or gravity wall that resists the pressure from the remainder of the mass by gravity. The design requires a detailed study of the joint patterns so that the blocks are securely restrained to form a unit that cannot separate.

REINFORCED EARTH / One of the more innovative developments in retaining structures has been *reinforced earth* (Fig. 9.18). The system consists of thin strips of metal at right angles to the wall face that provides tensile strength to the backfill and which also supports the *skin* or outer face of steel or precast concrete. It resembles a metal crib wall, but its behavior is far more complex and not fully understood. A reasonable rationale for design, however, was developed by K. Lee et al.[9:6]

The metal strips (or even plastic cloth) are laid horizontally (or sloping down to the outside a few degrees) with a spacing of S (center to center), sandwiched between backfill layers of thickness ΔH. Typical dimension are $S = 0.7$ m or 2 ft and $\Delta H = 0.25$ m to 0.3 m (10 to 12 in). Design requires that the skin resist the soil pressure from the layer, that the strip length, L, be great enough to support the skin and provide a stable mass, and that the strip be strong enough to resist the tension in it.

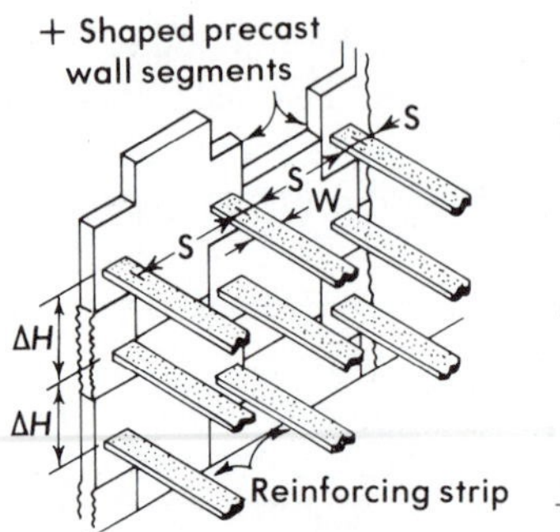

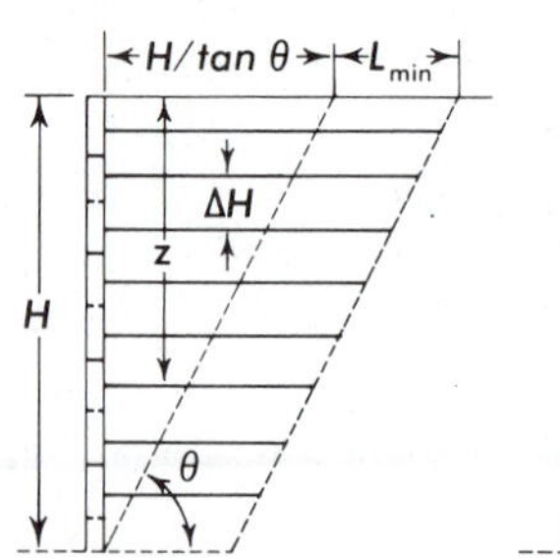

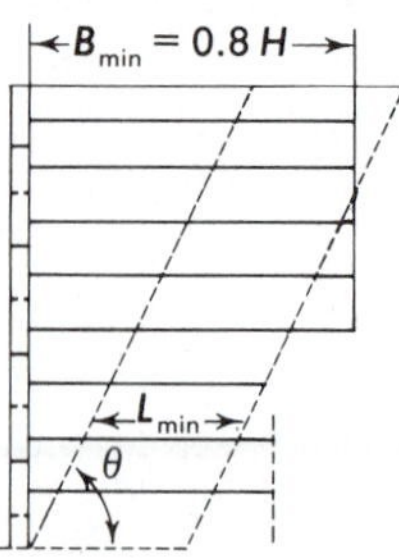

a. Reinforcing strip width and spacing behind + segmented precast wall b. Minimum reinforcing length, L_{min} c. Reinforcement for overall stability, $L_{avg} > L_{min}$

Figure 9.18 *Reinforced earth wall.*

At a depth of z below the wall top, the force, P, against the skin element defined by S and ΔH, with an earth pressure coefficient, K

$$P = K\gamma z S\,\Delta H \tag{9:19a}$$

The friction force, F, developed by the top and bottom faces of the metal strip of width W, length L, and angle of friction with the backfill of δ is

$$F = 2LW\gamma z \tan\delta \tag{9:19b}$$

The required length L_{min} is found by multiplying P by an appropriate safety factor F_s, (usually 1.5 or 2) and equating the expressions

$$F_sP = F$$

$$F_sK\gamma zS\,\Delta H = 2LW\gamma z \tan\delta$$

$$L_{min} = \frac{F_sKS\,\Delta H}{2W\tan\delta} \tag{9:19c}$$

It is significant that L_{min} is constant regardless of z. The strip thickness is computed from P [Equation (9:19a)], the width, W, usually about 75 mm (3 in.), and the metal working stress with allowances for corrosion. The wall skin structural design is also found from P and the dimensions ΔH and S.

The length, L, is considered to be measured beyond the zone of Rankine failure, as shown in Fig. 9.18. For overall stability a top width $B = 0.8H$ has been expedient (Fig. 9.18c). The uppermost strips may be shorter than L_{min}, but as long as the average L exceeds L_{min}, the wall will be satisfactory. Sometimes the length of the lower strips is made less than $0.8H$, but exceeding L_{min} (Fig. 9.18c).

When backfilled with compacted sand, experience has shown that movement is sufficient that $K = K_A$. For other backfills, the data are insufficient; for clays K probably approaches K_0 (although clay backfills are discouraged). Reinforced earth walls have been economical for many applications where good cohesionless backfill materials are available.

WALL FOUNDATION / Faulty foundations are a frequent cause of retaining wall failures. The combination of vertical and horizontal forces that are supported are not always considered in design. Furthermore, it has been pointed out that the deflection of the foundation can be a major factor in the magnitude of the earth pressure acting against the wall. The most economical wall design presumes active earth pressure, which requires movement of the wall. By way of contrast, a rigid wall on an unyielding foundation must resist at-rest pressure.

The design of foundations to resist combined vertical and lateral pressure is considered in Chapters 10 and 11. Generally, a spread footing design requires that the resultant force of the earth pressure, wall weight, and foundation weight pass through the middle third of the foundation width.

If the wall is supported on rock or very hard soil, the safety against overturning should be determined. Moments of earth pressure about the outside corner of the bottom of the foundation, the foundation *toe*, must be resisted by the moment of weight of the wall, foundation, and any vertical component of the backfill. The ratio of the resisting to the overturning moments should be at least 1.5.

If the foundation is very weak, the weight of the backfill plus the wall can cause a large-scale shear failure involving both the wall and the backfill (Fig. 9.19a). If there are external loads on the backfill, such as railroad tracks or structures, both the pressure against the wall as well as the load on the foundation are increased. A *relieving platform* supported on piles (Fig. 9.19b) can carry part of the backfill load. Platforms are particularly useful in wharf construction, where the foundation soils are likely to be poor and the loads on the backfill are as high as 50 kN/m^2 (1 kip/ft^2). Methods for analyzing the total stability of the mass are discussed in Chapter 12.

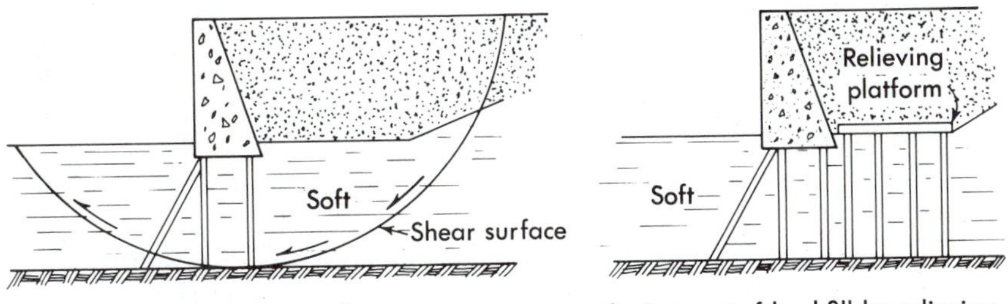

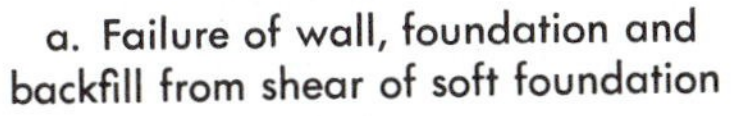

a. Failure of wall, foundation and backfill from shear of soft foundation

b. Support of backfill by relieving platform on piles

Figure 9.19 *Total wall stability.*

9:5 Excavation Bracing[9:7, 9:8, 9:9]

In many construction jobs deep excavations must be made before the structure can be built. Sometimes (unfortunately) the planning of the excavation is left to the excavation superintendent or to a shovel operator. When expensive excavations or those that may endanger lives or adjacent property are involved, however, bracing to support the soil must be designed as in any other important structure.

Bracing design involves many different factors; most of which are related to construction requirements and procedures, the hazards to nearby structures, as well as soil, rock, and ground water. Most are interdependent: the design earth pressure is a function of soil or rock strength and rigidity, ground water, and its changes produced by excavation, the soil movement tolerated by the surroundings, and construction procedure. The construction procedure, in turn, depends on the earth pressure, the tolerable soil movement and the excavation requirements. The designer and constructor must work together in both design and construction stages, with knowledge of conditions within and adjacent to the site.

Open excavations are those that require no bracing to support the soil or to control ground water. The soil is cut to the steepest slope on which it will safely stand, usually 1.5(H) to 1 (V) for sandy soils, and up to vertical slopes for shallow excavations in stiff clay or decomposed rock. The proper slopes are usually determined by trial and error or from past experience in similar soils, but are also governed by building and other safety codes. The deeper the excavation and the weaker the soil, the flatter are the slopes required. Flat slopes, small size, and great depths all require considerable excavation beyond that actually needed for the structure; therefore open excavations are usually limited to strong soils, large areas, or shallow depths [under 6 m or (20 ft)]. For analysis of stability and the safe slopes required in open cuts see Chapter 12.

CUT BRACING / When it is uneconomical, illegal, or impossible to use open excavations, bracing is employed to support the soil. Many systems have been devised, and some have even been standardized and prefabricated by organizations (such as sewer contractors) that must do considerable trench excavation work. Unfortunately, however, some excavation contractors spend little or no time designing their bracing, and as a result there have been numerous accidents. Almost every year workmen are crushed to death by the failures of inadequately designed bracing, which occur even in shallow excavations.

The simplest bracing is the strut (Fig. 9.20b), a horizontal timber whose ends are jacked against the soil. Struts are commonly applied in shallow cuts in cohesive soils where the soil can stand unbraced for a short time. The struts, applied near the top of the cut, relieve the tension in the soil that occurs above the depth $z = 2c/\gamma$ and prevent the formation of tension cracks

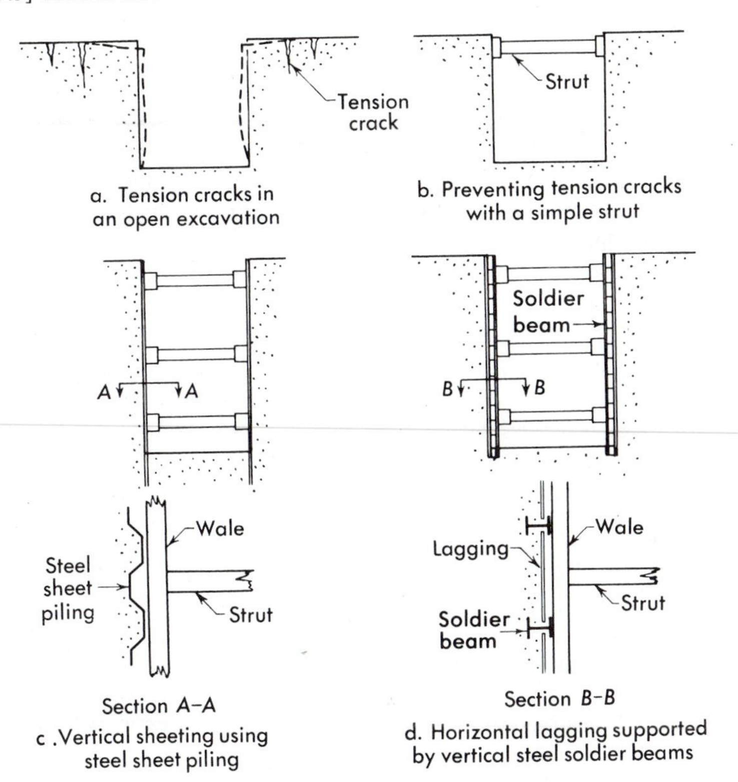

Figure 9.20 *Methods for bracing an excavation.*

(Fig. 9.20a) that would lead to a collapse of the walls of the excavation. If two rows of struts are required, they may be set against vertical timbers known as *soldier beams.*

When more extensive bracing is required, three methods are used.

1. Vertical sheeting.
2. Horizontal lagging.
3. Cast-in-place wall.

When the soil is very soft and runny, the time-honored method of *vertical sheeting* is employed, as shown in Fig. 9.20c. Wood or steel sheet piling is driven along the line of the excavation before the soil is removed. As excavation proceeds, horizontal members known as *wales* are placed along the inside of the excavation to support the sheeting and are braced with struts.

If the depth of excavation is greater than the length of the sheeting, a second row of sheeting is driven inside the first after excavation has extended near the bottom of the first row.

When the soil is not runny, horizontal lagging (Fig. 9.20d) may be used. If the cut will stand without bracing for a few hours, it is excavated without supports, and then horizontal boards known as *lagging* are placed against the soil. These are held in place by vertical *soldier beams* supported by struts. If the soil requires support at all times but does not run, the soldier beams, consisting of steel H-sections are driven into the soil, and the lagging is wedged tightly against the soil to prevent excessive movement. The soldier beams are supported by a wale and strut system in the same manner as vertical sheeting.

Cast-in-place concrete retaining walls can be installed where obstructions prevent pile driving or where the shock and vibration of pile driving are objectionable. One form consists of overlapping concrete cylinders (Fig. 9.21). These can be constructed in two ways, depending on the strength of the soil and on ground water. In firm clays above the water table a hole 0.4 m (16 in.) or more is drilled with a large auger. A reinforcing cage is inserted, and the hole filled with concrete. In less stable soils, the hole is drilled but kept filled with soil or mud during the drilling process. Cement mortar is injected through the hollow stem of the drill from the bottom of the hole up as the drill is withdrawn, so that the mortar replaces the soil and there is never a moment that the walls of the hole are not supported. (The same

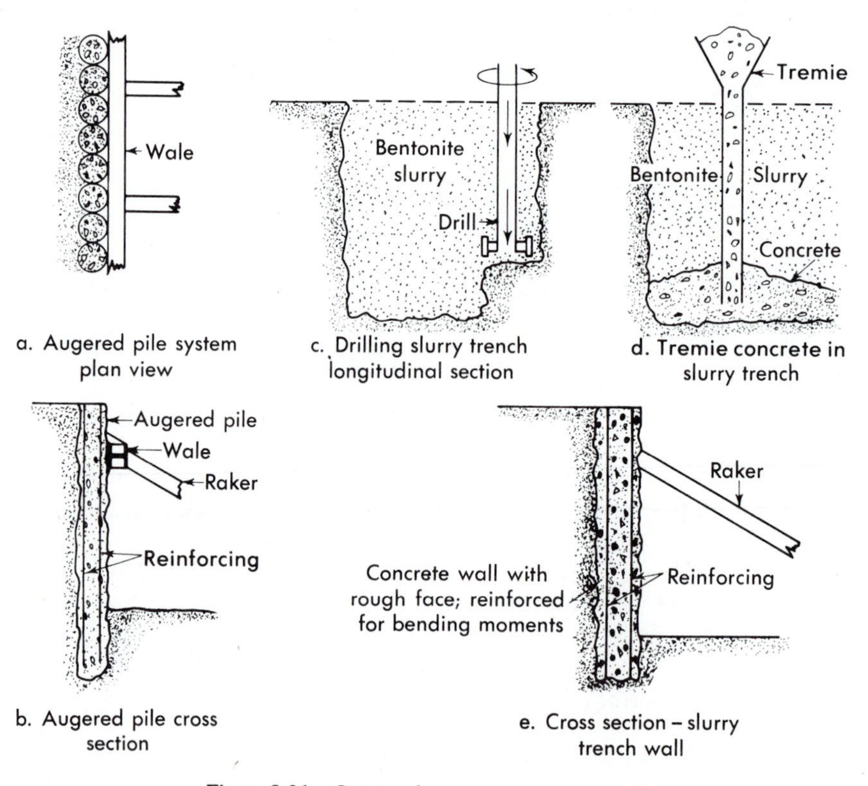

Figure 9.21 *Cast-in-place concrete bracing walls.*

method applied to foundations is described in Chapter 11.) Reinforcing steel is set by placing it through the hollow auger stem or by forcing it down through the mortar; however, accurate placement is impossible. Wales are placed against the cylinders to support the wall as required.

The second wall form uses a trench filled with a slurry of clay and water to support the soil during construction. The *slurry trench* is excavated in short sections, 2 to 10 m or (6 to 33 ft), and with an 0.8 to 1.5 m or (2.5 to 5 ft) wall thickness, by a variety of drilling or digging methods, depending on the soils and the equipment available. The essential feature of the process is that the excavation is kept full of a viscous mixture of water, clay (usually bentonite), and soil that supports the earth by fluid pressure. Gravel and even boulders can be removed by special clamshell buckets, and sand can be pumped out by circulating the slurry. After excavation, prefabricated reinforcing steel is lowered through the slurry and positioned by rollers or spacers against the walls of the hole. Concrete is tremied into the trench from the bottom up, displacing the slurry which is stored for reuse. Slurry trenches as deep as 30 m (100 ft) have been constructed in a variety of soils. Generally, the resulting wall is uniform enough that it can be used as a form for the permanent wall and sometimes as the permanent wall with only a thin facing.

SUPPORT SYSTEM / The bracing for very shallow excavations is supported by using the sheeting, soldier piles or cast-in-place walls as cantilevers. Excavations deeper than 3 m (10 ft) generally require support systems (Fig. 9.22).

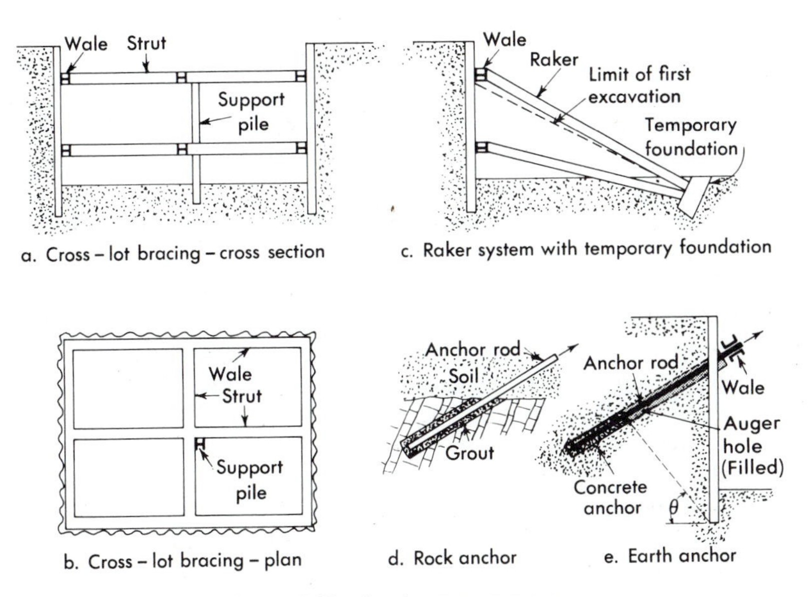

Figure 9.22 *Bracing support systems.*

Narrow excavations, such as trenches or small building excavations can be supported by horizontal columns or struts (Fig. 9.22a and b). If the excavation is wide, the system is tied together in both the vertical and horizontal directions to reduce the slenderness ratio (L/r) of the struts and to minimize buckling of a strut if excavating equipment should strike it. Sometimes the struts and vertical supports are tied together with diagonal members in the vertical plane to form trusses. In this way the upper struts can be used to support construction equipment, and the excavation bottom can be free of obstructions.

Cross-lot bracing obstructs the site, and if the excavation width is several times the depth, a system of diagonal supports or *rakers* is convenient (Fig. 9.22c). The excavation is completed to final grade at the center, leaving sloping banks to support the sheeting or soldier piles, as shown by the dotted line (Fig. 9.22c). The raker is then set in-place, reacting against a special foundation or against a completed portion of the foundation of the permanent structure. The next increment of excavation is made to the level of the second wale and raker, and these are installed. If the soil is very weak, and requires a very flat slope for support, the rakers are placed in trenches, leaving the greater part of the bank intact between them. Generally, the raker angles are not steeper than 35° with the horizontal to minimize the upward component of thrust on the sheet piles or soldier piles.

Anchors or *tiebacks* (Fig. 9.22d) eliminate obstructions in the excavation inherent in raker or struts. They consist of rods that extend well beyond any potential failure surface into firm undisturbed soil or rock. A number of systems are employed—some with high tensile cables grouted into rock and prestressed against a wale, and others utilizing ordinary reinforcing steel. Anchors produce a downward load on the bracing wall, which must be resisted by friction between the soil and wall or by the wall foundation. The design and construction of anchors is discussed in Chapter 11.

DESIGN OF A BRACING SYSTEM / A bracing system is a temporary structure that usually is removed when the job is completed. Actually, it is a dam to keep water and soil from the building site in order that construction can proceed in the dry, and consequently is often called a *land cofferdam* or simply a *cofferdam*. The latter term, however, is more often applied to temporary dams in open water. In most situations, since safety, ease of construction, and convenience are the most important considerations, and economy of materials is less important, refined, accurate methods of analysis are seldom justified. An understanding of the nature of earth pressure against bracing is necessary, however, for even the most approximate design.

DEFORMATION AND PRESSURE / The earth pressure on the bracing system depends on the soil rigidity and strength and the amount of deformation or yield of the bracing. Unlike the retaining wall, which is structurally a rigid unit against which the earth is placed after construction is completed, the bracing system is somewhat flexible; moreover, it supports earth pressure while it is being constructed. The result is irregular deformation and erratic variations in earth pressure with depth that cannot be calculated by simple theories.

The stresses on an element of intact soil before excavation (Fig. 9.23) are drastically changed by excavation. The lateral stress, σ_h, is reduced, and the element bulges outward and subsides vertically. The combined effect is bulging in the lower part and subsidence in the upper part of the bank, as shown in Fig. 9.23a. The soil is in tension near the surface, because of the downward and outward movement, and tension cracks may form. The crack location is typically between 0.4 and 0.7 times the face height, H, from the top. The first cracks to appear are close to the top; subsequent cracks are progressively further away.

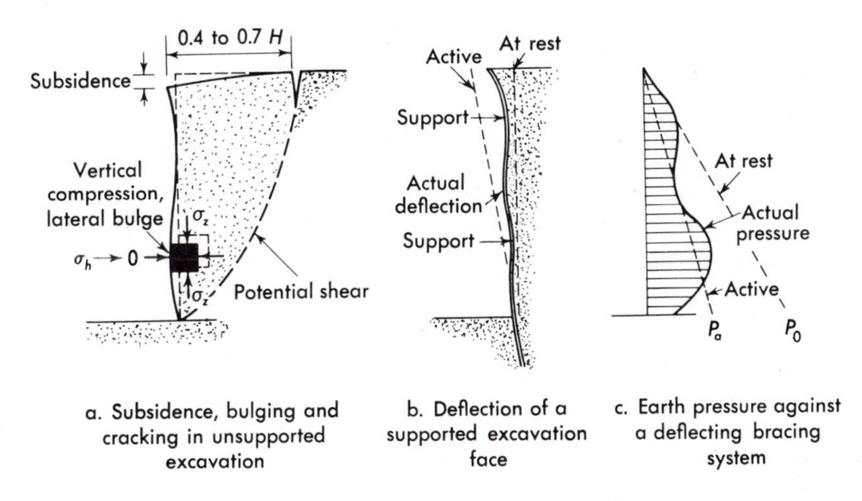

Figure 9.23 *Deflection and earth pressure against a bracing system.*

The excavation bracing system restrains the bulge, keeps the crack closed, and minimizes the surface subsidence. If the bracing system is so rigid that no lateral deformation is possible, the earth pressure will be at rest (Fig. 9.23c). Usually, the system deflects; this alters the total force and the pressure distribution. Excavation to the first support level allows the bracing to tilt (Fig. 9.23b) and the pressure to drop toward the active (Fig. 9.23c). The first support prevents appreciable further deflection at that point. Deeper excavation allows the system to bulge below the support and the earth pressure to drop. The pressure at the support increases correspondingly because the load is transferred by horizontal shear from the bulging zone. At the excavation bottom, the soil is restrained from bulging by the horizontal shear within the mass, and the pressure against the bracing is reduced.

The earth pressure diagram is irregular. The resultant force is somewhat more than the active, and less than the at-rest; the location of the resultant is above the third point associated with the active pressure in a cohesionless soil. Pressures on actual systems computed from strut loads and the bending moments in the vertical sheet piles or soldier piles, as well as finite element analyses, have generally verified the form of pressure diagram shown in Fig. 9.23c. The actual pressure, however, varies considerably from one point to another because of the differences in construction sequence and the displacements of the support system.

DESIGN EARTH PRESSURE FOR BRACING / Two approaches can be made to finding the design pressure for excavation bracing. The finite element analysis appears to be more rational; however, because the potential deflection of the system (a major factor in earth pressure) is not known until the design is complete, several cycles of trial and revision or design iteration will be necessary. The results may not be realistic because of the uncertainties in soil constants and the unknowns of bracing installation procedure and care.

The more realistic approach is to assume a simple earth-pressure distribution that envelopes much of the observed data. Several of such procedures have been proposed, and even revised slightly, by Terzaghi and

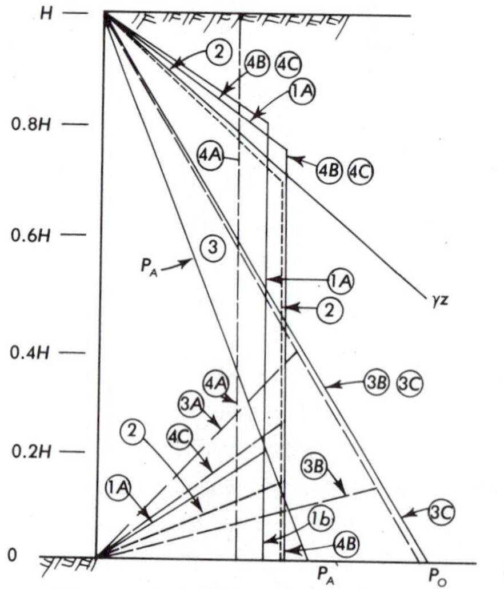

a. Composite of design recommendations after Terzaghi and Peck 1949[9.2] 1967[9.5] and Tschebotarioff, 1951[9.9]

b. Simplified design pressure P_D = design force/unit wall length

1A T & P '49 Md Sa	3A Tschebotarioff '51 Stf Cl	4A T & P '67 Sand
1B T & P '49 Lse Sa	3B Tschebotarioff '51 Fm Cl	4B T & P '67 So to fm Cl
2C T & P '49 So to fm Cl	3C Tschebotarioff '51 So Cl	4C T & P '67 Stf fsr Cl

Figure 9.24 *Earth pressure design for excavation bracing.*

Peck[9:8] and Tschebotarioff.[9:9] However, when all are normalized to permit comparison, they are remarkably similar (Fig. 9.24a).

It has been the author's experience that far too much attention has been paid to minute differences in design pressure, diverting effort from the more important tasks of specifying construction procedures compatible with the design assumptions and inspecting the work for compliance. Therefore, one design pressure diagram is recommended (Fig. 9.24b). It is a simple composite of the numerous others. The diagram has two breakpoints: the upper, $0.2H$ below the top and the lower, $0.1H$ above the bottom. If the upper or lower support points are within 1 m ($3\frac{1}{2}$ ft) of those levels, the breakpoint could be at the supports for ease in computation. The resultant force of the diagram, P_D, per unit length of wall length, is given in Table 9:4.

TABLE 9:4 / RESULTANT OF EARTH PRESSURE FOR EXCAVATION BRACING DESIGN

Soil	Design Resultant Force (Per Unit of Wall Length)
Loose sand, gravel	$P_D = 1.4P_A$
Dense sand, gravel	$P_D = 1.3P_A$
Soft clay	$P_D = 1.5P_A$ or P_0 (larger of the two)
Stiff clay	$P_D = 1.4P_A$ or $0.9P_0$ (larger of the two)
Partially saturated clay	$P_D = 1.3P_A$ or $0.8P_0$ (larger of the two)

The resultant can be computed by the Rankine approach or even Coulomb (with the effective c' one-half the measured value to allow for possible tension cracking in undrained shear). For long-term stability, drained shear parameters should be used; for short terms, undrained. These pressures assume some wall movement, but not enough to establish the active state.

$$p_{\max} = \frac{1.2P_D}{H} \tag{9:20a}$$

$$p_{\min} = \frac{0.6P_D}{H} \tag{9:20b}$$

Below the groundwater level, full water pressure, u, would be added to the design pressure in sand and gravel. It would also be added in clays if drained strength is used for P_A or drained K_0 for P_0.

COMPONENT DESIGN / The moments in structural members subject to bending are computed as if the members were simple beams between support points. Because yield in bending rarely endangers a bracing system, the working stresses in major members, such as steel sheeting or soldier piles, can be 10 to 20% higher than those used for permanent structures. For minor members in flexure, such as wood lagging, working stresses are 2 to 3 times the usual allowable, unless failure of a member would cause loss of life or serious damage.

Members in compression or tension such as struts and anchors must be designed conservatively because failure of one will transfer its load to the adjoining members. If the adjoining members are already overloaded, progressive or chain reaction failure will occur. For example, the author observed one strut in a braced sewer trench break, immediately transferring the load to the adjacent struts and increasing each by 50%. They broke a few seconds later, almost simultaneously transferring their load and overload to the next adjoining struts. Strut failure propagated, in opposite directions, from the initial break, until the entire 30 m (100 ft) of excavation had collapsed. The elapsed time was about 10 minutes and the cost was four lives! Compression members are often cross-braced to reduce the slenderness ratio, L/r. Horizontal compression members are subject to bending and curvature from their own weight, which sharply reduces their capacity. They require alignment and cross bracing. Cross-bracing is also recommended to prevent their buckling from impact of construction equipment. Joints require stiffening to prevent local crushing.

The design dimension tolerances are much greater than ordinarily used in structural design. It is seldom possible to drive piling and supports closer than 25 to 50 mm (2 to 4 in.) to the design position. In deep excavations the differences are much larger. Field fabrication of steel members by cutting and welding and cast-in-place concrete can minimize reconstruction. Designs must allow for such differences.

DEFLECTION AND ITS CONTROL / The recommended design pressures presume that the bracing system can move enough to mobilize a substantial part of the shearing resistance in the soil so that the resultant force approaches that of the active state. The necessary movement comes from the deflection of the bracing system: bending of the sheet piles, soldier piles, lagging and wales; shortening of struts and rakers stretching of anchors local readjustments of structural connections; and displacements of the raker supports.[9:10,9:11]

Often bracing systems move considerably more than is necessary to establish the resultant of active earth pressure. Figure 9.25 shows a range of observed vertical and horizontal movements at the face of the bracing and at distance x from the bracing. Two facts stand out: (1) The settlements related to bracing movement extend horizontally as far as 3 to 4 times the excavation depth, and (2) the greater the preload imposed on the bracing while it is constructed, the less the movement. Although the graphs represent maximum, and most well-constructed systems should move less, some movement is likely.

Deflection can be reduced materially by *prestressing* or *preloading* the system to remove the elastic deflection and slack in the bracing structure and possibly to regain some of the lateral soil strain that develops before the bracing can be tightened. This can be seen in Fig. 9.25b[9:7] based on anchored or tied-back walls with construction similar to conditions I and II of Fig. 9.25a. Loads on walls braced by rakers or struts are in the same range as Fig. 9.25b or somewhat smaller if the walls are structurally sound and preloaded immediately after thay are installed.

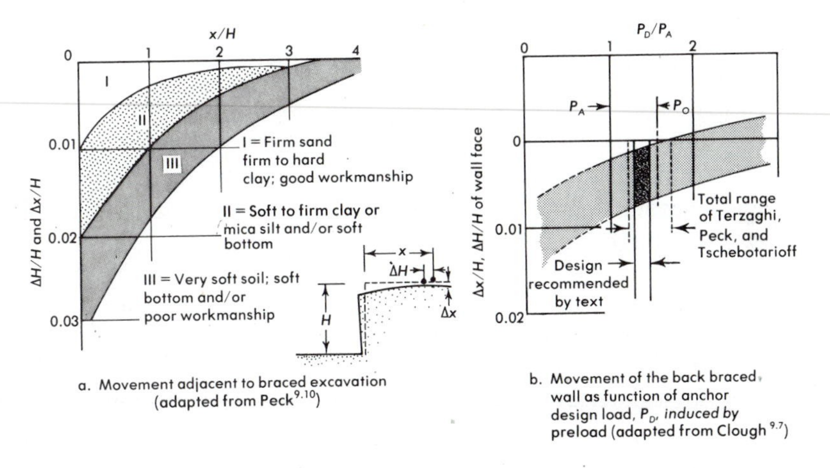

a. Movement adjacent to braced excavation (adapted from Peck[9.10])

b. Movement of the back braced wall as function of anchor design load, P_D, *induced by* preload (adapted from Clough[9.7])

Figure 9.25 *Settlement, displacement, and bracing prestress.*

It is technically possible (but not always practical) to prevent significant movement. In order to do this the structural members in flexure are designed for a limiting deflection. The system is based on the at-rest earth pressure, and the supporting members are prestressed to that magnitude.

STABILITY OF BOTTOM OF EXCAVATION / When the bottom of the excavation extends into soft clay, there is danger of a failure by bulging upward. The weight of the soil adjacent to the excavation bears on the soil stratum at the level of the bottom of the excavation, and if the bearing capacity of that soil is not great enough to support the weight, a failure will occur. The failure zone can be approximated by drawing a 45° line from one bottom corner of the excavation and connecting it to a circular arc whose center is the opposite bottom corner, as in Fig. 9.26. The zone of soil contributing to the failure has a width of 0.7 times the width of the excavation in this case. The downward force of this mass of soil is reduced by shear along the boundary of the mass, so its effective vertical force per linear foot of cut is $Q_L = 0.7B\gamma H - cH$. The pressure per square foot is $q = \gamma H - cH/0.7B$. Since the bearing capacity of clay is given approximately by $q_0 = 5c$ (see Chapter 10), the safety factor, F_S, of the bottom of the

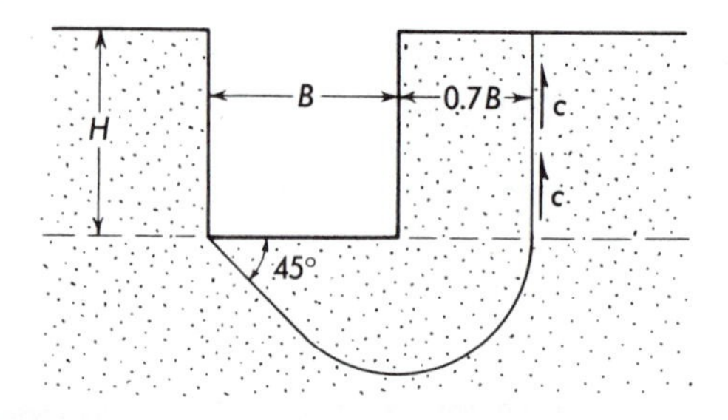

Figure 9.26 *Stability of the bottom of an excavation in soft clay.*

excavation can be expressed by the relation

$$F_s = \frac{5c}{\gamma H - cH/0.7B} \tag{9:21}$$

A safety factor of at least 1.3 should be used. If sheeting extends below the bottom of the cut, the effective load is reduced by the shear along both sides of the imbedded portion of the piling.

Excavation bottoms in sand are ordinarily stable so long as the water level inside the excavation is not lower than the groundwater level outside. As soon as the level inside is lowered by pumping, an upward seepage of water is created and heave or collapse is possible. Seepage erosion can cause subsidence of adjoining structures and is prevented by filters, as discussed in Chapter 3.

LOST GROUND / Both deformation of the bracing and heave of the excavation bottom are accompanied by a subsidence of the soil adjacent to the excavation (Figs. 9.23 and 9.25). This is known as *lost ground.* Although in many instances a moderate subsidence is of no consequence, in others even a slight movement of the soil will cause damage to adjacent buildings. In some cases lost ground is produced by sand flowing or *running* in a "quick" condition from excessive neutral stresses; in still other cases it may be caused by the slow, plastic creep of clays that are strong enough to stand in open excavations with no bracing at all.

Before any excavation that could cause damage to adjacent structures is begun, a survey should be made to determine the condition of those structures. The location, elevation, and size of all building cracks are recorded and photographed. This information can do much to prevent annoying and expensive lawsuits that often arise during excavation work.

During construction, level readings are made on points adjacent to the excavation to check for subsidences that might go unnoticed because of the usual din and confusion within the excavation. The bench mark should be located far enough from the excavation so that it will not subside and produce erratic level readings. A distance of at least 5 times the depth of the excavation from the excavation should be sufficient.

If subsidence is caused by deformation, it can be minimized by tightening the bracing system: preloading or prestressing it against the soil. If the bottom heaves, it can be prevented by driving the sheeting deeper and by loading the portions of the bottom of the excavation not actually involved in construction with excavation waste or berms of sand. If creep of unbraced soil is the cause, it can be prevented by bracing. If *sand flow* or *running sand* is the cause, it can be prevented by filtered drainage to relieve the neutral stress or by a water- and sand-tight bracing system.

9:6 Anchored Bulkheads[9:12–9:15]

An *anchored bulkhead* is a special form of retaining wall of sheet piling that is widely used in waterfront construction. Because it is built from the surface down by driving the sheets, it is adapted to sites where the water level is so high or the soil immediately beneath is so soft that the cost of constructing a retaining wall of masonry or concrete from the foundation up would be prohibitive.

BULKHEAD CONSTRUCTION / The components of an anchored bulkhead are shown in Fig. 9.27. The wall itself is of interlocking sheet piling. For very low walls, laminated creosoted timbers are occasionally used. Reinforced concrete sheets (often prestressed) 0.15 to 0.2 m (6 to 8 in.) thick and 0.3 to 0.75 m (12 to 30 in.) wide are sometimes employed in salt water because of their resistance to corrosion. The concrete must be dense and free of honeycombing in order to protect the reinforcement. Steel

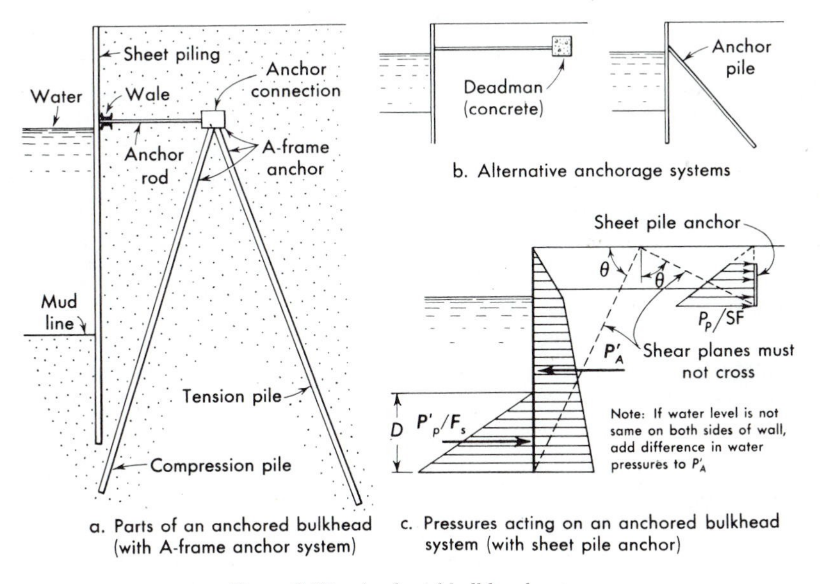

Figure 9.27 *Anchored bulkhead system.*

sheet piling is the most widely used because of its ease of handling and driving. Protection against corrosion, by coatings or by electrochemical methods (cathodic protection) is desirable, particularly in salt water.

The sheets are driven into the ground to provide lateral support at the bottom. The upper end is supported by the anchor system, consisting of the *wale, anchor rod*, and *anchor.* The wale is a continuous beam, usually a pair of channels or a H-section, that ties the sheets together and carries their load to the anchor rod. Structurally, it is simpler to place it on the outside of the wall. However, because it is vulnerable to damage from ships in that position, it is often placed on the inside. The anchor, or tie, rod connects the wale to the anchor. It is threaded for a nut at the bulkhead end or provided with a turnbuckle so its length can be varied and the sheeting aligned after installation. The anchor can take many forms, as shown in Fig. 9.27. The simple concrete *deadman* and the sheet pile deadman are used when the soils are strong enough and there is sufficient space. The A-frame and the single batter pile are employed when space is limited or when the upper soils are weak.

Two general methods of construction are used. When the bulkhead is built in open water and then the fill placed behind it, it is a *fill bulkhead.* When it is constructed in natural ground and then the earth removed from its face, it is a *dredged bulkhead.*

The wale and anchor rod are ordinarily placed as low as possible to minimize bending moments in the sheeting. Their depth is limited by low water; otherwise, their cost of installation becomes excessive.

BULKHEAD DESIGN[9:9, 9:12–9:14] / The forces acting on a bulkhead are shown in Fig. 9:27c. The inner face of the sheeting supports active earth pressure. This is resisted by the wale and anchor near the top and by passive earth pressure distributed along the outside of the sheeting at the bottom. Such a condition is termed *free earth support* because the embedded section of the pile is free to rotate as the unsupported section bulges outward under load. If the sheeting penetrates deeply into a rigid soil, it is partially fixed against rotation and the condition is termed *fixed earth support.* The free support analysis described below is satisfactory for many design problems. The more complex fixed earth support analysis is described elsewhere.

The pressure acting on the inner face of the bulkhead is essentially the effective active earth pressure. Although the pressure distribution is probably distorted by arching, as shown in Fig. 9.10, the magnitude and location of the resultant P'_A are probably not changed enough to affect the analysis.

In addition to the earth pressure there may be a difference in water pressure between one side of the sheeting and the other, caused by tidal fluctuations or rainfall infiltration. The unbalanced head can also cause a reduction in the passive pressure in the zone of embedment. Weep holes are installed to equalize the pressures but can lead to erosion of the backfill, unless protected by filter fabric or a graded filter.

The depth of embedment is determined by the passive pressure required to support the toe. The effective resultant pressure P'_p is divided by the safety factor, usually 2, to obtain a *mobilized* or *working* resistance P'_{pm}. The depth of embedment, D, is found, so that the algebraic sum of the moments of the active resultant and the mobilized passive resultant is zero.

The wale reaction unit length is the difference between the active resultant and the mobilized passive pressure. The wale is designed as a uniformly loaded beam with support at the anchor rods. The anchor rod pull is determined by the accumulated wale load. The rod is designed conservatively because corrosion and physical damage can reduce its capacity.

The moments in the sheeting are determined from the distributed loading (Fig. 9.27c). The working stresses in bending for design of the sheet piling are ordinarily 10 to 20% greater than for other structures, since the real bending moments are less than those computed because of arching.

The deadman anchor is essentially a second wall with passive pressure acting on its face. The working passive pressure is one-third to one-half the maximum. The anchor and the bulkhead must be separated a sufficient distance so that their shear zones do not interfere. The anchor wale is located at the depth of the resultant, which sometimes means a sloping anchor rod. If a concrete deadman is employed, its friction with the soil beneath provides added resistance. The A-frame anchors are designed on the assumption that the heads and tips of the piles are hinged. All anchor systems must be structurally flexible so that rotation and deflection of either the sheeting or the anchor will not develop secondary stresses that lead to failure.

Most bulkhead failures can be traced to inadequate depth of penetration of the sheeting, insufficient anchor resistance and poor structural connections; therefore special attention should be paid to these aspects of design.[9:15]

9:7 Underground Structures

Underground structures, including culverts, water and sewer pipes, tunnels and chambers such as powerhouses, must support both the horizontal and vertical pressures exerted by soil or rock. The analyses are similar to those for horizontal pressure alone: The limiting conditions can be mathematically approximated; the conditions between the limits must be evaluated by empirical adjustments based upon experience.

ELASTIC STATE / The at-rest state in a level mass of soil or rock was described in Section 9.1. The vertical pressures are defined by Equations (9:1a) and (9:1b). If the underground structure could be built by excavating the opening and immediately replacing the material removed with a perfectly rigid structure of exactly the same dimensions, the mass would remain in the same at-rest state of elastic equilibrium (Figs. 9.1 and 9.2).

Although it is physically possible to construct such a structure, it is far more likely that the mass will deform, and change the stresses in the vicinity of the opening. An approximate analysis for an unlined circular tunnel of radius R at a great depth z was developed by Kerisel (Fig. 9.28) assuming

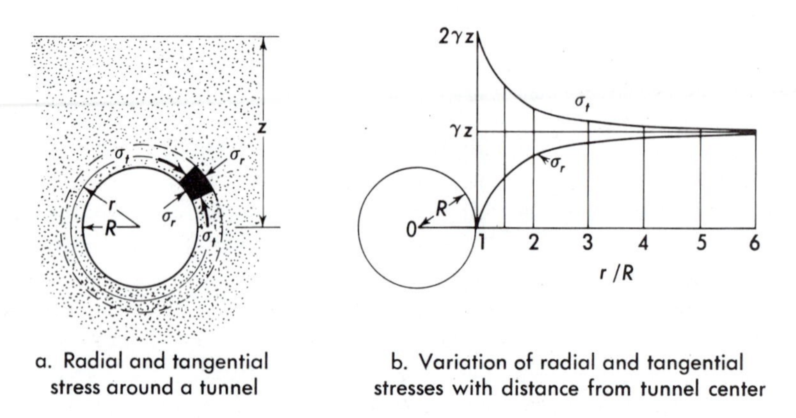

a. Radial and tangential stress around a tunnel

b. Variation of radial and tangential stresses with distance from tunnel center

Figure 9.28 *Stresses in an unlined tunnel in the elastic state.*

$K_0 = 1$[9:16,9:17] The stresses are expressed in cylindrical coordinates, σ_r, the radial normal stress and σ_τ, the tangential normal stress, with the origin at the tunnel center.

$$\sigma_r = \gamma z \left(1 - \frac{R^2}{r^2}\right) \tag{9:22a}$$

$$\sigma_\tau = \gamma z \left(1 + \frac{R^2}{r^2}\right) \tag{9:22b}$$

At the edge of the opening the radial stress is 0, and the compressive stress in the innermost ring of material surrounding the tunnel is $2\gamma z$. The reduction in the radial stress in the tunnel wall from the ambient γz is accompanied by inward deflection. If the tunnel wall could be entirely restrained, the stress would remain γz. Therefore, a support system that allows inward or radial deflection also causes σ_r to become smaller, and the tangential stress, σ_t to increase. The minimum radial stress, σ_r, is reached when the corresponding tangential stress reaches the compressive strength of the rock. At this point a state of plastic equilibrium develops; the radial deflection increases; and the stress, σ_r, is likely to increase. If the maximum tangential compressive stress, $2\gamma z$, is less than the rock compressive strength, theoretically no support will be required. Economical tunnel design requires that the soil or rock contribute to its own support in the elastic and that the deformations be controlled by adding sufficient radial restraint to prevent developing plastic equilibrium and substantially higher support loadings.[9:18]

More complex cases of elastic equilibrium have been analyzed and can be studied in the references.[9:16, 9:17] In addition, two-dimensional models and the finite element approximation have been utilized in solving problems of complex shapes, such as subway stations and underground power plants.

Both the temporary and permanent tunnel support or lining required for stability depends on the ability of the soil or rock to support itself or "arch" around the opening. The theoretical stress changes are shown in Fig. 9.28. Although the maximum tangential stress σ_τ is $2\gamma z$, actual measurements show it to be between $1.5\gamma z$ and $2\gamma z$ unless significant residual stresses are added to it. The radial stress at the inner face of an unlined tunnel, σ_r, is 0. The likelihood of failure can be determined from the Mohr circle, Fig. 9.29a, with $\sigma_1 = \sigma_\tau$ and $\sigma_3 = \sigma_r$. If the circle is below the envelope when $\sigma_r = 0$, the tunnel wall will stand unsupported. If not, sufficient radial stress will have to be supplied by temporary and permanent supports to prevent failure (Fig. 9.29b).

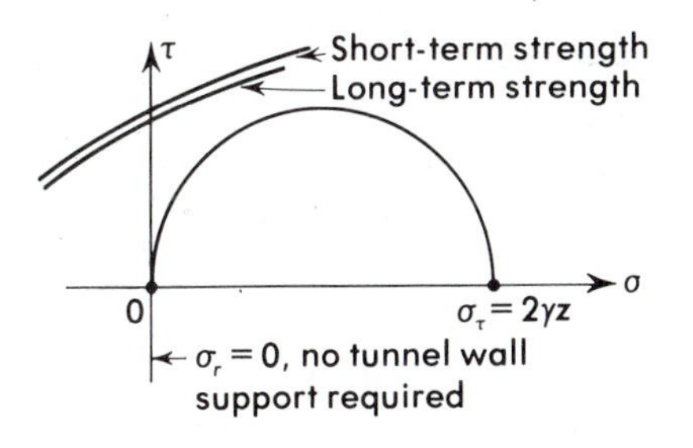

a. Mohr envelope, σ_r and σ_τ in tunnel wall of strong material

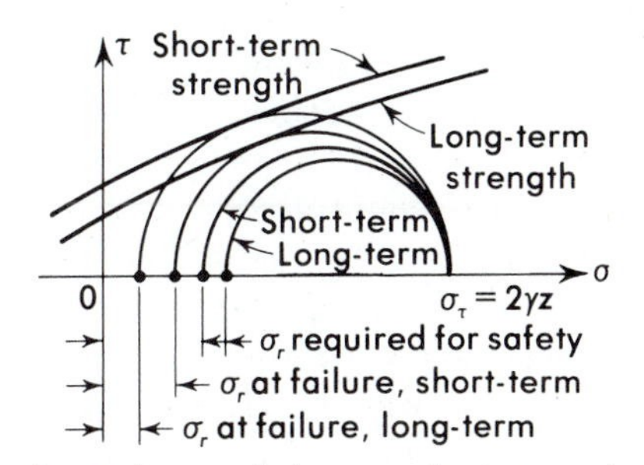

b. Mohr envelope, σ_r and σ_τ in tunnel wall of weak material

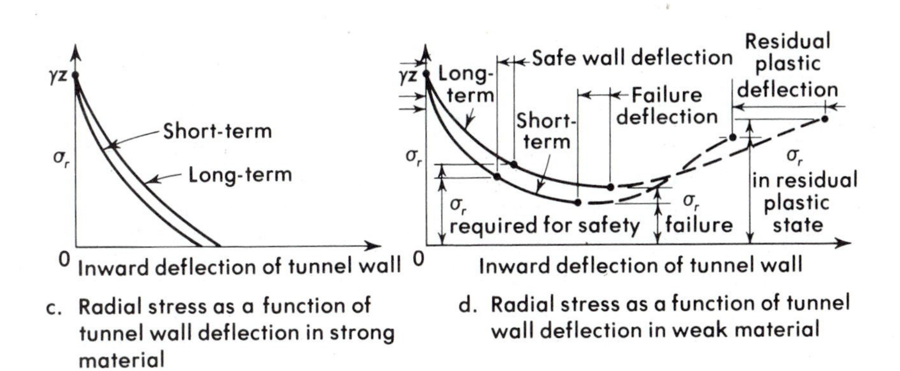

c. Radial stress as a function of tunnel wall deflection in strong material

d. Radial stress as a function of tunnel wall deflection in weak material

Figure 9.29 *Strength, failure, and tunnel wall deflection and support required in strong and weak materials.*

EXAMPLE 9:4

Compute the maximum unsupported tunnel depth in a granite having an unconfined compressive strength of 140,000 kN/m^2 (20,000 lb/in.^{2}) weighing 24.75 kN/m^3 (165 lb/ft^2) (a density of 2650 kg/m^3).

$$\sigma_T = 2\gamma z, \qquad \sigma_r = 0$$
$$q_u = (\sigma_T - \sigma_r)_f = 140{,}000 \text{ kN/m}^2$$
$$140{,}000 = 2 \times 24.75 z$$
$$z = 2828 \text{ m} = 9276 \text{ ft} \tag{9:21b}$$

Because both rock and soil undergo some strain before failure, σ_r will decrease with radial strain or inward deflection at the tunnel wall as shown in Fig. 9.29c, when the tunnel wall is sufficiently strong so that it will not fail. If the tunnel wall fails at a limiting σ_r, as shown in Fig. 9.29b, the radial strain and the corresponding support level are shown in Fig. 9.29d. Tunnel design involves establishing the amount of radial support required for safety. Tunnel construction engineering requires determining the corresponding strain or inward deflection of the tunnel wall and installing the support to insure that the limit is not passed. Time is critical because the long-term strain for a given radial stress exceeds the short-term strain and the long-term stress for a given strain can exceed that for a short-term strain. Although the soil or rock is strong enough to require no support, some may be desirable to prevent local spalling of the tunnel wall that might destroy the continuity of the natural arch.

If the radial strain exceeds the amount that produces failure [or if the radial stress required to prevent failure exceeds the resistance of the tunnel support, the tunnel wall fails in triaxial compression (shear)]. The state of stress becomes plastic and the stresses determined by the elastic state no longer apply. Because the residual strength of some soils and most rocks exceeds the peak or failure strength, shear failure of the tunnel wall is usually followed by an increase in radial stress and load on tunnel supports as shown by Fig. 9.29d. In this situation, the design of the tunnel support system and lining are related to plastic theories.

PLASTIC ANALYSIS OF STRESS / An analysis of pressures produced in a state of plastic equilibrium in the vicinity of the underground structure depends on the displacements along the soil–structure interface that define the extent of the shear zone. An approximate solution for both the horizontal and vertical pressures was developed by Terzaghi[9:2] based on the simple shear pattern proposed by Marston and Spangler[9:19] for computing the load on pipes in trenches. Although the real shear patterns are different, depending on the displacements on the soil–structure interface, the computed pressures and the significant factors defined by this approach are useful in the solution of many problems.

The zone of shear in a trench of width B is shown in Fig. 9.30. The trench is filled with soil above a level surface of a structure at depth H. The

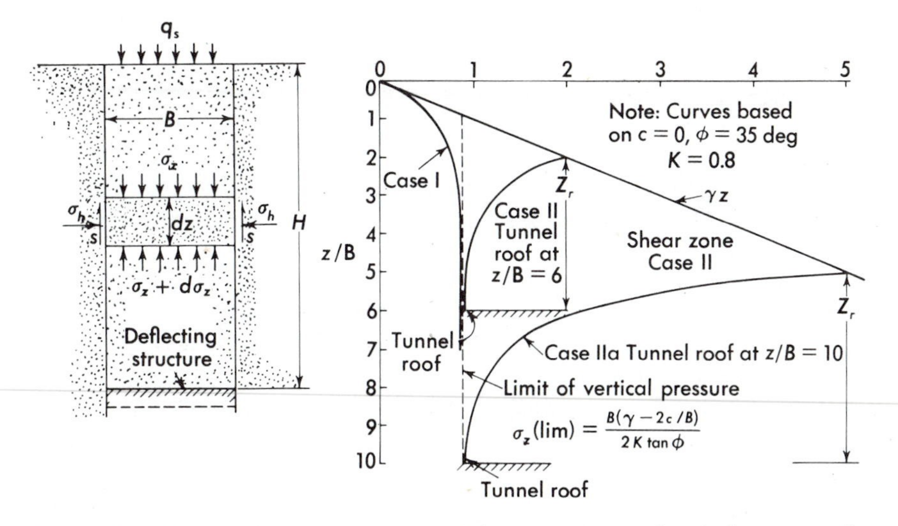

Figure 9.30 Plastic zone in a trench above a deflecting structure and vertical stresses on the structure roof.

structure and the soil above deflect downward, so that shear develops between the soil in the trench and the mass beyond. If the shear stress mobilized is equal to the shear strength of the soil, the equilibrium of a prism of soil at depth z can be described in terms of its weight dW, the vertical earth pressure on its upper and lower surfaces, and the shear strength produced by the lateral earth pressure $K\sigma_z$.

$$dW = \gamma B\,dz$$

$$dW + B\sigma_z = B(\sigma_z + d\sigma_z) + 2s\,dz$$

$$\tau_{ff} = c + p\tan\phi$$

$$\gamma B\,dz + B\sigma_z = B\sigma_z + B\,d\sigma_z + 2c\,dz + 2(\sigma_z K\tan\phi)\,dz$$

$$\frac{d\sigma_z}{dz} = \frac{\gamma - 2c}{B} \qquad (\text{if } \phi = 0)$$

At the upper surface $z = 0$ and the vertical stress is equal to any surcharge, q_q. Solving the differential equation with these boundary conditions, the vertical earth pressure σ_z is

$$\sigma_z = \frac{B(\gamma - 2c/B)}{2K\tan\phi}(1 - e^{-2K(z/B)\tan\phi}) + q_q\,e^{-2K(z)/B)\tan\phi} \qquad (9{:}23\text{a})$$

$$\sigma_z = \left(\frac{\gamma - 2c}{B}\right)z + q_q \qquad (\text{if } \phi = 0). \qquad (9{:}23\text{b})$$

The significance of the general equations (9:23a) and (9:23b) is illustrated by Examples 9:2 and 9:3 and Fig. 9.30.

In Case I, a shallow trench is backfilled over a deflecting structure, and Case II, a deep trench is first backfilled and then the structure installed by tunneling.

Example 9:2

CASE I—BACKFILL OVER STRUCTURE. For Case I, assume a trench of width B and no surcharge. The soil properties are $c = 0, \phi = 35°$, and $K = K_0 = 0.8$. The vertical stress is plotted in a dimensionless form, $\sigma_z/\gamma B$, and the depth in terms of z/B. The vertical stress (Fig. 9.29b) initially increases with depth at a rate of $\sigma_z = \gamma z$. The rate of increase in σ_z decreases rapidly as the weight of the prism of soil becomes supported by shear along its sides. Eventually, at a depth of $z/B = m$ (about 4 in this case), the vertical stress approaches a limit:

$$\sigma_z\ (\text{lim}) = \frac{B(\gamma - 2c/B)}{2\,K \tan\phi} \tag{9:23c}$$

Below this level, only enough soil shear is mobilized in each increment of depth to support the increased weight, and no additional vertical pressure develops, regardless of depth. Above that level, the full soil shear strength is mobilized.

Example 9:3

CASE II—TRENCH BACKFILLED, THEN TUNNEL EXCAVATED. If a trench should be backfilled, and then a tunnel excavated, the shear mechanism is somewhat different. Before the tunnel is excavated, the vertical stress is γz at all levels. After excavation and deflection of the tunnel roof the shear develops from the bottom up rather than the top down as previously described. The zone of full shear extends upward a distance of $m(z_r/B)$ to the point where the vertical stress, σ_z, is equal to the initial vertical stress at that level, γz. Above that level the mass is still in elastic equilibrium, $\sigma_z = \gamma z$, and there is no shear mobilized.

The height $m(z_r/B)$ of the shear zone in Case II is not sharply defined and is found by trial. This can be illustrated by Case II utilizing the same soil conditions as the first. Assume that the tunnel roof is at a depth of $z = 6B$. A preliminary estimate of the top of the shear zone can be made from the zone of full shear of Case I where $m = 4$. At that level the vertical pressure can be expressed by $\gamma z = \gamma(6B - 4B)$. This pressure is considered to be q_q in Equation (9:23a). Below that level the stresses are computed, using Equation (9:23a), considering that $z_r = 0$ at the top of the shear zone: $z_r = z - 2B$. The vertical pressure decreases rapidly through the shear zone until at $z = 6B$, the σ_z approaches a lower limit, σ_z(lim). This is the same limit as in Case I, and is defined by Equation (9:23c). If the assumed m were too small, the value of σ_z would not approach the limit; if the assumed m were too large, the limit would be approached well above the level of the tunnel. The height of the shear zone in Case II increases somewhat with the tunnel depth. If the tunnel depth should be greater, for example $z/B = 10$ (Case IIa), the limiting vertical stress immediately above the tunnel will be the same as for Case II. The height of the shear zone m, will be $z_r/B = 5$; the vertical stress at the upper limit of the shear zone will be larger. If the tunnel depth is less than the height of the zone of full shear as found by Case I, the vertical stress is always less that γz, and the pressure on the tunnel is computed as for Case I.

LATERAL PRESSURE / The simple shear zone for the approximation of the lateral stress is shown in Fig. 9.31. Above the structure, plastic equilibrium is established as described in Equations (9:23a) and (9:23b),

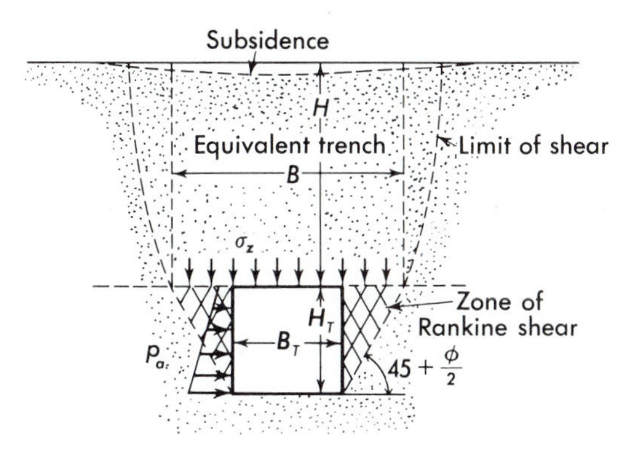

Figure 9.31 *Plastic wedges and lateral earth pressure on a deflecting buried structure.*

and the vertical stress σ_z acts on the plane of the top of the structure at a depth of $z = H$. The height of the structure is H_T and its width B_T. The inward deflection of the tunnel walls produces zones of plastic equilibrium in the mass on both sides, equivalent to the Rankine active state. The angle of the shear surfaces is $\theta = 45 + \phi/2$ and the width of each shear zone is $H_T/\tan(45 + \phi/2)$. The deflection of the shear zone also induces a vertical subsidence adjacent to the subsiding structure surface. The total width of the zone of greatest subsidence, which is equivalent to the trench width B_E of Equation (9:23a,b) is given by the following:

$$B_E = B_T + \frac{2H_T}{\tan[45 + (\phi/2)]} \tag{9:24}$$

This width is used to compute σ_z that acts on both the top of the structure and on the top of the Rankine shear zones as a surcharge. The lateral pressure is computed by the Rankine expressions including surcharge.

The real shear zone extends beyond the limits of the hypothetical trench, as shown in Fig. 9.30. The trench approximation, however, has been useful for solving real problems.

DITCH CONSTRUCTION / The load on a pipe or small culvert of diameter D laid in a narrow ditch. (Fig. 9.32a) is a simple case of the application of plastic equilibrium in a vertical direction. Generally, the trench backfill is not as well compacted as the surrounding soil, and settles under its own weight In addition the pipe subsides from its distortion under load and from settlement of the foundation. The movements approximate those assumed by Equations (9:23a) and (9:23b). The total load carried by the pipe per unit of length is $B\sigma_z$—thus the narrower the ditch the smaller the load. Sometimes a small subditch (Fig. 9.32b) is used to minimize the pipe load if the walls of the subditch can support the remaining load from above.

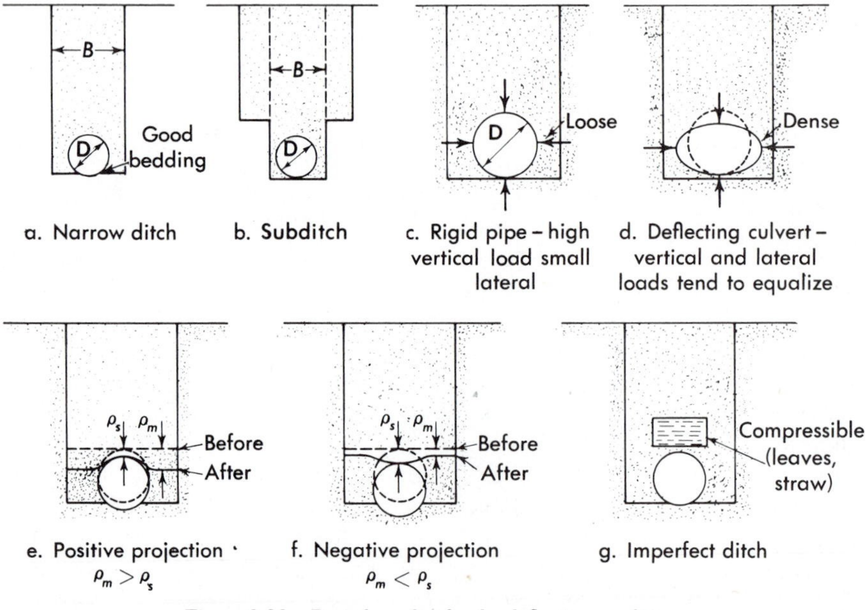

Figure 9.32 *Buried conduit load—deflection conditions.*

The lateral pressure depends on the deflection of the conduit as well as the compaction of the soil beside the pipe, if any. If the pipe is rigid and the soil beside it is poorly compacted, the lateral pressure will be negligible and the pipe will carry only the vertical load (Fig. 9.32c). If the soil beside the pipe is compacted and the pipe deflects, the lateral pressure will build up (Fig. 9.32d). In this way a thin flexible pipe can become structurally more stable in a deep ditch than a rigid pipe, provided its walls do not buckle.

Considerable empirical data on the soil properties to be used in Equations (9:23) have been published.[9:19] Usually, they suggest that K for ditches is close to K_A, probably due to lack of support of the ditch wall and the generally inadequate compaction of the backfill which allows enough movement for the active state to develop. (See Table 9:5.)

TABLE 9:5 / SOIL PROPERTIES FOR DITCH LOADS

Soil	$K \tan \phi$
Sand, gravel maximum	0.19
Sand, gravel minimum	0.16
Moist silts, clays	0.13
Saturated clay	0.11

PROJECTION-CUT AND COVER / Culverts in highway fills and shallow tunnels are frequently constructed on the ground surface or in a large excavation, and then the backfill placed around them. This has been termed the *projection condition* in culvert design. Large tunnels or underground structures are similarly constructed in place or prefabricated and lowered into relatively wide excavations. This technique has even been used in underwater tunnels where a trench is excavated in a river bottom and the tunnel sections, built on land are submerged and then connected in place.

The earth pressure on this form of construction is greatly dependent on the backfill compaction and on the relative deflection of the soil and the structure. The structure, and its foundation deflect downward, an amount ρ_s, while the adjoining mass and its foundation deflect ρ_m. Their difference compared to the mass deflection is termed the settlement ratio R_s.

$$R_s = \frac{\rho_m - \rho_s}{\rho_m} \tag{9:25}$$

If it is positive, termed *positive projection*, a plane initially level across the top of the structure is displaced as in Fig. 9.32c. If it is negative, the plane is displaced downward as in Fig. 9.32f. If they deflect equally, the value of R_s is 0 and the plane remains plane.

When $R_s = 0$, the vertical stress is σ_z by Equation (9:23) with B equal to the ditch width. The load on the pipe is $D\sigma_z$ and the lateral pressure is $K_0\sigma_z$. When the ratio is sufficiently negative, the condition is essentially that of a narrower ditch whose width is the pipe diameter, D. The vertical pressure is less than for the case of $R_s = 0$. If the ratio is sufficiently positive, the pressure will be equal to that of zero projection. However, the pipe supports the load of the full width of backfill, $B\sigma_z$.

bulkhead fitted with doors or portals that permit excavation of a limited portion of the face at one time. The permanent lining is constructed inside the shield as the excavation progresses. The shield is then jacked ahead, using the completed tunnel as the reaction. Lagging construction requires permanent support, ordinarily by a concrete lining. The liner plate temporary support sometimes is adequate for permanent support, but is ordinarily protected from corrosion by a concrete facing. If the liner plate is inadequate, a permanent reinforced concrete lining is necessary.

The elastic at-rest state is present in the undisturbed soil. This is altered by excavation, and if the soil is sufficiently strong, the compressive stress in the hypothetical ring around the tunnel will double [Equation (9:21b)], accompanied by inward deflection. Such a tunnel may not need support. An unlined tunnel through marl at Charleston, South Carolina, has been supplying raw water from a river diversion upstream for more than 50 years without serious distress. However, creep at high stress will cause the diameter to gradually reduce. In a few cases such tunnel *squeeze* has reduced the bore to a fraction of its original size. Supports or lining can prevent squeeze. However, because of creep, the pressures in softer clays approach the at-rest, with K_0 between 0.6 and 1.0 for total stress.

The pressures on tunnels in sands usually require support during construction as well as permanently. Some inward deflection is inevitable with excavation and the shear approaches the conditions described by Fig. 9.30 and Equations (9:23) and (9:24).

The art of tunneling described in References 9:17, 9:21, and 9:22 is fascinating. The techniques of excavation must be intimately adapted to the soil and groundwater conditions. Temporary liners, drainage, soil stabilization, and internal air pressure to partially balance water pressure are tools utilized by the tunnel builder or *mud hog* in boring through treacherous formations and are beyond the scope of this text.

ROCK TUNNELS / Rock tunnels include many of the features of those in soils. However, there are a number of significant differences. First, rock shear strengths at low confining stresses usually exceed those for soils despite joints. Second, the joints, bedding planes, and shear zones often destroy the continuity of the mass and cause local failure followed by local stress increases. Third, there are often significant residual stresses from tectonic movement or overburden erosion that obscure the stresses due to rock weight. Fourth, the vibration of machinery and the shock of explosives can reduce the rock strength.

The tunneling technique depends on the hardness of and uniformity of the rock, the presence of discontinuities such as shear zones or very weak seams, and on the state of stress. Homogeneous, sound rock from soft shale and limestone to gneiss can be drilled by continuous tunneling machines or *moles*. They are very complex, expensive devices, custom designed for each project and, despite their high production rate, are not justified economically without a uniform diameter several kilometers long. They incorporate provisions for installing temporary supports, for placing lining segments, and for continuous removal of the excavated material or *muck*.

Most rock tunneling requires drilling and blasting in advances of 2 to 6 m (7 to 20 ft). Horizontal holes are drilled in the tunnel end or *face* in patterns based on experience. Explosives in the holes are detonated in sequence starting at the inside and ending with the tunnel perimeter. The entire sequence or *round* usually requires only a few seconds. The broken rock, termed muck, is removed, temporary supports are installed as needed, and the process is repeated.

If the rock is sufficiently strong and intact, no temporary or permanent support may be needed. When the analyses of stress, strain, and strength (Fig. 9.29) indicate that supporting radial stress is necessary, a number of methods are possible depending on the stress level required and the time needed for installation. The simplest are steel dowels, 2 to 3 m apart, grouted in place perpendicular to the tunnel face. These hold the rock in place and develop some confining stress as the tunnel walls expand and the dowels stretch. Rock bolts $1\frac{1}{2}$ to 3 m apart and as long as $\frac{1}{4}$ to $\frac{1}{2}$ the tunnel diameter are pretensioned to the required restraining load or adjusted to allow the rock to expand to the appropriate strain level as indicated by Fig. 9.29d. If the potential fracture patterns are closer than the bolt spacing, reinforcing mesh is attached to the bolts and the tunnel wall protected by shotcrete 75 to 150 m thick.

An older system is to provide support by steel ribs with channels between them, termed *ribs* and *lagging*. Because the tunnel face is irregular, wood wedges are driven between the lagging and supports to provide the needed restraint. Such support systems are slower to install than dowels or bolts. Moreover, there is less control over the tunnel wall strain and restraining load.

If the rock is greatly weakened by joints and is allowed to expand without control beyond the minimum or failure restraint level, plastic conditions can develop in the vicinity of the tunnel.[9:21] In closely jointed or fractured rock such as in a shear zone, the mass resembles a cohesionless soil. The mass deflects inward, with a shear pattern that is approximated by Fig. 9.31 and Equation (9:23a). Even rather closely spaced joints (Fig. 9.34a) may resemble the simple model of Fig. 9.30 and the vertical tunnel load can be estimated in that way. The margin of uncertainty is greater, however, because K in rock with high internal residual stresses that are relieved by excavation can only be approximated by sophisticated analyses.

If the temporary support system is installed after a delay, the soil or rock expands, wedges or slabs drop from the roof, and even plastic theories may not apply. In such situations, the loads on supports are estimated from the fracture patterns.

When the joint spacing is less than half the tunnel width, the pressures on a tunnel lining can be estimated from the geometry of the joint blocks above the opening. The shear resistance of the joint surfaces must be estimated or determined by in-place tests. The equilibrium of each system of blocks is evaluated, utilizing the most realistic appraisal of the possible displacements. If the joints are tight and high internal stresses cause high friction across them, possibly no pressure will develop. This would be analogous to a strong arch ring around the tunnel (Fig. 9.34b) that supports the load. More commonly, the joint patterns, with some open joints, permit wedgelike masses to drop, or portions of the mass to slide down, partially resisted by friction. Typical shear zones and the loads to be resisted are shown in Fig. 9.34c, d, and e.

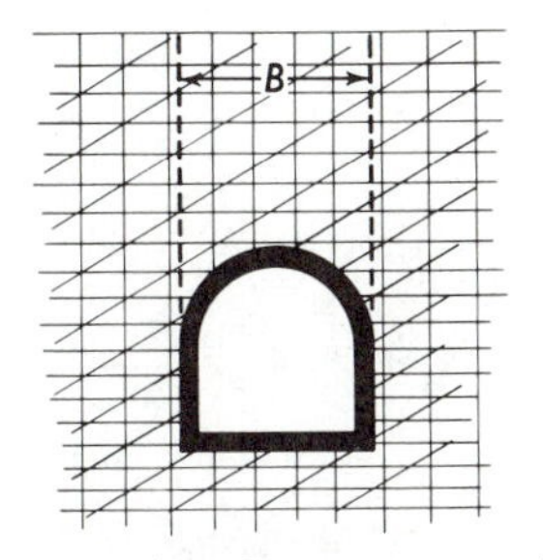

a. Close fractures – a cohesionless mass. Lining supports $\sigma_{z(lim)}$

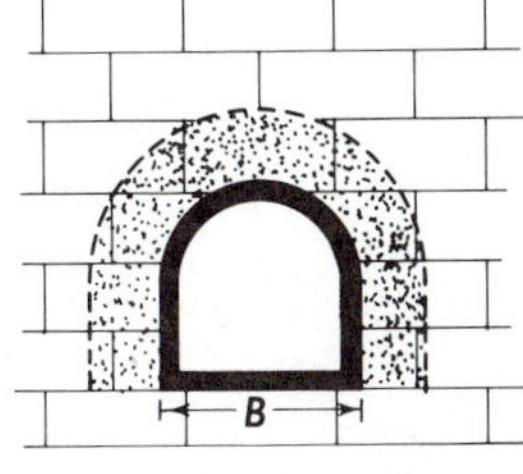

b. Wide-spaced joints under pressure allow arching No pressure on lining

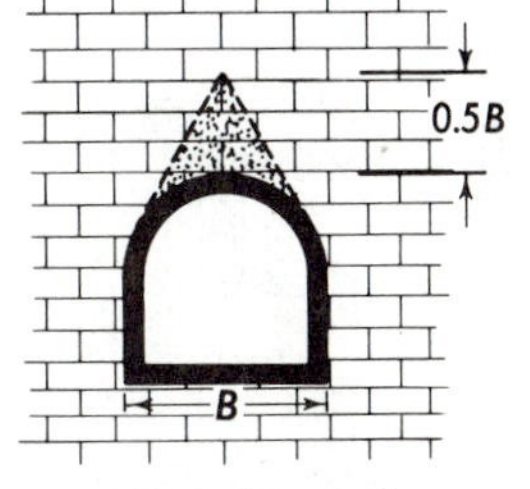

c. Staggered joints forming corbelled arch Lining supports wedge

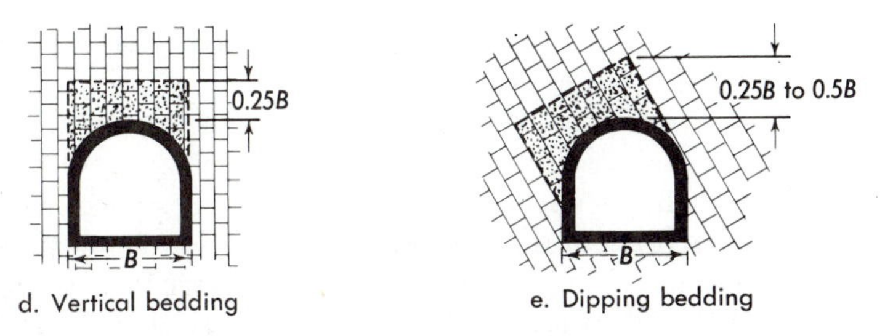

d. Vertical bedding

e. Dipping bedding

Figure 9.34 *Load of rock on tunnel lining after rock arch fails.*

In rocks with high residual internal stresses, local concentrations of tangential stress at the rock surface cause progressive *popping* of wedge-shaped segments of rock. The popping is aggravated in zones of poorer induration or where the rock has been weakened by blasting. Some weak rocks, such as shale, creep under load causing squeezing and a reduction of the tunnel bore. A few rocks, such as shale, expand on exposure to air, causing increased pressures, progressive failure, or squeeze.

ROCK BOLTS / Rock bolts have revolutionized underground construction, and to a lesser degree surface construction, in rock. They enhance the ability of the rock to support itself and reduce the need for tunnel linings, beams, and other supporting members.[9:22, 9:23]

A typical rock anchor is shown in Fig. 9.35a, installed in a hole drilled into the rock. The inside end may be fitted with an internal wedge that is forced against the hole bottom by driving the anchor against the rock or with an expansion shell that is forced against the hole walls by screwing the anchor rod into it. After the bolt is anchored, it is tensioned by a nut on the face end that squeezes the rock together, in simple precompression. Portland cement grouting after tensioning stabilizes the anchor as well as protects it from corrosion. Anchors cemented in place with rapid-setting plastic adhesives are very effective and are replacing other forms.

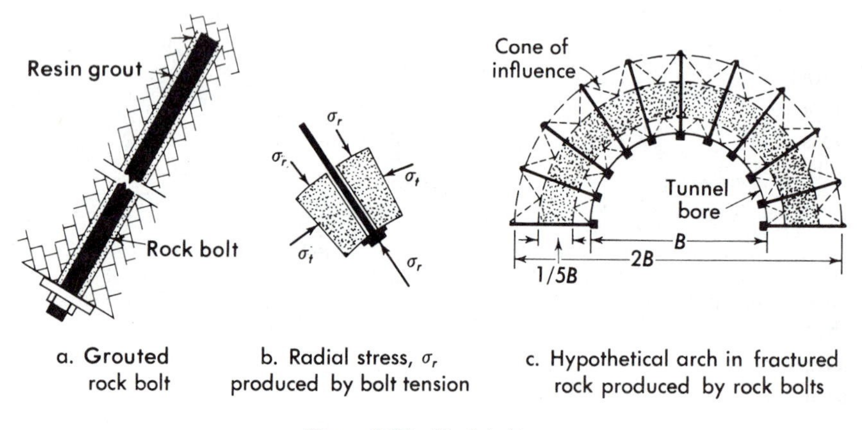

Figure 9.35 *Rock bolting.*

The rock bolts support the mass by a number of related mechanisms. Essentially, they impose a radial pressure, σ_r, or confining stress on the unrestrained surface. This confining stress increases the shear strength of the rock (Fig. 9.35b) and develops friction between blocks of jointed rock. If the rock is horizontally bedded, the increased stress on the bedding planes produces friction so that the laminated mass acts as a beam. If the strata are badly fractured, the blocks are wedged together to form a continuous arch ring (Fig. 9.35) Loose rock prisms can be held in place by suspending them from the mass above. Coherent masses are created by tying the joint blocks together.

The bolt length depends on the fracture patterns, the thickness of the beam or arch to be formed, and on the soundness of the rock at the point of wedging. Typical bolts are 1 to 3 m or (4 to 10 ft) long, but to form an arch, their length should be approximately half the tunnel width. The spacing and pattern depend on the fractures and the amount of stress to be imparted in the rock.

In closely fractured rock or in formations that spall because of high stress, the bolts serve a dual purpose. They support a protective cover termed *lagging*, which restrains the blocks that would otherwise drop free and destroy the arch. Small steel channels, wire fencing, or reinforced shotcrete are used for this purpose.

The installation of a rock bolt system and any auxiliary support, such as lagging, is largely based on experience. An all-over bolting plan is adopted based on initial studies of the fractures. The final plan is developed in the field when the fracture system and its behavior can be observed directly.

9:8 Pressure—A Construction-Related Design Load

Some textbooks on structural design as well as building codes give the impression that earth or rock pressure is a unique value that is entirely dependent on soil properties. Instead, there is a wide range of pressures possible for any given soil or rock formation. The magnitude and distribution depend largely on two factors:

1. The relative movement of the restrained soil or rock and the structure.
2. Groundwater pressure.

Both depend on construction procedures and on environmental changes that cannot always be determined in advance, but which can be controlled. Therefore, design and construction represent a cooperative and iterative process that must be integrated for economy and safety.

REFERENCES

9:1 A. Kezdi, *Handbook of Soil Mechanics*, Elsevier Scientific Publishing Co., Amsterdam, 1974, pp. 249–252.

9:2 K. Terzaghi, *Theoretical Soil Mechanics*, John Wiley & Sons, Inc., New York, 1943, p. 48.

9:3 K. Terzaghi, "General Wedge Theory of Earth Pressures," *Transactions, ASCE* **106**, 68, 1941.

9:4 M. G. Spangler, "Lateral Earth Pressures on Retaining Walls Caused by Superimposed Loads," *Proceedings, Highway Research Board*, **18**, 57–65, 1938.

9:5 K. Terzaghi and R. B. Peck, *Soil Mechanics in Engineering Practice*, 2nd ed., John Wiley & Sons, Inc., New York, 1967, pp. 363–368; 1st ed., 1948, pp. 348–350.

9:6 K. L. Lee, B. Dean, and J. M. J. Vageron, "Reinforced Earth Retaining Walls," *Journal of the Soil Mechanics and Foundations*

Division Proceedings ASCE **99**, SM10, pp. 745–764, October 1973.

9:7 G. W. Clough, "Deep Excavations and Retaining Structures," in *Analysis and Design of Building Foundations*, H. Y. Fang, ed., Envo Publishing Co., Lehigh Valley, Pa., 1976, Chapter 14, pp. 417–466.

9:8 Reference 9:5, pp. 394–413.

9:9 G. P. Tschebotarioff, *Soil Mechanics, Foundations and Earth Structures*, 2nd ed., McGraw-Hill Book Company, New York, 1973.

9:10 R. B. Peck, "Deep Excavations and Tunneling in Soft Ground: State of the Art Report," *Seventh International Conference on Soil Mechanics and Foundation Engineering, Mexico City, Mexico, 1969*, State of Art Vol., p. 266.

9:11 G. W. Clough and R. R. Davidson, "Effects of Construction on Geotechnical Performance," *Relationship Between Design and Construction in Soil Engineering*, Specialty Session 3, IX International Conference on Soil Mechanics & Foundation Engineering, Tokyo, 1977.

9:12 K. Terzaghi, "Anchored Bulkheads," *Transactions, ASCE*, **114**, 1243, 1954.

9:13 *U.S. Steel Sheet Piling*, Carnegie Illinois Steel Corp., Pittsburgh, Pa.

9:14 F. E. Richart, Jr., "Analysis for Sheet Pile Retaining Walls," *Transactions, ASCE*, **122**, 1113, 1957.

9:15 G. B. Sowers and G. F. Sowers, "Bulkhead and Excavation Bracing Failures," *Civil Engineering*, Jan. 1967.

9:16 J. Kerisel, *Proceedings of the Fourth International Conference on Soil Mechanics and Foundation Engineering, London, 1957.*

9:17 K. Szechy, *The Art of Tunneling*, Akademiai Kiado, Budapest, 1966.

9:18 J. Golser, "The New Austrian Tunneling Method," *Proceedings Conference on Shotcrete for Ground Support, ASCE*, New York, 1976, pp. 323–347.

9:19 M. G. Spangler, "Culverts and Conduits," *Foundation Engineering*, McGraw-Hill Book Company, New York, 1962, Chapter 11.

9:20 R. K. Watkins, "Buried Structures," *Foundation Engineering Handbook*, H. F. Winterkorn and H. Y. Fang, eds., Van Nostrand Reinhold Company, New York, 1975, pp. 649–672.

9:21 K. Terzaghi, "Rock Defects and Loads on Tunnel Supports," *Rock Tunneling with Steel Supports*, Commercial Shearing and Stamping Co., Youngstown, Ohio, 1946.

9:22 L. Obert and W. I. Duval, *Rock Mechanics and the Design of Structures in Rock*, John Wiley & Sons, Inc., New York, 1967.

9:23 H. K. Schmuck, "Theory and Practice of Rock Bolting," *Quarterly of the Colorado School of Mines* (Second Symposium on Rock Mechanics), Golden, Colorado, 1957.

SUGGESTIONS FOR FURTHER STUDY

J. Brinch Hansen, *Earth Pressure Calculation*, Danish Technical Press, Institution of Danish Civil Engineers, Copenhagen, 1953.

A. Kezdi, *Handbook of Soil Mechanics*, Elsevier Scientific Publishing Co., Amsterdam, 1974, Chapter 9.

R. V. Proctor and T. L. White, *Rock Tunneling with Steel Supports*, Commercial Shearing and Stamping Co., Youngstown, Ohio, 1946, 269 pp.

R. V. Procter and T. L. White *Soil Tunneling with Steel Supports*, Commercial Shearing and Stamping Co., Youngstown, Ohio, 1977.

K. Terzaghi, *Theoretical Soil Mechanics*, John Wiley & Sons, Inc., New York, 1943.

E. E. Wahlstrom, *Tunneling in Rock*, Elsevier Scientific Publishing Co., Amsterdam, 1973, 250 pp.

PROBLEMS

9:1 A vertical wall 6 m (20 ft) high has a backfill of sand whose $\phi = 35°$, that weighs 17 kN/m^3 (108 lb/ft^3) dry and 20 kN/m^3 (127 lb/ft^3) saturated. Backfill is horizontal.
  a. Compute the active earth pressure diagram and the resultant, assuming the backfill to be dry.
  b. Compute the active pressure, assuming the water table in the backfill rises to the top of the wall. Compare with (a).

9:2 A vertical wall 10 m (33 ft) high moves outward enough to establish the active state in a dry sand level backfill.
  a. Draw the pressure diagram and compute P_A if the sand $\phi = 33°$ and it weighs 16 kN/m^3 (102 lb/ft^3) dry.
  b. Compute the pressure and the resultant on the assumption that the wall does not move at all.

9:3 A wall 13 m (40 ft) high retains sand. In the loose state the sand has a void ratio of 0.67 and a ϕ of 34°. In the dense state the sand has a void ratio of 0.41 and a ϕ of 42°. Which would produce the lesser resultant of active pressure, the loose or the dense state? Which would produce the greater resultant of passive pressure? How much difference is there in the resultants for both conditions? Use either SI or English units.

9:4 A wall 8 m (26.2 ft) high retains sand weighing 15 kN/m^3 (96 lb/ft^3) dry and 19 kN/m^3 (121 lb/ft^3) saturated. The water table is permanently 3 m (10 ft) below the top of the wall. Estimate ϕ of 36°.
   a. Compute the effective and total active earth pressure diagrams assuming no capillary rise.
   b. Find the location of the resultant.
   c. How much reduction in overturning moment about the base of the wall would occur if the groundwater level could be lowered to the bottom of the wall?

9:5 A vertical wall 10 m (33 ft) high has a soft clay backfill. The clay weighs 17 kN/m^3 (105 lb/ft^3) and its strength c is 40 kN/m^2 (835 lb/in.^{2}).
   a. Compute the at-rest pressure, draw the pressure diagram, and find the resultant.
   b. Compute the active pressure and draw the pressure diagram. Find the resultant, neglecting the tension because of cracks.
   c. How much is the overturning moment caused by earth pressure reduced by allowing the wall to yield enough to establish the active state, neglecting tension?
   d. How much increase in overturning moment would occur if water got in the tension cracks?

9:6 Derive the expressions for active and passive pressures of a partially saturated clay whose shear strength is expressed by $\tau_{ff} = c + \sigma_{ff} \tan \phi$.

9:7 An anchor consisting of a sheet pile wall 4 m (13 ft) high is embedded in a partially saturated clay whose $\phi = 19°$, $c =$ 75 kN/m^2 (1570 lb/ft^2), and $\gamma = 18$ kN/m^3 (114.6 lb/ft^3). The top of the wall is at the ground surface.
   a. Compute the passive pressure, draw the pressure diagram for a 1-m (1-ft) width section of wall, and find the resultant force.
   b. Determine the depth below the top of the wall at which the anchor rod should be attached.

9:8 Compute the active, passive, and at-rest pressure for a sand having a unit weight of 17.5 kN/m^3 (111 lb/ft^3), and $\phi = 37°$. The sand is placed behind a bridge abutment that is 6 m (19.6 ft) high. Draw the pressure diagrams for each condition, and find the resultant.

9:9 A retaining wall 4 m (13 ft) high supports a level backfill of sand whose $\phi = 37°$ and $\gamma = 19$ kN/m^3 (119 lb/ft^3). The back of the wall has a batter of 1 (horizontal) to 4 (vertical). Trucks park with their rear wheels 1 m (3.28 ft) from the wall. Each truck's rear wheels carry 15 kN (3370 lb), and there is one truck for every 3 m

(9.8 ft) of wall. Compute the moment of earth pressure about the heel of the wall per meter or foot of wall.

9:10 A retaining wall 8 m (26.3 ft) high supports a dry sand fill whose $\phi = 34°$ and $\gamma = 19\ \mathrm{kN/m^3}$ ($121\ \mathrm{lb/ft^3}$). The top of the fill rises on a slope of 3 (horizontal) to 1 (vertical). The back of the wall slopes at an angle of 75° with the horizontal.

a. Compute the resultant of the active earth pressure and its direction and line of action.

b. Find the base width required to place the resultant at the third point of the base of a stone gravity wall with a top width of 42 cm (18 in.) and a masonry weight of $25\ \mathrm{kN/m^3}$ ($159\ \mathrm{lb/ft^3}$).

9:11 A gravity retaining wall 6 m (19.6 ft) high supports a level sand backfill. The top of the wall is 0.6 m (2 ft) wide and the front face has a batter of 1 (horizontal) to 6 (vertical). Find the base width required so that the resultant of the earth pressure and the weight of the wall will pass through the outside third point of the base. Assume that ϕ is 35°, the unit weight of the sand is $19\ \mathrm{kN/m^3}$ ($121\ \mathrm{lb/ft^3}$), and the unit weight of the concrete is $24\ \mathrm{kN/m^3}$ ($152.8\ \mathrm{lb/ft^3}$).

9:12 A concrete cantilever retaining wall is 5 m (16.4 ft) high and supports a sand backfill that rises on a slope of 2 (horizontal) to 1 (vertical). The sand $\phi = 32°$ and $\gamma = 18\ \mathrm{kN/m^3}$ ($114.6\ \mathrm{lb/ft^3}$). The wall is 0.45 m (1.5 ft) thick at the top and 0.75 m (30 in.) at the base; its back side is vertical. The wall foundation is 0.6 m (2 ft) back of the back face of the wall. The foundation soil is clay, with $c = 75\ \mathrm{kN/m^2}$ ($1566\ \mathrm{lb/ft^2}$). The concrete weighs $24\ \mathrm{kN/m^3}$ ($153\ \mathrm{lb/ft^3}$). The foundation is 0.75 m or 2.5 ft thick.

a. Compute active earth pressure diagram and the resultant.

b. Compute width of foundation so that resultant is within middle third.

c. Check foundation for safety against sliding. A sand cushion is placed between the foundation and the clay; $\phi = 34°$.

9:13 Recompute the resultants of Problems 9:10, 9:11, and 9:12, using the Coulomb analysis with $\delta = \frac{2}{3}\phi$ and assuming that the resultant meets the wall at a depth of $2H/3$ below the top. Find the base width required and compare with that required for the wall, based on the Rankine analysis.

9:14 An excavation 9 m (29.5 ft) deep and 12×12 m (39.4×39.4 ft) is to be made in sand with $\phi = 40°$ and $\gamma = 19\ \mathrm{kN/m^3} = 121\ \mathrm{lb/ft^3}$. The bracing system is to consist of horizontal wood lagging supported by vertical 200-mm deep (8-in.) H-pile soldier beams. The water table will be drained during construction.

a. Determine the pressure diagram.

b. Determine thickness of lagging if the soldier beams are 2 m (6.6 ft) apart. Wood is southern shortleaf pine (dense structural select).
c. The uppermost wale is at a depth of 1.5 m (4.9 ft) and the others are spaced 3 m (9.8 ft) apart. Design wales and struts, assuming only one strut in each direction at each elevation. Struts to be tied vertically at intersection in center of excavation.

9:15 A long excavation 11 m (36.1 ft) deep and 8 m (26.2 ft) wide in soft clay is supported by steel sheet piling; the clay $c = 30$ kN/m^2 (625 lb/ft^2), and $\gamma = 17$ kN/m^3 (108 lb/ft^3). Three sets of wales are used. The uppermost is 2 m (6.6 ft) below the ground and the other two are 3.5 m (11.5 ft) and 7 m (23 ft) below the first.
a. Compute the pressure diagram.
b. Select the steel sheet piling (see manufacturer's catalogs).
c. Determine the size wales and the strut size and spacing. The struts are braced vertically at their centers.
d. Check the stability of the bottom of the excavation if the sheet piling extends just below the bottom of the excavation. The clay extends 15 m (49.2 ft) below the bottom of the excavation.

9:16 An anchored bulkhead retains sand weighing 19 kN/m^3 = 121 lb/ft^3 saturated and 17 kN/m^3 = 108 lb/ft^3 damp, with an angle of internal friction of 36°. Below the dredge line is sand weighing 21 kN/m^3 (133.8 lb/ft^3), with an angle of internal friction of 39°. Low tide is 8 m (26.2 ft) above the dredge line, the tidal fluctuation is 2 m (6.6 ft), and the top of the bulkhead is 5 m (16.4 ft) above low tide.
a. Compute the active and working passive pressure diagrams, using a safety factor of 2 for passive pressure and assuming a maximum difference in water levels of 0.6 m (2 ft), front to back of the wall.
b. Compute the minimum embedment, assuming the wale to be 0.3 m (1 ft) above low tide.
c. Determine the wale reaction per meter or foot.
d. Determine the sheet pile section to be used if the maximum bending stress is 165.4 MN/m^2 (24,000 lb/in^2.).
e. Determine the wale size and anchor rod diameter if one rod is used for every six sheets.

9:17 Design a sheet pile anchor system for Problem 9:16, including the sheeting length, the anchor location, the anchor rod length, and the wale size.

9:18 An excavation 25 m (82 ft) deep, and all above ground water, is to be constructed in a site that does not restrict excavation slopes.

a. If the excavation is 8 m × 8 m or 26.3 × 26.3 ft square, compare the costs to brace the excavation or to excavate with 2(H) to 1(V) side slopes. (Use the unit prices for basement excavation, and backfill compaction around structures and the average values for steel sheet piling.)

b. If the excavation is 60 × 60 m (197 × 197 ft) square and the unit prices for earthwork and embankment compaction apply to excavation and later backfilling around the structure, compare the costs of a braced and open excavation.

9:19 A tunnel nominally 5 × 5 m (16.4 × 16.4 ft) square is driven through a mass of damp sand whose unit weight is 20 kN/m^3 (127 lb/ft^3), angle of internal friction is 36°, and whose capillary cohesion is 10 kN/m^2 (208 lb/ft^2). The tunnel is 20 m (65.6 ft) deep. Compute the vertical and horizontal pressures on the tunnel assuming $K = 0.6$.

CHAPTER 10
Foundations

A large retail store complex was being built of precast, prestressed concrete frames supported by individual square footings. After the building frame had been erected, the plumbing contractor excavated a trench adjacent to one of the footings. Suddenly, the soil supporting that footing sheared into the trench. The footing and the column dropped, and pulled the prestressed beam attached to the column head down and out. Each adjoining column and beam in succession was pulled out of line. The column-to-beam connections tore loose, the beams fell, and with them collapsed 5000 m^2 (50,000 ft^2) of roof. Three workmen were crushed to death and several injured, all because of the bearing capacity failure of one small foundation.

Square footings resting on young limestone supported the upper floors and roof of a three-story college building. The ground floor was supported by 1.5 m (5 ft) of fill required to raise the building above its low-lying site. Because of the limestone (possibly guided by the Biblical reference to wise men building on rock) the architect was not concerned about settlement. The construction sequence required building the foundations and columns to the level of the first structural floor and then placing the fill around the columns. No one could believe the settlement of 80 to 100 mm (3 to 4 in.), which occurred during filling. Even more surprising was the lack of settlement of a group of closely spaced columns and stairwells with the largest structural load but no fill.

An investigation disclosed that the limestone, 4 to 5 m (13 to 17 ft) thick, was underlain by extremely porous limestone (void ratio exceeding 2) and very loose sand. The fill weight caused the porous limestone and the very loose sand to consolidate. The stresses at the base of the limestone were less at the group of closely spaced columns and stairwells because the limestone spread the concentrated loads. The fill, continuous over the greater part of the building area, increased the stress in the porous limestone and loose sand about the same amount as the fill weight, about 30 kN/m^2 (625 lb/ft^2). The site was eventually stabilized by injecting portland cement–flyash grout into the loose and porous materials.

Several five-story reinforced concrete apartment buildings in Niigata, Japan, were supported on concrete mats that were continuous between the outside walls. The mats were founded on loose, saturated fine sand. After several settlement-free years, an earthquake of moderate severity shook the area. Several of the buildings suddenly sank into the sand; one tilted 35° from the vertical, another 10°, but in the opposite direction. The buildings, designed to resist earthquakes, remained intact. Nearby heavy buildings subsided; empty buried concrete tanks floated upward. The earthquake had liquefied the sand, momentarily destroying its strength and bearing capacity.

These cases illustrate the effect of foundation inadequacy on the behavior of the structure. In each case the foundation was structurally sound and did not suffer damage; instead the critical factor in the foundation performance was the soil below—the structural member that was present even before the building was built and the ultimate support of everything above.

Foundation construction is one of the oldest of man's arts. The prehistoric lake dwellers of Europe built their homes on long wood poles driven securely into soft lake bottoms; the ancient Egyptians built their monuments on mats of stone resting on bedrock. The Babylonians found only deep alluvium in their flood plains between the Tigris and Euphrates, which settled under the weight of their cities. Buildings and walls were supported on masonry mats; adjacent parts of structures were provided with sliding connections so that they could settle different amounts without cracking. The artisans of the Middle Ages supported their masterpieces on inverted tables or arches of stone, on rafts of timber, or on wood piling, following rules laid down by Roman builders before them. Until the twentieth century, however, the design of foundations was based entirely on past experience, ancient rules, and guesswork. Soil and rock mechanics have given foundation engineers powerful tools with which they can analyze stresses and strains in the substructure, in the same way as for the superstructure, and formulate a rational design to fit the structure to the capabilities of the soil and rock below.

10:1 Essentials of a Good Foundation

The *foundation* is the supporting part of a structure. The term is usually restricted to the structural member that transmits the superstructure load to the earth, but in a larger sense it includes the soil and rock below. It is a transition or structural connection whose design depends on the characteristics of both the structure and the soil and rock. A satisfactory foundation must meet three technical requirements:

1. It must be placed at an adequate depth to prevent frost damage, heave, undermining by scour, or damage from future construction nearby.
2. It must be safe against breaking into the ground.
3. It must not settle enough to disfigure or damage the structure.

These requirements are considered in the order named. The last two are capable of reasonably accurate determination through methods of soil and rock mechanics, but the first involves consideration of many possibilities, some far beyond the realm of engineering. During the long period of time the ground must support a structure, it may be changed by many man-made and natural forces. These should be evaluated in choosing the location for a structure and particularly in selecting the type of foundation and the minimum depth to which it must extend.

Once the technical requirements have been satisfied, there are economic considerations:

1. The quantity and quality of the materials must be minimized, compatible with the technical requirements.
2. The labor for construction, a major factor, must be minimized.
3. The impact of possible changes, such as caused by hidden geologic features, must be considered because small changes in some designs can cause large increases in construction or remedial cost. The designer must foresee the unforeseen in order to make allowances for it.
4. The persons responsible for design, construction, and maintenance must react promptly and together when the unforeseen occurs. Plans, specifications, and maintenance instructions should establish the chain of communication for reaction to prevent the unforeseen from becoming unmanageable.

FOUNDATION DEPTH / The surface zones of many soils change volume regularly with the seasons. In much of the United States frost action swells the ground in winter, which means foundations should be placed below the maximum depth of frost penetration. This can be determined by local experience or estimated from Fig. 3.30.

Clay soils, particularly those with high plasticity, shrink large amounts on drying as described in Section 4:5. Such clays and some shales expand when wet. In regions having pronounced wet and dry seasons, the soils close to the ground surface expand and contract. The outer walls move the most, while the inside, where the soil is protected from the sun and the rain, moves the least. In normally moist regions a prolonged dry spell can cause soil shrinking and foundation settlement. Accelerated drying and settlement can be brought on by certain types of vegetation that extract moisture from the soil or by boilers and kilns that heat the soil abnormally. In very dry regions the opposite occurs. Added moisture from leaking pipes, irrigation, and even watering lawns can cause a desiccated clay to expand and lift a structure. In most cases the volume change becomes less with increasing depth, and if possible, foundations are placed below the volume-change zone.[10:1] Other methods of handling high volume change are given in Section 10:8.

The scour of river bottoms, especially during periods of flood, has caused a number of bridge failures. This occurs in two ways: the normal scour of the riverbed and the outsides of bends due to the increased velocity of flow during floods, and the accelerated scour caused by the obstruction offered by the bridge pier to the flow. The first is a characteristic of the stream, and the amount can be estimated by correlating the height of water surface rise during a flood to the increase in depth of the river bottom. In many streams with a sand or gravel bottom a foot rise of the surface is accompanied by a foot or more of scour.[10:2] The accelerated scour can be minimized by good hydraulic design including streamlining the piers and aligning them with the direction of flow.[10:3]

Ice is a serious problem with bridge piers and marine structures such as docks and wharves. Impact of floating ice, carried by river or tidal currents, can cause serious damage. The weight of ice and frozen spray can become great enough to overturn light structures.

All foundations should be designed with allowances for future excavation and construction. Bridge piers must be located to provide clear navigation channels and must be placed deep enough to allow for future dredging of the channel. Construction operations in congested city areas often affect building foundations. The undermining of foundations by deep excavation and subway tunneling can result in the settlement and failure of buildings that have stood safely for years. In such cases the contractors responsible for the damage have been liable, but it is usually cheaper in the long run if such trouble is anticipated when the foundations are designed. Many building codes make a contractor liable for damage to adjacent property only if the excavation is deeper than 2 m (6 ft).

Ground water is a factor in several ways. First, excavation below the groundwater level is expensive and often hazardous because upward seepage loosens sands and tends to create a quick condition, and water standing over exposed clays softens them. Second, when the groundwater level is above the lowest floor, seepage into the structure and hydrostatic uplift become serious problems. Third, changes in the elevation of the water table have caused much trouble. In cities the water level drops because of drainage into sewers and deep excavations or because of pumping for water supply. This increases building settlements by increasing soil stresses, or it may cause rotting of timbers formerly submerged well below the water level. On the other hand, if the water level rises through flooding, protracted rainfall, or broken water mains, soil strength is decreased and failures occur.

In some cases watertight structures such as empty concrete swimming pools and buried tanks have floated out of their normal locations because of the high water table that normally occurs in late winter and spring.

Underground cavities such as mines, caves, and sewers are hazards to foundations because they sometimes collapse from overload or structural deterioration. Piping or internal erosion of soil into leaky sewers or cavities likewise can cause trouble. If possible, foundations should be moved from these defects or corrective measures taken to make them harmless.

10:2 Stability—Bearing Capacity

The *bearing capacity* of a soil, often termed its *stability*, is the ability of the soil to carry a load without failure within the soil. It is analogous to the ability of a beam to carry a load without breaking. The load-carrying capacity of soil varies not only with its strength but also with the magnitude and distribution of the load. When a load Q is applied to a soil in gradually increasing amounts, the soil deforms, making a load–settlement curve similar to a stress–strain curve. When the critical or failure load, Q_0, is reached, the rate of deformation increases. The load–settlement curve goes through a point of maximum curvature, indicating failure within the soil mass. Different curves (Fig. 10:1) are obtained, depending on the character of the soil that is loaded. Dense sand and insensitive clay usually show a sharp, sudden failure, whereas loose sand and sensitive clay show a more gradual transition associated with progressive failure.

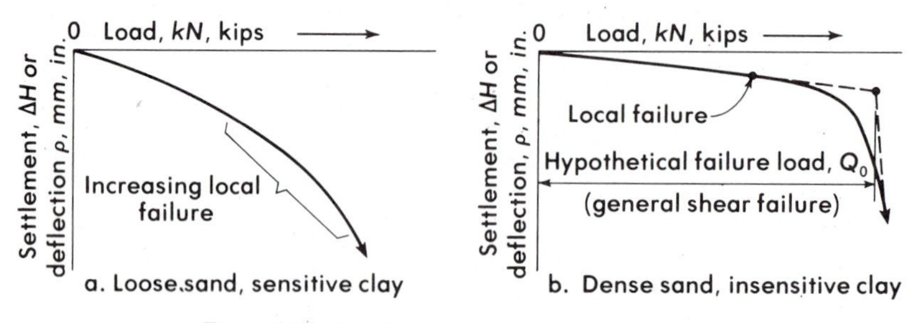

Figure 10.1 *Load settlement curves for foundations.*

If the soil is observed during loading by means of a glass-sided model or by an excavation adjacent to a full-size foundation, it will be seen that there are usually three stages in the development of failure. First, the soil beneath the foundation is forced downward in a cone or wedge (Fig. 10.2a). The soil below the wedge is forced downward and outward. Imaginary lines in the soil that were initially vertical now bulge outward like a barrel. Second, the soil around the foundation perimeter pulls away from the foundation, and surfaces of shear propagate outward from the tip of the cone or wedge (Fig.

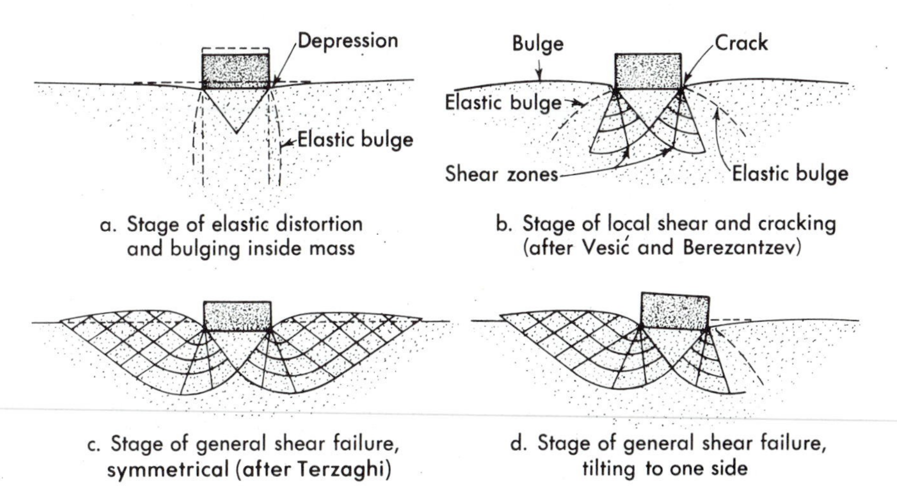

Figure 10.2 *Development of shear failure beneath a foundation.*

10.2b). If the soil is very compressible or can endure large strains without plastic flow, the failure is confined to fan-shaped zones of local shear. The foundation will displace downward with little load increase: one form of *bearing-capacity* failure. If the soil is more rigid, the shear zone propagates outward until a continuous surface of failure extends to the ground surface and the surface heaves (Fig. 10.2c). This is termed *general shear* failure. The failure can be symmetrical, if rotation is restricted by a column attached to the foundation, or it can tilt as in Fig. 10.2d. Such a *bearing-capacity failure* is not common, but it almost always results in a complete failure of the structure.

No exact mathematical analysis has been derived for evaluating such a failure. A number of approximate methods that have been developed are based on simplified representations of the complex failure surface and of the soil properties.

BEARING-CAPACITY ANALYSIS / A simple and conservative analysis was developed by Bell, extended by Terzaghi, and further modified by the author. The method approximates the curved failure surfaces with a set of straight lines, as shown in Fig. 10.3.

A foundation having a width of B and an infinite length is assumed, similar to a long wall footing. At the moment of failure the foundation exerts a pressure of q_0, which is the *ultimate bearing capacity*, or simply *bearing capacity*, of the soil. The soil immediately beneath the foundation is assumed to be in compression similar to a specimen in a triaxial shear test. The major principal stress on this zone, II, is equal to the foundation load q_0 if the weight of the soil beneath the footing is neglected. The minor principal stress on zone II is produced by the resistance of zone I to being compressed. Zone I is like a triaxial shear specimen lying on its side with the major principal

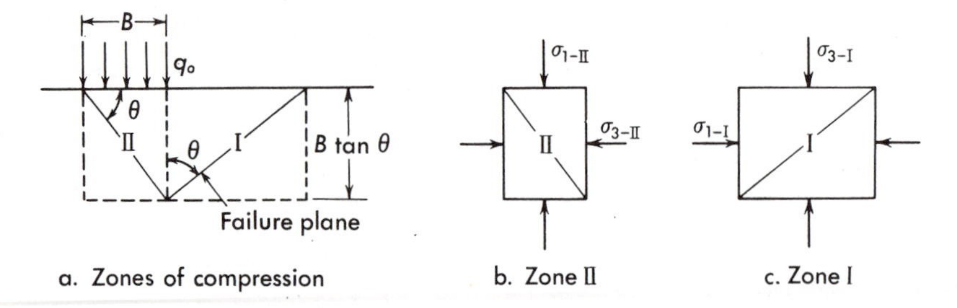

Figure 10.3 *Assumed straight-line failure plane and prismatic zones of triaxial compression and shear beneath a uniform load q_0 of width B.*

stress horizontal. At the moment of foundation failure both zones shear simultaneously, and the minor principal stress on zone II, σ_{3-II}, equals the major principal stress on zone I, σ_{1-I}.

The minor principal stress on zone I is provided by the average vertical stress caused by the soil's own dead weight and any surcharge q_q. The *surcharge* (Fig. 10.4) is any permanent confining pressure above the foun-

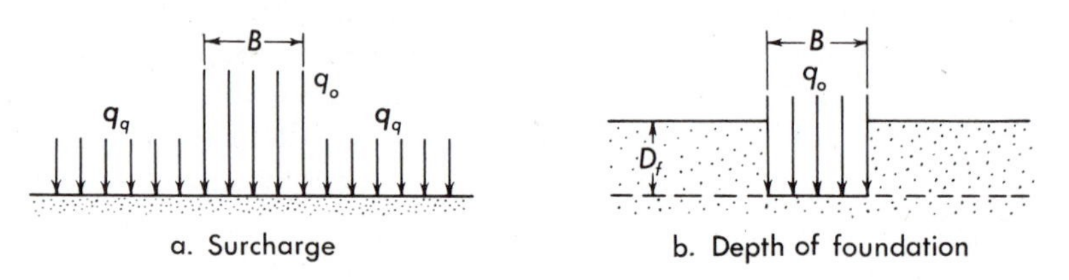

Figure 10.4 *Surcharge and depth of foundation.*

dation level such as the weight of a basement floor or the weight of soil above the foundation level:

$$q_q = \gamma D_f \tag{10:1}$$

The height of the failure zone is $B \tan \theta$, where θ is the angle of the failure zone, $\theta = 45 + (\phi/2)$. The average minor principal stress due to soil weight is therefore $(\gamma B/2) \tan \theta$. The total minor principal stress is

$$\sigma_{3-I} = q_q + \frac{\gamma B}{2} \tan \theta \tag{10:2a}$$

If the minor principal stress is known, the major principal stress on zone I can be found graphically by a Mohr circle (Fig. 10.5). This is essentially passive earth pressure and it resists the bulging of zone II. Since this is equal to the minor principal stress on zone II, a second Mohr circle will give the major principal stress on zone II, the ultimate bearing capacity:

$$\sigma_{1-II} = q_0 \tag{10:2b}$$

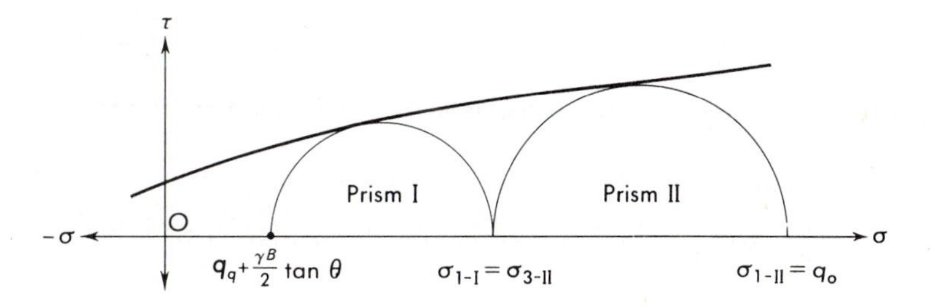

Figure 10.5 *Mohr circle analysis of bearing capacity based on straight-line failure planes and prismatic zones of triaxial compression and shear.*

The graphical analysis can be used in any soil, regardless of the shape of the Mohr envelope. If the Mohr envelope can be approximated by a straight line of the form

$$\tau_{ff} = c + \sigma_{ff} \tan \phi \tag{5:17b}$$

the ultimate bearing capacity can also be derived analytically from the trigonometry of the Mohr circle (Fig. 10.6):

$$\frac{\sigma_1 - \sigma_3}{2} = \left(\frac{c}{\tan \phi} + \frac{\sigma_1 + \sigma_3}{2}\right) \sin \phi$$

$$\sigma_1 = \sigma_3 \left(\frac{1 + \sin \phi}{1 - \sin \phi}\right) + 2c\left(\frac{\cos \phi}{1 - \sin \phi}\right)$$

$$\sigma_1 = \sigma_3 \tan^2 \theta + 2c + \tan \theta \tag{10:2c}$$

$$\sigma_{1\text{-I}} = \left(q_q + \frac{\gamma B}{2} \tan\right) \tan^2 \theta + 2c \tan \theta$$

$$q_0 = \sigma_{1\text{-II}} = \left[\left(q_q + \frac{\gamma B}{2} \tan \theta\right) \tan^2 \theta + 2c \tan \theta\right] \tan^2 \theta + 2c \tan \theta$$

$$q_0 = \sigma_{1\text{-II}} = \frac{\gamma B}{2} \tan^5 \theta + 2c (\tan \theta + \tan^3 \theta) + q_q \tan^4 \theta \tag{10:2d}$$

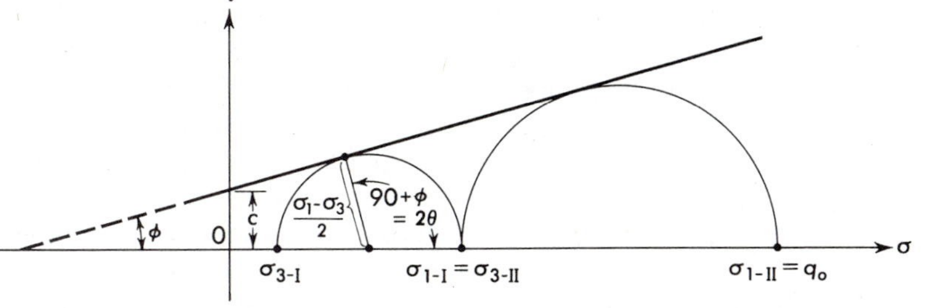

Figure 10.6 *Mohr's circle analysis based on straight-line Mohr envelope.*

This is a general expression for the ultimate bearing capacity for any soil with a straight-line Mohr envelope. It can be used for a cohesionless soil by setting $c = 0$ and for a saturated clay in undrained shear by setting $\phi = 0$, and $\tan \theta = 1$.

GENERAL BEARING-CAPACITY EQUATION—TERZAGHI-MEYERHOF / The equation for bearing capacity can be rewritten in a simple form:

$$q_0 = \frac{\gamma B}{2} N_\gamma + cN_c + q_q N_q \tag{10:3}$$

The symbols N_γ, N_c, and N_q are *bearing-capacity factors* that are functions of the angle of internal friction. The term containing factor N_γ shows the influence of soil weight and foundation width, that of N_c shows the influence of the cohesion, and that of N_q shows the influence of the surcharge. The values of these factors for different values of ϕ are given in Fig. 10.7.

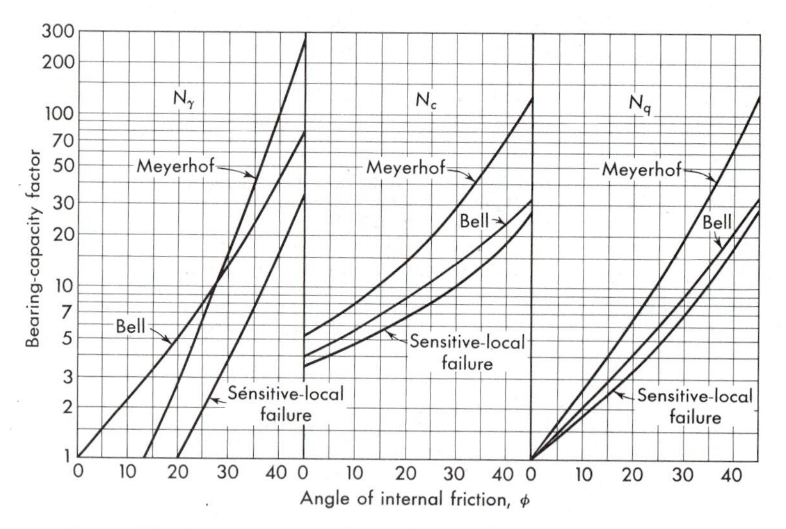

Figure 10.7 *Bearing-capacity factors for general bearing-capacity equation.*

This general expression was also derived by Terzaghi[10:4] from a more rigorous analysis of bearing capacity. It is based on approximating the surface of shear by a combination of straight lines and logarithmic spirals. The analysis was later improved by Meyerhof, but the results are expressed in the same form. Meyerhof's values for the bearing-capacity factors are given in Fig. 10.7.[10:5, 10:6]

Both the Terzaghi and the Meyerhof analyses assume the development of the full shear surface and the complete shear failure. However, loose sands and highly sensitive clays fail by *local* or progressive shear when local cracking develops around the foundation or when the cone or wedge of soil under the foundation forms. Terzaghi suggested an empirical reduction to the bearing-capacity factors for this condition. The reduced Meyerhof factors, which apply to sands having a relative density of less than 30 or to clays with a sensitivity of more than 10, are also shown in Fig. 10.7. The factors of the simplified Bell analysis fall between the limiting values of the more accurate ones. Vesic[10:7, 10:8] has proposed rational analyses of these conditions, but they have not been verified by full-scale tests.

RECTANGULAR AND CIRCULAR FOUNDATIONS / Both the Bell–Terzaghi and the Terzaghi–Meyerhof analyses assume an infinitely long foundation. When the foundation has a limited length, shear develops on surfaces at right angles to those previously described, and the bearing-capacity factors N_c and N_γ are changed. Correction factors to be multiplied by the bearing-capacity factors are given in Table 10:1, where L is the foundation length and B the width.

TABLE 10:1 / CORRECTION FACTORS FOR RECTANGULAR AND CIRCULAR FOUNDATIONS

Shape of Foundation	Correction for N_c	Correction for N_γ
Square	1.25	0.85
Rectangular $L/b = 2$	1.12	0.90
$L/b = 5$	1.05	0.95
Circular (use diameter D for B)	1.2	0.70

EFFECT OF SOIL PROPERTIES AND FOUNDATION DIMENSIONS / As can be seen by the general equation, the bearing capacity depends on the angle of internal friction ϕ, the soil unit weight γ, the foundation width B, the "cohesion" c, and the surcharge q_q. The angle of internal friction has the greatest influence because all three factors increase rapidly with only small increases in the angle.

If the angle of internal friction is zero, as for a saturated clay in undrained shear, the first and third terms become very small and only the cohesion contributes materially to the bearing capacity. Thus for all practical purposes in a saturated clay,

$$q_0 = cN_c \tag{10:4a}$$

$$q_0 = 5.2c \text{ (for long footings)} \tag{10:4b}$$

$$q_0 = 6.5c \text{ (for square footings)} \tag{10:4c}$$

Both the first term and the third term in the equation depend on the unit weight of the soil. When the shear zone is above the water table (the bottom of the footing a height of about B above the water), the full soil unit weight is used in computations. When the water table is at the base of the foundation, the submerged unit weight, $\gamma' = \gamma - \gamma_w$, must be used in the first term. The effect is to reduce that part of the bearing by about one-half. If the water table is above the bottom of the footing, the surcharge weight is also affected.

The first term of the equation varies in direct proportion to the foundation width. This means that in cohesionless soils such as sands, the bearing capacity of small foundations is low and that of large foundations is very high. Estimating the bearing capacity of sand by small-scale tests can be misleading because the bearing capacity of a full-sized foundation will be much greater. In saturated clays in undrained shear, foundation width has little effect on bearing capacity.

The third term is proportional to the surcharge q_q. For a saturated clay where $\phi = 0$ and $N_q = 1$, the contribution of surcharge to bearing capacity is small. In a soil with a high angle of internal friction, a small amount of surcharge produces a large amount of bearing capacity.

EFFECT OF ECCENTRICITY / If the load is not applied concentrically, the overturning moment reduces the bearing capacity. According to Meyerhof,[10:6] the eccentrically loaded foundation responds as if it had a reduced width, B_r:

$$B_r = B - 2e \tag{10:5}$$

In this expression e is the eccentricity of the resultant of the column load and the foundation weight (Fig. 10.8a). If there is eccentricity in two directions, both the length and width are reduced according to Equation (10:5). The value of q_0 computed from the reduced width is the average, and is used with the reduced width again in computing total capacity, Q.

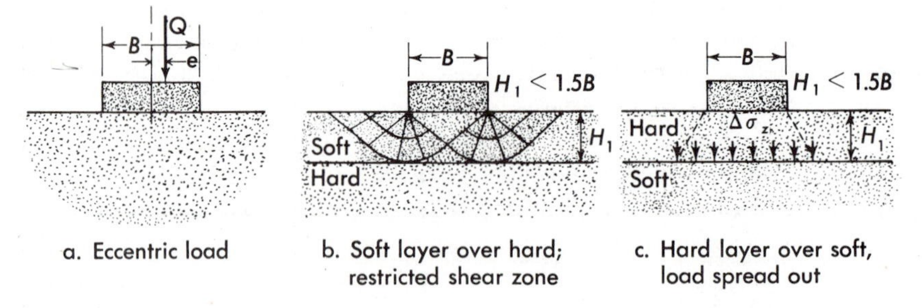

Figure 10.8 Bearing capacity with eccentric loads and stratified soils.

An older method has been to compute the pressure distribution on the foundation produced by eccentric loading by assuming a linear variation from one side to the other, similar to the stress distribution in an eccentrically loaded column. The maximum pressure is then used in computing the safety factor. This approximation is reasonable for small eccentricities, but the reduced width of Equation (10:5) is more realistic.

EFFECT OF INCLINED LOADING / If the loading is not vertical, the shear pattern is altered. The horizontal component of the load increases the lateral stress on the surrounding zone, leaving less resistance to support the lateral stress generated by the vertical component of the load. Meyerhof had proposed corrections to the bearing-capacity factors to be used in computing the ultimate capacity under the vertical component of load. These are given in Table 10:2.

TABLE 10:2 / CORRECTIONS FOR INCLINED LOAD[10:6]

		Inclination of Load from Vertical			
Factor	D_f	0	10°	20°	30°
N_γ	0	1.0	0.5	0.2	0
N_γ	B	1.0	0.6	0.4	0.25
N_c	0 to B	1.0	0.8	0.6	0.4

NONHOMOGENEOUS BEARING / If the soil is nonhomogeneous, the analyses are not directly applicable, but reasonable approximations can be made. When there are random variations in c and ϕ or thin repeating sequences of strata with different ϕ and c parameters within the hypothetical shear zone (Fig. 10.3), the mean of the c and ϕ can be used. If the range in variation is more than ±20% of the mean, somewhat higher safety factors should be used in design.

If a weak stratum overlies a strong one (Fig. 10.3b) the shear will be confined to the weaker material and the stronger will not be involved in the failure. The bearing capacity should be computed from the strength of the weaker stratum. Because the shear zone is restricted, the real bearing will exceed the computed value.

If a stronger layer overlies a weak stratum, the strong layer spreads the load, reducing the bearing pressure on the weaker material (Fig. 10.8c). Failure occurs by shear in the softer stratum as the stronger one bends down under load. The bearing capacity is computed from the strength of the weaker stratum using a reduced bearing pressure $q = \Delta\sigma_z$ computed by the approximation of Equation (10:6).

10:3 Stress and Settlement

When a load, such as the weight of a structure, is placed on the surface of a soil mass, the soil deflects, resulting in settlement of the structure. This is not a unique property of soils but one shared by all materials. In the same way that the deflection of a beam may be the limiting factor in structural design, the settlement of loads on soil is often the controlling factor in foundation design.

Settlement of soil or rock caused by load comes from two mechanisms: (1) distortion or change in shape of the prism of material beneath the load, and (2) the change in void ratio. The first is *distortion* or *contact* settlement; the second is *compression settlement,* sometimes termed *consolidation settlement.* Both mechanisms are related to the stresses produced by the foundation or other external loads as well as to the stresses already in the soil, due to its own weight or other body forces.

Both the compression and distortion settlements depend on the stresses produced in the soil by the foundation or other surface loads. By making simplifying assumptions about the physical properties of the soil, the stresses can be computed by the theories of elasticity. The settlements are then found from the stresses, using the physical properties of the soil determined by laboratory tests.

STRESSES DUE TO SOIL WEIGHT / The initial vertical effective stress in a soil mass before a structure is built is approximately the weight of the soil minus the neutral stress. At a depth of z in a homogeneous soil from (8:1b),

$$\sigma'_{z0} = \gamma z - u \tag{8:1b}$$

If the soil consists of different strata, each with a different unit weight, the vertical stress at any level is equal to the sum of their loads minus the neutral stress.

Changes in the neutral stress can play an important part in the settlement of a structure.

As described in Section 8:1, lowering of the water table can increase the effective stress and produce settlements comparable to those produced by the weight of a building. Typically lowering the water table 0.6 to 1 m (2 to 3 ft) is equivalent to the load of a one-story building (see Example 8:1). Serious settlements, extending far beyond the building site have been produced by construction drainage and must be included in any settlement evaluation.

STRESSES DUE TO SURFACE LOADS / When a load is applied to the surface of the soil mass, the vertical stress within the mass increases. If the soil were a series of independent columns, the load would be supported by the column immediately beneath it and the others would feel no change. The soil, however, is a coherent mass with the columns of soil interconnected elastically. Load at one point is transferred throughout the mass, spreading laterally with increasing depth.

As a very crude approximation, it can be assumed that load spreads through the soil as though it were supported by a flat-topped pyramid, as shown in Fig. 10.9. The sides of the pyramid are sloped $1(H)$ to $2(V)$, which means that the base of the pyramid becomes 1 m (3.3 ft) larger in length and

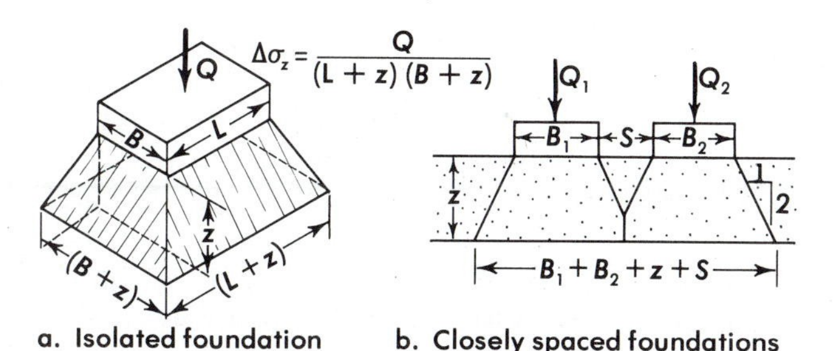

Figure 10.9 *Approximate method for computing the average vertical stress beneath a rectangular foundation with dimensions $L \times B$. The foundation is assumed to be supported by a pyramid of soil whose sides slope at $1(H)$ to $2(V)$.*

width for each meter increase in depth. The average stress increase in the soil at any depth z beneath a foundation whose dimensions are L and B and which has a load of Q and an average pressure of q is

$$\Delta\sigma_z = \frac{Q}{(L+z)(B+z)} = \frac{qLB}{(L+z)(B+z)} \tag{10:6}$$

This approximation is useful in preliminary studies of settlement. It can be misleading because it fails to show the variation in stress at a uniform depth, and it suggests no stresses beyond the pyramid.

A more accurate representation of the stress distribution can be obtained from various theories of elasticity. These show that a load applied to the soil increases the vertical stress throughout the entire mass. The increase is greatest directly under the load, as shown in Fig. 10.10, but extends infinitely far in all directions. As depth increases, the concentration of stress directly beneath the load decreases, but at any depth, if the

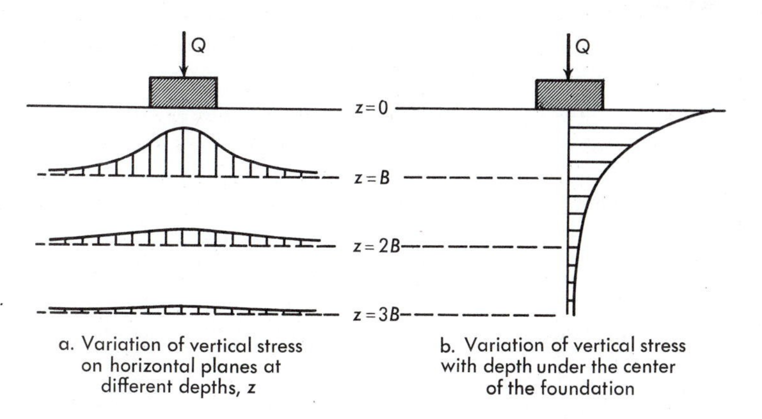

Figure 10.10 *Vertical stresses in a homogeneous formation due to a load Q applied to the ground surface by a square foundation of width B.*

increases in stress were to be integrated over the area to which they applied, the total force would equal the applied load Q. Near the surface the stress distribution depends on the size of the loaded area and on the contact pressure distribution, but at depths greater than about twice the width of the loaded area the stress distribution is practically independent of the way Q is applied.

Many formulas based on the theory of elasticity have been used to compute soil stresses. They are all similar and differ only in the assumptions made to represent the elastic conditions of the mass and its geometry. One of the most widely used formulas is that published by Boussinesq, a French mathematician, in 1885 and adapted to soil engineering by Jurgenson.[10:9] He assumed a homogeneous, elastic, isotropic body that extended infinitely in all directions below a level surface. A concentrated load of Q is applied to the surface of the body, and the increase in vertical stress, $\Delta\sigma_z$, at a depth z and at a horizontal distance of r from the point of application of Q is calculated by the formula:

$$\Delta\sigma_z = \frac{3Q}{2\pi}\frac{z^3}{(r^2+z^2)^{5/2}} \tag{10:7}$$

Westergaard published an analysis that more closely represents the elastic conditions of a stratified soil mass.[10:10] He assumed a homogeneous, elastic mass reinforced by thin, nonyielding, horizontal sheets of negligible thickness. The formula for the increase in vertical stresses produced by a concentrated surface load on a compressible soil (with Poisson's ratio = 0) is

$$\Delta\sigma_z = \frac{Q}{\pi z^2[1+2(r/z)^2]^{3/2}} \tag{10:8}$$

Both equations can be used to compute the stress increase caused by a footing if the depth z is greater than about twice the footing width B. For shallower depths the foundation pressure must be integrated over the foundation area to give the stress increase. The results of such integrations are presented in the charts of Figs. 10.11 to 10.14. The first two give contours of equal stress beneath foundations having widths of B and which exert a uniform pressure of q on the soil surface. The left side of each chart is for an infinitely long foundation and the right side for a square foundation. The depth and horizontal distances are expressed in terms of the foundation width B. The stress contours are expressed in fractions of the foundation pressure q. When the foundation is rectangular, the chart for a square foundation can be used with little error by assuming $B = \sqrt{A}$, where A is the foundation area.

Figures 10.13 and 10.14 are circular charts originally devised by Newmark.[10:11] The foundation is drawn on tracing paper to such a scale that the depth z, at which the stresses are to be computed, is equal (on the same

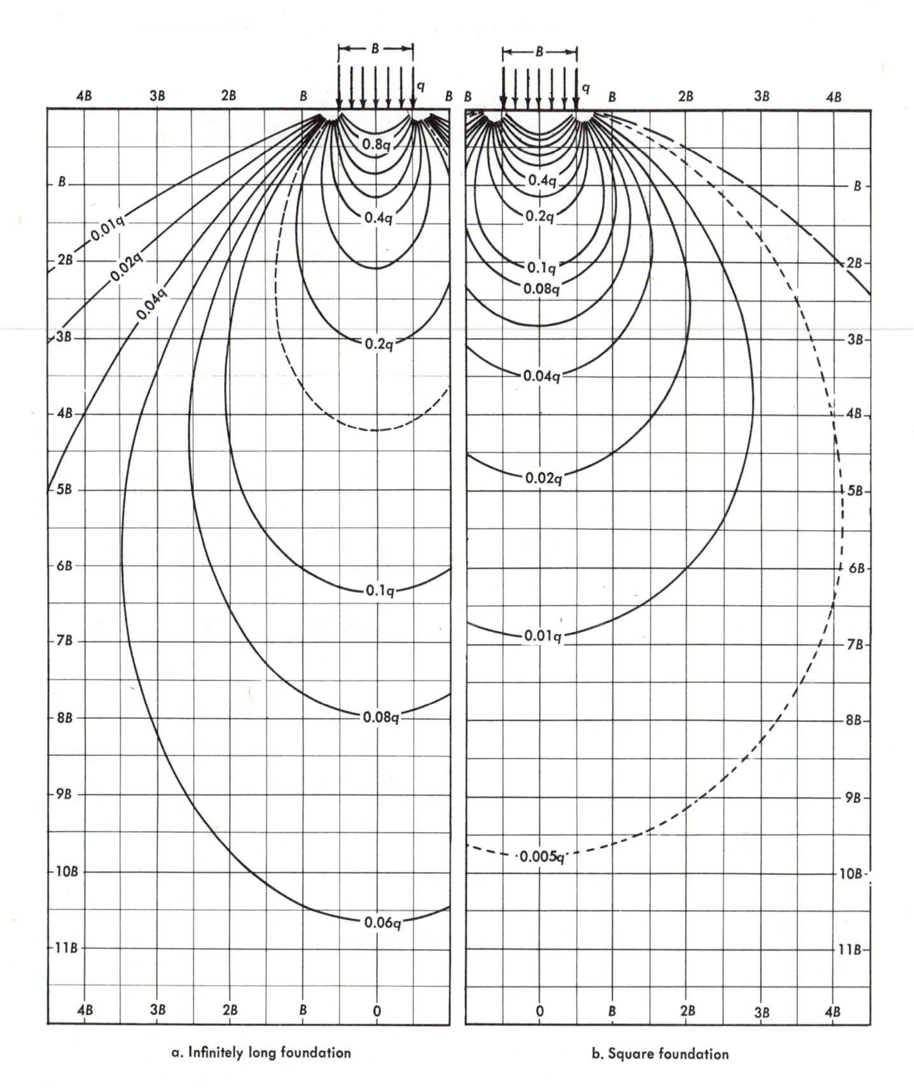

Figure 10.11 *Contours of equal vertical stress beneath a foundation on a semiinfinite, homogeneous, isotropic elastic solid—the Boussinesq analysis. Stresses are given as proportions of the uniform foundation pressure, q; distances and depths in terms of the foundation width, B.*

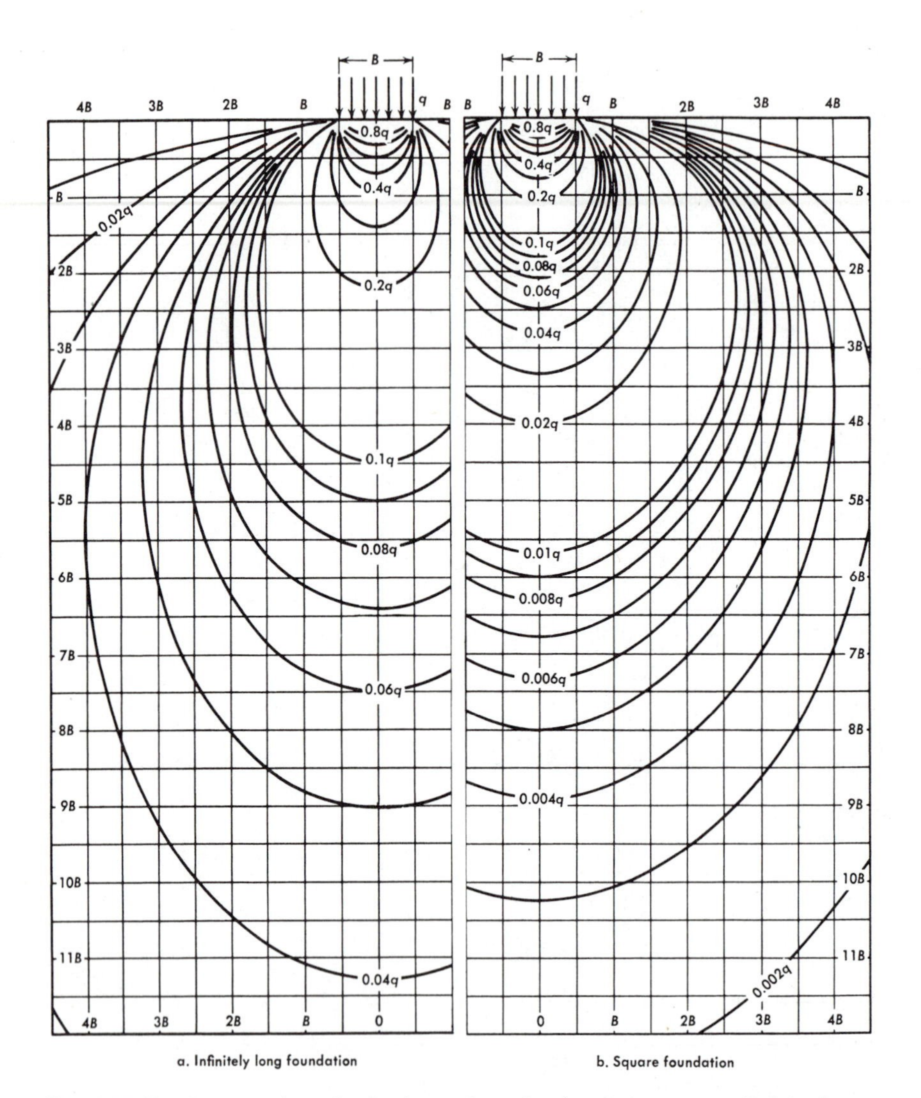

Figure 10.12 *Contours of equal vertical stress beneath a foundation on a semiinfinite, homogeneous, thinly stratified material—the Westergaard analysis. Stresses are given as proportions of the uniform surface pressure, q; distances and depths in terms of the foundation width, B.*

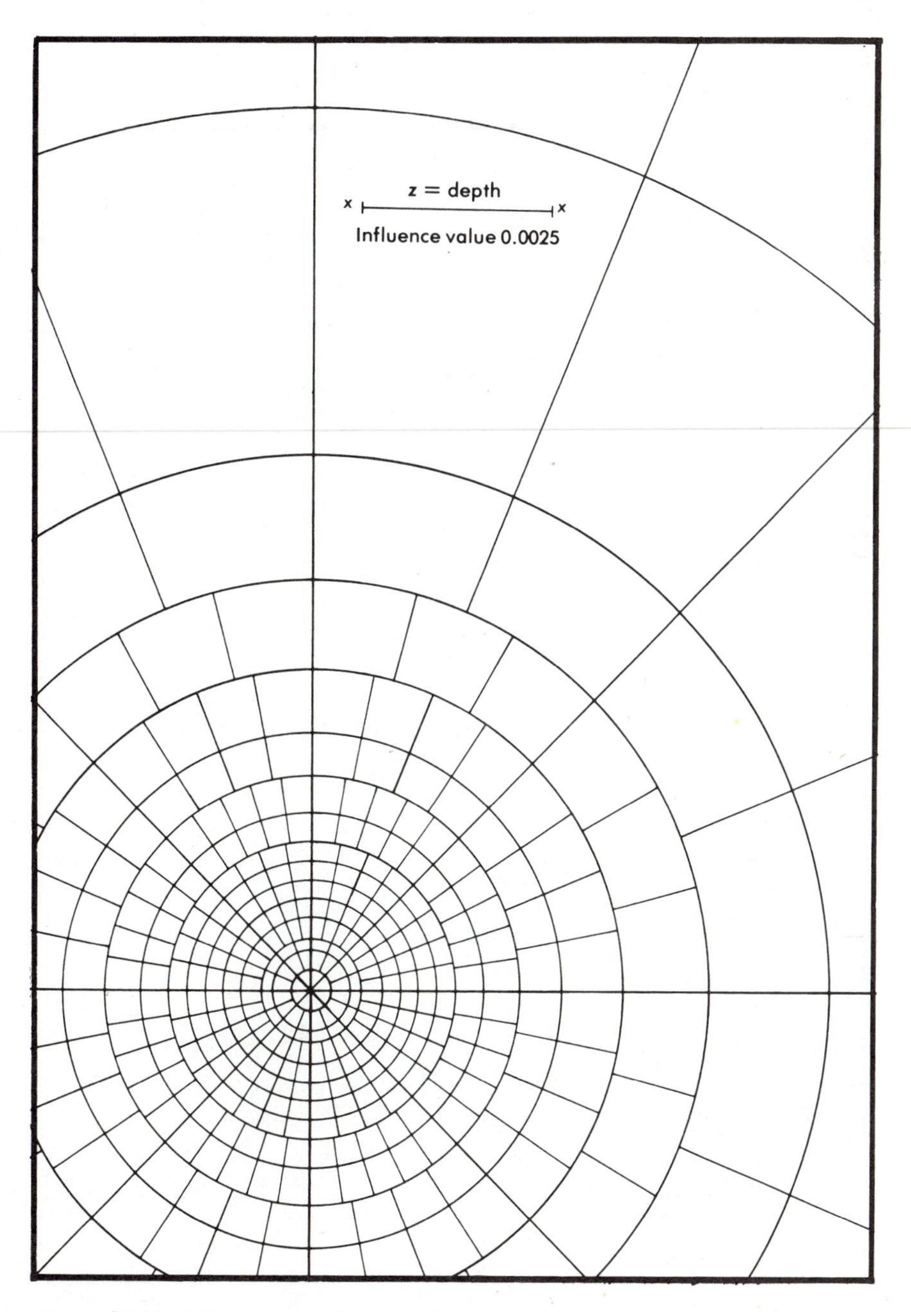

Figure 10.13 *Influence chart for computing vertical stress beneath a uniformly loaded foundation on a semiinfinite, homogeneous, isotropic elastic solid—the Boussinesq analysis.* (Adapted from N. Newmark.[10:11]).

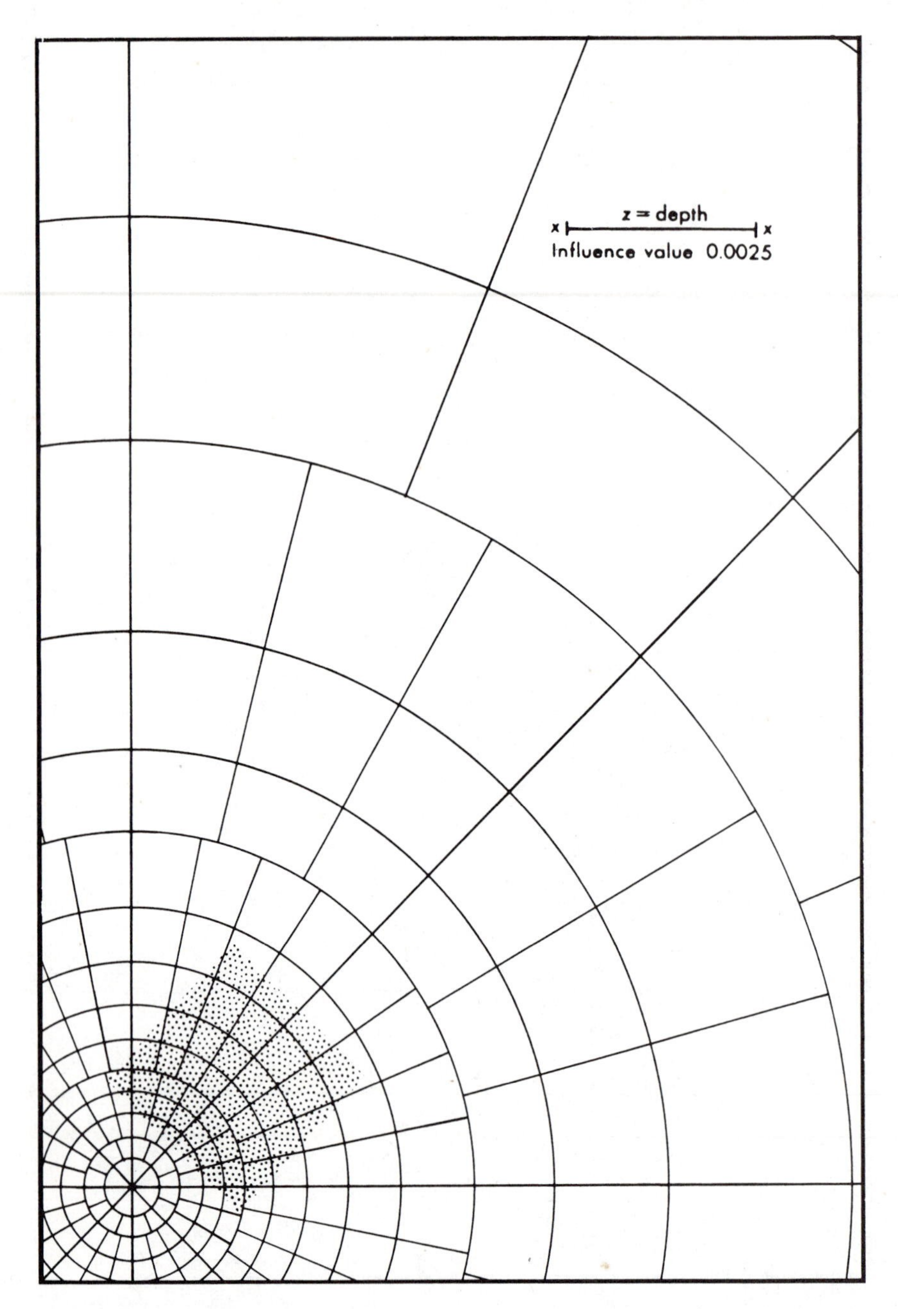

Figure 10.14 *Influence chart for computing vertical stress beneath a uniformly loaded foundation on a semiinfinite, homogeneous, thinly stratified material—the Westergaard analysis.* (Adapted from N. Newmark.)

scale) to the key line x—x on the chart. The paper is placed over the chart so that the point at which the stresses are to be computed is at the circle center. The number of squares covered by the foundation are counted. This number, multiplied by the foundation pressure and the chart influence value 0.0025, gives the stress increase at that depth and location.

Example 10:1

Compute the stress at a depth of 3 m and 2.4 m from the center of a footing that is 3 m × 3 m and which exerts a stress of 150 kN/m² (3.13 kips/ft²) on a stratified soil.

1. Using Fig. 10.12 (right side for a square foundation). The depth of the point $z = B$. The horizontal distance, 2.4 m is $r = 0.8B$. From the chart the contour is 0.09.

$$\Delta\sigma_z = 0.09 \times 150 = 13.5 \text{ kN/m}^2 \text{ (282 lb/ft}^2\text{)}$$

2. Using Fig. 10.14. The footing drawn to scale covers 36 "squares."

$$\Delta\sigma_z = 36 \times 0.0025 \times 150 = 13.5 \text{ kN/m}^2 \text{ (282 lb/ft}^2\text{)}$$

Example 10:2

Compare the increase in stress $\Delta\sigma_z$ in a soil stratum 1 m thick produced by a surface load of 150 kN/m² (3.13 kips/ft²) on a 0.5 × 0.5 m square footing with the stresses produced by a footing 3 m × 3 m square with the same load. The center of the soil stratum is 4 m beneath the ground surface. Use the pyramid approximation Equation (10:6).

1. For the 0.5 × 05 m footing

$$\Delta\sigma_z = \frac{0.5 \times 0.5 \times 150}{(4+0.5)(4+0.5)} = 1.85 \text{ kN/m}^2 \text{ (38 lb/ft}^2\text{)}$$

2. For the 3 × 3 m footing

$$\Delta\sigma_z = \frac{3 \times 3 \times 150}{(4+3)(4+3)} = 27.6 \text{ kN/m}^2 \text{ (576 lb/ft}^2\text{)}$$

The total load of the larger footing is 36 times that of the smaller and the stress increase beneath the larger foundation is 15 times greater than beneath the smaller although both foundations exert the same bearing pressure, q. Although the pyramid approximation is rough, it permits simple, quick comparison of the effect of foundation size and load differences.

Where several foundation loads, Q_1, Q_2, Q_3, . . . act simultaneously, such as adjacent footings of a building, the stress increase at the point of interest is calculated for each load independently, $\Delta\sigma_{z-1}$, $\Delta\sigma_{z-2}$, and $\Delta\sigma_{z-3}$.

The total stress increase is their algebraic sum:

$$\Delta\sigma_z = \Delta\sigma_{z\text{-}1} + \Delta\sigma_{z\text{-}2} + \Delta\sigma_{z\text{-}3} \cdots \tag{10:9}$$

The stress reduction produced by a small excavation is approximated by a similar computation. The weight of the soil is considered to be a negative load, $-Q$, acting at the level of the excavation bottom, and as if it were a foundation of the dimensions of the excavation. The stress decrease would be included in the algebraic sum of Equation (10:9) for deep foundations.

These stress analyses are based on loads placed at the surface of a very deep, wide (infinite) elastic body. If the load is placed below the surface, the stress can be evaluated by other elastic analyses or by a finite elements approximation. The stress increases for the same distances below the level of the load are smaller for loads beneath the ground surface. Thus stress increases and settlements computed by Equations (10:7) and (10:8) or Figs. 10.11 to 10.14 are conservative (too great).

The *average stress increase* over an area, such as the area directly under a foundation, is used to compute settlement of the soil strata beneath that foundation. As can be seen in Fig. 10.10, the stress immediately below a uniformly loaded foundation is greatest beneath the center and least beneath the edge. The average beneath the area is between the two limits; as an approximation it is about 10% less than the average of the center and edge stresses.

DISTORTION SETTLEMENT IN CLAYS / Distortion settlement occurs because the prism of soil immediately below the foundation shortens and bulges elastically, as shown in Fig. 10.15a. Saturated clays, many rocks,

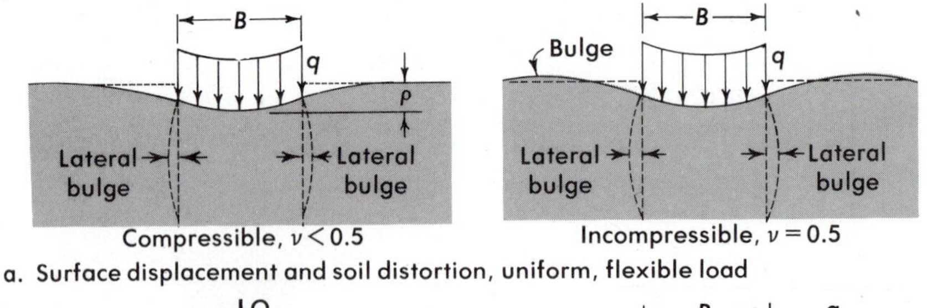

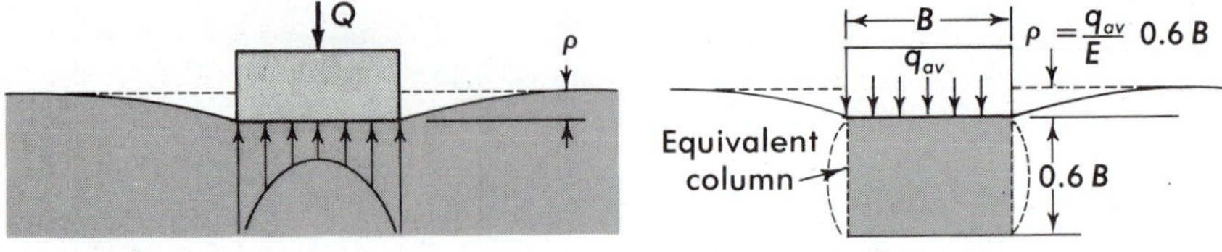

b. Contact pressure and settlement of rigid load

c. Equivalent soil or rock column

Figure 10.15 *Distortion settlement beneath a foundation on a homogeneous, elastic material such as saturated clay.*

and similar materials whose modulus of elasticity is reasonably constant within the stress range involved behave similar to a bowl of gelatin. If a uniform load of q per unit of area is applied to the surface, the loaded area and the adjacent surface deforms in a sagging profile. This is similar to the way a sponge mattress sags if one sits on it. The surface displacement from elastic distortion ρ_d can be computed from an integration of the Boussinesq equation for displacement. For a square loaded area of width B, the center and corner displacements $\rho_{d\text{cen}}$ and $\rho_{d\text{cor}}$ are:

$$\rho_{d\text{cen}} = \frac{0.84qB}{E} \tag{10:10a}$$

$$\rho_{d\text{cor}} = \frac{0.42qB}{E} \tag{10:10b}$$

The soil is assumed to have constant volume ($\nu = 0.5$) as the load is applied and is homogeneous to a depth of at least $2B$. If $\nu = 0$, the settlements are $\frac{1}{3}$ greater. For rectangular areas, an equivalent $B = \sqrt{A}$ is a reasonable approximation so long as $L < 3B$; for circles $B = \sqrt{\pi D^2/4}$. Uniform loads occur under shallow fills or metal tanks whose bottoms rest directly on the soil.

A rigid foundation, such as a concrete footing, cannot bend sufficiently to apply a uniform load on the elastic soil. Instead, the average pressure is redistributed as shown in Fig. 10.15b. The pressure between the ground and the foundation is greater at the outside edges where the uniform load produced less displacement and less at the center where the uniform load produced more displacement. The average pressure, $q_{\text{av}} = Q/A$. The distortion settlement is nearly an average of (10:10a) and (10:10b) for the uniform load (for constant E and $\nu = 0.5$).

$$\rho_d = \frac{0.6q_{\text{av}}B}{E} \tag{10:11a}$$

$$\rho_d = \frac{0.6Q}{E\sqrt{A}} \tag{10:11b}$$

The amount of settlement is the same as if the foundation were supported on an isolated soil or rock column whose height is $0.6B$ (Fig. 10.15c). Although the expression is derived for a square foundation, the assumption of an equivalent width $B = \sqrt{A}$ is reasonable if $L < 3B$. The Boussinesq analysis has been integrated for other conditions for which the reader should consult the references.

Equation (10:11b) implies that if the distortion settlement produced by different foundations on the same soil are to be equal,

$$\frac{q_{\text{av-1}}}{q_{\text{av-2}}} = \frac{Q_2}{Q_1} \tag{10:12a}$$

For larger loads, the average pressures must be reduced. For equal pressures,

$$\frac{\rho_{d\text{-}2}}{\rho_{d\text{-}1}} = \frac{Q_2}{Q_1} \qquad (10{:}12b)$$

In other words, for equal average pressures, the distortion settlement is proportional to the total foundation load.

The elastic distortion of clays is seldom critical for building foundations because a clay strong enough to support a footing will be rigid enough that the distortion settlement usually will not be serious. It occurs rapidly, during construction, and usually goes unnoticed. Therefore it is seldom included in settlement predictions. However, if building loads are suddenly increased, such as by an addition, the resulting distortion occurs immediately. Such sudden settlement is more likely to damage brittle materials than slowly developing consolidation. Therefore, the engineering decision to include clay distortion in evaluating potential settlement is based on construction sequence and potential structural changes.

COMPRESSION SETTLEMENT / The compression settlement for each compressible stratum (except within a depth of $2B$ for sands as subsequently described) is computed from the average initial effective stress in the stratum (8:1b), the stress increase due to the foundation load [Equations (10:6) to (10:8); Figs. 10.11 to 10.14], and the appropriate stress–void ratio curve or stress–strain curve for the stratum. The settlements for all the compressible strata ΔH_{com} are added to obtain the total.

$$\Delta H_{com} = \Delta H_1 + \Delta H_2 + \Delta H_3 \cdots \qquad (10{:}13)$$

The average initial effective stress in each stratum is the same as the initial stress at the middle of the stratum because the stress increases in direct proportion to the depth. The average increase in stress, however, is not the same as the stress at the stratum middle because the stress increase–depth relation is not linear. If the stratum is thin and relatively deep, it is sufficient to use the middle stress as the average. If the stratum is thicker than the footing width and if its depth is less than twice the footing width, it should be divided into thinner substrata and the average stresses and settlements computed for each.

Compression settlement is often a slow process requiring years to develop fully, as shown in Chapter 4. Estimates of the total settlement that will occur in any period can be made from the coefficient of consolidation of the soil. Because different strata are likely to have widely differing rates of consolidation, each stratum must be analyzed separately and the total settlement at any given time found from the sum of their individual settlements.

Consolidation settlement is adapted to a tabular form of computation. The stresses, changes in void ratio, or percentage of settlement, and settlements are computed for each compressible stratum or substratum; the sum of these at each point is the total settlement. Tabular solutions, with the aid of a pocket calculator, utilize the stress–void ratio curve directly. Computer programming requires approximating the stress–void ratio curve with a straight line or series of straight lines in the appropriate stress range. Programs are available that input column load and spacing, soil profile, and stress-compression strain; and output stress increase, void ratio change, and settlement for each stratum.

DISTORTION OF COHESIONLESS SOILS / The analysis of settlement distortion of a cohesionless soil is far more complex because its modulus of elasticity changes with σ_1 and σ_3 as shown in Fig. 5.15 and Equations (5:6a) and (5:6b). The cohesionless soil, confined beneath the center of a uniformly loaded foundation will have a higher E than the soil beneath the edge. Therefore, the edge settlement should exceed the center, as shown in Fig. 10.16. This is confirmed by field observations.

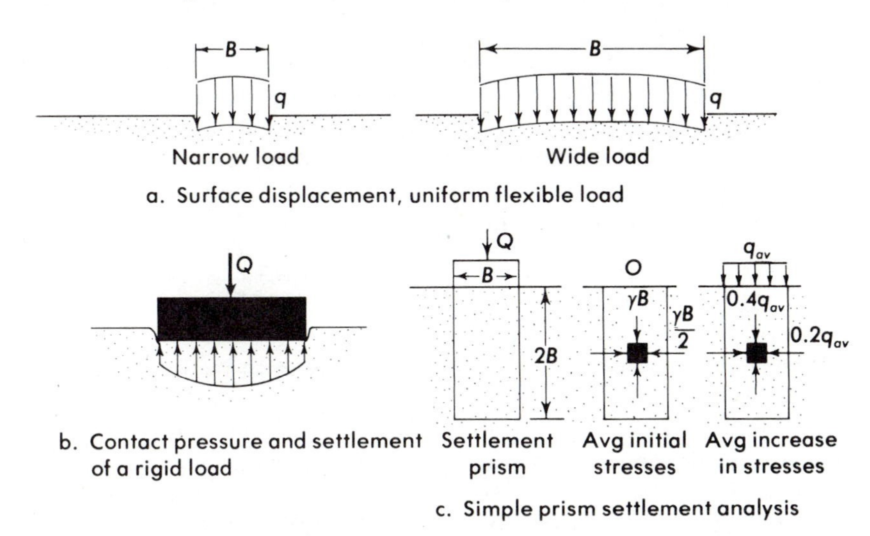

Figure 10.16 *Distortion settlement beneath a foundation on sand.*

The contact pressure between a rigid foundation and the cohesionless soil reflects the surface distortion. The pressure is somewhat greater than the average beneath the center of the foundation and falls off at the edge.

A simple direct equation for distortion of sand has not been written. It is possible to make a stress increment by increment, layer by layer evaluation, approximating the horizontal and vertical stresses by the Boussinesq analysis. Analyses by Schmertmann, using the finite element technique, show that 90% of the distortion settlement in sands occurs within a depth of

$z = 2B$ and more than $\frac{2}{3}$ within $z = B$. He utilized the Dutch cone and empirical correction factors to evaluate the distortion settlement.[10:12]

The author has used a simple approach based on the modulus of elasticity of sand determined from the laboratory, assuming that the settlement occurs in a prism whose height $z = 2B$. The initial average vertical stress, $\sigma_z = \gamma B$; the initial horizontal stress $\sigma_{h0} = 0.5\gamma B$. The average increase in vertical stress, $\Delta\sigma_z$, within the prism can be estimated from Equation (10:6) or from Fig. 10:11. It is about $0.4q$. The increase in horizontal stress would be $\Delta\sigma_h = 0.2q$.

Condition	Average Vertical Stress, $\sigma_z = \sigma_1$	Average Horizontal Stress, $\sigma_h = \sigma_3$
Before loading	γB	$0.5\gamma B$
After loading, q	$\gamma B + 0.4q$	$0.5\gamma B + 0.2q$

The two values of E for these conditions are found from laboratory tests. The distortion settlement is computed from the average of the two computed E values, E_1 and E_2.

The modulus of elasticity of sand from the usual drained laboratory test includes the effects of void ratio change and soil compression. Therefore the distortion settlement based on that E includes consolidation. The separate effects of the two mechanisms are not easily isolated because they occur simultaneously. Therefore, both distortion, ρ_d, and compression, ρ_c, within the depth $z = 2B$ are determined.

$$\rho_{d+c} = \frac{0.4qz}{E_{\text{avg}}} = \frac{0.4 \times 2Bq}{(E_1 + E_2)/2} = \frac{1.6Bq}{E_1 + E_2} \qquad (10:14)$$

The settlement occurs rapidly within a few hours to a few days after the load is applied.

A third approach to computing distortion settlement and associated shallow consolidation of cohesionless soils is extrapolation of plate load tests (Section 10:5).

DEEP COMPRESSION SETTLEMENT OF SAND / The contact settlement above $z = 2B$ includes compression. Deeper, there is not likely to be significant contact + compression settlement from surface loads because E increases with depth and confinement, while $\Delta\sigma_z$ from the foundation load decreases rapidly with increasing depth, as shown in Fig. 10.10.

Very wide loads such as a fill over a large area or stress increases from lowering the water table can cause sands to compress, especially if $D_r <$ 50%. The sand is tested in one-dimensional consolidation, as described in Chapter 4, using undisturbed samples if possible. Otherwise, disturbed sand at the same relative density as the virgin formation is tested. Stress increases and void ratio changes and settlements, ΔH for each stratum are computed as for clays. However, because of soil disturbance, estimates of compression settlement for sand are likely to be 2 to 3 times greater than measured values. Moreover, sand compression is often unnoticed because it is usually small and occurs within a few days of the load increase (often while the load is being placed). This is a mixed blessing. Because it is usually small and occurs early in construction, both design engineers and builders commonly ignore it. However, unexpected settlement can damage existing construction as well as play havoc with the elevation monuments, including those used to measure settlement.

10:4 Settlement Observations[10:14]

Whenever settlement exceeding 10 mm ($\frac{2}{5}$ in.) is predicted for a structure, careful observations should be made to check the accuracy of the estimates. So many assumptions are made in settlement studies that their accuracy is often poor, and the only way such studies can be improved is by correlating the actual, measured settlements with the predicted.

The first essential to accurate measurements is a stationary bench mark. This should be founded on bedrock if possible. In areas with deep soil such benchmarks can be constructed by driving a 100-mm (4-in.) pipe into the soil, through all compressible strata. This pipe is cleaned out and the hole extended into the firm soil and rock strata below by boring with earth drills or a diamond drill (Chapter 7). A 50-mm (2-in.) pipe is placed inside the larger outer casing and securely grouted into the firm stratum or rock. The inner pipe forms the benchmark, and the outer pipe acts as a sleeve to insulate the inner pipe from the settling soil. The second essential is permanent reference points at different key locations on the structure. Readings of settlement should be taken once every few days during the construction period and once or twice a year thereafter. Settlement can be read with a good engineer's level or special water-level devices made for the purpose.

Settlement observations during construction often can give warning of trouble from other sources. Landslides, underground subsidences, and bearing-capacity failures usually begin with slow but gradually increasing settlement rates. Usually, the trouble can be corrected before failure takes place if it is caught in time.

Excessive settlement usually leads to building cracks and in some cases structural failure. The engineer should recognize the causes of different types of cracking in order to correct them before structural failure results. Uniform settlement will produce no cracking except of water and sewer lines into the structure. Differential settlement can produce cracks, tipping of the structure, or both. The crack pattern depends on whether the center of the building or its edges settle more. The diagrams in Fig. 10.17 illustrate types of failure that occur.

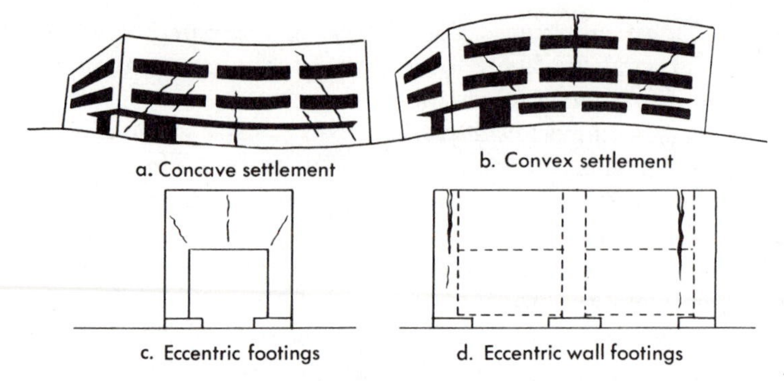

Figure 10.17 *Settlement crack patterns in buildings.*

The concave settlement is the usual pattern for a uniformly loaded structure on a compressible soil. The zone of settlement is saucer-shaped and extends well beyond the limits of the structure. Nearby buildings may be affected by the zone of settlement and develop new cracks. The convex pattern develops with wall-bearing structures or structures on loose sands.

Cracks often occur when footings are eccentrically loaded. This puts a bending moment into the base of the column or bearing wall and can cause failure. This is particularly true in outside bearing walls that are very close to the property line. The designer is tempted to make the footing wider inside the building than it is outside, and the result can be a crack, as shown in Fig. 10.17c and d.

Tipping is serious in narrow, tall structures such as chimneys and bridge piers. It is likely to occur when the soil compressibility is not uniform. It can also develop when the major cause of the settlement is a heavy load at some distance from the tall structure, as shown in Fig. 10.18a. The sagging setlement profile develops beneath the larger load, and the tall but lighter structure tilts in that direction. Such settlement caused by the weight of an approach fill for a bridge resulted in the abutment tilting backward against the fill, as shown in Fig. 10.18b.

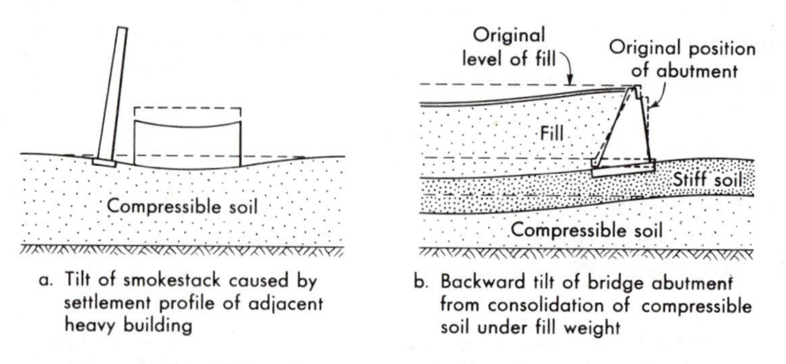

Figure 10.18 *Tilting of structures caused by adjacent heavy loads.*

Not all cracks in structures are caused by settlement. Shrinkage of mortar or of concrete blocks and similar masonry units is a common cause of cracking. Plaster is likely to shrink differently from the wood or masonry base which supports it. Shrinkage cracks are usually vertical and horizontal and are of uniform width or become narrow at both ends. Thermal expansion and contraction are important causes of cracks in exterior walls. Such cracks can be identified by their opening and closing with temperature changes. Vibration, shock, and earthquakes can cause cracking. Usually, these have an *x* pattern at the ends of walls and a ✳, or +, at the middle.

Gage marks on cracks can aid in studying their movement and in their identification. A straight pencil line is drawn across the crack at right angles to it, and gage points are set a definite distance apart [such as 50 mm (2 in.)] on each side of the crack. Measurement of the gage distance at regular intervals will show how much and in which direction the crack is moving.

OTHER CAUSES OF SETTLEMENT / There are many causes of structural settlement, all of which must be considered in design and which should be evaluated in studying the possible effects of settlement on the structure. The major causes, distortion and consolidation, are directly related to the foundation load and are controlled by the foundation design. Consolidation is also produced by loads induced by a changing environment that increases the effective stress by altering the neutral stress. Other forms of settlement are not directly produced by the structural load, although possibly aggravated by it. Neither the amount nor rate of these other causes of settlement can be predicted in advance, although the susceptibility to settlement may be estimated. Table 10:3 summarizes the principal features of each form.

10:5 Allowable or Design Pressure

After a foundation meets the requirements of location and minimum depth, two conditions remain that must be satisfied: First, there must be adequate safety against a failure within the soil mass; and second, the settlement of the foundation must not endanger the structure. It is obvious from the methods developed to analyze bearing capacity and settlement that these two conditions are independent of one another. For foundation design, however, it is desirable to know the maximum pressure that can be placed on soil without exceeding either of these two limits. This maximum is known as the *allowable pressure*, or *allowable loading* q_a.

PRESUMPTIVE BEARING PRESSURE / The oldest method of determining the allowable foundation pressure is to rely on past experience with similar materials in the region. Most engineers accumulate information on the success of their past designs, and these are used as a basis for future work. In many areas, such as the larger cities, the records of which design pressures were successful and which were not have been assembled and

TABLE 10:3 / CAUSES OF SETTLEMENT

Cause	Form of Mechanism		Amount of Settlement	Rate of Settlement
Structural load	Distortion (change in shape of soil mass)		Compute by elastic theory (partly included in consolidation)	Instantaneous
	Consolidation: Change in void ratio under stress	Initial	Stress–void ratio curve, time curve	Rapid
		Primary	Stress–void ratio curve	Compute from Terzaghi theory
		Secondary	Compute from log time–settlement	Compute from log time–settlement
Environmental load	Shrinkage (due to drying)		Estimate from stress–void ratio or moisture–void ratio and moisture loss limit–shrinkage limit	Equal to rate of drying. Seldom can be estimated
	Consolidation (due to water table lowering)		Compute from stress–void ratio and stress change	Compute from Terzaghi theory

Load independent (but may be aggravated by load; often environment related, but not dependent)	Reorientation of grains—shock and vibration	Estimate limit from relative density (up to 60–70%)	Erratic, depends on shock, relative density
	Structural collapse—loss of bonding (saturation thawing, etc.)	Estimate susceptibility and possibly limiting amount	Begins with environmental change, rate erratic
	Raveling—erosion into openings, cavities	Estimate susceptibility but not amount	Erratic, gradual or catastrophic, often increasing
	Biochemical decay	Estimate susceptibility, possible limits	Erratic, often decreases with time
	Chemical attack	Estimate susceptibility	Erratic
	Mass collapse—collapse of sewer, mine, cave	Estimate susceptibility	Likely to be catastrophic
	Mass distortion—shear-creep or landslide in slope	Compute susceptibility from stability analysis	Erratic, catastrophic to slow
	Expansion—frost, clay expansion, chemical attack (resembles settlement)	Estimate susceptibility sometimes limiting amount	Erratic, increases with wet weather

condensed in tabular form. These are called *presumptive bearing pressure* because it is presumed on the basis of past performance that the soil or rock can support such a pressure without a bearing-capacity failure or excessive settlement. Most *building codes* include such tables, and they are often a helpful guide to local practice.

Table 10:4 gives presumptive bearing pressures based on the author's experiences for simple structures up to four stories.

TABLE 10:4 / TYPICAL PRESUMPTIVE BEARING PRESSURES FOR PRELIMINARY DESIGN

Material*	N† (Standard Penetration Resistance)	kN/m^2	kips/ft^2	kg/cm^2
Loose sand, dry	5–10	70–140	1.5–3	0.75–1.5
Firm sand, dry	11–20	150–300	3–6	1.5–3
Dense sand, dry	31–50	400–600	8–12	4–6
Loose sand, inundated	5–10	40–80	0.8–1.6	0.4–0.8
Firm sand, inundated	11–20	80–170	1.6–3.5	0.8–1.8
Dense sand, inundated	31–50	240+	5+	2.5+
Soft clay	2–4	30–60	0.6–1.2	0.3–0.6
Firm clay	5–8	70–120	1.5–2.5	0.8–1.2
Stiff clay	9–15	150–200	3–4.5	1.5–2.2
Hard clay	30+	400+	8+	4+
Loose mica silty sand, damp	5–10	120–200	2.5–4.5	1.2–2.0
Firm mica silty sand, damp	11–20	200–350	4.5–7.5	2.2–3.8
Badly fractured or partially weathered rock	50+	500–1200	10–25	5–12
Hard rock, occasional soft seams	RQD† = 50%	1500–5000	30–100	15–50
Massive hard rock	RQD† = 90%	10,000+	200+	100+

*Section 2:10.
†Section 7:5.

Unfortunately, the use of presumptive bearing pressures often leads to trouble. Most of the tables are based on experiences going back to the nineteenth century and on entirely different types of structures than are built today. The soil and rock characteristics are defined by description, and often the most important properties are not mentioned. Sometimes the building code table is a copy of that from some other city and does not reflect local practice. For example, one municipal code includes a bearing pressure for granite, although there is none within 600 km (400 miles).

Finally, the values do not reflect the influence of building size and weight, excavation depth, or type of structure. As a result, presumptive pressures in design do not insure a safe, economical foundation. The engineer has just as much responsibility in computing the allowable foundation pressure for the structure as in determining the size of a beam or thickness of a floor slab; to rely only on experience can be disastrous. It is ironic that most of the foundation failures that the author has investigated have involved designs that nominally conformed to the local building code.

LOAD TEST / The *load test* or *test plate* method of determining allowable soil pressure (Fig. 10.19) was developed because of the failures of the design tables. Essentially, the load test is a model test of a foundation. A small plate, usually 0.3 m (1 ft sq) or 0.76 m (30 in.) in diameter, is placed on the undisturbed soil and is loaded in increments. The results of the test are presented in a load–settlement curve of the test plate. Properly conducted and correctly interpreted, the load test is a valuable *aid* to rational design, but, as usually conducted, it is a waste of time and money and often leads to a dangerous sense of false security.

A pit is dug to the bearing level. Its width is 5 times the plate width. Although 0.3-m (1-ft) plates are widely used, the larger the plate the more realistic the test. Full-scale tests are expensive but justified when soil bearing capacity or distortion settlement computations are in doubt. A load is placed on the plate, as shown in Fig. 10.19, by a loading platform weighted with pig

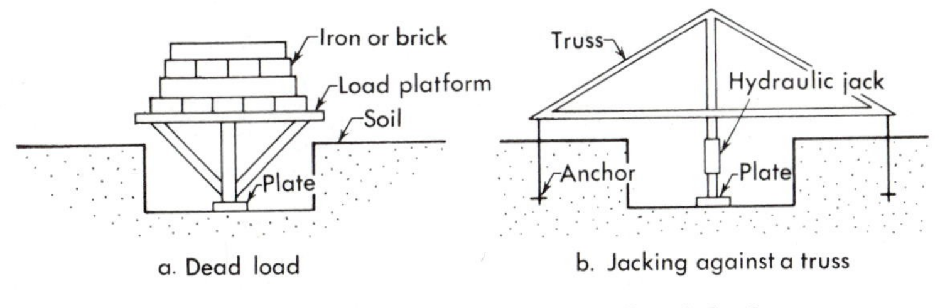

Figure 10.19 *Methods for making a small-scale load test.*

iron, lead, or concrete, or by jacking with calibrated hydraulic jacks against a beam or truss held down by earth anchors or weights. Loads are applied in increments of one-fourth the estimated allowable soil pressure and increased until 2 times the estimated allowable soil pressure is reached for sands and gravels and 2.5 times the estimated pressure for clays. Settlement under each increment should be read to 0.025 mm (0.001 in.), and must be referred to a benchmark beyond the limits of the possible settlement profile. Each increment should be maintained constant until the rate of settlement is less than 0.1 mm (0.005 in.) for 30 min, before adding another increment. The final increment should be maintained at least 1 hour before ending the test. A time–settlement curve (Fig. 10.20) should be plotted on semilogarithmic coordinates for each increment of load. The semilog plot will show a break into a straight, nearly flat line, point A. This point should be selected as the "ultimate settlement" under each load increment. The loads and their corresponding ultimate settlements should be plotted to form a load–

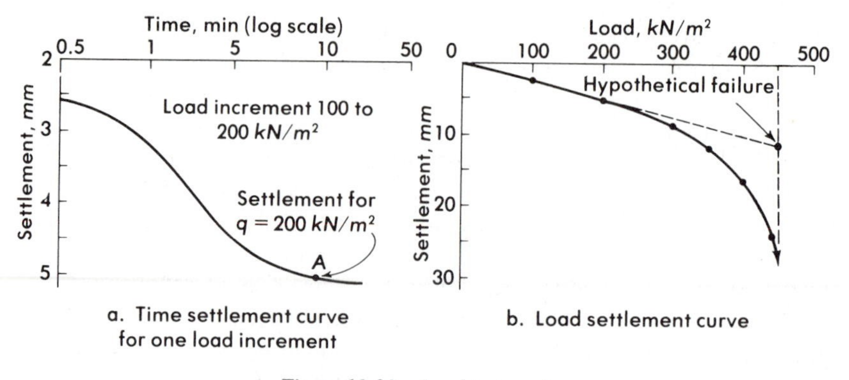

Figure 10.20 *Load test results.*

settlement curve (Fig. 10.20b). A definite break or the intersection of tangents to this curve represents the ultimate bearing capacity of a foundation the same *size*, *depth*, and *location* as the test load.

The test is useless unless its results are interpreted in terms of a full-size foundation. Interpretation must be based largely on theory because so few field observations have been reliable enough to correlate foundation performance with load test results. Because bearing capacity of footings on a deep homogeneous saturated clay soil is independent of the width of the loaded area, the critical pressure determined by the load test is the same for all footing sizes:

$$q_0 \text{ (foundation)} = q_0 \text{ (load test)} \tag{10:15a}$$

In sands the gravels the bearing capacity increases in direct proportion to the width of the loaded area, and so the following correction should be made:

$$q_0 \text{ (foundation)} = q_0 \text{ (test)} \times \frac{B \text{ (foundation)}}{B \text{ (test)}} \tag{10:15b}$$

If the soil is not homogeneous to a depth of at least the width of the proposed foundation, the load test results are meaningless as far as bearing capacity is concerned.

The load test can be used to determine distortion settlement in a homogeneous soil, provided the soil is uniform to a depth of twice the footing width. For saturated clays, the distortion settlement at a given pressure per square foot varies directly with the width of the loaded area (10:11a).

$$\rho_d \text{ (foundation)} = \rho_d \text{ (test plate)} \times \frac{B \text{ (foundation)}}{B \text{ (plate)}} \tag{10:16a}$$

The load test cannot predict compression settlement. Compression requires time, particularly in the more critical soils such as thick soft clays, and the few hours or days allotted to the load test can allow only a negligible part of the total volume change to take place. Furthermore, if the compressible soil extends to any depth, the stresses throughout the stratum resulting from the loaded test plate will be very small compared to the full-size foundation. The same pressure applied by a wide foundation will produce correspondingly greater settlement.

In cohesionless soils the compression and distortion settlement occur simultaneously. Most of the settlement in a thick deposit of sand or gravel occurs near the surface because the modulus of elasticity increases and the stress decreases rapidly with increasing depth. Based on limited field test data, as well as Equation (10:14) for distortion and consolidation of sand, the following applies to cohesionless soils where $z \geqq 1.5B$:

$$\rho\ (\text{foundation}) = \rho_{c+d}\ (\text{test plate})\left[\frac{B\ (\text{foundation})}{B\ (\text{plate})}\right]^n \qquad (10{:}16b)$$

The exponent n is typically between 0.5 and 0.7.

Rules for conducting load tests are included in many building codes and standard specifications. The most important rules are:

1. Use as large a load plate as practical.
2. Make the tests in both good and bad soils.
3. Correlate the results with accurate boring data.
4. Interpret the data with full understanding of the mechanics of bearing capacity and settlement.

The load test cannot be the entire answer to the question of allowable foundation pressure because it supplies only a part of the data (bearing capacity and distortion) for a model of the proposed structure.

10:6 Rational Procedure for Design Pressures

A rational determination of allowable foundation pressure is similar to the design of other parts of the structure. First, a trial pressure is assumed, based on experience. Second, the trial pressure is checked by a bearing-capacity analysis to determine the safety against soil failure. Third, if it is safe, an analysis is made of the settlement produced, to see if it is excessive. Fourth, the trial pressure is revised to increase the safety, reduce the settlement, or improve the economy, depending on the analytical results.

In order to do this, the engineer requires accurate data on the ground below and the structure above the foundation. The data include the depth and thickness of the soil and rock strata, the level of ground water, and the physical properties of each soil, including its strength and compressibility. If the soil deposit is uniform, the analyses are based on the average properties of each material; if it is variable, the analyses are based on the variations in strength and compressibility from one part of the site to another.

The data required on the structure include the use or purpose, the elevations of the lowest floors, particularly basements and pits, the type of structural framing and its sensitivity to deflection, and the possibility of future additions. The load data are the depth and extent of general excavation and filling and the dead loads and live loads on columns, including the amount of live load that is likely to be continuous. If the lowest floor is supported directly on the ground, its average sustained load is needed.

The *safety factor* of a foundation, F_s, is the ratio of the ultimate bearing capacity q_0 to the actual foundation pressure q. The *safe bearing capacity* q_s is the ultimate bearing capacity divided by the minimum permissible safety factor

$$F_s = \frac{q_0}{q} \tag{10:17a}$$

$$q_s = \frac{q_0}{F_s} \tag{10:17b}$$

$$q_a \leqq q_s \tag{10:17c}$$

The allowable foundation pressure q_a cannot exceed the safe bearing capacity, but it often is less.

SAFETY FACTOR / The safety factor required for design depends on how accurately the soil conditions and the structural loads are known and what hazards are involved in a bearing-capacity failure. Any future changes in the site, such as a rising water table or excavation adjacent to a footing that will reduce the surcharge, must be taken into account by the ultimate bearing-capacity equation or else included in the safety factor. For temporary construction work, where a failure would be inconvenient but not disastrous, a safety factor of 1.5 is prudent. For most cases of structural design where there is reasonably accurate data on the soil and loadings, a safety factor of 2.5 is employed with dead load plus full live load. If a large part of the live load is not likely to develop, a minimum safety factor of 2 is permissible. When the conditions are questionable, a safety factor of 4 is sometimes warranted.

PERMISSIBLE SETTLEMENT / The settlement is computed by using the assumed foundation design pressure (not the ultimate bearing capacity). For foundations on soils that settle slowly, such as saturated clays, only the dead load plus any sustained live load is used in the analysis; but for partially saturated clays, silts, and organic soils, which usually settle rapidly, the dead plus total live load is used. In some cases it is necessary to compute the settlement of every column or part of a structure. In many cases it is sufficient to know the settlement of the more critical parts, such as footings for delicate machinery, smoke stacks, and the heaviest columns.

The permissible amount of settlement depends on its uniformity and on the dimensions and nature of the structure. If all parts of the structure settle the same amount, the structure will not be damaged. Only access, drainage, and utility connections will be affected, and these can tolerate movements of more than $\frac{1}{2}$ m (2 ft). Differential settlement causing tilting is important for floors, crane rails, machinery, and tall narrow structures such as chimneys, and the limits are established by their operation. Differential settlement causing curvature affects the structure itself and is limited by the structural flexibility. Table 10:5 gives the maximum settlements that can be permitted. It is based on both theory and observations of structures that have suffered damage.[10:14, 10:15, 10:16]

TABLE 10:5 / MAXIMUM PERMISSIBLE SETTLEMENT*

Type of Movement	Limiting Factor	Maximum Settlement
Total settlement	Drainage and access	0.15 to 0.6 m (0.5 to 2 ft)
	Probability of differential settlement	
	Masonry walls	25 to 50 mm (1 to 2 in.)
	Framed buildings	50 to 100 mm (2 to 4 in.)
Tilting	Tower, stacks	$0.004B$†
	Rolling of trucks, stacking of goods	$0.01S$†
	Crane rails	$0.003S$†
Curvature	Brick walls in buildings	$0.0005S$ to $0.002S$†
	Reinforced concrete building frame	$0.003S$†
	Steel building frame, continuous	$0.002S$†
	Steel building frame, simple	$0.005S$†
Maximum permissible settlement	Front slab, 100 mm thick	$0.02S$†

* B is base width; S is column spacing.
† Differential settlement in distance B or S.

The allowable pressure for each footing must satisfy both the requirements of safety and settlement. For convenience in its determination for each foundation, the results of the bearing-capacity and settlement analyses are expressed graphically, as shown in Fig. 10.21. The safe bearing pressure

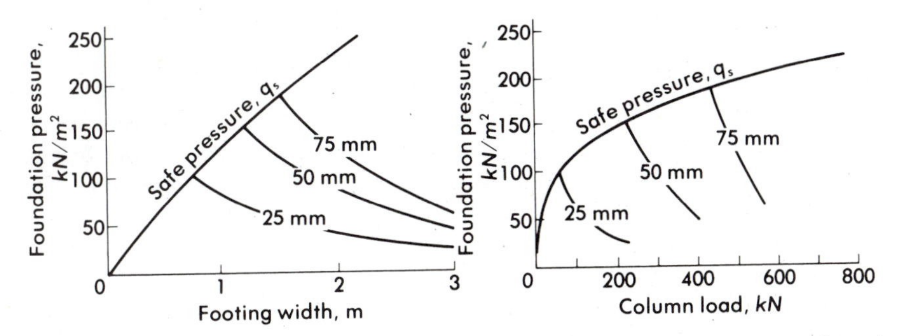

Figure 10.21 *Footing design charts showing safe bearing as upper limit and curves of equal settlement.*

represents the upper limit without regard to settlement. Curves of equal settlement for different column loads or foundation sizes are also shown. A number of such charts are prepared, each for a different set of soil conditions, different shape of foundation, or for varying ground water and depth of excavation. From these the designer can select the allowable pressure that will keep the total settlement and differential settlement within limits.

REDESIGN ALTERNATIVES[10:17] / When the analyses show that the assumed foundation pressure is not safe or that the settlement will be excessive, or when they indicate that the safety is so high and settlement so low that the foundation is uneconomical, a redesign is necessary. Usually, this is limited to the foundation, but it is often fruitful to extend the redesign to the structure and even to the soil.

The simplest procedure is to change the foundation. By reducing the design pressure, the safety factor against failure is increased. Reducing the pressure is not always effective in reducing settlement, however. If the compressible strata are at a shallow depth below the foundation, the settlement will be reduced almost in proportion to the pressure; but if the compressible layer is far below the foundation, a reduction in pressure may not reduce the settlement materially.

Example 10:3

What is the effect of reducing the foundation pressure from 300 kN/m^2 (6.26 kips/ft^2) to 150 kN/m^2 (3.13 kips/ft^2) on the stresses at a depth of 3 m (10 ft) beneath a column supporting a load of 600 kN (135 kips)?

1. For the higher pressure $B = \sqrt{\dfrac{600}{300}} = 1.414$ m.
2. $z/B = 3/1.414 = 2.12$. The factor from Fig. 10.12b is $0.07q$ at the center; $\Delta\sigma_z = 0.07 \times 300 = 21$ kN/m^2 $= 0.44$ kips/ft^2.
3. For the lower pressure $B = \sqrt{\dfrac{600}{150}} = 2$ m.
4. $z/B = 3/2 = 1.5$. The factor from Fig. 10.12 $= 0.13q$ at the center; $\Delta\sigma_z = 0.13 \times 150 = 19.5$ kN/m^2 $= 0.41$ kips/ft^2.

The stress reduction is negligible, not justifying the increased cost of the larger foundation. The limit in size is reached when the foundations touch and form a continuous footing. Such a continuous foundation can bridge over small soft areas, but it cannot reduce the dished-in settlement profile produced by thick compressible strata.

Increasing the foundation depth will increase the bearing capacity in homogeneous soils by increasing the surcharge, particularly if the soils have a high angle of internal friction. If the soil is stratified and becomes stronger with increasing depth, then depth will improve the bearing capacity; but if the soil deposit has a hard crust underlain by softer soil, an increase in depth will reduce the bearing. In most cases, increasing the depth will reduce settlement. However, if the compressible strata are deep, increasing the depth will bring the foundation closer to the source of trouble and will aggravate the settlement. Very deep foundations that transfer the load below the weak or compressible strata are considered in Chapter 11.

Sometimes, great benefits can be had by changes in the structure. The column loads can be reduced by reducing the spacing. This helps if the bearing capacity is limited, but it will usually not reduce settlement appreciably because the total structural load is not changed materially. Substituting lightweight construction for conventional forms will reduce the total weight and benefit both the safety against failure and settlement. If the site is sufficiently large, the structure may be spread over a larger area and the concentration of load reduced, which will also benefit safety and settlement. If extensive fill is planned to support the ground floor, its weight is a major factor in settlement. For example, 1 m (3 ft) of fill weigh as much as $2\frac{1}{2}$ stories of building. Eliminating such fill will reduce settlement materially. Changing the structure by making it so rigid that it will resist distortion is occasionally possible, as shown in Fig. 10.22. The floors and walls can be combined structurally to develop a box girder stiff enough to prevent differential settlement. Introducing trusses into the building frame can do the same thing. For very small buildings a continuous mat can be made rigid if its thickness is about one-tenth its span, but the weight of the mat itself becomes a major load on the soil.

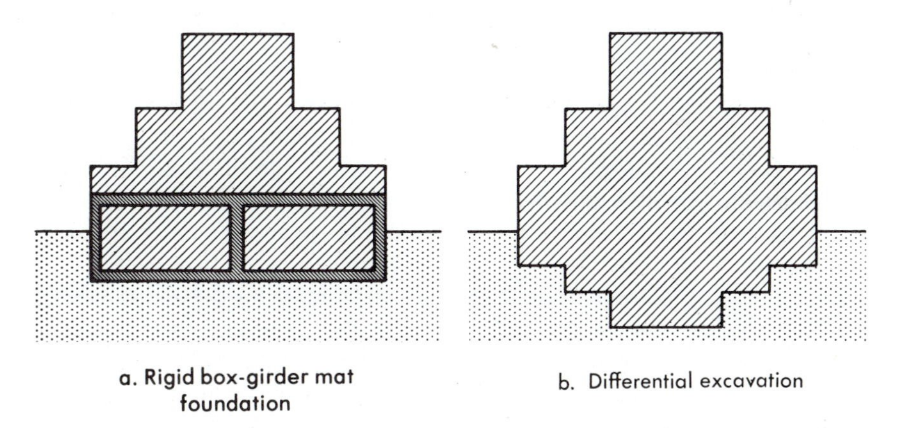

Figure 10.22 *Reducing differential settlement by changes in building rigidity or excavation.*

Making the structure flexible so that it can conform to the settlement is a simple method of preventing damage from differential movement. Simple structural framing, small wall panels that are not rigidly connected to the floors and columns, masonry walls with low-strength mortar, and ground floors that are reinforced, jointed, and keyed like a pavement will permit maximum movement with minimum damage. Flexible construction is best adapted to wide, low buildings where the structural framing would ordinarily be light.

Reducing the net load on the soil by excavation (Fig. 10.22b) is a very old method for minimizing settlement. If the weight of the soil excavated equals the structural weight, there will be no increase in the stresses in the soil below and therefore little settlement. Such a design is a *compensated foundation* or sometimes "floating the structure," since the structure appears to be buoyed up by the weight of the soil displaced. The soil, however, is a solid that expands when the load is relieved by excavation and which recompresses when reloaded; therefore there will be some settlement even with a good balance. A perfect balance is impossible because of the variable live load in the structure and the variations in the unit weight of the soil excavated.

If some areas of the structure are heavier than others, the balance can be improved by *differential excavation*: the excavation depth is increased in the areas of greatest load so that the net increase in stress in the compressible strata is relatively uniform across the total breadth of the structure. A changing water table can upset the balance. This method requires a very careful evaluation of the soil and loading conditions in order to be successful.

Example 10:4

A 10-story building weighs 7 kN/m^2 or 150 lb/ft^2 per floor, and the roof 5 kN/m^2 or 105 lb/ft^2 (dead load + full-time live load). Settlement analyses show that a clay stratum 3 m thick beginning 10 m below the ground surface will compress excessively. How can excavation offset this weight?

1. The building load is $10 \times 7 + 1 \times 5 = 75$ kN/m^2.
2. If the soil weighs 20 kN/m^3 (127 lb/ft^3), the required excavation for 0 net loading would be $75/20 = 3.75$ m $= 12.3$ ft.

Changing the soil to increase the safety or reduce the settlement includes drainage, densification, altering the soil by an additive, and changing to another site. Drainage improves the soil by reducing the neutral stress. In a cohesionless soil the bearing capacity is often doubled by lowering the water table below the foundation shear zone, but it must be permanent before it can be counted on in design. Reducing the neutral stress in a compressible soil will cause it to consolidate and become stronger and, of course, to settle. If the consolidation can be completed before the structure is built, the bearing capacity will be increased and the settlement reduced. Loss of drainage after construction is complete is not critical in the compressible soil and can be helpful because, as the neutral stress increases, the effective stress decreases.

Preconsolidating a compressible soil by a surcharge load is very effective in increasing strength and reducing settlement. In effect this is another form of drainage, by consolidation, and was discussed in Chapter 6. Stage loading makes it possible to improve the soil by the weight of the structure itself. For example, the soil at a riverside site was too weak to support with safety the full load of a grain elevator. The load was limited to half the capacity for the first year to maintain safety. The soil consolidated and become stronger and could then support the full load.

Densification of loose sands by shock and vibration is effective in increasing bearing capacity and reducing settlement. Altering the soil by injecting cementing agents, changing the chemistry of clays, and fusing it with heat are collectively termed *soil stabilization*. These are discussed in Chapter 6.

Sometimes it is better to move to a different site. It is physically possible to design a satisfactory foundation for any site if enough money can be spent. However, the cost of foundations added to the cost of the property can make cheap real estate very expensive. A complete study of all the economic factors involved is required to determine if special foundations are worth their cost.

10:7 Footings and Mats[10:17, 10:18]

FOOTING FOUNDATIONS / A footing is an enlargement of a column or wall in order to reduce the pressure on the soil to the maximum allowable. Beneath a wall the footing may be continuous, forming a long, rectangular, loaded area, called a *wall footing*. Beneath a column the footing may be any shape, but economy in construction favors the square. Rectangular shapes are often used if clearances prevent squares, or two or more square footings may be joined to form a single, rectangular footing under several columns. Occasionally, round or hexagonal footings are used, especially under chimneys or heavy machines, but the economy in materials of such shapes is usually overbalanced by the additional labor required for their construction.

Structurally, the footing is a wide beam, acted on by a distributed load (the soil pressure) and supported by a concentrated force (the column). Modern practice calls for reinforced concrete, which is designed by the usual methods. The soil pressure is usually assumed to be acting uniformly over the footing. This is a conservative assumption for footings on sand but may be somewhat unsafe for footings on saturated clays. The soil pressure under a rigid load on clay is high at the outer edges of the loaded area and decreases to about half the average at the center. The pressure under a concrete footing will not have such extreme variation, owing to some deflection of the footing edges, but it will not be uniform. Therefore it is well to be conservative in the structural design of footings on clay.

To avoid eccentricity, the centroid of the footing should coincide with the centroid of the load coming on it. Ordinarily, this is no problem, but along the exterior walls where the property line limits the extent of the footing or where elevator pits, machinery, and utilities obstruct the space, a concentrically loaded footing is impossible. If the eccentricity is small, it is sometimes possible to design the footing and column to absorb the unbalanced moment (Fig. 10.23a). A better method is to combine two adjacent footings into one large footing, as shown in Fig. 10.23b. The footing should be proportioned so that its center of gravity is at the same point as the center of gravity of the two column loads. If the column spacing is large, two adjacent footings can be connected by a small tie beam that absorbs the eccentric moment (Fig. 10.23c).

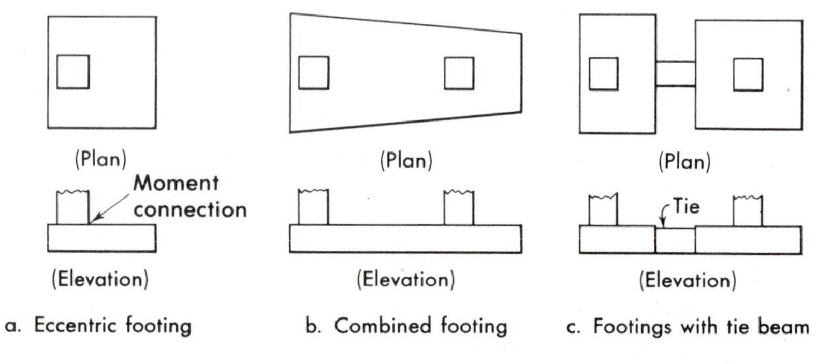

Figure 10.23 *Footing shapes to avoid eccentric moments.*

STRAP FOUNDATIONS / A *strap foundation* is a continuous footing that supports three or more columns in a straight line. It consists of a number of spread footings that have been connected together. Straps are employed for two reasons: to provide structural continuity and to achieve construction economy. When single footings are large and closely spaced in one direction, they can be combined to form a continuous shallow beam. This can bridge over small (less than half the column spacing) weak areas and achieve some economy through structural continuity. The strap foundation is often cheaper to build because the foundation excavation is a continuous trench rather than a series of isolated pits. The strap is designed like an inverted continuous beam with a uniform load on the bottom and with concentrated supports on the top.

MAT FOUNDATIONS / A *mat* or *raft* is a combined footing supporting more than three columns not in the same line. The mat provides the greatest total foundation area for a given space and the minimum foundation pressure, and therefore the maximum safety against soil failure. If the compressible strata are located at a shallow depth, the mat will minimize the settlement. However, if the compressible strata are deep, it will have little effect and in some cases, because of its weight, can increase the total settlement slightly.

A mat has other advantages. Like the strap it can bridge over small isolated soft areas in the foundation. It provides economy in design and construction by developing structural continuity and by permitting a uniform excavation depth. Cost comparisons between mats and large footings show that when the total area of the spread footings is more than from one-half to two-thirds the building area, a mat may be cheaper. Mats are employed when hydrostatic uplift must be resisted because the weight of the building is used to overcome the upward pressure.

MAT DESIGN PRESSURE / The pressure on the soil–mat interface depends on the stiffness of the structure–foundation system and the deflection and settlement of the soil under that load. If the structural system is very rigid (Fig. 10.24a,b) the pressures will resemble those for a perfectly rigid load—nearly uniform but falling off at the edges in a deep cohesionless soil and higher at the edges and less at the center in an elastic compressible material such as clay. If the structure–foundation is sufficiently flexible so that the total load for any segment of the mat defined by lines drawn midway between the columns is not changed by differential deflection of the structure, then the pressure on that portion of the mat will be equal to the column load divided by the area of the mat segment. For uniform column loads the pressure in this case will be approximately uniform (Fig. 10.24c). If the mat is so flexible that it deflects upward appreciably between columns, the pressure will be greatest at the columns and less between, as in Fig. 10.24d. The average pressure on each mat segment will be equal to the column load divided by the area of the corresponding mat segment.

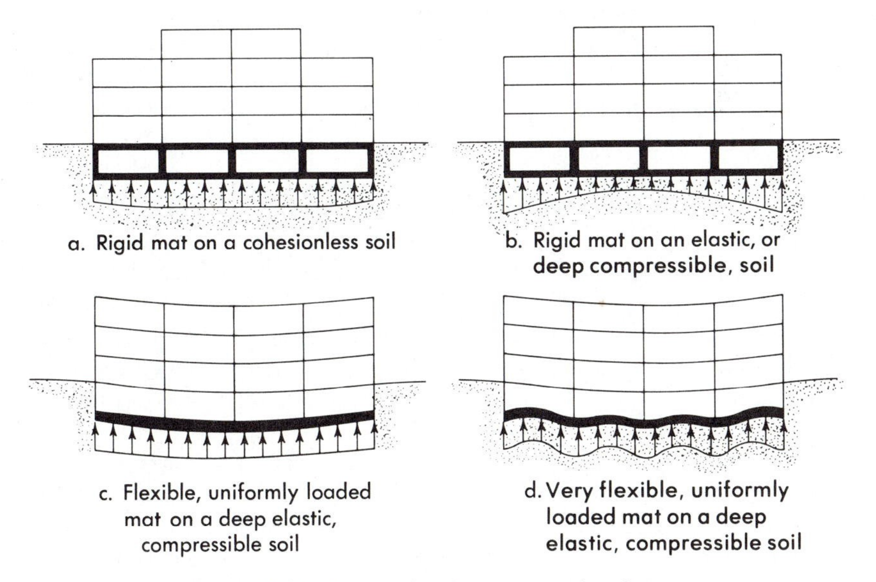

Figure 10.24 *Pressure distribution on mat foundations.*

Most mats are neither as rigid, as depicted by Fig. 10.24a and b, nor so flexible that the load is uniform. For deep sands the overall pressure is so nearly uniform that this can be assumed for design. For elastic soils the pressure distribution on the mat is statically indeterminant and must be compatible with the stress-deflection characteristics of both the soil mass and the structure. This can be analyzed in three ways: by successive approximations, by representing the soil by a simple mathematical model that can be solved directly, or by a finite element approximation of elastic response.

In the method of successive approximations, a pressure distribution is assumed, similar to Fig. 10.24b. The deflection of the structure and the settlement of the soil are then computed. If the pressure was correct, the two deflection curves should coincide. If not, the assumed pressure is revised and the process is repeated, until the deflections are compatible.

The second method makes use of an imaginary soil parameter, the *coefficient of subgrade reaction*, k_s, usually based on a plate load test, but also derived from settlement analyses:

$$k_s = \frac{q}{\rho_{c+d}} = \frac{q}{\Sigma \Delta H} \tag{10:18}$$

This implies that the settlement or deflection of any segment of the soil surface is proportional to the foundation pressure on that segment. The soil is thus assumed to react to load like a system of independent springs, instead of as an elastic mass. Generally, it is assumed that k_s is a constant, independent of B; however, Equations (10:11), (10:12), and (10:14) show it to be a function of B. Therefore, more than one k_s value must be used in analysis: (1) a maximum corresponding to the column spacing, and (2) a minimum corresponding to the total mat width. The value of k_s also varies with time. An initial k_s must be found corresponding to elastic or initial consolidation deflection and a final value corresponding to long-term consolidation.

The mat–structure system is represented by an equivalent elastic beam. The moments and deflections are then computed by the theories of beams on elastic foundations, utilizing four possible values for k_s: maximum and minimum and for long-term and short-term settlements. Although the assumption of a coefficient of subgrade reaction is not compatible with real soil behavior; model tests show that the moments and deflections in the structure computed in this manner are reasonably reliable if the structure is relatively flexible.[10:19]

The finite element analysis (Chapter 8) permits evaluation of the soil–structure interaction, assuming that the load-deflection behavior of both the soil and the structure can be defined mathematically.[10:20]

The structural design of a mat is similar to that of an inverted floor. Flat slab designs are used for small mats and have the advantage of unobstructed surface. If additional thickness is required at the columns, it is provided by lowering the bottom of the mat. Beam and slab designs are used for larger mats. To minimize thickness, the beams are wide and shallow. The space between the beams is filled with lightweight concrete to provide a flat floor surface, or a solid mat is used with the beams defined only by their reinforcing steel. Mats can be combined with the basement walls to form an inverted T-beam foundation or with the walls and upper floors to form a box girder (Fig. 10:24a). Such a rigid foundation–structure system can be designed to resist differential settlement.

10:8 Special Problems in Shallow Foundation Design

HYDROSTATIC UPLIFT / Structures below the groundwater level are acted on by uplift pressures. If the structure is weak, the pressure can break it and cause a blow-in of a basement floor or collapse of a basement wall. If the structure is strong but light, it may be forced upward or *floated* out of its original position. Uplift is taken care of by drainage or by resisting the upward force. Continuous drainage blankets, as described in Chapter 3, are very effective but must be designed with filters to function indefinitely without clogging. If possible the water should be disposed of by gravity because pumps sometimes fail when they are needed most.

The entire weight of the structure can resist uplift if a mat foundation is employed. In addition it is sometimes possible to increase the mat thickness to provide more weight. Because $\gamma_{conc} = 2.5\gamma_w$ then $\gamma'_{conc} = \gamma_w(2.5-1) = 1.5\gamma_w$. Thus, if the added mat thickness is gained by excavation, 1 meter of mat only balances 1.5 m of uplift. With still more effective weight and mat thickness required for some margin of safety, thickening the mat to gain uplift resistance is seldom economical. Anchors, grouted into bedrock or driven into hard soil, can provide uplift resistance. The anchor resistance is limited to the buoyant or effective weight of the soil or rock engaged and to the connection between the anchor and the earth, whichever is the smaller.

Structures designed to resist uplift must be waterproofed to prevent seepage and to minimize dampness. Bituminous membranes applied to the outside are useful if the head is only a few feet. If the heads are greater, the concrete structure itself must be made watertight. There are three essentials: (1) high-quality dense concrete placed without honeycombing; (2) water stops across all construction or expansion joints; and (3) masonry water-proofing (usually cement, powdered iron, and an agent to rust the iron) applied externally if possible, but internally if necessary to fill the hair cracks and to minimize capillary movement.

If the uplift occurs infrequently or if the value of the structure does not justify elaborate measures to overcome uplift, damage to structure can be prevented by intentional flooding to balance the uplift pressure. Relief holes with flap valves are sometimes installed in the bottom of sewage treatment tanks so they will flood automatically if they are empty at a time of groundwater rise. Flooding in buildings is seldom desirable because of the damage to the contents.

SEVERE VOLUME CHANGE[10:21] / Soils and rocks having high volume change were discussed in Section 4:5 and under depth in Section 10:1. In most cases the best way to avoid trouble from high volume change is to place the foundation below the level of severe movement. Movement of the upper soil strata along the column extending upward from the footing can break the column and lift the structure. This is prevented by placing a layer of weak, isolating material around the column. Mineral wool, vermiculite, and even sawdust have been used for this purpose. When the building is supported by deep footings, any grade beams must be separated from the soil so that they will not be damaged by soil or rock expansion. The space between the beams and the soil should be filled with vermiculite or mineral wool to keep out soil.

In arid regions where swelling is the most serious problem, it is sometimes possible to utilize a bearing pressure that exceeds the swell pressure. This is not always successful because the stress increase produced by the foundation decreases rapidly with increasing depth, while the swell pressure may not.

Small structures can be placed on relatively stiff mats that rise and fall with the volume change but do not deflect enough to cause trouble. Flexible foundations and structures that can deform without damage can be used if the volume change is not very irregular.

In arid regions special attention must be paid to drains, leaking pipes, and other sources of water that could cause heave. Piping under floor slabs can be placed in concrete troughs so that leakage can be kept out of the soil and so that soil movements will not cause pipes to leak.

LOESS / Loess soils are ordinary hard and incompressible from the partial cementing of clay and calcium carbonate. If they become wet, they soften and are very compressible. Surface water must be drained away from foundations on loess and piping must be routed so that leaks will not cause damage.

Similar settlements from structural collapse sometimes occur in the loose cemented soils of arid regions. The settlement can be minimized by prewetting the site, but at the cost of reduced bearing capacity.

COLD-STORAGE STRUCTURES / Frost action beneath cold-storage warehouses is a serious problem even in warm regions. If the temperatures are not very low, replacement of the frost-susceptible soil with a clean coarse sand or gravel is sometimes sufficient. This also provides some insulation. Insulation with cork or foam glass from 0.1 to 0.2 m (4 to 8 in.) thick is effective if the soil is warm. Isolation, by placing the cold area on piers so that air can circulate below is sometimes used for small structures.

When the cold is severe, it may be necessary to warm the soil by hot-air ducts or hot-water coils. Methods for analyzing the heat balance in such cases have been published.[10:22]

BOILERS AND FURNACES / Foundations for boilers, furnaces, and hot industrial processes can heat the soil and cause it to lose moisture and settle. The preventive measures are similar to those for cold-storage warehouses: soil replacement, insulation, and isolation.

FOUNDATIONS ON FILL / The ability of fill to support structures safely and with tolerable settlement depends on two factors: (1) the character of the fill, and (2) the settlement of the foundation of the fill. A properly compacted fill on a good foundation can be as good or better than virgin soil, although some archaic codes prohibit building on any fill. If the fill is uniformly compacted to the minimum standards given in Chapter 6, and if tests of the fill and foundation show them to have the required bearing and incompressibility, foundations on fill can be designed as any other foundation.

Old fills, and fills placed over low areas underlain by compressible or weak strata should be considered unsuitable unless extensive tests show them to be usable or unless the proposed structure can be adapted to low bearing and irregular settlement. Frequently, poorly compacted old fills continue to settle for years from secondary consolidation, aggravated by a breakdown of the soil structure by moisture. For example, a one-story masonry building on a 50-year-old 15-m (50-ft) deep fill settled near 0.1 m (4 in.) during the first 6 years after its construction, largely because of the additional weight of 1 m (3 ft) of new, well-compacted fill placed to provide a better foundation for the building.

Sanitary land fills and fills above organic debris and rubbish suffer from continuing organic decomposition and physicochemical breakdown.[10:23] One result is continuing erratic settlement. A more insidious effect is the production of methane and hydrogen sulfide gases that can be explosive and poisonous. A number of explosions, and some cases of illness and death, can be traced to gas accumulation in structures built on fills over organic matter.

The simplest solution, if the waste is shallow, is to remove it and replace it with compacted soil beneath buildings or critical structures. Preloading (Chapter 6) will reduce future consolidation and decay, but not eliminate gas. If wastefill remains below or adjacent to a building, gastight membranes between the fill and the building will reduce gas transfer. Venting the gas by gravel-filled gas drain trenches will dissipate it into the air.

Corrosion of underground piping is accelerated in debris-ladened fill. Anticorrosion protection or the replacement of the fill in the vicinity of the piping is usually necessary.

FLOORS SUPPORTED ON GROUND / Money can be saved in wide structures by supporting the lowest floor directly on the ground. This is feasible only when the building settlement (its exterior walls or grade beams and interior columns) caused by the structural live and dead loads and the floor settlement from its weight and live load are comparable. The maximum differential should be about 20 mm ($\frac{3}{4}$ in.). However, if the supporting strata for the building foundations and the floor are greatly different, serious differential settlement results. The typical example of trouble is with deep footings (or piles and piers, Chapter 11) supported on hard soil or rock and a heavily loaded floor supported either by poorly compacted fill or by fill above compressible soils. The floor edges restrained or supported by the walls and column footings cannot settle; between supports on the soil the floor may settle as much as 0.2 m (8 in.), with swales and cracking. There are three alternatives: (1) The floor is structurally supported by the walls and columns to minimize differential settlement; (2) the floor is supported on the soil, isolated from the walls and columns to settle independently (transition slabs may be required to accommodate differential movement between the settling floor and the rigid walls and columns); or (3) the building is supported on the same compressible soils as the floor and all are tied together structurally to minimize sharp settlement differences.

LATERALLY LOADED FOUNDATIONS / Foundations that support both vertical and horizontal loads are required for retaining walls, bridge abutments, and many kinds of industrial process machinery. The bearing capacity that supports the vertical component of the load is computed from the reduced bearing capacity factors of Section 10:2. Load inclinations exceeding 30° should not be analyzed by correcting the equations for vertical bearing but must be evaluated in terms of passive earth pressure on a sloping surface, which is beyond the scope of this text.

Resistance to the lateral load is provided two ways: (1) sliding resistance of the bottom of the foundation, and (2) passive earth pressure on the face of the foundation opposite to the thrust. The sliding resistance is either equal to the adhesion and friction between the foundation and the soil or the shear resistance of the soil (or rock), whichever is the smaller. Too often there is a thin layer of disturbed soil in the bottom of the footing excavation that acts as a plane of weakness. For maximum sliding resistance the soil surface is made irregular (but not notched as occasionally advocated) and cleaned just before pouring the foundation. To avoid disturbing the soil while placing reinforcing a 0.1 to 0.15 m (4 to 6 in.) working mat of plain concrete is poured on the freshly exposed soil surface. The steel is then placed on this surface which is left rough so as to bond with the remaining concrete.

If the sliding resistance is inadequate, it can be enhanced by a shear key on the edge of the foundation to which the thrust is directed. This adds to the passive pressure against the foundation face. The design should not rely on full passive pressure because considerable lateral movement is necessary for its development. Ordinarily, one-third of the potential passive pressure is assumed to be available.

Excavation adjacent to the foundation can destroy passive pressure. Saturation of the soil can reduce it. Therefore, designs relying on passive pressure should be conservative.

10:9 Foundations on Rock

Generally, there has been little concern for the bearing capacity and settlement of foundations on rock. Tradition, going back to the virgin stone platforms of the great Pyramid of Egypt, as well as the Biblical example of the wise man who built his house on rock to withstand rain, floods, and wind, has witnessesed the security of bedrock support. Facts, such as the compressive strengths of hard limestone, sandstone, and granite that exceed the capacity of good concrete, lull the modern engineer into a sense of false security. Foundations supported by rock do sometimes experience bearing capacity failures and settlement. Therefore, these possibilities must be investigated and any shortcomings minimized by design.

BEARING CAPACITY / The mechanics of bearing-capacity failure of homogeneous rock masses of great extent should be similar to that of soil because their Mohr envelopes are similar (Chapter 5). The major consideration in rock shear at low confining stresses is the drop of strength after failure and at low strains, termed *brittle failure*. From this it is deduced that the bearing capacity of rock is controlled by local shear, accompanied by cracking around the perimeter of the loaded area. This was confirmed by model tests by the author nearly 40 years ago. For homogeneous rock, Equation (10:3) and the bearing-capacity factors for local shear should be used (Fig. 10.7). For ordinary concrete foundations on sound rock, bearing capacity is no limitation because the rock is stronger than the concrete. For intensely loaded pile tips and the concentrated loads of steel supports in tunnels, bearing capacity of the softer homogeneous rocks, such as shale and sandstone, can be critical.

If the rock is jointed, the mechanism of potential failure is different, depending on the size of the loaded area, the joint spacing, joint opening, and the location of the load. Three simple possibilities can be analyzed, Fig. 10.25a, b, and c. In the first case where the joint spacing, S, is a fraction of B, and where the joints are open, the foundation is supported by unconfined rock columns. The ultimate bearing capacity approaches the sum of the unconfined compressive strengths of the rock prisms. The total capacity is somewhat less than the sum of the prism strengths because all do not have the same rigidity, and therefore some will fail before others reach their ultimate capacity.

If the joints are closed so that pressure can be transmitted across them without movement, the shear mechanism is essentially that described by the Bell–Terzaghi analysis (Section 10:2). The bearing capacity can be evaluated graphically by Fig. 10.5. If the joint spacing is much greater than width,

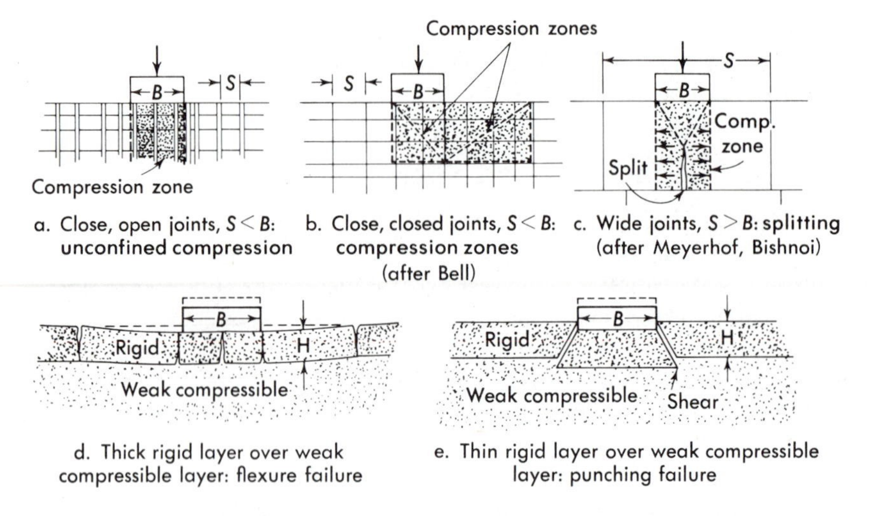

Figure 10.25 *Bearing-capacity failure modes on rock.*

$S \gg B$, the mechanism is different. The cone-shaped zone that forms below the foundation splits the block of rock formed by the joints. This condition has been analyzed first by Meyerhof,[10:24] and extended by Bishnoi and Sowers.[10:25] The results are approximated by a modification of Equation (10:4a), assuming that the load is centered on the joint block, and little pressure is transmitted across the joints.

$$q_0 = JcN_{cr} \tag{10:19}$$

Values of N_{cr} derived from models for splitting failure depend on the S/B ratio, and ϕ. These are given in Fig. 10.26 for circular footings. The values

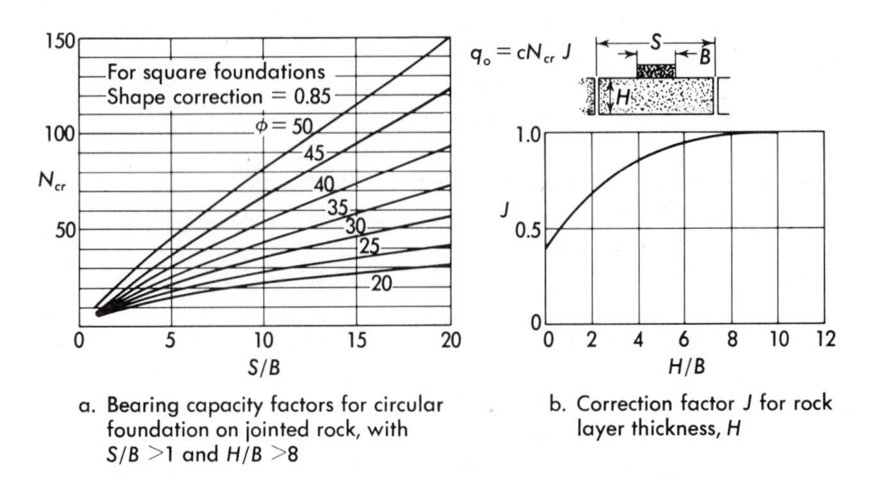

Figure 10.26 *Bearing-capacity factors for rock splitting.* (After Bishnoi.[10:25])

for square footings are 85% of the circular. The factor J considers the effect of the thickness of the upper layer: if $H/B > 8$, $J = 1$; if $H/B = 1$, $J = \frac{1}{2}$ approximately.

When the rock formation consists of an extensive hard seam underlain by a weak compressible stratum, two forms of failure occur, depending on the ratio H/B and S/B, and on the flexural strength of the rock stratum. If H/B is large and the flexural strength is small, the rock failure occurs by flexure (Fig. 10.25d). if the H/B ratio is small, punching is more likely (Fig. 10.25e).

Neither case has been adequately studied, and so only possible methods for analysis are suggested. A foundation over a rock cavity also fails by flexure or punching and could be analyzed approximately by those mechanisms. A major unknown factor in either case is the location of vertical joints and their effect on failure.

SETTLEMENT / Settlement of foundations on most rock formations is controlled by the joints. Settlement of hard rock with closed joints is negligible. If the joints are open or irregular, the observed settlement is comparable to the measured joint separations under the loaded area. Some porous limestones and weathered shales, poorly indurated sandstones, and earthlike rocks such as tuff consolidate similar to soils. The settlement potential can be evaluated by consolidation tests of undisturbed samples.

DESIGN / The design of foundations for rock is similar to foundations resting on soil. However, because much higher pressures are used, and because of unforeseen defects, special treatment of local defects is usually required. A narrow soft zone beneath the foundation (Fig. 10.27a) will not seriously reduce the bearing of the adjoining rock if the edges of the

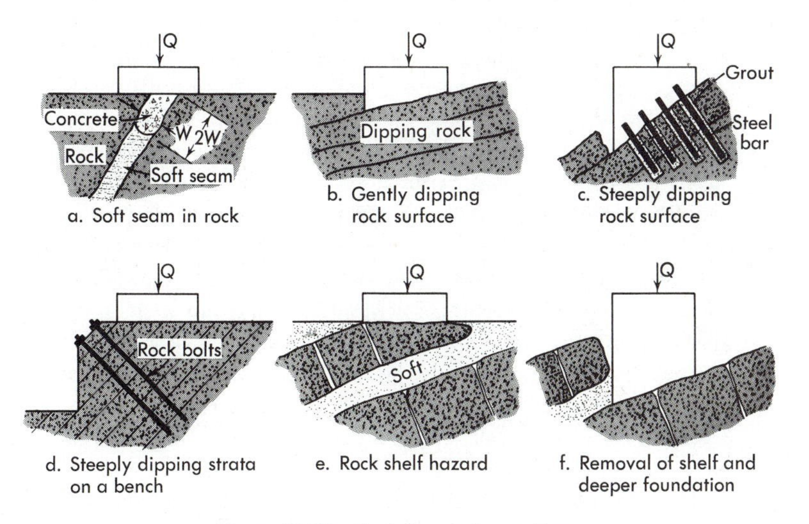

Figure 10.27 *Rock foundation problems.*

sound rock do not crumble. The fissure is cleaned out to a depth of twice its width and filled with concrete to support the rock corners above.

An irregular or sloping rock surface (Fig. 10.27b) is no problem if it can be sufficiently cleaned to obtain good bond and if the slope is less than the angle of internal friction of the rock or concrete. Excavation or blasting to level the rock will weaken it and do more harm than good. Resistance to sliding can be generated at little cost by drilling holes in the rock and inserting reinforcing bars as dowels (Fig. 10.27c).

Sloping seams of weakness (Fig. 10.27d and e) that define blocks that can slide easily can be corrected in two ways. The unstable mass is removed or reinforced by rock bolting.

Cavities and zones of high porosity that could lead to consolidation or bearing failure require grouting to strengthen the mass or to support the crust on which the foundation rests. Foundation design in rock is usually tentative. Ingenuous field decisions are required during construction to adapt the design to the rock formation as it is uncovered.

10:10 Foundations Subject to Vibrations[10:26]

Foundations are subject to vibrations from a number of sources, natural and artificial. Earthquakes, wind on tall, narrow buildings and towers, fast-flowing water, and waves pounding on marine structures produce both transient and near-continuous vibrations of variable frequencies. Machinery is the most common cause of vibration. Reciprocating engines, compressors, pumps, and oscillating machines are sources of continuous, low-frequency vibrations and are usually the most serious causes of trouble. Electric motors, rotary pumps, and turbines produce continuous high-frequency vibrations. Shock and transient vibrations often are caused by stamping machines, forges, pile drivers, moving vehicles, and blasting. Vibration consists of complex repeating motions and can include both rotation and translation in all three directions. Continuous vibrations usually have a constant frequency determined by the source but complicated by harmonics generated by the structure, while transient vibrations from shocks may have a variable frequency, depending on both the source and its supporting system.

Vibration occurs in different directions (Fig. 10.28a): vertical, longitudinal, and transverse linear motion (equivalent to an orthogonal coordinate system), and in three directions of rotation: rocking, pitching, and slewing, depending on the source of the vibration and the freedom to move. Many vibrations of structures are a complex combination of all six, which must be unraveled to understand the mechanism of motion and to devise corrective measures.

MECHANISM OF VIBRATIONS / If an impulse of short duration is applied to a body that is supported elastically, it will vibrate at its *natural*

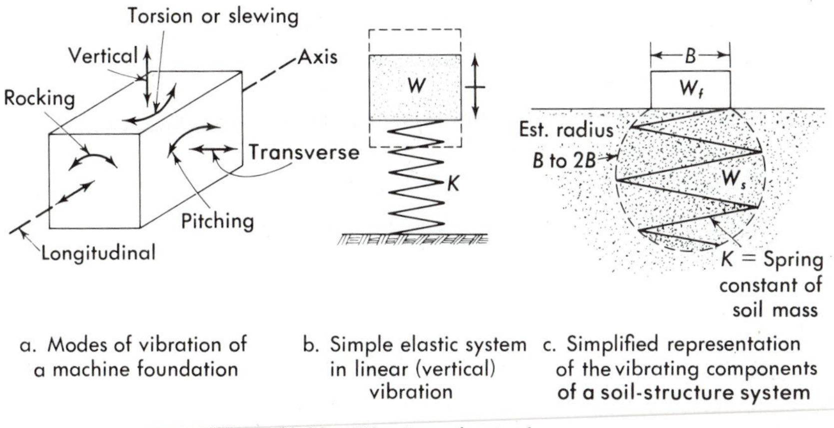

Figure 10.28 *Vibrations of a simple structure.*

frequency, which depends on its mass and elastic properties. For a perfectly elastic body (Fig. 10.28b) whose weight is W and whose resistance to deflection in force per unit of deflection (kN/m or lb/ft) is K, the natural frequency f_n is given by

$$f_n = \frac{1}{2\pi}\sqrt{\frac{Kg}{W}} \tag{10:20}$$

This means that the natural frequency increases as the square root of the rigidity and decreases with the square root of the weight of the body. When energy is lost in the process, the vibration is said to be *damped*, and the natural frequency is somewhat less.

The damping is described in terms of the *damping ratio*, C, which is an indication of the amount of vibration energy lost in each cycle: $C = 0$ denotes no loss; $C = 1$ indicates all the impulse energy is dissipated in one cycle of vibration.

The natural frequency of a foundation–soil system, as shown in Fig. 10.28c, is much more complex. The resistance per unit of deflection K can be estimated from the distortion settlement ρ. This depends on both the modulus of elasticity of the soil and the size of the foundation. The weight of the vibrating body, W, is the sum of the weight of the foundation, W_f, and the portion of the soil mass below the foundation which is vibrating, W_s. Therefore, the natural frequency of the soil is not a property of the soil alone but also depends on the weight and size of the foundation and the load it carries.

The intensity of the vibration is also a factor because the modulus of elasticity of most soils changes with confining pressure and with the strain. Tests of soil masses with vibrators having masses of from 10 to 30 kN (2 to 6 kips) and with square bases from 0.6 to 1 m (2 to 3 ft) wide indicate natural frequencies of from 12 Hz (720 cpm) for peat to 30 Hz (1800 cpm) for very dense sand. For heavier and wider foundations the natural frequency is less. The natural frequencies, unfortunately, are comparable to the vibrations or multiples of the vibrations generated by many reciprocating machines such as pumps and compressors. They are lower than the frequencies generated by turbines and high-speed motors, however.

FORCED VIBRATION-RESONANCE / If a periodic impulse is forced on a simple elastic system at a very low frequency, $f \ll f_n$, the mass and its elastic support will respond together, with the same amplitude of motion in *forced vibration*. The movement is neither *magnified* nor *attenuated*. If the impulse is forced on the system at a very high frequency, $f \gg f_n$, the elastic mass cannot respond because of its inertia, and the elasticity of the system absorbs the difference in movement. The vibration effect on the mass is attenuated. These effects are illustrated by the extreme ends of the graph in Fig. 10.29. If the vibration is applied at approximately the natural

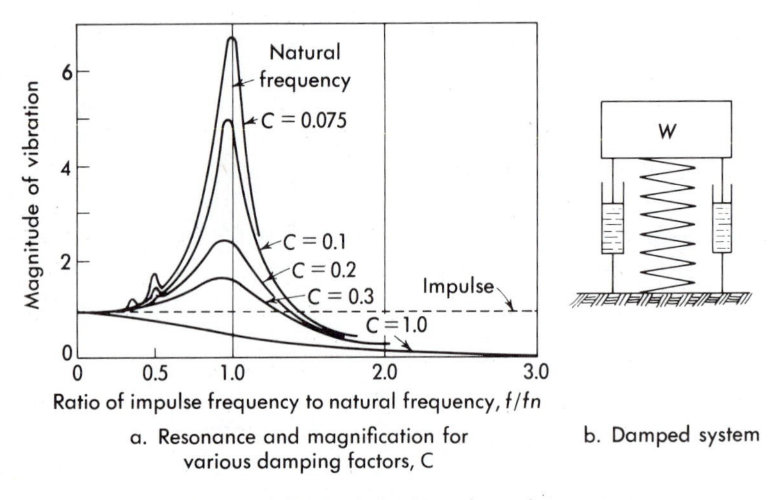

Figure 10.29 *Resonance in a damped system*

frequency, $f = f_n$, each recurring movement of the source applies additional impetus to the movement of the body. If there is no energy loss or damping, the amplitude of the movement of the elastic system is increased each cycle by the energy imparted by the impulse, and eventually becomes indefinitely large. This is termed *resonance*. At resonance with damping, the amplitude of the vibration of the elastic mass increases with each impulse until the energy lost in each cycle of vibration is equal to the energy input of the impulse during that cycle. The ratio of the magnitude of the vibration of the elastic system to that of the impulse is termed the *magnification*. For a continuous source of vibration, the magnification depends on the damping factor and how close the frequency of the impulse matches the natural frequency, as shown in Fig. 10.29. (Although high damping reduces the natural frequency somewhat, this does not change the concept of resonance.)

An impulse at half the natural frequency can cause a magnified vibration by adding energy every other cycle. The magnification is less for a given damping ratio because the energy input is only half that for one impulse per cycle. Similarly, resonance can occur with input frequencies of one-third and one-fourth the natural frequency. The magnification for these conditions is seldom great unless the damping ratio is very small.

EFFECT OF VIBRATIONS / Soil vibrations have a number of important effects. First, the vibration can be transmitted to other foundations and to other structures at some distance from the vibration source. These transmitted vibrations can be annoying and even damaging. If a foundation–soil system should be in resonance, amplification and severe damage could result. Second, the vibration can cause a reduction in the void ratio of cohesionless soils and result in severe settlement. Ordinarily, the settlement will be small if the relative density is greater than 60%, but if the vibration is severe, as in the case of resonance, settlements can occur until the relative density is nearly 80%. Third, vibration in loose, saturated cohesionless soil can bring about liquefaction, and failure. Soils with cohesion are resistant to vibration settlement and are not weakened appreciably.

Corrective measures include reducing the vibration of the source, changing the soil–foundation system to prevent resonance, and stabilizing the soil to prevent vibration damage. Vibration can be reduced by installing isolating systems such as damped spring mounts on the source or by cushioning them on vibration-absorbing materials. It can also be minimized by changing the type of machine, such as a substituting high frequency rotary for reciprocating compressors. The frequency of resonance is changed by altering one or more of the factors in Equation (10:20). Increasing the size and weight of the foundation will reduce the resonant frequency of the system. Increasing the modulus of elasticity of the soil by densification or stabilization (Chapter 6) will increase it. Changing the speed of the source can be helpful, of course, if it is possible mechanically. In general the natural frequency should be less than half or more than twice the vibration frequency. Stabilizing the soil by injecting a cementing agent or by densification can also prevent settlement or loss of strength.

EARTHQUAKES[10:27] / Much of the earth's surface is occasionally subject to short episodes of complex vibration, earthquakes from crustal fracture. As can be seen from Fig. 7.1, the severity of earthquakes varies greatly within the United States. Some areas such as South Florida have never felt them in 400 years of recorded history; others, such as southern California, feel them regularly. The duration is usually less than a minute, but several episodes of motion sometimes occur within minutes. The motion is random, with varying amplitudes and frequencies, although certain frequencies usually dominate in any single event.

The response of the ground is complex. The earth's crust can be visualized as a three-dimensional system of interconnected masses and springs, as in Fig. 10.28b, repeating in all directions. If the dominant frequency of the earthquake is different from the natural frequency of the soil–rock system, the ground responds in forced vibration. If the dominant and natural frequencies are comparable, the motion is amplified similar to that shown in Fig. 10.29. This explains the variable ground motion at adjacent sites as well as the difference in ground motion between different quakes of comparable magnitude. Similarly, a structure supported by the soil can respond in forced vibration or in amplified vibration depending on its natural frequency and that of the ground.

Three effects are considered in foundation design in regions subject to strong earthquakes: (1) the direct effect on the soil and rock beneath the structure, including liquefaction, strain, and fracture; (2) the direct effect of the motion transmitted to the structure; and (3) indirect effects such as *tsunamis* (tidal waves) and landslides generated by the motion. The reader should consult the references for a more detailed discussion.

10:11 Simple Construction—Complex Behavior

Shallow foundations, immediately beneath the lowest part of the structure, are functionally simple: enlargements of the columns or walls of the structure above. Because they are a part of the structure, there is a temptation to treat them as simple structural components whose capacity increases in proportion to their size or whose deflection is reduced in proportion to reduced stress or bearing pressure. This simplistic view ignores the ultimate support: the soil and rock. These formations are massive solids whose response to loading is far different from simple beams, columns, walls, and floors. Moreover, the soil and structure interact: Excessive settlement of the soil beneath a column of a building will cause a redistribution of load within the structure, increasing the load of adjacent columns and reducing that of the settling column. The reduced load retards further settlement.

The analyses of bearing capacity and settlement in this text go far beyond the simplistic structural component sizing approach and are adequate for many problems of design. The complex interaction, as well as the environmental changes that cannot be evaluated quantitatively (Table 10:3) produces surprises. Designs must allow for them as far as possible. More important, the building designer, the owner, and the public should be warned of the uncertainties. Even unforeseen problems need not be disastrous: The settlement problem of the belfry of the Cathedral of Pisa in Italy is one of the nation's assets.

REFERENCES

10:1 "Theoretical and Practical Treatment of Expansive Clays," *Quaterly Colorado School of Mines*, **54**, 4, October 1959.

10:2 E. W. Lane and W. M. Borland, "River Bed Scour During Floods," *Transactions, ASCE*, **119**, 1072, 1954.

10:3 E. M. Laursen and A. Toch, "Scour Around Bridge Piers and Abutments," *Bulletin 4, Iowa Highway Research Board, Ames*, 1956.

10:4 K. Terzaghi, *Theoretical Soil Mechanics*, John Wiley & Sons, Inc., New York, 1943.

10:5 G. G. Meyerhof, "The Influence of Roughness of Base and Ground Water on the Ultimate Bearing Capacity of Foundations," *Geotechnique 5*, **3**, 227, September 1955.

10:6 G. G. Meyerhof, "The Bearing Capacity of Footings Under Eccentric and Inclined Loads," *Proceedings of the Third International Conference on Soil Mechanics and Foundation Engineering, 1, Zurich, 1953.*

10:7 A. S. Vesić, "Analysis of Ultimate Loads on Shallow Foundations," *Journal of the Soil Mechanics and Foundations Division, Proceedings, ASCE*, **99**, SM1, 4572, January 1973.

10:8 A. S. Vesić, "Bearing Capacity of Shallow Foundations," *Foundation Engineering Handbook*, H. F. Winterkorn and H. Y. Fang, eds., Van Nostrand Reinhold Company, New York, 1975, Chapter 3, pp. 121–147.

10:9 L. Jurgenson, "The Application of Theories of Elasticity and Plasticity to Foundation Problems," *Journal of the Boston Society of Civil Engineers*, July 1954.

10:10 H. M. Westergaard, "A Problem of Elasticity Suggested by a Problem of Soil Mechanics: Soft Material Reinforced by Numerous Strong Horizontal Sheets," in *Contributions to Mechanics of Solids*, Macmillan Publishing Co., Inc., New York, 1938.

10:11 N. M. Newmark, "Influence Charts for Computation of Stresses in Elastic Soils," *Bulletin 38, University of Illinois Engineering Experiment Station*, Urbana, 1942.

10:12 J. H. Schmertmann, "Static Cone to Compute Static Settlement Over Sand," *Journal of the Soil Mechanics and Foundations Division, Proceedings, ASCE*, **96**, SM3, 1011–1043, May 1970.

10:13 F. Baguelin, J. F. Jezequel, and D. H. Shields, *The Pressuremeter and Foundation Engineering*, Trans. Tech. Publications, Aedermannsdorf, Switzerland, 1978, Chapter 3.

10:14 K. Terzaghi, "Settlement of Structures in Europe and Methods of Observation," *Transactions, ASCE*, 1938, p. 1432.

10:15 A. W. Skempton and D. H. McDonald, "The Allowable Settlement of Buildings," *Proceedings of the Institute of Civil Engineers*, **5**, 3, London, December 1956, p. 727.

10:16 D. E. Polshin and R. A. Tokar, "Maximum Allowable Differential Settlement of Structures," *Proceedings of the Fourth International Conference on Soil Mechanics and Foundation Engineering*, 1, London, 1957, p. 402.

10:17 G. F. Sowers, "Analysis of Lightly Loaded Foundations," in *Analysis and Design in Geotechnical Engineering*, Vol. 2, ASCE, New York, 1975, pp. 49–78.

10:18 J. Bowles, *Foundation Analysis and Design*, McGraw-Hill Book Company, New York, 1968.

10:19 A. B. Vesić, "Beams on Elastic Subgrade and Winklers Hypothesis," *Proceedings of the Fifth International Conference on Soil Mechanics and Foundation Engineering*, 1, Paris, 1961, p. 845.

10:20 W. F. Swiger, "Evaluation of Soil Moduli," in *Analysis and Design in Geotechnical Engineering*, Vol. 2, ASCE, New York, 1975, pp. 79–92.

10:21 F. H. Chen, *Foundations and Expansive Soils*, Elsevier Scientific Publishing Co., Amsterdam, 1975.

10:22 W. Ward and E. C. Sewell, "Protection of the Ground From Thermal Effects of Industrial Plants," *Geotechnique*, 2, 1, June 1950, p. 64.

10:23 G. F. Sowers, "Foundations on Sanitary Land Fill," *Journal of the Sanitary Engineering Division, Proceedings, ASCE*, **94**, S1, February 1968.

10:24 G. G. Meyerhof, "Bearing Capacity on Rock," *Magazine of Concrete Research*, April 1953.

10:25 B. W. Bishnoi, *Bearing Capacity of Jointed Rock*, A thesis presented to the Georgia Institute of Technology in partial fulfillment for the Ph.D. in Civil Engineering, 1968.

10:26 R. V. Whitman and F. E. Richart, Jr., "Design Procedures for Dynamically Loaded Foundations," *Journal of the Soil Mechanics and Foundations Division, Proceedings, ASCE*, **93**, SM6, 169, November 1967.

10:27 L. Zeevaert, "Foundation Problems in Earthquake Regions," *Analysis and Design of Building Foundations*, H. Y. Fang, ed., Envo Publishing Co., 1976, Chapter 27, pp. 753–770.

SUGGESTIONS FOR FURTHER STUDY

C. W. Dunham, *Foundations of Structures*, McGraw-Hill Book Company, New York, 1950.

Okamoto, *Introduction to Earthquake Engineering*, John Wiley & Sons, New York, 1973, 571 pp.

R. B. Peck, W. E. Hansen, and T. H. Thornburn, *Foundation Engineering*, 2nd ed., John Wiley & Sons, Inc., New York, 1974, p. 514.

F. E. Richart, Jr., J. R. Hall, Jr., and R. D. Woods, *Vibrations of Soils and Foundations*, Prentice-Hall, Englewood Cliffs, N.J., 1970.

N. A. Tsytovich, *The Mechanics of Frozen Ground*, McGraw-Hill Book Company, New York, 1975, 426 pp. (USSR Publication 1973.)

A. S. Vesić, ed., *Bearing Capacity and Settlement of Foundations, A Symposium*, Dept. of Civil Engineering, Duke University, Durham, N.C., 1967.

R. L. Wiegel, ed., *Earthquake Engineering*, Prentice-Hall, Inc., Englewood Cliffs, N.J., 1970, p. 518.

L. Zeevaert, *Foundation Engineering for Difficult Subsoil Conditions*, Van Nostrand Reinhold Company, New York, 1972, 652 pp.

PROBLEMS

10:1 A long footing is 1 m (3.28 ft) wide. Its base is 0.75 m (2.5 ft) below the ground surface. Find the safe bearing capacity if the soil is saturated clay having a unit weight of 17.5 kN/m^3 (111.5 lb/ft^3) and a strength c, of 150 kN/m^2 (3130 lb/ft^2). The safety factor is 3. Use Mohr's circle and compare with bearing capacity computed by the general formula, using both the Bell–Terzaghi and Meyerhof factors.

10:2 A square footing is 2.5 m (8.2 ft) wide with its base 1 m (3.3 ft) below the ground surface. The soil is saturated clay having a unit weight of 19 kN/m^3 (121 lb/ft^3) and a cohesion of 200 kN/m^2 (4175 lb/ft^2). Find the safe bearing capacity by the Meyerhof factors if a minimum safety factor of 2.5 is required.

10:3 A long footing 1.5 m (4.9 ft) wide is 0.9 m (3 ft) below the surface of a sand weighing 20.5 kN/m^3 (130.6 lb/ft^3) saturated and 17.5 kN/m^3 (111 lb/ft^3) dry, and having an angle of internal friction of 37°. Compute the safe bearing capacity for a safety factor of 2.5, using (1) the graphical method; (2) the general equation with the Bell–Terzaghi factors; and (3) the Meyerhof factors. For each case find the capacity of (a) the water table 3 m (9.8 ft) below the footing, (b) the water table at the base of the footing, and (c) the water table at the ground surface.

10:4 A column carries 900 kN (202 kips). The soil is a dry sand weighing 18 kN/m^3 (114.6 lb/ft^3) and having an angle of internal friction of 40°. A minimum safety factor of 2.5 is required, and the Meyerhof factors are to be used in computation.

a. Find the size of square footing required if it is placed at the ground surface.
b. Find the size of square footing required if it is placed 1 m (3.3 ft) below the ground surface.
c. Find the size of footing required for (b) if the water table rises to the ground surface, increasing the soil weight to 21 kN/m^3 (133.8 lb/ft^3).

10:5 A small steam turbine whose base is 6 m × 3.7 m (19.7 ft × 12.1 ft) weighs 10,000 kN or 2220 kips. It is to be placed on a clay soil with $c = 145$ kN/m^2 (3030 lb/ft^2). Find the size of foundation required if the minimum safety factor is 3. The foundation is to be 0.6 m (2 ft) below the ground surface.

10:6 A column carries 1600 kN (360 kips). It is to rest on a square footing on sand with $\phi = 38°$ and $\gamma = 19$ kN/m^3 (121 lb/ft^3). The safety factor is 2.5.

a. Find size of square footing if it is at ground surface.
b. Find size of square footing if it is 1.2 m (4 ft) below the ground surface.
c. Would it be cheaper to lower the footing as in (b) if the column is 0.5 m (1.64 ft) square and the footing is 0.6 m (2 ft) thick than to place it at the ground surface? Concrete costs \$90.00/m^3 (\$69.00/yd^3) in place, and excavation costs \$2.75/m^3 (\$2.10/yd^3).

10:7 A column carries 2110 kN (475 kips) to a square footing that rests 1 m (3.3 ft) below the surface of a partially saturated clay. Then if $\phi = 15°$, $c = 50$ kN/m^2 (1044 lb/ft^2), and $\gamma = 18$ kN/m^3 (114.5 lb/ft^3), find the footing size required for a safety factor of 2.5.

10:8 A smoke stack foundation 12 m × 12 m (39 ft × 39 ft) exerts a pressure of 250 kN/m^2 (5200 lb/ft^2) at the surface of a sand that weighs 17 kN/m^3 (108.3 lb/ft^3) dry and 20 kN/m^3 (127 lb/ft^3) saturated. Below the sand at a depth of 10 m (32.8 ft) is an organic silt that is 2 m (6.6 ft) thick, weighing 16 kN/m^3 (102 lb/ft^3) saturated.

a. Construct a diagram showing the variation of vertical stress increase under the center of the foundation as a function of depth.
b. Construct a similar diagram showing the initial effective stress in the soil as a function of depth. The water table is at a depth of 3 m (9.8 ft).

c. Construct a diagram showing the increase in stress at the center of the clay stratum [a depth of 11 m (36 ft)] as a function of the horizontal distance from the footing centerline. What is the average stress increase directly beneath the foundation?

10:9 The clay of Problem 10:8 has the stress–void ratio curve of Problem 4:5.

a. Find the settlement caused by the initial load and the average stress increase in the stratum.

b. Below the silt is more sand. If the coefficient of consolidation is 1.2×10^{-3} m^2/day (1.3×10^{-2} ft^2/day), compute the time required for (1) 25%, (2) 50%, and (3) 75% consolidation.

10:10 Find the additional increase in stress and increased settlement of problem 10:9 caused by permanently lowering the water table to a depth of 6 m (20 ft). This occurs after the settlement due to the foundation load is completed.

10:11 An elevated water tank weighing 4000 kN (900 kips) rests on four square footings 6 m (19.7 ft) apart (center to center). The allowable soil pressure is 250 kN/m^2 (5.2 kips/ft^2). The soil consists of 8 m (25 ft) of gravel underlain by 2.5 m (8 ft) of clay underlain by more gravel. The water table is below the clay. Both clay and gravel weight 18 kN/m^3 (115 lb/ft^3). The clay has the stress–void ratio curve of Problem 4:2.

a. Compute the average effective stress in the clay before and after construction. Use the Westergaard chart.

b. Compute the tank settlement.

c. Compute the tank settlement if the footing area is doubled. How much was the settlement reduced? (Express as a percentage of the original sattlement.)

10:12 A monument has a base 12 m $\times$ 18 m (39.4 ft $\times$ 59 ft). It weighs 14,000 kN (3150 kips). It rests on a stratum of sand 12 m (39 ft) thick underlain by a stratum of soft clay 2 m (6.56 ft) thick. The clay rests on bedrock. The water table is at the ground surface and the γ of the sand is 20.5 kN/m^3 (130.6 lb/ft^3) saturated and the γ of the clay is 17 kN/m^3 (108.3 lb/ft^3).

a. Compute the average effective stress in the clay before and after construction. Use the Westergaard chart.

b. Compute the settlement of the monument if, for the clay $e_0 = 1.13$ and $C_c = 0.31$.

c. Compute the time required for 80% of the ultimate settlement to take place if $k = 2 \times 10^{-7}$ mm/sec.

d. Recompute the stress and settlement, using the approximate method.

10:13 A turbine foundation mat 8 m × 20 m (26.3 ft × 65.6 ft) carries a load of 80 kN/m^2 (1670 lb/ft^2). The soil consists of 8 m (26.3 ft) of clay overlying dense sand. The water table is at the ground surface. The clay has the following characteristics: $c = 96$ kN/m^2 or 2000 lb/ft^2, $\gamma = 17$ kN/m^3 (108.3 lb/ft^3), $E = 43{,}000$ kN/m^2 (897,000 lb/ft^2), $c_v = 1.5 \times 10^{-4}$ m^2/day or 1.6×10^{-3} ft^2/day. The stress–void ratio curve is given in Problem 4:2.

a. Is the foundation safe if a minimum factor of safety of 2.5 is necessary?
b. What is the distortion settlement?
c. What is the average stress increase in the upper 2 m (6.6 ft) substratum of clay, the next 3 m (9.8 ft), and the bottom 3 m (9.8 ft) of clay?
d. What is the total compression settlement?
e. How much settlement will take place in one year if a thin layer of sand is placed between the clay and the concrete mat?

10:14 Prepare a report describing the failure of excessive settlement of a structure due to faulty foundations. Include the following items:

a. Soil conditions.
b. Foundation.
c. Description of failure.
d. Cause of failure.
e. Corrective measures, if any.

10:15 A building 30 m × 24 m (98.5 ft × 79 ft) of reinforced concrete weighs 33.5 kN/m^2 (700 lb/ft^2) of gross area. Columns are to be spaced 6 m or 19.7 ft apart. The soil is deep, dry sand weighing 19 kN/m^3 (121 lb/ft^3). The following results were obtained with a load test made at the soil surface, using a 0.3 m (1 ft) square plate.

Load		Settlement	
kN	lb	mm	in.
4.45	1000	1.3	0.05
8.90	2000	2.5	0.10
13.35	3000	3.8	0.15
17.8	4000	5.1	0.20
22.25	5000	7.6	0.30
26.7	6000	15.2	0.60

The footings are to be placed 1 m (3.28 ft) below the ground surface. The minimum safety factor is 2.5.

a. Plot load test and get failure q_c.
b. Compute ϕ.
c. Find minimum square-footing sizes.
d. Find footing settlements for (c). [Assume that the settlements at higher pressures can be found by extending the straight-line portion of the load settlement curve and that Formula (10:15b) applies.]
e. If the settlement is excessive, what should be done?

10:16 A bridge pier has a total load of 10,000 kN (2250 kips). It is located in a flood plain in which the water velocities are nominal. The soil profile consists of:
1. 10 m (32.8 ft) of firm sand weighing 19.6 kN/m^3 (124.2 lb/ft^3), $\phi = 35°$.
2. 6 m (19.7 ft) of dense sand weighing 21 kN/m^3 (133.8 lb/ft^3) and $\phi = 42°$.
3. 2 m (6.6 ft) of clay, normally consolidated with $e = 1.25$ and $c_c = 0.48$.
4. 30 m (98.4 ft) of dense gravelly sand.

Ground water is at a depth of 1 m (3.3 ft). Assume that the designer wishes a spread footing foundation and presumes a depth of 3.5 m (11.5 ft) will provide safety against scour.
a. What is the minimum width of a square foundation required at a depth of 3.5 m if the land can flood and if the foundation has an eccentricity of load of 1 m (3.28 ft) perpendicular to the roadway? A safety factor of 3 is required.
b. Find the depth of penetration of a steel sheet piling cofferdam required to provide a safety factor of 1.5 against heave if the contractor excavates to the foundation level under water and then pumps the excavation from a ramp so that the water level is 0.6 m (2 ft) below the excavation bottom. What will be the approximate rate of pumping if the firm sand has a permeability coefficient of 3×10^{-4} mm/sec and the dense sand is relatively impervious? Assume that 30 m (98.4 ft) from the excavation the groundwater level remains unchanged and that a two-dimensional flow net is a reasonable approximation of the seepage conditions.
c. What stress changes and settlement (or expansion) are produced by (1) excavation followed by (2) dewatering? Assume c_c is the same for both increase and decrease in stress in the clay.
d. How much settlement is produced by the weight of a concrete footing 3 m (9.85 ft) thick poured in the excavation, assuming the groundwater level regains its level 0.6 m (2 ft) below the ground surface?

e. How much stress change and additional settlement takes place when the 10,000 kN (2250 kips) pier load and 5000 kN (1125 kips) dead load of the bridge are added to the foundation?

10:17 A chimney weighing 20,000 kN (4500 kips) is placed on a circular foundation with a bearing pressure of 200 kN/m^2 (4175 lb/ft^2) (including the foundation weight). The soil profile consists of sand from the surface to a depth of 12 m (39.4 ft) underlain by normally consolidated, soft organic clay 2 m (6.6 ft) thick. Under the soft clay is dense sand to a depth of 25 m (82 ft) underlain by rock. The upper sand weighs 16 kN/m^3 (102 lb/ft^3) dry and 195 kN/m^3 (124 lb/ft^3) saturated, and has an angle of internal friction of 32°. The soft clay weighs 16.5 kN/m^3 (105 lb/ft^3) saturated, has a void ratio of 1.5, and a compression index of 0.71. The groundwater table before construction is 1 m (3.3 ft) below the ground surface and is lowered permanently to 4 m (13.1 ft) below the ground surface before foundation construction is begun. The base of the foundation will be 3 m (9.8 ft) below the ground surface.

a. Compute the effective stress in the clay before construction.
b. Compute the effective stress increase in the clay caused by lowering the water table, and the settlement ultimately produced by the unwatering.
c. Compute the stress change in the clay caused by excavating the foundation and building the chimney, and the settlement resulting from these operations, assuming that the time interval between excavation and the pouring of concrete is so short that there is no time for decompression of the clay.
d. If the coefficient of consolidation of the clay is 2×10^{-3} m^2/day (2×10^{-2} ft^2/day), how long will it take for 90% of the settlement to develop?
e. Is the foundation safe from bearing-capacity failure under static load?
f. Is the foundation safe from bearing-capacity failure if it must undergo a wind of 125 km/hr (75 mph)? The chimney is 6 m (19.6 ft) in diameter and 75 m (246 ft) high.
g. If the chimney foundation is marginal, what should be done to remedy the situation?
h. If the settlement is considered excessive, what alternate foundations might be considered and why?

CHAPTER 11
Deep Foundations

The site for a 20-story hospital was underlain by a recent alluvial deposit more than 30 m (100 ft) thick. The formation was predominantly soft, compressible clay. However, there were several strata of firm-to-dense sand within the clay; one about 12 m (40 ft) below the surface was about 3 m (10 ft) thick. Because the soft clay had insufficient bearing capacity for spread footings, the decision was made to support the structure on piles.

Timber piles were driven at several locations on the site. None could penetrate deeper than about 14 m (46 ft) below the surface. Several of the piles were load-tested and none settled more than 10 mm ($\frac{3}{8}$ in.) at the design load of 267 kN (30 tons). On the basis of the tests the structure was supported on more than 10,000 piles with their tips embedded in the dense sand. Within a year after the completion of construction the building had settled in a saucer-shaped depression nearly 30 cm (1 ft) below its original level at the center. Within 20 years the depression was more than 120 m (400 ft) in diameter and nearly 1 m (3 ft) deep at the center. Major repairs were necessary to keep the building serviceable. Today the settlement is increasing slowly in the secondary compression range, and, although only minor repairs are now necessary, the movement will continue indefinitely.

A single preliminary test boring for a three-story building near the seacoast found loose to firm sand 13 m (43 ft) thick, underlain by cemented sand with lenses of shells. Without further study, the structural engineer concluded that settlement of the upper sands might be excessive with spread footings and that piles driven to the cemented sand would provide a trouble-free foundation. For economy, creosoted wood piles with a design load of 350 kN (40 tons) were specified.

Pile driving went rapidly. Although the piles encountered considerable resistance when the tips reached a depth of about 7 m (23 ft), about $\frac{1}{2}$ m ($1\frac{1}{2}$ ft) of hard driving brought more rapid penetration, but occasional levels of high resistances. Hard driving was reached when 13 m (43 ft) had been driven into the ground, confirming the level of the cemented sand. Driving was stopped at this point, although the driving resistance was not as great as the engineer had specified.

The contractor expressed growing concern with the failure to develop high driving resistance. At his expense he load-tested a pile. It settled 5 cm (about 2 in.) at a load 267 kN (30 tons). The engineer, now alarmed, required that driving continue when the length in the ground reached 13 m (43 ft), hoping to embed the pile tip 1 or 2 m ($3\frac{1}{2}$ to $6\frac{1}{2}$ ft) in the cemented sand and develop more capacity. The driving resistance did not increase with increased pile length. A second longer pile was tested. It also settled more than 50 mm (2 in.) with a load of 270 kN (30 tons).

A geotechnical specialist, after examining the construction records, suggested that the test pile be pulled from the ground. A 7-m (23-ft) long piece emerged, split and broomed at the lower end. More piles were pulled; all were broken 7 m (23 ft) below the ground surface. One 13-m (43-ft) long piece was recovered with crushing fractures at two points, one of which was 7 m (23 ft) below the upper end of the pile.

New soil borings disclosed a continuous seam of dense cemented sand about $\frac{1}{2}$ m ($1\frac{1}{2}$ft) thick 7 m (23 ft) below the ground surface. The hard driving at this level broke the piles, apparently splitting some and making **z** shapes of others. The uppermost fracture was 7 m (23 ft) below the upper end of the woodpile where the cross-sectional area of the wood was great enough to resist further fracture. However, the split, fractured 7-m (23-ft) long piece tip crushed against the cemented seam at a load of 300 kN (34 tons). Additional load tests found no piles capable of supporting more than 300 kN (34 tons) without deflections of from 50 to 100 mm (2 to 4 in.). The wood piles were abandoned. The building plan was shifted enough that new piles of a different type could be driven through the hard seam at 7 m (23 ft) to the cemented sand at 13 m (43 ft). Ironically, the building probably could have been supported on lightly loaded spread footings or 7-m (23-ft) long piles at much less cost than the original design. The project ended with multiple lawsuits between the architect, the engineer, the contractor, and the owner.

These two examples illustrate some of the problems associated with deep foundations. Many architects and engineers believe that deep foundations are a simple foolproof solution to all problems of weak or compressible soils and therefore foundation analyses and construction surveillance are not needed. In the first example, the piles bypassed one compressible layer, but concentrated the building load above a second one just as bad. The test of a single pile was misleading because the stress increase in the clay was not sufficient to cause appreciable consolidation. In the second example, a simplistic design based on one boring might have been satisfactory except for unforeseen factors. First, the cheap preliminary boring had sampling intervals of 1.5 m (5 ft) which failed to disclose the dense cemented sand layer at 7 m (23 ft). Second, the hard driving required to develop a safe pile capacity of 250 kN (40 tons) broke the slender lower ends of woodpiles whose strengths were about 40% of the average strength of that wood species. The inadequate investigation, a premature design decision, and faulty specifications cost far more money than was saved by not completing the site studies.

Two types of deep foundations are recognized: *piles* and *piers*. Piles are long slender shafts that are forced into the ground. Although there is no specific limit, most are less than 0.6 m (2 ft) in diameter. Piers are larger, constructed by excavation, and usually permit visual examination of the soil or rock on which they rest. In effect they are deep spread footings or mats. A sharp distinction between piles and piers is impossible because some foundations combine features of both.

11:1 Development and Use of Piles

Piles are older than history. The Neolithic inhabitants of Switzerland 12,000 years ago drove wooden poles in the soft bottoms of shallow lakes and on them erected their homes, high above marauding animals and warring neighbors. Similar structures are in use today in jungle areas of Southeast Asia and South America. Venice was built on woodpiling in the marshy delta of the Po River to protect early Italians from the invaders from Eastern Europe and at the same time enable them to be close to the sea and their source of livelihood. Venezuela was given its name (meaning little Venice) by the Spanish explorers who found the Indians living in pile-supported huts in lagoons around the shores of Lake Maricaibo. Today, pile foundations serve the same purpose: to make it possible to build in areas where the soil conditions are unfavorable for shallow foundations.

USES OF PILES / Piles are used in many ways, as shown in Fig. 11.1. *Bearing piles* that support foundation loads are the most common form. They do this either by transferring the load of the structure through soft strata into stronger, incompressible soils or rock below, or by distributing the load through soft strata that are not capable of supporting the concentrated loading of shallow footings. Bearing piles are also used when there is danger of the upper soil strata being scoured away by current or wave action, or when wharves and bridges are built in water.

Tension piles are used to resist upward forces. These are used in structures subject to uplift, such as buildings with basements below the groundwater level, aprons of dams, or buried tanks. They are also used to resist overturning of walls and dams and for anchors of guy wires, bulkheads, and towers.

Laterally loaded piles support loads applied perpendicular to the axis of the pile in foundations subject to horizontal forces, such as retaining walls, bridges, dams, and wharves, and as fenders and dolphins for harbor

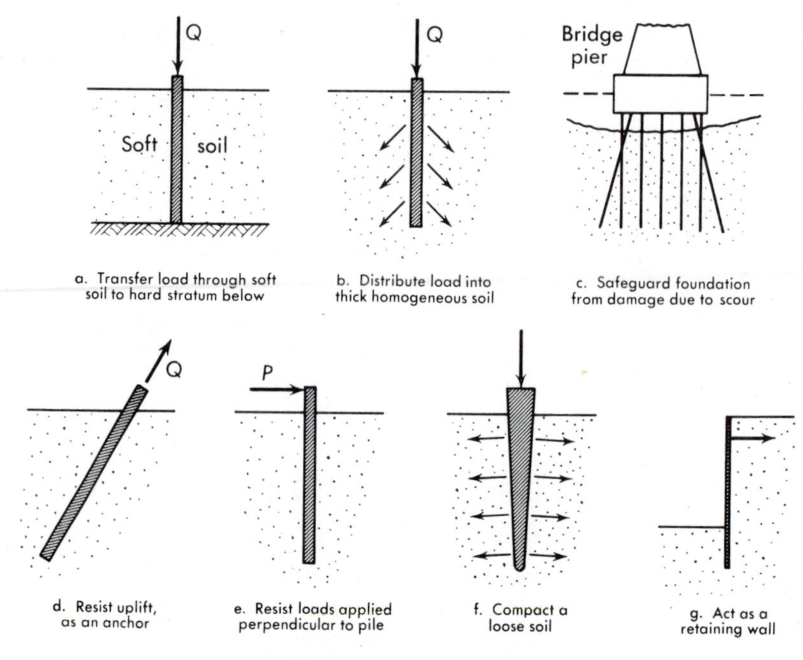

Figure 11.1 *Uses of piles.*

construction. If the lateral loads are great, they can be resisted more effectively by *batter piles* driven at an angle. Frequently, a combination of vertical and batter piles is used, as in Fig. 11.1c. Piles are sometimes employed to compact soils or to serve as vertical drains through strata of low permeability. Closely spaced piles and wide, thin *sheet piles* that interlock together are used as retaining walls, temporary dams, or seepage cutoffs.

11:2 Pile Driving

The operation of forcing the pile into the ground is *pile driving.* Like many construction operations it is an art, the success of which is dependent on the skill and ingenuity of the workers. However, also like many construction operations, it is increasingly dependent on engineering science for greatest effectiveness. Even more important both the art and engineering mechanics involved in the construction are major factors in the ultimate ability of the pile foundation to fulfill its function. Therefore, the foundation engineer must be involved in construction and the construction engineer in the design. The oldest method and the one most widely used today is by means of a hammer. Oriental constructors have used a stone block as a hammer for centuries. It is lifted by ropes held taut by laborers arranged in a star pattern around the pile head. The rhythmic pulling and stretching of the ropes throws the stone up in the air and guides the downward blow on the pile head. The Romans used a stone block hoisted by an A-frame derrick with slave or horse power, and guided in its fall by vertical poles.

PILE DRIVING EQUIPMENT / While the simple A-frame, pile-driving rig of the Romans is still in use today (with mechanical power), the more common machine is essentially a crawler-mounted crane (Figs. 11.2 and 11.3). Attached to the boom are the *leads*: two parallel steel channels fastened together by U-shaped spacers and stiffened by trussing. These serve as guides for the *hammer*, which is fitted with lugs so as to slide between them. The leads are braced against the crane with a *stay*, which usually is adjustable to permit driving batter piles. A steam generator or air compressor is required for steam hammers.

The pile is placed between the sides of the leads under the hammer. Lateral support to steady the pile is sometimes provided by sliding guides placed in the leads at the midpoint or thirdpoints of the pile.

Some large pile-driving rigs are mounted on I-beam bases that are supported by steel beams and timber cribbing. They are moved by skidding them along the beams or on rollers. Rubber-tired motor crane rigs are used for highway work, and even forklift trucks mount hammers for working inside buildings. Barge-mounted rigs are available for marine construction and compact rail-mounted rigs for work on tracks. Sometimes, small *swinging leads* suspended from cables are used when there is insufficient room for a crane.

The most important feature of the driving rig, from the engineer's point of view, is its ability to guide the pile accurately. It must be rugged and rigid enough to keep the pile and hammer in alignment and plumb in spite of wind, underground obstructions, and the movement of the pile hammer.

PILE HAMMERS / The simplest is the *drop hammer*, a block of cast steel commonly weighing from 2 to 13.3 kN (0.5 to 3 kips) is lifted 2 to 3 m

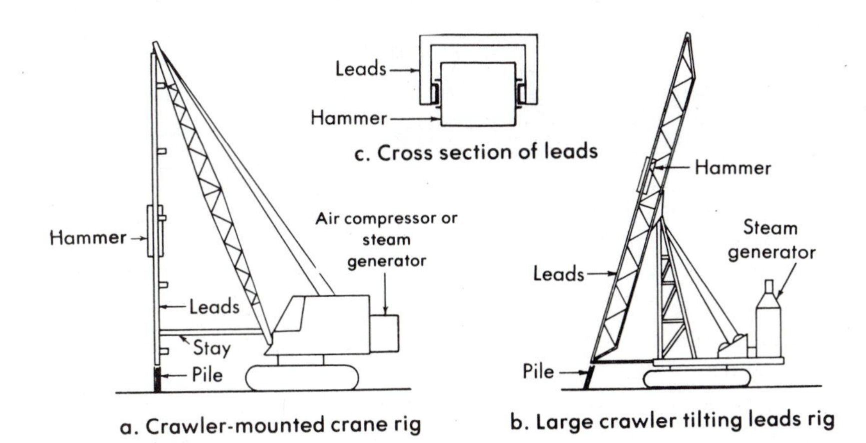

Figure 11.2 *Essential parts of pile-driving rigs.*

Figure 11.3 *Pile-driving rig mounted on crawler treads with leads that can be tilted to drive batter piles.*

(6 to 10 ft) above the pile by a winch and then released and allowed to drop. The drop hammer is simple but very slow and is used only on small jobs where the contractor must improvise the equipment or where the cost of bringing in special machinery is not justified.

The *single-acting* steam hammer employs a heavy cast-steel block known as the *ram*, a piston, and a cylinder (Fig. 11.4a). Steam or compressed

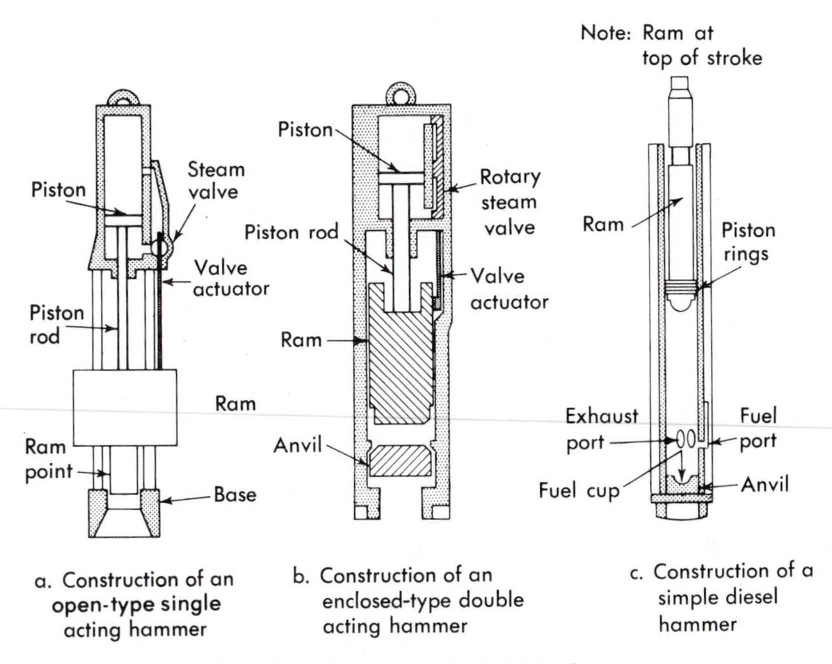

Figure 11.4 *Construction of pile-driving hammers.*

air is introduced into the cylinder to lift the ram 0.6 to 1 m (2 to 3 ft) and then is released to allow the ram to fall on the head of the pile. These hammers are simple and rugged, and they deliver a low-velocity blow whose energy is relatively constant in spite of wear, adjustment, or small variations in steam pressure. Their characteristics are given in Table 11:1.

The *double-acting* or *differential-acting* hammer (Fig. 11.4b) employs steam or air pressure to lift the ram and then accelerate it downward. The blows are more rapid, from 95 to 240 blows per minute, thus reducing the time required to drive the pile and making the driving easier in saturated loose sands. They can lose some of their effectiveness with wear or poor valve adjustment. The amount of energy delivered in each blow varies greatly with the steam or air pressure. Careful inspection is necessary to ensure that it is the amount specified. If the number of hammer blows per minute is approximately the rated value, as given in Table 11:1, the steam pressure is probably correct.

Steam hammers can operate on both steam and compressed air. Steam operation is more efficient, particularly with circulating steam generators. If the hammer is to be operated underwater, as can be done with enclosed double-acting types, air is required.

Diesel pile hammers are available in an increasing range of sizes. They consist of a solid-bottom cylinder and an enclosed-piston ram. The ram is raised upward mechanically and then allowed to fall. Fuel is injected into the

TABLE 11:1 / CHARACTERISTICS OF TYPICAL PILE-DRIVING HAMMERS[11:1]

Hammer	Type	Wt. Ram kN	kips	10^3 kg	Blows per Min.	Energy per Blow kNm	ft-kips	kgfm
Vulcan, Raymond 1	Single-acting	22	5	2.3	60	20	15	2,080
Vulcan, Raymond 0	Single-acting	33	7.5	3.4	50	33	24.4	3,380
Vulcan 040	Single-acting	178	40	18	60	163	120	16,630
Vulcan 3150 CT	Single-acting	668	150	68	60	610	450	62,360
MKT S-5	Single-acting	22	5	2.3	60	22	16.3	2,260
MKT S-14	Single-acting	62	14	6.4	60	51	37.5	5,200
MKT C-5	Double-acting	22	5	2.3	110	22	16	2,220
MKT 11B-3	Double-acting	22	5	2.3	95	26	19.1	2,650
MKT C-826	Compound	36	8	3.6	95	33	24	3,330
Vulcan 80C	Differential	36	8	3.6	111	33	24.5	3,400
Vulcan 140C	Differential	62	14	6.4	103	49	36	5,000
Raymond 0–30	Hydraulic	29	6.5	3.0	130	26	19.5	2,700
MKT DE20	Diesel	9	2	0.9	48	22*	16*	2,220*
MKT DE40	Diesel	18	4	1.8	48	43*	32*	4,430*
Link Belt 520	Diesel	23	5.1	2.3	80–84	41*	30*	4,160*
Kobe K-32	Diesel	32	7.1	3.2	40–60	81*	60.1*	8,330*
Delmag D-39	Diesel	29	6.6	3.0	39–60	74*	54.3*	7,520*

*Energy variable: maximum value for very hard driving, far less for easy driving; maximum energy at minimum blows per minute.

cylinder while the hammer drops and is ignited by the heat of the air compressed by the ram. The impact and the explosion forces the cylinder down against the pile and the ram up, to repeat the cycle automatically. The important advantages of the diesel hammers are that they are self-contained, economical, and simple to service. The energy per hammer blow is high, considering the weight of the hammer, but it is developed by a high-velocity blow from a middle-weight ram. The biggest disadvantage is that the energy per hammer blow varies with the resistance offered by the pile and is extremely difficult to evaluate in the field. In some types of diesel hammers the length of the ram stroke can be observed visually and the available energy approximated by the product of the stroke and weight. In others the hammer energy can be estimated from the air pressure generated in a recoil chamber above the hammer. Because of the variable energy the diesel hammer is best adapted to conditions where controlled energy is not critical, or where it can be closely monitored at critical times.

A double-acting hammer actuated by hydraulic pressure is somewhat faster and lighter than equivalent steam hammers because the operating pressure is much greater. The compact hydraulic pump system is easier to move than the bulky air compressor or steam generator, although the higher pressures do involve more critical mechanical problems. The light double-acting hammer will develop the same foot-pounds of energy at the instant of contact with the pile head as the heavy ram of a single-acting steam hammer falling 1 m (3 ft). The effects of the blows, however, are different, owing to the greatly different velocities of the falling rams at the instant of striking. Consider the driving of a railroad spike, using first a tack hammer and striking a hard, fast blow. Then use a heavy iron sledgehammer that drops a few inches so as to develop the same amount of energy. The slow, heavy blow drives the spike, whereas the tack hammer bounces. The same difference in effect can be observed in the driving of piles. Experience has shown that the weight of the driving ram should be from $\frac{1}{3}$ to 2 times that of the pile.

Most pile hammers require the use of *driving heads, helmets,* or *caps* that distribute the force of the hammer blow over the butt of the pile. The head is made of cast steel and contains a renewable wood, fiber, or laminated metal and rubber or plastic cushion block on which the hammer strikes. Heads for driving reinforced concrete piles may also provide for a wood cushion between the driving head and the pile.

BEHAVIOR OF THE PILE DURING DRIVING / Pile driving is a fascinating operation that never fails to attract crowds of onlookers. Clouds of steam and the repeated hammering are arresting, but they often obscure what merits much attention from the engineer—the behavior of the pile during driving. In very soft soils the first few blows of the hammer may drive the pile several feet; in fact, the pile may "run" into the ground under the static weight of the hammer. In harder soils, however, each blow of the hammer is accompanied by definite distortion of the pile and consequent losses of energy. If a piece of chalk is held against a pile and is moved with a steady horizontal motion while the driving is progressing, a graph will be traced on the pile that represents the vertical movement of the pile with time. A typical example of such a graph is shown in Fig. 11.5. The blow of the hammer produces an initial downward movement of the pile, but this is followed by a partial rebound or *bounce* that represents the temporary elastic compression of the pile and the soil surrounding it. The initial movement minus the bounce is called the *set* and is the net movement of the pile into the soil under one hammer blow. The average set for several hammer blows can be found from the driving resistance, which is the number of blows necessary to drive the pile a specified distance, usually 25, 100, or 300 mm (1, 4, or 12 in.).

When the pile is very long and the driving hard, the pile behavior is more complex. At the instant of impact the top part of the pile moves downward. The section of pile immediately below is compressed elastically but the tip of the pile momentarily remains fixed. The zone of compression travels swiftly down the pile, reaching the tip a fraction of a second after the initial impact. As a result of this compression wave the entire pile does not

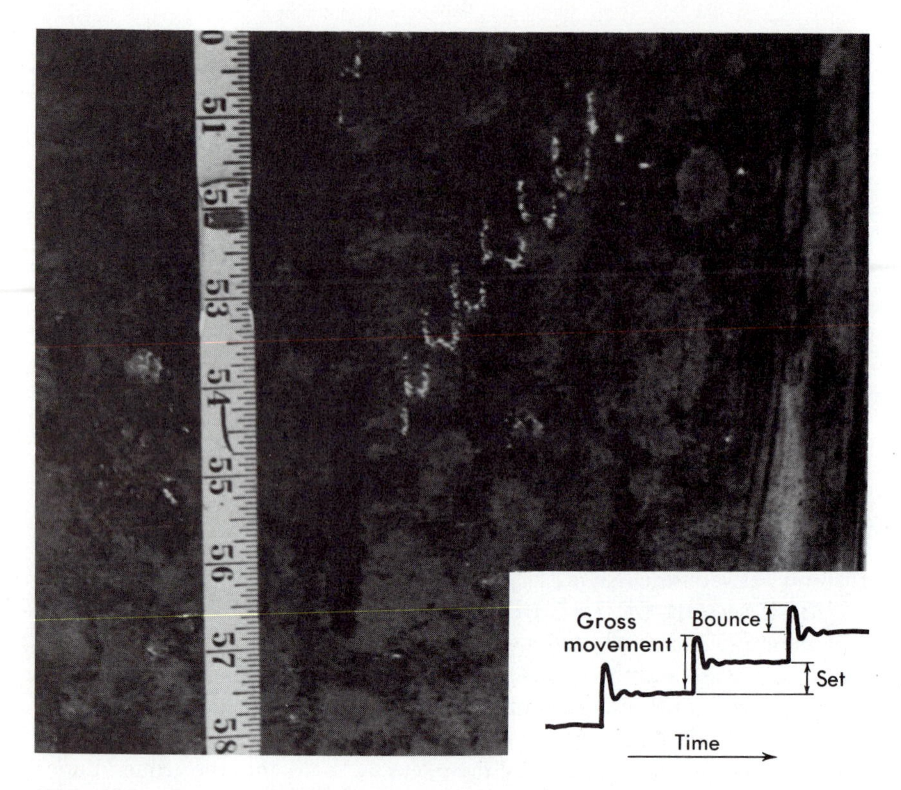

Figure 11.5 *Movement of head of pile during driving. Photograph of chalk trace on a pipe pile.* (Inset): *Bounce and set.*

move downward at any one instant but instead moves in shorter segments or waves.

OTHER METHODS OF PILE DRIVING / In cohesionless soils *jetting* can be used to place short, lightly loaded piles in their final position or as an aid to driving long, heavily loaded piles. The jet consists of a 40 to 50 mm ($1\frac{1}{2}$ to 2-in.) pipe, with a nozzle half that diameter, that is supplied with water at from 1000 to 2000 kN/m^2 (150 to 300 lb/in.^{2}). It can be used to wash a hole in the sand before driving or can be fastened to the pipe, singly or in pairs (or even embedded in concrete piles), so that driving and jetting can proceed simultaneously. Since jetting loosens the soil, it is usually stopped before the pile reaches its final position, and the last few feet of penetration secured only by hammering. If too much water is used, the jetting can loosen previously driven piles. It is of greatest benefit in dense sands and of little help in clays.

Where stiff clays or soft rock must be penetrated at high levels in order to reach the bearing stratum, time and expense can be saved by *preboring*. In dry soils this is done with an auger, and the pile is dropped into the open hole. If the soil is continuously stiff, a concrete pile can be cast in the open hole, forming a *bored pile* (discussed later in the chapter). If the soil contains soft seams, the hole can be made with a rotary well drill and kept open by a slurry of soil and water. The pile is driven through the slurry to bearing in firm strata below.

Spudding is the driving of a heavy steel H section into the soil to punch through obstructions or to break up hard seams that could damage or even prevent penetration of small piles. The spud is withdrawn before the pile is driven.

Jacking is employed to drive piles when the vibration of hammering is not permissible or when the head room is too small to permit use of a pile hammer. It is used principally in underpinning where the piles are jacked in short sections, using the existing structure as a reaction.

Vibrators have been effective in driving piles in silty and sandy soils. The driver consists of a pair of counter-rotating weights oriented to provide up-down motion. Vibration frequencies of 12 to 30 Hz (720 to 1800 cycles/min) are available with dynamic forces of from 71 kN (16 kips) to 912 kN (205 kips). The vibrators weigh from 6.7 kN (1.5 kips) to 169 kN (38 kips) and are driven by electric motors from 12 to 24 hp. Several vibrators, synchronized to provide simultaneous impulses have been used to drive 1.5-m (5-ft)-diameter piles.

If the vibration frequency equals the natural frequency of the pile-soil system, resonance increases the effectiveness greatly. This requires variable speed motors, typically at frequencies exceeding 30 Hz. Although resonant or *sonic* pile driving has been spectacularly successful, the high dynamic forces also damage the vibrators; therefore, the method is seldom used.

11:3 Pile Capacity

The ability of a pile foundation to support loads without failure or excessive settlement depends on a number of factors: the pile cap, the pile shaft, the transfer of the pile load to the soil, and the soil and underlying rock strata which ultimately support the load. The pile-cap analysis and design is essentially a structural problem and is covered adequately in textbooks on reinforced concrete design. It is rarely a critical problem or a source of trouble. The analysis and design of the pile shaft involves both the pile and the soil. Usually, the shaft capacity is dictated by construction needs and is far more than is needed for the ultimate load, but it can be critical with heavily loaded slender piles or when construction difficulties are encountered. The transfer of the pile load to the soil is termed the *pile-bearing capacity*. It is a frequent source of trouble in pile foundations. The ability of the underlying strata to carry the load depends on the combined effect of all the piles acting together. Although the capacity of the underlying strata seldom receives attention, it is a frequent source of trouble in pile foundations.

PILE SHAFT / The pile shaft is a structural column that is fixed at the point and usually restrained at the top. The elastic stability of piles, their resistance against buckling, has been investigated both theoretically and by load tests.[11:3] The buckling of a pile depends on its straightness, length, moment of inertia, and modulus of elasticity, and the elastic resistance of the soil that surrounds it. Both theory and experience demonstrate that the lateral support of the soil is so effective that buckling will occur only in extremely slender piles in very soft clays or in piles that extend through open air or water. Therefore, the ordinary pile in sand or soft clay can be designed as though it were fully braced or were a short column. This is substantiated by load tests of 30-m (100-ft) long piles in soft clay at a Michigan site. H-piles failed above the ground surface at the yield point of their steel, and concrete piles failed by crushing at the compressive strength of their concrete (Fig. 11.23).

The most important consideration in limiting the capacity of the shaft is faulty construction, particularly of connections between two sections of pile. This can lead to deflection of the lower part of the pile and the development of a dog-leg and to a reduction of the pile cross section and a loss of strength as a short column. Tests of dog-legged piles shows that their capacity is not materially reduced, provided the surrounding soil is firm. The reduction in column strength can be prevented by careful control of the construction procedures.

EFFECT OF PILE ON SOIL[11:4, 11:5, 11:6] / The stress pattern, settlement, and ultimate capacity of the pile foundation depend on the effect of the pile on the soil. The pile, represented by a cylinder of length L and diameter D (Fig. 11.6a) is a discontinuity in the soil mass that either replaces or displaces soil, depending on whether it is installed by excavation like a pier, or by driving.

Excavation disturbs the soil by changing the stress pattern; the soil may squeeze inward (Fig. 11.6b) disrupting the structure of clays and reducing the density of sands. Forcing a pile into the hole or placing wet concrete in the hole partially forces the soil back, creating more disturbance.

Driving the pile creates even greater disturbance. The tip acts as a small footing that accumulates a cone of soil and punches down, forcing the soil aside in successive bearing capacity failures (Fig. 11.6c). A zone of disturbed or remolded soil is formed around the pile, with a width of from D to $2D$. If the driving is aided by jetting or predrilling a small hole, the disturbed zone is smaller. Within the disturbed zone there is a reduction of cohesive strength in saturated clays and cemented soils. In most cohesionless soils there is an increase in density and angle of internal friction.[11:5] However, in a very dense soil there might be a density reduction in the immediate vicinity of the pile due to shear and a slight local reduction in the angle of internal friction. Evidence of high tip forces is seen in Fig. 11.7a.

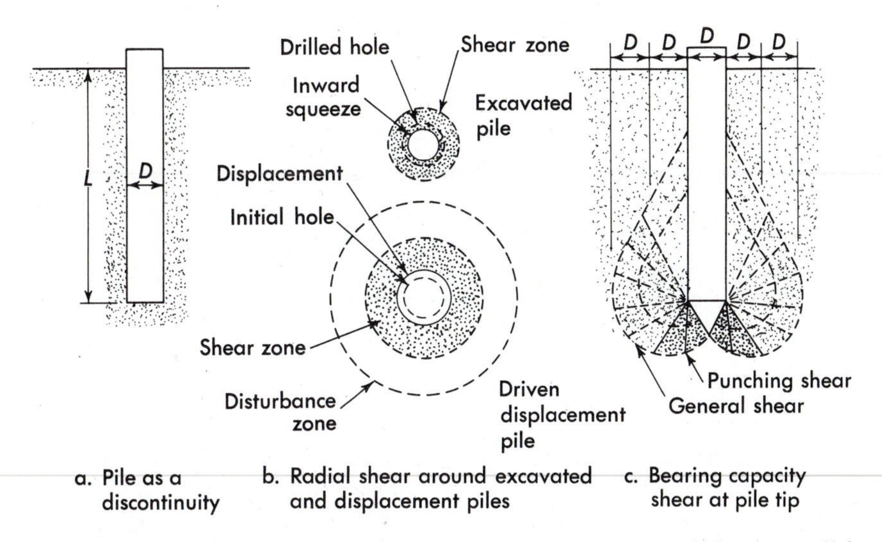

Figure 11.6 *Effect of pile on soil during driving.* (Adapted from Meyerhof[11:5] and Vesić.[11:6])

The displacement of driven piles has two effects. First, there is heave of the ground in saturated clays and dense cohesionless soils. The heave sometimes pushes previously driven piles laterally as much as 0.3 to 0.6 m (1 to 2 ft) or raises the ground surface an amount equivalent to the volume

Figure 11.7 *Effect of driving on piles.* (Left): *Tip of 273 mm ($10\frac{3}{4}$ in.) OD steel pipe pile wrinkled by continuing driving by a very heavy hammer after tip had reached refusal in dense soil.* (Right): *Collapse of 355 mm (14 in.) ID corrugated steel tubular pile from pressure developed by displacement of adjacent piles in stiff clay.*

of soil displaced. Second, high lateral pressures are set up in the soil. The limited data available indicate that the total lateral pressure in saturated clay can be as much as twice the total vertical overburden pressure; and in sands, the effective lateral pressure can be from 0.5 to 3 times the vertical effective stress.[11:6] In saturated clays even higher pressures have been indirectly indicated by the collapse of cofferdams and thin-walled open pipe or steel shell piles and the heaving of structures near piles being driven.

The tubular pile collapse [Fig. 11.7 (right)] occurred in a group of 36 piles driven 1 m ($\frac{1}{2}$ ft) apart into stiff clay. The displacement of piles driven in clay on one project broke spread footings and raised a wall of an adjoining building 75 mm (3 in.).

In saturated clays the pressure increase is largely in the neutral stress phase. This is dissipated into the surrounding soil with time, causing the lateral pressure to drop toward its original value, somewhat less than the overburden pressure. The reduction in neutral stress in the clay is accompanied by a regain in strength, which in some cases eventually exceeds the original strength of the undisturbed soil.

Driving piles with hammers produces shock and vibration that are transmitted through the ground to adjoining structures. This can annoy the occupants, and when sufficiently severe, cause physical damage. If there are very loose saturated fine sands present, the vibration may cause temporary liquefaction, loss of bearing capacity and severe damage. This is a rare occurrence, however. More commonly, the vibration in loose sand deposits causes a subsidence of the ground surface in spite of the pile displacement. The subsidence may extend as far as 30 m (100 ft) from the structure, depending on the pile length and severity of driving, and cause settlement and damage to nearby buildings.

TRANSFER OF LOAD / The pile transfers the load into the soil in two ways, as shown in Fig. 11.8:[11:7, 11:8, 11:9] first, by the tip in compression,

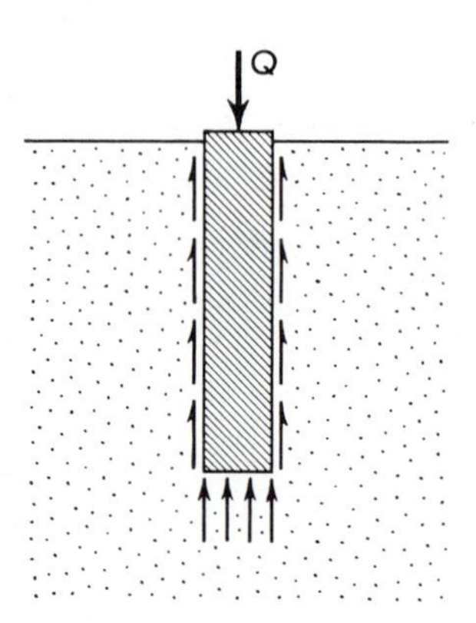

Figure 11.8 *Transfer of load from pile to soil by end-bearing and skin friction.*

termed *end bearing*, and second, by shear along the surface commonly termed *skin friction* (although true friction does not develop in all cases). Piles driven through weak strata until their tips rest on a hard stratum transfer the greater part of their load by end bearing and are sometimes called *end bearing piles*. Piles in homogeneous soils transfer the greater part of their load by skin friction and are thus called *friction piles*. Nearly all piles develop both end bearing and skin friction, however.

STRESS FIELD AROUND PILE / The initial stress field around a pile installed by drilling or jetting is probably close to the at-rest, depending on any stress reduction accompanying squeeze and any stress increase produced by displacement.

Upon loading the pile, the stress field is changed by the transfer of the pile load to the soil.

The analysis of stresses produced by a vertical load introduced below the surface of a semiinfinite isotropic elastic solid was developed by Mindlin (Fig. 11.9). It is the equivalent of the Boussinesq analysis for surface loads. The vertical stress increment, $\Delta\sigma_z$, produced by a load Q at a depth of z, and pile length L is given by the expression

$$\Delta\sigma_z = \frac{Q}{L^2} I_p \tag{11:1a}$$

$$I_p = f\left(\frac{z}{L}; \frac{x}{L}\right) \tag{11:1b}$$

Contours of equal vertical stress in terms of I_p for unit depths of $z/L = m$ and load Q are shown in Fig. 11.9. The right half shows the vertical stresses at a radius of $n = x/L$ and $m = z/L$ for end bearing only at depth L; the left half shows the vertical stress for a uniform distribution of Q by skin friction along the length of the pile (Fig. 11.8). The vertical stress computed by the Boussinesq analysis assuming that the pile end bearing is at the surface of an elastic mass is shown by a dotted line on the right half of the diagram for comparison. It shows that the increased stress due to end bearing inside the elastic mass near the pile tip is approximately half of that found by the Boussinesq analysis for surface loads.

Above the pile tip, within a cylindrical zone whose radius is about half the pile length, the end bearing produces a negative $\Delta\sigma_z$, or a reduction in vertical stress in the soil. The radial stresses (in the lateral direction) are similarly influenced by the vertical stress transferred into the soil by the pile. Above the point of loading the radial stress is reduced; below the point of loading, it is increased.

The combined effect of end bearing and skin friction on the stress field depends on their relative magnitudes as well as the distribution of skin friction along the pile length.

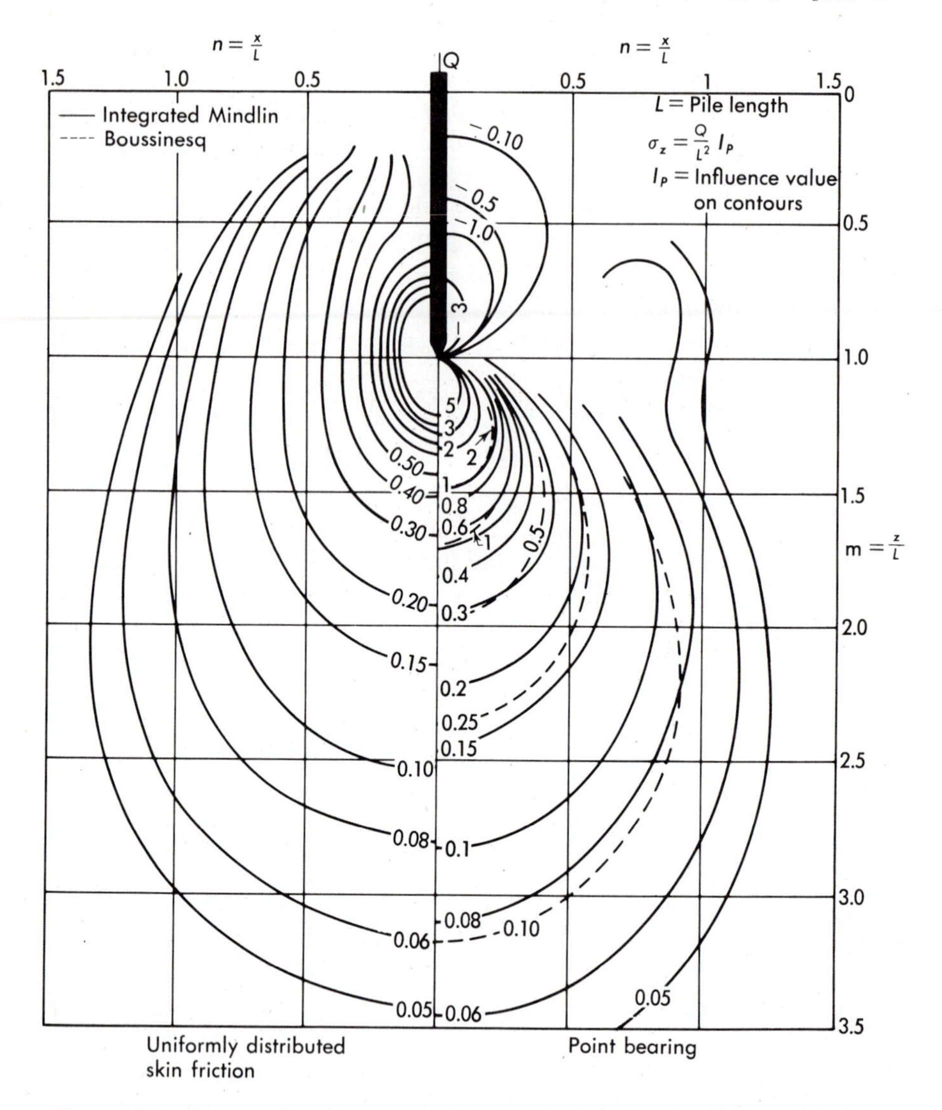

Figure 11.9 *Contours of equal stress around and below the point of a pile in a semiinfinite, homogeneous elastic solid—the Mindlin analysis.* (After O. Grillo.[11:9])

Estimates of the vertical stress produced by a pile at loads below failure can be made with the use of Fig. 11.9b. Limited field observations of piles in homogeneous materials suggest that for lengths exceeding 20 diameters, the end bearing is about $\frac{1}{4}$ to $\frac{1}{3}$ the total; for shorter piles the proportion carried by end bearing increases in proportion to D/L. If the soil or rock at the pile tip is more rigid than along the shaft, the end bearing would be greater. As the load approaches failure the proportion of load transfer in end bearing depends on the ultimate or limiting shear in end bearing compared to the limiting shear in skin friction.

STRESSES ADJACENT TO A PILE / The vertical stress, σ_{zp}, immediately adjacent to an unloaded drilled pile is γz (Fig. 11.10). As the pile load is increased, there is a reduction in the vertical stress immediately adjacent to the lower part of the pile due to the load transferred in end bearing as can be seen from the tension zone of Fig. 11.9. Although this may be partially offset by the increase in vertical stress caused by load transfer in skin friction above, the net effect for long slender piles will be a stress reduction. In addition the subsidence of the soil mass surrounding the pile produces a vertical stress reduction similar to that in a backfilled trench (Fig. 9.29). As a result the vertical stress adjacent to a loaded pile is less than γz below a critical depth z_c (Fig. 11.10b). Below the depth of z_c the vertical

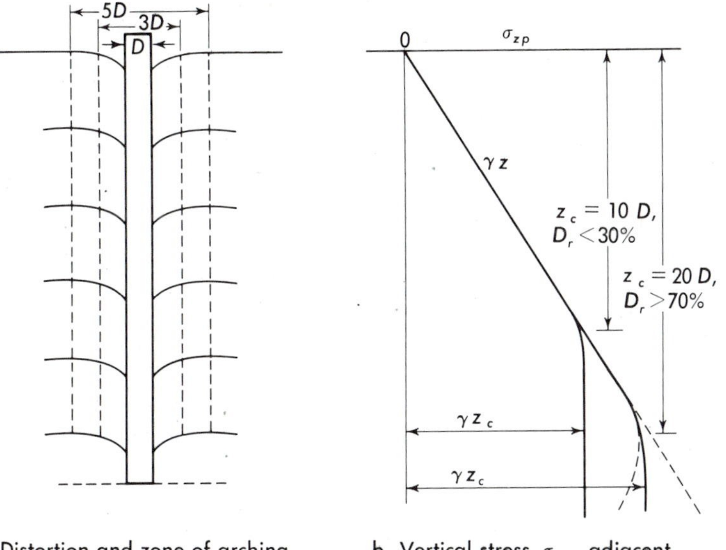

a. Distortion and zone of arching around a pile approaching failure

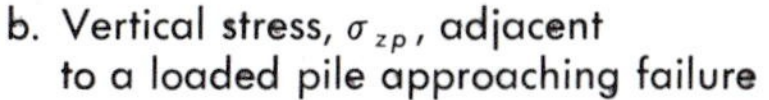
b. Vertical stress, σ_{zp}, adjacent to a loaded pile approaching failure

Figure 11.10 *Soil distortion and vertical stress adjacent to a loaded pile.* (Adapted from Vesić.[11:6])

stress immediately adjacent to a drilled or jetted pile depends on the pile load. At failure, tests indicate that it is approximately γz_c. At smaller loads it is somewhat higher. For driven piles (which have already "failed" successively during driving) the vertical pressure below z_c is apparently close to γz_c, regardless of load. Large-scale tests in soils by Vesic[11:6] at the Georgia Institute of Technology and by Kerisel[11:11] in France indicate that the critical depth z_c is a function of relative density. For $D_r < 30\%$, $z_c = 10D$; for

$D_r > 70\%$, $z_c = 20D$; for intermediate densities it is approximately proportional to relative density.

The lateral pressure of soil against the pile surface can be expressed by the equation above the depth z_c,

$$\sigma'_h = K_s(\gamma z - u) \tag{11:2a}$$

Below z_c, σ'_h remains constant,

$$\sigma'_h = K_s(\gamma z_c - u) \tag{11:2b}$$

The earth pressure coefficient K_s depends on the displacement of the pile and the density or compressibility of the soil. (See Table 11:2.) For the jetted or drilled piles the values of K_s increase with load; the maximum occurs at failure.

TABLE 11:2 / COEFFICIENT OF LATERAL EARTH PRESSURE IN COHESIONLESS SOIL ADJACENT TO PILE AT FAILURE

Soil	Displacement Condition	K_s
Loose sand,	Jetted	0.5 to 0.75
$D_r < 30\%$	Drilled pile	0.75 to 1.5
	Driven pile	2 to 3
Dense sand,	Jetted	0.5 to 1
$D_r > 70\%$	Drilled pile	1 to 2
	Driven pile	3 to 4

STATIC ANALYSIS OF BEARING CAPACITY / The ultimate bearing capacity of the pile or pier is the sum of end bearing and skin friction at the instant of maximum load:

$$Q_0 = Q_{EB} + Q_{SF} \tag{11:3}$$

The ultimate values for Q_{EB} and Q_{SF} can be analyzed separately. Both are based upon the state of stress around the pile (or any deep foundation) and on the shear patterns that develop at failure.

In end bearing the pile tip resembles a deeply buried spread footing. When the pile is loaded, a cone of intact soil adheres to the tip. As the tip penetrates deeper with increasing load, the cone forces the soil aside, shearing the mass along a curved surface (Fig. 11.6c). If the soil is loose or compressible or has a low modulus of elasticity, the mass beyond the shear zone compresses or deforms, allowing the cone to penetrate further. This is a form of local shear, similar to that described for shallow foundation (Chapter 10). If the soil or rock is very rigid, the shear zone expands until the total displacement allows the cone to punch downward. Various forms of shear zone have been proposed to evaluate end bearing. Like the results of shallow foundation analyses these can be expressed in the general form

$$q_0 = \frac{B\gamma}{2} N_\gamma + cN_c + q_q N_q \tag{11:3}$$

For piles where B is small, the first term is often omitted:

$$q_0 = cN_c + q_q N_q \tag{11:4}$$

Although many different bearing-capacity factors have been derived for deep foundations, the range that has been verified to some extent by full size test piles is depicted in Fig. 11.11. The lower curves are the Meyerhof factors for shallow foundations, corrected for a circular or square shape. The upper curves are for general shear failure, adapted from Meyerhof,[11:5] and require the complete development of the shear zone that can only occur in a rigid-plastic solid or a dense sand. The middle curves are adapted from Berezantzev's work in sands; they fit the results of load tests of driven large-scale models and full-sized piles.[11:12]

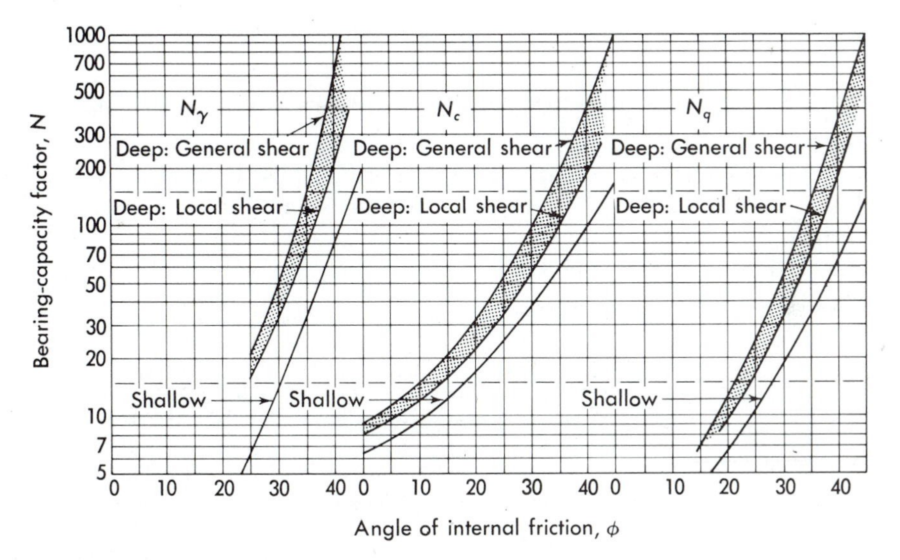

Figure 11.11 *Bearing-capacity factors for shallow and deep square or cylindrical foundations.* (Adapted from Meyerhof[11:5] and Berezantzev.[11:11])

Figure 11.11 shows a considerable range in N for deep foundations, the shaded zone. It represents the uncertainties caused by changes in soil density with pile driving as well as the reduction in vertical stress with depth below z_c. The lower shallow curves apply to depths of $z/2D$. They also apply to a pile or pier extending through a weak stratum and end bearing on a hard one. Below $z = 4D$ the factors increase in proportion to increasing depth to the shaded zone between the two upper curves, at $z = z_c/2$. The upper curve applies to hard clays or very dense sands. At depths approaching z_c, the end bearing appears to become constant. According to Meyerhof[11:4] that part produced by surcharge q_{0q}, has been found by full-scale tests to reach a limit,

$$q_{0q} = 50N_q \tan\phi \text{ (kN/m}^2\text{)} \tag{11:5a}$$

$$q_{0q} = N_q \tan\phi \text{ (kips/ft}^2\text{)} \tag{11:5b}$$

In these limiting expressions, N_q is found from the upper side of the shaded range.

If the piles are driven into the ground, the appropriate angle of friction is that obtained after driving. In sands the increase is from 2° to 5° above the value before driving, according to Meyerhof.[11:5] If the foundation is installed by jetting or drilling, the angle remains virtually unchanged.

The skin friction that acts along the pile shaft is equal to either the sum of friction plus adhesion on the pile face or to the shear strength of the soil immediately adjacent to the pile, whichever is smaller. If f_0 is the skin friction

$$f_0 = \begin{cases} c + \sigma'_h \tan\phi \\ c_a + \sigma'_h \tan\delta \end{cases} \tag{11:6}$$

where c_a is the adhesion and δ is the angle of friction of soil against the pile face.

The values of c_a and $\tan\delta$ can be determined by direct shear tests in which one-half of the shear box is replaced by the same material as the pile surface.

Typical values of $\tan\delta$ are given in Table 11:3, based on limited test data. The effective lateral stress is computed from Equations (11:2a) and

TABLE 11:3 / SKIN FRICTION OR SLIDING RESISTANCE SOILS AGAINST PILES AND SIMILAR STRUCTURES
(From Laboratory Tests)

	Friction, Cohesionless Soil		Adhesion, Cohesive Soil (Saturated, Undrained)
	$\tan\delta$	δ	
Wood	0.4	22°	$0.9c$ to c
Rough concrete, cast against soil	$\tan\phi$	ϕ	c
Smooth formed concrete	0.3–0.4	17	$0.8c$ to c
Clean steel	0.2	11	$0.5c$ to $0.9c$
Rusty steel	0.4	22	c
Corrugated metal	$\tan\phi$	ϕ	c

(11:2b), using the coefficients estimated from Table 11:2. The value of σ'_h does not increase below z_c.

The adhesion, c_a, and undrained shear strength, c, adjacent to instrumented piles, vary considerably from the values obtained from laboratory tests, according to Meyerhof,[11:4] from less than $\frac{1}{2}$ to 2 times the laboratory values for driven piles with displacement and from $\frac{1}{2}$ to 1.5 times the laboratory values for drilled piles with no displacement. The variations are probably the combined effects of soil displacement and disturbance followed by reconsolidation of the soil after driving. The measured average is about equal the laboratory values for driven piles and $\frac{2}{3}$ the laboratory values for bored piles. For preliminary design the averages are appropriate; they should be verified by load tests.

STATIC CAPACITY-PENETRATION TESTS / Empirical correlations have been developed for end bearing and skin friction of piles in cohesionless soils based on the penetration resistance measured during site exploration. Such relationships should be expected because a penetrometer is a miniature pile.

The point resistance of the Dutch cone (Fig. 7.9) is proportional to end bearing. Because of the great disparity in sizes, the cone resistance in force per unit of area often exceeds that for a full-sized pile. The pile skin friction is approximately double that of the cone friction sleeve. Because of the size effects, the relation between the cone resistance and pile capacity for each site and pile type should be established by full-scale load tests.

For the Standard Penetration Test, N (Chapter 7) Meyerhof suggests the following:[11:4, 11:13]

$$q_0 = 40N\frac{L}{D} \leq 400N \text{ (kN/m}^2\text{)} \tag{11:7a}$$

$$q_0 = 0.8N\frac{L}{D} \leq 8N \text{ (kips ft}^2\text{)} \tag{11:7b}$$

$$f_0 = 2N \text{ (kN/m}^2\text{)} \tag{11:7c}$$

$$f_0 = 0.04N \text{ (kips/ft}^2\text{)} \tag{11:7d}$$

For the nondisplacement piles in Equations (11:7c) and (11.7d), divide the values by 1.5 to 2.

TIME-CAPACITY / The bearing capacity of piles changes with time. Driving a pile into loose saturated sand causes soil densification and a temporary increase in pore pressure, and a temporary loss of both driving resistance and bearing. Neighboring piles already driven, including those of old structures that have been safely supporting their loads for years can suffer temporary loss of bearing and even damage. The excess eventually dissipates, in minutes in coarse sands but sometimes a few days in silty sands. The reverse occurs in dense sands. Driving produces dilation and temporary negative pore pressures that sometimes prevents driving the pile to its required depth. When the pore pressure is relieved by seepage, the resistance drops. Occasionally, engineers and contractors are fooled by this effect and stop driving before the proper bearing level is reached. Again the time for return to normal ranges from minutes to days. Ultimately, the bearing of piles in loose sand will be enhanced by the densification because of increased displacement.

In clays the time effect is more complex. Positive pore pressures and loss of strength develop from structural breakdown of sensitive clays; dilation causes negative pore pressures and increased strength in some overconsolidated clays. The effect is compounded by the increased lateral stress from pile displacement that ultimately produces soil consolidation and increased strength. Moreover, some clays gain strength by reestablishing interparticle bonds: thixotropic hardening. Therefore, although many clays lose strength from remolding and pore pressure increase during driving, they ultimately become stronger than in their undisturbed state. The time required varies from a few days for thixotropic hardening to years for consolidation. Overconsolidated clays retain part of the strength gain generated by dilation.

Piles pulled out of clay frequently are covered with a skin of soil several inches thick that adheres tightly to their surface, verifying the strength gain.

The time effects in clays are not well understood. Load tests should be delayed at least a week and preferably longer to evaluate at least part of the changes that occur.

The total pile capacity is nominally the sum of the mobilized end bearing and the product of the mobilized skin friction and the surface area of the pile. The ultimate or failure load, Q_0, however, is not necessarily the sum of the ultimate end bearing and the ultimate skin friction. First, the end bearing and the skin friction along different sections of the shaft may not be mobilized simultaneously. Consider a pile whose shaft is in a weak, nonrigid soil but whose point rests on a rigid stratum. A relatively small downward movement of the pile would be sufficient to produce bearing-capacity failure, but the same movement would not be great enough to produce skin-friction failure. Therefore only part of the skin friction would be mobilized at the instant of failure. The deflection of the pile shaft under load (which is greatest at the ground surface but less at the point), the different rigidities of different strata in contact with the pile, and the compression of soil beneath the pile point also contribute to the unequal mobilization of end bearing and skin friction. As a result the actual pile capacity can be materially less than the sum of the ultimate values. The difference is aggravated in ultrasensitive soils where failure brings about a loss of strength. For these reasons the skin frictions of weaker strata are often neglected in analysis.

A driven pile usually has a higher ultimate capacity than one placed by excavation or jetting, because both skin friction and end bearing reach their ultimate values during driving. Moreover, the displacement of driving increases ϕ and ultimately c.

A second cause of difference between the computed and the actual ultimate capacity of piles arises from negative skin friction. The stresses introduced into the soils by the pile and by any surface loads, such as fill, not supported by piles, will cause the soils to consolidate. If there is a highly compressible stratum at some level above the pile point, its consolidation will cause the soils above to move downward with respect to the pile. Instead of supporting the pile, these strata, by their downward movement, now add load. This negative skin friction has been great enough to cause failure of pile foundations in a few cases and must be considered in design.

Example 11.1

Compute the ultimate bearing capacity of a 0.274 m (10.75 in.) OD steel pipe 20 m or 67 ft long driven into a homogeneous stiff insensitive clay with $c = 60\ \text{kN/m}^2$ or 1.25 kips/ft^2. The pipe has a flat-plate closed end. The soil and unit weight = 18 kN/m^3 or 115 lb/ft^2. The water table is at the ground surface.

1. End bearing

$$\frac{z_p}{D} = \frac{20}{0.274} = 73 > 20$$

From Fig. 11.11, $N_c = 7$ to 9, $N_q = 1$:

$$q_0 = 8 \times 60 + 1 \times 20 \times (18 - 9.81) = 480 + 164 = 644\ \text{kN/m}^2$$

$$Q_0 = 644 \times \frac{\pi \times 0.274^2}{4} = 0.0598 \times 644 = 38\ \text{kN} = 9\ \text{kips}$$

2. Skin friction. Estimate $c_a = 0.9c$ for clean steel (Table 11.3)

$$f_0 = 20 \times \pi \times 0.274 \times 0.9 \times 60 = 930\ \text{kN} = 209\ \text{kips}$$

3. $Q_0 = 9309 + 38 = 968\ \text{kN} = 217$ kips.

LOAD TESTING / The most reliable method of determining pile capacity for most sites is a load test. Pile load tests are made to determine the ultimate failure load of a pile or group of piles or to determine if the pile or pile group is capable of supporting a load without excessive or continuous settlement.

The bearing capacity of all piles except those driven to rock does not reach the ultimate until after a period of rest. Load tests are not a good indication of performance unless made after this period of adjustment. For piles in permeable soils this period is two or three days, but for piles partly or wholly surrounded by silt or clay it can be more than a month.

Pile load tests can be made by constructing a loading platform or box on top of the pile or group of piles (Fig. 11.12a) on which the load is applied, using sand, pig iron, concrete blocks, or water. A safer and more easily controlled test uses a large, accurately calibrated hydraulic jack to apply the load (Fig. 11.12b). The resistance above the jack can be secured with a loaded platform or by a beam held down by piles in tension. An added advantage of the jacking method is that the load on the pile can be varied rapidly and cheaply. Settlements are measured by a precision level or preferably by micrometer dial gages mounted on an independent support.

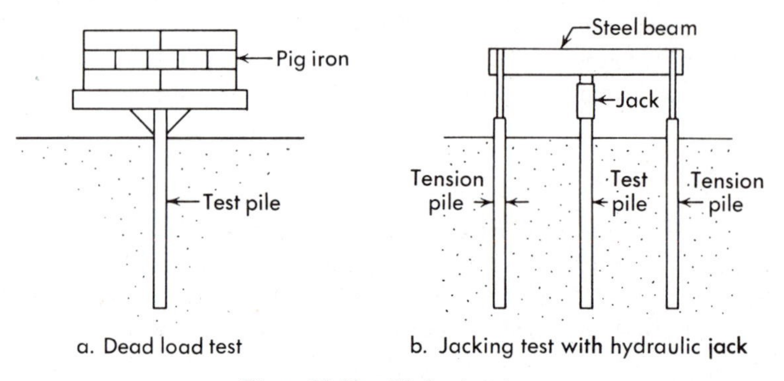

Figure 11.12 *Pile load testing.*

The loads are applied in increments of one-fifth or one-fourth the design load until failure or two times the design load is reached; the load is then reduced to zero by similar increments. Each load is maintained constant and the settlement is measured at regular intervals until the rate of movement is less than 0.3 mm (0.001 ft) per hr. A load final–settlement curve is plotted similar to that for the plate load test.

Many different criteria for working load have been proposed, but the best is the same as for any other foundation: the load having an adequate safety factor (1.5 to 2 when a test is made) or the load giving the greatest permissible total settlement (as described in Chapter 10), whichever is the smaller.

SETTLEMENT OF A SINGLE PILE / Settlement of a single isolated pile comes from the elastic shortening of the pile shaft and the distortion of the soil around the pile. These can be determined best from a load test. Settlement can be computed from the static analysis of pile capacity by computing the elastic shortening of each section of the pile shaft from that portion of the total load remaining in that section.

The elastic distortion of the soil immediately surrounding the pile can be estimated from the displacement in a direct shear test. A sample 20 to 30 mm (0.8 to 1.3 in.) thick produces shear displacements that are comparable to short-term soil distortions as measured in load tests.

The major settlement of all piles except those that are end bearing on rock comes from the consolidation of the underlying soil by stresses developed by the pile group. This is considered in Section 11.5.

TENSION PILES / Tension piles can be analyzed by the static method (with no end bearing) or by tension load tests. The resistance of tension piles with enlarged bases can be determined best by load tests.

11:4 Dynamic Analysis of Pile Capacity[11:9]

Because the driving of a pile produces successive pile-bearing failures with each hammer blow it should be possible theoretically to develop some relationship between pile capacity and the resistance offered to driving. For foundation design this implicitly presumes that the long-term capacity, Q_0, equals the driving resistance, R_0, a dynamic force. While this is valid in some situations, it may not be in others.

The pile loading and "failure" produced by driving with a hammer occurs in a fraction of a second, whereas in the structure the load is applied over a period ranging from hours to years. A fixed relation between the dynamic and long-term capacity can exist only in a soil whose shear strength is independent of the rate of loading. This is approximately true in a dry cohesionless soil, and in wet cohesionless soils that are of intermediate density or so coarse-grained that shear does not generate appreciable neutral stresses. In clays and in both very loose or dense fine-grained saturated cohesionless soils, the strength depends on the rate of shear; in such soils a dynamic analysis may have no validity.

WAVE ANALYSIS / The dynamic process of pile driving is analogous to the impact of a concentrated mass on an elastic rod. The rod is partially restrained along the surface by skin friction and at the tip by end bearing. This system can be approximated by a lumped mass elastic model, Fig. 11.13a.[11:14–11:18] The distributed mass of the pile is represented by a series of small concentrated masses, W, linked by springs that simulate the longitudinal resistance of the pile. The skin resistance can be represented by a rheologic model of surface restraint that includes friction, elastic distortion, and damping.

When the hammer W_r strikes the pile cap, a force R_c is generated that accelerates the cap (W_c) and compresses the cap. This transfers a force, R_1, to the top segment of the pile, W_1, and causes it to accelerate, slightly after the acceleration of W_c. The compressive force induced in the top of the pile, R_1, induces acceleration in the next segment of pile, W_2. A wave of compression moves down the pile. The vertical force at any instant, t, is equivalent to the spring compression. The force R travels down the pile in a wave, $R_1 \rightarrow R_2 \rightarrow R_3$ .... In a pile of uniform cross section, the force is diminished slightly in each successive length increment by side friction and damping, f. At the tip, however, R_T can decrease if the soil below is less rigid

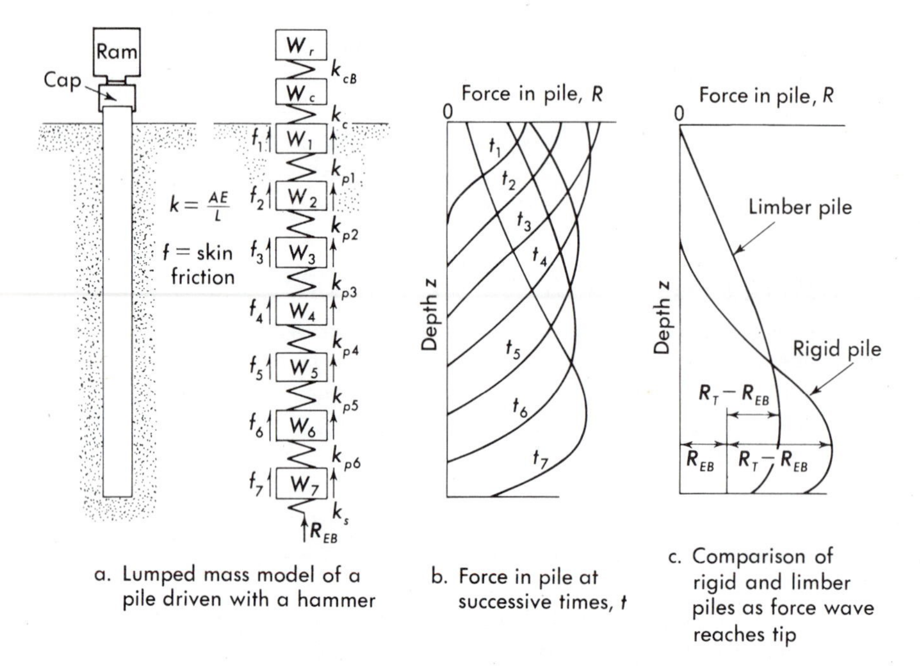

Figure 11.13 *Wave analysis of a driven pile.*

than the pile or even increase if it is more rigid. The force, R_T is dissipated at the tip in overcoming the dynamic end bearing R_{EB} and in temporarily distorting the soil. If $R_T < R_{EB}$ the pile will not move, this is termed *refusal.* All the force is dissipated in temporary soil distortion, sometimes termed *quake*, s_s. If $R_T > R_{EB}$, the pile moves into the soil permanently a distance s_t approximately the set of Fig. 11.5.

The shape of the force wave depends on the pile rigidity; a rigid pile (stiff springs) exhibits a sharper force wave with a higher peak than a limber pile. The force available for overcoming end bearing, $R_T - R_{EB}$ is greater for the rigid pile (Fig. 11.13c). The peak force is also a function of hammer energy and efficiency: the higher energy produces a higher force. The force, divided by the cross-sectional area of the pile equals the stress produced during driving. If the maximum stress exceeds the pile strength, the pile will be damaged, a condition known as *overdriving.*

In order to use the wave equation, the following data are required: (1) hammer weight W_r; (2) hammer velocity at impact, V_r; (3) area and rigidity of pile cap block or helmet, A_cE_c; (4) coefficient of restitution of cap block or helmet, n_c; (5) pile weight (in increments, $W_{p\text{-}1} + W_{p\text{-}2} \ldots$); (6) rigidity of the pile, A_pE_p; (7) distribution of skin friction and damping along the pile ($f_1, f_2, \ldots$); (8) end bearing, R_{EB}; (9) soil distortion at tip, s_s, and tip damping, J_t. Some of these can be measured: (1) (2) (3) (4) (5) (6); the others must be estimated. The progress of R down the pile is computed for various times, t, using short time intervals, Δt, with the aid of a digital computer.[11:17] Generally, a value for R is assumed and the unknowns, s_T, per blow, and R_{max} at different levels within the pile are computed.

The wave analysis provides a picture of the mechanics of the process and a realistic relation between R_T and s_s, when the input data, enumerated above, can be evaluated. This requires judgment as well as a reasonably accurate evaluation of ΣfA compared to R_T. Load tests of piles instrumented to measure the distribution of skin friction and R_T are necessary for realistic computations.

Hammer velocity and the cap data can be estimated. A dynamic instrument package plus a computer to evaluate the wave dynamics have been developed in an integrated system by Rausche and Goble.[11:16] This makes it easier to use the analysis to obtain specified capacities, R_0, during routine construction.

In situations where the input data are incomplete or uncertain, the wave analysis is still of value. It indicates the effect of changing pile hammer weight, cap block material, and pile length and type without making load tests for each different condition.

APPROXIMATE METHODS / Approximate methods of dynamic analysis or "pile formulas" have been used for more than a century and are still useful in estimating ultimate capacity from simple observations of driving resistance, bounce, and set.

All are based on the transfer of the kinetic energy of the falling pile hammer to the pile and the soil. This accomplishes useful work by forcing the pile into the soil. Energy is wasted in the mechanical friction of the hammer, in the transfer of energy from the hammer to the pile by impact, and in temporary compression of the pile, pile cap (if any), and of the soil. The basic relationship, therefore, will be

$$(R_0 \times s) + \text{losses} = E_r \times (\text{efficiency}) \tag{11:8}$$

where R_0 is the resistance of the pile to driving; s, the net distance it moves into the ground from one hammer blow (the *set*); and E_r, the hypothetical hammer energy at impact (Table 11:1). The relation is solved for R_0, which is then assumed to be equal to the capacity of the pile under sustained loading, Q_0.

The major uncertainty in this approach and the basic difference between all pile formulas is the way in which the energy losses and the mechanical efficiency of the process are computed. The most elegant is that of Hiley as described by Chellis.[11:19] The mechanical efficiency of the hammer is expressed by e, a coefficient that ranges from 0.75 for drop hammers operated by a drum winch, or most steam pile hammers or hydraulic hammers. The energy available from the hammer after impact is approximated by impulse and momentum concepts. This considers the coefficient of restitution, n, which ranges from 0.9 for aluminum-plastic laminates to 0.25 for a hammer striking on the head of a woodpile or a wood cushion block in a pile helmet or cap. In addition it involves the weight of the hammer, W_r, and the pile weight, W_p. The available energy after impact is the hammer energy multiplied by

$$\frac{W_r + n^2 W_p}{W_r + W_p}$$

This shows that as the weight of the pile increases with respect to that of the hammer, the relative inertia increases and there is less energy available for useful work. For long piles this is not strictly valid because the pile moves in a wave rather than as a rigid body, as previously discussed.

The energy lost by elastic compression of the pile, any helmet, and the soil can be approximated by assuming a linear increase in the stress acting from 0 to R_0 while the compression develops. The energy loss will therefore be

$$\frac{R_0 s_c}{2} + \frac{R_0 s_p}{2} + \frac{R_0 s_s}{2}$$

where s_c, s_p, s_s are, respectively, the temporary compression of the helmet, pile, and soil (s_s is comparable to s_s of the wave analysis). The value of $s_p + s_s$ is the bounce of the pile with each hammer blow (Fig. 11.5) and is easily measured during driving. The value of s_c must be estimated from the value of R_0 and the dimensions and material of the cap. The energy distribution per blow is

$$\overbrace{R_0 s + R_0\left(\frac{s_c}{2} + \frac{s_p}{2} + \frac{s_s}{2}\right)}^{\text{work done on pile}} = \overbrace{E_r e\left(\frac{W_r + n^2 W_p}{W_r + W_p}\right)}^{\text{energy available to pile}} \qquad (11{:}9a)$$

$$Q_0 = R_0 = \frac{E_r e}{s + \frac{1}{2}(s_c + s_p + s_s)}\left(\frac{W_r + n^2 W_p}{W_r + W_p}\right) \qquad (11{:}9b)$$

Note: The formula is dimensionally homogeneous and both E and s must be in the same units.

Detailed tables of the constants for use in the Hiley formula have been published by Chellis. Although the values of e, n, and (for long piles) W_p must be estimated, which requires considerable experience, the method is reasonably accurate for piles driven in cohesionless soils. A safety factor of from 2 to 2.5 is ordinarily employed to obtain the safe load.

For long piles and very rigid piles, the Hiley formula is overly conservative because only a fraction of the total pile weight is accelerated at one time, as is demonstrated by the wave analysis. The approximation is more realistic if the moving mass, W_p, is taken to be the weight of the pile cap plus the weight of the uppermost part of the pile. The proper length depends on the pile rigidity and weight per unit of length: For heavy steel mandrels and precast concrete piles it is 9 to 15 m (30 to 50 ft).

Because the wave equation requires computer analysis, it is inconvenient to use as a pile-by-pile criterion for driving. Instead, pile load results can be correlated with driving resistance for the driving equipment, pile type and range of lengths on the project, using the Hiley formula as a base. Arbitrary constants can be intoduced for the hammer efficiency and impact factor, while s and $s_p + s_s$ are measured. The custom formula is used as a guide to construction.

ENGINEERING NEWS FORMULA / The Hiley formula can be further simplified by substituting arbitrary constants for the different factors in the equation. The *Engineering News* formula was derived from observations of the driving of wood piles in sand with a free-falling drop hammer. The value of $s_c + s_p + s_s$ is assumed to be 50 mm (2 in.) and both the hammer efficiency and impact factor are assumed to be 1. In its familiar form in English dimensions, with $E_r = W_r h$

$$R_0 = \frac{W_r h}{s+1} = \frac{E_r}{s+1} \qquad (s \text{ in in. } E_r \text{ in in.-lb}) \tag{11:10a}$$

A safety factor of 6 was introduced to make up for the inaccuracies arising from the use of the arbitrary constants. Since the height of fall of drop hammers is usually measured in feet and s is measured in inches, a factor of 12 was added to make it possible to use the mixed units. This reduced to the familiar form of the equation:

$$R_s = \frac{1}{6} R_0 = \frac{W_r(h \times 12)}{6(s+1)} = R_s = \frac{2 W_r h}{s+1} \tag{11:10b}$$

In this expression, h is the hammer drop in feet W_r the hammer weight and s is in inches. R_s is the safe pile dynamic resistance, including the built-in "safety factor." The formula was later modified for steam hammers by substituting 0.2 in. for the temporary compression to give

$$R_s = \frac{2 W_r h}{s+0.1} \tag{11:10c}$$

Numerous pile-load tests show that the real safety factor of the *Engineering News* formula averages 2 instead of its apparent 6, and that the safety factor can be as low as $\frac{2}{3}$ and as high as 20. For wood piles driven with free-falling drop hammers and for lightly loaded short piles driven with a steam hammer, the *Engineering News* formulas are a crude indication of pile capacity. For other conditions they can be very misleading.

11:5 Pile Groups

Since piles are ordinarily closely spaced beneath footings or foundations, the action of the entire pile group must be considered. This is particularly important when purely "friction" piles are used and when the hard stratum on which the points of end-bearing piles rest is underlain by more compressible soils.

GROUP BEARING CAPACITY / The group bearing capacity is computed by assuming that the piles form a giant foundation whose base is at the level of the pile points and whose length and width are the length and width of the group (Fig. 11.14a).[11:20] The group capacity is the sum of the bearing capacity developed by the base of the "foundation" and the shear developed along the vertical sides of the group "foundation."

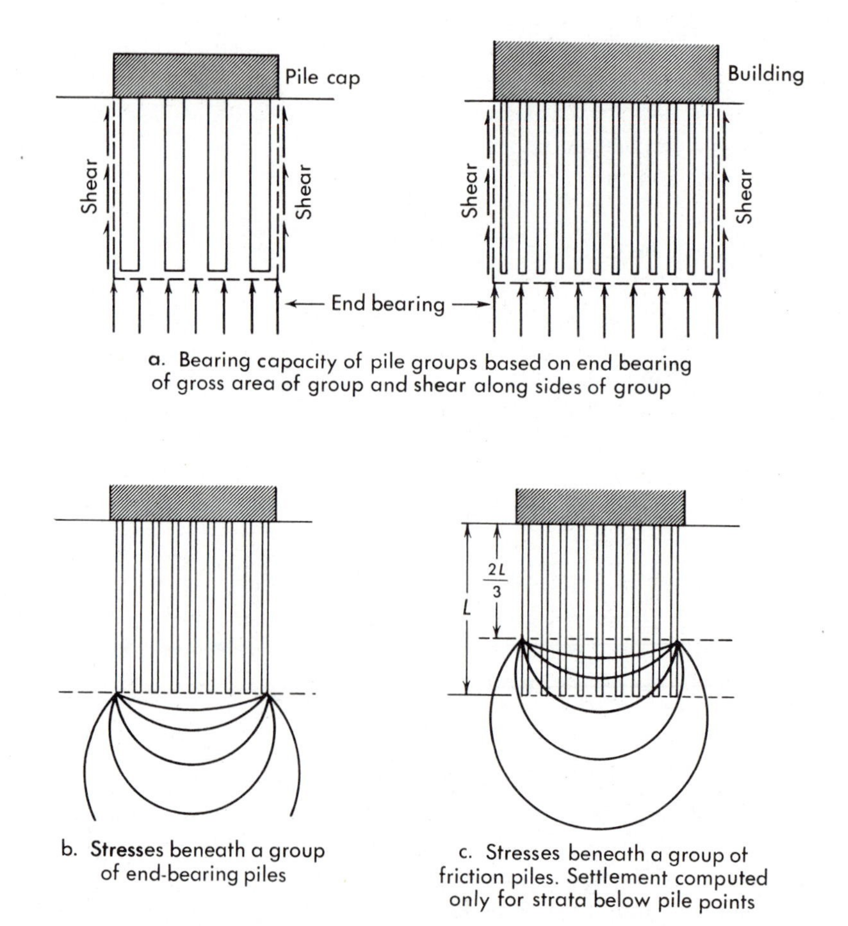

Figure 11.14 *Approximate method for analyzing the bearing capacity and settlement of pile groups, by assuming them to act as a single foundation unit.*

The bearing is computed by using the general bearing-capacity Equation (10:3). The factors for deep foundations are used when the pile length is at least 10 times the group width and when the soil is homogeneous; in all other cases the factors for shallow foundations are used. The shear around the group perimeter is the soil strength, determined without any increase in lateral pressure from pile displacement, multiplied by the surface area of the group. While model tests show that the actual group capacity is always slightly less than the computed values, the difference is well within a safety factor of 2.

PILE EFFICIENCY / The efficiency of the pile group e_g is the ratio of the group capacity, Q_g, to the sums of the capacities of the number of piles, n, in the group:

$$e_g = \frac{Q_g}{nQ_0} \qquad (11:11)$$

Although a number of empirical formulas have been derived for group efficiency, none have been shown to be realistic. Instead the efficiency should be evaluated from the group capacity using the definition [Equation (11:11)]. The group capacity increases with spacing, but the individual capacity of piles in clay does not. A plot of theoretical efficiency versus spacing (Fig. 11.15a) shows the group capacity equals the sums of the

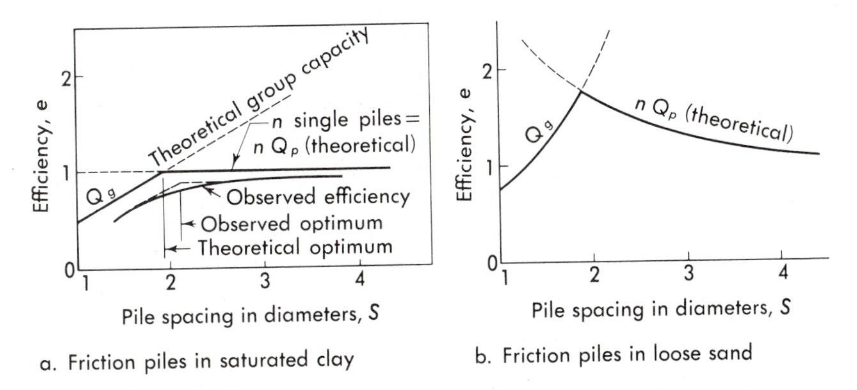

Figure 11.15 *Efficiency of groups of long friction piles.*

individual capacities at an optimum spacing, and an efficiency of 1. The optimum spacing S_0 for long piles in clay and group efficiency at the optimum spacing is given by the following:

$$S_0 = 1.1 + 0.4n^{0.4} \qquad (11:12a)$$

$$e_g = 0.5 + \frac{0.4}{(n-0.9)^{0.1}} \qquad (11:12b)$$

Typically, S_0 is 2 to 3 pile diameters center to center. Model tests in clay[11:19] indicate that the actual efficiency at the optimum spacing is somewhat less than 1 (0.85 to 0.9) and increases slowly at greater spacings. For design, utilizing the common safety factor of 2, the error in assuming the real efficiency to be 1 at the optimum spacing is inconsequential.

For piles in cohesionless soil the capacity of the individual piles increases with reduced pile spacing because of increased soil strength due to densification.[11:21] The optimum spacing (Fig. 11.15b) is very small, and has an efficiency greater than 1. Unfortunately, it is impossible to drive the piles that close. Practical spacings are 2.5 to 4 diameters, center to center.

GROUP SETTLEMENT / The settlement of a group of piles results from consolidation of the soil strata beneath the pile points. Such settlement will exceed that of an isolated pile that carries the same load as each pile in the group unless the piles are end bearing on rock or on a thick stratum of incompressible soil. The group settlement may be analyzed by again considering the group to represent a giant foundation. When the piles are end bearing, the base of the imaginary footing is assumed to be at the level of the pile tips, as shown in Fig. 11.14b, and the stresses are computed on that basis. When the piles are supported by friction, the stresses beneath the footing are computed by assuming that the entire group load is introduced in the soil at a depth of from one-half to two-thirds the pile length. The load is distributed at this level over the gross area of the group. Settlements of the soil strata beneath the pile points are computed from these stresses. The actual increased stress will be less because the load is distributed within the mass. Typically, the stress will be about $\frac{2}{3}$ the computed increase. Alternatively, the relative load supported by end bearing and skin friction of the group can be found from the ultimate computed side friction and end bearing of the group. The total load is assumed to skin friction and end bearing in proportion. The stress increase produced by each is found from Equation (11:1a) and Fig. 11.9. At best, the computations are approximate and can indicate the source of settlement and the order of magnitude.

Example 11.2

Compute the increase in stress due to piles at a depth of 3 m (10 ft) below the tips of a group of 9 friction piles 20 m (66 ft) long. The outer dimensions of the group are $2\text{ m} \times 2\text{ m}$ or $6.6\text{ ft} \times 6.6\text{ ft}$. The total load is 2500 kN (562 kips).

1. The group load is assumed to act at a depth of $2/3 \times 20 = 13.33$ m. The area over which it acts is $2 \times 2 = 4\text{ m}^2$.
2. The depth at which the stress is found is $3 + 20/3 = 9.67$ m below the 2/3 level. Thus $z = 9.67/2 = 4.83B$ for Fig. 10.11.

3. From Fig. 10.11, the surface loading in a homogeneous mass at a depth of $4.83B$, the average stress increase is $0.021q$. The area load $q = 2500/4 = 625\ \text{kN/m}^2 = 13\ \text{kip/ft}^2$.
4. The stress increase from the pile group computed from the surface loading chart is

$$\Delta\sigma_z = 0.021 \times 625 = 13.1\ \text{kN/m}^2 = 0.27\ \text{kip/ft}^2$$

5. Because the loading is introduced well below the ground surface, the stress increase can be estimated by Fig. 11.9. The depth at which stresses are computed $z/L = (20+3)/20 = 1.15$, the width, 2 m, $X/L = 2/20 = 0.1$.
6. Assume that 1800 kN will be supported by side friction and 700 kN by end bearing of the group. (The exact distribution of load would be found from computing the relative amount of ultimate friction and end bearing of the group.)
7. For end bearing $I_p = 5$; for friction $I_p = 1$. $\Delta\sigma_z = (Q/L^2)I_p$ [Equation (11:1a)]. $\Delta\sigma_z$ end bearing $= (700 \div 400) \times 5 = 8.8\ \text{kN/m}^2$. $\Delta\sigma_z$ side friction $= (1800 \div 400) \times 1 = 4.5\ \text{kN/m}^2$. Total stress increase $= 13.3\ \text{kN/m}^2 = 0.28\ \text{kip/ft}^2$. These stresses are about the same as those computed by assuming that the load is concentrated at a depth of $2L/3$.

11:6 Lateral Loads

VERTICAL PILES / A vertical pile loaded laterally deflects as a partially retrained cantilever beam (Fig. 11.16); when the loads are small, the resistance of the soil is reasonably elastic. The soil can be represented by

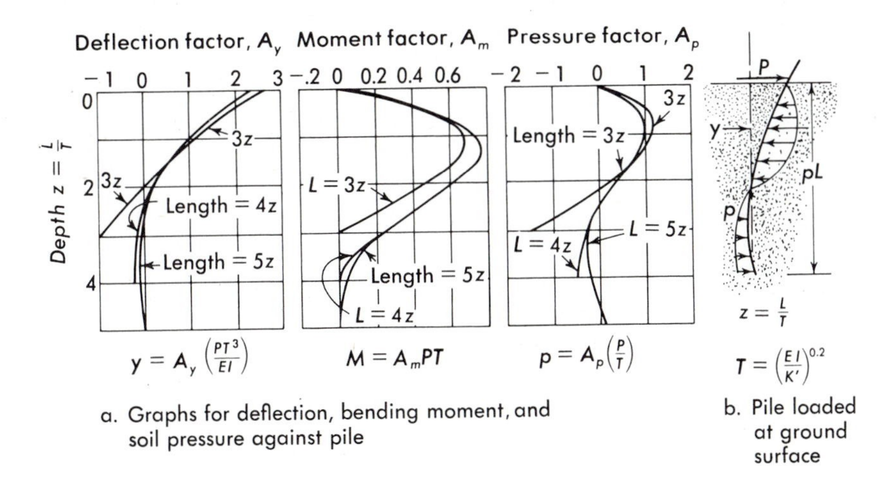

Figure 11.16 *Behavior of a laterally loaded vertical pile.* (After Reese and Matlock.)

a series of horizontal springs whose resistance to displacement is expressed by a modulus of subgrade reaction k, in force per unit length of pile per unit displacement. The differential equation of bending of the pile can be solved for deflection bending and pressure by successive approximations. The elastic restraint of the soil can also be evaluated by a finite element analysis.

Solutions are available in nondimensional form, as developed by Reese and Matlock.[11:22, 11:23] An example is given in Fig. 11.16 in which the soil rigidity is expressed by a coefficient of subgrade reaction, k, that increases linearly with depth z: $k = k_1 z$. The pile stiffness is uniform and expressed in terms of the moment of inertia of its cross section I, and modulus of elasticity E. The values of k_1 can be estimated from Table 11:4. A lateral load test in which load and deflection are measured to obtain deflection can be used to obtain more reliable k, using the analysis in reverse, E.

TABLE 11:4 / COEFFICIENT OF SUBGRADE REACTION, k_1, FOR LATERALLY LOADED PILES

Soil	k_1 (kN m^{-2} m^{-1})	k_1 (lb $in.^{-2}$ $in.^{-1}$)
Soft clay	270–1,360	1–5
Stiff clay	2,700–5,400	10–20
Loose sand	1,360–2,700	5–10
Dense sand	6,800–13,600	25–50

The curves of Fig. 11.16 are expressed in terms of the relative stiffness of the pile and soil, T, with the dimensions of a length.

$$T = \left(\frac{EI}{k_1}\right)^{0.2} \tag{11:13a}$$

The depth is expressed by the dimensionless coefficient Z, the ratio of pile length L to T.

$$Z = \frac{L}{T} \tag{11:13b}$$

Similar elastic analyses have been presented by Poulos.[11:24]

If the lateral load is great enough, the soil pressure will exceed the soil strength and the pile will fail. The failure resistance is sometimes computed to be the passive earth pressure against the upper part of the pile. This is unrealistic because the usual theories of passive pressure assume two-dimensional or plane strain shear, while the laterally loaded pile will fail in three-dimensional shear at a pressure exceeding the passive. Moreover, the deflection accompanying failure is so great that a structure supported by a laterally loaded pile would be in distress long before failure of the pile is reached.

Typical load test results suggest fully embedded vertical piles can support lateral loads of only $\frac{1}{10}$ to $\frac{1}{5}$ of their vertical capacity without excessive [more than 25 mm (1 in.)] deflection. If greater rigidity or greater lateral resistance is needed, batter piles are required.

BATTER PILES IN GROUPS / Batter piles combined with vertical piles are the most effective device for resisting horizontal thrusts. Anchors for wharves and bulkheads that combine a vertical pile in tension and a batter pile in compression, as shown in Fig. 11.17, have proved to be compact and economical. Batter piles combined with vertical piles have been utilized to support retaining walls and similar structures that develop horizontal loads. A rational analysis of batter-pile loading is impossible because the problem is statically indeterminate to a high degree. A simple useful approximation assumes the piles to be hinged at their points and at their butts (Fig. 11.17b).

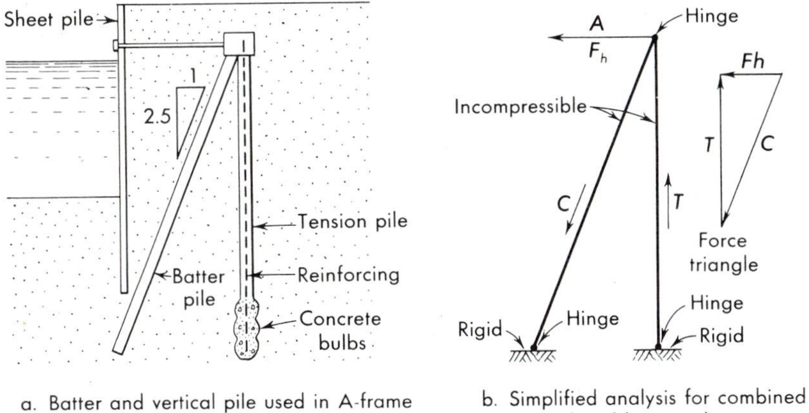

Figure 11.17 *Batter and vertical piles in combination.*

11:7 Types of Piles and Their Construction

PILE SHAPES / Constructors through the ages have tried and used with varying degrees of success many shapes and types of piling. Each shape has probably been successful under certain conditions. However, the use of a certain type or shape of pile that has proved successful in one job may not meet with success in a different situation. In the United States the establishment of large and well-equipped pile-driving organizations has led to the general use of a few types and shapes of piles.

Four basic shapes are commonly used: first, uniform cross section throughout the length; second, enlarged base; third, tapered; and fourth, sheet. These are shown in Fig. 11.18.

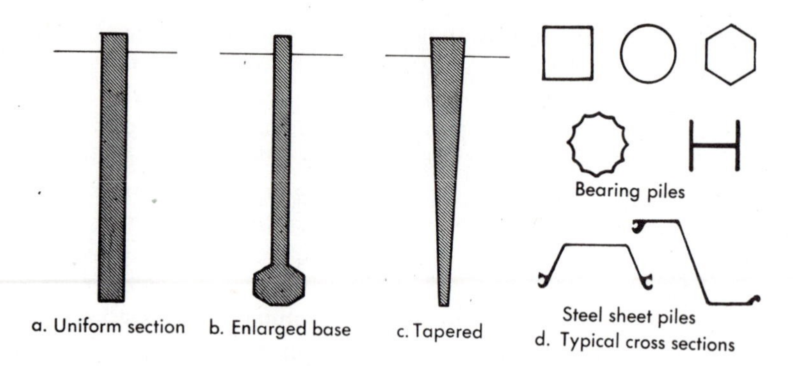

Figure 11.18 *Basic pile shapes.*

The uniform-section pile comes in a variety of forms: cylindrical, square, octagonal, fluted, and H-section. The uniform section provides uniform column strength from the point to the butt, and "skin friction" is well distributed over the entire shaft. It is well adapted to splicing and cutting, since each section of the pile is identical.

In order to increase the end bearing and the friction on the lower portion of the pile, different forms of enlarged points have been used. In one form a large, precast point is attached to a cylindrical pile, while in another form a bubble of concrete is forced into the soil at the pile point. Piles of this shape have proved very effective in developing end bearing on firm, cohesive soils and even in loose sands. They are of little value as friction piles and have little advantage over uniform-section piles when used as end bearing on rock.

The tapered shape originated with wood piles that conform to the natural shape of the tree. However, the taper has been imitated in concrete and steel in order to permit easier construction. Tapered piles are useful in compacting loose sands because of their wedge action but in other cases may be less effective than uniform-section piles. Both the point bearing and the skin friction on the lower portions of the tapered piles are low, since both the point area and the surface area of the pile are small. The result is that end-bearing, tapered piles require greater lengths than uniform-section piles in order to support a given load. Tapered piles that depend on friction for support may transmit a large portion of their load to the upper, weaker soil strata and produce objectionable settlement. Unexplainably, the skin friction of tapered piles is greater in both cohesionless and cohesive soils than displacement or wedge effects suggest. Thus tapered friction piles are often shorter than cylindrical piles of the same surface area.

Sheet piles are relatively flat and wide in cross section, so that when they are driven side by side they form a wall. Many different forms of wood, concrete, and steel sheet piles have been developed for special purposes such as cofferdams, wharves, retaining walls, and cutoffs. Some have arch-shaped and Z-shaped cross sections to provide rigidity, and most all types are made to interlock with adjacent sheet piles to form a soiltight wall.

Piles that are hollow have a distinct advantage over those that are not, for it is possible to inspect the entire length of the piles after driving. Piles may deviate from the vertical, develop sharp bending or "dog-legs," or may be damaged from overdriving. Hollow piles may be inspected by dropping a burning flare into them or reflecting the sun's rays into them with a mirror, but other forms must be assumed satisfactory without any check. Therefore higher safety factors should be used with piles that cannot be inspected. Hollow piles that are driven with open ends and then cleaned out make possible an examination of the soil beneath the pile point. When an open-end pile rests on an irregular rock surface, the rock may be smoothed by drilling; and if the pile is found to be hung on a large boulder above the supporting stratum, the boulder may be drilled or dynamited to permit the pile to reach its full depth.

WOOD PILES / Wood is one of the most commonly used pile materials because it is cheap, readily available, and easy to handle. Some kind of timber suitable for piling will be found available in nearly every section of the world. Spruce, fir, and pine up to 30 m (100 ft) long; oak and mixed hardwood piles up to 15 m (50 ft); southern pine up to 20 m (65 ft); and palmetto are commonly used for piling. Untreated timber piles completely embedded in soil below the water level will remain sound and durable indefinitely. When the campanile of St. Mark's in Venice fell in 1902, it was found that the 1000-year-old piles were in such a good state of preservation that they were left in place and used to support the new tower. Sound timber piles that have been water-soaked for many years should not be allowed to dry before redriving, since in drying the wood fiber becomes brittle.

Above the water table untreated timber is subject to decay and damage from termites and other insects. In salt water timber is susceptible to marine borer attack.[10:19, 10:25] Many types of marine borers are found, but most are related to the lobster and crab or to the clam and oyster families. The crablike limnora destroy the wood from the outside in, leaving the pile as a slender spindle of wood (Fig. 11.19a). The clamlike toredo destroy the wood from the inside out; they enter the pile through a small opening, destroy the inside of the pile, and leave it a hollow shell (Fig. 11.19b). Timber piles can be made to last longer through treatment with zinc chloride, copper sulfate, or numerous patented chemicals. Creosote impregnation has proved to be one of the most efficient and long-lasting means of protecting timber piles. From 0.19 to 0.4 g/ml (12 to 25 lb/ft^3) of creosote are forced into the wood by vacuum- and pressure-treatment processes. In areas of very severe marine borer attack combination treatment of copper arsenate followed by a coal tar–creosote mixture, both applied under heat and pressure, are necessary to assure a life of 15 to 25 years in salt water.

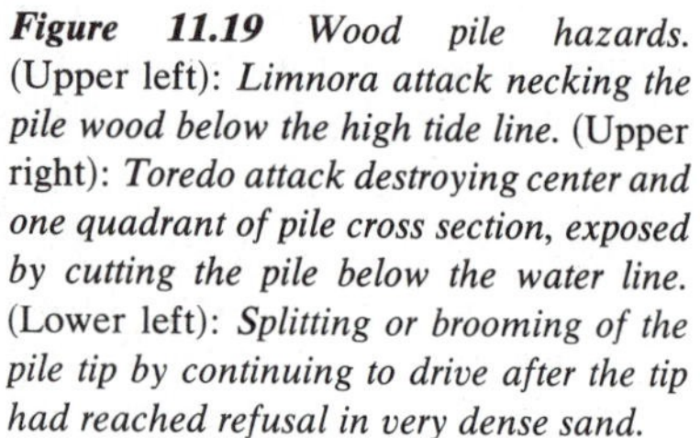

Figure 11.19 *Wood pile hazards.* (Upper left): *Limnora attack necking the pile wood below the high tide line.* (Upper right): *Toredo attack destroying center and one quadrant of pile cross section, exposed by cutting the pile below the water line.* (Lower left): *Splitting or brooming of the pile tip by continuing to drive after the tip had reached refusal in very dense sand.*

Timber piles tend to suffer badly from overdriving. The tops of the piles become "broomed," and the shafts are very likely to split or break, as shown in Fig. 11.19c, when stiff resistance to driving is encountered. Such a situation was discussed in the chapter introduction.

Timber piles are ordinarily capable of supporting safely from 150 to 300 kN (15 to 30 tons) per pile. Timber piles have been utilized for design loads exceeding 500 kN (50 tons) in friction, and the safety of these designs have been demonstrated by load tests. The major problem of such designs is ensuring that the structural quality of the wood piles is uniformly high so that there is no danger of the piles breaking during driving. The very low cost for materials and for driving often makes timber the cheapest pile foundation per ton supported.

PRECAST CONCRETE PILES[11.19, 11:26] / Precast concrete piles are uniform-section circular, square, or octagonal shafts with sufficient reinforcing to enable them to withstand handling stresses. The smaller sizes 0.2 to 0.3 m (8 to 12 in.) wide and are usually solid. Larger sizes are solid or are hollow so as to reduce their weight. Prestressing makes it possible to secure adequate strength with relatively thin concrete walls; diameters up to 1.5 m (60 in.) with walls 0.1 to 0.15 m (4 to 6 in.) thick, similar to concrete pipe, have been used where great stiffness and high bearing capacity are required. At least 50 mm (2 in.) of cover over the reinforcing is needed to prevent corrosion in marine work.

Precast piles are used principally in marine construction and bridges where durability under extreme exposure is important and where the pile extends above the earth as an unsupported column. In the latter case the reinforcing is increased as dictated by the column requirements. Typical lengths of the small, solid piles are 15 to 20 m (50 to 65 ft), and of the larger hollow piles, up to 60 m (200 ft). Typical loads for the small piles are 400 to 500 kN (40 to 50 tons), and for the larger piles, over 2000 kN (200 tons).

Two factors limit the use of precast piles. First, they are relatively heavy compared with other piles of comparable size. Second, it is difficult to cut them off if they prove to be too long and even more difficult to splice them to increase their length.

CAST-IN-PLACE CONCRETE PILES / Concrete piles that are cast in the ground are the most widely used types of piles for 300 to 600 kN (30 to 60 tons) loads. These can be divided into two groups: *cased piles* in which a thin metal casing is driven into the ground to serve as a form, and *uncased* piles where the concrete is placed directly against the soil. Many types of each have been developed, and the engineer will find it enlightening to study the catalogs of pile contractors to see the different methods of construction.

The *Raymond Standard* pile (Fig. 11.20a) was one of the earliest cased types. A thin metal shell with a 0.2 m (8 in.) diameter tip and with a taper of 33 mm/m (0.4 in./ft) is driven into the ground on a close-fitting steel core or mandrel. After the mandrel is withdrawn, the tapered hole, supported by the shell, is filled with concrete. This pile is employed for lengths up to 12 m (40 ft) and for loads of 300 to 500 kN (30 to 50 tons).

The *Raymond–Step-Taper* pile (Fig. 11.20b) consists of a series of cylindrical, corrugated sheet-metal sections, each 2.4 m (8 ft) long and 25 mm (1 in.) in diameter larger than the one below and screwed together to form a continuous tube. The minimum tip diameter is 0.2 m ($8\frac{5}{8}$ in.), but larger tips up to 0.35 m ($13\frac{3}{8}$ in.) can be used by starting the pile with the

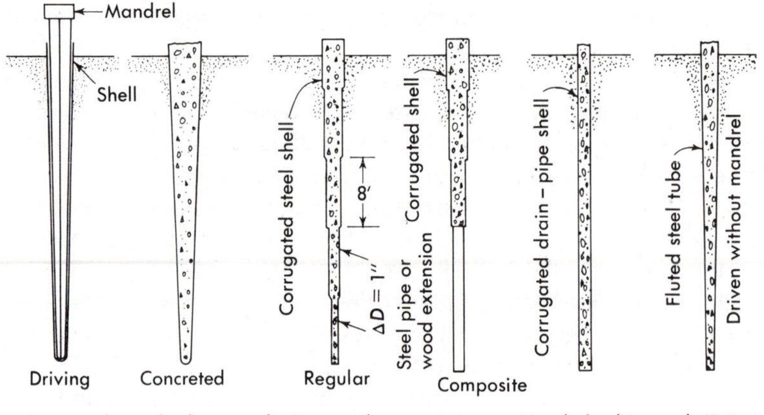

Figure 11.20 *Cast-in-place concrete piles.*

larger cylindrical sections. The pile is driven by a loosely fitting mandrel that drives agains the tip and the shoulders or *plow rings* of each larger section. Lengths up to 29 m (96 ft) and loads of from 400 to 750 kN (40 to 75 tons), depending on the tip diameter, are used.

The *Cobi pile* and the *Hercules pile* employ a cylindrical corrugated sheet-metal shell, similar to drainage pipe, from 0.2 to 0.5 m (8 to 12 in.) inside diameter. The shell is closed at the lower end with a flat or cone-shaped boot. It is driven with a cylindrical steel core that expands to grip the inside of the pipe and its corrugations tightly. The Cobi-type core expands by air pressure in a rubber tube, whereas the Hercules type expands by mechanical wedging. Lengths up to 30 m (100 ft) are possible.

The *Union Monotube* consists of a thin-walled, fluted steel tube that is driven into the soil without a mandrel or core. The fluting makes the thin steel shell capable of withstanding the driving stresses without buckling. Monotubes as long as 37 m (125 ft) carrying loads of 300 to 600 kN (30 to 60 tons), are used. They are particularly suited to small jobs because they require no special driving equipment such as the mandrel.

Many variations of the thin shell cast-in-place pile are used. The *Button Bottom* employs an 0.45 m (18 in.) diameter precast concrete pedestal at the bottom of the 0.3 m (12 in.) corrugated pipe shell. After the mandrel is withdrawn, the corrugated shell is filled with concrete to make a continuous pile. This form develops unusually high end bearing but reduced skin friction because the tip punches a larger hole than the pile shaft.

The thin shell piles have a number of features in common. Ordinarily, they are not reinforced, because they are in compression when supporting vertical loads. Reinforcing can be added before pouring if the piles are to resist tension or flexure. The thin shell is usually not considered a part of that reinforcement, because of possible corrosion. It is easy to cut them off if too long or to increase their length during driving by welding on more shell. They can be inspected after driving to check their straightness. The shell keeps water and soil away from the wet concrete and develops a shaft of uniform quality. The thin shells sometimes are damaged by obstructions that tear them or smooth out their corrugations and reduce their strength, or they may collapse because of the extremely high lateral pressures that develop in stiff clays and dense sands.

Uncased piles formed in a casing are suitable for loose sands and firm clays where the lateral pressures developed will not squeeze the unprotected fresh concrete. Lengths of 18 m (60 ft) and loads of from 300 to 750 kN (30 to 75 tons) are the usual limits for such piles. Uncased piles require heavy driving rigs and are economical where the size of the job justifies the initial expense of equipment.

The *Franki* uncased concrete pile (Fig. 11.21a) is formed by ramming a charge of dry concrete in the bottom of a 0.5 m (20 in.) steel casing so that the concrete grips the walls of the pipe and forms a plug. A 32-kN (3.5-ton) ram falling 3 to 6 m (10 to 20 ft) inside the casing forces the plug into the ground, dragging the casing downward by friction. Alternatively, the casing is driven with an expendable cap which is forced out by the concrete bulb. At the bearing level, the casing is anchored to the driving rig, and the concrete plug is driven out its bottom to form a bulb over 1 m (3 ft) in diameter. The casing is then raised while successive charges of concrete are rammed in place to form a rough shaft above the pedestal. Lengths up to 30 m (100 ft)

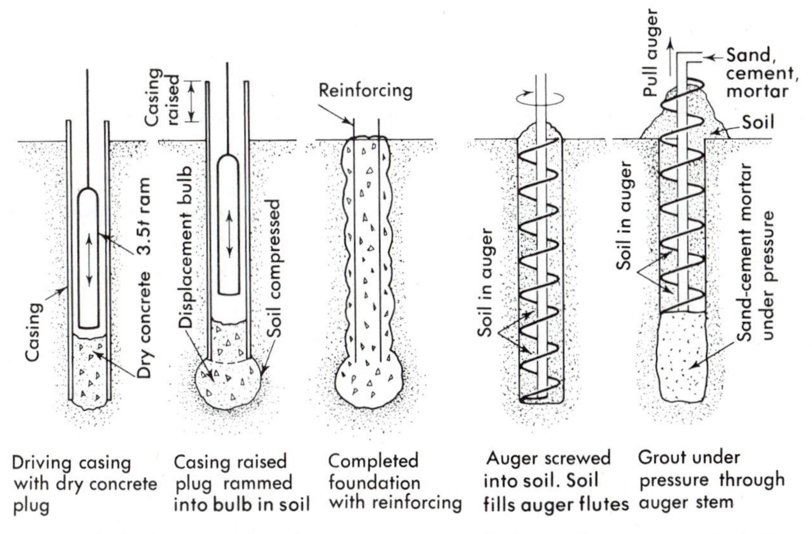

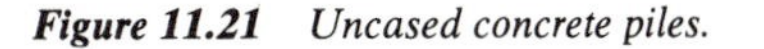
Figure 11.21 *Uncased concrete piles.*

with capacities of from 1000 to 10,000 kN (100 to 1000 tons) are typical. With reinforcement they make excellent tension piles.

A number of forms of augered uncased piles are available. The *Auger-cast* pile (Fig. 11.21b) is drilled with a continuous auger with a hollow central stem. The rate of drilling is such that the auger screws itself into the ground rather than expelling the soil. The hole, therefore, remains full of intact soil, until it reaches the bearing stratum. At that point the augur is slowly withdrawn. At the same time sand cement grout is pumped down the hollow stem. The rate of auger pulling is controlled so that there is always a positive pressure on the grout to fill the hole, prevent hole collapse, and to force the grout a few inches into loose sands and gravels. The resulting pile develops both end bearing and skin friction from its irregular grout penetration. The process is economical and free of vibrations, an advantage in building additions and underpinning. Lengths of 18 m (60 ft) and diameters of 0.35 to 0.45 m (14 to 18 in.) are commonly used.

Bored piles consisting of augered holes filled with concrete can be used where the soils are firm enough to stand without support. Diameters from 0.15 m (6 in.) up to lengths of more than 15 m (50 ft) are commonly constructed. [Diameters larger than 0.6 m (2 ft) are considered as piers and will be discussed later.]

It is difficult to measure the quality of the augered piles except by load testing. Test piles should be installed before the design is final. After their strength has been verified by load tests, continuous inspection is essential. If there is any change in construction procedure, new load tests must be made to check the effect of change.

STRUCTURAL STEEL PILES / Structural steel shapes, particularly the H-pile sections, are widely used for bearing piles, usually when high end bearing is required on soil or rock. The cross-sectional area is small compared with the strength and makes driving much easier through obstructions such as hard, cemented seams, old timbers, and even thin layers of partially weathered rock. The sections can be obtained in pieces and can be easily cut off or spliced. The sections ordinarily driven are 8BP36 to 14BP117, with working loads of 400 to 1500 kN (40 to 150 tons). Wide-flange sections as deep as 0.9 m (36 in.) have been driven, and built-up piles from channels and railroad rails are occasionally used. The lengths are limited only by driving; 14-in. H-sections over 100 m (330 ft) long have been installed.

H-piles driven onto rock have demonstrated their capability of supporting loads up to the yield point of the steel. Figure 11.22 shows the local buckling of a 30-m (100-ft) long H-pile driven through soft clay and then load-tested to 4000 kN (400 tons), approximately the yield point. Apparently, the pile bites into the rock and establishes full bearing in spite of irregularities in the surface. In very hard rock, the point end is sometimes reinforced by gusset plates or steel tips to prevent local buckling. The H-pile

Figure 11.22 *Failure of 30-m long H-piles above ground at the yield point of steel. The piles were driven through soft clay to end bearing on rock.*

penetrates soil with minimum displacement and development of heave and lateral pressure. When H-piles are used in friction, the surface area between the flanges is so great that friction failure occurs by shear parallel to the web across the outer edges of the flanges, and by friction against metal along the outer faces of the flanges.

The structural sections have three disadvantages. First, they are relatively flexible and easily deflected or twisted by boulders. In fact, a few H-piles have been deflected so far that their points skidded along the bearing stratum instead of biting in. Second, the soil packs between the flanges so that the friction area is equal to the rectangle which encompasses the pile rather than to the total area of the pile surface. Third, corrosion reduces the effective cross section. In most soil a corrosion allowance of 1 to 2 mm (0.05 to 0.1 in.) is realistic because a heavy film of rust protects the pile from further attack. In strongly acid soils such as fill and organic matter, and in seawater, corrosion is more serious. Cathodic protection or jacketing with concrete is necessary to prevent deterioration.

STEEP PIPE PILES / Steel pipes filled with concrete make excellent piles. In most cases they are driven with the tip closed by a flat plate or a conical point. The flat plates are cheapest and tend to form a conical point of soil ahead of them during driving. An X shape of plate welded to the tip helps the pile to break through gravel and cemented layers and to bite into bed rock. Open-ended pipes are used where minimum displacement is essential. The plug of soil that pushes up into the pipe must be removed at intervals to prevent its packing and causing the pile to drive as though it had a closed end.

Both the closed-end and open-ended pipes are filled with concrete after driving (after cleaning the open pipe, of course). This increases the shaft load capacity because both the strength of the steel and of the concrete contribute to the column strength.

Pipes from 27 cm (10.75 in.) OD×4.8 mm (0.188 in.) wall to 1 m (39 in.) OD×15 mm (0.6 in.) wall have been driven with capacities of from 500 kN (50 tons) to more than 2000 kN (200 tons). Lengths are limited by the driving equipment; pipe piles 120 m (400 ft) long have been installed.

Pipes are light, easy to handle and to drive, and can be cut off and spliced readily. They are stiffer than H-piles and not so likely to deflect when they strike an obstruction. They have the distinct advantage that they can be inspected internally after driving and before concreting.

In driving steel piles, the hammer must strike squarely over the centroid of the section. An off-center or wobbling hammer will "accordion" pipes and batter structural sections, which destroys the effectiveness of the blow. The carbon content of the pile is important, for if it is too high, the pile will split; if too low, the steel will yield. Pipe specifically made for pile driving should be used instead of gas or oil line pipe, which is sometimes cheaper.

COMPOSITE PILES / Composite piles are a combination of a steel or timber lower section with a cast-in-place concrete upper section. In this way it is possible to combine the economy of a wood pile below the groundwater level with a durability of concrete above water, or to combine the low cost of cast-in-place concrete with the great length or relatively greater driving strength of pipes or H-sections.

The design and construction of the splice between the two sections is the key to a successful composite pile. The head of the lower section must be protected against damage; a tight joint must be maintained to exclude water and soil from the shell; good alignment between the sections must be maintained to prevent doglegging; and the splice must be as strong as the weakest member it connects.

Two methods are used. In one the lower section is driven its full length. The metal shell, on its driving core, is attached to the lower section, and the whole assembly is driven to the final penetration. The core is then withdrawn and the shell concreted. A second method consists in driving a steel casing first. The core is withdrawn and the lower section is placed in the casing like a projectile in a gun. The lower section is then driven out the casing by the core. The thin steel shell for the cast-in-place section is lowered into the casing and locked on the lower section, after which the casing is withdrawn and the pile concreted.

Composite wood–concrete piles 43 m (140 ft) long and steel pipe–concrete piles 55 m (180 ft) long have been used with loads of up to 300 and 600 kN (30 and 60 tons), respectively.

SAND PILES / Holes rammed into soil and filled with sand or crushed slag for the purpose of compacting and draining a soil are known as *sand* or *wick* piles. They have little structural strength other than that of the compacted sand. They are constructed in the same way as uncased cast-in-place concrete piles, but a free-draining material is used instead of concrete.

PIN OR ROOT PILES / Small-diameter piles, 50 to 150 mm (2 to 6 in.) diameter have been used in a number of special applications where available space, soil conditions, and economics are unfavorable for larger piles. Steel tubes, driven with small vibrators or pneumatic hammers, are termed *pin piles*. They can be driven in short sections and spliced by welded or threaded couplings. The *root piles* are augered holes drilled with casing where required. Steel reinforcing is added and the shaft grouted as the casing is withdrawn. They are a variant of the auger-cast pile (Fig. 11.21). Lengths of 20 m (66 ft) are typical, but they have been installed 50 m (170 ft) long.

In bearing, there appears to be little danger of buckling despite their slenderness because of the lateral restraint of the soil. Load tests, however, should be made to establish safe bearing capacity as well as their feasibility in a particular site. Their ability to safely reinforce soil has not been evaluated analytically; local experience is the best guide.

These small piles can be placed where clearances are limited and when time precludes obtaining large pile-driving equipment. They are often employed as corrective measures when spread footing excavations disclose localized weak soils, or in underpinning. A more unique application is in a closely spaced XX in three dimensions to strengthen a weak soil over a tunnel or below a failing foundation.

STONE COLUMNS / Cylinders of compacted, crushed stone installed through soft clays and similar weak materials to end bearing on firmer materials are *stone columns*. A hole is jetted through the weak deposit to firm bearing by the same cylindrical probe-vibrator used for vibroflotation (Chapter 7). The diameter is 0.6 to 1 m (2 to 3.3 ft). Angular crushed stone is added to the annular space between the vibrator and the soil; it falls through the upflowing jet water to the hole bottom. The probe vibrates as it is withdrawn compacting the stone. One or more columns provide support to footings that are similar to pile caps.

The column owes its support to the passive pressure of the soil developed by the vibration–compaction. Its capacity is analyzed in that way.

$$\sigma_3 = \gamma z + 2c - u$$

$$\sigma_1 = q_0 = (\gamma z + 2c - u) \tan^2\left(45 + \frac{\phi}{2}\right)$$

In this expression ϕ is the angle of internal friction of the compacted crushed stone, sometimes exceeding 40°.

It is not certain how much passive restraint will be offered by any soil formation. Therefore, full-scale load tests of the columns are essential to determine their safe bearing and deflection under load. A second uncertainty is the use of undrained shear parameters in evaluating long-term bearing. However, lateral pressure of the column eventually will cause consolidation of the soil and an increased strength, probably offsetting any loss of apparent cohesion of the undrained state.

11:8 Design of Pile Foundations

The design of a pile foundation is similar to the design of any other part of a structure. It consists of assuming a design, then checking the proposed design for safety and revising it until it is satisfactory. Several such designs are then compared, and the final one is selected on the basis of cost and time required for construction.

Piles in a foundation may be valueless in some locations, and under some conditions their use actually may be very harmful. For example, a layer of reasonably firm soil over a deep bed of soft soil might act as a natural mat to distribute the load of a shallow-footing foundation. The driving of piles into the firm layer might break it up or remold it. The result would be a concentration of load in the soft soil strata, with excessive settlement likely to take place.

SELECTION OF PILE LENGTH / The selection of the approximate pile length is made from a study of the soil profile and the strength and compressibility of each soil stratum. Such studies may be made by using the method of pile group analysis discussed in Section 11:5. End-bearing piles must reach a stratum that is capable of supporting the entire foundation without undue settlement or failure, and friction piles must be long enough to distribute the stresses through the soil mass so as to minimize settlement and obtain adequate safety of the entire group of piles.

SELECTION OF POSSIBLE PILE TYPES / The pile type and the material from which it can be made must be carefully chosen to fit:

1. The superimposed load.
2. The time available for the completion of the job.
3. The characteristics of the soil strata through which it penetrates as well as that of the strata to which the load must be transferred.
4. The groundwater conditions.
5. The size of the job.
6. The availability of equipment and getting it onto the site.
7. The availability of material for the piles.
8. The building code requirements.

If the structure is a bridge abutment or a wharf, the depth of the water, its velocity, ice condition, and the possibility of marine borers or chemicals in the water attacking the pile material must be given full consideration. Scour is quite likely to take place around new bridge piers and abutments because of the increased water velocity; the piling in such cases should be protected by concrete and the structure should be braced by batter piles.

If the foundation loads are low and scattered, a pile of low cost per foot and per pile may be the most economical. If the loads are high and concentrated within small areas, a pile having a high load-supporting value will probably be the lowest in cost per ton of load. If the load is a single load exceeding 3000 kN (300 tons), and there are several such load points, some type of pier may be more economical.

Project logistics are a major factor. Cost of moving equipment is a significant factor in small projects. Because many pile types require special equipment, some of which is proprietary or patent, the choice of pile types will be limited to those driven by local contractors for structures like small bridges that require only a few dozen piles. Availability of materials is a limitation; areas that do not produce steel favor concrete; forest regions favor wood piles. Handling is a factor. Most wood piles float, an advantage in marine construction; reinforced concrete is difficult to splice or cut off, a factor where variable pile lengths are expected.

Building codes usually define maximum pile loads and other design limitations. Sometimes these limits are rational, or can be waived if a sound engineering analysis is made of site requirements. However, in some jurisdictions, the limits are arbitrary and unchangeable, favoring local contractors or trade association pressure groups. The design must be safe, despite the code; unfortunately, the code may not permit the best design from the technical point of view.

PILE DESIGN LOADS / The design of the pile shaft is governed by stresses produced by driving. During driving the actual or working load on the shaft equals the failure load between the pile and the soil, R_0. The driving stresses can be estimated from Equation (11:9b) or the wave analysis, regardless of the type of soil, because only the dynamic resistance is concerned. The shaft should have a safety factor of at least 1.2 with respect to R_0, which means that the safety factor with respect to the design load, Q_a, is larger than for other short columns. The safe load on the pile is governed by the soil-to-pile connection and the group capacity. These are analyzed as described in Sections 11:3, 11:4, and 11:9, and the appropriate safety factor applied depending on the reliability of the analyses and the structural loading data.

SPACING / The final pile spacing is based on the analysis of pile group action. The piles are placed so that the capacity of the pile group acting as a unit is equal to the sum of the capacities of the individual piles. Greater spacing may be necessary in dense sands and stiff clays to minimize lateral pressures from displacement.

TOLERANCES / It is impossible to construct piles in exactly the required location or angle because they tend to drift out of line when they encounter hard or soft spots in the ground. The designs and specifications should allow for 50 to 75 mm (2 to 3 in.) of tolerance in the top of small piles in soil and 0.15 m (6 in.) (and sometimes more) for piles driven through water. Out of plumbness and doglegging of 2 to 3% of the pile length generally does not affect pile capacity and should be permitted by the design and specifications. More is usually permitted if load tests show the capacities are adequate.

INSPECTION AND RECORDS / Similar to earth filling, the geotechnical engineer plays a larger role in decision making during pile construction than for the other work. Each pile is different, although its neighbors are only 2 to 4 diameters away. The driving resistance, bounce, and other behavior of the pile are significant throughout driving, indicating unforeseen obstructions and possible pile damage. Suspension in driving allows pore water pressure dissipation and an increase in resistance in loose sand, but a decrease in dense sand. The equipment condition and performance, compared with manufacturer's specifications, indicate if hammer energy is that expected. Changes in the soil (heave, shrinking, or lateral movement) can indicate negative skin friction, pile bending, or even collapse of hollow piles. Finally, the elevation of the pile tip and top are found for contractor payment and verification of design.

PILE CAP / The load of a wall or column must be transferred to the pile by means of a footing or pile cap. In designing the cap, consideration must be given to the fact that the pile butts may be from 50 to 150 mm (2 to 6 in.) out of their required position. In some cases the piles may be pulled or jacked into position, depending on the rigidity of the pile and the soil, but it is cheaper if the pile cap is designed with allowances for some misalignment. In dock structures the pile must resist horizontal forces and sometimes rotation. In such cases the pile must be anchored to the cap by adequate embedment and in some cases with reinforcing steel. The structural design of the cap is similar to the design of a footing foundation. Care must be taken to see that the footing is rigid enough to transfer the load to the outermost piles in the group.

11:9 Pier Foundations

The pier foundation is a relatively large, deep footing. Its function is to transfer a foundation load through soft soil to hard soil or rock or to transfer a load through soils that may be scoured away by rivers or tidal currents. The chief differences between piles and piers are size (piles larger than 0.6 m (2ft) in diameter are sometimes called *piers*) and the method of construction. Piles are ordinarily forced into the ground without previous excavation, whereas piers usually require soil excavation ahead of or during their construction.

Piers are divided into two classes, *open shafts* or wells, and *caissons*, depending on the method of construction. Open shafts are merely deep excavations that are provided with bracing or lining when needed as the work progresses. A caisson is a box or chamber that excludes water and soil from the excavation. It is usually prefabricated above ground and then sunk to the required level by excavation from within. The word *caisson* is often applied to any pier, but, strictly speaking, it refers only to those that employ the box or chamber that is lowered as excavation proceeds.

The materials employed and the type of structure are dependent on the load, ground water conditions, depth of load-supporting strata, building code requirements, and the availability of materials and equipment. If the pier is to be in water, the velocity of the water, its maximum depth under scouring conditions, and the effect of ice and debris must be provided for in the design.

BEARING CAPACITY AND SETTLEMENT / A pier is actually a deep footing foundation that is supported by end bearing and by shear of friction along its sides. The end bearing is computed by the general bearing-capacity Equation (10:3), using the appropriate bearing-capacity factors in Fig. 11.11. The skin friction on the pier is either friction plus adhesion or shear strength, whichever is the smaller. In computing friction and shear, no allowance is made for any increase in lateral pressure from displacement because the piers are constructed by excavation.

For piers resting on hard rock, with soil above, skin friction is neglected. For piers entirely in a homogeneous soil, the skin friction in the soil is included. If the rock is weak, and sufficient end bearing cannot be obtained, the pier can be extended below the rock surface. The skin friction or shear below that level augments the total bearing of the pier.

It is essential that the soil or rock immediately beneath a very heavily loaded pier or caisson be examined because local soft or compressible seams could seriously impair the bearing or settlement. If the caisson is large enough and can be unwatered, one to three 50-mm (2-in.) holes are drilled in the bottom, 1.5B to 3B (pier) deep. These are probed with a hooked rod to find soft seams or cavities. It is also possible to make core borings in the bottom, with the drill rig on the ground surface, particularly if the pier cannot be unwatered. Boring in a deep open shaft is difficult, and only the best workmanship is permissible. It is essential that loose material be removed from the bottom—every millimeter of loose soil means virtually a millimeter of settlement.

The settlement of piers is often the governing factor in their design. Settlement can be estimated by the methods described in Chapter 10 by assuming the end of the pier to be a footing. The stresses in the soil below such a "footing" are considerably less than those computed by the Boussinesq or Westergaard methods because those analyses are based on loads at the ground surface. Based on the Mindlin analyses (Fig. 11.9), the stresses beneath a narrow deep end-bearing pier can be as little as half those computed by surface load methods.

OPEN SHAFTS / The simplest form of the open shaft is an open excavation similar to a dug well (Fig. 11.23a and b). Shallow wells in firm soil can be dug by hand. Large power augers are capable of drilling open excavations as large as 3 m (10 ft) in diameter and more than 30 m (100 ft) deep. Special underreaming drills can be used to enlarge or *bell out* the bottom of the shaft to nearly double the shaft diameter. After it is drilled, the shaft is filled with concrete to form the pier.

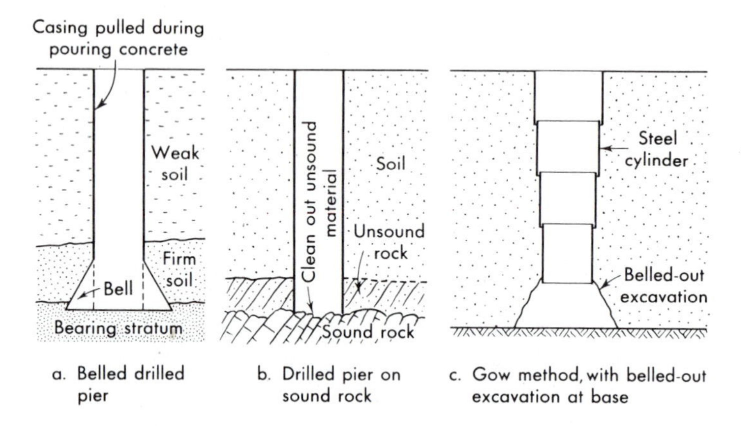

Figure 11.23 *Construction of open shaft piers; holes to be filled with concrete.*

When the pier must extend below the groundwater level, or when the soil is not strong enough to stand without support, some form of bracing is required. The simplest is a metal cylinder that is lowered into the shaft immediately after drilling to hold it open until the concrete can be placed. The cylinder is lifted out as the concrete is poured, because the pressure of the wet concrete is ordinarily capable of supporting the soil and keeping out water. In very soft or wet soils it is sometimes necessary to drill and install lining cylinders successively in short sections 2.6 to 5 m (8 to 16 ft) long. The cylinders telescope, forming a tapering shaft known as the *Gow caisson* (Fig. 11.23c). Piers as deep as 9 m (30 ft) with belled bottoms have been installed by this method, using both hand and auger excavation.

If the pier rests on rock, a laborer is lowered into the hole to clean the surface so that there is no soil between the rock and the concrete. This is hazardous because of gas and danger of blow ins. If the rock is seamy or the surface is badly weathered, it is sometimes necessary to remove the unsound materials by air hammers or even blasting. A level bottom is unnecessary. If the surface is steep, the pier should be keyed to it by a socket into the rock or by steel dowels placed in drilled holes.

The *Benoto* system makes it possible to excavate pier foundations to great depths through soil that is not easily drilled with augers. The Benoto drill is a specialized crane equipped with a grab bucket similar to a clamshell, but with four blades that can remove boulders and soft rock. A heavy steel churn drill is used to break up boulders and penetrate soft rock and a long cylindrical bucket can bail the soil–water slurry from the hole. The machine is always equipped with a chute to aid in concreting. An auxiliary machine aids in driving casing, when needed, by rotating the steel tube back and forth to reduce driving friction. Ordinarily, the casing is removed during concreting and the completed foundation is similar to the drilled pier (Fig. 11.23a and b).

The *Chicago well* is an open shaft lined with vertical wood sheeting held in place by steel hoops inside the shaft. It is dug by hand and the sheeting installed in 1.2- to 2-m (4- to 6-ft) lengths. Shafts as deep as 60 m (200 ft) and 4 m (13 ft) in diameter have been placed by this method.

The wet excavation method is sometimes used in soils that are too soft to permit open excavation without support. A large rotary well drill bores a hole that is the diameter of the finished pier. The hole is kept full of a mixture of clay, water, and heavy minerals that has the same unit weight as the soil and which provides an internal pressure that keeps the hole open. After drilling, a cylindrical steel shell is lowered into the excavation, and then the mud is replaced by clean water. Concrete is tremied through the water to form the pier. Such methods do not permit thorough cleaning of the excavation or an inspection of the stratum on which the pier rests.

The balance of pressure is often augmented by extending a casing a few feet above the ground surface and keeping the mud level at its top. The coarse cuttings can be removed by pumping them up a hollow drill stem, a process termed *reverse circulation.* In this way the high velocity of fluid through the stem can keep the gravel-sized particles in suspension. Cylindrical buckets with bottom valves or internal pistons also are used to suck up the coarse cuttings, but they disturb the hydrostatic balance. Concrete is tremied into the mud-filled hole to concrete the shaft from the bottom up. A casing can be placed after drilling is complete, and the hole flushed out and unwatered for *final cleanup.*

CAISSONS / Three forms of caisson are used in the United States; the caisson pile, the open caisson, and the pneumatic caisson. The caisson pile is a large diameter pipe. 0.6 to 1.5 m (2 to 5 ft) in diameter, which is driven with open ends by a very large pile-driving rig. The soil within the pipe is excavated, after which the pipe may be entered for inspection or for cleaning the surface on which the caisson rests. It is then filled with concrete to form the pier.

In the process known as the *drilled-in-caisson* (Fig. 11.24a) the pipe is fitted with a tool-steel cutting edge that can be driven into rock. The soil inside can be removed with a small-size bucket, by blowing out with a jet of

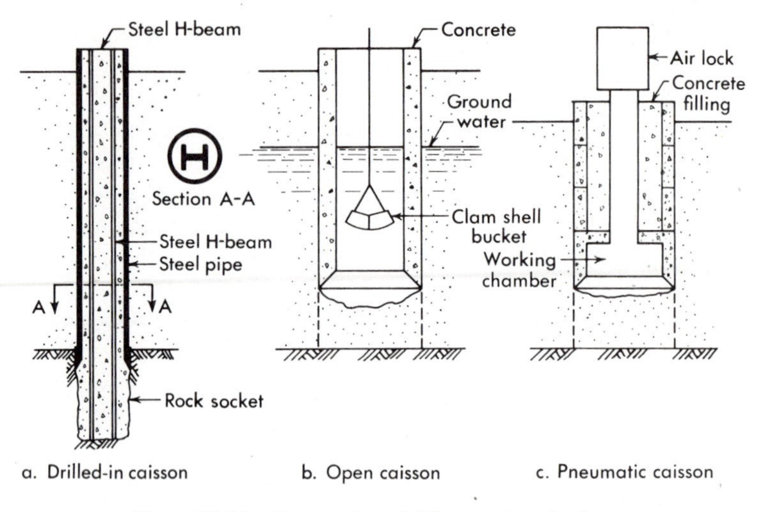

Figure 11.24 *Construction of different types of caissons.*

compressed air, by driving a coring tube into the soil and then pulling the tube, or by adding water and churning the soil into a slurry, after which it can be bailed out. After cleaning, a large well drill bit is placed inside the pipe and used to drill a socket in the rock. The pipe is then driven until the cutting edge develops a watertight seal against the rock. The drilling continues until a socket from 0.5 to 3 m ($1\frac{1}{2}$ to 10 ft) deep is formed in the solid rock. The caissons may be unwatered and the rock sockets inspected before concreting. In order to increase their column capacity, they may be provided with a steel H-beam core. Drilled-in caissons have been built with diameters from 0.6 to 1 m (2 to 3 ft), as long as 80 m (250 ft), and with capacities as high as 20,000 kN (2000 tons) each. They have the advantages that any types of soil may be penetrated and obstructions such as boulders can be drilled out, they can be extended through partially decomposed rock until sound rock is encountered, and visual inspection is ordinarily possible before concreting. They can be placed in a batter of 1 (H) to 6 (V) and in this way can be combined with vertical caissons to form A-frames that resist lateral loads.

The open caisson (Fig. 11.24b) is an open box that has a cutting shoe on its lower edge. As the soil is excavated from inside, the box is forced down by weights until it comes to rest on the desired bearing stratum. Open caissons are often used for constructing bridge piers in deep water. The caisson is prefabricated on land, floated into position with pontoons, and then lowered into place. No attempt is made to unwater the caisson until it is seated on the bearing stratum. Sometimes it is necessary to seal the bottom with concrete placed under water before unwatering is possible.

The width of the opening or *dredging well* must be great enough to permit the use of a clamshell bucket for excavation [at least 3 m (10 ft) but preferably more]. Water jets are sometimes placed in the outside walls to reduce friction during sinking.

Large caissons for bridge piers are made up of a number of small caissons, or *cells*, each with an independent dredging well but all opening into a common bottom chamber. The depth at which open caissons may be extended is limited by skin friction which overcomes the effects of weighting. Boulders and other obstructions that catch under the cutting edge may limit the depth, since it is difficult to remove them.

PNEUMATIC CAISSONS / Compressed air or pneumatic caissons must be used when an acceptable bearing stratum cannot be reached by open caisson methods because of water conditions. The high cost is justified only where the loads to be supported are very high.

The pneumatic caisson (Fig. 11.24c) is like an inverted tumbler lowered into water. It is a box with an open bottom and an airtight roof or cover that is filled with compressed air to keep the water and mud from coming into the box. The lower section is a working chamber in which the excavating is carried on and the pier is constructed. Above the working chamber is the air lock, which permits the workmen and materials to enter and leave without loss of air pressure in the working chamber.

The pneumatic caisson process provides a better means for controlling the sinking of the caisson and makes possible the removal of boulders, logs, and debris from under the cutting shoe. Also the foundation bed upon the rock or bearing strata can be better prepared and inspected.

The workmen have to work under air pressure sufficient to balance the pressure of the surrounding mud and water. This working pressure limits the depth to which pneumatic caissons may be used and greatly slows down the progress of sinking the caisson. There are numerous hazards in the use of the pneumatic caisson and only experienced engineers and contractors should undertake their design and use.

11:10 Anchors

An *anchor* is a special form of deep foundation designed to resist a lateral or upward force. It is used to resist hydrostatic uplift or provide support for anchored bulkheads, excavation bracing, or tied retaining walls.

GENERAL FEATURES / A number of different forms are used: anchor piles and blocks or thrust walls in soil and anchor rods in rock. The anchor blocks and thrust walls are analogous to footings or retaining walls. Their capacity is controlled by bearing capacity or earth pressure against the anchor, and by the resistance of the mass, whichever is smaller. The anchor pile and the anchor rod in rock are analogous to friction piles. Their capacity is controlled by the skin friction or shear developed along the anchor shaft and by the resistance of the mass, whichever of the two is the smaller.

ANCHOR BLOCKS / The mechanisms of shear resistance depend on the depth below the surface (Fig. 11.25). For shallow blocks, $z_2 < 4$ $(z_2 - z_1)$, the shear pattern with horizontal loading resemble passive pressure (Figs. 9.6 and 9.7). The resultant of earth pressure against the anchor face approaches that of passive pressure for the entire depth z_2, minus the resultant of active earth pressure on the opposite face. (Tension in cohesive soils should be neglected.) For small anchors, with horizontal thrust, wall friction is neglected and the Rankine analysis applies. For anchors with an upward thrust or pull, the wall friction acts upward, and the passive pressure is less than the Rankine value. The Coulomb analysis with $-\delta$ (Fig. 9.16c) is an adequate approximation.

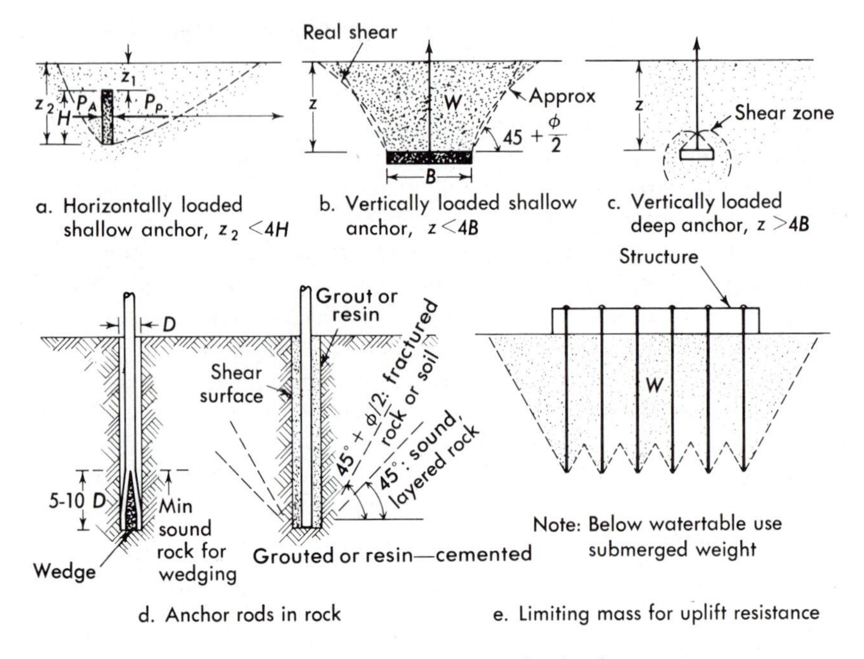

Figure 11.25 *Anchors in soil and rock.*

For vertical uplift blocks the shear pattern is that of Fig. 11.25b. This can be approximated by a truncated cone or pyramid whose sides slope at $45 + \phi/2$ with the vertical, using the slow or drained ϕ_D. The uplift resistance is equal to the weight of the block and the soil it engages. The vertical component of c (in undrained shear) can be added for dynamic uplift resistance but only c_D is included for static loading. Below the groundwater level the submerged unit weight applies.

If the anchor depth z_2 exceeds about 4 times its height, $z_2 - z_1$, for horizontal loads or 4 times its width B for vertical loads, then the anchor shears the soil like a small inverted deep footing (Fig. 11.25c). Limited data from model tests of square anchors in saturated clay in undrained shear as well as full-scale pullout tests[11:27, 11:28] indicate resistances, r_0, per unit of area perpendicular to the thrust is similar to that for footings.

$$r_0 = q_0 = cN_c + q_q N_q \tag{11:4}$$

For typical anchors $N_c = 5$ to 7, between the values for shallow and deep foundations. This suggests that for other soils the general bearing capacity [Equations (11:3)], with factors for shallow foundations, may be a reasonable but conservative approximation.

Because of the technical problems of obtaining an enlarged base anchor, as in Fig. 11.25b, cylindrical, concrete-filled shafts are preferred, as described in Fig. 11.21b and Section 11:7. Their capacity is evaluated as friction piles.

ANCHOR PILES AND RODS / Anchor piles in soil and grouted rods in rock transfer their loads by shear along their surface. The uplift resistance for the anchor pile can be computed in the same way as skin friction for a bearing pile. Of course, end bearing is neglected.

The anchor rod is placed deep in a drilled hole (Fig. 11.25d). If the rock is strong and sound, it can be secured by wedges or expansion sleeves like rock bolts. In softer rocks it is held by filling the hole with cement grout or quick-hardening plastic. A number of forms of anchor rods are used: plain bars with an expansion sleeve or an enlargement held by grout, high strength steel tendons similarly anchored, or ordinary reinforcing bars grouted the greater part of their length. The capacity of the anchor is limited by two factors: the pullout resistance of the bar, and the resistance of the rock mass engaged by the bar.

The pull-out of anchor wedges and expansion sleeves is dependent on the rock, the bolt size, and design. In hard rock, 25 mm (1 in.) in diameter, sleeve bolts have ultimate capacities of 100 to 150 kN (10 to 15 tons). The capacity of grouted bars depends on the shear between the grout and the ground as well as on bond or bearing between the bar and the grout. In soft rock, grout to rock shear may be as low as 200 kN/m^2 (30 lb/in^2), in hard rock as high as 2000 kN/m^2 (300 $lb/in.^2$). Grouted rods can also be used in coarse sand, gravel, and very fractured rock. The grout-to-soil shear is equal to the soil shear at the perimeter of the cylinder of grout.

Both the anchor rod and the pile are limited to the resistance of the mass surrounding the rod. In rock the geometry depends on the joint pattern. For cohesionless soils and closely jointed rock a cone with its angle $45 - \phi/2$ with the rod is conservative. For widespread rock joints that overlap, a cone angle of 45° is more realistic. For uplift anchors the limit of capacity is the weight of the cone (submerged weight below the water table).

TESTS / Because of the uncertainties in anchor wedging, bond, and the resistance of the soil or rock surrounding the anchor, pullout tests are essential for safe design. Moreover, it is prudent to proof-test all anchors to

1.3 to 1.5 times their working loads, because minor peculiarities in soil or rock structure have a great influence on anchor capacity.

After the overload test the full working load (or a large fraction of it) is retained permanently as a *prestress*. This minimizes anchor deflection, which can be damaging for structures subject to occasional uplift. The prestress also induces compressive stresses in the soil or rock and increases their strength (which is one of the functions of rock bolting).

ANCHOR SYSTEM / The total capacity of a system of anchors is limited by the weight of the portion of the mass that might break loose when all are loaded simultaneously (Fig. 11.25e). For uplift anchors closely spaced the limit is the weight of the soil mass penetrated by all the anchors. For lateral anchors the limit is the passive earth pressure or sliding resistance of the mass. For masses of complex shape, the resistance of the mass is analyzed as for slope stability (Chapter 12).

11:11 Underpinning[11:29]

Underpinning is the construction of new foundations under existing structures. The underpinning of structures is a highly specialized type of foundation engineering involving construction methods adapted to very limited working space and the handling of soils already under load. The work is done by only a few highly skilled contractors whose many years of experience qualify them for such exacting and critical work.

Underpinning is necessary when the foundations of a structure prove incapable of supporting the structure with adequate safety or without undue settlement. Underpinning also is required when changing conditions, such as the construction of a nearby building with a deep basement or the excavation for a subway, making existing foundations inadequate.

Two procedures are commonly used: first, installing new foundations in small pits excavated beneath the existing foundations; and second, installing new foundations adjacent to the old and transferring the load between new and old with steel beams.

The pit method (Fig. 11.26a and c) requires excavating a small hole beneath part of the existing foundation. A new, deep footing is poured into this hole, or pipe piles are forced into the soil by jacking against the existing foundation. The pipes, in sections about 0.6 m (2 ft) long, are jacked into the soil and then excavated by small buckets, a steam jet, or augers. The new foundation is placed section by section so that the old foundation is never without some support.

The second method (Fig. 11.26b) involves driving piles or constructing a new foundation as close as possible to the old. This is necessary when the old foundation is so small or weak that a pit beneath it is impossible. It is often cheaper than the pit method, for more room in which to work is available. The load is transferred from the old to the new foundation by

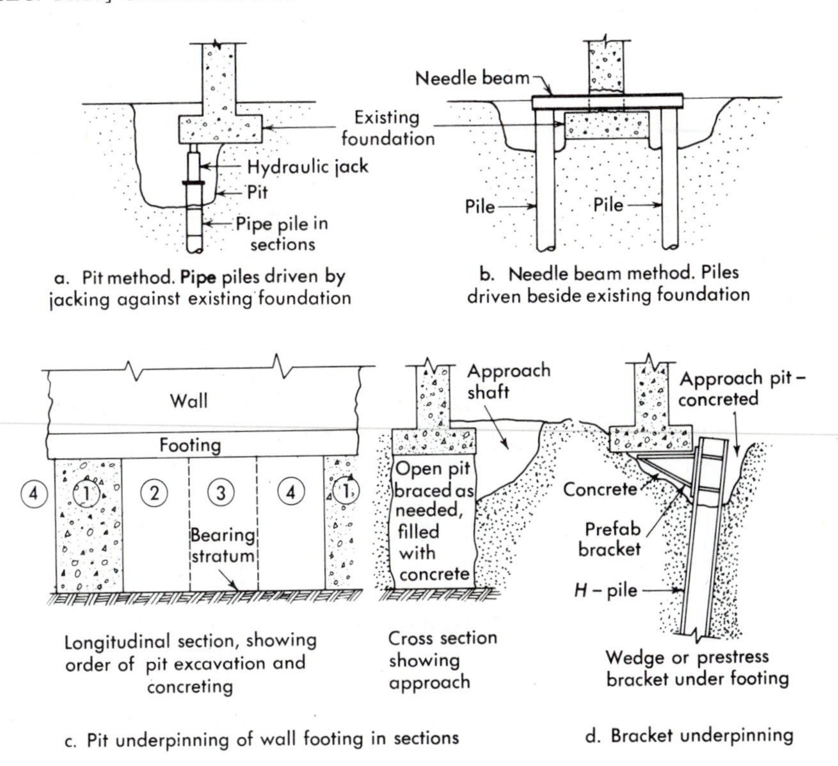

Figure 11.26 *Methods for underpinning an existing foundation.*

horizontal *needle beams* which are placed beneath the old footing or through it.

Clamps tightly bolted to a notched concrete or welded to a steel column make it possible to place the needle beam above the footing.

A bracket system (Fig. 11.26d) employs a small drilled pier or pile reinforced to support bending moments. The pile or pier is placed immediately adjacent to the foundation and terminated about 0.6 m (2 ft) below it. A bracket of steel or reinforced concete is then constructed under the old foundation and either wedged or prestressed to pick up the load. Steel members are encased in concrete, after prestressing, for corrosion protection.

The transfer of the load from the old to the new foundation is accompanied by some settlement. This can be minimized by jacking the new foundation against the old until the new carries the load. The settlement is prevented by extending the jack as the new foundation deflects under load. After the new foundation has stopped settling, the jack is replaced by steel wedges, which are then encased in concrete. This is known as the *Pretest method* of underpinning.[11:29] It was developed by Lazarus White in constructing subways in New York, making it possible to build beneath the largest structures without damaging them and to preserve monumental structures despite the ravages of time or the demands of deep new construction next to old.

11:12 Deep Foundations: Value and Uncertainties

The object of a deep foundation is to transfer loads past unsuitable materials to strong, rigid formation. However, merely driving piling or excavating caissons to the first hard stratum that is encountered may not solve the problem as the case histories at the opening of the chapter illustrate. Instead, deep foundations bring greater stresses to the deeper strata than shallow foundations. Moreover, constructing a slender column from the surface down, is difficult, sometimes hazardous, and always involves some uncertainties.

Analyzing deep foundation behavior during construction and predicting bearing capacity and settlement are more complex than surface foundations. Therefore, analysis and design require more technical effort and sound judgment than shallow foundations. However, some engineers persist in employing deep foundations as the answer to all difficult site conditions. They often are, but the exceptions have been exceptional problems.

REFERENCES

11:1 "Pile Hammers," *Conmaco, Inc.*, Kansas City.

11:2 D. D. Barkan, "Foundation Engineering and Drilling by the Vibration Method," *Proceedings of the Fourth International Conference on Soil Mechanics and Foundation Engineering*, 2, London, 1957, p. 3.

11:3 L. Bjerrum, "Norwegian Experience with Steel Piles to Rock," *Geotechnique*, **7**, 2, June 1957, p. 3.

11:4 G. G. Meyerhof, "Bearing Capacity and Settlement of Pile Foundations," *Journal of the Geotechnical Engineering Division, Proceedings, ASCE*, GT3, March 1976, pp. 195–227.

11:5 G. G. Meyerhof, "Compaction of Sands and Bearing Capacity of Piles," *Journal of the Soil Mechanics and Foundations Division, Proceedings, ASCE*, **85**, SM6, December 1959.

11:6 A. S. Vesic, "Ultimate Loads and Settlements of Deep Foundations," *Symposium of Bearing Capacity and Settlement of Foundations*, Duke University, Durham, N.C., 1967, p. 53.

11:7 L. C. Reese and H. B. Seed, "Pressure Distribution Along Friction Piles," *Proceedings, ASTM*, 1955.

11:8 H. M. Coyle and L. C. Reese, "Load Transfer for Axially Loaded Piles in Clay," *Journal of the Soil Mechanics and Foundation Division, Proceedings, ASCE*, **92**, SM2, March 1966.

11:9 A. Kezdi, "Pile Foundations," *Foundation Engineering Handbook*, H. F. Winterkorn and H. Y. Fang, eds., Van Nostrand Reinhold Company, New York, 1975, pp. 556–600.

11:10 O. Grillo, "Influence Scale and Chart for Computation of Stresses Due to Point Load and Pile Load," *Proceedings of the Second International Conference on Soil Mechanics and Foundation Engineering*, Rotterdam, 1947.

11:11 J. Kerisel, "Deep Foundations—Basic Experimental Facts," *Proceedings of the Conference on Deep Foundations*, Mexico City, 1964, p. 5; also, J. L. Kerisel, "Vertical and Horizontal Bearing Capacity of Deep Foundations in Clay," *Symposium on Bearing Capacity and Settlement of Foundations*, Duke University, Durham, N.C., 1967, p. 45.

11:12 V. G. Berezantzev, V. S. Khristoforov, and V. M. Golubkov, "Load-Bearing Capacity and Deformation of Piled Foundations," *Proceedings of the Fifth International Conference on Soil Mechanics Foundation Engineering*, 2, Paris, 1961, p. 11.

11:13 G. G. Meyerhof, "Penetration Tests and the Bearing Capacity of Cohesionless Soils," *Journal of the Soil Mechanics and Foundation Division, Proceedings, ASCE*, **82**, SM1 (Separate 866, 1956).

11:14 E. A. Smith, "Pile Driving Analysis by the Wave Equation," *Transactions, ASCE*, **127**, Part 1, 1145, 1962.

11:15 T. Ramot, "Analysis of Pile Driving by the Wave Equation," *Foundation Facts*, Raymond Concrete Pile Co., New York, 3, 1, Spring 1967.

11:16 (a) F. Rausche and G. G. Goble, "Soil Resistance Predictions from Pile Dynamics," *Journal of the Soil Mechanics and Foundation Division, Proceedings, ASCE*, **98**, SM9, September 1972. (b) F. Rausche and G. G. Goble, "Performance of Pile Driving Hammers," *Journal of the Construction Division, Proceeding, ASCE*, **98**, CO2, 201–218, September 1972.

11:17 P. W. Forehand and J. L. Reese, Jr., "Predicting Pile Capacity by the Wave Equation," *Journal of the Soil Mechanics and Foundation Division, Proceedings, ASCE*, **90**, SM2, 1–26, March 1964.

11:18 J. E. Bowles, *Foundation Analyses and Design*, McGraw-Hill Book Company, New York, 1968, pp. 470–479; 642–645.

11:19 R. D. Chellis, *Pile Foundations*, 2nd ed., McGraw-Hill Book Company, New York, 1961.

11:20 G. F. Sowers, L. Wilson, B. Martin, and M. Fausold, "Model Tests of Friction Pile Groups in Homogeneous Clay," *Proceedings of the Fifth International Conference on Soil Mechanics and Foundation Engineering*, 2, Paris, 1961, p. 155; 3, pp. 261 and 279.

11:21 H. Kishida and G. G. Meyerhof, "Bearing Capacity of Pile Groups Under Eccentric Load in Sand," *Proceedings of the Sixth International Conference on Soil Mechanics and Foundation Engineering*, 2, Montreal, 1965, p. 270.

11:22 L. C. Reese and Hudson Matlock, "Non-Dimensional Analysis for Laterally Loaded Piles with Soil Modulus Assumed Proportional to Depth," *Proceedings of the Eighth Texas Conference on Soil Mechanics and Foundation Engineering*, University of Texas, Austin, 1956.

11:23 H. Matlock and L. C. Reese, "Foundation Analysis of Offshore Structures," *Proceedings, Fifth International Conference on Soil Mechanics and Foundation Engineering*, Vol. 2, Paris, 1961, p. 91.

11:24 H. G. Poulos, "Behavior of Laterally Loaded Piles," *Journal of the Soil Mechanics and Foundation Division, Proceedings, ASCE*, **97**, SM5, 711–751, May 1971.

11:25 R. D. Chellis, "Finding and Fighting Marine Borers," *Engineering News Record*, March 4, March 18, April 1, April 15, 1948.

11:26 *Concrete Piles*, Portland Cement Association, Chicago.

11:27 G. P. Tschebotarioff, "Retaining Structures," *Foundation Engineering*, McGraw-Hill Book Company, New York, 1962, Chapter 5.

11:28 G. G. Meyerhof and J. I. Adams, "The Ultimate Uplift Capacity of Foundations," *Canadian Geotechnical Journal*, **5**, 4, November 1968.

11:29 E. A. Prentis and L. White, *Underpinning*, 2nd ed., Columbia University Press, New York, 1950.

SUGGESTIONS FOR ADDITIONAL STUDY

Behavior of Piles, Institution of Civil Engineers, London, 1971, 222 pp. (A conference collection of 18 papers.)

J. B. Burland, B. B. Broms, and V. F. B. DeMello, "Behavior of Foundations and Structures," *Proceedings, Ninth International Conference on Soil Mechanics and Foundation Engineering*, Vol. 2, Tokyo, 1977, pp. 495–546.

A. B. Carson, *Foundation Construction*, McGraw-Hill Book Company, New York, 1965.

V. F. B. DeMello, "Foundations in Clay," *Proceedings, Seventh International Conference on Soil Mechanics and Foundation Engineering*, Mexico, 1969, State of Art Volume, pp. 49–137.

"Ground Anchors," *Proceedings Specialty Conference 4, Ninth International Conference on Soil Mechanics and Foundation Engineering, Tokyo, 1977.* (Published by LeBaitment, Paris, 1977.)

S. M. Johnson and T. C. Kavanagh, *The Design of Foundations of Buildings*, McGraw-Hill Book Company, New York, 1968, 393 pp.

G. A. Leonards, et al., *Foundation Engineering*, McGraw-Hill Book Company, New York, 1961.

B. J. McClelland, "Design of Deep Penetration Piles for Ocean Structures," *Journal of Geotechnical Engineering Division, Proceedings, ASCE*, **100**, GT7, 705–747, July 1974.

J. D. Parsons and S. D. Wilson, "Safe Loads on Dog-Leg Piles," *Transactions, ASCE*, **121**, 1956, p. 695.

W. C. Teng, *Foundation Design*, Prentice-Hall, Inc., Englewood Cliffs, N.J., 1962.

R. J. Woodward, W. S. Gardner, and D. M. Greer, *Drilled Pier Foundations*, McGraw-Hill Book Company, New York, 1972, 278 pp.

L. Zeevaert, *Foundation Engineering for Difficult Subsoil Conditions*, Van Nostrand Reinhold Company, New York, 1972, 652 pp.

PROBLEMS

11:1 Prepare a table showing the point diameters, point areas, and surface areas of the lower 3 m (10 ft) of different sizes of concrete, steel, and wood piles. Consult the catalogs of different pile-driving contractors and manufacturers of wood and steel piling.

11:2 A wood pile with a 0.2-m (8-in.)-diameter tip and a 0.36-m (14-in.)-diameter butt 11 m (36 ft) long is driven into a dry loose sand weighing 17 kN/m^3 (108 lb/ft^3); angle of internal friction of 32°. Compute its bearing capacity.

11:3 A 10H57 steel pile 12 m (40 ft) long is driven into saturated clay weighing 17 kN/m^3 (108 lb/ft^3) and having an undisturbed c of 100 kN/m^2 (2090 lb/ft^2) and a remolded c of 50 kN/m^2 (1040 lb/ft^2). Compute its skin friction, assuming (1) the skin friction is developed over the entire pile surface, and (2) the skin friction is developed along the surface of a rectangle that encloses the pile. Depending on which governs, compute the end bearing for either the gross area of the enclosed rectangle or the net area of the pile. (Use the undisturbed c for end bearing and the remolded c for skin friction.)

11:4 A 0.35-m (14-in.) OD pipe pile 50 m (164 ft) long with a flat end is driven into a deep deposit of clay having the following characteristics:

Depth		c		Soil Weight	
m	ft	kN/m^2	lb/ft^2	kN/m^3	lb/ft^3
0–12	0–39	75	1570	19	121
12–33	39–108	25	520	16	102
33–45	108–148	48	1000	17	108
45–55	148–180	175	3650	21	134

Water table is at ground surface.

a. Compute the capacity in kN and kips.
b. If 25 of these piles are to be driven in a group, determine the minimum spacing to ensure that the group capacity will not be less than the sum of the capacities of the individual piles.

11:5 A steam hammer weighing 13.35 kN (3000 lb) and falling 0.76 m (30 in.) is used to drive a precast concrete pile 0.3 m (12 in.) square and 12 m (40 ft) long. The total bounce is 0.8 cm (0.3 in.) per blow, and the driving resistance is 4 blows/25 mm (1 in.). The coefficient of restitution is estimated to be 0.40 and the value of s_c is 5 mm (0.2 in.).

a. Compute the safe capacity by the Hiley formula, using a safety factor of 2.
b. Compute the safe capacity by the *Engineering News* formula.
c. How do they differ and why?

11:6 A steam hammer weighing 4.45 kN (1000 lb) that falls 0.9 m (3 ft) is used to drive wood piling. Compute the dynamic resistance according to the *Engineering News* formula if the pile penetrates 12 mm ($\frac{1}{2}$ in.) under each of the last few hammer blows.

11:7 Compute the safe bearing capacity of a steel 225 mm ($10\frac{3}{4}$ in.) OD by 8 mm (0.315 in.) wall pipe pile 15 m (49 ft) long driven into sand by a steam hammer with a 22.25-kN (5000-lb) ram and 0.9 m (35.4 in.) stroke. The "bounce" is measured and found to be 8 mm (0.3 in.), and the net penetration or "set" is 10 mm (0.4 in.) per blow. The hammer efficiency is 75% and the coefficient of restitution of the pile hammer on the pile cap is estimated to be 0.80; s_c is 2.5 mm (0.1 in.).

a. Compute the safe load, using the Hiley formula and including a safety factor of 2.
b. Compute the safe load, using the *Engineering News* formula.
c. Which shows the greater safe load and why?

11:8 A pile load test produces the following data. The pile is 26 × 26 cm (10 × 10 in.), 26-kg (57-lb) H-section.

Load		Deflection		Load		Deflection	
kN	tons	mm	in.	kN	tons	mm	in.
200	22.5	3	0.12	1000	112	20	0.79
400	45	6	0.24	1200	135	37	1.45
600	67	9	0.35	600	67	28	1.10
800	90	13	0.51	0	0	22	0.87

a. Find the safe pile load, using a safety factor of 2.
b. How much of the above settlement can be attributed to elastic shortening of the pile if the pile is assumed to be end bearing? The length of pile is 10 m (33 ft).

11:9 A machine weighing 12,000 kN (2700 kips) is to be supported on piles. The untreated wood piles at 180 kN (20.2 tons) must be 15 m (49 ft) long, and steel pipe piles loaded to 450 kN (50.6 tons) must be 20 m (66 ft) long. Which will be cheaper? The wood piles will cost $16.00/m ($4.88/ft) driven; the pipe piles $36.00/m ($11.00/ft).

11:10 An anchor consists of a vertical tension pile and a pile driven on a batter of 1 (horizontal) to 3 (vertical). Both piles are 25 m (82 ft) long. If the horizontal load is 100 kN (22.5 kips), compute the load in each pile. Assume the piles hinged at both ends.

11:11 Prepare a report describing a pile or caisson construction job. Include the following items:
1. Soil profile.
2. Reason for selection of piles or caissons.
3. Equipment.
4. Description of features of piles and their construction.
5. Tests, if any.

11:12 A caisson pile 0.9 m (3 ft) in diameter, of concrete with a compressive strength of 42 MN/m^2 (6100 lb/in^2) is drilled 0.6 m (2 ft) into thin-bedded close-jointed sandstone with q_u = 35 MN/m^2 (5080 lb/in.^{2}). The angle of the sandstone test failure plane θ was 65°.
a. Determine the safe load on the caisson.
b. If the joint spacing were 2 m × 2 m (6.6 ft × 6.6 ft), what would the safe load be?

11:13 For the soil profile and loadings described in Problem 10:17, suggest alternative types of deep foundations including the advantages and disadvantages of each. Which would be your choice and why do you think it best?

11:14 A closed end 0.275 m ($10\frac{3}{4}$ in.) OD pipe pile 20 m (65.6 ft) long is driven into a homogeneous sand formation 26 m (85.3 ft) thick underlain by a compressible clay stratum 3 m (10 ft) thick. The sand weighs 18 kN/m^3 (115 lb/ft^3) above the water table and 21 kN/m^3 (134 lb/ft^3) below. The angle of internal friction of the sand is 38° and the relative density 62%. The groundwater table is at a depth of 4 m (13.1 ft). The clay weighs 16.8 kN/m^3 (107 lb/ft^3) saturated, has a void ratio of 1.38, and a compression index of 0.63. It is normally consolidated. Below the clay is more sand.

a. Compute the safe bearing capacity of the pile assuming that it is jetted to a depth of 12 m (39 ft) and then driven.
b. Compute the safe bearing capacity assuming that the pile is driven for its full length.
c. Compute the stress increase below the single pile, and the settlement, assuming that the pile was driven. Pile load is 200 kN (45 kips).
d. Compute the stress increase due to 9 of these piles driven 0.9 m (3 ft) on centers, and the resulting settlement in the clay. Each pile carries 200 kN (45 kips).

CHAPTER 12
Stability of Soil and Rock Slopes

A highway embankment more than 40 m (130 ft) high crossing a narrow mountain valley was constructed of fresh, hard broken shale and sandstone blasted from deep cuts through the adjoining ridges. The pieces of rock from gravel to small boulder size, 25 mm (1 in.) to 1 m (3 ft), were compacted by heavy hauling machines, and topped with fined crushed stone for road base and by topsoil for landscaping. Fifteen years after the divided-lane freeway had been placed in service, arc-shaped cracks appeared in the southernmost pavement. Cracking was followed by slow subsidence of the pavement by about 0.3 m (1 ft). During the ensuing year, the subsidence increased to nearly 1 m (3 ft); the traffic lane was maintained by filling the subsidence zone and repaving. Boring and sampling were underway when suddenly the moving segment slid, followed by new arcuate scarps and subsidences extending across both traffic lanes (Fig. 12.1).

Additional borings found two surprises: (1) The fill was now a mass of soft clay containing occasional sandstone boulders floating in the clay matrix, and (2) the groundwater level within the embankment was only 6 m (20 ft) below the pavement, or 34 m (110 ft) above the original ground surface. Ground water had been flowing into the ends of the embankment from springs in the steep, interstratified sandstone–shale formations of the adjoining ridges. The fluctuating ground water plus air within the fill caused the shale (geologically considered to be sound and stable) to slowly slake into 1-mm thick plates of soft shale coated with clay. These expanded to fill the large voids in the rock fill, and continued to melt into clay. The rock fill with an initial angle of internal friction of nearly 45° had been slowly transformed into a clay fill with a residual drained angle of friction of about 12°. The embankment slopes of 1.5(H) to 1(V) or 33° could not withstand such a strength reduction and failed. Further movement was minimized by installing deep wells to lower the water level and reduce further slaking. In addition, the weakened portion of the fill was replaced with sound sandstone.

Figure 12.1 *Slide in highway embankment 40 m high caused by the deterioration of shale in the rock fill.*

Design for a two-lane highway in the Andes Mountains of Bolivia required a 10-m (33-ft) wide horizontal cut in the steeply sloping mountainside. The rock was hard quartzite sandstone dipping downward parallel to the mountainside. The cut design was based on past experience with a smaller road in the same vicinity; fill was avoided because it slid on the steep bare rock face. The cut was excavated by blasting the rock in 3-m (10-ft) deep benches with a nearly vertical slope in the hard rock uphill. The cut was already 7 m (20 ft) wide and drilling was underway for the last 3 m (10 ft) of cut when the mountainside above suddenly dropped down into the cut. The dipping sandstone layers separated on their bedding surfaces; the upper 6 m (20 ft) continued across the cut like a giant guillotine blade, slicing men and equipment apart and then grinding them into small pieces as the rock broke apart on its way down. More than 60,000 m^3 of rock moved below an inverted V-shaped crack and nearly half did not stop until it reached the bottom, 1000 m (3300 ft) below.

A subsequent international investigation disclosed a number of factors which had not been foreseen in design. The formations in the mountainside were warped. The dip of the bedding in the rockslide section was nearly 45°, compared to the 30° dip in the section upon which the design was based. The formations were deeply jointed, in a system of X cracks perpendicular to the bedding, which had opened sufficiently that small shrubs rooted in them, forming narrow ridges of vegetation on the bare rock. Temperature changes of 30°C from day to night in the clear tropical air generated creep in the steeply sloping slabs of rock whose support had been removed by the cut. Failure finally occurred when the accumulated creep strains caused the rock strength between the undulating bedding planes to drop from the peak ϕ of 47°, defined by friction plus shelving (Section 5:9), to the residual $\phi = 30°$ of the friction alone.

These failures illustrate some of the interrelated engineering changes and physical, physicochemical, and environmental factors of earth movements. In both cases, engineering construction started the succession of changes that culminated in failure; however, in both cases, there was a significant delay between the initial construction and eventual movements: 15 years in the first case and 3 months in the second. The succeeding changes that led to failure illustrate the effects of geologic structure, alteration of soil and rock strength by water, physicochemical changes, and even temperature. The changes were environment related; all might have been foreseen if those involved had investigated the properties of the formations. Although the extent of change might not be predicted quantitatively, the direction of change could have been foreseen and corrective measures undertaken before failure occurred.

12:1 Analysis of Stress and Stability

The safety of a body of earth against failure or movement is termed its *stability*. It must be considered not only in the design of earth structures but also in the repair and correction of failures. The design of open-cut slopes and embankment, levee, and earth dam cross sections is based primarily on stability studies—unless the project is so small that occasional failures can be tolerated. When failures such as landslides and subsidence do occur, their correction requires stability studies to determine the cause of failure and the best method of preventing future trouble.

CAUSES OF EARTH MOVEMENTS / Earth failures have one feature in common: There is a movement of a large body of soil along a more or less definite surface, as shown in Fig. 12.2. In most cases the earth remains intact during the first stages of the movement, but finally it becomes distorted and broken up as movement progresses. Some failures occur suddenly with little or no warning, while most take place leisurely after announcing their intentions by slow settlement or by the formation of cracks.[12:1]

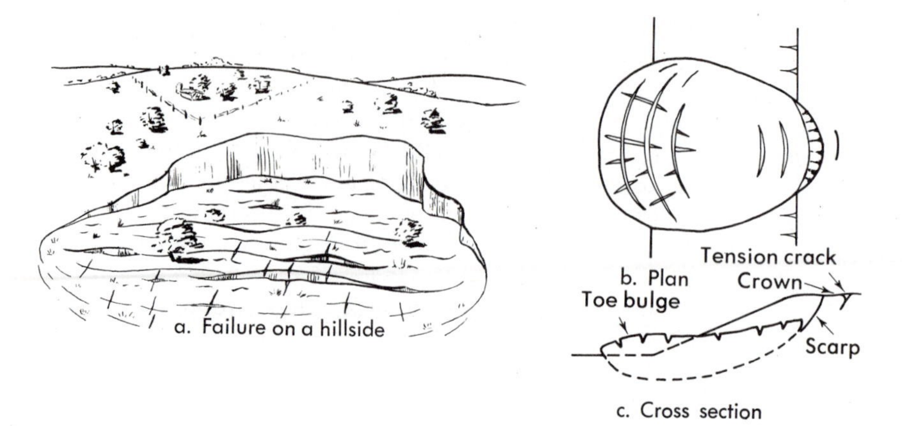

Figure 12.2 *Failure of a homogeneous earth mass by rotational sliding.*

Movements occur when the shear strength of the soil is exceeded by the shear stresses over a relatively continuous surface. Failure at a single point in the mass does not necessarily mean that the entire soil-rock body is unstable. Instability results only when shear failure has occurred at enough points to define a surface along which movement can take place. It is hard to determine the cause of many earth movements. Actually, anything that produces a decrease in soil strength or an increase in soil stress contributes to instability and should be considered in both the design of earth structures and in the correction of failures. Table 12:1 serves as a guide in analyzing for stability.

Failure is generated by any one or any combination of these factors. Most are independent, but some may be interdependent. For example, progressive strain in loose saturated sands causes a neutral stress increase that leads to a loss of strength and possibly a breakdown or collapse of soil structure. The possible number of combinations of factors leading to stability is staggering: 16!, or 2.09×10^{13}. The effect of water is vital: Water pressure or changes produced by water are involved in 12 of the 16 listed.

In cases of failures that involve property damage or loss of life, the engineer is often called upon to determine *the* cause of the failure. In most cases, several "causes" exist simultaneously; therefore, attempting to decide which one finally produced failure is not only difficult but also technically incorrect. Often the final factor is nothing more than a trigger that sets a body of earth in motion that was already on the verge of failure. Calling the final factor *the cause* is like calling the match that lit the fuse that detonated the dynamite that destroyed the building *the* cause of the disaster.

STRESSES IN A SLOPE / The stresses in a slope that has not reached the point of impending failure can be analyzed by elastic theories or by two- (and in some cases three-) dimensional finite element studies.

TABLE 12:1 / FACTORS IN INSTABILITY

Increased Stresses	Decreased Strength
*1. External loads such as buildings, water, or snow	*1 Swelling of clays by adsorption of water
*2. Increase in unit weight by increased water content	*2. Pore water pressure (neutral stress)
3. Removal of part of slope by excavation	*3 Breakdown of loose or honeycombed soil structure with shock, vibration, or seismic activity
*4. Undermining, caused by tunnelling, collapse of underground caverns, or seepage erosion	*4. Hair cracking from alternate swelling and shrinking or from tension
5. Shock, caused by earthquake or blasting	5. Strain and progressive failure in sensitive soils and brittle rocks
6. Tension cracks	*6. Thawing of frozen soil or frost lenses
*7. Water pressure in cracks	*7. Deterioration of cementing material
	*8. Loss of capillary tension on drying
	*9. Weathering—chemical or biochemical deterioration

* Water involved.

Although they do not define the state of stress at failure, they demonstrate (1) the stresses are dependent on construction sequence, particularly whether the slope was produced by cutting or filling; and (2) the locations within the soil body at which shear stresses and tensile stresses are greatest and therefore where failure is likely to commence and propagate outward.

Figure 12.3 shows two examples of such analyses. The vertical normal stress, σ_z, horizontal normal stress, σ_x, and shear stresses τ_{xz} are shown in terms of γH for a homogeneous, isotropic elastic body resting on a rigid, rough base, as in Fig. 12.3 adapted from Poulos.[12:2] The vertical normal stress, σ_z is approximately equal to γz at any point. The horizontal shear stress is virtually zero beneath the top of the slope; it is greatest along the foundation about midway between the top and toe. The shear stress, also, is greatest beneath the midpoint of the slope.

Figure 12.3b, adapted from Dunlop and Duncan,[12:3] shows the propagation of maximum shear stress in a homogeneous elastic solid while excavation progresses deeper forming a slope. The solid's shear strength increases with the original depth but does not change during excavation, similar to the behavior of a normally consolidated saturated clay in undrained shear. As the excavation progresses, the maximum shear stress increases until it reaches the limit, τ_{ff}. There is local failure beneath the slope face, but the slope remains stable. The zone of failure or plastic equilibrium increases in extent as the excavation becomes deeper. Eventually, the zone of plastic equilibrium is large enough so that a continuous surface of failure

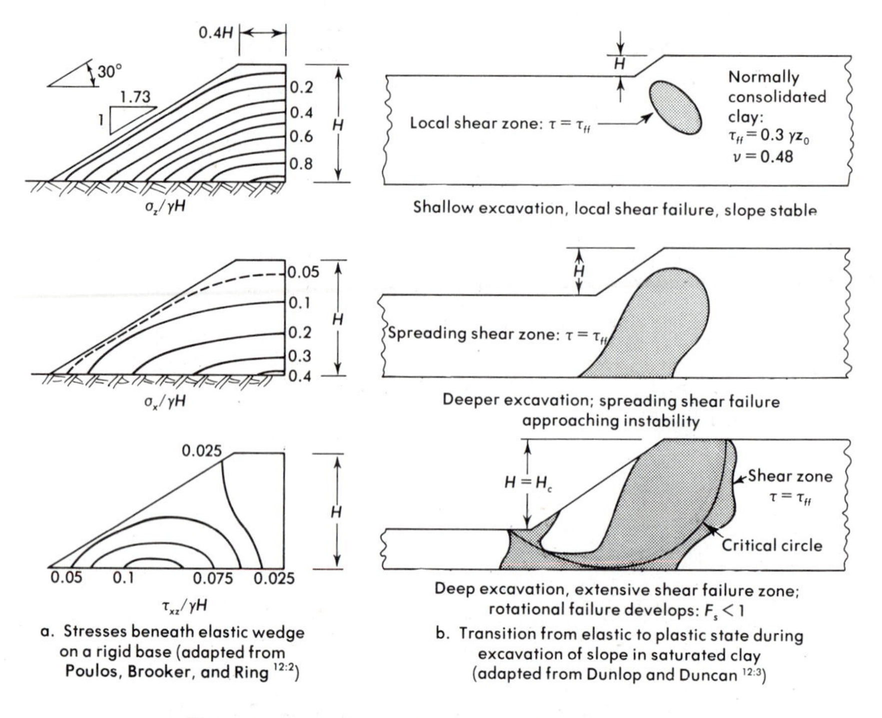

Figure 12.3 *Stresses in slopes, and failure propagation.*

compatible with slope movement can occur. At that point the slope fails in rotation, similar to Fig. 12.2b and c. If there are pronounced surfaces of weakness or anisotropy, shear stress would be concentrated on them. The plastic zone would be distorted and the failure surface elongated following the weaker materials.

STABILITY OF SLOPES / Among the most common of earth-mass failures are those resulting from unstable slopes in reasonably isotropic and homogeneous formations. Gravity, in the form of the weight of the soil mass and of any water above it, is the major force causing failure, while the shearing resistance of the soil is the major resisting force. The failure surface has the shape of the bowl of a teaspoon or half an egg sliced lengthwise, with the smaller end at the top of the slope and the wider end at the bottom as shown in Fig. 12.2b and c.

The intact soil immediately above the slide is the *crown*. The arc-shaped *scarp* with near-vertical displacement defines the upper limit of significant sliding. Tension cracks in the crown, however, can develop into new scarps because the unsupported scarp presents a new steeper slope. The upper portion of the sliding mass subsides, and moves outward, tilting trees toward the scarp. The lower portion bulges upward, forming tension cracks and causing trees to tilt downhill close to the toe. Sometimes the cracked bulge zone absorbs water from rainfall or springs and becomes so weak that it eventually flows viscously outward in narrow tongues.

Failure in isotropic materials initially occurs in one of three forms (Fig. 12.4). The *base failure* develops in soft clays and soils with numerous soft seams. The top of the slope drops, leaving a vertical scarp, while the level ground beyond the toe of the slope bulges upward. *Toe failures* occur in steeper slopes and in soils having appreciable internal friction. The top of the slope drops, often forming a series of steps, while the soil near the bottom of the slope bulges outward, covering the toe. The *slope* or *face* failures are special cases of toe failures in which hard strata limit the extent of the failure surface. Zones or surfaces of pronounced weakness, define part or most of the shear surface, elongating it and even causing linear motion. If there are large external forces, they also distort the shear surface.

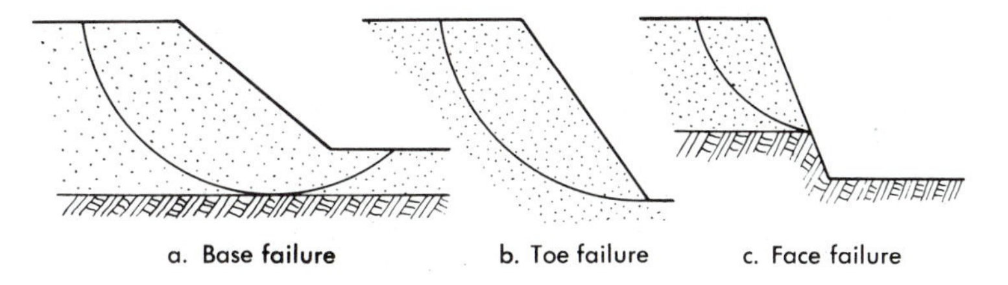

Figure 12.4 *Types of failure: Cross sections approximated by circular arcs.*

STABILITY BY TRIAL / Stability analysis is a problem in plastic equilibrium. When the mass is on the verge of failure, the forces causing movement have become equal to the resistance of the mass to being moved. A slight increase in forces is sufficient to produce continuing strain, as described in Chapter 5. Because of the irregular geometry of the soil and rock formations and the complex force systems in any real problem, the methods of direct analysis, such as used for earth pressure, are seldom applicable. Instead, trial of a very large (theoretically infinite) number of possibilities is the most useful approach to determining the safety factor of a tentative design or the potential failure of an existing slope.

First, a potential failure surface is assumed, and the shearing resistance acting along the surface is calculated. The forces acting on the segment of soil within the failure surfaces are determined, and then the safety factor of the segment is calculated as follows:

Safety against rotation

$$F_{sr} = \frac{\Sigma \text{ resisting moments}}{\Sigma \text{ moments causing failure}} \tag{12:1a}$$

Safety against translation (straight-line movement)

$$F_{st} = \frac{\Sigma \text{ forces opposing motion}}{\Sigma \text{ forces causing motion}} \tag{12:1b}$$

When all possible failure surfaces have been evaluated, the smallest safety factor found will be the actual safety factor. In practice, however, the smallest safety factor found by analyzing a limited number of well-chosen, possible failure segments will be sufficiently accurate. The computations are repetitive, making the analyses well adapted to computer programming and computation.[12:4, 12:5]

Generally, the forces causing motion are considered to include inertia, gravity, and all external loads. Those that resist motion include the soil strength and other forces along the potential surface of movement. There are different opinions as to which forces belong in the numerator and denominator of Equations (12:1a) and (12:1b) although all produce $F_s = 1$ under the same set of conditions. For example, an external resisting moment could be added to the numerator instead of subtracting it from the other external moments producing overturning in the denominator. The effect is a large number for F_r if $F_r > 1$. This may generate a feeling of greater safety and confidence from the public, but it blinds geotechnical engineers to their reliance on the soil shear parameters.

CIRCULAR ARC ANALYSIS / The general method for analysis of the stability of slopes was first suggested by the Swedish engineer K. E. Petterson[12:6] as a result of studies of a landslide in the harbor of Gothenburg. The actual surface is approximated by a segment of a cylinder which in cross section is an arc of a circle (Fig. 12.5). The overturning moment M_0 per unit

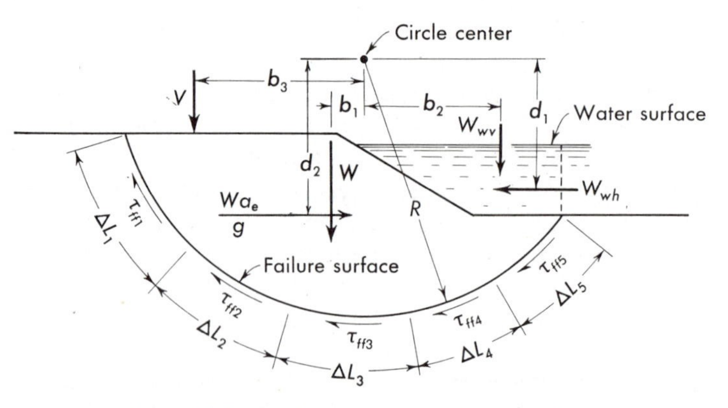

Figure 12.5 *Simple circular arc analysis of failure.*

of width about the circle center is the algebraic sum of the moments due to the weight of the mass, W, the horizontal and vertical components of water pressure (if the slope is inundated) acting on the surface of the slope, W_{wh} and W_{wv}, and any other external forces acting on the mass V

$$M_0 = Wb_1 - W_{wh}d_1 - W_{wv}b_2 + Vb_3 + \frac{Wa_e d_2}{g} \qquad (12:2a)$$

In this expression, b_1, b_2, b_3, d_1, d_2 are the respective moment arms of the centroids of the weights or action lines of the forces about the circle center. The resisting moment is provided by the soil strength. If the shear strength is τ_{ff} along each segment of the arc ΔL, whose radius is R, then the resisting moment for each foot of width is

$$M_r = R \sum \tau_{ff} \Delta L = R(\tau_{ff1}\, \Delta L_1 + \tau_{ff2}\, \Delta L_2 + \cdots) \tag{12:2b}$$

The safety factor of the circular segment is found by

$$F_{sr} = \frac{M_r}{M_0} \tag{12:3}$$

The basic circular analysis can be applied to any slope and combination of forces where the shear strengths of the soils are independent of the normal stresses on the failure surface, such as saturated clays in which failure occurs so rapidly that there are no changes in water content or soil strength.

Example 12:1

Calculate the safety of the following assumed segment (Fig. 12.6) if the crack, which is 1.5 m deep, is filled with water. The arc radius is 9.9 m.

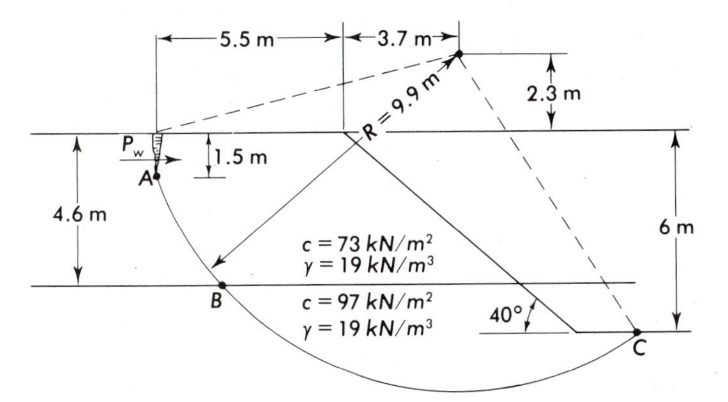

Figure 12.6 *Simple circular arc analysis in Example 12:1 of a slope in thick strata of saturated clays under conditions of undrained shear.*

1. Divide the arc into two segments, AB and BC. Determine the length of each.

$$AB = 3.7 \text{ m} \qquad BC = 13.4 \text{ m}$$

2. Calculate the resisting moment.

$$M_r = 73 \times 3.7 \times 9.9 + 97 \times 13.4 \times 9.9 = 15.542 \text{ kN-m}$$

3. Calculate the weight of the segment and find the centroid by methods of statics.

$$W = 1109 \text{ kN} \qquad a = 3.1 \text{ m}$$

4. The moment caused by the weight is $1109 \times 3.1 = 3439$ kN-m.
5. The resultant force of water pressure in the crack, P is $\frac{1.5 \times 9.8 \times 1.5}{2} = 11$ kN. It acts horizontally at a distance of 3.3 m from the center of the arc; $M = 36$ kN-m.
6. The total moment tending to cause overturning is $M_0 = 3438 + 36 = 3474$ kN-m.
7. The safety factor F_s is given by

$$F_s = \frac{15542}{3474} = 4.5$$

METHODS OF SLICES / In order to compute the stability of slopes in soils whose strength depends on the normal stress, it is necessary to determine the effective normal stress along the failure surface. The method of slices developed by Fellenius[12:7] has proved to be a workable approximation.

The failure zone is divided into vertical slices, as shown in Fig. 12.7. They need not be of equal width, and for convenience in computation, the boundaries should coincide with the intersections of the strata with the circle and with the slope face. In the basic analysis, it is assumed that each slice acts independently of its neighbor: There is no shear developed between them, and the normal pressures on each side of a slice produced by the adjoining slices are equal.

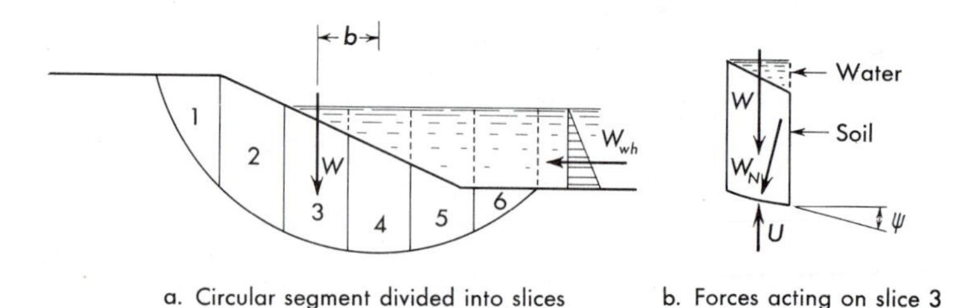

Figure 12.7 *Method of slices for circular arc analysis of slopes in soils whose strength depends on stress.*

The vertical force acting on each slice, W, is the weight of the soil in the slice plus the weight of water directly above the slice. The weight of any external load on the slice, such as a structure, is also included. The net or effective downward force acting on the curved bottom of the slice is the total weight minus the upward force due to neutral stress, $W' = W - U$. The upward force U is found by multiplying the neutral stress u (computed from the flow net as described in Chapter 3) by the slice width.*

If the slice is sufficiently narrow, the curved boundary can be approximated by a straight line that makes an angle of ψ with the horizontal axis. The component of the vertical force normal to the plane, $W'_{N'}$ is computed by $W'_N = W' \cos \psi$. The shear strength along that segment of arc can be expressed as follows:

$$\tau_{ff} = c + \sigma'_{ff} \tan \phi_D \tag{5:12b}$$

$$\tau_{ff} = c + \frac{W'_N}{\Delta L} \tan \phi_D \tag{12:4}$$

The total resisting moment for all the arc segments is found as before by Equation (12:2b).

The overturning moment can be found, as previously described, by Equation (12:2a). The moment due to the vertical forces is the algebraic sum of the moment of the total weight W of each slice about the circle center Wb. To this must be added algebraically the total moments due to the horizontal component of water pressure on the slope, and water pressure in cracks.

Many variations and refinements of this basic method have been developed. Although none are rigorous, they have proved to be sufficiently reliable for analysis and design. The differences in the more refined methods lie largely in the assumptions made regarding the shear and normal forces on the sides of the slices.

The analysis requires trial of a large number of assumed failure surfaces. That which has the smallest safety factor is the most critical surface—the one on which failure is most likely to occur. The analysis is well adapted to tabular or digital computer solution.[12:5]

Various automatic search programs have been incorporated in computer analyses to locate the minimum safety factor. The author has found their indiscriminate use misleading and sometimes unsafe: (1) The computed minimum is often unrealistic or geometrically impossible, diverting attention from the real answer. (2) In some stratified formations, there is more than one minimum; most programs stop at the first, although a second could be lower. A grid of circle centers, each with varying radii, with the minimum for each center plotted, as in Fig. 12.8, is far more useful. Contours can define the location of the most critical circle centers. These circles should be plotted to ensure that they are geometrically reasonable. At this point, an automatic search program could further refine the location of the most critical circle. More elegant, rigorous analyses, such as those of

* An alternate approach taken by some is to compute U_N by ΔL and subtract from W_N. However, this overcompensates for the effective weight of the soil $W' = W - U$. The author's experience demonstrates that the method proposed produces reasonably reliable safety factors that are consistent with the results of more sophisticated analyses.

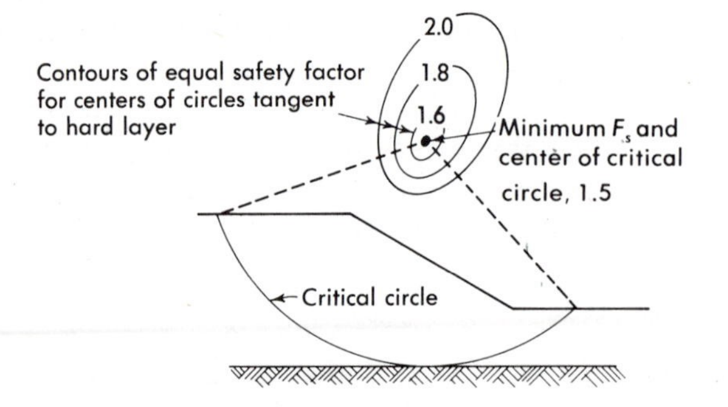

Figure 12.8 *Contours of equal safety factor for locating center of critical circle.*

Bishop,[12:8] Janbu,[12:9] and Morganstern and Price,[12:10] consider the forces acting between slices and their effect on the stresses on the failure surface, as well as the equilibrium of both forces and moments. Some, such as Janbu, Morgenstern and Price, include noncircular surfaces of rupture. All require assumptions of the distribution of forces and shears between slices, none of which have been verified by field measurements. Moreover, as shown by Wright and Kulhawy,[12:11] and in the author's experience, the differences between the safety factors are usually well within the accuracy of the shear parameters used in the analyses. The real test of the validity of any method is its ability to predict movement when the computed safety factor is 1. There is little evidence that any one is much more reliable than the other, except when the quality of the input shear parameters and data on geologic nonhomogeneity and anisotropy justify the effort.

EARTHQUAKE STRESSES / The horizontal acceleration of an earthquake, a_e, imposes a transient force of Wa_e/g on the mass. A simple approximation of its effect is to add this inertia force, multiplied by its moment arm, to the overturning moments. The real effect is less critical because the acceleration is momentary. More realistic analyses, such as that proposed by Newmark[12:12] consider the strains produced and their effects. For earthquakes with accelerations exceeding 0.15 g, dynamic analyses of the earth response are made using earthquake time histories and a finite element analysis of changing displacements and stresses.[12:13] The simple analysis described is not adequate for such conditions.

CRACKING AT THE TOP OF THE SLOPE / As described in Chapter 9, the upper part of a steep slope in a cohesive soil is in a state of tension. Under continued tension, vertical cracks develop which destroy part of the shear strength and which can contribute to failure if they fill with water. The depth d of the tension cracks can be approximated by

$$d = \frac{2c}{\gamma} \tag{12:5}$$

The soil above this level does not contribute to the resisting moment of the failure arc, as shown in Example 12:1. If the crack fills with water, water pressure contributes to the overturning moment.

EFFECT OF SUBMERGENCE AND SEEPAGE / Submergence of a slope has three effects. First, the weight of the circular segment is increased by the weight of the water above the slope and the greater soil unit weight, which increases the overturning moment. Second, the increase is more than offset by the resisting moment due to horizontal water pressure. Third, neutral stress increases on the failure surface, depending on the seepage flow net that develops, and offsets some of the gain in strength produced by the additional weights of the soil and water. The result is that the submerged slope usually has a higher safety factor than the same slope without submergence.

When the level of submergence is reduced so rapidly that the neutral stress within the slope cannot adjust itself to the new water level, the condition is termed *sudden drawdown.* The helpful moment due to horizontal water pressure is reduced. The weight of the soil and water is also reduced, but the neutral stress is not greatly changed. As a result the safety factor drops sharply, usually below that for the nonsubmerged condition. This is often the most critical condition in the design of the upstream face of an earth dam.

Seepage through the soil toward the face of a slope is caused by excess neutral stress within the soil mass and results in lower strength and a smaller safety factor compared with that of the same slope without seepage. This condition is often critical in deep excavations, highway and railroad cuts, the downstream face of earth dams, and in natural slopes.

SLOPES IN HOMOGENEOUS, SOFT CLAY / The special case of a uniform slope in a homogeneous, soft clay whose shearing resistance in undrained shear is approximated by $\tau_{ff} = c$ can be solved analytically, and the results are presented in the form of a dimensionless number, m, termed the *stability number.* The stability number depends only on the angle of the slope, β, and on the *depth factor*, n_d, which is the ratio of the depth of a hard, dense stratum measured from the top of the slope to the height of the slope. The height of slope, H_c, at which a failure will occur is given by the relation:

$$H_c = \frac{c}{m\gamma} \tag{12:6a}$$

and the safety factor of a slope of height H is given by

$$F_s = \frac{c}{mH\gamma}* \tag{12:6b}$$

* This is *not* the same safety factor as found by Equation (12:1) but instead is the safety factor with respect to cohesion.

A chart (Fig. 12.9a) showing the relation of the stability number to the slope angle and to the depth factor has been prepared from the results of a study by D. W. Taylor.[12:14]

From the chart it can be seen that toe failures occur for all slopes steeper than 53°. The location of the center of the failure arc can be found from a chart (Fig. 12.9b and c) developed by W. Fellenius,[12:7] a Swedish engineer

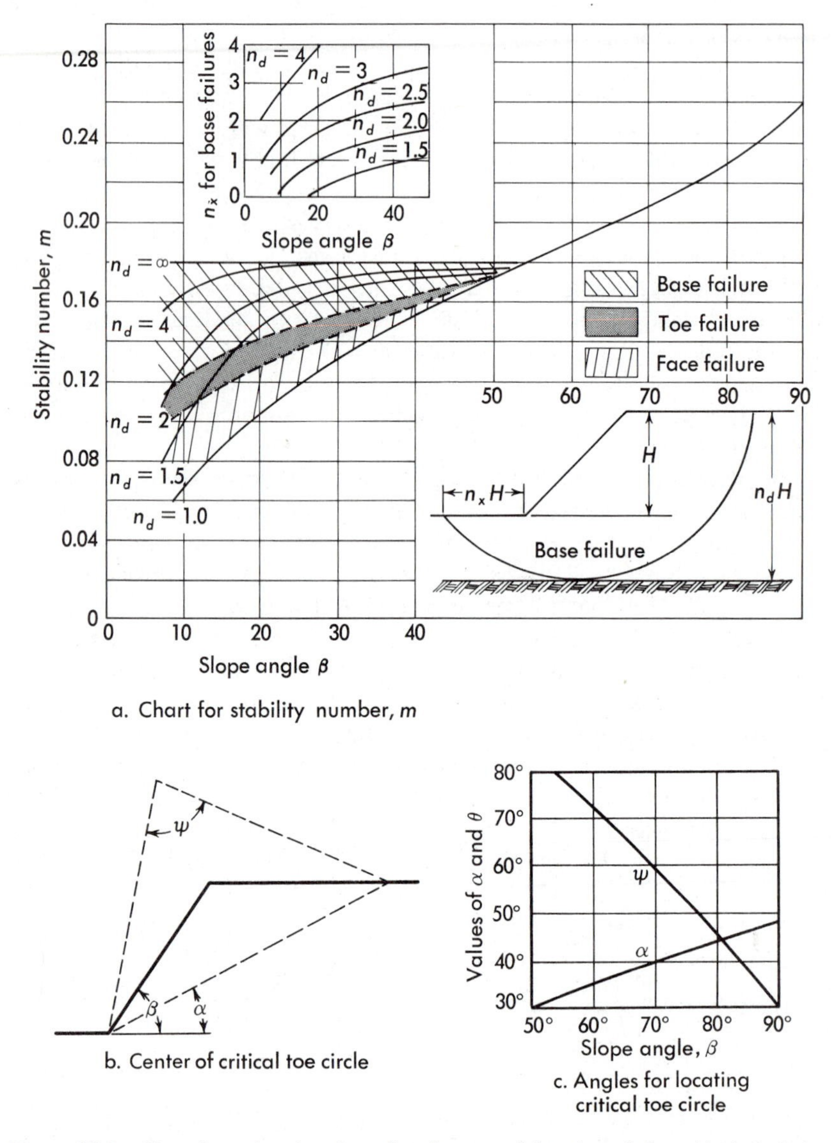

Figure 12.9 *Chart for estimating the safety factor and location of the critical circle in a homogeneous saturated clay in undrained shear.* (After D. W. Taylor[12:14] and W. Fellenius.[12:7])

Stability problems in cracked soils or jointed rocks involve multiple planes of weakness with different orientations. These cannot be analyzed by the simple representation of a two-dimensional cross section, such as shown in Fig. 12.10. Instead, sliding takes place on two intersecting planes in a direction parallel to their intersection (Fig. 12.11). The forces in such a three-dimensional system are most readily handled by vectors, as described by Deere et al.[12:16] and Goodman.[12:17]

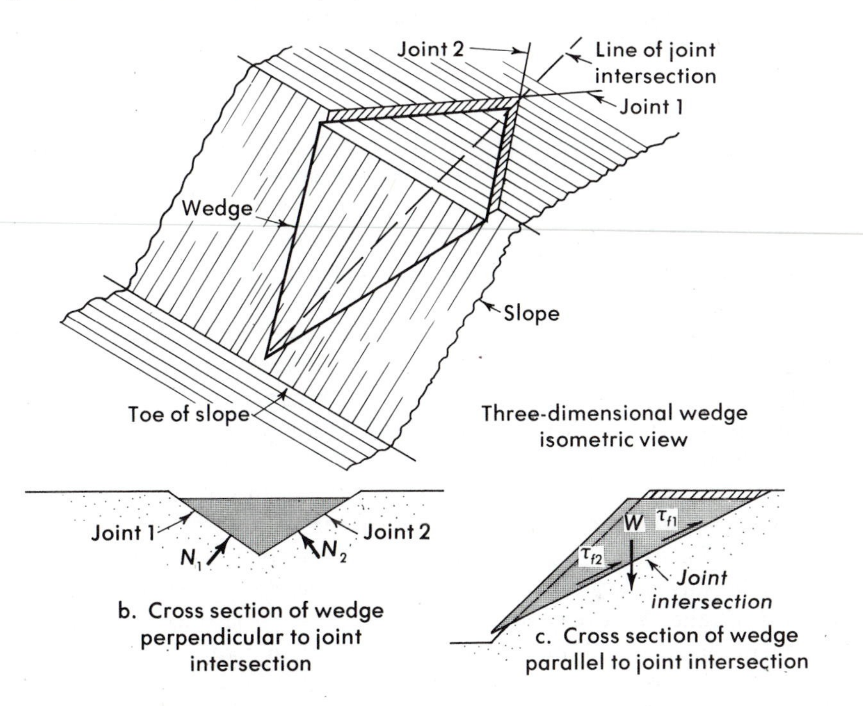

Figure 12.11 *Wedge failure in jointed rock.*

SHEAR STRENGTH FOR ANALYSIS / The shear strength of rock or soil varies greatly, depending on the environment and particularly on the degree of saturation, the effective stress and its change due to neutral stress variations, and the effects of progressive strain. Saturation destroys capillary tension and causes the buildup of neutral stresses under quickly applied loads. Therefore, if the environmental conditions indicate any possibility of saturation, the shear strength in that state is employed in analysis.

The effective stress depends on the rate of loading compared with the rate of drainage within the soil or rock. If the loads will be placed on the soil quickly, such as by rapid construction of a large embankment, the undrained strength is used. If the embankment load is applied slowly, the undrained strength should be safe, but the drained strength with proper consideration of neutral stress is more realistic and should produce a more economical design. For excavated or natural slopes that are exposed for long periods of time, it is necessary to use the drained strength because the unloading produced by erosion or excavation eventually reduces the effective stress on the soil and thereby the strength. The long-term effects of weathering upon exposure should be included for permanent excavations.

The consolidated-undrained (CU or R) shear test is often a realistic model of shear in slopes where the mass develops equilibrium under static loads, but failure is caused by a sudden change in load such as produced by flood or heavy rainfall. The drained (D, CD, or S) test is better only if pore water pressures can be fully evaluated.

Temporary strength produced by capillary tension is misleading in utilizing the results of undrained or consolidated-undrained tests to analyze natural slopes and excavation slopes. Although the slope may be temporarily stable, supported by capillary tension, either saturation or drying can eliminate the tension and cause a loss of shear strength or failure.

Progressive failure brings about failure at one point before the adjoining point is highly stressed and a transfer of load from one point of the soil or rock mass to the next. In such cases the average strength on the failure surface is not the peak strength, but between the peak and residual.

Judgment and experience are necessary in interpreting the laboratory test results and selecting the strengths to be used in stability analysis. The best shear strength for design is that found by analyzing actual failures in that same soil. This is particularly true when the problem is to correct a past failure. When a new design must be made for a situation for which there is no record of failures, it may be expedient to make an artificial—a full-scale model of the proposed slope, embankment, or dam, that can be made to fail under careful observation and control. The results of such a full-scale test, when correlated with the laboratory data for the same soil, furnish the engineer with the effective strength for design of similar structures.

SAFETY FACTORS FOR DESIGN / When existing slopes and embankments have been analyzed for safety, it has been found they have relatively small factors of safety when compared with those of other structures. Although a safety factor of 2 or 2.5 is not uncommon in building design, the same factors applied to embankments would make the cost so high that they could not be constructed. Some earth structures having a *computed* safety factor as low as 0.9 have not failed!

The following table gives the significance of different values of the safety factor for soil masses. The safety factors in Table 12:2 apply to the most critical combination of forces, loss of strength, and neutral stresses to which the structure will be subjected. Under ordinary conditions of loading, an earth dam should have a *minimum* safety factor of 1.5. However, under extraordinary loading conditions, such as a design superflood followed by

TABLE 12:2 / SIGNIFICANCE OF SAFETY FACTORS FOR DESIGN

Safety Factor	Significance
Less than 1.0	Unsafe
1.0–1.2	Questionable safety
1.3–1.4	Satisfactory for cuts, fills, questionable for dams
1.5–1.75	Safe for dams

sudden drawdown, a minimum safety factor of 1.1 to 1.25 is often considered adequate.

The safety factors for use in design include an allowance for the differences between laboratory tests results and the real shear strength of the soil. If the design is based on an analysis of a failure, somewhat lower values are acceptable.

12:2 Open Cuts

Open cuts are excavations in which no bracing is used to support the soil. They are used in constructing excavations when hard soil is encountered that needs no support and in highway, railroad, and canal cuts where the cost of long lines of bracing would be great. Cuts less than 15 m (50 ft) deep whose failure would not endanger life are ordinarily designed on the basis of experience. Railroad design manuals and standard highway specifications give $1\frac{1}{2}(H)$ to $1(V)$ as a standard slope for most conditions and 2 to 1 for weak soils. This is based upon both soil strength and the observation that it is difficult to maintain protective vegetation on slopes steeper than $1.5(H)$ to $1(V)$ or $1.75(H)$ to $1(V)$ in erosion susceptible silts.

DEEP CUTS / Deep cuts are investigated on the basis of a geologic and soil study and a slope-stability analysis utilizing the drained shear strength. If the soil is a swelling or fissured clay or subject to unusual seepage, further investigation is necessary. The condition of nearby cuts in similar soil is a guide to cut performance. In critical or unusual situations a trial cut excavated at a slope steep enough to cause failure. The soil strength, determined by an analysis of the failure and correlated with the laboratory data on the soils in the cut, is used to determine the safe slope. In extreme cases, where accurate analysis is impossible because of erratic soils, it may be prudent to place benchmarks at the top of the finished slopes to warn of any unusual movements that could lead to failure.

IMPROVEMENT OF CUT STABILITY / The stability of a cut can be improved by decreasing the soil stress or by increasing its strength. Soil stress can be reduced in most cases by making the slope flatter. If the sections that require improved stability are short, the slope can be partially supported by a small retaining wall or by cribbing. Water pressure in cracks in cohesive soils can be relieved by surface drains above the slope to intercept water and by horizontal drains driven into the face of the slope.

Piles driven through the potential shear plane can increase the resisting moment; however, the additional shear resistance offered by piles is usually small compared to the shear forces or overturning moments. Better still, piles driven near the top of the slope can support part of the potentially sliding mass. Piles are most effective in increasing marginal stability; if the slope is very unsafe, enough piles to be effective are usually uneconomical.

Strength of cohesionless, slightly cohesive, or fissured soils and fractured rocks can be increased by relieving neutral stresses with surface drains and horizontal drains in the face of the slope (Chapter 3). Good drainage has always been the most effective measure for improving slope stability where water is a factor in instability. The strength of cohesive soils is difficult to improve permanently. In some instances large ventilation ducts, driven into the slope, have been able to reduce the soil's water content and increase its strength, but the method is very expensive.

CUTS IN LOESS AND TUFF / True loess, a cemented soil, has high shear strength in spite of its loose structure. It will stand vertically in cuts as deep as 20 m (65 ft). Sloping cuts are stable only until rain falls. The bare, porous soil absorbs the water, which seeps downward rapidly because of the high vertical permeability. The cementing of the soil breaks down in water, and the slope disintegrates by gullying and slumping until it becomes vertical. Vertical cuts (Fig. 12.12) will stand for many years with only occasional slumping or scaling along the vertical cleavage planes. Cuts are made wider than is necessary in order to allow room for the debris that collects at the toe.

Cemented volcanic ash or tuff also stands on very steep slopes until the cementing is destroyed by saturation or rapid tropical weathering.

12:3 Embankments

An embankment is an artificial mound of soil or broken rock that supports railroads, highways, airfields, and large industrial sites in low areas, or impounds water. Because embankments are constructed of filled-in material, they are often termed *fills*, but this term applies also to other earth construction.

HIGHWAY AND RAILROAD FILLS / Highway and railroad fills are usually designed on the basis of experience unless heights greater than 15 m (50 ft) are involved. The standard slopes are usually $1\frac{1}{2}(H)$ to $1(V)$ or $2(H)$ to $1(V)$ unless the embankment is subject to flooding. Highway fills are carefully constructed of selected soils compacted to prevent settlement and a rough surface (see Chapter 6). Railroad fills are not always highly compacted because settlement can be offset by maintenance of the ballast.

Figure 12.12 *Near-vertical cut in loess.*

HIGH FILLS AND FILLS SUBJECT TO FLOODING / High fills and those subject to flooding require careful analysis and design based on the shear strength and compressibility of soils or fractured rock to be used in the fill construction. The different soils and rocks that could be used should be selected and tested as described in Chapter 6, and their shear strengths, drainage and consolidation for future environmental conditions are made available to the design engineer. The slopes required by each different material to provide a safe embankment are found by stability analyses. From these data, trial designs are made, and the cost of each is estimated. The best material is the one that provides the required performance at the lowest cost.

When fine-grained soils are compacted at water contents several percent wetter than optimum, temporary pore water pressures develop. If the body of fine-grained soil is very wet, and the neutral stresses do not dissipate as rapidly as fill load is added, the soil is weakened and shallow toe or face failures will occur in the wet areas. Undrained shear tests with pore pressure measurements can diagnose this possibility in advance and help set limits of moisture, or control the rate of construction to match the rate of drainage.

Fills subject to flooding are especially critical. Railroad fills that have stood the pounding of heavy trains for years sometimes collapse after periods of inundation. They should be designed on the basis of the shear strength determined after soaking samples of the soil in water as described in Chapter 6. Typical slopes for such fills may be as flat as 3(H) to 1(V) or even 4 to 1 when made of soils that soften readily on absorbing water.

LEVEES / Levees are small, long earth dams that protect low areas of cities and towns, industrial plants, and expensive farmland from flooding during periods of high water. Unlike highway and railroad embankments, small settlement is not an important consideration; and unlike earth dams, levees must often be placed on poor foundations. Since levees usually extend for many miles, the cost of borrow materials and of construction is extremely important. Where land is inexpensive and space is available, the embankment is seldom compacted except by the weight of hauling equipment. Where the water table is high, the embankment in constructed by casting from a dragline working on top of the newly completed fill, which interferes with compaction. Typical slopes for uncompacted or indifferently compacted levees are as flat as 5(H) to 1(V) on the outer or water-covered slope and 3(H) to 1(V) on the inner.

High levees, levees in restricted spaces where flat slopes cannot be used, or levees protecting critical areas such as power plants are designed on the basis of soil tests and stability analyses. In such cases controlled compaction is required, but because steeper slopes are used, the saving in soil volume compensates, to some extent, for the added cost of construction.

12:4 Embankment Foundations

Most difficulties with embankments come from faulty foundations (Fig. 12.13). It is not difficult to construct a fill that is strong, free from volume

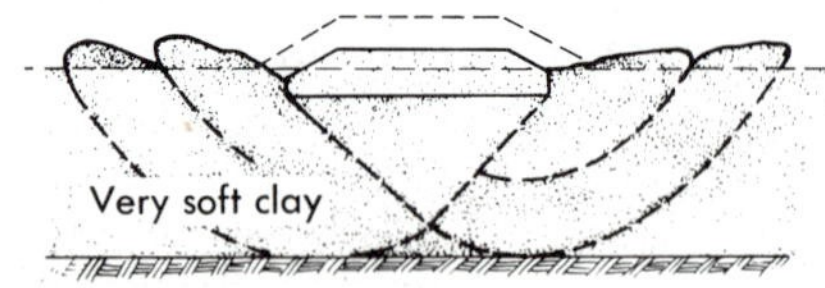

a. Mud waving and embankment subsidence from shear in thick soft clay

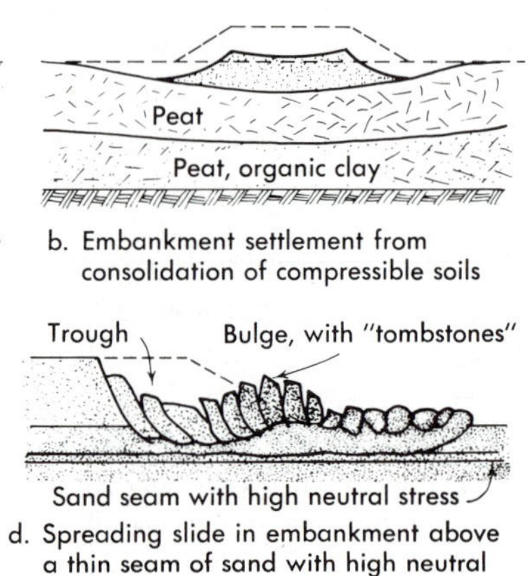

b. Embankment settlement from consolidation of compressible soils

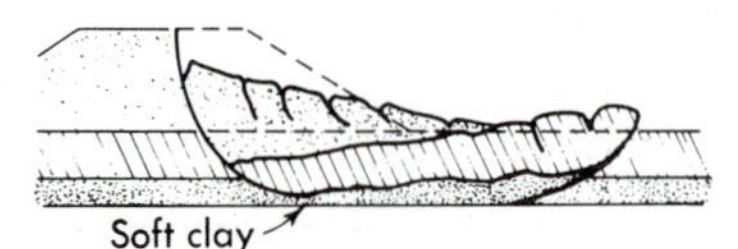

c. Elongated shear slide in a stratum of soft clay

d. Spreading slide in embankment above a thin seam of sand with high neutral stress

Figure 12.13 *Embankment foundation problems.*

change, and incompressible; but if the soil below it is poor, failure can occur in spite of careful construction. The failure commences below the fill and in some cases spreads into the fill itself, obscuring the actual cause of the trouble.

FILLS ON THICK STRATA OF WEAK SOIL / Fills on deep soils of little strength fail because of inadequate bearing capacity (Fig. 12.13a). The analyses for bearing capacity (Chapter 10) can be used for such failures if the weak stratum is at least half as thick as the base of the fill is wide. Otherwise the possibility of failure is determined by trial and error, using the circle method of analysis. If a hard crust overlies the soft soil, its strength can not always be relied on to support the load. In one case a levee 12 m (40 ft) high was built on top of a thin crust of hard clay above a thick stratum of soft clay. Twelve hours after the levee was completed, it had sunk until it was only 1 or 2 ($3\frac{1}{2}$ or $6\frac{1}{2}$ ft) above the ground surface. The hard clay, which held up the partially completed fill, broke under the full load and allowed the fill to drop. Bulges or mud waves appeared in the ground surface adjacent to the toe of the fill, and smaller waves were pushed up as far as 300 m (1000 ft) from the levee.

Failures of this type can be prevented in a number of ways. Lightweight fill materials, such as slag, expanded shale, volcanic ash, and even foam plastic, can reduce the stresses beneath the fill to a safe amount. The overturning moment can be reduced by flat slopes or a berm at the toe of the slope that acts as a counterweight. If the soil is normally consolidated, its strength can be improved by consolidation under the weight of the fill (Chapter 6). Construction must proceed slowly, however, in order to give the soil time to consolidate. Vertical sand drains or sand piles can decrease the length of the drainage paths and increases the rate of consolidation.

If the soft stratum is relatively thin [2 to 3 m (6 to 10 ft)] and shallow, it may be cheaper to excavate the soil beneath the fill area and replace it with a stronger material, such as broken rock. If the soft soil is less than 10 m (33 ft) thick, it can sometimes be replaced by *displacement.* In this method the fill is constructed on the top of the soft soil as high and with as steep slopes as possible. In most cases it is expedient to allow it to sink under its own weight, displacing the soft soil in a bearing-capacity failure. In a few cases, it is feasible to remove the soft soil from beneath the fill by blasting.[12:18] Explosives are distributed in the virgin soil before the fill is placed. An appropriate fill volume is placed above to confine the explosive force and to fill the void created by the subsequent explosive excavation. The explosives below the fill toe are detonated a fraction of a second before the explosives beneath the fill center. The sequential detonation forces the weak soil outward and the fill subsides into the gap. Additional fill height is added as necessary. Although the weak soil is not entirely removed, the technique produces a better foundation than the soil it replaced.

Where the fill load is heavy and the weak soil too thick to be displaced or blasted out, a pile-relieving platform supports the fill weight. Untreated woodpiles, entirely below ground water, spaced less than 1 m ($3\frac{1}{2}$ ft) apart, can support well-compacted fill by arching between their wide butts. Heavy fill, concentrated surface loads, and very deep, weak soils may require a reinforced concrete slab across the piles, termed a *relieving platform* (Fig. 9.19b).

FILLS ON COMPRESSIBLE SOILS / In some cases the soil may be strong enough to support the fill without failure but is so compressible that the fill settles excessively (Fig. 12.13b). This is typical of organic silts and organic clays and occurs to an extreme degree in peat. Highways and airfield runways across marsh areas often assume wavy profiles because of irregular settlement.

Excessive settlement due to compression can be minimized by preconsolidating the soil, by slow construction, by sandpiles, or excavating the compressible soil. Any procedures involving consolidation require evaluation of the maximum and minimum rates of primary consolidation, as well as continuing secondary compression, to determine if the time required is available and if the cost of correcting postconstruction settlement is less than that of removal.

FILLS ON THIN STRATA OF SOFT CLAY / Fills on relatively thin strata of soft clay fail by sliding horizontally along a complex failure surface that extends upward through the fill, as is shown in Fig. 12.13c. Failures of this type usually occur during or shortly after construction, before the weak clay stratum can consolidate under load. The safety against this type of failure can be determined by the sliding-block analysis, using the undrained shear strength. If the safety factor is marginal, and no loss of life or other serious consequences of failure are likely, the low safety factor can be considered to be a reasonable risk. The weak stratum will eventually consolidate under the fill load and become stronger, increasing the safety. Increased safety against this type of failure can be provided by a lightweight fill, flat slopes, and slow construction that permits the soil to consolidate and gain strength under the load. Vertical sand or fiber drains can accelerate consolidation. Their installation weakens sensitive clays, sometimes to the degree that their benefit is not worth their cost. If the stratum is close to the ground surface, it may be most economical to remove it completely.

FILLS ABOVE THIN, COHESIONLESS STRATA SUBJECTED TO NEUTRAL STRESS / When water pressure builds up in thin, cohesionless strata beneath a fill, failure may take place suddenly without warning. Although the neutral stress is generally greater under the center of the embankment, so is the confining stress due to the weight of the fill. Near the toe the vertical stress decreases faster than u, and $\sigma' \rightarrow 0$. Failure starts with outward movement of a wedge of soil adjacent to the toe, followed by toe failure of the steep faces left by the moving wedge (Fig. 12.13d). The secondary toe failures sometimes form a trough or *graben* in the fill surface with the intact fragments of sliding soil looking like rows of tilted tombstones.

Fills on hillsides sometimes seal natural outlets for seepage and cause pressures to build up in the embankment foundation and in the natural slope above. Dams and levees create high pressures by the reservoirs they impound. Temporary high water pressures sometimes develop during construction in sand or silt lenses from which water cannot drain readily.

The safety against failure due to water pressure in cohesionless seams is analyzed by the sliding-block method, considering the neutral stress and its effect on strength along the surface of sliding. This is easily done where the head or neutral stress is known, such as when it is produced by a man-made lake or reservoir. In other cases it must be estimated from groundwater observations or from the maximum weight of the embankment. Piezometers sealed into critical strata warn of excessive neutral stresses and permit corrective measures to be taken. These include drainage or a temporary halt of construction.

12:5 Earth and Rock-Fill Dams[12:19—12:24]

Earth and rock-fill dams are special embankments designed to impound water more or less permanently. Dams are the most critical of all engineeering structures, for their failure can cause property damage and loss of life for great distances downstream, involving people who may not be aware of the dam's existence. Earth dams, if not properly designed and constructed, are particularly vulnerable because the very material of which they are constructed can be weakened or disrupted by the water they are supposed to retain. In spite of these dangers, earth dams have proved to be among the most enduring of structures. Dams built in India and Ceylon over 2000 years ago are still storing water for irrigation. The largest structure ever built is the Garrison Dam in South Dakota with nearly 100 million m^3 (130 million yd^3) of embankment. Many earth dams over 120 m (400 ft) high have been built on all the continents and rock fill–earth dams 300 m (1000 ft) high are now in use.

CRITERIA FOR USE / Dams of earth or rock fill are employed for a number of reasons. First, materials suitable for earth construction are available at a large proportion of dam sites. Second, earth and broken rock are easily handled, either by hand in remote areas with cheap labor or by great machines. Third, earth and rock-fill dams are often suited to sites where the foundations are not strong enough or sufficiently rigid for masonry dams. Finally, the earth or rock-fill dam is frequently cheaper than any other type. Of course, there are some disadvantages: Suitable materials are not always available; greater maintenance is usually necessary; and a separate spillway is required.

A successful earth or rock-fill dam must satisfy two technical requirements: safety against hydraulic failure and safety against structural failure. These include failure of the foundation and the embankment, acting together as a unit, and any part of either.

Hydraulic failure can be external or internal. Externally, the dam must have enough spillway to be safe against overtopping, and both the upstream and downstream faces must be protected against surface erosion. Safety against internal hydraulic (seepage) failures was discussed in Chapter 3. The dam and the foundation must be sufficiently impervious that the loss of water is not objectionable, and they must be resistant to seepage erosion or piping. The latter has been an important cause of failure of earth dams, and provisions to prevent seepage erosion are essential features of design.

Structurally, the dam and the foundation must support the weight of the embankment and the load of the water under the worst possible combinations of maximum reservoir, seepage forces, changes in reservoir level, and earthquake acceleration. The embankment is analyzed in the same manner as other earth structures, but the required safety factors are higher, as given in Table 12.2. Settlement of the embankment can create cracks through which seepage erosion could develop. Ordinarily, settlement of the foundation of 1% of the dam height and settlement in the embankment of 1 or 2% of the height will not be serious, provided there are no sharp differences between adjacent parts of the dam.

COMPONENTS OF AN EARTH DAM / The components of an earth dam are shown in Fig. 12.14a. The basic parts are: (1) the foundation, (2) the cutoff and core, (3) the shell, and (4) the drainage system. The

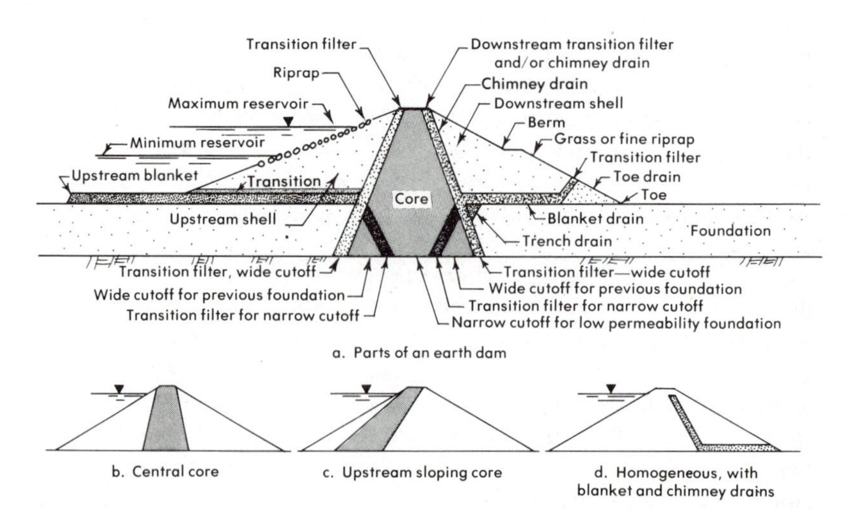

Figure 12.14 *Basic earth and rock fill dam cross sections. Note that all parts are not necessarily incorporated in any one dam.*

foundation, either earth or rock, provides support for the embankment and also resists seepage beneath the dam. The *core* holds back the water, and if the foundation is pervious, the core is extended downward to form a *cutoff*. The *shell* provides structural support for the core and distributes the loads into the foundation. The internal drains carry away any seepage that passes through the core and cutoff and prevent the buildup of neutral stress in the downstream part of the embankment. The internal drains take many forms, depending on the anticipated seepage; trench drains and sloping chimney drains just downstream from the core, blanket drains between the dam and the downstream foundation, the toe drains at the downstream toe. All must be provided with filters, as described in Chapter 3, to prevent internal erosion and clogging.

Transition filters are necessary between the core and upstream shell to prevent migration of the fine-grained core into the coarser shell. They are sometimes omitted if the grain sizes of the core and shell are not greatly different or if the seepage gradient through the core is very small. *Riprap* is required on the upstream face to prevent erosion and wash by waves. Sod or fine riprap is required on the downstream face to prevent rain wash. Berms are often provided to permit access to the face of the dam during construction or for maintenance afterward. They also help prevent rain wash by breaking the long continuous slope. The parts of a rock-fill dam are similar, but the rock fill itself is a drain.

CROSS SECTION / The three basic cross sections are the central core, the upstream core, and the homogeneous, as shown in Fig. 12.14b, c, and d. The *central core* provides equal support for the core and is most stable during sudden drawdown. It utilizes the minimum amount of core material. The *upstream core* or sloping core is more stable with a full reservoir and provides the cheapest design when there is little or no sudden drawdown. The upstream and central core are common for rock fill dams. The *homogeneous* cross section results when both the core and shell are made of similar materials. Unless a *core zone* is defined by internal drainage (Fig. 12.14d), a homogeneous cross section requires somewhat flatter slopes than a core type and for that reason is seen more frequently in dams less than 15 m (50 ft) high.

CUTOFF AND CORE DESIGN / Cutoffs and cores are made of earth, steel sheet piling, concrete, or a curtain of grout injected into the soil. Earth is cheap and can be made sufficiently flexible so that it will remain watertight in spite of small movements in the dam or foundation. Almost any soil can be used if its permeability is sufficiently low (less than 10^{-3} mm/sec) and provided that it is protected from internal erosion or piping by filters and it does not develop swelling pressures. Earth is used for cutoffs if sufficient volumes are available and if it is possible to cut a trench in the foundation without damaging the foundation or creating undue construction difficulties. Earth is always used for the dam core unless no suitable material is present.

The minimum thickness of the core or cutoff depends on the soil; clay cores as thin as 5% of the head have been used, but better practice calls for the minimum thickness of clay cores to be 15 to 25% of the head and for silty cores to be 30 to 50% of the head, in order to minimize erosive gradients. In many dams the cutoff is constructed of the same material as the core. Yet the cutoff is often wedge-shaped, narrowing with increasing depth, as shown in Fig. 12.14a. This construction expediency can be hazardous because the greatest hydraulic gradient through the core-cutoff is at the contact with the impervious foundation. An irregular foundation is smoothed with mortar or "slush grout" to provide a continuous tight surface against which the cutoff can be compacted. Slots or cracks in the contact zone are cleaned out to a depth of 2 to 3 times their width and then filled with concrete or mortar so that any water percolating through the cracks will not contact the core and erode it.

A trench filled with a clay slurry can be used in a cutoff where open excavation is impractical because of a high groundwater table or where sheet piling is impossible because of boulders. The trench is excavated with a dragline but is maintained full of a viscous highly plastic clay–water slurry that supports the foundation soil. The trench is backfilled with a pasty clay–gravel that displaces the viscous clurry, and that contains enough gravel that will not consolidate appreciably under its own weight. Cutoffs more than 30 m (100 ft) deep have been made in this way. They are usually placed upstream from the dam toe so as not to weaken the foundation unduly, and are connected to the dam core by an upstream blanket.

Steel sheet piling can be an effective cutoff in sand. Gravel, however, can cause the interlocks to open. The cutoff, in this case, generates such high local gradients that piping can occur. Grouting can extend a cutoff into fractured but otherwise impervious materials. However, unless the grout holes are so closely spaced that they form a continuous wall, the high gradients across the grout curtain can generate seepage erosion. Some designers reduce the hypothetical gradients by several parallel grout lines; others by concentrating on a single line, with very closely spaced holes. Piezometer readings upstream and downstream of both types show that neither is perfect. Therefore, filtered foundation drains downstream of a grout curtain are essential.

SHELL DESIGN / The design of the shell consists of selecting the material on the basis of its strength and availability for construction, as outlined in Chapter 6 and then determining the slopes necessary to provide stability for the embankment. The design process is essentially trial and revision: Tentative slopes are selected, the stability is analyzed, and the design is then revised to provide greater economy or greater stability.

Typical upstream slopes range from 2.5(H) to 1(V) for gravels and sandy gravels to 3.5(H) to 1(V) for micaceous sandy silts. Typical downstream slopes for the same soils are 2(H) to 1(V) to 3(H) to 1(V). A seepage analysis is made of the trial design to determine the flow net and the neutral stresses within the embankment and foundation. Safety against seepage erosion and the amount of leakage through the dam are computed.

Stability analyses are made of both faces of the dam, using the method of slices previously described. The upstream face is usually analyzed for three conditions: full reservoir, sudden drawdown, and reservoir empty before filling. The downstream face is analyzed for full reservoir and minimum tailwater and also for sudden drawdown of tailwater from maximum to minimum if that condition can develop.

Shells of broken rock, both dumped and compacted make excellent dams. Typical downstream slopes are often steeper than $2(H)$ to $1(V)$ even in very high dams. The entire dam shell is a drain, with no pore pressure above the tailwater level. Because of the sharp contrast between the grain sizes of the shell and core, multilayer transition filters are essential.

If the stability is insufficient, there are several possibilities for improvement. First, the slope can be flattened. Frequently, only the lower half or third need be changed, making a composite slope that is steep at the top and flatter at the bottom. Berms at the toe of the slope serve the same purpose. Second, the soil strengths can be improved by increasing the required density or by using different materials. Third, weak zones in the foundation can be removed. Fourth, the position of the core and cutoff can be shifted. Finally, internal drainage can be designed to reduce neutral stresses in the downstream foundation and shell.

12:6 Earth Movements in Nature

CLASSIFICATION OF NATURAL EARTH MOVEMENTS / Earth movements are commonplace geologic phenomena that are part of the process of *mass wasting*. Large quantities of fractured or weathered rock and soil are constantly moving down slopes and into streams where they are carried away to be deposited elsewhere. The impelling force for all these movements is gravity, assisted at times by rain erosion, water pressure, expansion and contraction forces, earthquake shock, and man's interference with nature.

There are innumerable classes of natural movements that have been proposed: by the mechanisms responsible (there are 2.09×10^{13}), the geometry of the moving mass, or the features that can be readily observed, as suggested by Varnes.[12:1] The author considers four major categories:

1. Creep—the slow, relatively steady movement of soil down slopes.
2. Landslides—fairly rapid movements of soil or rock masses in combined horizontal and vertical direction.
3. Subsidences—movement of earth masses vertically downward.
4. Rockfalls—steep or vertical superficial rock movements.

CREEP / Creep is a slow, nearly continuous movement of soil, resembling the creep of metals under small stresses or the plastic flow of concrete. It is manifested by the tipping of fence posts and similar rigid objects embedded in the soil. The best indication of creep is the gentle curving of trees, with the convex side point downhill in the direction of movement, as shown in Fig. 12.15. (Trees in areas subject to landslides exhibit an abrupt change in trunk tilt that corresponds to each movement; trees along roads tilt toward the road.)

Figure 12.15 *Tree trunks bowed downhill by creep.*

The mechanism of creep is not fully understood. On slopes in which the safety factor is low, the movement is probably true creep (Section 5:10) at stresses close to shear failure. On flatter slopes, seldom less than 4°, creep may be the result of alternate shrinking and swelling from seasonal changes in moisture coupled with the continuing downhill force of gravity. Generally, creep is confined to the upper 5 to 6 m (16 to 20 ft) of the soil or broken rock mass and it is most rapid close to the ground surface. It is an indication of potential trouble—a quasi-equilibrium state that can be easily upset and turned into a landslide by engineering construction, such as a deep cut or a heavy fill.

Creep cannot be stopped, but its rate of movement can be decreased materially by drainage to increase the strength of the soil and to reduce periodic swelling and shrinking. In most cases the best method of preventing trouble is to make allowances for it. Benchmarks should be set in solid rock in level areas not subject to movement by creep. Pipelines should be made with flexible joints when laid in slopes on which creep is taking place. Building foundations must be strong enough and deep enough to resist movement or should be tied together so the entire structure can move. The latter procedure, of course, would be practical only for very small buildings.

LANDSLIDES[12:25] / Natural landslides are difficult to analyze because of their complex nature. The strength of natural deposits is so variable and the number of different forces acting is so great that theoretical studies are at best only indications of what is likely to occur. Most landslides do not occur spontaneously. The slope is usually unstable for years and gives warning of its instability from time to time by continuing creep, and a hummocky topography.

Finally, some event increases the stress or decreases the strength to the level that failure takes place. This event triggers the failure, although it actually may be insignificant by itself. Loud noise has been blamed for slides in loose debris in mountainous areas; the added weight of water from a hard rainfall is often responsible for the start of slides in humid regions.

Although many landslide classification schemes have been proposed to aid in deducing the mechanisms and designing control measures[12:1], every slide is unique: Each must be evaluated in detail if control is to be effective. However, the author has found simple classification of the following paragraphs useful for describing similar mechanisms,

FLOW SLIDES / If a soil or rock mass suddenly loses its strength, it becomes liquid and flows downhill, spreading out over the flat land below. The basic cause of failure is ordinarily neutral stress that builds up until the strength becomes virtually zero, as in a liquid. If the material is loose, structural collapse, as described in Chapters 1 to 5, contributes to the water pressure increase and loss of strength.

Loose, saturated cohesionless soils are particularly vulnerable to sudden flow of a large portion of the mass caused by localized shear or shock. A local failure produces a pore pressure buildup that propagates further failure and more pore pressure. Pile driving in a shore deposit of loose fine sand triggered a landslide that carried houses and people into a shallow lake. An explosion triggered the flow of a saturated fine-grained industrial waste that engulfed the processing plant that produced the waste that destroyed it.

Water pressure from rainfall and snow melt produces flow slides in coarse-grained cohesionless deposits like talus and rock debris and in closely jointed rock. Heavy rain and snow melt at Gros Ventre, Wyoming saturated a fractured conglomerate on a mountainside, and built up enough neutral stress that several million cubic meters flowed down to destroy a small village and dam up a river in the valley. The lake has remained for 50 years, covering farms and houses. Neutral stress accumulating in fragmented rock from a coal mine waste dump in Wales brought on a flow slide that buried a school and killed more than 100 children in a few seconds.

Severe cyclical loading from earthquakes has caused liquefaction of sand deposits and devastating flow slides. According to Seed,[12:26] the widespread damage that accompanied many past earthquakes was the result of flow slides. One of the disastrous landslides at Anchorage, Alaska in 1963 has been attributed to flow in a sand seam underlying a firm clay. The water in the collapsing voids in a flow slide is forced to the surface in small sand–water geysers that add to the terror of the phenomenon.

FLOW SLIDES IN SENSITIVE CLAY / Flow slides also occur in the ultrasensitive marine clays such as the marine clays of eastern Quebec, Canada. Although the undisturbed strength of the clay is moderately high, the remolded strength is extremely low. Failure at an isolated point in the mass, induced by local shear, brings about progressive failure and flow in which large blocks of intact clay float like icebergs on a stream of viscous liquid remolded clay.

LINEAR SHEAR SLIDES / Linear shear slides occur along well-defined plane surfaces. Although crescent-shaped segments of soil break loose in the upper end of the slide zone, and bulges and mud waves occur at the lower end, the failure zone is elongated, and most of the movement is linear.

The movements take several forms, as shown in Fig. 12.16. The simplest occurs on a sloping plane of weakness such as a seam of weathered shale (Fig. 12.16a). A common variant is a wedge of loose fill whose plane of contact with the virgin soil becomes saturated by seepage from pervious strata. Another common variant is a cohesionless seam dammed up by an impervious fill (Fig. 12.16b). Failure in linear slides can be triggered by excavation that reduces support at the toe, by local softening of the soil or water pressure in pervious strata or joints.

Movements on a horizontal plane can occur by squeezing of a clay or weak shale seam (Fig. 12.16c). The blocks above move out as if on a belt conveyor. Water pressure in joints aggravates such movement.

Another form is shear in a sloping stratum of weak material, usually a weathered zone, underlain by a hard boundary. This is a rotational slide that is severely elongated and flattened by the solid boundary below (Fig. 12.16d). These slides are common in residual soils overlying rock. They also occur in hard, highly plastic clays that gradually expand and weaken from the ground surface down, creating a blanket of soft soil above a harder unweathered base. Such slides have developed in slopes as flat as 4(H) to 1(V) or 14° with the horizontal where the clays have been exposed to rainfall by highway cuts.

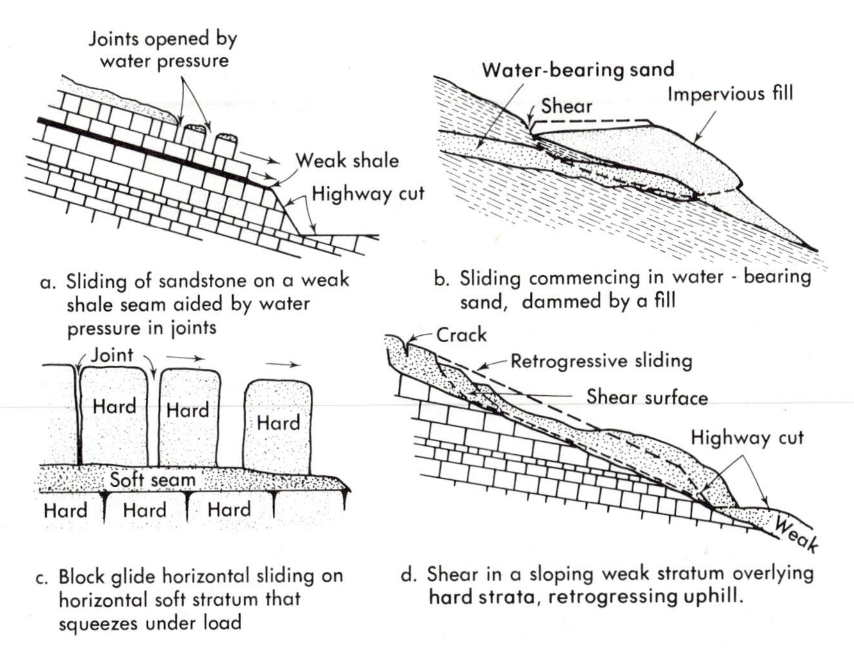

Figure 12.16 *Linear shear slides.*

Progressive failure or *retrogression* is characteristic of many landslides but particularly of linear shear forms. The slide progresses up the hillside, with each successive movement leaving a steep slope that is unstable and which subsequently fails as the earlier slide continues to move down. Progressive failure is also characteristic of soils which lose strength upon continuing shear strain. Even some stiff clays and shales can fail by progressive movement and loss of strength.

Linear slides in weathered rocks including shales are often triggered by the slow expansion or elastic rebound of the material exposed by excavation or earlier sliding. Whole hillsides sometimes begin to move although tests show that materials are strong enough that a conventional stability analysis indicates adequate safety.

ROTATIONAL SLIDES / Rotational slides take place in homogeneous soils, particularly clays, and in any thick deposit where there are numerous noncontinuous planes of weakness. The classic examples of the rotational slides are the deep base failures that occur in the soft clays of the coastal areas of Norway and Sweden. Failure is triggered by undercutting of the toe by dredging or erosion and by external loads on the upper part of the slope.

A major slide occurred in the edge of a valley in a deep deposit of glacial clay. It was caused by filling at the top of the slope to gain a parking area for a manufacturing plant. The underlying clay sheared under the weight of the fill. The bulge at the toe of the slope was 4 m (14 ft) high. The force exerted by the moving wall of clay against an adjacent railroad trestle was great enough to bend the steel legs of the support towers and to shear their concrete foundation pedestals.

Steep slopes in stiff clays fail in rotational toe slides (Fig. 12.17a). The clay cliffs of the Great Lakes suffer from such movements when they are undercut by wave action. Several successive slides are often seen at the same point, forming narrow, arc-shaped terraces or steps leading down the slope, their surfaces intact and with trees and shrubs growing on them.

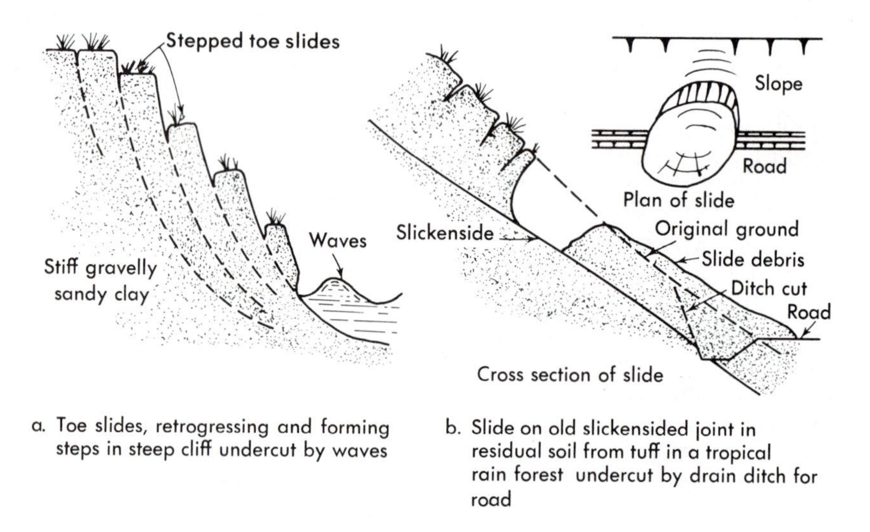

a. Toe slides, retrogressing and forming steps in steep cliff undercut by waves

b. Slide on old slickensided joint in residual soil from tuff in a tropical rain forest undercut by drain ditch for road

Figure 12.17 *Slides in stiff soils initiated by undercutting.*

MODIFIED ROTATIONAL / Modified rotational slides occur when multiple surfaces of weakness distort the shear zone. One typical form occurs in clays with seams or lenses of cohesionless soils. When water pressure (neutral stress) builds up, the strength drops until shear failure occurs. Failure begins with a horizontal movement, but this leads to an elongated rotational slide that often cuts across the cohesionless seams and allows them to drain. An example of such a slide is shown in Fig. 12.18. It took place after snow melt, spring rainfall, and finally a leaking water pipe augmented the normal groundwater pressure in numerous thin sand seams in a thick clay deposit.

A second form occurs in residual soils that retain the structural defects, such as joints, of the original rock. The joints can sustain no tension. Often they are surfaces of advanced weathering that are weaker in shear than the remainder of the mass. Failure may start with a plane tension crack at the top of the slope or a plane bulge at the toe, followed by shear in the remainder of

Figure 12.18 *Landslide caused by water pressure in thin seams of cohesionless silt and fine sand in a thick deposit of clay. Note the undamaged brick chimney despite a toe bulge rise of 6 m (20 ft).* (Photograph by G. B. Sowers.)

the mass along a curved surface. Such slides are common in tropical regions with deeply weathered rock. The weathered formation is generally strong and therefore will stand safely on steep slopes. However, local failures occur in areas where the joints are favorably oriented, and particularly after equilibrium has been upset by some construction operation such as excavating for a highway (Fig. 12.17b).

The slides in the weathered volcanic formations of the continental divide in Panama are more complex. The Gaillard Cut (originally aptly named the Culebra Cut—"snake" in Spanish) was commenced in 1907 with slopes initially about 45°. Although the intricately interlayered volcanic rocks were so hard that 25,000 tons of dynamite were required to loosen them, slides of nearly 1,000,000 m^3 occurred several times during construction. The total cut volume was about 75,000,000 m^3, which was 3 times the original design. Since completion there have been numerous new slides ranging in volume from a few hundred m^3 to 10^6 m^3. In October 1974, following heavy rain, the canal was partially blocked by a slide of 750,000 m^3 (1,000,000 yd^3), Fig. 12.19. The fractured partially weathered volcanic and tuffaceous sedimentary rocks slowly continue to deteriorate in the moist tropical environment, and move in elongated base, rotation, and linear shear slides, as the cut faces are exposed and strains open cracks.

DIAGNOSING LANDSLIDE SUSCEPTIBILITY / The foregoing discussions show that natural landslides are related to soil and rock types,

Figure 12.19 *Landslide of 750,000 m^3 (1,000,000 yd^3) in east bank of the Gaillard (Culebra) Cut in the Panama Canal, triggered by heavy rains in weathered complex volcanic mudstones, October 1974.* (Courtesy of George Berman and the Panama Canal Co.)

geologic structure, topography, and environment. Therefore, if landslides have occurred in one location, similar slides are likely in areas in which these features are similar. Landslides and old landslide-related topography can be identified in stereoscopic air photographs and other forms of remote sensing. Maps of past sliding correlated with maps of the other relevant factors can define areas that are most susceptible to future movements. Such landslide studies and their accompanying maps have been prepared for local areas of frequent movements that are also areas of intense land development for housing, industry, and transportation. They can be helpful in land-use planning and in restricting development in the most susceptible areas.[12:28, 12:29]

CORRECTING LANDSLIDES / Most landslides are caused by a combination of factors. Before any remedial action can be taken, a detailed study must be made to determine which factors are the significant ones for that particular situation. The structure of the soil and rock formations and the physical properties of the different materials must be established. The groundwater levels and the pressures in cracks and fissures are particularly important.

Correction involves controlling as many of the factors as possible. Often the best method is found by trial. Drainage, flattening the slope, stabilization of the soil by grouting, removing external loads, erosion protection at the toe, and providing support with piling and retaining walls—all have proved to be successful under certain circumstances, but no method is of value unless it fits the specific needs of the particular slide.

SUBSIDENCES / Subsidences are actually vertical earth movements. They are of two types: rapid, caused by undermining or failure of the underlying strata; and slow, caused by consolidation. Rapid subsidence occurs frequently in areas of abandoned mines. Disintegration of old timbering in shallow workings causes caving of the rock above and the formation of a cavity beneath the soil. Sooner or later the soil bridging the cavity will break apart until an intact mass slides vertically downward. The same phenomenon occurs in areas underlaid by cavernous limestone. The countless sinkholes that dot the landscape in many parts of the eastern United States as well as other parts of the world are the results of small subsidences. In a few cases subsidences have been caused by the underground erosion of cohesionless strata by artesian water.

Rapid subsidence can be caused by excavation for sewers, tunnels, and buildings. If more soil is removed from an excavation than the finished volume of the excavation, it indicates that the soil is squeezing into the hole as it is being removed. This phenomenon, known as *lost ground*, is particularly troublesome in soft clays such as are found in Chicago, Detroit, Cleveland, and other cities in glaciated areas. The lost ground results in subsidence of the surrounding ground surface and often causes damage to adjacent buildings. Careful bracing of excavations and checks to determine possible building settlement are required to prevent such troubles.

Slow subsidence caused by consolidation occurs in areas in which the soil stresses increase materially. The Long Beach area of California subsided at a rate of 0.25 m (10 in.) per year for the period 1941 to 1945 and is still sinking. The excessive pumping from the many oil wells in the area is reducing the neutral stresses in the oil-bearing rocks and is increasing the effective stresses. The rocks, therefore, consolidate as the oil is removed and the ground surface sinks correspondingly. Local areas of Houston,, Texas, and Mexico City are subsiding from pumping water out of the sand strata that are interbbedded with the soft clays. The only remedy for such subsidences is to make allowances for them in design of structures or control the withdrawal of water or oil. They cannot be prevented without correcting the causes.

Subsidences may induce stresses in the soil or rock strata that aggravate failure by another unrelated mechanism. The failure of a water reservoir in Baldwin Hills, Los Angeles, is an example. Subsidence over a wide area, blamed on oil well pumping, caused an old fault to reactivate. One limb of the fault cut across the earth dike foundation and the reservoir bottom; its movement ruptured the reservoir lining. Water seeping from the leak softened the foundation, aggravating the leak. The reservoir eventually failed by piping followed by collapse of a portion of the dike into the hole eroded by the seepage.

ROCKFALLS / Rockfalls are movements of detached rock fragments down steep slopes. They often occur in cuts in badly jointed rock and in cuts where all the materials loosened by blasting were not excavated. Periodic checks should be made on the condition of rock cuts or other steep slopes in rock and all unstable pieces removed. Some roads must be closed during periods of heavy rain or freezing and thawing when water pressure or frost wedging set loose rock in motion. In some cases it has been practical to anchor loose rocks with rods and cables to prevent their movement.

Rock bolts can be used to prestress jointed formations and bind them together in a coherent mass. Concrete or shotcrete facings, supported by rock bolts, can prevent the movement of small, loose pieces, and at the same time protect the rock from weathering. The facing is drained to prevent a buildup of neutral stress in pervious seams or joints.

Deep cuts in rock are frequently provided with horizontal berms to catch falling rock and prevent them from tumbling down the full height and gaining enough momentum to create serious damage. Sand cushions on the berms have been effective in minimizing the bouncing of such rocks. Rock fills or fences are used to catch the rocks and keep them from rolling beyond the slope toe and endangering people and structures below. In mountainous regions, avalanche sheds of heavy timber or concrete are sometimes built over railroads and highways to protect them from rockfalls and snow slides. Any preventive or corrective program must start with a complete picture of the joints, fissures and bedding planes upon which the failures focus. Control of the movements on these surfaces is the essence of the design.

12:7 Instability and the Geotechnical Engineering Challenge

Instability in the form of creep, landslides, rockfalls, and subsidence are major geologic phenomena that help shape the earth's surface. They are also major geotechnical problems for two reasons: (1) Naturally occurring movements interfere with human activities and well-being, and (2) activities of humans (usually for their own well-being) aggravate, accelerate, or precipitate instability and subsequent motion. Unlike those involved in many of the other geotechnical problems in this text, people are sometimes only innocent bystanders, dwarfed by the earth in motion. In other cases, they are helpless victims of the forces they helped unleash. By a careful study of the geology and mechanics of the process, engineers can prevent many of the movements, control others to the extent they can live with them, and define areas of trouble that should be avoided by the public. These alternatives summarize the human relation to the earth and the ultimate challenge of geotechnical engineering.

REFERENCES

12:1 D. J. Varnes, "Classification of Landslides, Chapter 2," *Landslide Analysis and Control, Special Report 176*, Transportation Research Board, Washington, D.C., 1978.

12:2 H. G. Poulos, J. R. Baker, and G. J. Ring, "Simplified Calculation of Embankment Deformations," *Soils and Foundations*, **12**, 4, December 1972 (Japanese Society of Soil Mechanics and Foundation Engineering, Tokyo), pp. 1–18.

12:3 P. Dunlop and J. M. Duncan, "Failure Around Slopes," *Journal of Soil Mechanics and Foundations Division, Proceedings, ASCE*, **96**, SM2, March 1970, pp. 471–493.

12:4 R. V. Whitman and W. A. Bailey, "Use of Computers in Slope Stability Analysis," *Journal of the Soil Mechanics and Foundations Division, Proceedings, ASCE*, **93**, SM4, July 1967.

12:5 J. E. Bowles, *Analytic and Computer Methods in Foundation Engineering*, McGraw-Hill Book Company, New York, 1974, p. 518.

12:6 L. Bjerrum and Nils Flodin, "The Development of Soil Mechanics in Sweden, 1900–1925," *Geotechnique*, **X**, 1, March 1960.

12:7 W. Fellenius, *Erdstatische Berechnungen*, rev. ed., W. Ernst u. Sohn, Berlin, 1939.

12:8 A. W. Bishop, "The Use of the Slip Circle in the Stability Analysis of Slopes," *Geotechnique*, **5**, 1, pp. 7–17, 1955.

12:9 N. Janbu, "Application of Composite Slip Surfaces for Stability Analysis," *Proceedings, European Conference on Stability Analysis of Earth Slopes, Stockholm, Sweden*, 3, pp. 43–49, 1954.

12:10 N. R. Morgenstern and V. E. Price, "The Analysis of Stability of Slopes," *Geotechnique*, **15**, 1, pp. 79–93, 1965.

12:11 S. G. Wright and F. H. Kulhawy, "Accuracy of Equilibrium Slope Stability Methods," *Journal of Soil Mechanics and Foundations Division, Proceedings, ASCE*, **99**, SM10, October 1973, pp. 783–791.

12:12 N. Newmark, "Effects of Earthquakes on Dams and Embankments," *Geotechnique*, **15**, September 1965, p. 140.

12:13 H. B. Seed, I. M. Idriss, K. L. Lee, and F. I. Makdisi, "Dynamic Analysis of the Slide of the Lower San Fernando Dam During the Earthquake of February 9, 1971," *Journal of Geotechnical Engineering Division, Proceedings, ASCE*, **109**, GT9, September 1975, pp. 889–911.

12:14 D. W. Taylor, "Stability of Earth Slopes," *Journal of the Boston Society of Civil Engineers*, July 1937.

12:15 John M. Lowe III, "Stability Analysis of Embankments," *Journal of the Soil Mechanics and Foundations Division, Proceedings, ASCE*, **93**, SM4, July 1967.

12:16 D. U. Deere, A. J. Hendron, F. D. Patton, and E. J. Cording, "Design of Surface and Near-Surface Construction in Rock," *Proceedings, 8th Symposium on Rock Mechanics*, American Institute of Mining and Metallurgical Engineers, 1967.

12:17 R. E. Goodman, *Methods of Geological Engineering in Discontinuous Rocks*, West Publishing Co., St. Paul, Minn., 1976, pp. 213–244.

12:18 *Blasters Handbook*, 6th ed., E. I. du Pont de Nemours and Co., Wilmington, Del., 1978.

12:19 A. Casagrande, "Notes on the Design of Earth Dams," *Journal of the Boston Society of Civil Engineers*, **37**, 1950.

12:20 J. L. Sherrard, R. J. Woodward, S. F. Gizienski, and W. A. Clevenger, *Earth and Earth-Rock Dams*, John Wiley & Sons, Inc., New York, 1963.

12:21 G. F. Sowers, *Earth and Rockfill Dam Engineering*, Asia Publishing House, Bombay, 1961.

12:22 "Problems in the Design and Construction of Earth and Rockfill Dams," *Journal of the Soil Mechanics and Foundations Division, Proceedings, ASCE*, **93**, SM3, May 1967, p. 129.

12:23 "Symposium on Rockfill Dams," *Transactions, ASCE*, **125**, Part II, 1960.

12:24 H. B. Seed, "Earthquake Resistant Design of Dams," *Journal of the Soil Mechanics and Foundations Division, Proceedings, ASCE*, **92**, SM1, January 1966.

12:25 K. Terzaghi, "Mechanism of Landslides," *Application of Geology to Engineering Practice, Berkey Volume*, Geological Society of America, 1950.

12:26 H. B. Seed, "Landslides During Earthquakes due to Liquefaction," *Journal of the Soil Mechanics and Foundations Division, Proceedings, ASCE*, **94**, SM5, September 1968, p. 1053.

12:27 L. Bjerrum, "Progressive Failure in Slopes of Overconsolidated Plastic Clay and Clay Shales," *Journal of the Soil Mechanics and Foundations Division, Proceedings, ASCE*, **93**, SM5, September 1967, pp. 1–49.

12:28 D. F. Radbruch-Hall, R. B. Colton, B. A. Skipp, I. Lucchitta, and D. J. Varnes, *Preliminary Landslides Overview Map of the Conterminous United States*, U.S. Geological Survey, Misc. Field Studies Map MF-771, USGS, Denver, 1977.

12:29 T. H. Nilsen, Fred A. Taylor, and R. M. Dean, *Natural Conditions That Control Landsliding in the San Francisco Bay Region*—U.S. Geological Survey Bulletin 1424, Washington, D.C., 1976.

SUGGESTIONS FOR FURTHER STUDY

Design of Small Dams, 2nd ed., U.S. Bureau of Reclamation, Denver, 1974, 816 pp.

A. R. Golze, ed., *Handbook of Dam Engineering*, Van Nostrand Reinhold Company, New York, 1977, 793 pp. (A joint effort of 27 authors.)

D. P. Krynine and W. R. Judd, *Principles of Engineering Geology and Geotechnics*, McGraw-Hill Book Company, New York, 1957.

P. Londe, G. Vigier, and R. Vormeringer, "Stability of Rock Slopes, A Three-Dimensional Study," *Journal of the Soil Mechanics and Foundations Division, Proceedings, ASCE*, **95**, SM1, 1969, pp. 235–262.

N. Morganstern, G. E. Blight, N. Janbu, and D. Resendiz, "Slopes and Excavations," *General Report, Ninth International Conference on Soil Mechanics and Foundation Engineering*, Tokyo, 1977, 2, pp. 547–604.

"Proceedings of the 1966 Slope Stability Conference, ASCE," *Journal of the Soil Mechanics and Foundations Division, Proceedings, ASCE*, **93**, SM4, July 1967; 30 papers on a wide variety of stability problems.

A. Richards, ed., "Marine Slope Stability," *Marine Geotechnology*, 2, 1977, 392 pp. (A symposium of 23 papers on submarine slopes.)

R. L. Schuster and R. J. Krizek, eds., *Landslide Analysis and Control, Special Report 176*, Transportation Research Board, Washington, 1978. This is a revision of the HRB Report, 29, *Landslides and Engineering Practice*, of 1959. It is a comprehensive manual on earth and rock slope investigation, analysis, and design (excepting dams) by a group of engineers and geologists.

W. J. Turnbull and M. J. Hvorslev, "Special Problem in Slope Stability," *Journal of the Soil Mechanics and Foundations Division, Proceedings, ASCE*, **93**, SM4, July 1967, p. 499.

Q. Zaruba and V. Mencl *Landslides and Their Control*, Elsevier Publishing Co., Amsterdam and New York; Academia Prague, 1969, 205 pp.

PROBLEMS

12:1 A slope of 2 (horizontal) to 1 (vertical) is cut in homogeneous, saturated clay whose $c = 53\ \text{kN/m}^2 = 1.107\ \text{kip/ft}^2$ and whose $\gamma = 17.5\ \text{kN/m}^3$ or $111.6\ \text{lb/ft}^3$. The cut is 13 m (42.7 ft) deep and the clay deposit extends 5 m (16.4 ft) below the bottom of the cut. The clay rests on rock.

a. Compute the safety of this slope, using the charts for homogeneous soils.
b. Check, using the circular arc analysis by computer.

12:2 A cut is excavated at an angle of 45°. It is 15 m (49.2 ft) deep. The soil profile from the surface down is

Depth			Shear Strength, c		Unit Weight	
m	ft	Soil	kN/m^2	kip/ft^2	kN/m^3	lb/ft^3
0–3	0–10	Stiff clay	75 =	1.57	19 =	121
3–10	10–32.8	Stiff clay	60 =	1.25	18 =	114.6
10–20	32.8–65.6	Firm clay	55 =	1.15	17 =	108.3
20	65.6	Shale (rock)				

a. Find the safety factor with respect to base failure, assuming the center of the failure circle to be above the midpoint of the slope. Verify by various centers and radii.
b. Find the safety with respect to toe failure.
c. Check the results, using the stability factor chart and assuming that the effective c is the weighed average c of all the strata.

12:3 An excavation 10 m (32.3 ft) deep and 20 m (65.6 ft) wide at the bottom is to be made in clay with $c = 40\ \text{kN/m}^2$ or $835\ \text{lb/ft}^2$ and $\gamma = 17.3\ \text{kN/m}^3$ or $110\ \text{lb/ft}^3$). How wide should the top of the excavation be if the minimum safety factor is 1.3? (Use the chart of stability number for saturated clay.)

12:4 An embankment of sand is 20 m (66 ft) high, 10 m (33 ft) wide at the top, and has side slopes of $1\frac{1}{2}$ (horizontal) to 1 (vertical). The embankment soil $\phi = 43°$ and $\gamma = 20\ \text{kN/m}^3$ or $127\ \text{lb/ft}^3$. The foundation soil consists of clay with $c = 40\ \text{kN/m}^2$ or $0.83\ \text{kip/ft}^2$. Compute the safety of the embankment against a sliding-block failure along the line of contact of the foundation and the fill, or a sloping plane within the foundation.

12:5 A canal is dug in a soil whose characteristics when saturated are $c = 30$ kN/m^2 or 0.63 kips/ft^2, $\phi = 16°$, and $\gamma = 19$ kN/m^3 or 121 lb/ft^3 saturated. The canal is to be 6.7 m (22 ft) deep and the slopes are 2 (horizontal) to 1 (vertical).

a. Compute the safety factor when the canal is full of water.
b. Compute the safety factor if the canal is suddenly drained, leaving the soil saturated.
c. Which condition is worse?

HINT: When the canal is full of water, the unit weight is reduced by buoyancy, an amount of 9.8 kN/m^3 (62.4 lb/ft^3) effectively.

12:6 An embankment 25 m (82 ft) high is 13 m (42.7 ft) wide at the top is constructed with a slope of 50°. The soil is partially saturated clay. The uppermost 15 m (49.2 ft) of soil has $c =$ 100 kN/m^2 or 2.09 kip/ft^2, $\phi = 15°$, $\gamma = 18$ kN/m^3 or 114.6 lb/ft^3. The remaining 10 m (32.8 ft) has $c = 45$ kN/m^2 or 0.94 kip/ft^2, and $\gamma = 17$ kN/m^3 or 108.3 lb/ft^3. The foundation is rock. Compute the safety factor with respect to toe failure. Use the method of slices. Assume a toe failure and find the critical circle center and the safety factor.

12:7 A highway fill 8 m (26.2 ft) high and 10 m (32.8 ft) wide at the top is constructed across an area of soft compressible soil. The fill weighs 21 kN/m^3 (133.8 lb/ft^3) as compacted. The foundation soil is deep saturated clay with $c = 30$ kN/m^2 (626 lb/ft^2) and $\gamma = 16.5$ kN/m^3 or 105 lb/ft^3.

a. Compute the safety against a bearing-capacity failure, assuming that the foundation soil is deep enough that the bearing-capacity factors of Chapter 10 apply.
b. If the foundation soil is only 5 m (16.4 ft) thick, compute the safety against a circular arc failure using the method of slices. The angle of friction of the fill is 20° and $c = 50$ kN/m^2 (1044 lb/ft^2).
c. Compute the safety against a sliding block failure, assuming a tension crack depth of $2c/\gamma$ in the fill.

12:8 An earth dam is 30 m (100 ft) high and 6 m (19.7 ft) wide at the crest; the upstream slope is 3(H) to 1(V) and the downstream slope 2(H) to 1(V). The dam has a central clay core 6 m (19.7 ft) wide at the dam crest and 9 m (29.5 ft) wide at the base with $c = 25$ kN/m^2 or 520 lb/ft^2 weighing 16 kN/m^3 (102 lb/ft^3). The shell of the dam is sand with $\phi = 40°$ and $\gamma = 19$ kN/m^3 or 121 lb/ft^3. The foundation is clay 10 m (33 ft) thick with $c =$ 125 kN/m^2 or 2600 lb/ft^2, on rock. A thin sand seam with $\phi = 40°$ extends from the reservoir, and is 4 ft below the dam, ending a few feet from the downstream toe. The maximum head on the dam is 26 m (85.3 ft).

a. Compute the safety of the downstream half of the dam against sliding by a circular arc failure through the foundation, neglecting the sand seam.
b. Compute the safety of the core and downstream shell sliding on the sand seam, assuming full reservoir head in the sand seam.

12:9 A cut 12 m (39.4 ft) deep has a slope of $2(H)$ to $1(V)$. The soil is saturated clay with a unit weight of 20 kN/m^3 (127 lb/ft^3) and an undrained shear strength of 50 kN/m^2 (1043 lb/ft^2). Below the slope is a 4-m (14-ft) thick stratum of softer clay with a unit weight of 17 kN/m^3 (108 lb/ft^3) and a shear strength, c, of 40 kN/m^2 (835 lb/ft^2). Below is bedrock. Find the critical circle by trial, with each student assuming a different circle center and assuming base failure tangent to the rock. Each student will try two circles, each of whose center is in a grid with centers commencing with (0, 0) at the top of the slope and increasing in 4-m (14-ft) increments above the top of the slope and 4-m (14-ft) increments outward from the top of the slope. The area searched will be $[0(H), 9(V)]$ to $[0(H), 24(V)]$ to $[24(H), 0(V)]$, $[24(H), 24(V)]$ (in meters).

a. Draw contours of safety factor, interchanging safety factors among students or use computer solutions.
b. Compute by the Taylor Factor chart using an average of the strengths and unit weights.

12:10 Prepare a discussion of a landslide or slope failure from the published description in an engineering journal or magazine. Include the following points:

1. Description of failure.
2. Chain of events leading to failure.
3. Probable cause.
4. Corrective measures, if any.

APPENDIX 1
Unit Costs

Engineering analysis and design cannot be divorced from cost of construction. Science and technology have made it possible to do many remarkable things; whether or not they will be of use depends on their ultimate value compared with their cost. The following table will give the student some concept of the cost of soil and foundation work. It is based on typical costs in the United States in 1979 as published in *Engineering News Record* and similar publications. Generally, costs have increased 75 to 100% in the last decade.

	$	$
FOUNDATIONS		
Basement excavation in soil	1.00–4.00/yd^3	1.30–5.00/m^3
Basement excavation in rock	10.00–20.00/yd^3	13.00–25.00/m^3
Footing excavation in soil	4.00–20.00/yd^3	5.00–25.00/m^3
Footing excavation in rock	10.00–60.00/yd^3	13.00–80.00/m^3
Footing concrete, including any forming	75.00–120.00/yd^3	100.00–160.00/m^3
Drilled pier foundations, including concrete	100.00–150.00/yd^3	130.00–200.00/m^3
Wood piles, untreated, including driving	3.00–5.00/ft	10.00–16.00/m
Wood piles, treated, including driving	4.00–6.00/ft	13.00–20.00/m
Concrete 50-ton precast, including driving	10.00–15.00/ft	33.00–50.00/m
Concrete 50-ton cast-in-place, including driving	8.00–12.00/ft	25.00–40.00/m
Steel H-pile, 10 in., 50 ton, including driving	10.00–15.00/ft	35.00–50.00/m
Steel pipe pile, $10\frac{3}{4}$, 50 ton, including driving	12.00–15.00/ft	40.00–50.00/m
EXCAVATION BRACING		
Wood sheeting (20 ft deep)	8.00–12.00/ft^2	80.00–120.00/m^2
Steel sheeting, or H-beam and lagging, (40 ft deep)	10.00–14.00/ft^2	100.00–140.00/m^2
Slurry wall	12.00–16.00/ft^2	120.00–160.00/m^2

DRAINAGE		
Wellpoint installation and operation, 30 days	25.00–60.00/ft*	80.00–200.00/m
Trench excavation in soil	4.00–10.00/yd^3	5.00–13.00/m^3
Pumping, 20-ft head	0.10–2.00/1000 gal	
Drain pipe, 8-in. concrete	4.00–7.00/ft	13.00–25.00/m
EARTHWORK (Dams, embankments, large fills)		
Soil excavation, placement, compaction in embankments	1.00–1.75/yd^3	1.30–2.50/m^3
Rock excavation, placement, compaction	1.50–4.00/yd^3	2.00–5.00/m^3
Backfill, around structures, compacted	3.00–8.00/yd^3	4.00–10.00/m^3
SOIL STABILIZATION		
Soil-cement, cement, mixing and compaction	10.00–12.00/yd^3	13.00–16.00/m^3
Cement, grouting, including injection (sacks of cement)	4.00–6.00/	— —
Chemical grouting, including injection (vol. soil)	100.00–300.00/yd^3	130.00–400.00/m^3
EXPLORATION		
Auger boring	3.50–5.00/ft	12.00–16.00/m
Test boring, including split barrel samples	6.00–9.00/ft	20.00–30.00/m
Diamond-core drilling	15.00–25.00/ft	50.00–80.00/m
Undisturbed sampling	40.00–600.00/sample	— —

* Of header pipe.

APPENDIX 2
Age of Geologic Formations

Era	Period	Epoch	Age Range*
Cenozoic	Quarternary	Holocene (Recent)	0–11,000 years
		Pleistocene	11,000–2 million
	Tertiary	Pliocene	2–5 million years
		Miocene	5–25
		Oligocene	25-35
		Eocene	35–55
		Paleocene	55–65
Mesozoic	Cretaceous	Upper	65–90
		Lower	90–140
	Jurassic		140–190
	Triassic		190–230
Paleozoic	Permian		230–280
	Pennsylvanian	Carboniferous	280–320
	Mississippian	Carboniferous	320–350
	Devonian		350–400
	Silurian		400–430
	Ordovician		430–500
	Cambrian		500–600
Precambrian	Proterozoic		600+
	Archeozoic		from 4600?

SOURCES: J. L. Kulp, "Geologic Time Scale," *Science*, **133**, 1105–1114, 1961; D. L. Eicher, *Geologic Time*, Prentice-Hall, Inc., Englewood Cliffs, N.J., 1968; and W. A. Berggren, "Cenozoic Chronostratic Planktonic Forminiferal Zonation and the Radiometric Time Scale," *Nature*, **224**, 1072–1075, December 1969.

* There is considerable controversy in ages in the Paleozoic Era and in the beginning of the Recent and Pleistocene epochs. The Pleistocene may have begun 1 million years ago and ended as late as 5000 years ago according to some authorities.

Index